International Conference on Chemical and Environmental Engineering (ICEE-11)

Journal of Physics: Conference Series Volume 2305

Online
29-31 March 2022

ISBN: 978-1-7138-6032-7
ISSN: 1742-6588

Printed from e-media with permission by:

Curran Associates, Inc.
57 Morehouse Lane
Red Hook, NY 12571

Some format issues inherent in the e-media version may also appear in this print version.

Institute of Physics
Dirac House, Temple Back
Bristol BS1 6BE UK

Phone: 44 1 17 929 7481
Fax: 44 1 17 920 0979

techtracking@iop.org

Additional copies of this publication are available from:

Curran Associates, Inc.
57 Morehouse Lane
Red Hook, NY 12571 USA
Phone: 845-758-0400
Fax: 845-758-2633
Email: curran@proceedings.com
Web: www.proceedings.com

TABLE OF CONTENTS

CHEMICAL & RADIATION DETECTION IN ENVIRONMENT

INDUSTRIAL AND ENVIRONMENTAL ENGINEERING

NANOTECHNOLOGY AND CHEMICAL APPLICATIONS

ENERGETIC MATERIALS AND ENVIRONMENTAL SCIENCE

POLLUTION AND ENVIRONMENTAL PROTECTION

RENEWABLE ENERGY AND SUSTAINABILITY

ENVIRONMENTAL AND CHEMICAL REACTION ENGINEERING

Journal of Physics: Conference Series **2305** (2022) 011001 doi:10.1088/1742-6596/2305/1/011001

Ministry of Defense
Military Technical College

The 11th International Conference on Chemical & Environmental Engineering

I.C.E.E.2022

29 - 31 March, 2022
Cairo, Egypt

Military Technical College
Kobry El-Kobbah,
Cairo, Egypt

I.C.E.E.2022

11th International Conference
on
Chemical & Environmental
Engineering
29– 31 March, 2022.

PREFACE

The military Technical College is honored to organize the 11th International Conference on Chemical and Environmental Engineering (ICEE) in the period 29-31 March, 2022 sponsored by the Egyptian Ministry of Defense. The conference provides a valuable opportunity to exchange, update the knowledge and stimulate discussions on recent developments and research activities.

Accepted papers are arranged in the following sessions:

1. Chemical and Radiation Detection in Environment (CRDE-I)
2. Chemical and Radiation Detection in Environment (CRDE-II)
3. Industrial and Environmental Engineering (IEE)
4. Nanotechnology and Chemical Applications (NCA-I)
5. Nanotechnology and Chemical Applications (NCA-II)
6. Energetic Materials and Environmental Science (EME-I)
7. Energetic Materials and Environmental Science (EME-II)
8. Pollution and Environmental Protection (PEP-I)
9. Pollution and Environmental Protection (PEP-II)
10. Pollution and Environmental Protection (PEP-III)
11. Renewable Energy and Sustainability (RES)
12. Environmental and Chemical Reaction Engineering (ECRE-I)
13. Environmental and Chemical Reaction Engineering (ECRE-II)

We are highly indebted to all conference contributors. Also; the effort of the scientific committee is highly appreciated.

Journal of Physics: Conference Series **2305** (2022) 011001 doi:10.1088/1742-6596/2305/1/011001

Military Technical College
Kobry El-Kobbah,
Cairo, Egypt

I.C.E.E.2022

11th International Conference
on
Chemical & Environmental
Engineering
29– 31 March, 2022.

Conference Committees

Higher Committee

- Maj. Gen. Prof. Dr. Ismail Kamal — *Chairman*
- Maj. Gen. Prof. Dr. Motaz E. AboNoor — *Vice - Chairman*
- Maj. Gen. (R) Prof. Dr./Ahmed E. Abdalla — *General Rapporteur*

Organizing Committee:

- Maj. Gen. (R) Prof. Dr. Hosam E. Mostafa — *Rapporteur*
- Brig. Gen. Assoc. Prof. Dr. Tamer Z. Wafy — *Assistant Rapporteur*
- Col. Dr. Mahmoud Abdelhafiz Mahmoud — *Assistant Rapporteur*

Scientific Committee:

- Maj. Gen. (R) Prof. Dr. Mohmed A. El-Shaheer
- Maj. Gen. (R) Assoc. Prof. Dr. Said M. A. Hassanein
- Maj. General (R) Assoc. Prof. Dr/ Mohamed Kassem
- Brig. Gen. Prof. Dr. Mohamed Abo El fotoh Elsayed
- Brig. Gen. Prof. Dr. Mohammed M. Gobbara
- Brig. Gen. Dr. Mohamed I. Awaad
- Brig. Gen. Assoc. Prof. Dr. Ahmad Elbeih
- Brig. Gen. Assoc. Prof. Dr./ Mohamed Abdel-Rahman
- Brig. Gen. Assoc. Prof. Dr./ Tamer Z. Wafy
- Brig. Gen. Assoc. Prof. Dr. Sherif Shaaban Elbasuney
- Brig. Gen. Assoc. Prof. Dr. Hisham Gad El Kareem Tantawy
- Col. Dr. Ahmed Farghal Tawfik
- Col. Dr. Mohamed M. Kotb
- Col. Dr. Ibraheem Samir Saad
- Col. Dr. Mahmoud Abdelhafiz
- Lt. Col. Dr. Mohamed Maraden
- Lt. Col. Dr. Ahmed K. Hussein
- Lt. Col. Dr. Ahmed M. Omar
- Lt. Col. Dr. Basem Zaghloul
- Lt. Col. Dr. Osama abo Zalat

Military Technical College
Kobry El-Kobbah,
Cairo, Egypt

I.C.E.E.2022

11th International Conference
on
Chemical & Environmental
Engineering
29– 31 March, 2022.

CONTRIBUTORS

I Egyptian Universities and Organizations

1. Ain Shams University, Faculty of Engineering, Cairo, Egypt.
2. Air Pollution Department, Environment and Climate change Research Institute, National Research Centre, 33 EL Bohouth St., Dokki, Giza, P.O.12622, Egypt.
3. Advanced Materials/Solar Energy and Environmental Sustainability (AMSEES) Laboratory, Basic Engineering Science Department, Faculty of Engineering, Menoufia University, Shebin El-Kom, Egypt.
4. British University in Egypt, Department of Chemical Engineering, Egypt, 11837.
5. Cairo University, Department of Chemical Engineering, Egypt
6. Chemical Engineering Department, British University in Egypt, Egypt
7. Chemical Engineering Department, Faculty of Engineering, Cairo University, Egypt
8. Chemical Engineering Department, Military Technical College, Kobry Elkobbah, Cairo, Egypt.
9. Chemical Engineering Department, Menia University, El-Menia, Egypt.
10. Chemistry Department, Faculty of Science, Assiut University, Assiut, 71516 Egypt.
11. Chemistry Department, Faculty of Science, Menoufia University, Shebin El-Kom, Egypt.
12. Canal High Institute for Engineering and Technology.
13. Chemistry Department, Faculty of Science, Al-Azhar University, Egypt.
14. Civil Engineering Department, Faculty of Engineering, Kafrelsheikh University, Egypt.
15. Defense Systems Studies Center, Cairo, Egypt
16. Department of Uranium Ores Processing, Production Sector, Nuclear Materials Authority, Cairo, Egypt.
17. Department of Environmental Studies, Institute of Graduate Studies and Research, Alexandria University, Postal code 21545 El-Shatby, Alexandria, Egypt.
18. Department of Soil & Water Science, Faculty of Agriculture, Alexandria University, Postal code 21545 El-Shatby, Alexandria, Egypt.
19. Department of Botany, Faculty of Science, Zagazig University, Egypt
20. Department of Chemistry, Faculty of Science, Ain Shams University, Cairo, Egypt.
21. Department of Chemistry, Faculty of Science, Mansoura University, Egypt
22. Department of Chemistry, Faculty of Science, Menoufia University, Egypt
23. Department of Chemistry, Faculty of Science, Al-Azher University, Cairo, Egypt.
24. Environmental and Climate changes Institute.
25. Egyptian Organization for Standardization and Quality (EOS), Egypt
26. Egyptian Armed Forces, Cairo, Egypt.
27. Faculty of Science, Al-Azhar University, Cairo, P.O. Box 11884, Egypt
28. Faculty of Engineering, King Salman International University, Sinai Egypt.
29. Faculty of Science, Geology, Geology Department, Tanta University, Egypt
30. Factually of Science, Ain Shams University, Cairo, Egypt.
31. Geology Department, Faculty of Sciences, Suez Canal University, Egypt
32. Geology Department, Faculty of Science, Fayoum University, Fayoum, Egypt

33. Head of Nanotechnology Research Center, Military Technical College (MTC), Egypt.
34. Medical Laboratory Department, Faculty of Applied Health Sciences Technology, Pharos University in Alexandria, Alexandria, Egypt.
35. Mechanical Engineering Department, Faculty of Engineering, Kafrelsheikh University, Kafrelsheikh 33516, Egypt.
36. Microwaves Physics and Dielectrics Department, National Research Centre, Elbuhoth Street 33, Dokki, Cairo, 12622, Egypt.
37. Main Laboratories, Chemical Ware Fare, Egyptian Army, Egypt.
38. Nuclear Power Plants Authority, 4 El Nasr Avenu-Nasr City, Cairo.
39. Nuclear Materials Authority, El Maadi P.O. Box 530, Cairo, Egypt
40. Nuclear and Radiological Regularity Authority, Cairo- Egypt
41. Nuclear Safety Research Center, Egyptian Atomic Energy Authority, Cairo, Egypt
42. Nuclear Research Center, Egyptian Atomic Energy Authority, Cairo, Egypt
43. Nanomaterials Investigation Laboratory, central Laboratories Network, National Research Centre, Elbuhoth Street 33, Dokki, Cairo, 12622, Egypt
44. Nuclear Engineering Department, Military Technical College, Cairo, Egypt
45. Nuclear Materials Authority, P.O. Box 530 El-Maadi, Cairo, Egypt
46. Nuclear and Radiological Safety Research Center (NRSRC). Egyptian Atomic Energy Authority (EAEA).
47. Physics Department, Faculty of Science / Al Azhar University, Girls' branch, Egypt)
48. Packaging Materials Department, National Research Centre, Elbuhoth Street 33, Dokki, Cairo, 12622, Egypt
49. Polymer chemistry department, National Centre for Radiation Research and Technology, Egyptian Atomic Energy Authority, Cairo, Egypt.
50. Polymeric Materials Research Department, City of Scientific Research and Technological Applications (SRTA-City), Borg El-Arab City, Alexandria 21934, Egypt.
51. Port Said University, Department of Chemical Engineering, Egypt, 42526
52. Physics Department, Faculty of Science / Al Azhar University, Girls' branch, Egypt
53. Radiation Engineering Department, EAEA.
54. Radiation Safety Department, EAEA.
55. Radiation measurements Branch, Main Defense Chemical Laboratory, Cairo-Suez Road - Almaza, Cairo- Egypt.
56. Radiation Research of Polymer Chemistry Department, National Center for Radiation Research and Technology (NCRRT), Egyptian Atomic Energy Authority (EAEA).
57. Radiation safety department, nuclear and radiological safety research center, Egyptian atomic energy authority.
58. Space technology Centre, Cairo, Egypt.
59. Science and Technology Center of Excellence (STCE), Ministry of Military Production, Cairo, Egypt.
60. Technical Research Center, Cairo, Egypt
61. Water Pollution Research Department, Environmental Sciences Division, National Research Centre (NRC), Egypt, P.O. 12622.

Military Technical College
Kobry El-Kobbah,
Cairo, Egypt

I.C.E.E.2022

11th International Conference
on
Chemical & Environmental
Engineering
29– 31 March, 2022.

II Arab and Foreign Universities and Organizations

1. Department of Chemistry, College of Science and Humanities, Prince Sattam bin Abdulaziz University, Alkharj 11942, **Saudi Arabia**.
2. University of Jeddah, College of Science, Department of Chemistry, P.O. Box 80327, Jeddah 21589, **Saudi Arabia**.
3. «Water Supply and Drainage», Academy of Construction and Architecture of the Don State Technical University, **Russia**.
4. Ministry of Electricity and Water, **Kuwait**, P.O.12010
5. Collage of Applied Science, Department of Engineering, Suhar, 331, **Sultanate of Oman**.
6. Department of Chemical and Environmental Engineering, University of Nottingham, Nottingham, NG7 2RD, **United Kingdom.**
7. School of Chemical and Process Engineering, University of Leeds, Leeds, LS2 9JT, **United Kingdom.**
8. Liquid crystal Institute, chemical Physics Interdisciplinary Program, Kent State University, Kent, Ohio, 44242.
9. Water Resources Research Center, DPRI, Kyoto University, **Japan**.
10. Université de Perpignan Via Domitia, **France**.
11. Department of Science Laboratory Technology, **Nigeria**.

Military Technical College
Kobry El-Kobbah,
Cairo, Egypt

I.C.E.E.2022

11th International Conference
on
Chemical & Environmental
Engineering
29– 31 March, 2022.

REVIEWERS

1	Amr M.I. Kany	Faculty of Science, Al-Azhar University
2	Ahmed A. Baraka	Military Technical College (MTC)
3	Ahmed Farghal Tawfik	Military Technical College (MTC)
4	Ahmed M. Omar	Military Technical College (MTC)
5	Ahmad Elbeih	Military Technical College (MTC)
6	Ahmed K. Hussein	Military Technical College (MTC)
7	Ahmed Shalaby	Science and Technology Center of Excellence, Military production
8	Ahmed El Sayed Azazy	Science and Technology Center of Excellence, Military production
9	A. Salah El-Din El-Kholany	Pharos University in Alexandria, Egypt
10	Ahmed Saleh	Science and Technology Center of Excellence, Military production
11	Basem Zaghloul	Military Technical College (MTC)
12	Ghareb S. El-Sayyad	Egyptian Atomic Energy Authority
13	Hosam E. Mostafa	Military Technical College (MTC)
14	Hisham Gad El Kareem Tantawy	Military Technical College (MTC)
15	Hany A. Elazab	British University in Egypt (BUE)
16	Hussein S. Hegazy	National Research Centre, Dokki, Giza, Egypt
17	Ibrahim Samir Saad	Military Technical College (MTC)
18	M. Mahmoud El-Oquer	Faculty of Science, Al-Azhar University
19	Mohamed S. El-Deab	Faculty of Science, Cairo University
20	Mohamed A. Gobbara	Military Technical College (MTC)
21	M. Kassem Abdel-Magid	Military Technical College (MTC)
22	Mohamed I. Awaad	Military Technical College (MTC)
23	Mahmoud El-Ahdal	Nuclear Research Centre, Atomic Energy Authority
24	Manar Fathy Omran	Atomic Energy Authority, Cairo Egypt
25	M. Amin El-Sahir Sadek	British University in Egypt
26	Mahmoud Yossery	Military Technical College (MTC)
27	Mona Mostafa Mahmoud	Faculty of physics Minia university
28	Mostafa Radwan	British University in Egypt
29	M. Mohamed El Dosokey	British University in Egypt
30	Mahmoud Abdelhafiz	Military Technical College (MTC)
31	Mohamed Y. Hassaan	Faculty of science Azahar University
32	Mohammad Husain Sammour	Military Technical College (MTC)
33	Magdy M. Hassan Senna	Atomic Energy Authority, Cairo Egypt.
34	Mohamed Abo El fotoh Elsayed	Military Technical College (MTC)
35	Mohamed M. Kotb	Military Technical College (MTC)
36	Mohamed Maraden	Military Technical College (MTC)

Journal of Physics: Conference Series **2305** (2022) 011001 doi:10.1088/1742-6596/2305/1/011001

Military Technical College
Kobry El-Kobbah,
Cairo, Egypt

I.C.E.E.2022

11th International Conference
on
Chemical & Environmental
Engineering
29– 31 March, 2022.

37	Nabeel M. Abdel-Monem	Cairo University of Engineering
38	Sherif Fahmy Mahmoud	Faculty of Medicine, Al-Azhar University
39	Sherif Sh. El-Bassuny	Military Technical College (MTC)
40	Said A. Hassanein	Military Technical College (MTC)
41	Salah El Mofty	Cairo University, Mining Engineering Department
42	Samah Ibraheem Radwan	Atomic Energy Authority, Cairo Egypt.
43	Sayed Ali El-Mongy	Nuclear Regulatory Authority (ENRRA)
44	Salwa Abo El kassem Maqbol	Faculty of science Minia university
45	R. M. Megahid	Nuclear Research Centre, Atomic Energy Authority
46	Osama Ali Abuzalat	Military Technical College (MTC)
47	Waleed Fekry	Egyptian Atomic Energy Authority
48	Zaynab Abdel-Hameed	Central Metallurgical R & D Institute, Cairo, Egypt

Journal of Physics: Conference Series **2305** (2022) 011001 doi:10.1088/1742-6596/2305/1/011001

Military Technical College
Kobry El-Kobbah,
Cairo, Egypt

I.C.E.E.2022

11th International Conference
on
Chemical & Environmental
Engineering
29– 31 March, 2022.

Paper Code, Authors, and affiliation

Session Code	Paper Code	Authors	Affiliation (Corresponding Author)
CRDE-I	ICEE-0001	**El Tohamy, Amira*** - Kamar, Mohamed - Mira, Hamed - Mossad, Ahmed.	Nuclear Materials Authority
	ICEE-0002	**Hassan, Mohamed*** - El Shafey, A. M. - bayoumi, moustafa bayoumi - Ebyan, Osama M	Nuclear Materials Authority
	ICEE-0003	El-Dabour, Sameh E. - Harpy, Nareman M. - **Khattab, Mahmoud R.***	Nuclear Materials Authority
CRDE-II	ICEE-0004	Sallam, Asmaa M.* - Faheim, A. A. - El-Elshafiey, Z. A. - M. Amer, S. A. - El Feky, Mohamed G	Nuclear Materials Authority
	ICEE-0005	**El gamel, amal abd el razik***	Egyptian Atomic Energy Authority
	ICEE-0006	**Omar, Omar*** - Elmongy, Sayed Ali	Military Technical College
	ICEE-0007	**Mubarak, Fawzia Mubarak Abd ElKader*** - Ali, Abdallah	Egyptian Atomic Energy Authority
IEE	ICEE-0008	**Sorour, Faisal Hassan*** - Abd E-lmonem, Nabil - Mohamed, Ahmed Fathy - El Sayed, Waleed Abd El Azim	Canal High Institute for Engineering and Technology
	ICEE-0009	**Azab, Rana Adel*** - Abd E-lmonem, Nabil - Soliman, Moustafa Aly - Ismail, Ibrahim - Refaat, Ahmed Ahmed	British University in Egypt
	ICEE-0010	**Sheashea, Mohamed*** - Gobara, Mohamed - zaghloul, basem	Military Technical College

Military Technical College
Kobry El-Kobbah,
Cairo, Egypt

I.C.E.E.2022

11th International Conference
on
Chemical & Environmental
Engineering
29– 31 March, 2022.

NCA-I	ICEE-0011	**Hamed, Abdelaziz*** - Gobara, Mohamed - Mokhtar, Mohamed - Elbasuney, Sherif	Military Technical College
	ICEE-0012	**Abu Elezz, Mariam*** - abd elmonem, Nabil - Ismail, Ibrahim - Refay, Amr	Cairo University
	ICEE-0013	**Ibrahim, Saber Ahmed*** - Labeeb, Ahmad Mohamed	National Research Center
	ICEE-0014	**ismael, shukri*** - maraden, Ahmed - deif, Ahmed - Elbasuney, Sherif	Military Technical College
NCA-II	ICEE-0015	Ahmed Khalil, **waleed Fekry*** - Rayan, Rafat Ahmed - Agha, Ahmed Rabaa	Egyptian Atomic Energy Authority
	ICEE-0016	**Salah Soliman Asser, Mohamed*** - El-Arnaouty, M. B. - Eid, E. - Soliman, El-Sayed - Hegazy, El-Sayed A.	Egyptian Atomic Energy Authority
	ICEE-0017	**Ali, Sayeda Sadek*** - Abdel-moneam, Yasser Kamal - Hegazy, Mohammad	Egyptian Organizations for Standardization and Quality (EOS)
	ICEE-0018	**Elsayed, Mohamed*** - Tantawy, Hesham - Zalat, osama - Elmowafy, Mohamed	Military Technical College
EME-I	ICEE-0019	Hussein, Ahmed - Elbeih, Ahmed - Abdelhafiz, Mahmoud - Salah, hesham - shalaby, Ahmed - Azzazy, Ahmed - **Saleh, Ahmed***	Science and Technology Center of Excellence
	ICEE-0020	Elbeih, Ahmed - elnogomy, mohamed - Shenawy, Tamer - **Saleh, Ahmed*** - Azzazy, Ahmeed - Hussein, Ahmed - gaber, mohamed - Salah, hesham	Science and Technology Center of Excellence
	ICEE-0021	**Fawzy, Mona*** - Abu El Ghar, Mohamed - Gaafar, Ibrahim - El Shafey, Abdel Hay - Diab, Mohamed - Hussein, Ahmed	Nuclear Materials Authority
	ICEE-0022	**Abdelhafiz, Mahmoud*** - Mostafa, Hosam E.	Military Technical College

Journal of Physics: Conference Series **2305** (2022) 011001 doi:10.1088/1742-6596/2305/1/011001

Military Technical College
Kobry El-Kobbah,
Cairo, Egypt

I.C.E.E.2022

11th International Conference on Chemical & Environmental Engineering
29– 31 March, 2022.

EME-II	ICEE-0023	**Saber, ahmed talaat*** - Elbeih, Ahmed	Military Technical College
	ICEE-0024	**Eissa, Mohamed F.*** - Maraden, Ahmed - M. Elhedery, Tarek - M. Abd elall, Ali - E. Mostafa, Hosam	Military Technical College
PEP-I	ICEE-0025	**Abdelghany, Amr*** - Elmansy, Asmaa - Abdelmonem, Nabil - Shaaban, Ahmed	Cairo University
	ICEE-0026	Al Barawy, **Reham Samy*** - Abd E-lmonem, Nabil - Ismail, Ibrahim - Abdelghany, Amr	Cairo University
	ICEE-0027	**Khalil, Ahmed Sobhy*** - Sharshir, Swellam	Kafrelsheikh University
PEP-II	ICEE-0028	**Abdelghany, Amr*** - El Bouraie, Mohamed - Abdelmonem, Nabil - hanafy, mohamed - Samy, Reham	Cairo University
	ICEE-0029	**Tony, Maha Abdel Baseet***	Menoufia University
	ICEE-0030	**Mohamed, Ghada Osama*** - Saleh, Maher Elsayed - Shalaby, Elsayed Ahmed - Elsafty, Ahmed Samir	Alexandria University
	ICEE-0031	**Osman, Asmaa Mohamed *** - Khalifa, Ibrahim Hussein - El- Sayed, Rehab Abd El-Halim - Ibrahim, Ashraf El-Azab - Attia, Reda Mohamed	Suez Canal University
PEP-III	ICEE-0032	**Farouk, elham***	National Research Centre
	ICEE-0033	**Elhelece, wael abdalla***	Egyptian Atomic Energy Authority
	ICEE-0034	**Tawfic, Ahmed *** - Omar, Ahmed - Galal, Moemen - El Minyawi, S. M. - Taha, Samia - Galal, Mohamed - salama, Hanna - El-Afandy, Adel H.	Military Technical College

Military Technical College
Kobry El-Kobbah,
Cairo, Egypt

I.C.E.E.2022

11th International Conference
on
Chemical & Environmental
Engineering
29– 31 March, 2022.

RES	ICEE-0035	**MOHAMED SHABAAN ATREES SEMIDA SELIM*** - Ali, Mohamed Eid - Elmutire, Mohamed Semri - Salah, Mohamed Mahmoud - Ebrahiem, Ebrahiem Ismaiel - khawassek, Yasser mahoud	Nuclear Materials Authority
	ICEE-0036	**Mohamed, Nourhan*** - Mohamed, Marwa - El Azab, Hany - Gadalla, Mamdouh - Abo el azayem, Omar	British University in Egypt
	ICEE-0037	**Thabet, Rahma Hussien*** - Fouad, Mai - El-Sherbiney, Shakinaz - Tony, Maha	Menoufia University
ECRE-I	ICEE-0038	**Mosa, Mostafa Mohamed*** - Kotb, Mohamed - Fouda, Hany - Gobara, Mohamed	Military Technical College
	ICEE-0039	Radwan, Mostafa - Mankarious, **Remon Ashraf Sabry*** - Roushdy, Mai	British University in Egypt
	ICEE-0040	**El-Kholany, Ahmed*** - Hafith, Hagar - Gaber, Nourhan Mohamed	Pharos University in Alex.
ECRE-II	ICEE-0041	**Abdelwahab, Fatma*** - El Terb, Refat - Elgendy, Nader - barakat, Moatz - Wael, Wael - Saad, Mostafa - Mohammed El Tohamy, Amira Mohammed El Tohamy - Galal, Mohamed	Nuclear Materials Authority
	ICEE-0042	**Mohamed Shabaan Atrees Semida Selim ***	Nuclear Materials Authority
	ICEE-0043	**Khaled, Esraa*** - Gadalla, Mamdouh - Amin, Mohamed - Ashour, Fatma - Kamel, Dina	British University in Egypt
	ICEE-0044	**Farrag, Nessren*** & **Ghallab, Ayat Ossama ***	British University in Egypt Cairo University

SCIENTIFIC SESSIONS

Journal of Physics: Conference Series **2305** (2022) 011001 doi:10.1088/1742-6596/2305/1/011001

MilitaryTechnical College
Kobry El-Kobbah,
Cairo, Egypt

I.C.E.E.2022

11th International Conference
on
Chemical & Environmental
Engineering
29 - 31 March, 2022.

Session No 1

Chemical and Radiation Detection in Environment (CRDE-I)

Room: **A8**
Date : Tuesday, 29/03/2022
Time : **12:00 – 13:30**
Board: Prof. Dr. / Tarek Fahmy Youssef NMA
Col. Dr. / Ahmed Farghal Tawfik MTC

0001	**Geology and Environmental Impact Assessment of Psammitic Gneiss and Lamprophyre Dykes at Wadi Sikait, South Eastern Desert, Egypt.** Amira El Tohamy*; Mohamed Kamar; Hamed Mira; Ahmed Mossad.
0002	**Mineralogical Studies and Radioactivity on Stream Sediments Around Wadi Aidib Area, South Eastern Desert, Egypt.** Mohamed Hassan*; A. M. El Shafey; moustafa bayoumi Bayoumi; Osama M. Ebyan
0003	**Radionuclides and Toxic Elements Contamination during the Mining and Processing of Ore and Waste Samples.** Sameh E. El-Dabour; Nareman M. Harpy; Mahmoud R. Khattab *

MilitaryTechnical College
Kobry El-Kobbah,
Cairo, Egypt

I.C.E.E.2022

11th International Conference
on
Chemical & Environmental
Engineering
29 - 31 March, 2022.

Session No. 2

Chemical and Radiation Detection in Environment (CRDA-2)

Room : **A8**
Date : Tuesday 29-03-2022
Time : **13:45 – 15:15**
Board: Prof. Dr. / Farag Mohamed Khalil NMA
Lt. Col. Dr. / Ahmed Mohamed Omar MTC

0004	**Natural Radionuclides Characterization on Baltim-El Burullus Coastal Sediments and Their Environmental Risk Assessments** Sallam, Asmaa M. * - Faheim, A. A. - El-Elshafiey, Z. A. - M. Amer, S. A. - El Feky, Mohamed G
0005	**Simulation of Ozone Concentrations Inside ^{60}CO Industrial Irradiator** El gamel, amal abd el razik *
0006	**Study Matrix Densities Variation on Assay of Cs-137 & Co-60 using Gamma Spectrometric Analysis and Correction Equation.** Omar, Omar * - Elmongy, Sayed Ali
0007	**Isotopic ratio of radionuclides in natural materials** Mubarak, Fawzia Mubarak Abd ElKader * - Ali, Abdallah

Journal of Physics: Conference Series **2305** (2022) 011001 doi:10.1088/1742-6596/2305/1/011001

MilitaryTechnical College **Kobry El-Kobbah,** **Cairo, Egypt**	**11th International Conference** **on** **Chemical & Environmental Engineering** **29 - 31 March, 2022.**

I.C.E.E.2022

Session No 3

Industrial and Environmental Engineering (IEE)

Room: **A8**
Date : Wednesday, 30/03/2022
Time : **08:30 – 10:00**
Board: Prof. Dr./ Gehan Ali Mohamed Ali NMA
Maj. Dr. / Moamen Galal Elsamra MTC

0008	**Process Simulation and Optimization of Dodecyl Benzene Sulfonic Acid Production Plant** Sorour, Faisal Hassan * - Abd E-lmonem, Nabil - Mohamed, Ahmed Fathy - El Sayed, Waleed Abd El Azim
0009	**Hybrid Hydro-pyrometallurgical Process for Zn Recovery from Electric Arc Furnace Dust** Azab, Rana Adel * - Abd E-lmonem, Nabil - Soliman, Moustafa Aly - Ismail, Ibrahim - Refaat, Ahmed Ahmed
0010	**Improve the Corrosion Resistance of Mild Steel in a Dichloroisocyanuric Acid Solution** Sheashea, Mohamed * - Gobara, Mohamed - zaghloul, basem

MilitaryTechnical College
Kobry El-Kobbah,
Cairo, Egypt

I.C.E.E.2022

11th International Conference on Chemical & Environmental Engineering
29 - 31 March, 2022.

Session No 4

Nanotechnology and Chemical Applications (NCA-I)

Room: **A8**
Date : Wednesday, 30/03/2022
Time : **10:15 – 11:45**
Board: Brig. Gen. Prof. Dr. / Mohamed Gobara MTC
Brig. Gen. Prof. Dr. / Sherif S. Elbassiouny MTC

0011	**Kinetic Decomposition Models for Ammonium Perchlorate Hybrid Catalyzed with Nano Thermite Particles** Hamed, Abdelaziz * - Gobara, Mohamed - Mokhtar, Mohamed - Elbasuney, Sherif
0012	**Utilization of Nano Thin Films for Solar Cell self-cleaning** Abu Elezz, Mariam * - abdelmonem, Nabil - Ismail, Ibrahim - Refay, Amr
0013	**Smart, Self-Cleaning Coating Based Ultra-Hydrophobic Copolymer Nanocomposites** Ibrahim, Saber Ahmed * - Labeeb, Ahmad Mohamed
0014	**The Excellent Catalytic Activity for Thermal Decomposition of Ammonium Perchlorate Using Copper Chromite Nanoparticles Synthesized by Hydrothermal Process** Ismael, shukri * - maraden, Ahmed - deif, Ahmed - Elbasuney, Sherif

MilitaryTechnical College
Kobry El-Kobbah,
Cairo, Egypt

I.C.E.E.2022

11th International Conference on Chemical & Environmental Engineering
29 - 31 March, 2022.

Session No. 5

Nanotechnology and Chemical Applications (NCA-II)

Room : **A8**
Date : Wednesday 30-03-2022
Time : 12:00 – 13:30
Board: Prof. Dr / Mohamed Amin El-shaheer BUE
Brig. Gen. Assoc.Prof. Dr. / Hisham Tantawy MTC

0015	**Removal of Co(II) from Water Using Graphene Oxide Functionalized with TOPO** Ahmed Khalil, waleed Fekry * - Rayan, Rafat Ahmed - Agha, Ahmed Rabaa
0016	**Gamma Radiation Synthesis and Characterization of Poly (Polyaniline /Glycidyl methacrylate) Containing TiO2 Nanoparticles and Studying Their Physicochemical Properties.** Salah Soliman Asser, Mohamed * - El-Arnaouty, M. B. - Eid, E. - Soliman, E Sayed - Hegazy, El-Sayed A.
0017	**Preparation and Application of Titanium Dioxide Nanoparticles in Offset Ink** Ali, Sayeda Sadek * - Abdel-moneam, Yasser Kamal - Hegazy, Mohammad
0018	**A Review on Recent Developments in Metal-Organic Frameworks (MOFs) Derived Carbon as a Microwave Absorption Material (MAMs)** Elsayed, Mohamed * - Tantawy, Hesham - Zalat, osama - Elmowafy, Moham

MilitaryTechnical College **Kobry El-Kobbah,** **Cairo, Egypt**	**11th International Conference** **on** **Chemical & Environmental Engineering** **29 - 31 March, 2022.**

I.C.E.E.2022

Session No 6

Energetic Materials and Environmental Science (EME-I)

Room: **A8**
Date : Thursday, 31/03/2022
Time : 13:45 – 15:15
Board: Prof. Dr / Mohamed Amin Elshaheer BUE
Col. Dr. / Ahmed Mohamed Maraden MTC

0019	**Sensitivity and Performance of Dinitroanisole (DNAN) in comparison with TNT** Hussein, Ahmed - Elbeih, Ahmed - Abdelhafiz, Mahmoud - Salah, hesham - shalaby, Ahmed - Azzazy, Ahmed - Saleh, Ahmed *
0020	**High Performance Sheet Explosive based on NTO** Elbeih, Ahmed - elnogomy, mohamed - Shenawy, Tamer - Saleh, Ahmed * - Azzazy, Ahmeed - Hussein, Ahmed - gaber, mohamed - Salah, hesham
0021	**Recovery of Valuable Heavy Minerals Via Gravity and Magnetic Separation Operations from Diit Quaternary Stream Sediments, Southern Coast of the Red Sea, Egypt** Fawzy, Mona * - Abu El Ghar, Mohamed - Gaafar, Ibrahim - El Shafey, Abdel Hay - Diab, Mohamed - Hussein, Ahmed
0022	**A Novel Approach for Synthesis of Low Sensitive FOX-7 with a high % yield** Abdelhafiz, Mahmoud * - Mostafa, Hosam E.

MilitaryTechnical College
Kobry El-Kobbah,
Cairo, Egypt

I.C.E.E.2022

11th International Conference
on
Chemical & Environmental Engineering
29 - 31 March, 2022.

Session No 7

Energetic Materials and Environmental Science (EME-II)

Room: **A8**
Date : Thursday, 31/03/2022
Time : **08:30-10:00**
Board: Brig. Gen. Prof. Dr. / Mohamed Gobara MTC
Lt. Col. Dr./ Basem Zaghloul MTC

0023	**Thermal Behaviour and Explosive Properties of TNT/RGO** Ahmed Talaat*[1], Ahmed Elbeih[1], Hesham Tantawy[1], Ahmed K. Hussein[1]
0024	**Aziridine type bonding agents for composite solid rocket propellant** Eissa, Mohamed F. *, Maraden Ahmed, Elhedery Tarek, Ali.A. Elall, Hosam E. Mostafa

MilitaryTechnical College
Kobry El-Kobbah,
Cairo, Egypt

11th International Conference on Chemical & Environmental Engineering
29 - 31 March, 2022.

I.C.E.E.2022

Session No 8

Pollution and Environmental Protection (PEP-I)

Room: **A8**
Date : Thursday, 31/03/2022
Time : **10:15-11:45**
Board: General (R) Assoc. Prof. Dr/ Mohamed Kassem MTC
Col.Dr./ Ibrahim Samir Saad Naeem MTC

0025	**Process Engineering Design of Tobacco wastes Incinerator with Utilization of Heat Energy from Combustion Gases** Abdelghany, Amr[*1], Elmansy, Asmaa[1], Abdelmonem, Nabil[1], Shaaban, Ahmed[2]
0026	**Removal Techniques of Ammonia from Drinking Water and Wastewater in Egypt and Developing Countries: a Review Study** Reham Samy[1], Nabil Abdelmonem[1], Ibrahim Ismail[1], and Amr Abdelghany[1]
0027	**Improving the Performance of Biological Tanks in Wastewater Treatment Plants by Modifying the Conventional Ceramic Aerator** Ahmed Khalil[1], Chenskij Ilia[2], Serpokrylov Nikolaj[2], Swellam W. Sharshir[3]

Journal of Physics: Conference Series **2305** (2022) 011001 doi:10.1088/1742-6596/2305/1/011001

MilitaryTechnical College
Kobry El-Kobbah,
Cairo, Egypt

I.C.E.E.2022

11th International Conference
on
Chemical & Environmental
Engineering
29 - 31 March, 2022.

Session No 9

Pollution and Environmental Protection (PEP-II)

Room: **A8**
Date : Thursday, 31/03/2022
Time : **12:00-13:30**
Board: General (R) Assoc. Prof. Dr/ Mohamed Kassem MTC
Col.Dr./ Ibrahim Samir Saad Naeem MTC

0028	**Experimental and Process Design of a Treatment System to Remove Aluminium Ions from lean Sludge in an Egyptian Water Treatment Plant** Abdelghany, Amr[*1], El Bouraie, Mohamed[2], Abdelmonem, Nabil[1], hanafy, Mohamed[1], Samy, Reham[1]
0029	**Compendium Data and Preliminary Results Based on Alum Sludge and Gravel Continuous Fixed-Packed-Bed Column Substrate as Sustainable Practice for Constructed Wetland Development in Northern Egypt** Maha A. Tony
0030	**Using Biochar to Control Nitric Oxide Air Pollution** Ghada O Mohamed[1], Maher E Saleh[2], Elsayed A Shalaby[1] and Ahmed S Elsafty[1]
0031	**Radiological and Mineralogical Studies on the Coastal Area Between Port-Said and Damietta** A.M. Osman[1], R.A. Seif[1], R.M. Attia[2], A. EL-Azab[2]. &, I.H. Khalifa[1]

MilitaryTechnical College
Kobry El-Kobbah,
Cairo, Egypt

I.C.E.E.2022

11th International Conference
on
Chemical & Environmental
Engineering
29 - 31 March, 2022.

Session No 10

Pollution and Environmental Protection (PEP-III)

Room: **A8**
Date : Thursday, 31/03/2022
Time : **13:45-15:15**
Board: Brig. Gen. Prof. Dr. / Mohamed Aboelfottoh MTC
Lt. Col. Dr. Osama abo Zalat MTC

0032	**Air Pollution Remediation Using Synergetic Coupling of AC/TiO_2 Filter for Benzene Gas Removal: Regeneration and Reuse of Spent Activated Carbon After the Treatment Process** E. F. Mohamed
0033	**Influence of Chemical Treatment on the Spectral and Thermal Behavior of Activated Carbon Obtained from Olive Oil Solid Waste** Wael A A elhelece*
0034	**Radioactivity and Environmental Impacts of Granites from Um Ara, Southeastern Desert, Egypt** El-Afandy A H[1], Mohamed G El-Feky[1], Samia H Taha[*2], S M El Minyawi[2], H A Sallam[2], A F Tawfic[3], A Omar[3], M G El-Samrah[3]

Journal of Physics: Conference Series **2305** (2022) 011001 doi:10.1088/1742-6596/2305/1/011001

MilitaryTechnical College
Kobry El-Kobbah,
Cairo, Egypt

I.C.E.E.2022

11th International Conference
on
Chemical & Environmental
Engineering
29 - 31 March, 2022.

Session No 11

Renewable Energy and Sustainability (RES)

Room: **A9**
Date : Thursday, 31/03/2022
Time : **08:30-10:00**
Board: General (R) Assoc. Prof. Dr/ Ahmed Baraka MTC
Col. Dr./Mohamed Mokhtar Kotb MTC

0035	**Comparative Studies on Hydrometallurgy and Pyrometullargy to the Extraction of Beryllium from Decrepitated Beryl from Wadi El Gemal Region, Eastern Desert, Egypt** Mohamed Shabaan Atrees[1], Hamed I. E. Mira[1], Yasser M Kawassik[1] and Mohamed E. M. Ali[2]
0036	**Waste Cooking Oil Management in Egypt Production of Biodiesel-Development of Rapid Test Method** Marwa Mohamed[1], Nourhan Sherif[*1], Omar Aboelazayem[2], Hany A. Elazab[1,3], Mamdouh Gadalla[1,4], Basudeb Saha[5]
0037	**Multivariate-Parameters Effect of Dewatered Aluminium-Based Sludge in Catalytic Oxidation into Wtreatment with a Sstrategy for Solar Energy Application** Rahma H. Thabet[1], Mai K. Fouad[2], Shakinaz A. El Sherbiny[2], Maha A. Tony[1*]

MilitaryTechnical College
Kobry El-Kobbah,
Cairo, Egypt

11th International Conference on Chemical & Environmental Engineering
29 - 31 March, 2022.

Session No 12

Environmental and Chemical Reaction Engineering (ECRE-I)

Room: **A9**
Date : Thursday, 31/03/2022
Time : **10:15-11:45**
Board: General (R) Assoc. Prof. Dr/ Ahmed Baraka MTC
Col. Dr./Mohamed Mokhtar Kotb MTC

0038	**Study of Elastomeric Heat Shielding Materials for Solid Rocket Motor Insulation** M Mosa[1], M Mokhtar Kotb[1], H Fouda[2] and Mohamed Gobara[1]
0039	**Implementing Nano composite Shear Thickening Fluid to Improve the Impact Resistance of Polyamide Fabrics Against Intermediate Velocity Projectiles** Remon A Mankarious[1*] and Mostafa A Radwan[2] and M H Roushdy[3]
0040	**Spectrophotometric and Biological Studies for the Interaction of Rifaximin with Co, Cu, Fe and Vitamin C** A S El-Kholany[*1], H A Hafith[1] and N M Gaber[1]

Journal of Physics: Conference Series **2305** (2022) 011001 doi:10.1088/1742-6596/2305/1/011001

MilitaryTechnical College
Kobry El-Kobbah,
Cairo, Egypt

I.C.E.E.2022

11th International Conference on Chemical & Environmental Engineering
29 - 31 March, 2022.

Session No 13

Environmental and Chemical Reaction Engineering (ECRE-II)

Room: **A9**
Date : Thursday, 31/03/2022
Time : **12:00-13:30**
Board: General (R) Assoc. Prof. Dr/ Said Hassanin MTC
Brig. Gen. Prof. Dr. / Ahmed Elbeih MTC

0041	**Radionuclide Content and Radiation Dose rates of Nraniferous and Mineralized Granites at GV Occurrence in Gabal Gattar, Northeastern Desert, Egypt.** F M Abdelwahab[1], N H El Gendy[2], M Kh Barakat[2], M S Mohamed[1], A M El-Tohamy[1], R A Rl Terb[1] and M G El Feky[1]
0042	
0043	**Economic and Environmental Analysis of Cumene Production in Egypt: Process simulation and Improvement** Esraa Khaled[1], Mamdouh Gadalla[1], Mohamed Amin Sadek[1], Fatma Ashour[2], Dina Ahmed[1]
0044	**Implementation of Mass Integration Design Concept for Material Preservation in Acrylonitrile Industry** Nessren M Farrag[1] and Ayat O Ghallab[2]

Journal of Physics: Conference Series **2305** (2022) 011002 doi:10.1088/1742-6596/2305/1/011002

Peer Review Statement

All papers published in this volume have been reviewed through processes administered by the Editors. Reviews were conducted by expert referees to the professional and scientific standards expected of a proceedings journal published by IOP Publishing.

- **Type of peer review:** Double Anonymous
- **Conference submission management system:** https://icee.conferences.ekb.eg/
- **Number of submissions received:** 84
- **Number of submissions sent for review:** 84
- **Number of submissions accepted:** 44
- **Acceptance Rate (Submissions Accepted / Submissions Received × 100):** 52.4
- **Average number of reviews per paper:** 2
- **Total number of reviewers involved:** 48
- **Contact person for queries:**
  Name: Prof. Hosam Mostafa
  Email: h.mostafa@mtc.edu.eg
  Affiliation: Conference Rapporteur

Geology and Environmental Impact Assessment of Psammitic Gneiss and Lamprophyre Dykes at Wadi Sikait, South Eastern Desert, Egypt

M. S. Kamar, A. M. El Tohamy*, H. I. Mira and A. M. Ismail

Nuclear Materials Authority, P.O. Box 530 El Maadi, Cairo, Egypt

*amira_eltohamy2011@yahoo.com

Abstract. The present study was carried out on promising lamprophyre dykes at Wadi Sikait, through the mineralogical and petrography studies with emphasis on the radiological hazard's indices along Wadi Sikait. So, the radioelement contents in the studied rocks were measured radiometrically by using gamma-spectrometry (NaI-detector). The psammitic gneisses formed an elongated zone striking NNE-SSW and are silicified, banded, jointed, sheared, mylonitized, and hematized. The lamprophyre dykes strike NNW-SSE varies in thickness from 0.5 m to 1.5 m, and extend more than one kilometer. The averages of the eU/eTh ratio are 2.02, indicating uranium accumulation, which confirmed positive disequilibrium (P-factor). Many of the radiological hazard parameters in both psammitic gneisses and lamprophyre dykes are higher than the international limits, suggesting probable hazard. ^{40}K plays the main and most important role in dose rate contribution. The high activity and the elevated radiological hazard parameters in the lamprophyre are related to the presence of radioactive and radioelements bearing minerals like kasolite, thorite, uranothorite, xenotime, cheralite and baddeleyite. Other mineralization was also recorded as gold and base metal minerals.

Keywords: Egypt; Eastern Desert; Wadi Sikait; Lamprophyre dyke; Psammitic gneiss; Radiological hazards indices

1. Introduction

[1] developed the phrase Pan-African based on a review of available Rb-Sr and K-Ar ages in Africa, which was later expanded to include the time range between 900 and 450 Ma [2]. It is presently used to define Neoproterozoic to Paleozoic tectonic, magmatic, and metamorphic activity, particularly for crust that was once part of Gondwana.

Egypt's Eastern Desert was divided into three provinces based on differences in exposed lithologies and notable differences in physiographic features: Northern Eastern Desert (NED), Central Eastern Desert (CED), and Southern Eastern Desert (SED) (SED)(figure 1) [3] and [4]. The provinces are separated by two important structural elements: the Qena-Safaga Shear Zone separates the NED from the CED, and the Idfu-Marsa Alam Shear Zone separates the CED from the SED [5]. The border dividing the CED and the SED, according to [6], is the northerly dipping Sha'it–Nugrus shear zone.

Nugrus – Sikait area is located on the Arabian Nubian Shield, bordering the Nugrus thrust fault [7], or Nugrus strike-slip fault [8] and the Sha'it-Nugrus shear zone, and encompasses around 320 km^2. This shear zone separates the Hafafit complex (high-temperature metamorphic rocks) in the southwest from the Ghadir group (low-grade ophiolitic and arc volcanic assemblages) in the northeast [6]. The Nugrus thrust travels in a northwesterly direction along the upper section of W. Sikait until it reaches the southern edge of G. Ras Sha'it, then swings southwest along

W. Sha'it west of G. Migif. The footwall of the Nugrus thrust is represented by the Migif-Hafafit gneisses, while the hanging wall is represented by the Ghadir group [9].
In Egypt's Eastern Desert, the W. Sikait area is about 95 kilometres southwest of Marsa Alam City (ED). This is one of the most promising places in the Eastern Desert. In the middle of the wadi, archaeologists discovered Sikait temple, which is now a tourist destination. In portions of W. Sikait and Gabal (G.) Sikait, there are numerous historic beryl and emerald workings. In the middle of W. Sikait, there are also gneiss rocks that include radioactive minerals including thorite and uranothorite, as well as base metals. Metamorphosed sandstone existed in two portions in the northern part of the wadi, each with a significant percentage of uranium and molybdenum. W. Abu Rusheid is located west of W. Sikait, and the Sikait- Abu Rusheid pluton is elongated in the NW-SE (12 km) and thinning in the NE-SW directions (3 km in width). Many authors had looked at the W. Sikait area [e.g., 10,11,12, 13].

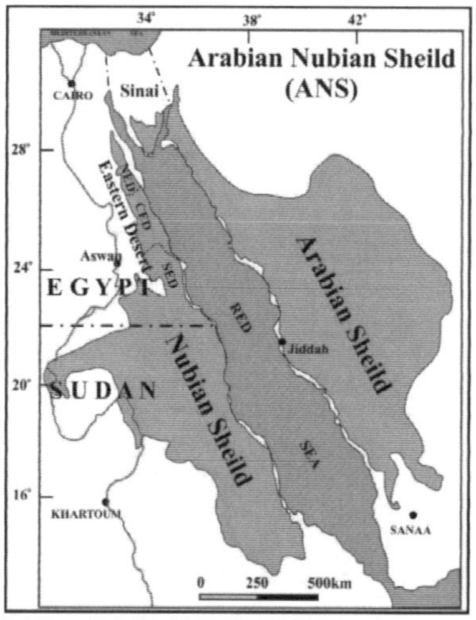

Figure 1. The Arabian Nubian Shield (ANS), [14] North Eastern Desert (NED), Central Eastern Desert (CED), and Southern Eastern Desert (SED).

Ionizing radiation of natural origin, such as terrestrial and interplanetary radiation, has always been a part of human life **[15, 16]**. Extraterrestrial radiation is emitted by high-energy cosmic ray particles and is about 30 $nGyh^{-1}$ at sea level, whereas terrestrial radiation is emitted by naturally occurring radionuclides, primarily potassium (^{40}K), rubidium, and radionuclides in the decay chains of thorium (^{232}Th) and uranium (^{238}U) **[17]**. The goal of this study is to conduct a survey and measure natural terrestrial radionuclides such as ^{40}K, ^{238}U, ^{226}Ra, and ^{232}Th, as well as the mineralogical contents of W. Sikait lamprophyre dykes, in order to better understand the relationship between natural radioactivity and radioactive mineral content. In addition, the geologic setting, spectrometric prospecting, and risk assessment for psammitic gneisses and lamprophyre dykes on humans and the environment will be discussed.

2. Materials and Methods

W. Sikait lamprophyre dykes yielded a total of 8 representative samples. Crushed, powdered, quartered, and sieved each sample were used in this study. The heavy fraction of the size fraction 0.125–0.250 mm was separated using bromoform and subjected to magnetic fractionation utilising the Frantz Isodynamic Magnetic Separator (Model LB 1). The binocular microscope was used to select the desired minerals in order to obtain mono-mineralic fractions for identification and analysis. An environmental scanning electron microscope (ESEM model Philips XL 30) with an energy dispersive spectrometer (EDX) instrument was used to identify minerals.

In addition, radiometric analysis was performed on ten (10) and eight (8) samples of psammitic gneiss and lamprophyre dykes, respectively, after drying at 80°C to remove moisture, which would otherwise result in radioactive self-absorption during radioassay. The dried samples were then packed in labelled cylindrical plastic containers with a base diameter of 9.5 cm and a height of 3 cm, weighing 300 g by mass. For the achievement of secular equilibrium between ^{238}U and ^{232}Th and their progenies, the plastic containers were carefully covered, sealed, and left for 30 days before counting [18].

A Bicron scintillation detector NaI (Tl) 76x76 mm, hermetically sealed with the photomultiplier tube in aluminum housing, was employed in the analysis of the four radioactive radionuclides. A copper cylinder with a thickness of 0.6 cm protects the tube from induced X-rays, while a chamber of lead bricks protects it from external radiation.

^{234}Th, ^{212}Pb, ^{214}Pb, and ^{40}K for uranium, thorium, radium, and potassium are measured utilizing four energy areas at 93 kV, 239 kV, 352 kV, and 1460 kV for uranium, thorium, radium, and potassium, respectively.

2.1. Calculations of radiological hazards indices

Using the conversion factors provided by the Polish Central Laboratory for Radiological Protection, the measurement values of U, Th, and Ra in parts per million (ppm), as well as K, measured as a percent value, were converted to activity concentrations in Becquerel per kilogram ($Bq.kg^{-1}$) [19, 20].

The radiological indices are: radium equivalent activity (Ra_{eq}), absorbed gamma dose rate (Da), annual effective dose (HE), internal hazard index (H_{in}), external hazard index (H_{ex}), and gamma-ray index (I_{γ}).

The radium equivalent activity is a weighted sum of activities of the ^{238}U, ^{232}Th, and ^{40}K radionuclides based on the assumption that $1 Bqkg^{-1}$ of ^{226}Ra, 0.7 $Bq.kg^{-1}$ of ^{232}Th, and 13 $Bq.kg^{-1}$ of ^{40}K produces the same gamma-ray dose rate [21]. The radium equivalent activity (Ra_{eq}) was calculated by the equation described by [22, 23] as indicated by the following equation 1;

$$Ra_{eq} = (10)/130\ C_k + 10/7\ C_{Th} + C_U \qquad (1)$$

Where C_U, C_{Th}, and C_K were the activity concentrations in Bqkg-1 of ^{238}U, ^{232}Th, and ^{40}K respectively.

Absorbed gamma dose rate (Da) is an index that calculates the absorbed dose rate in the air for external gamma radiation at about 1.0m above the ground from the natural radionuclides Da($nGyhr^{-1}$) given by [17].

$$Da\ (nGyhr^{-1}) = 0.0417C_k + 0.604C_{Th} + 0.462C_U \qquad (2)$$

The absorbed dose rate in the air was used to estimate the annual outdoor effective dose equivalent for individuals and determine the biological impact of the radiation exposure. The effective dose conversion factor was taken to be 0.7 Sv. Gyr^{-1} and an outdoor occupancy factor of 0.2 with the annual occupancy time approximately 8760 $hr.y^{-1}$. Hence the annual effective dose equivalent was estimated using the equation 5:

$$H_E\ (mSvyr^{-1}) = D_a\ (nGyhr^{-1}) \times 8760\ (hryr^{-1}) \times 0.2 \times 0.7 (SvGy^{-1}) \times 10^{-6} \qquad (3)$$

As described by [24], the external hazard index (Hex) was commonly employed to quantify radiation dose rate owing to external exposure to gamma radiation from natural radionuclides in soil samples. The calculated average external hazard index was found to be less than unity for the radiation hazard to be insignificant.

$$H_{ex} = \frac{C_k}{4810} + \frac{C_{Th}}{259} + \frac{C_{Ra}}{370} \leq 1 \qquad (4)$$

The internal hazard index governs the internal exposure to ^{222}Rn and its radioactive offspring (Hin). The internal hazard index (Hin) is a parameter used to estimate whether radioactive materials are harmful to the lungs or not. The risk of internal exposure to the natural radionuclides ^{40}K, ^{226}Ra, and ^{232}Th can be calculated using equation 3 [25], which is as follows:

$$H_{in} = \frac{C_k}{4810} + \frac{C_{Th}}{259} + \frac{C_{Ra}}{185} \qquad (5)$$

The Gamma-ray index (I) suggested by the European Commission is substantially connected to the yearly effective dose [26]. A common room model with dimensions of 4 m x 5 m x 2.8 m and 20 cm thick walls is used to compute the gamma-ray index. The formula is used to compute the index factor associated with external exposure.:

$$I\gamma = \frac{C_{Ra}}{300} + \frac{C_{Th}}{200} + \frac{C_K}{3000} \quad (6)$$

2.2. Geologic Setting

The study area lies between latitudes 24°38′ 36″ to 24° 39′ 21″N and longitudes 34° 47′ 03″ to 34° 48′ 06″E (figure 2). The tectonic sequence begins from the oldest as follows; psammitic gneisses, ophiolitic mélange, monzogranites, and lamprophyre dyke. At the contact with the ophiolitic mélange on the eastern side (figure 3a) and the monzogranites on the other side, the psammitic gneiss formed an elongated zone striking NNE-SSW. Jointed, sheared, mylonitized, and hematized rocks. With a NNW-W dipping of 50°-69°, the rocks show silicification and banding (figure 3b).

The ophiolitic mélange is characterized by tectonically mixed fragments and blocks of ultramafic rocks within fine to coarse-grained matrix and is abundant in the eastern portion of the studied area (figure 3c). On the western side of W. Sikait, there are acute contacts with the mineralized zone of the psammitic gneiss. It is mostly made of a mélange matrix (primarily schists) that encases numerous fragments of meta-peridotites, meta-pyroxenites, and metagabbros of various sizes and dimensions [27].

The mafic-ultramafic fragments are mostly serpentinized, and talc-carbonate rocks have formed in a number of places (creamy in colors). These fragments come in a variety of sizes, ranging from large slabs to massive mountains. These matrixes are characterized by strongly foliated layering and the presence of folds on a regular basis. The matrices are expressed in a wide range of ways (talc schist, quartzo-feldspathic schist, tourmaline-garnetiferous-biotite schist, graphite schist, and sillimanite schist).

Lamprophyre dykes are the most common dykes in the research region. The ophiolitic mélange, psammitic gneiss, and monzogranites are intruded by these dykes, which have a nearly vertical dip and strike NNW-SSE. It ranges in thickness from 0.5 to 1.5 m and is more than one km long, with right-lateral strike-slip faults cutting across it (figure 3d).

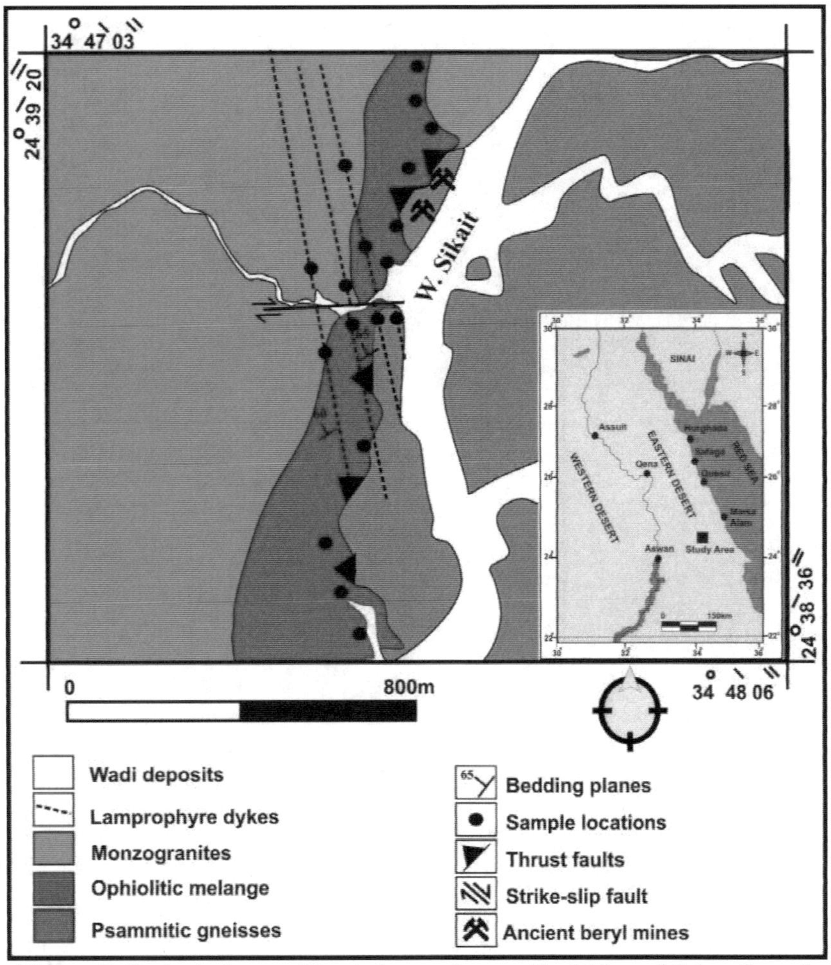

Figure. 2. Geological map of the study area (mapped by Mohamed Salem Kamar).

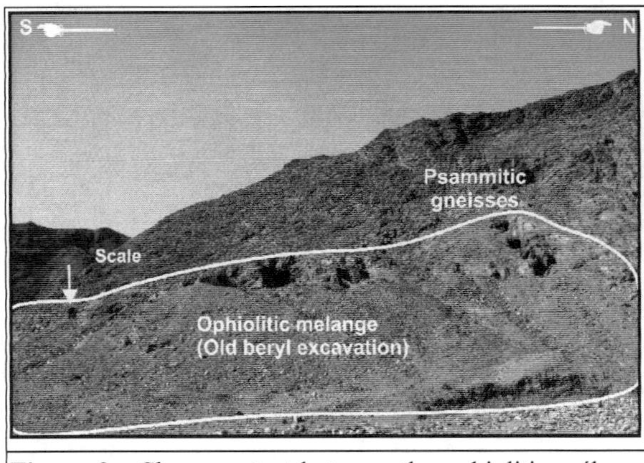

Figure 3a. Sharp contact between the ophiolitic mélange and the psammitic gneisses.

Figure 3b. Field photograph showing silicification and banding in the psammitic gneisses with dipping 65° toward NW direction.

Figure 3c. Ophiolitic mélange occupies the eastern part of the study area.

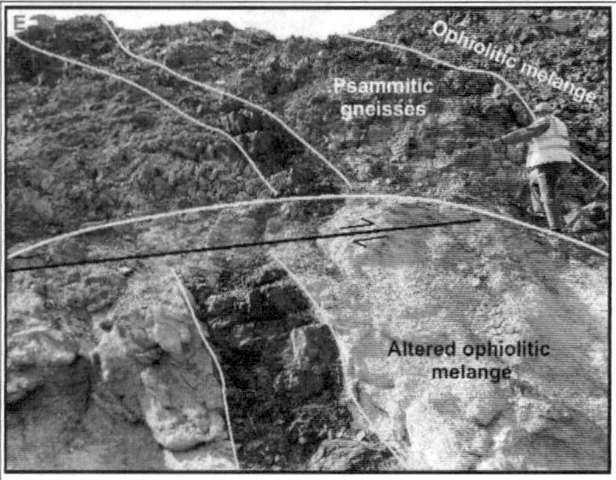

Figure 3d. Field photograph showing lamprophyre dykes cut both ophiolitic mélange and psammitic gneisses, then cut by right strike-slip fault.

3. Result and Discussion

3.1. Petrographic studies

Lamprophyres are phenocryst-bearing rocks that contain biotite, amphibole (with bright cleavage surfaces), and pyroxene but not feldspar. They differ from porphyries and porphyrites in which the feldspar crystallized in two generations. They're basically dyke rocks that can be found as dykes, narrow sills, and as peripheral facies of plutonic intrusions. W. Sikait Lamprophyre (SL) is primarily a melanocratic rock with an equigranular, holocrystalline panidiomorphic texture in hand specimens. W. Sikait lamprophyre has phenocrysts of brown amphibole, mica, plagioclase, and carbonate mineral (calcite) embedded in fine-grained groundmass, as seen under the microscope. Brown amphibole minerals are euhedral–subhedral (hornblende) and account for up to 40% of phenocrysts, with elongate and prismatic crystals up to 5 mm in length (figure 4a). Plagioclase (2-5%) is found in the form of euhedral crystals with simple twinning (figure 4b). Accessories include opacity, apatite, carbonate, and quartz (figure 4c, 4d).

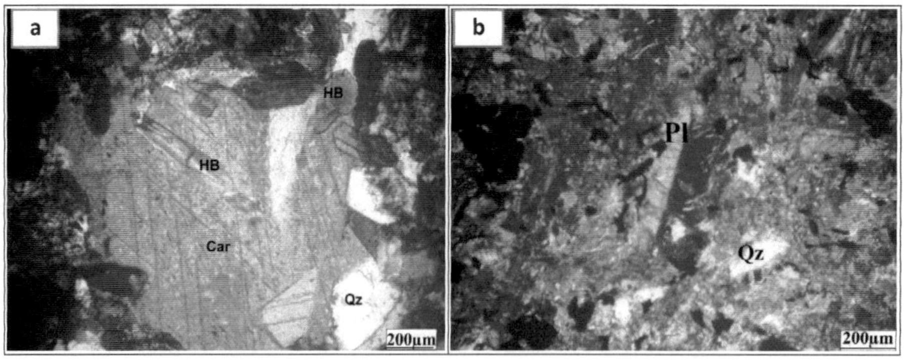

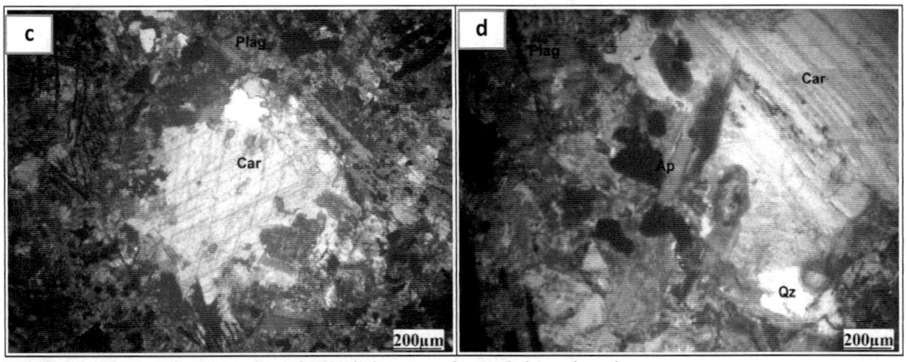

Figure 4. Photomicrographs of Sikait lamprophyre dykes showing:
a) Anhedral crystal of carbonate associated with hornblende (HB).
b) Euhedral crystal of plagioclase showing simple twinning
c) Anhedral crystal of carbonate associated with quartz (Qz) and surrounded by plagioclase
d) Elongated crystal of apatite.

3.2. Mineralogical Investigation

The observed mineralogical data revealed that W. Sikait lamprophyre dykes include a diverse mineral assemblage, including radioactive minerals, rare earth minerals, and base metals containing minerals, as well as a variety of accessory species **(Table 1)**.

3.3. Radioactive minerals

Kasolite ($Pb(UO_2)$ (SiO_4). H_2O)

Pitchblende oxidation in the zone of supergene alteration produced the hexavalent minerals, primarily kasolite and autunite. Kasolite is the only uranyl silicate that has a significant amount of lead. It is a secondary mineral that forms when meteoric water containing silica reacts with already created secondary uranium minerals. According to [28], uranium released during pitchblende alteration in the oxidation zone produces several generations of secondary uranium minerals, the first of which are lead-rich, such as kasolite and masuyite, and the latter of which are lead-deficient. The presence of kasolite was confirmed using an ESEM analysis. Kasolite is found as scattered crystals (100 m) primarily consisting of U (41 wt %) and Pb (36 wt %), with minor quantities of Fe, K, and Al (figure 5a and Table 1).

Thorite ($ThSiO_4$) was recorded in lamprophyre samples having a composition of Th (52 wt %), U (8 wt %), and Si (14 wt %) together with traces of Ca and Fe (figure 5b). Also, **Uranothorite (Th, U) SiO_4** was recorded and the EDX data of investigated mineral shows that Th-content is (64 wt %) and U (15.5 wt %) as the main constituents (figure 5c).

3.4. Rare earth-bearing minerals (ABO_4 Minerals with Monazite Structure)

Monazite is a REEs phosphate mineral that contains around 70% REE oxides and is structurally similar to huttonite ($ThSiO_4$), cheralite ($CaTh(PO_4)2$), gasparite, and roosevelite, with varying degrees of solid solution among the endmembers **[29]**. Monazite contains 1-28 wt. percent ThO_2 and lesser amounts of UO2 (0-6.5 wt. %) **[30, 31]**. Huttonite preferentially incorporates the high Th content in monazite, which is a severe impediment to the development of REE deposits. $REE^{3+} + P^{5+} \leftrightarrow (Th, U)^{4+} + Si^{4+}$ or by cheralite substitution

$$2REE^{3+} \leftrightarrow (Th, U)^{4+} + Ca^{2+}.$$

Cheralite is a phosphate mineral that is anhydrous and has the chemical formula $CaTh(PO_4)_2$ **[32].** Huttonite and monazite are isomorphous with it. It can be thought of as the result of the system's complete cationic substitution

$$2LREE^{3+} \leftrightarrow Ca^{2+} + Th^{4+}.$$

Monazite, like huttonite and cheralite, is isostructural, resulting in varied solid solutions among the endmembers **[33, 31, 34, 29]**. The BSE imaging and EDX analysis of cheralite grains (figure 5d and 5e) revealed that cheralite grains are primarily made up of P (9-6 wt%), Th+U (50-40 wt %), Ca (5.5-4 wt %), and LREEs (15-9 wt %). LREEs (71 wt %) are found in the rare earth concentration, along with traces of Y and Ca **(**figure 5f**)**.

3.5. Precious and base metals

Gold (Au) was chosen as the precious metal because it occurs in fine crystals (10 m) with Au (44 wt %), Cu (5 wt %), and minor levels of Fe and Si, as determined by EDX **(**figure 5g**)**. **[13]** used fire assay to establish the occurrence of gold in the study region, with gold concentrations ranging from 0.48 g t^{-1} to 0.52 g t^{-1}. Pyrite (FeS_2) exists as the primary base metal in lamprophyre samples having a composition of S (52 wt %), Fe (46 wt %) **(**figure 5h**)**.
Many nonferrous metals that aren't precious or noble metals make up base metal alloys. Copper, nickel, cobalt, tin, aluminium, and zinc are the most common basic metals. The base metal concentrations have a varied composition (figure 5i and 5j).

3.6. Accessory minerals

Xenotime (YPO_4) is Yttrium phosphate mineral. Due to uranium and thorium impurities, some xenotime specimens may be mildly to extremely radioactive. The EDX results demonstrate that xenotime has a chemical makeup that includes a considerable number of HREEs (30 wt %) **(**figure 5k).
Apatite (Ca PO_4) The most prevalent phosphate mineral, apatite, is the primary source of phosphorus. The colour intensity increases as the Mn content of the apatite increases **[35]**. Although apatite is not a rare earth mineral, REEs are used to replace Ca ions in the crystal lattice **[36]**. HREEs (30 wt %) are obtained from the EDX analyses **(**figure 5l).
Baddeleyite (ZrO_2) is a rare zirconium oxide mineral that come in several of monoclinic prismatic crystal shapes. Its hue ranges from colorless to yellow, green, and dark brown, and it is transparent to translucent with high refraction indices. Baddeleyite is a refractory mineral with a melting point of 2700°C. Hafnium is a replacement contaminant that can be present in concentrations ranging from 0.1 to a few percent. Baddeleyite is infrequently observed with zircon because it forms in silica-undersaturated rocks, such as mafic rocks ($ZrSiO_4$). When silica is free in the solution (silica-saturated/oversaturated) **[37]**, zircon, not baddeleyite, is the dominating phase. Baddeleyite, which contains Zr, was discovered using ESEM at the study site (80 wt. %) (figure 5m).

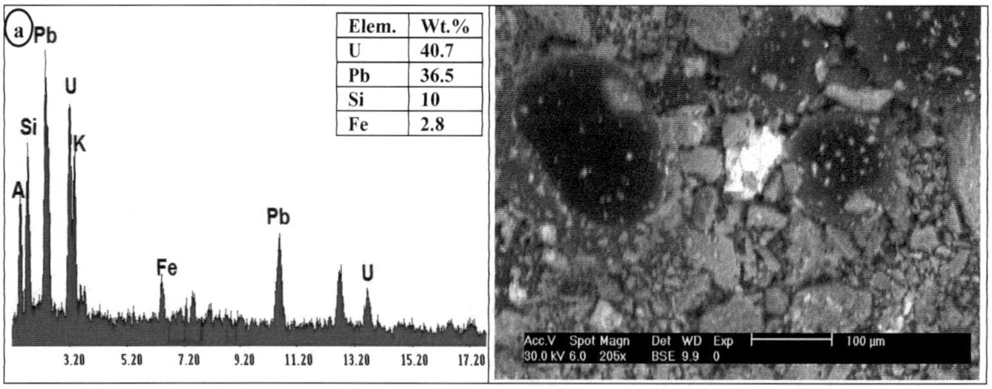

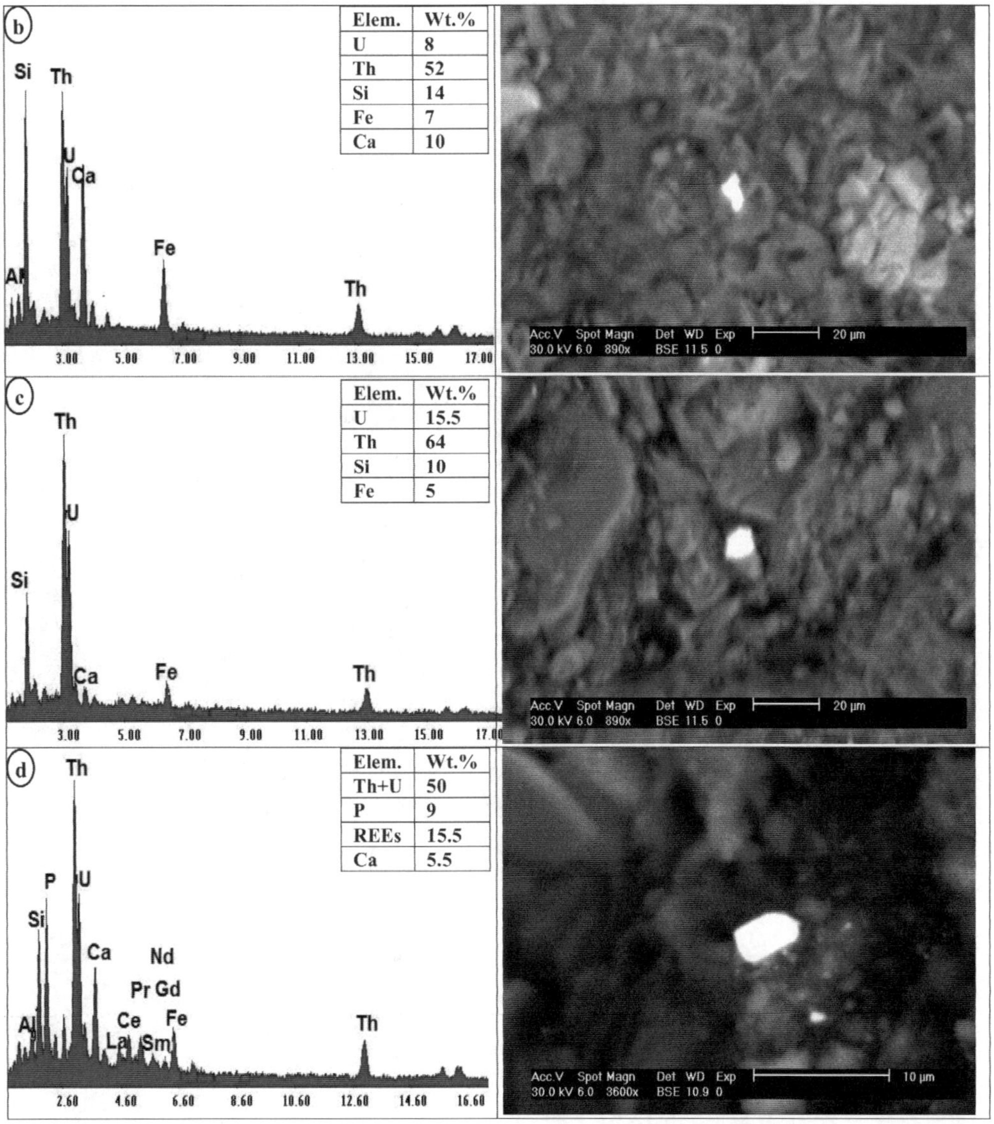

Elem.	Wt.%
U	8
Th	52
Si	14
Fe	7
Ca	10

Elem.	Wt.%
U	15.5
Th	64
Si	10
Fe	5

Elem.	Wt.%
Th+U	50
P	9
REEs	15.5
Ca	5.5

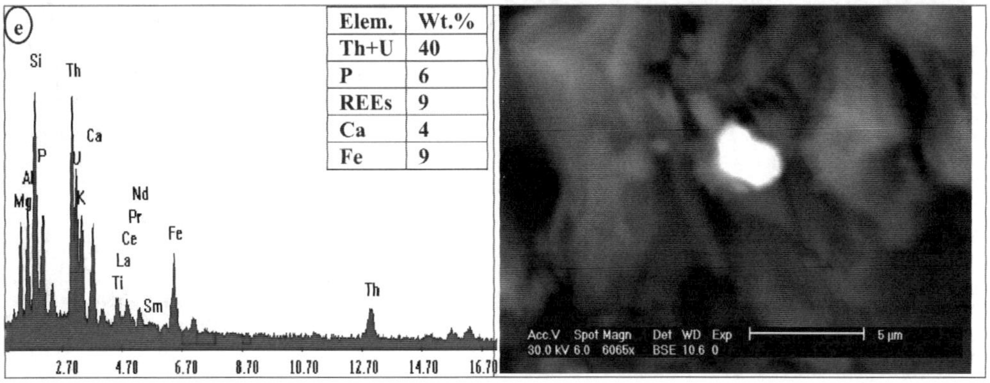

Elem.	Wt.%
Th+U	40
P	6
REEs	9
Ca	4
Fe	9

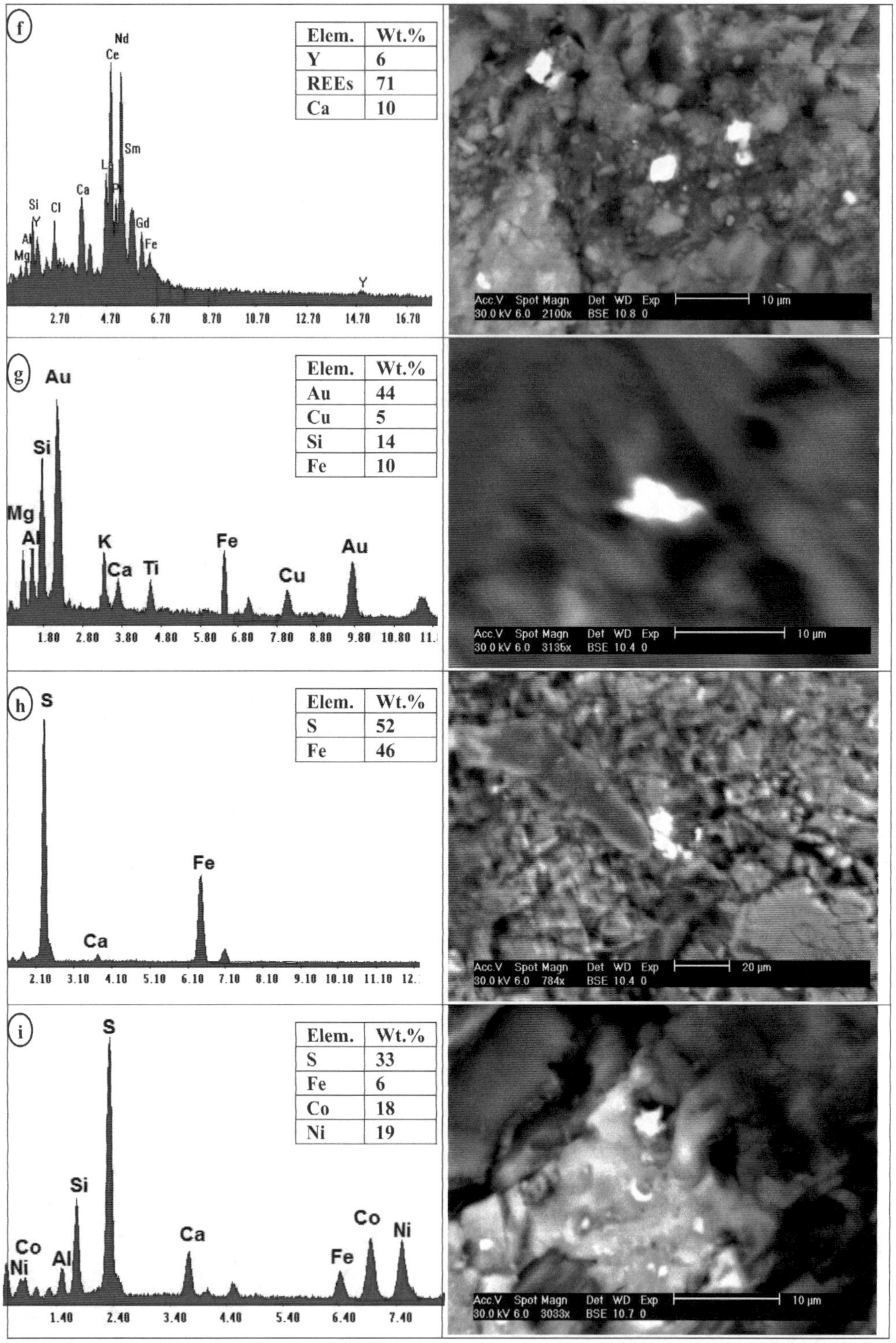
f
Nd
Ce
Sm
La
Pr
Ca
Si
Cl
Y
Al
Mg
Gd
Fe
Y
2.70 4.70 6.70 8.70 10.70 12.70 14.70 16.70
Elem. Wt.%
Y 6
REEs 71
Ca 10
Acc.V Spot Magn Det WD Exp 10 µm
30.0 kV 6.0 2100x BSE 10.8 0
g
Au
Si
Mg
Al
K
Ca
Ti
Fe
Cu
Au
1.80 2.80 3.80 4.80 5.80 6.80 7.80 8.80 9.80 10.80
Elem. Wt.%
Au 44
Cu 5
Si 14
Fe 10
Acc.V Spot Magn Det WD Exp 10 µm
30.0 kV 6.0 3135x BSE 10.4 0
h
S
Fe
Ca
2.10 3.10 4.10 5.10 6.10 7.10 8.10 9.10 10.10 11.10
Elem. Wt.%
S 52
Fe 46
Acc.V Spot Magn Det WD Exp 20 µm
30.0 kV 6.0 784x BSE 10.4 0
i
S
Si
Co
Ni
Al
Ca
Co
Ni
Fe
1.40 2.40 3.40 4.40 5.40 6.40 7.40
Elem. Wt.%
S 33
Fe 6
Co 18
Ni 19
Acc.V Spot Magn Det WD Exp 10 µm
30.0 kV 6.0 3033x BSE 10.7 0

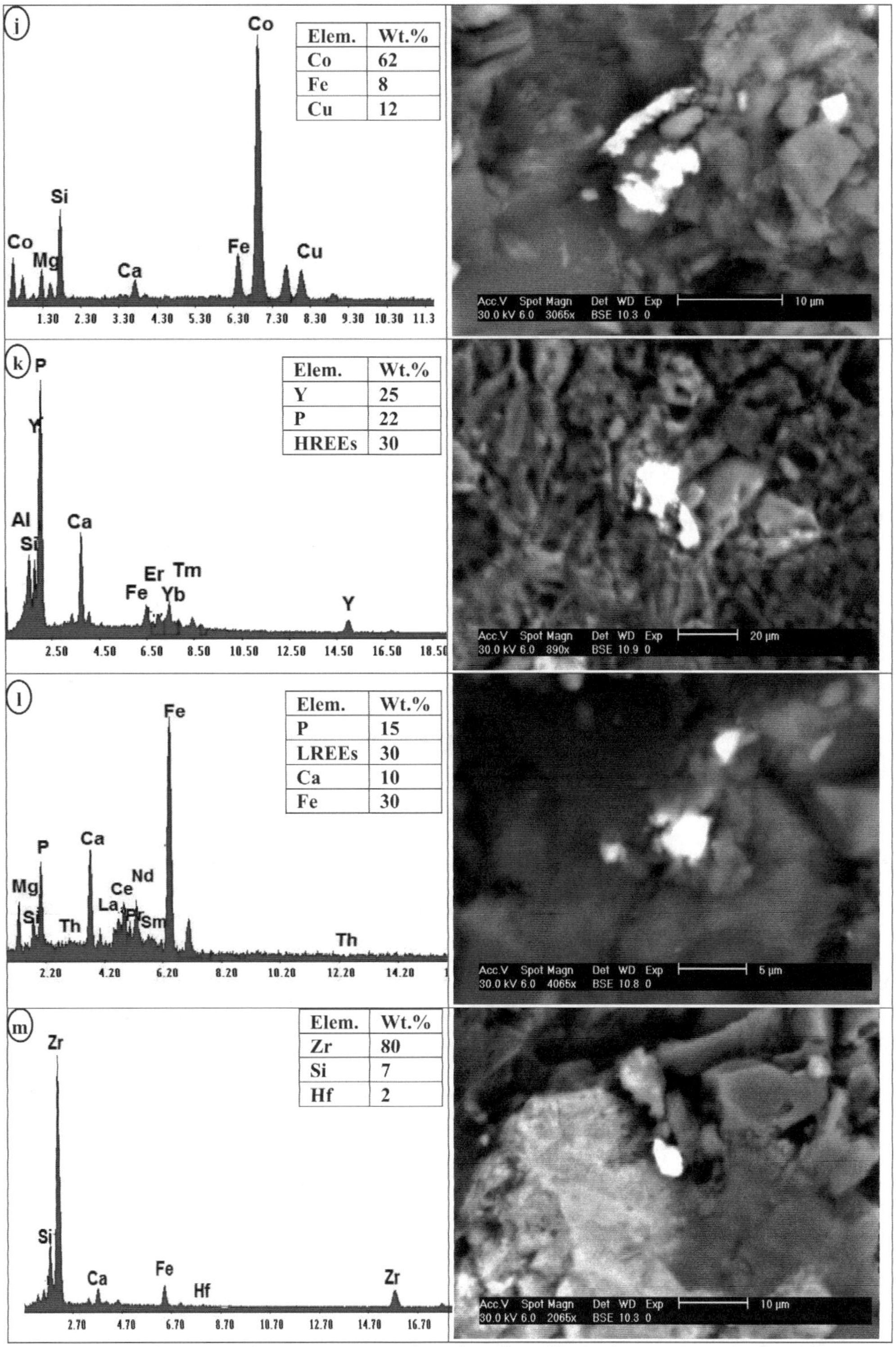

Figure 5. BSE images and EDX patterns of (a) kasolite, (b) thorite, (c) uranothorite, (d) cheralite, (e) cheralite, (f) rare earth concentration, (g) gold, (h) pyrite, (i) Co-Ni –Fe alloy, (j)

Co- Cu- Fe alloy, (k) xenotime, (l) apatite, (m) baddeleyite minerals recorded in the Sikait lamprophyre dykes.

Table 1. The recorded minerals in the Sikait lamprophyre dykes.

Area	Uranium minerals	Thorium minerals	Nb-Ta minerals	REEs-minerals	Precious and base metals Primary	Precious and base metals Secondary	Accessory minerals
SL	Kasolite	Uranothorite Thorite	--	Cheralite	Gold Pyrite	Base metal concentrations	Xenotime Apatite Baddeleyite
W. Sikait**	Soddyite Zippeite Ni-Zippeite Kasolite Uranophane Autunite	Thorite Uranothorite	Plumbobetafite Ishikawaite Columbite Yttrocolumbite	Allanite Monazite	Gold Pyrite Native Ni	Chromite Chrysocolla Atacamite Willemite Mimetite	Zircon Fluorite

**[13]

3.7. Radiometric Prospecting

In nature, thorium is three times more prevalent in rocks than uranium **[38]**. Chemically measured uranium (14 ppm) and eU (19 ppm) are present in the study samples on average **(Table 2)**. (Authigenic U) = U-Th/3.5 is the formula for calculating authigenic uranium content. The oxic condition is assumed to be represented by authigenic uranium values below 5, while suboxic to anoxic conditions are represented by values over 5. The average value of 16.3 in the SL samples indicates that suboxic conditions prevailed during mineralization processes.

The lamprophyre dykes are not a source of uranium or thorium, but they play an important role in uranium mineralization as a source of heat, causing remobilization and leaching of uranium from the hosting psammitic gneisses to be entrapped and precipitated along the dyke fractures, or as a geochemical barrier, causing a local reduction zone that aids in uranium precept.

Because of its geochemical behavior and the composition of its host rocks, uranium is the most likely element to be mobilized. For granitic rocks, the most typically documented eU/eTh ratio is around 0.33. **[39, 38, 40, 41].** This ratio is used as a radiometric indicator to determine the destiny of the U-mineralization and its vicinity. However, an increase in this ratio beyond 0.33 could imply uranium enrichment, whilst a decrease in this ratio below 0.33 could indicate depleted or initially uranium-poor granites.

In most productive uraniferous rocks, the eU/eTh ratio is around 1.**[42]**. The eU/eTh ratio on average is 2.02, indicating uranium buildup. The D-factor **[43]** and the activity ratio **(Table 2)** were used to determine the equilibrium/disequilibrium states for the SL samples. Table 2 reveals that eU is approximately higher in all samples than those evaluated chemically, indicating a disequilibrium condition with leaching processes. Also, if the P-factor is greater than unity, it implies that uranium was added, whereas a P-factor of 1 suggests that uranium was removed recently. According to the aforementioned remarks, the SL samples have a P factor of 2.4 but a D-factor of less than one, indicating that old uranium deposits (P-factor 1) were recently leached (D-factor 1).

Table 2. Chemical analysis and radiometric measurements (in ppm) in W. Sikait lamprophyre samples

Sample No	eU	eTh	Ra	K%	U_{ch}	Th_{ch}	P - factor eU/Ra	D-factor eU/U_{ch}	U/Th	Authigenic U
1	31	13	11	0.89	15	13	2.8	0.48	1.15	27.3
2	21	13	14	3.3	16	14	1.5	0.76	1.14	17.3
3	15	4	18	2.1	5	15	0.8	0.33	0.33	13.9
4	8	19	5	0.5	15	11	1.6	1.88	1.36	2.6
5	10	5	6	٠,٧	١٤	١٥	١,٧	١,٤٠	٠,٩٣	٨,٦

6	15	6	8	0.98	19	17	1.9	1.27	1.12	13.3
7	22	7	10	1.2	8	20	2.2	0.36	0.40	20.0
8	32	10	9	1.2	14	16	3.6	0.44	0.88	29.1
Av.	19	9.44	9.33	1.23	14	14.78	2.41	0.90	0.9	16.30

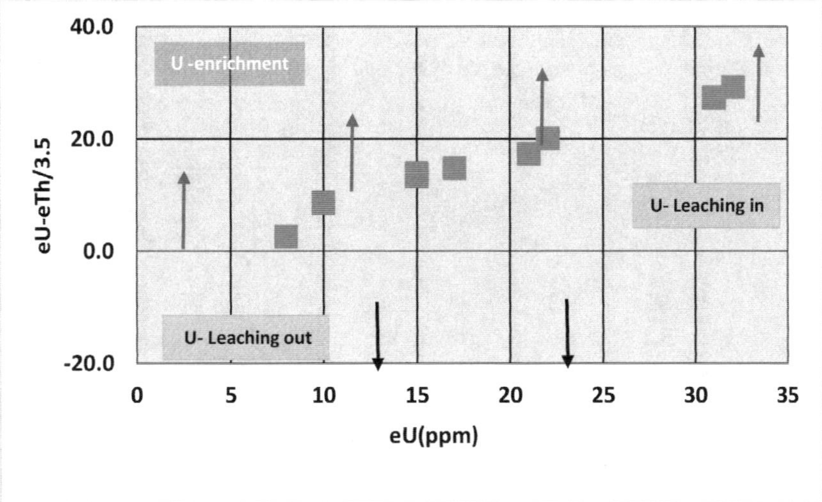

Figure 6. Uranium mobilization in the SL samples, SED, Egypt

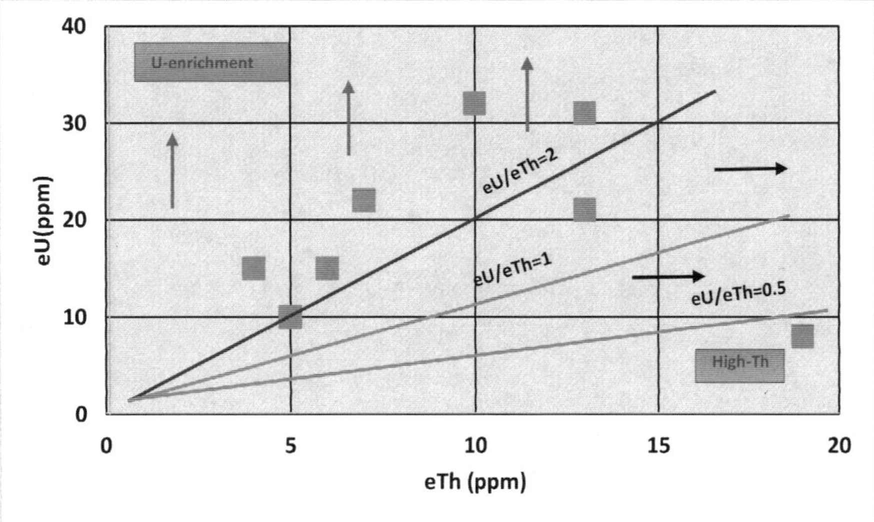

Figure 7. Variation diagrams between eU and eTh, in the SL samples, SED, Egypt

3.8. Radiological Hazard Parameters

Aiming to protect the public from excessive exposure to radioactivity, various radiological indices have been proposed in order to assess the exposure to natural radioactivity of building materials and environment, especially the presence of the Roman Sikait temple nearby the study area (southeast).

The activity concentrations (Bq.kg^{-1}) of ^{238}U, ^{232}Th, ^{226}Ra, and ^{40}K along with average values for ten samples for psammitic gneiss and eight samples of lamprophyre dykes are given in **table 3**. Data of table displaying the specific activity concentration values ranged from 238.7 to 275 Bq.kg^{-1} for ^{238}U; ^{232}Th ranged from 39 to 105 Bq.kg^{-1} as far as, the specific activities of ^{40}K from 425 to 1142 Bq.kg^{-1} and ^{226}Ra ranged from 112 to 261 Bq.kg^{-1} for lamprophyre dykes and psammitic gneiss respectively. The radionuclide-specific activity concentrations of W. Sikait samples are compared to those of the international average as shown in table 3 and figure 8. Generally, there is a noticeable increase in radionuclide concentrations concerning the international levels.

Table 3. Results of radionuclide concentrations, the dose rate (D), the annual effective dose equivalent (AEDE), radium equivalent activity (Ra_{eq}), external (H_{ex}), internal (H_{in}) hazard indices, and gamma index (I_γ) for lamprophyre dyke and psammitic gneiss samples.

Sample No	Rock type	eU $Bq.kg^{-1}$	eTh $Bq.kg^{-1}$	Ra $Bq.kg^{-1}$	K $Bq.kg^{-1}$	Abs. Dose (nGy/h)	Eff. Dose (mSv)	Ra_{eq}	H_{ex}	H_{in}	I_γ
1	Psammitic gneiss	967.2	68.68	1343.1	219.1	497.47	0.61	1082.17	2.92	5.54	7.28
2		136.4	44.44	255.3	1064.2	134.24	0.16	281.75	0.76	1.13	2.06
3		210.8	72.72	144.3	907.7	179.16	0.22	384.51	1.04	1.61	2.74
4		89.28	92.92	99.9	719.9	127.39	0.16	277.40	0.75	0.99	2.00
5		235.6	250.48	66.6	1502.4	322.79	0.40	709.00	1.92	2.55	5.08
6		161.2	125.24	210.9	1565	215.38	0.26	460.50	1.24	1.68	3.37
7		136.4	113.12	166.5	1565	196.60	0.24	418.38	1.13	1.50	3.08
8		105.4	72.72	144.3	1158.1	140.91	0.17	298.37	0.81	1.09	2.20
9		198.4	133.32	99.9	1752.8	245.28	0.30	523.69	1.42	1.95	3.82
10		508.4	76.76	77.7	970.3	321.71	0.39	692.70	1.87	3.25	4.80
Av.		274.9	105.04	260.85	1142.45	238.09	0.29	512.84	1.38	2.12	3.6
1	Lamprophyre dykes	384.4	52.52	122.1	278.57	220.93	0.27	480.86	1.30	2.34	3.27
2		260.4	52.52	155.4	1032.9	195.10	0.24	414.88	1.12	1.83	2.95
3		186	16.16	199.8	657.3	123.10	0.15	259.65	0.70	1.20	1.84
4		99.2	76.76	55.5	156.5	98.72	0.12	220.90	0.60	0.87	1.53
5		124	20.2	66.6	219.1	78.63	0.10	169.71	0.46	0.79	1.17
6		186	24.24	88.8	306.74	113.36	0.14	244.22	0.66	1.16	1.69
7		272.8	28.28	111	375.6	158.78	0.19	342.09	0.92	1.66	2.35
8		396.8	40.4	99.9	375.6	223.39	0.27	483.41	1.31	2.38	3.30
Av.		238.70	38.89	112.39	425.29	151.50	0.18	327	0.88	1.52	2.26
International average		33	45	32	420	57	0.48	370	1	1	0.5

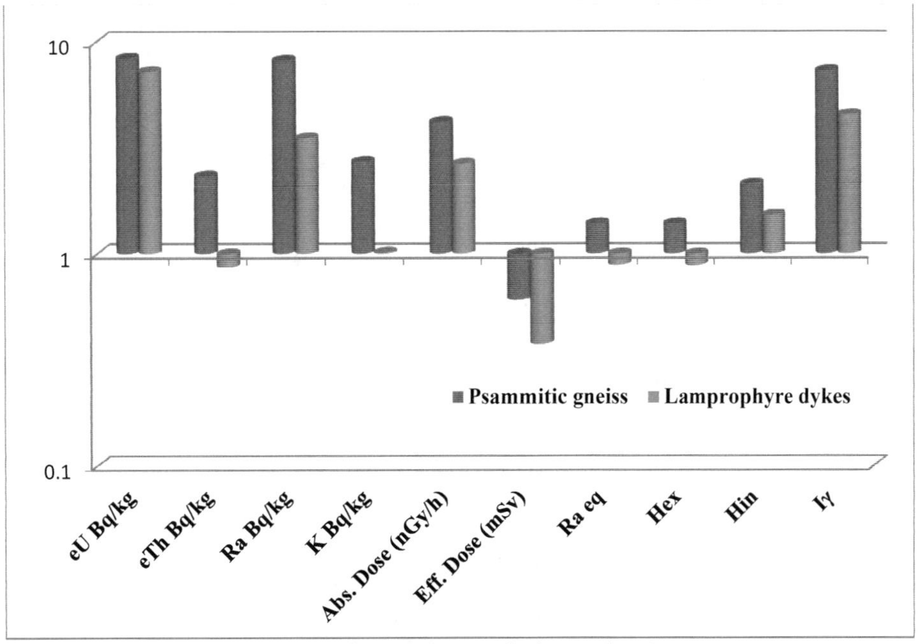

Figure 8. Enrichment – depletion diagram for W. Sikait samples, SED, Egypt

For safe use and to keep the external dosage below 1 mSvy-1, the maximum value of Ra_{eq} in building materials must be less than the limit value of 370 Bq.kg^{-1} suggested by the International Commission on Radiological Protection **[44]**. The results in table 3 show that Ra_{eq} values for lamprophyre dykes and psammitic gneiss vary from 327 to 512.8 Bq.kg^{-1}, respectively. These readings are lower than the maximum permissible value for lamprophyre samples, which is 370 Bq.kg^{-1}. Lamprophyre samples, on the other hand, are regarded safe when used as building materials, whereas psammitic gneiss samples are not.

The absorbed dose rate for lamprophyre dykes and psammitic gneiss, as indicated in table 3, varied between 151.5 nGy.h^{-1} and 238 nGy.h^{-1}, respectively. We found that all of the samples above the worldwide dosage rate of 24 to 85 nGy.h^{-1}, with an arithmetic mean of 55 nGy.h^{-1} **[45]**. The annual outdoor effective dose equivalent ranged between 0.18 and 0.29 mSvyr^{-1} which is below the limit of 1 mSvyr^{-1} **[17].**

For the samples analysed, the internal hazard index (H_{in}) ranged from 1.52 to 2.12. These values are greater above the European Commission's suggested safety limit of one **[26]**. In order to keep the radiation threat to a minimum, the external hazard index (H_{ex}) must also be smaller than unity **[46]**. The estimated external hazard index for the analyzed samples was approximately unity, as shown in table 3.

The annual dosage rate due to gamma radiation is associated with the gamma-ray index (external exposure). The annual effective dosage would be increased by 0.3 mSv for materials with $I\gamma < 2$, whereas the gamma-ray index for $2 < I\gamma < 6$ equates to an increase in effective dose of 1 mSvy^{-1}. Building materials that are utilized only on the surface rather than in bulk (tiles, boards, etc.) should be excluded from all radioactivity limits if the excess gamma radiation emitted by them increases the annual effective dosage of a member of the public by no more than 0.3 mSvy^{-1}. Dose rates greater than 1 mSvy^{-1}, on the other hand, are permitted only in extreme circumstances, such as when materials are utilized locally. Finally, samples with $I > 6$ are not suitable for use in structures.

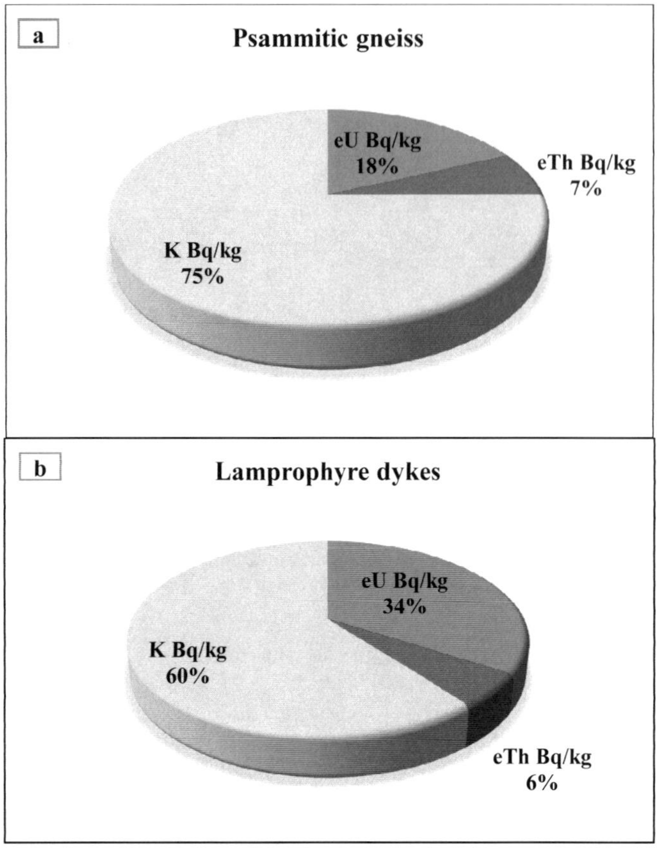

Figure 9. Contribution of the radionuclides for the absorbed dose rate within the studied rock samples: a) psammitic gneisses and b) lamprophyre dykes

The contribution of ^{238}U, ^{232}Th, and ^{40}K radionuclides for the absorbed dose rate within the studied samples are plotted in figure 9. The contribution of ^{238}U, ^{232}Th, and ^{40}K are different for most samples. ^{40}K plays the main and most important role in dose rate contribution.

4. Conclusion

Lamprophyre dykes that intrude the psammitic gneisses, ophiolitic mélange, and monzogranites, vary in thickness from 0.5 m to 1.5 m, ex more than one-kilometer and tend cut by right-lateral strike-slip faults. eU/eTh, P-and D factors indicate the presence of old uranium accumulation in the past but recently, there are leaching processes. The activities of 238U, 226Ra, 232Th, and 40K for most of the studied samples (especially the psammitic gneiss) exceeded the average level of these radionuclides, according to the average values of the radioelements worldwide. 40K plays the main and most important role in dose rate contribution. The high activity and the elevated radiological hazard parameters in the lamprophyre are related to the presence of radioactive and radioelements bearing minerals like kasolite, thorite, uranothorite, xenotime, cheralite and baddeleyite. Other mineralization was also recorded as gold and base metal minerals. According to the environmental parameters like Absorbed Dose Rate (D), annual effective dose equivalent (AEDE), radium equivalent activity (Raeq), external (Hex), and internal (Hin) hazard indices, the majority of the investigated samples are posing a hazard on the surrounding environment.

Acknowledgment

The authors sincerely thank the late prof. Mohamed E. Ibrahim for suggesting the research point. Also, thanks to the Professors Gehad M. Saleh and Mohamed G. El Feky for their revisions and critical comments.

References

[1] Kennedy WQ 1964 The structural differentiation of Africa in the Pan African (+ 500 million years) tectonic episode, Eighth Annu. Rep. Res. *Inst. African Geol*. Univ. Leeds, 48-49.

[2] Kröner A 1984 Late Precambrian plate tectonics and orogeny: a need to redefine the term Pan-African. In:Klerkx, J., Michot, J. (Eds.), *African Geology. Royal Museum for Central Africa, Tervuren, Belgium*, p 23-28.

[3] Stern R J and Hedge C E 1985 Geochronological and isotopic constrains on late Precambrian crustal evolution in the Eastern Desert of Egypt, *Am. Jour. Sci*., 285, p 97-127.

[4] El - Gaby S, List FK and Tehrani R 1988 Geology, evolution and metallogenesis of the Pan-African belt in Egypt. In: S. El Gaby & R. O. Greiling (eds), The Pan-African Belt of northeast African and adjacent areas, Fried. Vieweg and Shon, Braun Schweig, Wiesbaden, p 17-68.

[5] El Rakaiby ML 1988 The tectonic lineaments of the basement belt of the Eastern Desert, Egypt. *Egypt. J. Geol*., 32, p 77-95.

[6] Fowler AR and Osman AF 2009 The Sha'it-Nugrus shear zone separating Central and South Eastern Deserts, Egypt: a post-arc collision low-angle normal ductile shear zone. *J Afr Earth Sci*., 53, p16–32.

[7] Greiling RO, Kröner A, El Ramly MF, Rashwan AA 1988 Structural relationships between the southern and central parts of the Eastern Desert of Egypt: details of a fold and thrust belt. In: El Gaby, S., Greiling, R.O. (Eds.). The Pan-African Belt of Northeast Africa and Adjacent Areas. Vieweg and Sohn, Weisbaden, p 121– 145.

[8] Fritz H, Wallbrecher E, Khudier AA, Abu El Ela F, Dallmeyer RD 1996 Formation of Neoproterozoic metamorphic core complexes during oblique convergence, Eastern Desert, *Egypt. J Afr Earth Sci* 23,p 311–329

[9] El Bayoumi RM 1980 Ophiolites and associated rocks of Wadi Ghadir, east of Gabal Zabara, Eastern Desert, Egypt. Unpublished Ph.D. Thesis, Univ. Cairo, 227 pp.

[10] Ibrahim ME and Ragab AA 2011 Geochemistry of lamprophyre dykes, Wadi Sikait area, South Eastern Desert, Egypt. *Chin. J. Geochem.,* 30, p 323–331. https://doi.org/10.1007/s11631-011-0516-4.

[11] Saleh GM, Abdallah SA, Abbas AA, Dawood NA and Rashed MA 2012 Uranium mineralizations of Wadi Sikait mylonites, Southeastern Desert, *Egypt. J. of Geo. and Min. Res.* 3(5), p 86-104. https://doi.org/10.5897/JGMR10.035.

[12] Ibrahim IH 2013 Abu Rusheid-Sikait Mylonitic Rocks, South Eastern Desert, Egypt: Origin and Mode of Occurrence. International Conference on Natural Hazards and Geomatics -ICNHG, Tunisia.

[13] El Tohamy AM 2019 Mineralogy and geochemistry of the Sikait U-Au-bearing lamprophyre dyke, South Eastern Desert, Egypt. *Arab. J. of Geo.,* 12, p 237. https://doi.org/10.1007/s12517-019-4394-0

[14] Johnson PR, Andresen A, Collins, AS, Fowler, AR, Fritz, H, Ghebreab, W,Kusky, T, and Stern, RJ 2011 Late Cryogeniane Ediacaran history of the Arabian-Nubian Shield: a review of depositional, plutonic, structural, and tectonicevents in the closing stages of the northern East African Orogen. *Journal of African Earth Sciences* 61 (3), p 167-232.

[15] Sonkawade R, Kant K, Muralithar S, Kumar R and Ramola RC 2008 Natural radioactivity in common building construction and radiation shielding materials. *Atmos. Environ*. 42, p 2254–2259.

[16] Zakaly HM, Uosif MA, Madkour H, Tammam M, Issa S, Elsaman R and El-Taher A 2019 Assessment of natural radionuclides and heavy metal concentrations in marine sediments

in view of tourism activities in Hurghada city, northern Red Sea, Egypt. *J. Phys. Sci.* 30, p 21–47.
[17] UNSCEAR 2000 Report to General Assembly. Annex B: Exposure from Natural Radiation Sources. Report to General Assembly. with Scientific Annexes. Sources and Effects of Ionizing Radiation. United Nations Sales Publications No. E.00.Ix.3 Volume I: Sources and No. E.00.Ix.4 (Volume Ii: Effects). United Nations, New York, p. 1220.
[18] Papaefthymiou H, Papatheodorou G, Moustakli A, Christodoulou D and Geraga M 2007 Natural Radionuclides and 137Cs Distributions and Their Relationship with Sedimentological Processes in Patras Harbour, Greece *Journal of Environmental Radioactivity* 94 (2): p 55-74.
[19] Malczewski D, Taper L and Dorda J 2004 Assessment of natural and anthropogenic radioactivity levels in rooks and soils in the environs of Swieradow Zdroj in Sudetes, Poland by in situ gamma-ray spectrometry. *J.Environ. Radioact.* 73, p 233-245.
[20] Said AF, Salama AM, Hassan SF and Mohamed WS 2010 Assessment of the Environmental radioactivity impacts and health hazards indices at Wadi Sahu Area, Sinai, Egypt. 10th Radiation Physics & Protection Conf., Nasr City, Cairo, Egypt.
[21] Krisiuk E M, Tarasov S I, Shamov V P, Shlak N I, Lisachenko E P and Gomelsky L G 1971 A Study of Radioactivity in Building Materials (Leningrad: Research Institute for Radiation Hygiene). Ministry of Public Health of the RFSR.
[22] Beretka J and Mathew PJ 1985 Natural radioactivity of Australian building materials, industrial wastes and by-products. *Health Phys*. 48, p 87–95.
[23] Yang YX, Wu XM, Jiang ZY, Wang WX, Lu JG and Lin J 2005 Radioactivity concentration in soil of the Xinzhuang Granite Area, China. *Appl. Radiat. Isotop*. 63: p 225-259.
[24] Hamzah Z, Ahmad S, Noor H M and She D E 2008 Surface Radiation Dose and Radionuclide Measurement in Ex-Tin Mining Area, Kg Gajah, Perak, The Malaysian *Journal of Analytical Sciences* 12 (2): p 419-431.
[25] Nada A 2003 Evaluation of natural radionuclides at Um-Greifat area, Eastern Desert of Egypt. *Appl. Radiat. Isot*. 58, p 275–280.
[26] European Commission (E. C) 1999 Radiation Protection 112: Radiological Protection Principles Concerning the Natural Radioactivity of Building Materials Directorate – General Environment, Nuclear Safety and Civil Protection.
[27] Assaf HS, Ibrahim ME, Zalata AA, El-Metwally AA and Saleh GM 2000 Polyphase folding in Nugrus-Sikait area south Eastern Desert Egypt. *JKAW: Earth Sci*., 12, p1-16.
[28] Rismasite J 1982 The leaching of radionuclides and other ions during alteration and replacement of accessory minerals in radioactive rocks. Paper *Geol. Surv. Can*., 15: p 253-253.
[29] Förster HJ 2006 Composition and origin of intermediate solid solutions in the system thoritexenotime-zircon-coffinite. *Lithos* 88, p 35–55.

[30] Förster HJ 1998 The chemical composition of REE-Y-Th-U-rich accessory minerals in peraluminous granites of the Erzgebirge-Fichtelgebirge region, Germany, part I: the monazite XE "monazite" -(Ce)-brabantite solid solution series. *Am. Mineral*. 83, p 259–272.
[31] Zhu XK and O'Nions RK 1999a Zonation of monazite in metamorphic rocks and its implications for high temperature thermochronology: a case study from the Lewisian terrain, *Earth Planet. Sci. Lett*. 171,p 209–220.
[32] Kees Linthout 2007 "Tripartite division of the system 2REEPO4-CaTh (PO4)2-2ThSiO4, discreditation of brabantite, and recognition of cheralite as the name for members dominated by CaTh(PO4)2, Canadian Mineralogist 45 (2007) 503-508" (PDF). Retrieved May 11, 2018.
[33] Petrık I and Konecˇny´ P 2009 Metasomatic replacement of inherited metamorphic monazite in a biotite-garnet granite from the Nı´zke Tatry Mountains, Western

Carpathians, Slovakia: chemical dating and evidence for disequilibrium melting. *Am. Mineral*. 94, p 957–974.

[34] Zhu XK and O'Nions RK 1999b Monazite chemical composition: some implications for monazite geochronology. *Contrib. Mineral. Petrol*. 137, p 351–363.

[35] Deer W A, Howie R A and Zussman J 1992 An introduction to the rock-forming minerals; Longman Scientific and Technical, Hong Kong, 558p.

[36] Gupta CK and Krishnamurthy N 2005 Extractive Metallurgy of Rare Earths. CRC Press, Boca Raton, FL. 508p

[37] Bayanova TB 2006 "Baddeleyite: A promising geochronometer for alkaline and basic magmatism". *Petrology*. 14 (2): p 187. *doi*:*10.1134/S0869591106020032*.

[38] Rogers J W and Adams J A 1969 Uranium and thorium, In: K. H. Wedepohl (ed.), Handbook of geochemistry. Berlin, *Springier-Verlag*, 11 3, 92-B-1 to 92-0-8 and 90-Bb-1 to 90-00-5, 201 P.

[39] Clark S P, Petrman Z E and Heier K 1966 Abundance of uranium, thorium and potassium, In: S.P. Clark, Jr., (Editor), Handbook of Physical constraints, *Geol. Soc. Am. Mem*. 97, Section 24, P. 521 - 541.

[40] Stuckless J S, Nkomo I T, Wenner DB and Van Trump G 1984 Geochemistry and uranium favorability of the postorogenic granites of the northwestern Arabian Shield. Kingdom of Saudi Arabia. In: Pan-African crustal evolution in the Arabian-Nubian Shield (Convenor: A. M. Al Shanti), Bull. Fac. Earth Sci., king Abdelaziz Univ., Jeddah, Pergamon Press, Oxford, p.195-210.

[41] Boyle RW 1982 Geochemical prospecting for thorium and uranium deposits. Develop. *Economic, geol*., 16, El Sevier, Amsterdam. p.189.

[42] Darenely AG, Ford KL 1989 Regional airborne gamma-ray survey; a review. In: Proceedings of 3rd International Conference of Geophysical and Geochemical Exploration for Minerals and Ground Water l.3. *Geological Survey, Ontario*, p. 229–240.

[43] Hansink J D 1976 Equilibrium analyses of sandstone rollfront uranium deposits. *International Atomic Energy Agency*, Vienna, p.683-693.

[44] ICRP 1992 International Commission on Radiological Protection. The 1990-91 Recommendations of the International Commission on Radiological Protection, Publication 60, Ann. ICRP 21.

[45] UNSCEAR 1988 United Nations. Sources and Effects of Ionizing Radiation, Report to Ionizing Radiation: Sources and Biological Effects. And, Report to The General Assembly, With Scientific Annexes. United Nations Sales Publication E.94.Ix.2. United Nations, New York.

[46] Xinwei L 2004 Natural radioactivity in some building materials and by-products of Shaanxi, China. *J.Radioanal.Nucl. Chem*. 262, p 775-777.

Simulation of ozone concentrations inside ^{60}CO industrial irradiator

Ali Noaman[1]& Amal A. El gamel[2]

Radiation Engineering Department, EAEA[1]
Radiation Safety Department, EAEA[2]

amal.gamel@yahoo.com amal A. El gamel

Abstract. The motivation of this work is to study the relationship between gamma radiation and also ozone concentration in ^{60}CO industrial irradiators. Because ozone is considered to be one of the most important safety issues, because it has a destructive effect on the human body and has a corrosive effect on metal parts in the irradiated area. Moreover, due to its high reactivity. So first using ozone monogram method to measure ozone concentration, and also the effect of the factor on ozone concentration was determined. MATLAB computer code was used to calculate the ozone concentration-response with completely different statistics. Among them, ozone concentration in the ^{60}CO irradiation area is in a completely different state. These conditions include: different gamma emitters, completely different irradiators, airflow correction, and engine failure. It is suggested that ozone formation may be a direct effect of gamma radiations. the result show also that the health hazards due to the ozone produced during irradiation are significant and should not be overlooked. It was observed that in case of 1Ci motor failure, the detection time is 67.5 minutes. While, for 0.1Ci motor failure, the detection time is 32.2 minutes. To fulfill the employee safety limits. it is recommended to have a second control system with ventilation

Keyword: Gamma radiation and ozone concentration, monogram method, MATLAB, ozone generation, Influence of ozone concentration parameters.

1.Introduction

Industrial irradiators are used for many items, such as food, clinical disposables, electrical connections, gemstones, and different applications [1]. The purpose of this process is to improve performance or to clean

and purify these products. In industrial irradiators, electromagnetic radiation (gamma rays and X-rays) or electrons, a layer of air is encountered before interacting with the product being processed. When radiation interacts with this layer, it produces radiolytic effects on atoms in the surrounding atmosphere [2]. Molecular nitrogen comprises more than 70% of the Earth's atmosphere and is bombarded by energetic particles from cosmic rays and natural radiation from the Sun. Free electrons in air can interact with nitrogen molecules via electron impact dissociation [3]. This electron impact produces two nitrogen atoms, one in the ground state and the other in the doublet excited state. propagating TGF photons will naturally ionize air to produce free electrons that go into breaking apart N_2 in this reaction.
Once nitrogen molecules are dissociated, the resulting atoms can undergo chemical reactions with other species in the atmosphere. In a study of optical emission from sprites, [4] listed the important reactions for nitrogen atoms. the main interaction is with oxygen atoms. According to the Chapman cycle [5], this interaction is similar to what occurs in the stratosphere. When diatomic oxygen (O2) molecules absorb solar ultraviolet radiation with a wavelength less than 240 nanometers, they disconnect them and divide them in two. Active atoms that produce ozone (O3) when highly combined with another oxygen molecule. In the troposphere, ozone acts as a protective layer, protecting the earth's surface from the sun's high-energy ultraviolet rays [6]. Due to its high reactivity, ozone on the earth's surface is harmful to plants, animals and humans [7]. In devices that use gamma radiation, X-rays, or electron accelerators, the ozone generated interacts with the surface of the processed product and the structural elements of the radiator. Therefore, the high concentration of such harmful gases is an occupational safety issue in the irradiation room, especially when operators are required to enter for product maintenance or positioning. Therefore, it is necessary to install a ventilation system to reduce the concentration and subsequent release within the allowable gas level in the environment. In the ^{60}Co industrial irradiator, the air layer between the source and the processed product interacts with gamma radiation and produces ozone gas. [8] As a highly reactive gas, ozone is harmful to the human body. Figure (1) shows the relationship between ozone concentration and decreased lung function in different age groups. [9]

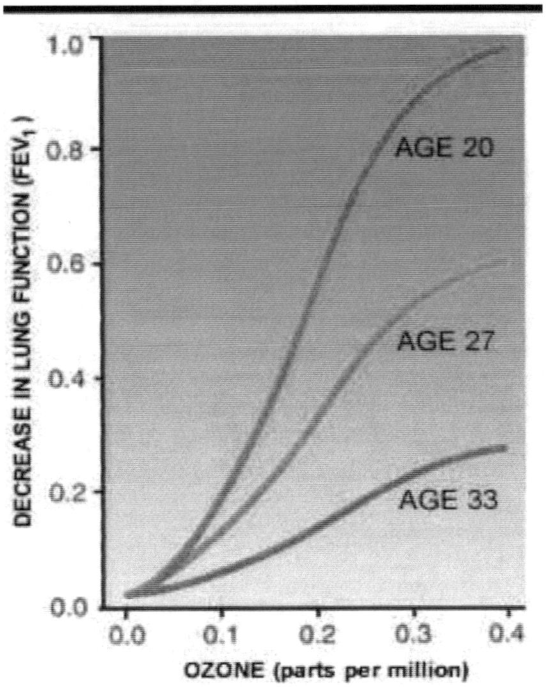

Figure 1. Relation between ozone concentrations and decreasing in lung function for different ages: Source*: Devlin et al. (1997)*

Also, ozone is considered a highly corrosive gas and should not be used with metallic materials unless it is anti-corrosive. [10] In order to avoid any potential harm to workers and metal parts in the irradiation room, a ventilation system was installed to reduce and control the ozone concentration and the amount of release into the environment to levels allowed. The most famous allowable level of ozone concentration for is 75 ppb (0.075 ppm). But even at a low concentration of 10 ppb (0.01 ppm), this gas can be identified by its characteristic odor. [11], [12]

2.Materials and methods

To determine the ozone concentration in the irradiated area, it will be used monogram method measured Figure (2) is shown that associates O3 concentration (in ppm) with supply activity (in Ci), irradiated area (in cubic feet), and air modification rate (per min). The abbreviations C, S, V and A .

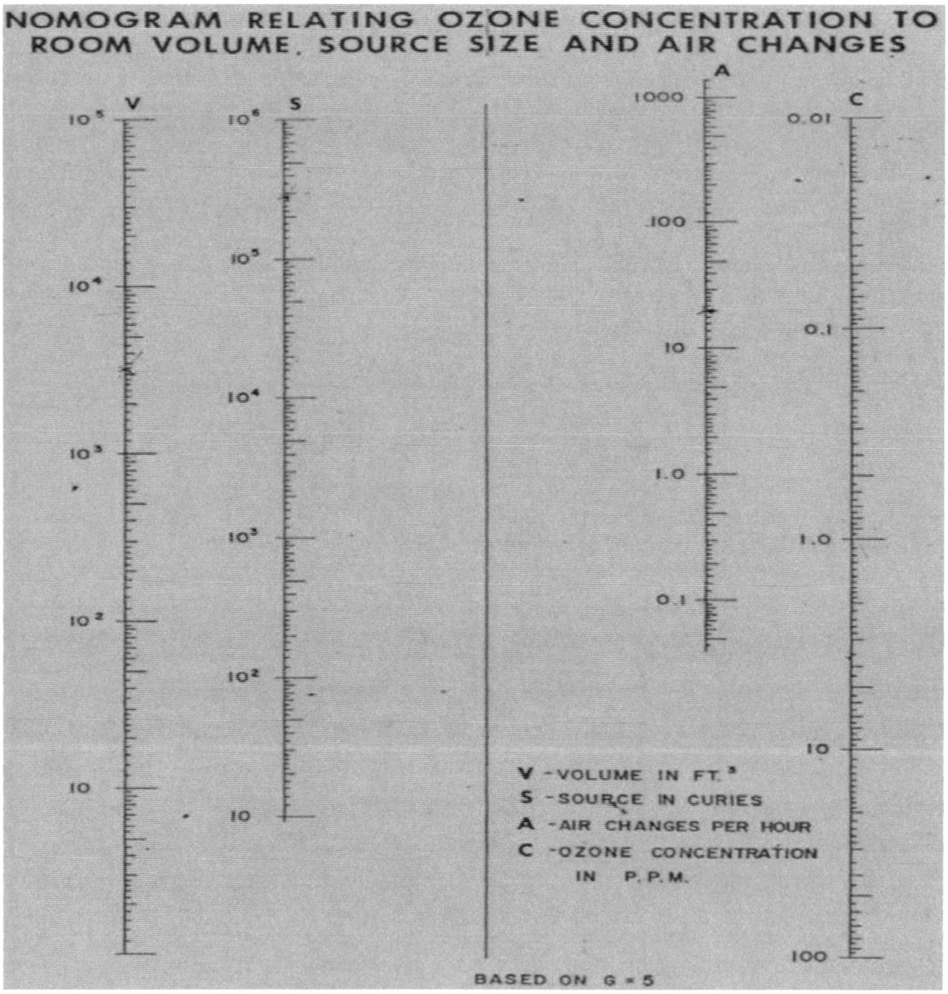

Figure 2. Ozone monogram method measured

MATLAB calculations are created by equations that provide a bet on the concentration of irradiation time or the development letter combination that provides the highest ozone concentration in a given area, betting on supply activity, air modification rate, and volume of the area. Due to reduced supply, source activity may be a variable. The volume of the irradiation chamber varies from one irradiator. In addition, the air modification rate is modified according to the engine power in the operator and similar irradiators. By creating 2 of the 3 installed variables, we can obtain the relationship between the ozone concentration and the last variable. In addition, there are some formulas that can express the ozone concentration in the irradiated area multiple times. These equations are used to calculate the time required to deplete the ozone concentration in the irradiated area in the event of damage to the air purification engine or similar conditions. These equations are from Technical Report Series No. 188, printed by the International Atomic Energy Agency [13]. The purpose of this article is to apply the correlation equation to obtain the equilibrium concentration of O3, as it can be similar to the equation mentioned in the reference, no matter where the concentration of O3, C (t) can be inferred, the equation can be obtained as follows:

$$C\ (t) = \frac{pT}{V}\ [1 - e^{\frac{-1}{T}}] \quad (1)$$

$$P = A\ .\ \kappa \quad (2)$$

$$T = \frac{T\ (vent).T1/2}{T\ (vent)+T1/2} \quad (3)$$

$$T\ (vent) = \frac{V}{Flow} \quad (4)$$

We substitute from equation (2) into the equation (1)
And also substitute from equation (4) into the equation (3)
Where:
t = irradiation time (min);
p = O3production rate (ppb/min);
V = irradiation chamber volume (64 m^3);
T1/2= the average time for the decay of O3in the installation
Flow = an extraction rate of air through the exhaust
A = exposed activity source (Ci)
K = proportionality constant (ppb/PBq.min)

2.1 Methodology

Gamma photons will naturally ionize air to produce free electrons that go into breaking apart No_2 in this reaction .Ozone concentration response as function of gamma radiation due to Chemical Coupling of O_3, NO and NO_2 was considered .It is well established that the inter-conversion of O_3, NO and NO_2 under atmospheric conditions is generally dominated by the following chemical reactions [14] :

$$NO_2 \quad + \quad h\gamma \quad \rightarrow \quad NO \quad (5)$$

$$O + O_2 = O_3 + M \quad (6)$$

$NO + O_3 = NO_2 + O_2$ (7)

M (usually N_2 or O_2) represents a molecule that absorbs the excess vibrational energy and thereby stabilizes the O3 molecule formed. hɣ represents the energy of a photon due to gamma radiation and O is an active monoatomic oxygen molecule. These equations form a cycle with no net chemistry i.e. the overall effect of reaction (2) is the reverse of reaction (1). These reactions therefore represent a closed system in which the NOx (NO and NO_2) components and the OX (O_3 and NO_2) components relate separately. During daytime hours, NO, NO_2 and O_3 are typically equilibrated on a timescale of a few minutes. This is known as a photo stationary state. NO, NO_2 and O_3 concentrations increases with an increase in $[NO_2]/[NO]$.

3.Results and discussion

Using monogram method measured the ozone concentration in the irradiation room. Dynamic volume (V), source activity (S) or air exchange rate (A) can change the ozone concentration (C) in the irradiation chamber. Active ozone concentrations with completely different parameters are usually shown in the following table (1).

Table 1. The relations between each parameter and the other (C, A, S and V)

V(m^3)	**6**				**9**				**15**				**30**				**S(Ci)**
A(m^3/min)	**10**	**20**	**40**	**50**	**10**	**20**	**40**	**50**	**10**	**20**	**40**	**50**	**10**	**20**	**40**	**50**	
	0.45	0.25	0.11	0.09	0.34	0.17	0.09	0.07	0.30	0.14	0.07	0.06	0.15	0.08	0.04	0.03	**0.1**
C(ppm)	1.30	0.60	0.31	0.27	1.00	0.50	0.26	0.20	0.80	0.40	0.20	0.18	0.50	0.26	0.12	0.09	**0.3**
	2.00	1.09	0.60	0.40	1.60	0.90	0.40	0.35	1.40	0.65	0.34	0.30	0.80	0.36	0.20	0.16	**0.5**
	3.50	2.00	1.20	0.85	3.00	1.60	0.80	0.70	2.60	1.30	0.70	0.60	1.60	0.80	0.40	0.33	**1.0**

According to the latest schedule, the ozone stages in the irradiated area are divided as follows:

3.1 Normal operation

3.1.1 Ozone concentration response as function of source activity (Linearity)

a) Correlation of V and A constant

The first stage deals with changes in the activity of the source in the same irradiator. (V, A) Constant. The font decays over time is the most common case. As shown in Figure 3. This is a natural relationship, where the air exchange rate is constant and the engine efficiency is a high value of 50 m3 / min at the same irradiator. A typical first-order relationship of can be written as:

$$Y = A + B * X \ (5)$$ (8)

where: B is the slope, A is the intercept The physical meaning of the variables x and y describes the ozone concentration (ppm) The function for the source activity is: the independent variable x represents the source activity, and the dependent variable y is the ozone concentration (ppm)

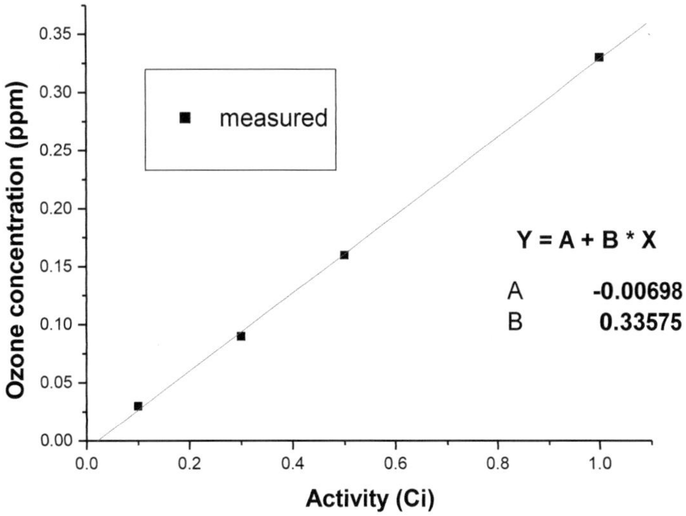

Figure 3. The relation between ozone concentration and source activity volume V_1 and air flow 50 m^3/min

As in the previous figure, using the linear regression method, the gradient of the curve shows good linearity (R = 0.99972). It shows that when the activity of the source increases, the concentration of ozone increases. Using this equation

$$\text{Ozone concentration (ppm)} = A + \text{source activity (Ci)} \times B \tag{9}$$

The conversion factor C.F must be determined as follows:
C.F that must be less than 1 under normal conditions for the ozone concentration value to be safe.

$$B = \text{slope} = \text{ozone concentration} / \text{source activity}$$

$$\text{ozone concentration (ppm)} = \text{source activity (Ci)} * 0.33575 \tag{10}$$

b) Different irradiators: (A) constant

The second phase is about a variation in room volume (different irradiators). As shown in Fig (4). This is a natural relation with different room volume ($V_3 < V_2 < V_1$) and air change rate is constant and is the highest value is 50 m^3/min in the same irradiator. With different source activity.

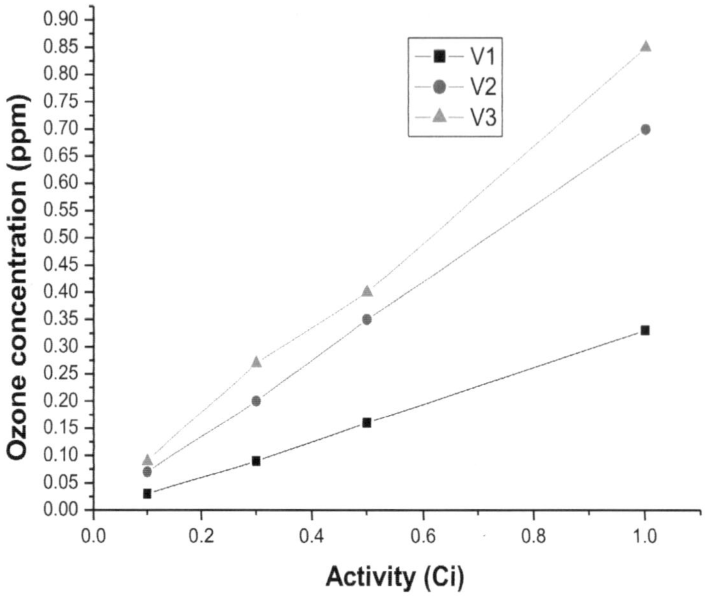

Figure 4. The relation between ozone concentration and source activity with different volume ($V_3<V_2<V_1$) and air flow of 50 m^3/min

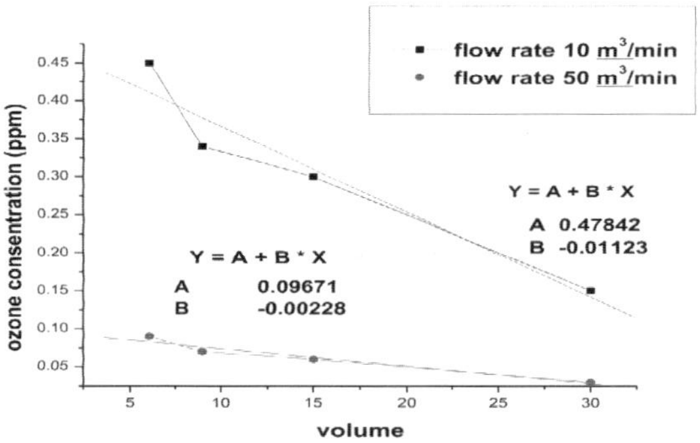

Figure 5. The relation between ozone concentration and room volume with different flow rate

As in the previous figure (5), the relation between measured line and linear fitting of data is small that it can be linear. Based on Fig (4,5) It indicates that ozone concentration decreased when volume is increasing.

c) Variation of change in air change rate: the constant V and S

The third phase of correlations deals with a variation in air change rate And ozone concentration figure 6.

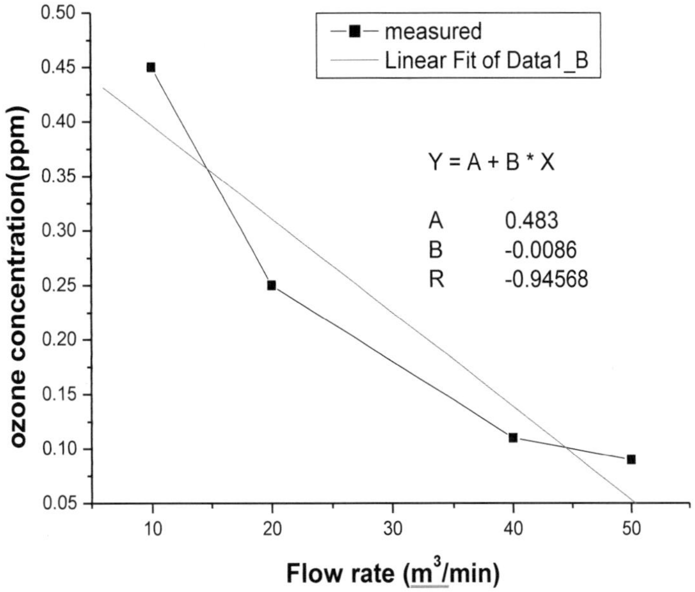

Figure 6. The relation between ozone concentration and air change rate with (S=0.1 Ci) & volume V_1

As in the previous figure (6), there is a variation between the line represents the ozone concentration and flow rate and line fit of data. The line fit of the ozone concentration is decreasing with increasing flow rate. When flow rate is almost zero, the ozone concentration is the highest value.

3.2 MATLAB simulation

3.2.1 Calculation the ozone concentration as time function for different source activities.

Figure (7) shows the temporal evolution of the O3 concentration, as time function, for different activities. It was observed that the systems tend to show a region of O3 equilibrium concentration. Which Measurements by monogram were taken from the exhaust pipe, in normal operation.

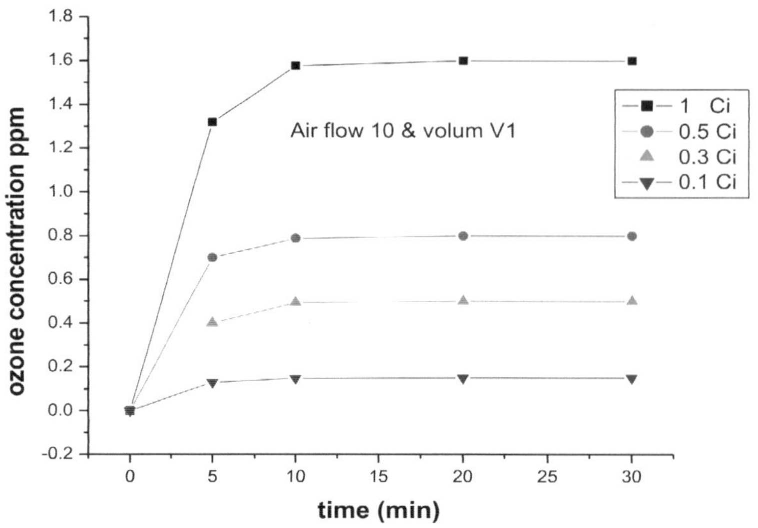

Figure 7. Measurements of O3 concentration, taken in the exhaust system pipe for different activities, until equilibrium condition.

The motor that blows air out of the radiation chamber may be damaged by any failure of any part of the motor. This means that the flow rate = 0. According to figure (8&9) the value of measured O_3 concentration in exhaust pipe for activity source S=0.1Ci and 1 Ci at volumeV3&V1 when ventilation is off as follows. So, figure (8) shows the value of measured O_3 concentration in exhaust pipe for activity source S=0.1Ci and 1 Ci when ventilation is off with volume V3. While figure (9) shows the value of measured O_3 concentration in exhaust pipe for activity source S=0.1Ci and 1 Ci when ventilation is off with volume V1 .and then from equation. (1) and estimate the half life time is 12 min to calculate the irradiation time which is needed that the ozone concentration be within safety limit (0.075ppm).

Figure (10) shows the O3 decay curve in the exhaust pipe, with the exhaust system on, V1 and after gathering sources of cobalt-60. It could be seen that after 50 min, the O3 concentration is already lower than the limit for workers (0.075ppm).as sown in table 2 and 3

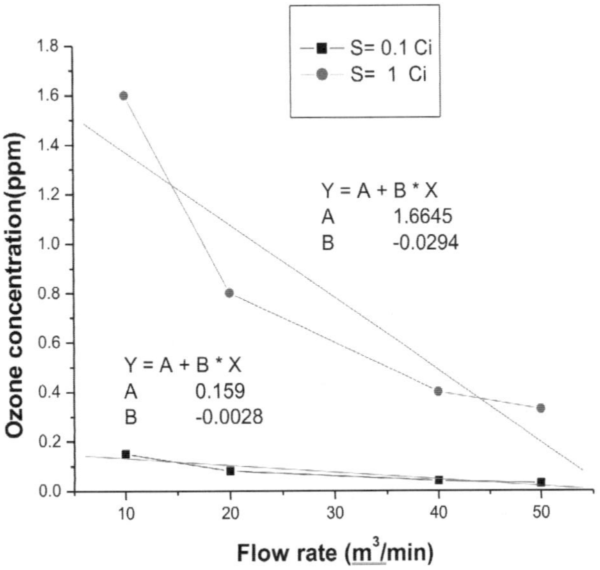

Figure 8. The relation between ozone concentration and air change rate with different activity source (0.1 & 1) Ci & volume V3

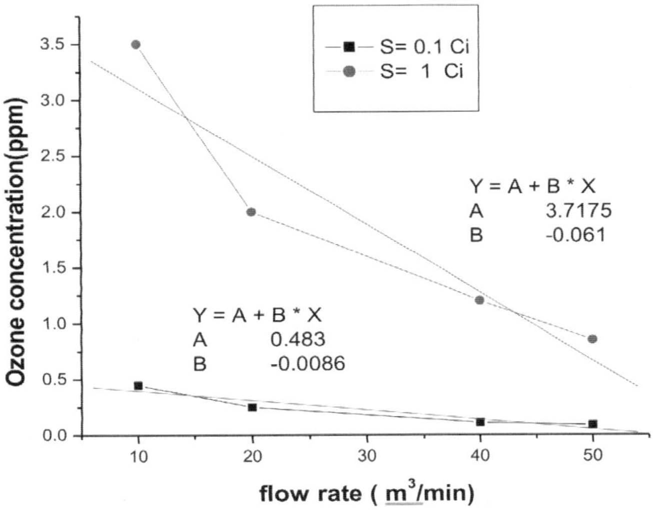

Figure 9. The relation between ozone concentration and air change rate with different activity source (0.1 & 1) Ci& volume V1

Table 2. the activity with V1 and max ozone concentration and time to be safety limits (0.075ppm)

Activity (Ci)	Maximum concentration (ppm)	Time (min)
1	3.71	67.5 min
0.1	0.483	32.2 min

Table 3. the activity with V3 and max ozone concentration and time to be safety limits (0.075ppm)

Activity (Ci)	Maximum concentration (ppm)	time (min)
1	1.66	53.6 min
0.1	0.159	13 min

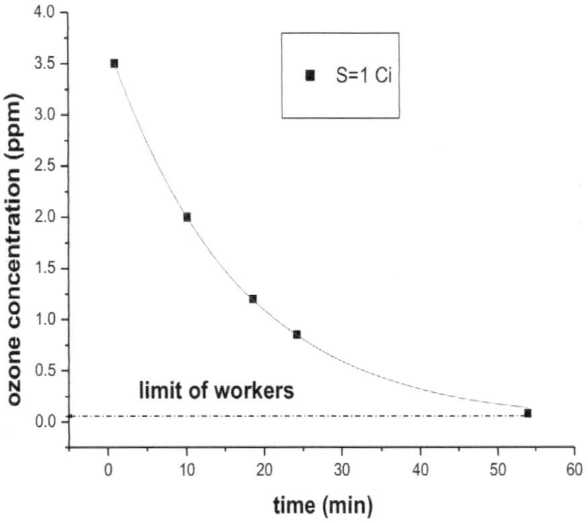

Figure 10. O3 decay curve in the exhaust pipe, with the exhaust system is not working, In ^{60}Co industrial irradiator with activity source 1 Ci and V1

as shown in Figure (11a). The relationship between the flow rate and the ozone concentration. Using an active source of 1 Ci, an ozone concentration was observed to be 50% after 10 minutes at a flow rate of 10 m^3/min. while, as shown in Figure (11b) This is a natural correlation where the air exchange rate is constant and the motor efficiency is a high value of 50 m3/min in the same irradiator. Ozone concentration percentage is 54.1% at 1 Ci activity source and 4.92% at 0.1 Ci.

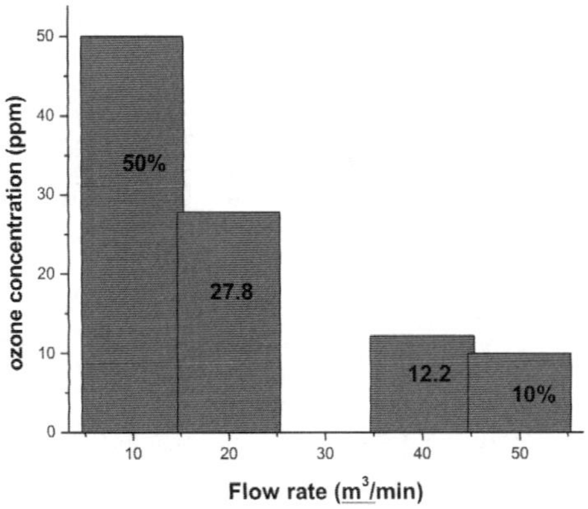

Figure (11-a)

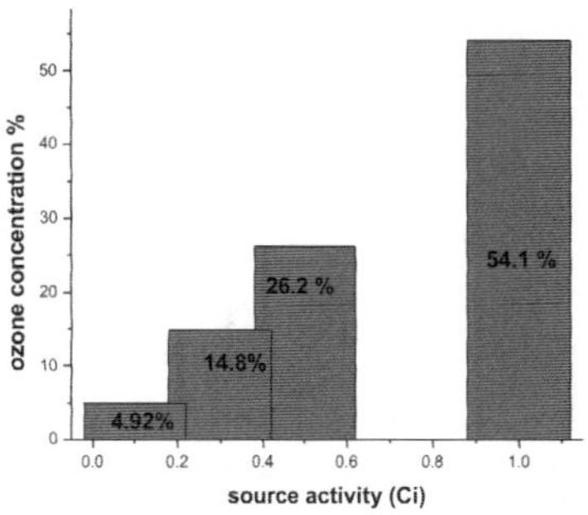

Figure (11-b)

Fig (11-a & 11-b) the flow rate and source activity vs. the percentage of ozone concentration

Table (4) Information from National Ambient Air Quality Objectives for Ground-Level Ozone, section 8 of the Science Assessment Document published by Health Canada, 1999

Exposure limits of ozone	
Exposure limits of ozone ppm	Exposure limits of ozone
0.05	Maximum allowable concentration averaged over an eight-hour period for heavy work
0.08	Maximum allowable concentration averaged over an eight-hour period for moderate work
0.1	Maximum allowable concentration averaged over an eight-hour period for moderate work
0.2	Maximum (short-term) exposure limit for light, moderate or heavy work, for less than or equal to 2 hours
1	Immediately Dangerous to Life and Health (IDLH)

4.Conclusions

- Ozone production from NOx pollutants, Oxygen atoms freed from nitrogen dioxide by the action of gamma radiation attack oxygen molecules to make ozone.
- So, the ozone concentration is directly proportional to the gamma activity.
-In traditional operation, ozone concentration (ppm) = source activity (Ci) * 0.33575,
The conversion factor CF = ozone concentration/source activity, indicates the safe level of ozone.that should be less than unity (< 1)
- This relationship between ozone and gamma radiation, obviously can be used as an indicator of gamma radiation level and intensity.
- in the case of engine failure, the hazardous ozone concentration which is obtained due to lack of environmental circulation in the irradiator area.
- This analysis indicates the time required to reach a stable ozone concentration, and describes the time required to successfully reach a stable gas concentration. 1 Ci motor failure detection time is 67.5 minutes, 0.1 Ci motor failure detection time is 32.2 minutes.to fulfill the employee safety limits.
-it is recommended to have a second control system with ventilation. These procedures should take precautions to protect all employees who may be exposed to hazardous substances. Exposure to high concentrations of ozone can have serious health effects and pose a risk to workers.
- the exposure limit for employee in the industrial irradiator must be less than or equal to 2 hours.

These procedures require that precautions be taken to protect all employees who may be exposed to hazardous materials. Such exposure to the highest concentrations of ozone can cause serious health effects and pose risks to people.
-Ozone odor can be detected at 0.1 ppm or less, so any room without a characteristic odor can be considered safe and unaffected by this toxic gas.
-On the other hand, if the odor is strong or detected frequently, Monitoring tools should be used for evaluation and detection the elevated levels.

5. References

[1] INTERNATIONAL ATOMIC ENERGY AGENCY, Manual of Food Irradiation Dosimetry, Technical Reports Series No. 178, IAEA, Vienna (1977).
[2] Akinloye, M.K.; Isola, G.A.; Olasunkanmi, S.K. and Okunade, D.A. (2015). Irradiation as a Food Preservation Method in Nigeria: Prospects and Problems. International Journal for Research in Applied Science and Engineering Technology (IJRASET). 3(11): 85-96
[3] Cosby, P. (1993), Electron-impact dissociation of nitrogen, *J. Chem. Phys.*, 98(12), 9544– 9553
[4] Liu, N., and V. P. Pasko (2007), Modeling studies of NO-γ emissions of sprites, *Geophys.* Res. Lett., 34, L16103, doi:10.1029/2007GL030352.
[5] A. Pretto, O estudo do comportamento dos gases-traço O3, NOX, CO, SO2 e de COVs na atmosfera da cidade de São Paulo. Tese (Doutorado em Ciências). Instituto de Pesquisas Energéticas e Nucleares, São Paulo. 2005.

[6] T. Peter (1994) “The stratospheric ozone layer –An overview” , Environmental Pollution, v. 83, pp. 69-79

[7] C. D. Klaassen, Casarett and Doull’s(2008): Toxicology -The basic science of poisons. 7. ed. McGraw-Hill

[8] D. H. Uzueli1(2013) "Ozone Concentration inside IPEN Industrial Irradiation Facility", Journal of Radiation Physics and Chemistry.

[9] R. J. Strittmatter, B. Yang, and D. A. Johnson,(1992) "Application of Ozone in Cooling Water Systems", National Association of Corrosion Engineers Corrosion, Nashville, Tennessee.

[10] IAEA, "Gamma Irradiators for Radiation Processing", Vienna, 2005.

[11] D. H. Uzueli, P. R. Rela, P. A. S. Vasquez, M. M. Hamada, and F. E. da Costa (2013) "Study on The Formation of Ozone Gas in Industrial Irradiation Process", International Nuclear Atlantic Conference – INAC, Brazil.

[12] J. H. Ruth(1986) “Odor thresholds and irritation levels of several chemical substances: A review”,Am. Ind. Hyg. Assoc. J.,v. 47, pp. 142-151.

[13] INTERNATIONAL ATOMIC ENERGY AGENCY. Toxic Gas Production at Electron Linear Accelerators. Technical Reports Series No. 188, Vienna ,1988

[14] Kenty, K.L., Poor, N.D., Kronmiller, K.G., McClenny, W., King, K., Atkeson, T. and Campbell, S.W. (2007). Application of CALINE4 to Roadside NO/NO2 Transformations. Atmos. Environ. 41: 4270–4280.

Study Matrix Densities Variation on Assay of ^{137}Cs & ^{60}Co using Gamma Spectrometric Analysis and Correction Equation

Omar Abo-Bakr Omar*[1], Sayed A. El-mongy[2]

[1] Radiation measurements Branch, Main Defense Chemical Laboratory, Cairo-Suez Road - Almaza, Cairo- Egypt
[2] Nuclear and Radiological Regularity Authority, Cairo- Egypt

* omar805805805@gmail.com

Abstract. There are many factors affecting the results of specific activity determination by using HPGe detection system. One of these important factors is the incompatibility and dissimilarity between the density of the certified standard source for efficiency calibration of the system and the samples to be measured. In this work, five standard samples containing 137Cs at 661.6 keV and 60Co at 1173.2and 1332.5 keV energies, with the same geometry and different densities were used to investigate and validate the percentage of error. It is scientifically known that the low energies are attenuated more than the high energies. The densities of the used standard samples are 0.7, 1, 1.2, 1.3, 1.6 g/cm3 respectively. The detection system was well calibrated before carrying out the measurements. The calculated error percentages between the measured activities (in Bq) and the certified values for 137Cs and 60Co in the standard samples were found to be in wide range from 9.8% to 19.2% for 137Cs and from 7.77% to 16.1% for 60Co respectively. A correction equation taking into consideration the densities variation was suggested and formulated to overcome and compensate these errors.

Keywords: Matrix densities ,^{137}Cs & ^{60}Co, correction equation, HPGe spectrometer

1. Introduction

The absolute photo-peak efficiency of the HPGe detection system is the main influencer on analysis results and it depends on many parameters like; the source and detector geometry– gamma energy - density – distance between the source and detector - specifications of the used HPGe detector [1].
One of the most important factors that may be affects the absolute photo-peak efficiency calculations is the density, and to clarify this point when using HPGe detector in analysis of different samples, the most important step in set-up of the HPGe detection system before measurements is the efficiency calculations. Usually, it is calculated by using a certified standard source having the same geometry and same density with respect to the samples to be measured. The use of efficiency curve established using a standard source matrix different from that of the samples to be measured leads to errors in the estimated values of the measured activity, so it is necessary to make corrections in order to consider the differences in densities and compositions between the measured and calibration conditions [2, 3].
Also, the dependence of absolute photo-peak efficiency of the HPGe detection system on samples densities is linked to another factor which is the energy range, by other words the low energy gamma lines are affected much more than high energy gamma lines by density change when they are used in radioisotopes activity calculations [4]. The effect of density change on radioisotopes activity calculations when using HPGe detection system mainly due to absorption processes which is a function

of energy and described by the exponential attenuation equation [5, 6]. Hence the use of certified standard source having the same geometry and same density with respect to the samples to be measured to create the efficiency curve of the HPGe detection system is the main method to overcome the densities incompatibility problem [7]. In many cases, because of a lack of standard radioactive sources researchers are confronted with a problem of incompatibility between the density of the certified standard source and the samples to be measured [8].
In this work, we are trying to use of the relative method in accompanied with HPGe detection system for determining the ^{137}Cs,^{60}Co activity (Bq) in five certified standard samples; S1, S2, S3, S4, S5, which have the same geometry and radioisotopic composition but different in their densities; 0.7,1,1.2,1.3,1.6 g/cm^3 respectively. Based on the error percentage values due to the density variation, Finally Acorrection equation was suggested and formulated to overcome and compensate these errors.

2. Experimental work

2.1. Measurements Arrangement and HPGe detection system set-up:
The five standard samples were analyzed using a p-type HPGe detector of ~ 60 % efficiency. The used HPGe detection system, the cooling of the HPGe crystal by using liquid nitrogen at (-77^{0}C) thermally isolated under vacuum in cryostat to reduce the noise of leakage current. The output signals have been treated with 16384 channels CANBERRA multi-channel analyzer (DSA-LX model). The used HPGe detection system shield was lead with approximately 10 cm thick with an inner layer of 1 mm tin and 1.6 mm copper to minimize the participation from Pb X-ray florescence and to inhibit the effect of x-rays peaks [9, 10].

The used five standard samples spectrum were analyzed using GENIE 2000 software. And some specifications of the used HPGe detector are given in table (1).

Table 1. Used HPGe specifications

Detector model	GX6020
Relative efficiency	≥ 60%
Resolution	≤ 2@ 1.33 MeV
Operating voltage	(+) 3500 V
Peak : Compton ratio	≥ 66:1

2.2. Energy Calibration of the system
Energy calibration of the used HPGe detector system were done by using standard material S1 which has been counted for (2hrs) to obtain high enough statistics, and then correlation between the channel number and the energy (keV) was obtained as shown in Fig. (1) with its equation number (4)[11].

$$\text{Energy (keV)} = -0.1032 + (0.3549 \times \text{channel number}) \quad (1)$$

2.3. The standard samples specifications and relative method
Selected photo peaks were used for calculations of ^{137}Cs and ^{60}Co activity (Bq). The 661.6 keV gamma line has been used for determination and validation of ^{137}Cs activity in the analyzed five standard samples (S1,S2, S3, S4, S5) which have the same geometry but different densities. Also, the 1173.24 keV and 1332.5 keV gamma lines have been used for determination and validation of ^{60}Co activity in the investigated five standard samples.
The overall uncertainty of the used five standard sample (S1,S2, S3, S4, S5) are calculated at the 99% confidence level and its main specifications are also given in table (2). The shapes of the used standard samples are also shown in Fig. (2).

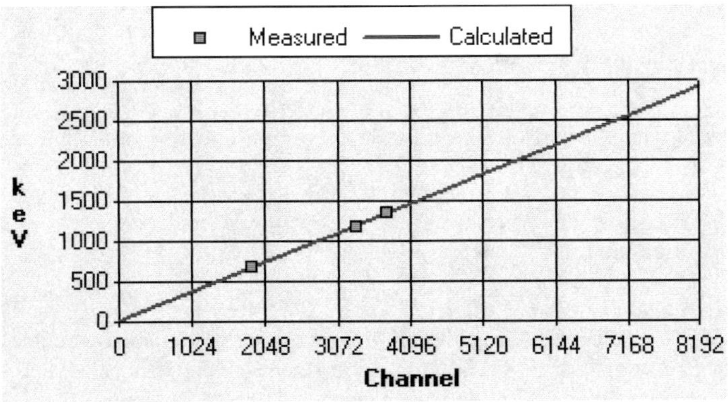

Figure 1. Energy calibration curve

Table 2. The specifications of the five standard samples

Parameter	S1		S2		S3		S4		S5	
Density (g/cm³)	0.7		1		1.2		1.3		1.6	
Radionuclides activity (Bq)	**^{137}Cs**	**^{60}Co**	**^{137}Cs**	**^{60}Co**	**^{137}Cs**	**^{60}Co**	**^{137}Cs**	**^{60}Co**	**^{137}Cs**	**^{60}Co**
	221	**2538.1**	**221.32**	**2542.4**	**216.64**	**2489**	**219.7**	**2525.3**	**221.32**	**2542.4**
Capsule type	Epoxy in plastic container									
Active volume	240 cm³									
Backing	Plastic									
Cover	Plastic									
Nature of active deposit	Multinuclear uniformly distributed in epoxy matrix									

Figure 2. A photo of the used standard samples

Every standard sample was counted three times for 2hrs.Then by using the relative method which is simply; Hit Interactive method, that is based on certain conditions; the five measured standard samples have the same geometry and radioisotopic composition with different densities. Where, they were counted for the same time and the photo peak efficiency of the detection system and the emission probabilities for the selected photo peaks are the same. Hence we can use the" Hit Interactive method " between the number of counts in the selected photo peaks (counts) and the activity (Bq) to calculate the values of ^{137}Cs,^{60}Co activities (Bq) for each standard sample and also validating and comparing all the calculated results to the certificates values (Ce_A).

3. Results and Discussion:

3.1. Activity calculation results of ^{137}Cs and ^{60}Co

By using (S1) as the a reference standard sample and the relative method to validate the activities of the other four standard samples (S2, S3, S4,S5),the error percentages were calculated and compared with respect to the certificates values (Ce_A).The calculations results for the samples S2,S3,S4 and S5 are given in table (3,4,5 and 6) respectively. The background counts and true coincidence correction due to summation peak in case of ^{60}Co activity calculations were also considered.

The calculated activity (Ca_A) in (Bq) were evaluated using the measured Region of interest counts (ROI_C) in the gamma line 661.6 KeV for ^{137}Cs, while the average of 1173.24 and 1332.5 KeV gamma lines were used in case of ^{60}Co activity calculations. The (% Err) was then calculated.

Table 3. Results of calculations for the used standard sample S2

Radioisotope	Energy (keV)	S1 (0.7 g/cm^3)		S2 (1 g/cm^3)			
		ROI_C	Ce_A	ROI_C	Ce_A	Ca_A	%Err
^{137}Cs	661.6	284404±323	2538.1	256940±314	2542.4	2293	9.8
^{60}Co	1173.24 1332.5	18481±73 16569±45	221	17024±65 15282±34	221.32	204.1	7.77

Table 4. Results of calculations for the used standard sample S3

Radioisotope	Energy (keV)	S1 (0.7 g/cm^3)		S3 (1.2 g/cm^3)			
		ROI_C	Ce_A	ROI_C	Ce_A	Ca_A	%Err
^{137}Cs	661.6	284404±323	2538.1	241466±298	2489	2154.9	13.4
^{60}Co	1173.24 1332.5	18481±73 16569±45	221	15804±55 14219±42	216.64	189.25	12.6

Table 5. Results of calculations for the used standard sample S4

Radioisotope	Energy (keV)	S1 (0.7 g/cm^3)		S4 (1.3 g/cm^3)			
		ROI_C	Ce_A	ROI_C	Ce_A	Ca_A	%Err
^{137}Cs	661.6	284404±323	2538.1	237563±246	2525.3	2120.1	16
^{60}Co	1173.24 1332.5	18481±73 16569±45	221	15843±47 14399±33	219.7	190.8	13.2

Table 6. Results of calculations for the used standard sample S5

Radioisotope	Energy (keV)	S1 (0.7 g/cm^3)		S5 (1.6 g/cm^3)			
		ROI_C	Ce_A	ROI_C	Ce_A	Ca_A	%Err
^{137}Cs	661.6	284404±323	2538.1	230222±221	2542.4	2054.6	19.2
^{60}Co	1173.24 1332.5	18481±73 16569±45	221	15387±61 14028±52	221.32	185.6	16.1

3.2The suggested correction equation

By using the results obtained in tables (3,4,5,6), a correlation between the calculated activity (Ca_A) and the certificate activity (Ce_A) using the ratio between the two densities in each table was obtained then by using the trial-and-error fitting method and curve expert software the following equation was deduced as given in equation (1). This correction equation is needed when there is a density difference between the standard sources used for efficiency calibration process of any HPGe detection system and the samples to be analysed; for assay of ^{137}Cs and ^{60}Co radionuclides activity (Bq).

$$Aco = \frac{1.5\times(\rho m/\rho s)}{0.5+(\rho m/\rho s)} \times Aca \qquad (2)$$

Where the abbreviations in the suggested correction equation are as follow:
A_{co} is the corrected activity (Bq).
A_{ca} is the calculated activity (Bq).
ρ_m is the density of the measured sample in (g/cm^3).
ρ_s is the density of the standard used for efficiency calibration process in (g/cm^3).
1.5 is equation coefficient.
0.5 is equation constant.

It should be mentioned that the suggested correction equation should be applied to correct the results obtained when there is a different in densities between the standard used for efficiency calibration process or in relative method as mentioned above and the samples to be analysed for ^{137}Cs and ^{60}Co activity calculations, finally the suggested equation was applied to the results in tables (3,4,5,6) and then the new calculated error percent values for the corrected activity (A_{co}) did not exceed 3.3%.

4. Conclusion

Based on the analysis carried out for the five standard samples; S1, S2, S3, S4 and S5, using HPGe detection system to calculate and validate the activities of ^{137}Cs and ^{60}Co, it was observed that the error percentages between the calculated and certificate activity (in Bq) are from 9.8% to 19.2% for ^{137}Cs activity and from 7.77% to 16.1% for ^{60}Co respectively. The densities of the used standard samples (S1, S2,S3, S4, S5) are 0.7,1,1.2,1.3,1.6 g/cm^3 respectively. Also, it was observed that the error percentage values increase as the difference in densities between the reference standard sample (S1) and the other standard samples increase. It can be concluded that in case of gamma spectroscopy analysis using HPGe detection system, it should carefully take into consideration the difference in densities between the standard sources to be used in efficiency calibration and the samples to be measured, because this may be considered a major source of error in the obtained results.
Finally, a correction equation was suggested to be used by researchers in case that there were a differences in densities between the standards and samples to be analysed for ^{137}Cs and ^{60}Co activity calculations.

References

[1] G. Xhixha, M. Alberi, M. Baldoncini, K. Bode, E. Bylyku, F. Cfarku, *et al.*, "Calibration of HPGe detectors using certified reference materials of natural origin," *Journal of Radioanalytical and Nuclear chemistry,* vol. 307, pp. 1507-1517, 2016.

[2] M. K. Hossain, S. M. Hossain, R. Azim, and A. M. H. Meaze, "Assessment of radiological contamination of soils due to shipbreaking using HPGe digital gamma-ray spectrometry system," *Journal of Environmental Protection,* vol. 1, p. 10, 2010.

[3] M. A. Abdel-Rahman and S. A. El-Mongy, "Study of Some Parameters Affecting Efficiency of HpGe Detectors for Accurate Radionuclides Analysis," in *The International Conference on Chemical and Environmental Engineering*, 2018, pp. 371-388.

[4] S. Harb, K. Salahel Din, and A. Abbady, "Study of efficiency calibrations of HPGe detectors for radioactivity measurements of environmental samples," 2008.

[5] A. Yusuf, "Gamma-ray spectrometry for linear attenuation coefficients and selfattenuation correction factors of the skimmed milk powder," *Sakarya Üniversitesi Fen Bilimleri Enstitüsü Dergisi,* vol. 21, pp. 426-430, 2017.

[6] J. Lee, J. Byun, and J. Yun, "Mass attenuation coefficients of environmental samples for gamma-ray energies from 46.5 keV to 1408 keV," *International Journal of Radiation Research,* vol. 18, pp. 201-207, 2020.

[7] O. Sierra, G. Parrado, Y. Cañón, A. Porras, D. Alonso, D. Herrera*, et al.*, "Characterization of HPGe gamma spectrometric detectors systems for Instrumental Neutron Activation Analysis (INAA) at the Colombian Geological Survey," in *AIP Conference Proceedings*, 2016, p. 080011.

[8] G. Gilmore, *Practical gamma-ray spectroscopy*: John Wiley & Sons, 2011.

[9] E. J. M. Nguelem, M. M. Ndontchueng, and O. Motapon, "Determination of 226Ra, 232Th, 40K, 235U and 238U activity concentration and public dose assessment in soil samples from bauxite core deposits in Western Cameroon," *SpringerPlus,* vol. 5, p. 1253, 2016.

[10] N. Marković, P. Roos, and S. P. Nielsen, "Digital gamma-gamma coincidence HPGe system for environmental analysis," *Applied Radiation and Isotopes,* vol. 126, pp. 194-196, 2017.

[11] O. A. B. Omar, M. A. Abdel-Rahman, and S. A. El-mongy, "Analysis of naturally occurring radioactive materials in environmental samples using gamma spectrometry," in *The International Conference on Chemical and Environmental Engineering*, 2018, pp. 356-370.

Mineralogical Studies and Radioactivity on Stream Sediments Around Wadi Aidib Area, South Eastern Desert, Egypt

M Hassan [1*], M El Shafey [1], Bayoumi, M B [1] and O A Ebyan [1]

[1] Nuclear Materials Authority, P.O. Box 530, El Maadi, Cairo, (EGYPT)

* Corresponding Author: M., Hassan (mohamed.hassan50@yahoo.com)

Abstract. Wadi Aidib lie at southern part of the Eastern Desert of Egypt, between 22° 15' - 22° 21' N and Longitudes 36° 20' - 36° 30' E, covering an area of about 30km2. It is covered by basement rocks dissected by several wadies, Wadi Aidib as a main wadi with several tributaries. Constant of channel maintenance showed the dissection of the studied basin by lineaments, fractures and joints, dissection is minimized in wadi Aidib. Landsat ETM+ false color composite image (bands 3, 2, 1) The rock types wide spread out at Wadi Aidib area contains metavolcanics, alkali feldspar granites, quartz diorites, biotite granites metasediments. Heavy liquid separation and magnetic separation techniques in addition to microscopic examination for the studied stream sediment samples were applied. The mineral resource tonnage for each concerned economic mineral 171.315 ilmenite; 93.150, magnetite; 25.110, zircon; 9.720, total rutile; 7.695, garnet; 1.215, monazite and 810.000 green silicates, in thousands of tons. The studied stream sediments are measured radiometrically using gamma-ray spectrometer for determination the radionuclides contents. U concentration varies between 1 and 2 ppm with 1.69 as an average, Th from 3 to 7 ppm with an average 4.92 ppm.

Keywords: Wadi Aidib, Heavy minerals, gamma-ray, Uraniferous rocks and Dose Rate.

1. Introduction

The study area is located in the Halaib area in the southeastern desert along the Red Sea coast. This area is located in the town of Abu Ramad on the coastal road. The source of river sediments is usually of medium-high relief, where the highest peak is Gabal Elba (557 m.a.s.l). The main Wadis that divide the region are the Aidid Wadi and the Yoider Wadi. All are drained east to the Red Sea. The geology of the Halaib area has been studied by many authors [1] and [2], and mineral deposits in the region have been studied [3]. On the other hand, several mining studies have been conducted on Egyptian sediments in the Red Sea coastal plain. [4] The concentration of heavy minerals on the Red Sea coast is evident in the area between Halaib and Shalitin. [5] Investigate the heavy minerals content of beach deposits in the area from Hurgada in the north to El Qusier in the south. This study examines the sediments that make up one of the important Wadi of Gabal Elba region in the southern part of the Egyptian Red Sea, with a focus on the content of heavy minerals. Many authors have studied the radioactivity of the Egyptian Eastern Desert [[6], [7], [8], [9], [10] and [11]. Various authors have applied remote sensing to the geology of Eastern Desert in Egypt in different ways e.g., [12], [13], [14], [15] and [16]. On the other hand, soil acts as a migration medium for transporting radionuclides into the biological system and is therefore the main indicator of radioactive pollution of the environment. Natural radionuclides in river deposits produce important elements of exposure to back radiation of the population [17]. The study area is located in the extreme sub-eastern desert of Egypt, between 22° 15' - 22° 21' N and Longitudes 36° 20' - 36° 30' E, (Figure 1).

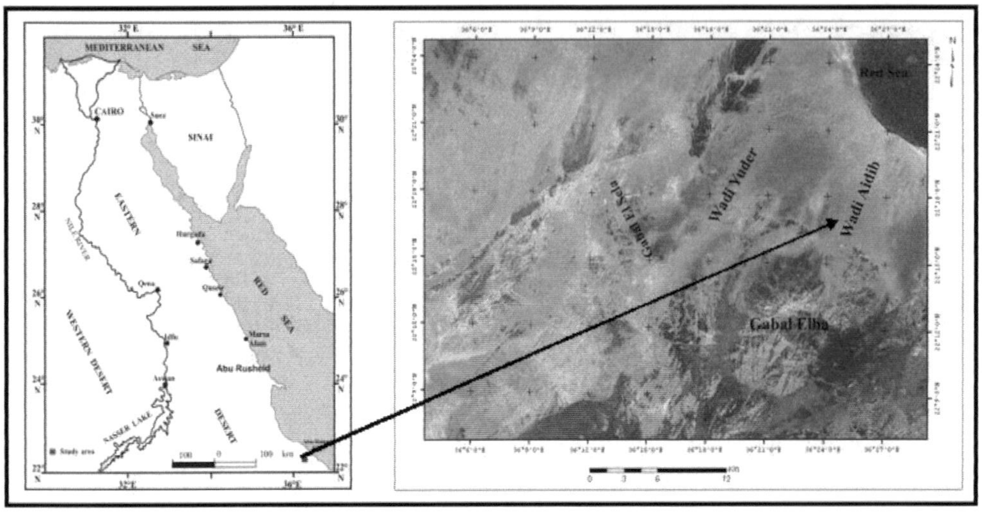

Figure 1. Location map of study are.

2. Remote sensing processing

The studied river deposits accumulate in the form of alluvial fans. They are mainly composed of sand, gravel and rock and contain heavy and valuable minerals from metavolcanics, metasediments, quartz diorite, biotite and alkaline granites. The main purpose of this study is to determine the natural radioactivity and environmental impact in various stations in the Wadi Aidib region. The morphological characteristics of the current sediments in the study area were known from Landsat ETM images, with the exception of some field observations. Landsat ETM+ (bands 3, 2, 1) are used for regional discrimination of different types of rocks and provide a base map of survey areas. Due to deformation, these rocks have the property of cutting, dissolution, and decreasing grains. In addition, the RGB green to yellowish-colored poles have a more reddish color than granite, so deformed fragments were transported from wadi to the sea in water drainage and recognized by Landsat ETM+ images and field verification. Figure 2 shows the difference between rock compounds. The basic rocks are reddish-colored. metavolcanic rocks are displayed in blue and granite is marked in violet. Wadi and alluvial deposits are displayed in blue RGB.

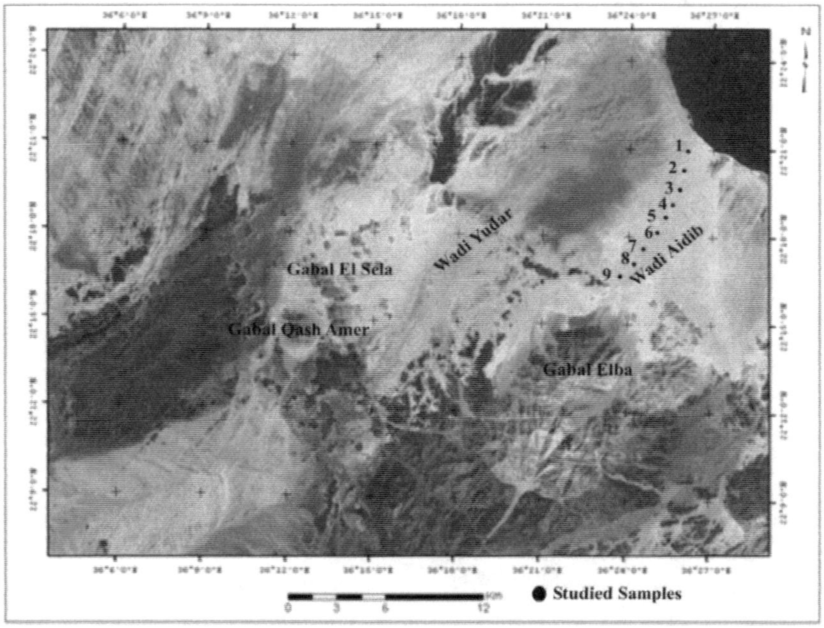

Figure 2. False color composite image FCC, (band 3, 2, 1) in RGB of study area.

3. Sampling and Techniques of Work

The study area is 10 km long and 3 km wide with 9 samples at a depth of one meter at a distance between the samples is about 1000 m almost (Figure 3). Representative samples were weighed and poured into a calibrated cylinder and well compacted by stirring to be analogous to a field deposit. The weight of the sand was divided by the volume to obtain the apparent specific gravity. John's splinter divided each field monster into two halves. One of them was re-divided to obtain a representative sample of about 50 grams for various analyses, while the rest was used as a mixed sample that was retained as a reference sample. Each mixed sample was quartered by a rotary dispenser of the correct weight (about 50 g). The quartered sample was weighed and washed with water to get rid of slimes and salts by settling. To decompose clay particles, about 5 ml of ammonium hydroxide was added. Samples of organic material were treated with a few drops of 30% hydrogen peroxide. After that, the samples were sieved and washed several times with a 50-μ screen to determine the amount of mucus. The undersized fraction and excess were dried and weighed. The slimes -free fraction was divided into heavy and light fractions using bromoform (Sp. gr. 2.88 g/cm3). Methylene iodide (Sp.gr 3.3 g/cm3) was used to reduce the size of the heavy fraction obtained by separating light silicates into floating subfractions, while isolating dark silicates and heavy economic minerals. A percentage of the mass of each heavy fraction of the corresponding original raw sample was calculated. A hand magnet was used to separate magnetite from the heavy fraction of methylene iodide. Magnetite was treated with a laboratory isodynamic magnetic Separator model (L-1) from Frantz with a longitudinal inclination of 20o, a lateral inclination of 5o and ampere values of 0.2, 0.5, 1.0 and 1.5. Subsequently, five factions were won. Each fraction was weighed and about 1000 grains were counted under a binocular microscope. With a comparison [18]. The weight percentage of each mineral in each fraction was calculated. The content of each mineral in each original composite sample were determined by adding its contents to various calculated conclusions binocular microscope and optics, the environmental scanning electron microscope (ESEM) conducted a detailed mineralogical examination and identification. Philips (XL30) ESEM, equipped with an energy dispersion X-ray spectrometer (EDX), was used to scan electron microscope images of the surface properties of economic minerals. In addition, semi-quantitative (EDX). were performed for determination of the chemical composition of the economic minerals. All of the analyses were carried out in the laboratories of the Nuclear Materials Authority, Egypt.

Figure 3. Photographs shows manuals procedures for sampling at the study area.

4. Drainage patterns

Geologically, the area has four main categories: the Red Sea Mountain range, the gentle isolated hills, the low conical hills, the Wadi deposits, the dunes, the coastal plain (fans) and the sand of the beach. The most important mountains within the area are El Sela, Qash Amer, Elba and Shendib. The area is traversed by many wades, the most important of them are, wadi Eishimhai, wadi EiKwan and wadi Yoider, Diit, Sarmati and Wadi Aidib. Drainage systems follow patterns of water flow that is influenced mainly by geologic structures, terrain properties, surficial processes and rainfall behavior. These patterns reflect a stable condition of the terrain surface on which they developed. However, drainage patterns exhibit anomalies in their orientation as a result of subsequent events and processes, (Figure 4).

5. Geologic setting

Precambrian substrates devolved during the Pan-African tectonic-thermal event 950-550 million years ago. Geological studies have shown that the study area is covered with muscovite granites containing ultramafic rocks of ophiolite Sol Hamid, intermediate metavolcanics, uninvited gabbro, granodiorites, biotite granites, monzogranites and garnets. These rocks are pumped from pegmatitic and quartz veins and intersect with acidic and basic ones. Granodiorites are less common and are found in the southeastern part of the study area and form separate hills with adverse weather conditions. They are characterized by a flaky texture and contain xenoliths of different shapes and compositions. Biotite granites are distributed in the eastern and central parts of the study area. These granites are the main component of the study area and are considered the main source of heavy minerals in alluvial lands. They penetrate monzogranites and intersect with acids and alkalis. Monzogranites are common in the northeastern part of the study area, the monzogranites are characterized by medium-large granular crystals, red color, weather-resistant surfaces and the presence of multiple joints and cracks. These granites mainly include feldspar, biotite, and quartz. Additional minerals are rutile, zirconium and opaque. Sericite, chlorite, and epidote are secondary minerals. Moscow grains containing pomegranates are rare and are found only in the southwestern part of the study area. These granites are located on a terraced hill that forms the Gabal Elba (1437 m.a.s.l.). They are characterized by the fact that they have a moderate to coarse-grained, light pink color, are weakly affected by bad weather and joints and fractures. This area is crossed by several wadis, the most important of which are: Wadi Serimtai, Wadi Baashoia, Wadi Elba, Wadi Hibro, Wadi Oqhoq, wadi Aideib and Wadi Ei kwan. These wadis are the most favorable places for rainwater recipients and for the search for fresh water in figure 6 (a), (b), (c) and (d).

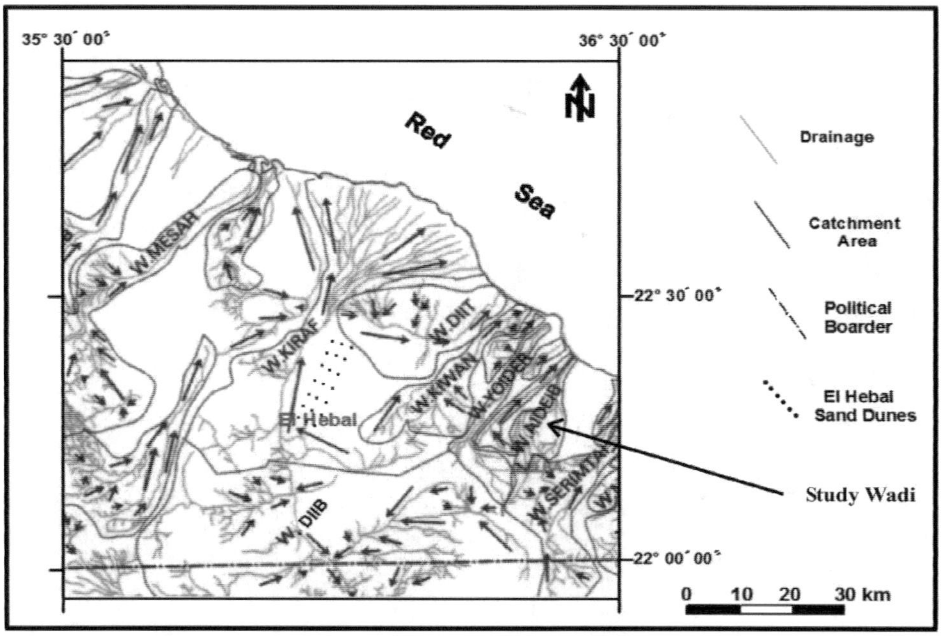

Figure 4. Drainage patterns image for wadies at study area.

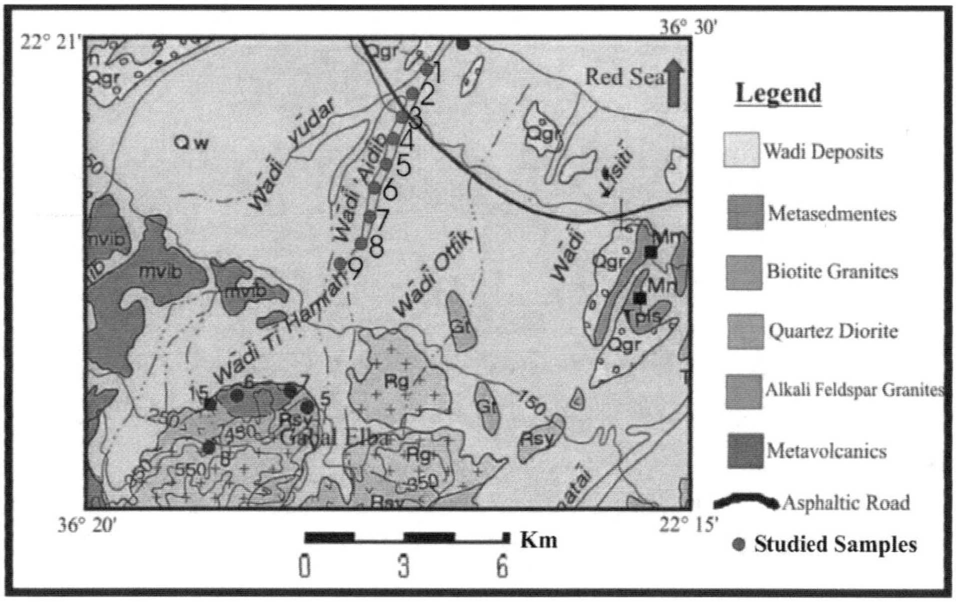

Figure 5. Geological map with Samples location of the study area.

Figure 6. Field Photographs shows Gabal Elba and Wadi Adieb, South Eastern Desert, Egypt.

6. Radiometric investigations

Ten samples from the studied stream sediments are analyzed radiometrically using gamma-ray spectrometer (NaI-detector) for determination the radionuclides contents. The results of the studied radioelements (eU eTh, RaeU and K) and their ratios are reported in table 1.

Table 1. Radionuclides contents (eU, eTh, RaeU and K in ppm) and their ratios

Samples No.	eU *ppm*	eTh *ppm*	Ra (eU) *ppm*	K%	Uc	eTh/U	eU-eTh/3.5	eU/RaeU	Uc/eU
1	2	5	2	2.12	8	2.50	0.57	1.00	4.00
2	1.8	4	3	1.84	6.8	2.22	0.66	0.60	3.78
3	1	3	2	2.08	7.5	3.00	0.14	0.50	7.50
4	1.5	5	2	2.08	8.3	3.33	0.07	0.75	5.53
5	2.3	4	3	2.07	7.5	1.74	1.16	0.77	3.26
6	2	4	3	1.87	7.9	2.00	0.86	0.67	3.95
7	1	7	2	1.88	7.7	7.00	-1.00	0.50	7.70
8	2	5	2	1.63	7.3	2.50	0.57	1.00	3.65
9	1.6	6	2	1.79	6.8	3.75	-0.11	0.80	4.25
10	1.78	6	2	1.88	6.9	3.37	0.07	0.89	3.88
Min.	1.00	3.00	2.00	1.63	1.74	0.50	-1.00	0.50	3.26
Max.	2.30	7.00	3.00	2.12	1.74	7.00	1.16	1.00	7.70
Av.	1.69	4.92	2.33	1.92	1.74	3.24	0.26	0.75	4.87

6.1. Distribution of radionuclides and their interrelations in the studied sediments

Uranium concentration varies between 1 and 2 ppm with 1.69 as an average. Radium content in the study area falls between 2 ppm and 3 ppm with an average of 2.33 ppm. Table 1. Thorium contents range from 3 to 7 ppm with an average 4.92 ppm. Also, eTh-eU show ill-defined trend and with eTh/eU, meaning that they are mobilize apart each other figure 7(a). Potassium varies between 1.63 wt.% and

2.12 wt.% with 1.92 wt.% as an average, also it has ill-defined correlation with both eU and eTh, indicating they aren't adsorbed on clay minerals (Figure 7(b) and (c)). Radium is the equivalent of uranium, though it has weak correlation with U (r = 0.29). The authigenic uranium content is calculated as follows: (Uthirojan U) = U-Th/. 3.5 l Athenian uranium values below 5 are considered to be an oxic state, while values above 5 indicate suboxic for the anoxic state. All authigenic uranium is less than 5 (-1-1,16), which means the proliferation of oxidative conditions Table 1.

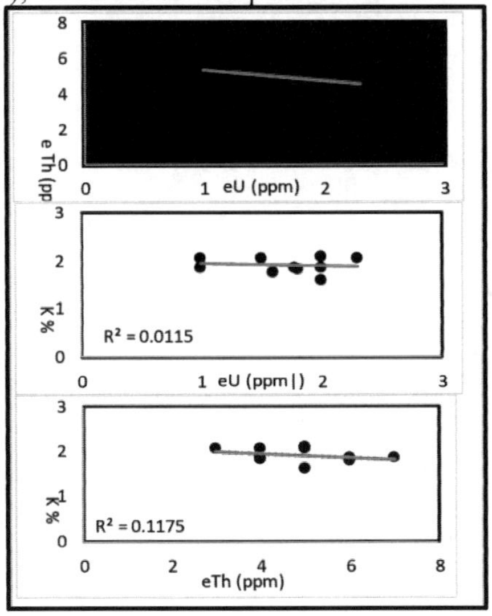

Figure 7: Inter element relations between radionuclides: A- eU-eTh, B- eU- K, C- eTh-K.

6.2. Radioactive Equilibrium

The radioactive balance in the study area can be determined by two methods, the first method depends on the calculation of the equilibrium coefficient (P) [20]. The second equilibrium method is carried out using chemically determined uranium (Uc) data [20]. The percolation and accumulation of uranium are studied in certain geological processes, such as the change that causes imbalance. In addition, groundwater can operate in some uranium deposits and lead to the separation of uranium from its original location and relocation to another location. Another factor controlling the stable state is the loss of radon as one of the daughters of uranium. This loss easily occurs due to the solubility of radon in water and its leakage through porous spaces, as well as along breakdowns and other cracks. P- and D-factors of the studied sediments were estimated Table 1. It is clear from the table that the average value of the P-factor of the studied stream sediment samples is less than or equal one (P = 0.75) referring to possible gain of radon or radium. There is a clear difference between uranium contents determined chemically and radiometrically. This may be resulted from the fact that when uranium is too recently deposited. Uranium concentrations are seldom. On this base, two types of uranium deposits are recorded, old and recent uranium deposits.

7. Analytical procedure

7.1. Apparent specific gravity

The apparent density of the samples varies from 1.50 gm/cm³ to 1.72 gm/cm³, and an average value of 1.62 gm/cm, shown in Table 2. Figure 8 (A) and (B).

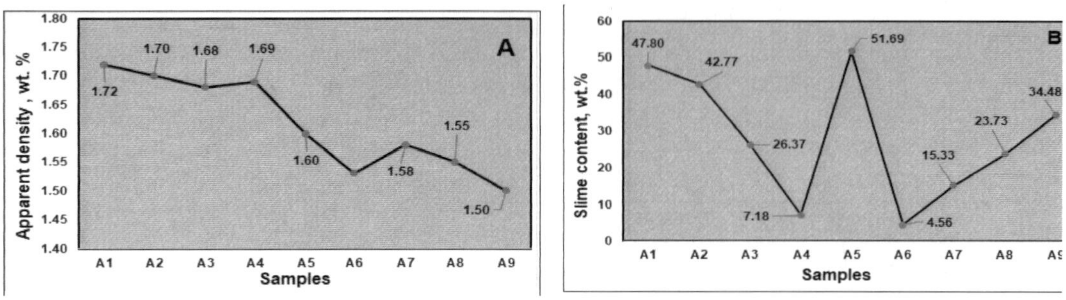

Figure 8. Scatter plot diagram showing A- the measured average apparent density for samples, B- The slime contents of samples along the studied area; wt. % = weight percent.

7.2. Determination of slime content

The contents of slimes along samples in the studied area are shown in figure 8. It was calculated that the slime contents for samples vary from 0.32 to 4.24 wt. %. The results are shown in Table 2.

Table 2. The measured average apparent density and the slime contents for the studied samples.

Samples No.	Apparent density *(gm/cm^3)*	Slime *wt. %*
A1	1.72	47.80
A2	1.70	42.77
A3	1.68	26.37
A4	1.69	7.18
A5	1.60	51.69
A6	1.53	4.56
A7	1.58	15.33
A8	1.55	23.73
A9	1.50	34.48
Average	1.62	28.21

7.3. Heavy liquid separation

A representative sample weighing about 50 gm from each collected drilled sample was taken and subjected to a process of heavy liquid separation using bromoform solution (sp. gr.=2.88 gm/cm^3). The heavy fraction of bromoform was taken on a filter paper and the light fraction was taken on another filter paper. The two fractions were left to complete the filtration. The remaining bromoform in the bores and the thin film coating the particles of both light and the heavy fractions was washed with acetone several times till the two fractions becomes clean from any bromoform traces. Both light and heavy fractions were dried, weighed and the obtained heavy fractions during the bromoform separation were subjected to Methylene Iodide (3.3 gm/cm^3). The separation was carried out to reduce the size of the heavy fraction by the separation of the light-colored silicates in the float layer, while the darker colored heavy economic minerals were sinking. Each bromoform heavy fraction was put in a separating funnel and a suitable amount from Methylene Iodide was added to permit each particle to freely sink or float. Both heavy and light fractions were washed with acetone, dried, weighted and their percentages were calculated relative to the original sample weight. The results are shown in Table 2 and Figure 9 (A) and (B). The obtained individual heavy Methylene Iodide fractions were subjected to a process of magnetite separated using a standard hand magnet. The treated sample is spread on a large sheet of clean paper. A small hand magnet was dripped in the spread sample and passes across it, the separated magnetite was then transferred to another clean paper. The step was repeated several times until the treated sample becomes completely free of magnetite. The collected magnetite was weighed and its corresponding original Wt. % was determined.

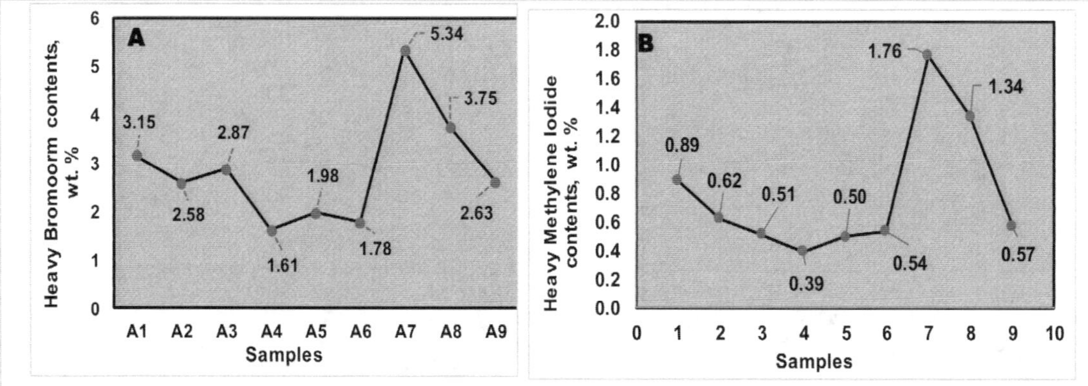

Figure 9. Scatter plot diagram showing A- The heavy bromoform contents for samples, B-The heavy methylene iodide contents of samples along the studied area; wt. % = weight percent.

7.4. The magnetic fractionation of Methylene Iodide fractions

This step was carried out to differentiate between the various heavy minerals into different magnetic fractions and hence facilitate their identification and counting under the microscope. The individual magnetite free samples were treated by laboratory Frantz Isodynamic Magnetic Separator Model (L-1) with longitudinal slope of 20°, side slope 5o and at ampere values of 0.2, 0.5, 1.0 and 1.5. Then, five fractions were obtained. Each fraction was weighted and about 1000 grains were counted under the binocular microscope. The weight percent of each mineral in every fraction was calculated using the equation of [18]. The contents of each mineral in each original composite sample were determined by adding its contents in the different counted subtractions, from different microscopic fields covering the entire prepared slide. The ilmenite particles and garnet were obtained in the highly magnetic fraction (0.2 Mag.), the leucoxene, the rest of monazite and garnet in the next magnetic fraction (0.5 Mag.), the rest of monazite in the third fraction (1.0 Mag.), the rest of opaque particles represent brown rutile (1.5 Mag.), and the last fraction represents the rest of reddish rutile and zircon (1.5 non-Mag.) [24]. In addition to Furthermore, the very low feeding rate in order to give a complete chance for each particle to move in the appropriate magnetic path. The frequency of each mineral in the original sample was calculated and the average values of economic minerals along 9 samples were calculated and given in Table 3.

Tables 3. The results of heavy liquid separation and average percent of the total heavy economic minerals, wt. % = weight percent; Br. % = Bromoform %; M.I. % = Methylene iodide.

Samples	Samples *Wt. g*	Slime *Wt. %*	Conc. *Wt. %*	Heavy. Br. *Wt. %*	Light Br. *Wt.%*	Heavy. M.I *Wt.%*	Light M.I *Wt. %*	Magnetite *Wt. %*	Ilmenite *Wt. %*	Zircon *Wt. %*	Rutile *Wt. %*	Garnet *Wt. %*	Monazite *Wt. %*
A1	7030	47.8	52.2	3.15	49.05	0.89	2.26	0.215	0.498	0.093	0.044	0.036	0.005
A2	6325	42.77	57.23	2.58	54.65	0.62	1.96	0.162	0.344	0.062	0.028	0.022	0.003
A3	5100	26.37	73.63	2.87	70.75	0.51	2.36	0.144	0.281	0.046	0.015	0.023	0.003
A4	6480	7.18	92.82	1.61	91.21	0.39	1.22	0.106	0.213	0.035	0.016	0.02	0.002
A5	8715	51.69	48.31	1.98	46.33	0.5	1.48	0.152	0.27	0.039	0.017	0.015	0.002
A6	5810	4.56	95.44	1.78	93.66	0.54	1.24	0.161	0.295	0.043	0.019	0.016	0.003
A7	6295	15.33	84.67	5.34	79.34	1.76	3.57	0.605	0.937	0.131	0.054	0.026	0.009
A8	5395	23.73	76.27	3.75	72.52	1.34	2.42	0.435	0.741	0.088	0.04	0.027	0.007
A9	6250	34.48	65.52	2.63	62.89	0.57	2.06	0.168	0.336	0.034	0.014	0.017	0.003
Average		28.21	71.79	2.85	68.93	0.79	2.06	0.239	0.435	0.064	0.027	0.022	0.004

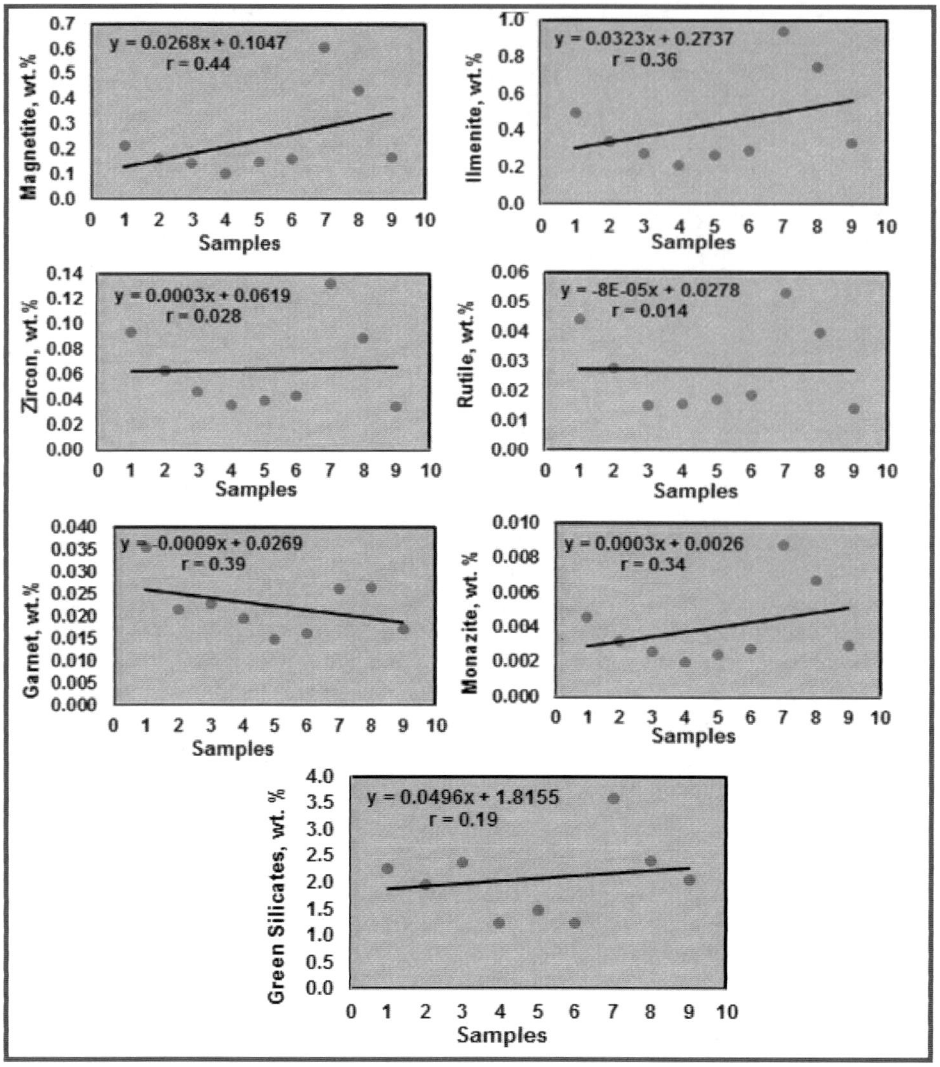

Figure 10. Scatter plot diagram showing the individual economic minerals for the drilled samples of the area of study: magnetite, ilmenite, zircon, rutile, garnet, monazite and green silicates; wt. % = weight percent

7.5. Distribution of Economic minerals

The heavy Methylene Iodide fraction contains the six economic minerals; magnetite, ilmenite, zircon, rutile, garnet and monazite in addition to leucoxene and green silicate minerals. In these concentrated black sands, the distribution of economic minerals is correlated by their specific gravities, their grain sizes or even their grain shapes and also the location of their occurrences. Investigating the contents of the different individual economic minerals and green silicates in Table 3. ensures that the content for each of these minerals increase in samples A1, A7 and A8 respectively through the area of study are shown in figure 10. The calculated different economic minerals and green silicates of the various drilled samples in the studied area are shown in Tables 3. It was detected that the correlation between the heavy mineral's contents and each of the other individual economic minerals are shown in figure 11.

Investigating the relationships between magnetite mineral content and the content for each of ilmenite, zircon, rutile, garnet, monazite, and green silicate minerals shows positive correlations.

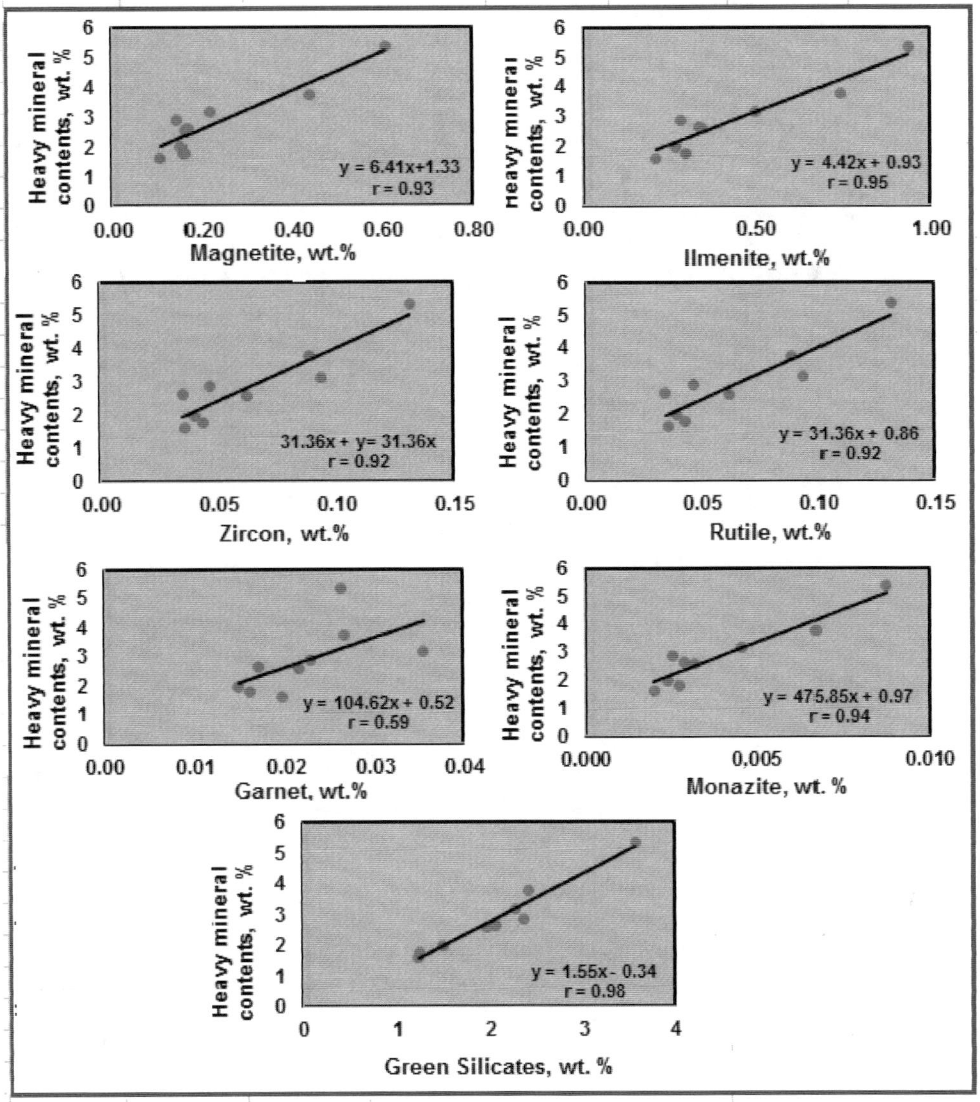

Figure 11. Scatter plot diagram showing the relationships between heavy mineral contents and each of the individual economic minerals for the drilled samples of the area of study: magnetite, ilmenite, zircon, rutile, garnet, Monazite and green silicates; wt. % = weight percent.

The correlation coefficient values equal 0.98, 0.89, 0.85, 0.44, 0.99 and 0.84, respectively (Figure 12). The correlation coefficients are 0.93, 0.95, 0.92, 0.92, 0.59, 0.94 and 0.98 for magnetite, ilmenite, zircon, rutile, garnet, monazite and green silicates. Due to the relatively lower specific gravity and coarser grain sizes, garnet content may be variable according to the extent of erosion. Hence, garnet has the lowest correlation coefficient value with heavy Methylene Iodide fraction content. It is obvious that there are direct relations between the heavy mineral's contents and each of the individual economic minerals and green silicates. In general, as the economic (heavy) mineral component content increases in the heavy mineral's contents, as its correlation coefficient increases. Then, it can be concluded that as magnetite mineral content increases as the content for each of zircon, rutile, monazite, or green silicate minerals increases, garnet has the lowest correlation coefficient value with magnetite mineral content.

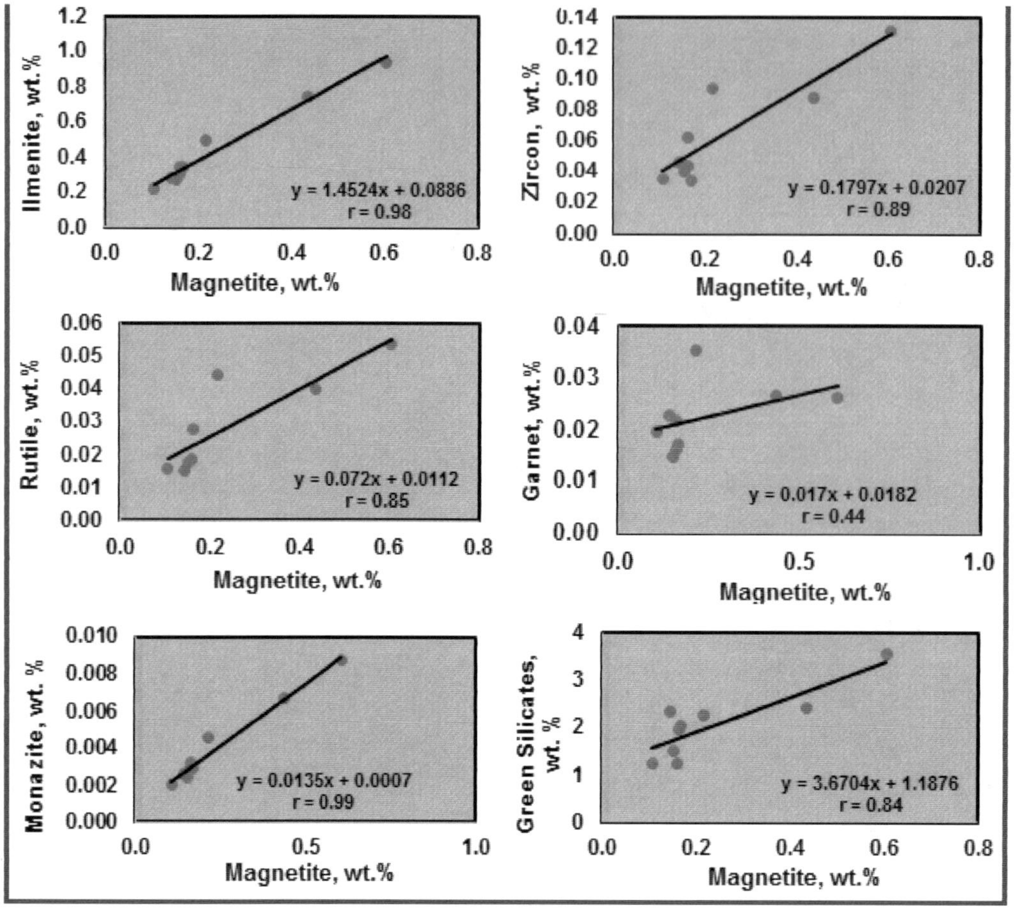

Figure 12. Scatter plot diagram showing the relationships between magnetite mineral content with: Ilmenite, zircon, rutile, garnet, Monazite and green silicates for the drilled samples of the area of study; wt. % = weight percent.

8. Ore dressing concentration

The concentration and separation of economic heavy metals using ore dressing techniques is important for the concentration and separation of economic heavy minerals on an industrial scale using an appropriate technological system based on the physical and chemical properties of mineral grains. Representative samples weighing approximately 2.500 kg have been subjected to a strong concentration of mineral and separation by means of a simple technological scheme involving gravitational and magnetic methods, (Figure 13). Each representative sample has dried out and left using a 0,5 mm screen to remove large shell residues and fragments. The gravitational concentration process is one of the oldest methods of enrichment using a medium-sized Wilfley moving table, a two-stage scheme for placing sand has been divided into three fractures: concentrate, a middle and tail fractions. The resulting concentration was again addressed to achieve the final economic concentration of heavy metals. Show the focus of mixing to separate the magnetite. The separation was carried out by the Laboratory Franz Isodynamic Magnetic Separator Model (L-1). The weigh percent of magnetism is separated; the percentage of the initial mass of the sample was calculated in Table 4. There is a relationship between the distribution of magnetite and common and economic heavy minerals is observed.

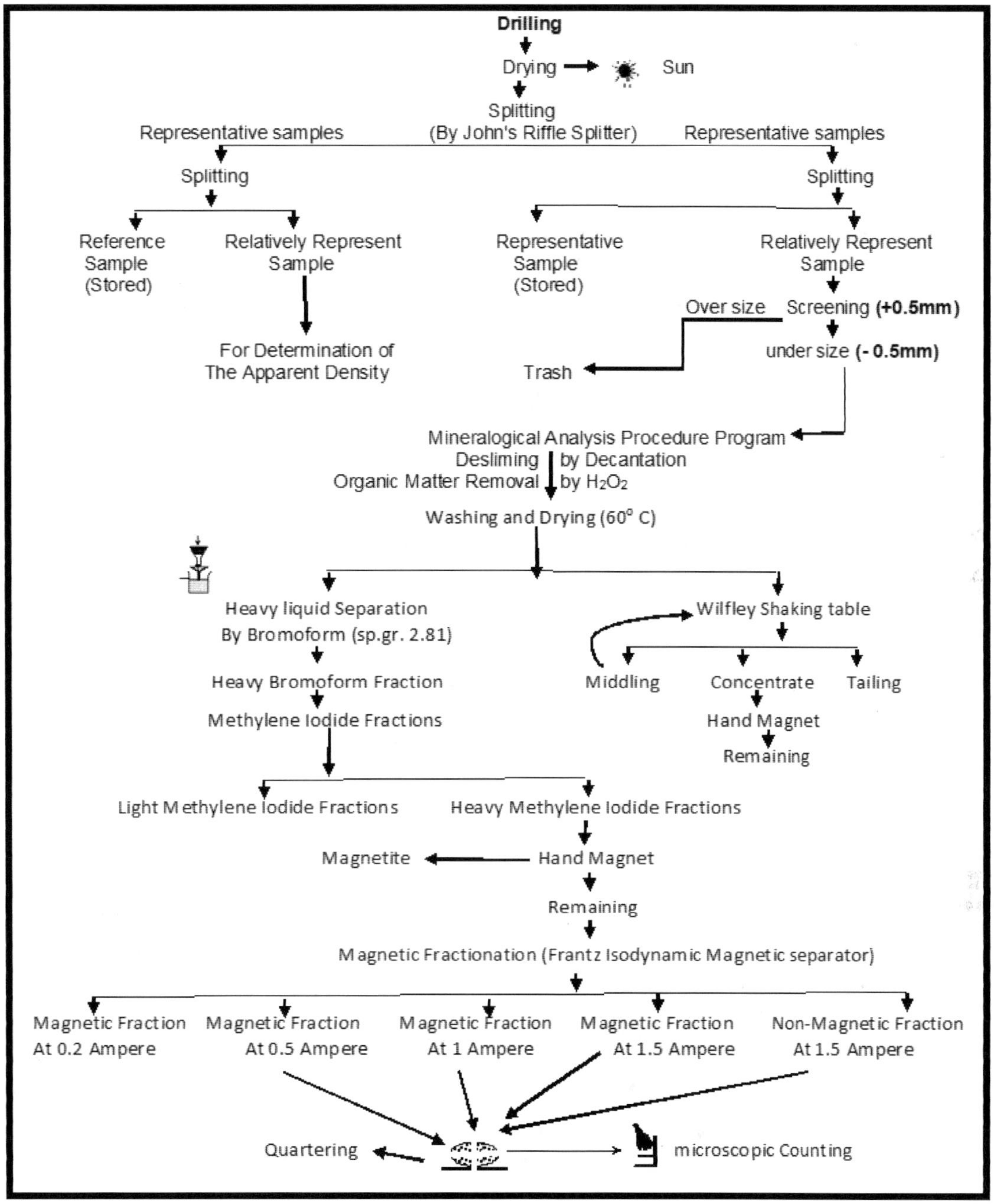

Figure 13. A flowsheet showing the different stages for collecting sample and various treatments of the mineralogical analysis procedure program.

Table 4: Frequency of the heavy economic minerals obtained by ore dressing techniques.

Mineral	Ilmenite	Magnetite	Zircon	Rutile	Garnet	Monazite	Green Silicates	Total
wt. %	0.423	0.230	0.062	0.024	0.019	0.0035	2.0	2.76

8.1. Estimation of the economic mineral reserves

The calculated average weight percentages of economic minerals and green silicates for the study area are presented in Tables 5. The total reserve of economic minerals and green silicates in the study area amounts to 1117.8 thousand tons of study area; Tables 6. The number of mineral resources for each relevant economic mineral in another zone is calculated in accordance with Table 5. From the table contains: 171,315 ilmenites; 93,150, magnetite; 25.110, zircon; 9,720, total rutile; 7,695, garnet; 1,215, monazite and 810,000, green silicates in thousands of tons.

Table 5: Volume, apparent density and tonnage of the raw sand of studied area.

Volume *(m3) (Length x width x depth)*	Apparent Density *(ton/m3)*	Tonnage of raw sand *(tons)*
10000 x2500 x1= 25,000,000	1.62	40,500,000

Table 6. The average content and resource for each economic mineral and Green Silicates along the studied area.

The Economic mineral	The average content *wt.%*	Resource *tons*
Ilmenite	0.423	171,315
Magnetite	0.230	93,150
Zircon	0.062	25,110
Total Rutile	0.024	9,720
Garnet	0.019	7,695
Monazite	0.003	1,215
Economic Minerals Content	0.76	308,407
Green Silicates	2.0	810,000
Total Heavy Minerals Content	2.76	1,118,407

9. Mineralogical Investigations

Correlation between soil radioactivity and its mineralogical content was investigated. The high radionuclides contents are mainly according to the presence of zirconium - and rare earth bearing monazite which is source of thorium, and rare earth elements. In the studied stream, monazite is the principal source of both uranium and thorium [25]. Uranium high concentrations in the stream sediments samples are mainly related to the metamict zircon and this indicated from the fact that most zircon grains vacant from any radioelements. The essential minerals in the studied sediments include opaque minerals (ilmenite, magnetite, and leucoxene), in addition to rutile, zircon, garnet, monazite, titanite and colored silicate minerals (amphibole, pyroxene, apatite, epidote, sturolite, kyanite, etc…). Magnetite reaching up to 29.25 % of the total heavy minerals of study area Table 2 occurs as octahedral angular to sub-rounded, sub-metallic grains. Magnetite is usually deep reddish brown to black color, with metallic to dull luster. The EDX/BSE images of magnetite of study area ESEM were shown in figure 14 (A). [26] suggested granitic source for magnetite based on its euhedral form

Ilmenite reaching up to 55 % of the total heavy minerals of study area contains 40 - 65% TiO_2 depending on its geological history. Under the binocular microscope, ilmenite colors are deep blue to black, ilmenite and leucoxene are subangular to rounded. The EDX/BSE images of Ilmenite of study area ESEM were shown in figure 14 (B). Zircon is found as accessory mineral in most igneous rocks and some metamorphic rocks, Zircon reaching up to 8.25% of the total heavy minerals of study area Zircon grain habits show euhedral crystals, irregular, ovoid, spherical and elongated grains. It is usually colorless, red, yellow, brown, brown, pink and green. The EDX and ESB images show differences in the chemical composition and prismatic morphology of bipyramidal zircon shown in figure 14 (C), D, E and (F), and figure 15 (A).

Apatite is one of the auxiliary minerals found in crustal rocks, and its presence and distribution are often used to create a model of geological processes such as mantle melting, hydrothermal processes, etc. Apatite is represented by Ca10(PO4)6(OH, F, Cl)2, with many possible substitutions of Ca, P, and F. A minor replacement of Ca in natural apatite is mainly Na, Sr, Mn and REE. Apatite grains usually green, yellow, less often colorless, and blue to violet, pink and brown. The EDX/BSE images of Apatite of study area ESEM were shown in figures. 15 (B) and (C).

Rutile grains are prismatic, elongated, tabular and of irregular shape. Reaching up to 3.5 % of the total heavy minerals of study area. The recorded varieties are translucent foxy red prismatic, the color of rutile is yellow, brown, red, black, straw yellow and green. Rutile concentrates with about 94 wt. % TiO2 can be used in pigment industry, welding rods coating and metallurgical uses ESEM were shown in figure 15 (D).

Secondary rutile derived from the alteration of Fe-Ti oxides especially ilmenite. this is the end product of the alteration of Fe-Ti oxides and is characterized by its black opaque appearance under microscope. ESEM were shown in figure 15 (E). Sphene is a calcium titanium silicate mineral, $CaTiSiO_5$. Trace impurities of iron and aluminum are typically present. Most of sphene grains are detected in green silicate fraction obtained during gravity concentration using shaking table. EDX semi-quantitative analyses and BSE images of euhedral sphene grains subjected to elemental chemical analysis (Figure 15 (F)).

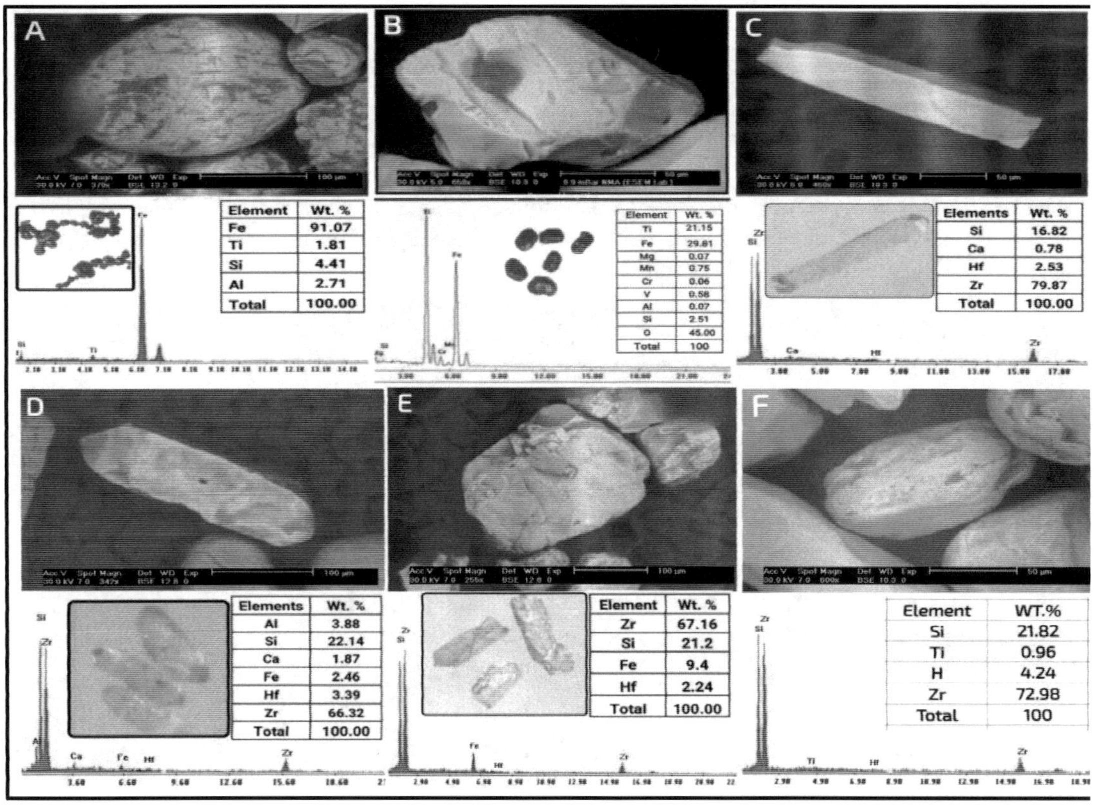

Figure 14 A, B, (C, D, E and F). EDX semi-quantitative analyses and BSE images of octahedral magnetite crystal grain, ilmenite Euhedral grain, prismatic bipyramidal zircon grains, euhedral zircon grain.

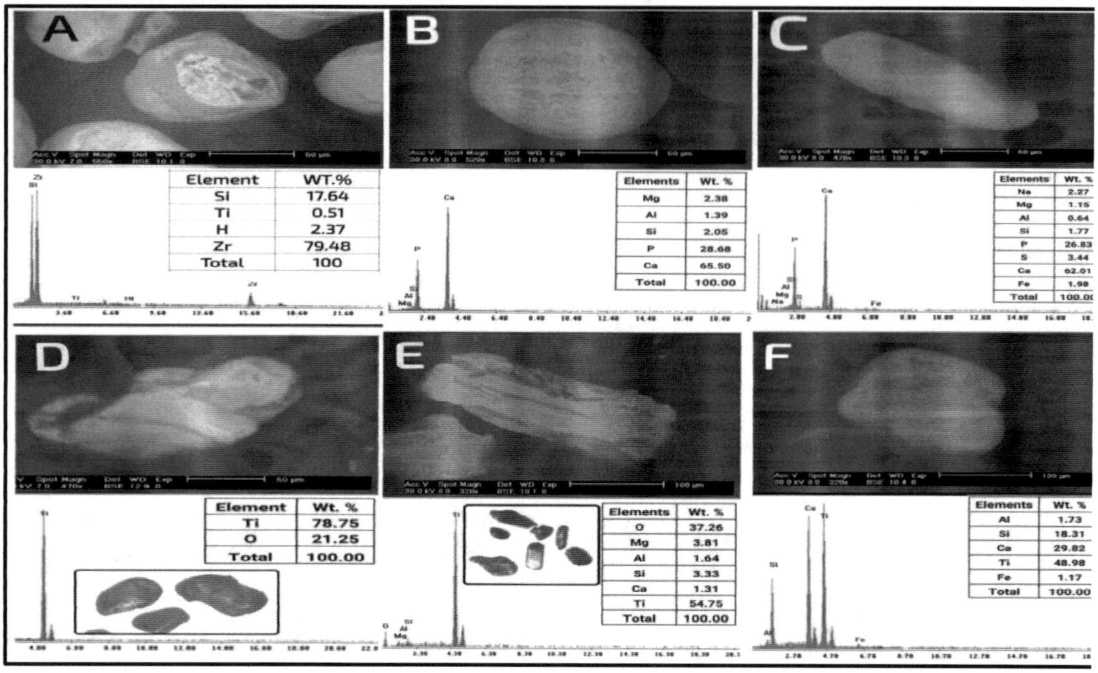

Figure 15 A, (B and C), D, E and F. EDX semi-quantitative analyses and BSE images of spherical zircon grain, rounded and prismatic apatite grains, spherical rutile grains, prismatic and spherical secondary rutile grains and euhedral sphene grains

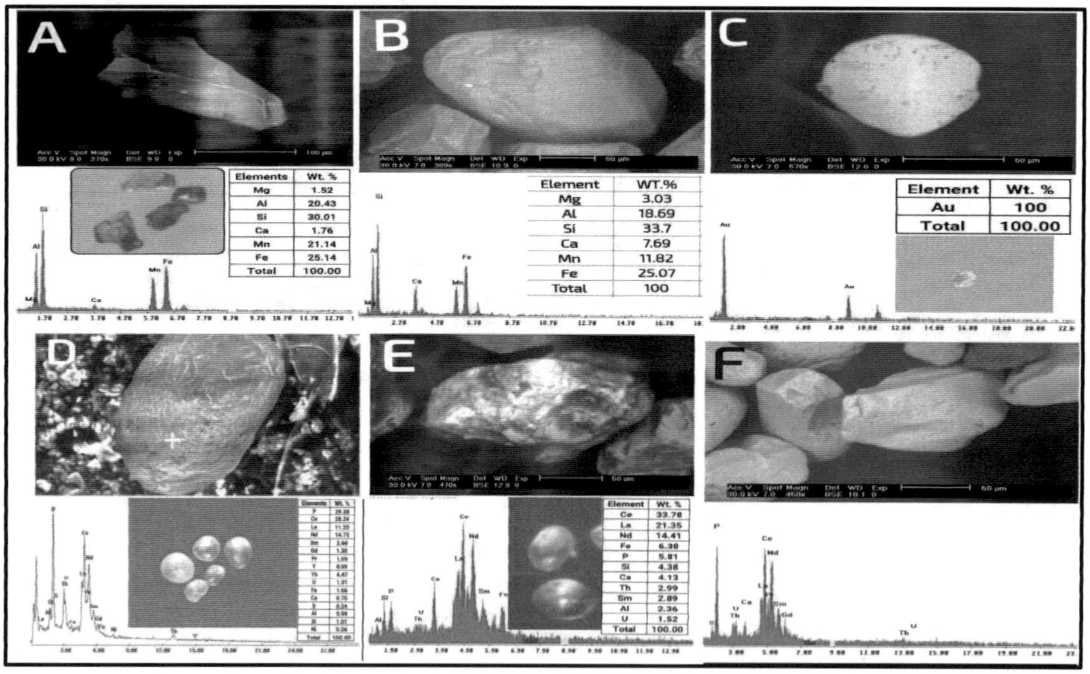

Figure 16 A, (B and C), (D, E and F). EDX semi-quantitative analyses and BSE images of Euhedral and rounded garnet crystal grain, discoidal gold grains, subrounded monazite grains with contaminated surfaces picked (D), rare earth elements bearing monazite (E) and subrounded monazite grains (F).

Garnet has a general formula A3B2 $(SiO_4)_3$, where A, may be calcium, iron, magnesium or manganese, and B may be iron, aluminum, or less commonly, chromium and vanadium. Because of increase of iron, manganese, calcium and magnesium content. Garnet is reaching up to 3.25 % of the total heavy minerals of study area. Garnet is a common mineral of metamorphic rocks. ESEM were shown in figures 16 (A) and (B). Native gold grains represented by discoidal, elongated, wire like, spherical, massive, platy and ornamented particles. A few grains of gold were picked from top concentrate of shaking table and from heavy Methylene Iodide fraction contains, by using binocular stereomicroscope and subjected to oxide analyses using ESEM gold source derived from ophiolitic ultramafites and serpentinite rock [19]. The oxide chemical compositions of the analyzed grains and the EDX and the back scattered electron image were graphically represented in figure 16 (C). Monazite is a phosphate of light lanthanides. Thorium and uranium on a very small scale are usually associated with lanthanides, where their tests reach 5-6% and <1, respectively. In other words, in addition to thorium and uranium, the mineral monazite consists to a much lesser extent of rare earth phosphates (Ce, La, Nd) PO4. Monazite seeds have different degrees of roundness, but sometimes maintain a prismatic and bipyramidal habit. Monazite reaches up to 3.25% of the total amount of heavy minerals in the study area. According to [25]. It was concluded that monazite contains 61.5% ∑REEs, 5.75% ThO2, 0.65% U3O8, 28.0% P2O5. Monazite grains have a light to deep canary color and lemon yellow, and the traces have a colorless color. Most grains are rounded to subrounded. The Monazite images of the EDX/BSE ESEM study area were presented in figure 16(D), (E) and (F).

10. Summary and Conclusion

Wadi Aidib lie at the extreme southern part of the Eastern Desert, Egypt, between 22o 15' - 22o 21' N and Longitudes 36o 20' - 36o 30' E. The studied stream sediments accumulate in the form of alluvial fans and mainly composed of sands, gravels, boulders and contain some heavy valuable minerals which are derived from metavolcanics, metasediments, quartz diorite, biotite and alkali feldspar granites. Heavy liquid separation and magnetic separation techniques in addition to microscopic examination for the studied stream sediment samples were applied to identify the heavy minerals in these sediments. These minerals are represented by magnetite, hematite, ilmenite, rutile, garnet, zircon, monazite and green silicates. U concentration varies between 1 and 2 ppm with 1.69 as an average, Th from 3 to 7 ppm with an average 4.92 ppm, RaeU between 2 ppm and 3 ppm with an average of 2.33 ppm and potassium between 1.63 wt.% and 2.12 wt.% with 1.92 wt.% as an average. eTh show ill-defined trend with eU and both eU, e Th haven't any correlation with eTh/eU, meaning their different behavior. Also, K has ill-defined correlation with both eU and eTh, indicating they are mobilized to adjacent rocks and not adsorbed on clay minerals or found on other minerals. All authigenic uranium values are lower than 5, meaning the prevailing of oxic conditions. P- and D- equilibrium factors of the studied sediments illustrated disequilibrium by leaching (P-factor) and addition (D-factor), suggesting the presence of old and young uranium deposits. eTh/eU ratio values of the studied sediments indicate that they lie in the poor weathering and rapid deposition of rock detritus.

References

[1] Hume W F 1937 *Geology of Egypt Survey of Egypt, Cairo II* **(3)**: 689-990.

[2] El Rammly M F 1972 *A new geological map for the basement rocks in the Eastern and Southwestern Desert of Egypt. Ann Geol Surv Egypt* **2**: 1-18.

[3] El Shazly et al 1976 *Regional prospecting for iron ores in Bahariya Oasis, El aiyum area, Egypt. Academy of Scientific Research and landsat1*.**iII**.24p.

[4] Dabbour G A , Morsy M A and El Hadary A F 1999 *Occurrence of black sands with potential nuclear elements along the southern part of the Egyptian Red Sea Coast. 1st Semin. Nuc. Raw Mater. and their Tech., NMA, Cai*ro (Abstract).

[5] Yousria M S. And El Gohary A M 2005 *Mineralogical composition of the coastal sediments between Hurghada and El Qusier, Red Sea Coast. Sed. Egypt,* Vol. **13**, pp. 115 – 132.

[6] El Ghawaby M A 1973 *Structures and radioactive mineralization of Wadi Zeidun area, Eastern Desert, Egypt. Ph. D. Thesis, Ain Shams Univ., Cairo, Egypt.*

[7] El Shazly E M , and El Ghawaby M A 1974 *Tectonic analysis of Wadi Zeidun area and its application in localizing radioactive mineralization in the Central Eastern Desert, Egypt. 2nd Arb. Conf. Miner. Resour., Jeddah, Saudi Arabia.* pp. 20-43.

[8] El Kassas I A 1974 *Radioactivity and geology of Wadi Atalla area. Eastern Desert of Egypt, Arab Republic of Egypt. Unpublished Ph.D. Thesis, Fac. of Sci., Ein Shams University, Geiza, Egypt.*

[9] Ammar A A, El Kattan E M, and Elsirafe A M 1991 *Aerial gamma-ray spectrometric exploration of the Um Naggat granite pluton, Eastern Desert, Egypt. Ann. Geol. Surv. Egypt, Vol.* ***XVII***, pp. 283-302.

[10] Ammar S E 1997 *Uranium and thorium mineralization at northern part of Gabal Um Naggat stock, Central Eastern Desert, Egypt. Mans. Sci. Bull., Vol.***24**, No. 2, pp. 11-44.

[11] Ramadan T M, and Sultan A S 2004 *Integration of remote sensing and geophysical data for the identification of massive sulphide zones at Wadi Allaqi area, South Eastern Desert,Egypt. M.E.R.C. Ain Shams Univ., Earth Sci. Ser., Vol.* **18**, pp. 165-174.

[12] Bayoumi M B 2011 *Remote sensing and petrochemistry studies for some high radioactive zones at Halaib Area, South Eastern Desert, Egypt. Unpublished Ph.D. Thesis, Fac. of Sci., Zagazig Univ., Egypt.* 174 p.

[13] Abu Halawa A, Masoud S M, Shahin H, and Embaby A 2010 *Economic and radioactive heavy minerals of El Hebal sand dunes, Southern Eastern desert, Egypt: Geologic Survey of Egypt.*

[14] El Afandy A H, Abu Halawa A, and Kandil M K 2011 *Heavy mineral potentiality of beach placer deposits of south Ras Banas, Red Sea coast, Egypt: Sedimentology of Egypt*, vol., **19**.

[15] El-Afandy A , Sakr S , and Abu Halawa A 2012 *Mineral exploration and geochemical signature of the ore mineral constituents, Ras Baghday black beach sand, red sea, Egypt. Al-Azhar Bull.*

[16] Tarek Ibrahim et al 2019 *Potential Heavy Mineral-Enriched Black Sand Deposits South Ras Banas Red Sea Coast (Egypt) © Springer Nature Switzerland AG 2019 D. M. Doronzo et al. (eds.), Petrogenesis and Exploration of the Earth's Interior Advances in Science, Technology & Innovation*, https://doi.org/10.1007/978-3-030-01575-6_31

[17] Suresh, Ramasamy, Meenakshisundaram, Venkatachalapathy, and Ponnusamy, 2011b *A relationship between the natural radioactivity and mineralogical composition of the ponnaiyar river sediments, India Journal of Environmental Radioactivity,* ***102*** pp. 370-377.

[18] Strakhov N M, Bushinskii G I and Pustovalov L V 1957 *Metody izocheniya ocadochnykh porod, tom I" (Methods of studying sedimentary rocks, vol.* ***I****) Moskva, Gosgeoitekhiz dat,* 611 pp.

[19] Kroner A 1984 *Late Precambrian plate tectonics and orogeny, a need to redefine the term Pan-African. In: Klerkx, J. and Michot, J. (eds.), African Geology, Tervuren,* pp. 23-28.

[20] Hussein A H 1978 *Lecture course in nuclear geology [M].* pp.101, *NMA, Egypt*

[21] Hansink J P 1976 *Equilibrium analysis of sandstone roll front uranium deposits. Proceedings Inter. Symposium on exploration of uranium deposits. Int. Atomic Energy Agency, Vienna.* pp. 683 - 693.

[22] Adams J A S, and Weaver C E 1958 *Thorium to Uranium Ratios as Indicator for Sedimentary Processes; an Example of Geochemical Facies. Am. Assoc. Pet. Geol.* Bull., **42**.

[23] Abed 2019 *Role of Alteration Processes in Mineralogical and Chemical Changes in Wadi El-Reddah Stream Sediments and Their Associating Granitic Rocks with Special Emphasis on Uranium Mobilization. Faculty of Science, Ain Shams university*, 205 p.

[24] Dabbour G A 1995 *Estimation of the economic minerals reserves in Rosetta beach sands, Egypt. Mineral.* Vol. **7**, pp. 1532 – 166.

[25] El-Shafey A M 2016 *Upgrading and mineralogical studies for some Economic and nuclear Elements Bearing Minerals of Rosetta black deposits, Egypt, with especial emphasis on Monazite. Ph. D. thesis: Fac. Sci. Zagazig University.*

[26] El Kammar A A, Arafa I H and El Sheltami O R 2007 *Mineral composition and environmental geochemistry of the beach sediments along the eastern side of the Gulf of Suez, Egypt. Journal of African Earth Sciences* Volume **49**, Issue **3**, October 2007, pp. 103-114

Natural Radionuclide characterization on Baltim-El Burullus coastal sediments and their risk assessments

Sallam A M[1], Faheim , A A[2] El-Elshafiey, Z A[2], Shimaa AMAmer[3], El Feky MG[3]

[1]Nuclear Power Plants Authority, 4 El Nasr Avenu-Nasr City, Cairo.
[2]Faculty of Science, Al-Azhar University, Cairo, P.O. Box 11884, Egypt
[3]Nuclear Materials Authority of Egypt, Kattamiya, Cairo

Abstract. The Egyptian black sand occurs especially along the beaches of northern part of the Nile Delta. Baltim-El Burullus coastal sediments lie between Longitudes 30° 10′ and 34° 15′E, and Latitudes 31° 10′and 31° 35′N. Sixty two samples from Baltim-El Burullus coastal plain were measured radio metrically by using Na-I detector for determination of ^{238}U, 232Th, 226Ra eU and 40K concentrations. 238U concentration ranged from 1 to 5 ppm with 1.63 as an average, ^{232}Th from 1 to 10 ppm with average 3.66 ppm, 226Ra (eU) from 1 to 6 with 1.92 as an average and 40K from 0.10 to 2.55 with average (0.76). The studied stream sediments could be intimated to different sources, leading to variability in contribution of the radioelements, thus the activity concentration of 40K > 232Th > 226Ra> 238U. eU / Ra eU ratios average (0.66), is lower than unity indicate the disequilibrium in the studied coastal sediments, suggesting disequilibrium with uranium migration out. The inter-element relations show the presence of U and Th in discrete minerals or their separate migration. The radiological hazard parameters, like the dose rate (D), the annual effective dose equivalent (AEDE), radium equivalent activity (Raeq), external (Hex), internal (Hin) hazard indices and gamma index (Iγ) were estimated for these sediments and compared with the permissible international limits. Nearly, all the studied samples are within the recommended permissible levels for most environmental parameters, with the exception of some samples lie out of the limits.

Keywords: Radio nuclides, Sodium iodide detector, Black sands, Baltim Elbrulus region.

1.Introduction:

The origin of the black sand deposits along the Delta and Sinai coasts has discussed; by Shukri (1950), he concluded that, the River Nile sediments outpoured into the Mediterranean Sea are the main source of these deposits. When the water of the river and its load enter the Mediterranean Sea, its velocity decreases abruptly to the laws of deposition. Though, the heavy and coarse grains have been deposited very near to the outpouring, while the fine and light mineral grains were transported long distance to be deposited far from the shore. According to Kuenan (1962), the Nile River carries about million m^3 of sediments to the Mediterranean Sea every year. The solid matter constitutes about 15-45% of the sand fractions; the rest are silt and clay. This enormous quantity of sediments have been outpoured through the present Rashid and Domiat mouths, as well as through the outpourings of the seven old extint Nile branches during the last century (Ball, 1942). The geology of the beach sands along the Mediterranean coast is directly related to the development of the Nile Delta and its past branches. Heavy minerals occurred mainly in the aeolian sand dune belts that extended from Gamsa at the east to El Burullus at the west, as well as sand dunes of Idku, and those along the northern coast of Sinai.

The Egyptian black sand occurs especially along the beaches of northern part of the Nile Delta. The loaded sands are concentrated as deposits at both of the mouths of the two Nile branches, near Rosetta and Damietta at the northern coast of Egypt, where the condition are most favorable for their accumulation as detrital beach deposits. This beach sand contains heavy minerals such as garnet and monazite. Many of these heavy minerals, zircon and monazite, contain radionuclides of the ^{238}U and ^{232}Th series. The concentration of ^{238}U and ^{232}Th in these minerals are much greater than the worldwide average concentration in soils and rocks (Abdel-Razek and Bakhit, 2009). Estimating the exposure of humans to various sources of radiation is very significant. Human beings have always been exposed to radiation from natural sources (Adewumi, 2016).

The area of study is located (Fig. 1). The accessibility of the area is relatively good, the entrance to the area is mainly from the Alexandria-Rafah highway. The climate is cold in winter and warm in summer, and the rain falling are highly frequency and density in winter season. Vegetation is moderate dense, while the habitants have been used the inter dunes space for cultivation. The coastal plain extends from Abu Qir to Burg El Burullus is slightly undulated to flat, its elevation ranges between -1 to 5 m, while the coastal plain between Burg El Burullus to Rafah is slightly rugged and characterized by dense, differentially elevated coastal dunes, with elevation ranges between 5 to 35 m.

This work is mainly aims to study natural radionuclides distribution and the probable conditions of sediments formation related to radioelements, in addition to the determination of their risk assessments.

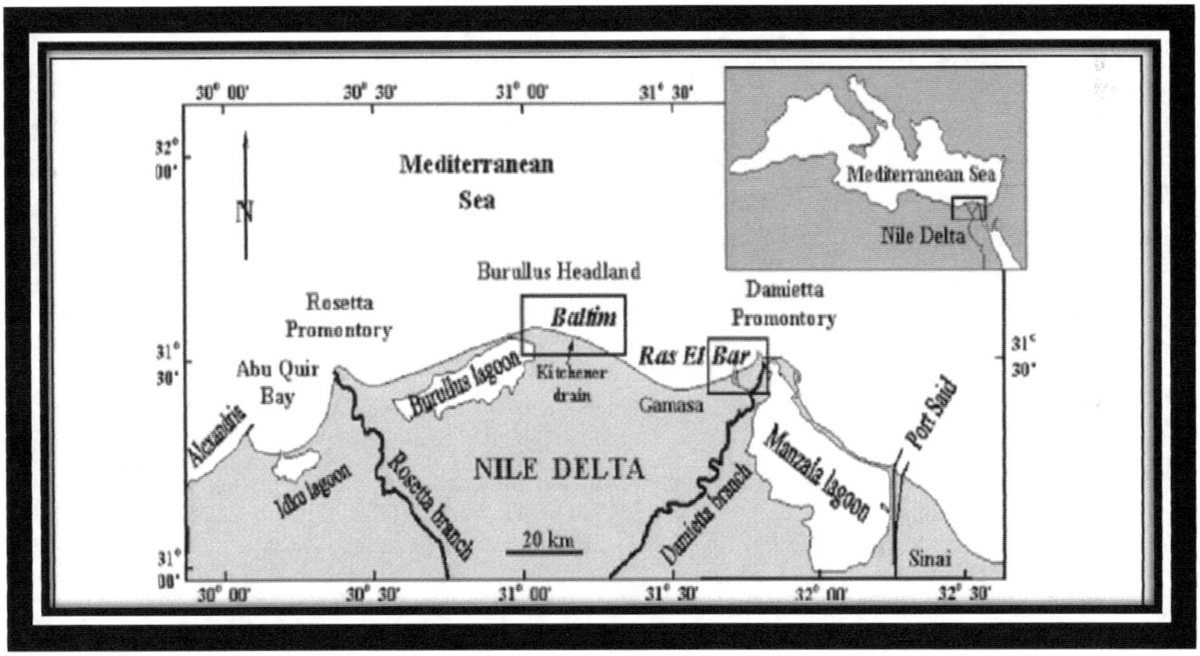

Fig. (1) : Map of the Nile delta coast showing the location of the studied area.

2. Analytical technique

The samples were crushed using a mixer Mill type MM-2 into fine powder, and then sieved using a 2 mm mesh to obtain a uniform particle size. All collected samples were dried in an oven at 110 °C for 24 hours to remove the moisture content from the samples. The sample plastic containers were cylindrical in shape, 212.6 cm^3 volumes with 9.5 cm average diameter and 3 cm height. The container was filled with about 300-400 g of the crushed sample, in plastic containers, sealed with PVC (polyvinyl chloride) insulation tape and left for at least 21 days to accumulate free radon to attain radioactive equilibrium. The relation between the percentages of ^{222}Rn accumulations and time increased till reaching the steady stage after about 38 days **(Matolin, 1991)**.

^{238}U, ^{232}Th, ^{226}Ra, and ^{40}K contents were determined radiometrically using NaI-Tl gamma-ray spectrometers. The instrument used in determination of the four radioactive elements consists of a Bicron scintillation detector NaI-Tl 76x76 mm, hermetically sealed with the photomultiplier tube in aluminum housing. The tube was protected by a copper cylinder protection of thickness 0.6 cm against induced X-ray and a chamber of lead bricks against environmental radiation. The detector was connected to an amplifier, model NE, 4658, Tennelec high voltage power supply, model TC 952 to the high voltage display. The detector was also connected to Nuclease PCA-8000 computer based, 8192 multichannel analyzer with color graphical display of spectra and high level technical operation features and an Hp-laser printer.

Calibration of the instrument was carried out using gamma emitting sources ^{57}Co (122.1 keV, set up in the channel 122) and ^{137}Cs (661.6 keV, set up in the channel 662) as follows; i) Calibration is started with the ^{137}Cs source (coarse adjustment) and then with ^{57}Co source (fine adjustment). ii) The ^{137}Cs source is repeatedly used as minimum procedure. iii) Since the electrical noise of the photomultiplier in the detector unit generated low amplitude pulses, the lower level discrimination of the instrument was set up to the level of channel 40 in order to avoid such pulses as well as natural cosmic and x-rays from the registered signals.

^{238}U, ^{232}Th, ^{226}Ra, and ^{40}K were measured by four energy regions representing ^{234}Th, ^{212}Pb, ^{214}Pb, and ^{40}K at 93 KeV, 239 KeV, 352 KeV, and 1460 KeV for ^{238}U, ^{232}Th, ^{226}Ra, and ^{40}K, respectively. The samples were measured two times 1000 second each, we have done the measurement twice to get the average and the accurate results of the samples.

3. Results and discussion:

Sixty two samples from Baltim-El Burullus coastal plain were measured radiometrically by using Na-I detector for determination of ^{238}U, ^{232}Th, ^{226}Ra eU and ^{40}K concentrations. The radioelement contents in the studied coastal sediments are registered in table(1).^{238}U concentration ranged from 1 to 5 ppm with 1.63 as an average, ^{232}Th from 1 to 10 ppm with average 3.66 ppm, ^{226}Ra (eU) from 1 to 6 with 1.92 as an average and ^{40}K from 0.10 to 2.55 with average (0.76). The studied stream sediments could be intimated to different sources, leading to variability in contribution of the radioelements, thus the activity concenteration of ^{40}K > ^{232}Th > ^{226}Ra> ^{238}U (Fig. 2).

It is indicated from table (1) that the average of uranium, thorium, radium and potassium contents are lower than the international averages (**UNSCEAR, 2000; IAEA, 1979; Boyl, 1982; Abu Elatta et al., 2014;** El-Kammar et al., 2018**). eTh /eU ratio of the** studied sediments (2.3) is lower than that of the Upper Continental Crust reported by McLennan and Taylor (1980; 1991); Rudnick and Gao (2004). The low Th / U ratio of the studied sediments may be related to U-bearing minerals like zircon, apatite and sphene (De Putter et al., 1999; 2002 El-Kammar et al., 2018). The eTh/eU ratio of the studied sediments indicate poor weathering and rapid deposition and the dominance of detrital radioactive minerals such as xenotime, monazite and zircon (Adams and Weaver, 1958; El Galy et al. 2007; El-Afandy et al. 2016; Mira et al. 2020)

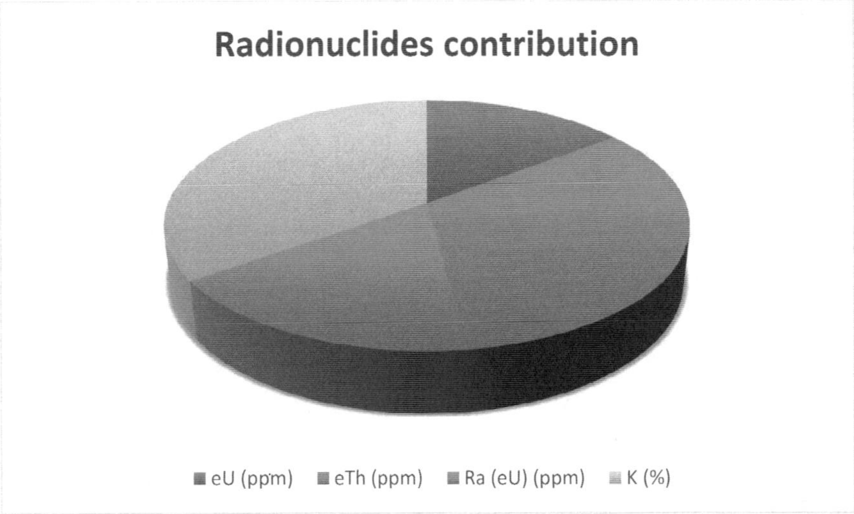

Fig. (2) : Radioelements contribution in the studied sediments.

4. Inter-element relations:

eTh-eU binary relation clarify that there is no any attachment between the two elements in these coastal sediments, suggesting their presence in discrete minerals or uranium mobilization separately apart from thorium or the reverse (Fig. 3a). The relation between eTh/eU and both eU and eTh gave the same vision (Fig. 3b & c). The crosscuts between eTh-K and eU-K indicate the absence any relation between U and Th with the alteration processes (Fig. 3d & e). eTh-K-eTh-U binary relation show the presence of uranium in fixed-U sector, confirming its presence in stream sediment refractory minerals (Fig. 3f). Low eTh/K ratios are considered an indication of potassium instability and lead to abrupt changes in this ratio in various sites with rising and lowering which may illustrate vigorous changes in physico-chemical conditions (Shives et al., 1995; El Kammar et al. 2018).

K/U ratio vary between 500 and 13000, with an average of 5509.52 ppm, which is lower than UCC (9475 ppm), (Rudnick & Gao, 2003). Potassium and uranium are considered incompatible elements and in the upper continental crust they have approximately constant K/U ratio, though ^{40}K plenty. K/U ratios variation could be clarify by the significant mobility of ^{40}K, while the immobility of ^{40}K in reduction conditions (McLennan, 2009), (Fig. 3g).

5. Radioactive equilibrium/disequilibrium

The study of radioactive equilibrium/ disequilibrium sheds some lights on the surface weathering, alteration and movement of uranium by groundwater. It also helps to explain the isotopic variation of some radioactive nuclides and consequently can be used as a tool for U-exploration processes. The equilibrium state means a closed system (for physical and chemical exchanges of the radionuclides) with the surrounding environments whereas disequilibrium indicates an open system, i.e., the radioactive nuclides may migrate out from their parents. The liberated radionuclides can be accumulated through several chemical and geological processes, including; chemical weathering, precipitation of minerals from aqueous solutions by biological and inorganic processes, adsorption on suspended clay minerals, etc.

Measurement of both the equivalent Uranium (eU) and Radium (RaeU) concentrations (in ppm) radio metrically. The equilibrium factor (P) is given as P = eU /Ra eU **(El Galy et al., 2008: El Feky et al. 2021)**. The equilibrium is achieved if the activity ratio1between the parent/daughter or between daughter/daughter is equal to unity.

In the present study, the radioactive equilibrium / disequilibrium states of the studied sediments are discussed through the equilibrium factor (P), i.e. P= eU/RaeU. The obtained average (0.66), is lower than unity and indicate the disequilibrium in the studied coastal sediments **(Fig. 3 h and Table(1))**, suggesting disequilibrium with uranium migration out.

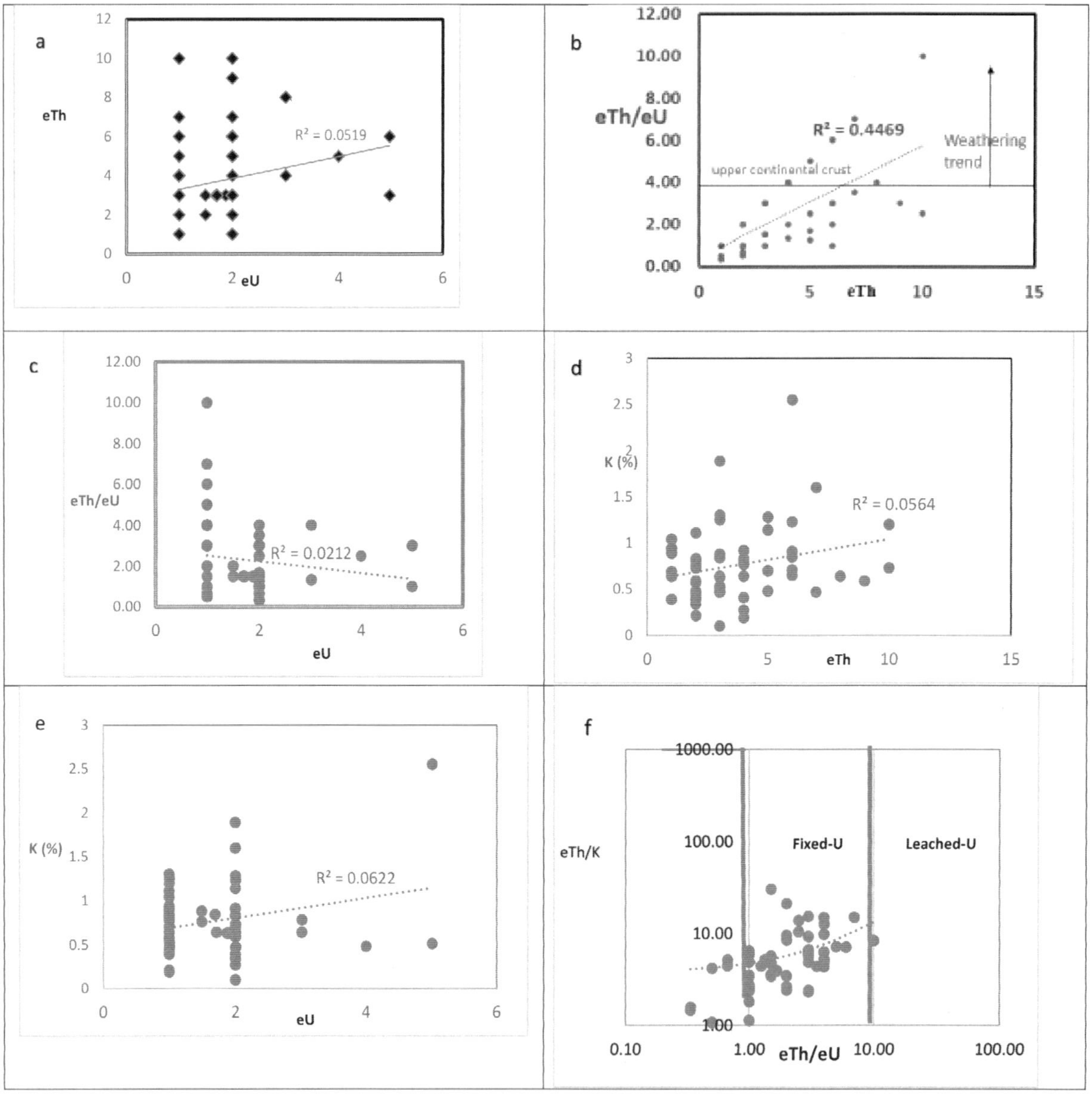

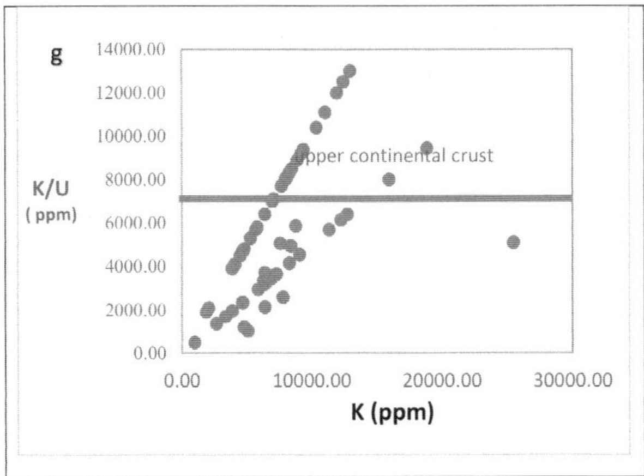

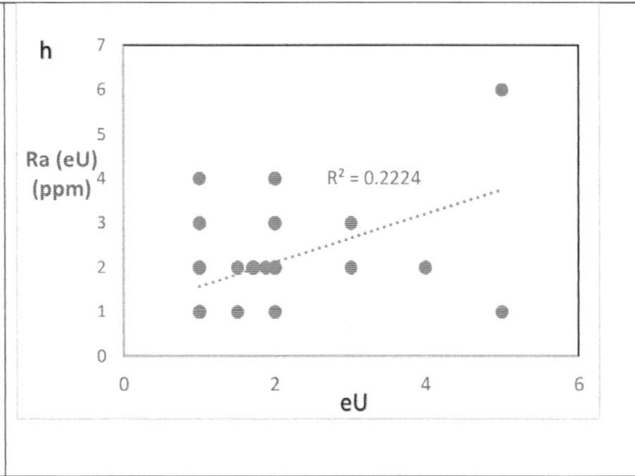

Fig. (3) : (a) eU-eTh, (b) eTh-eTh/eU, (c) eU-eTh/eU, (d) eTh-K, (e) eU-K, (f) eTh/eU- eTh/K (g) K-K/U and (h) eU-Ra eU of the studied sediments.

6. Authogenic U

Enrichment of such authogenic U can be calculated by the formula of Wignall and Meyers (1988):

(Ua) = (U) – (Th)/3, where the square brackets represent concentration expressed in ppm. The authogenic content, Ua, is extremely low. Most values of authogenic Uranium are positive, suggesting the presence of authogenic-U, though, the negative values indicate Uranium migration out.

7. Environmental Impacts:

7.1 Distribution of radionuclides in the studied granitic rocks:

Activity concentrations values of eU and eTh in ppm, as well as K, in %, were converted to activity concentration, Bq kg^{-1}, using the conversion factors given by the Polish Central Laboratory for Radiological Protection (Malczewski et al. 2004), where A_U, A_{Th}, and A_K are the average specific activities of ^{238}U, ^{232}Th, and ^{40}K in Bq kg^{-1}, respectively. The ^{238}U activity concentration ranges between 12.00 and 62.00, averaging 20.73 Bq kg^{-1}, ^{232}Th activity concentration varies between 4.04 and 40.40 Bq kg^{-1} with 15.01 Bq kg^{-1} as an average. ^{226}Ra range from 11.1 to 66.60 Bq kg^{-1} with an average 21.85 Bq kg^{-1}.^{40}K ranges between 406.90 and 1201.92 Bq kg^{-1} with average 913.96 Bq kg^{-1}. The studied sediments have lower average activity concentration than the world concentration limits for ^{238}U, ^{232}Th, ^{226}Ra and ^{40}K. The permissible levels are 33, 45, 32 and 412 Bq.Kg-1Bq kg^{-1} for ^{238}U, ^{232}Th, ^{226}Ra and ^{40}K, respectively, (UNSCEAR, 2008).

7.2 Absorbed gamma dose rates

According to Örgün et al. (2007), the absorbed gamma dose rates in air (D), at 1 m above the ground surface, for the uniform distribution of radionuclides (238U, 232Th, and 40K), were estimated on the basis of guidelines provided by UNSCEAR (2000) by the following equation: D (nGy h−1) = 0.462AU + 0.604ATh + 0.0417AK. Table clarifies the estimated indoor gamma dose rate values for the studied sediments. The indoor D values for the coastal sediments range from 12.66 to 78.69 nGy h−1 with a mean of 29.78 nGy h−1. The mean indoor D values for all the studied sediments are lesser than the worldwide average value (55 nGy h−1, Yang et al. 2005), (Table 2). It is worthy to mention that some sediment samples exceed the permissible levels, due to the presence of radioelement-bearing minerals such as zircon and monazite.

7.3 Annual effective dose equivalent

It was calculated from the absorbed dose by applying the dose conversion factor of 0.7 Sv/Gy and the outdoor occupancy factor of 0.2 (Örgün et al. 2007). The mean values of the studied sediments is 0.04, which are to a great extent less than the recommended worldwide average of the annual effective dose (0.48 mSvy−1) (Table 2).

Table (1): radioelement contents (ppm and Bq/kg) of the studied sediments and their ratios.

Serial No.	S. No.	eU (ppm)	eTh (ppm)	Ra(eU) (ppm)	K (%)	K (ppm)	K/U (ppm)	eTh/eU	eU/RaeU	eTh/K	eU-eTh/3.5
1	6 m	1.5	2	1	0.76	7600.00	5066.67	2.00	1.00	2.63	0.93
2	8 m	2	3	1	0.47	4700.00	2350.00	3.00	1.00	6.38	1.14
3	7 m	1	3	1	0.47	4700.00	4700.00	3.00	1.00	6.38	0.14
4	5 m	2	1	3	0.64	6400.00	3200.00	0.33	0.33	1.56	1.71
5	-1 m	1	6	3	0.71	7100.00	7100.00	2.00	0.33	8.45	-0.71
6	4 m	2	3	3	0.47	4700.00	2350.00	1.00	0.33	6.38	1.14
7	2 m	3	8	2	0.64	6400.00	2133.33	4.00	0.50	12.50	0.71
8	3 m	2	6	2	0.91	9100.00	4550.00	3.00	0.50	6.59	0.29
9	M8	1	2	1	0.58	5800.00	5800.00	2.00	1.00	3.45	0.43
10	sm	1.5	3	2	0.88	8800.00	5866.67	1.50	0.50	3.41	0.64
11	W28.3m	2	4	1	0.27	2700.00	1350.00	4.00	1.00	14.81	0.86
12	2 m	1	4	1	0.83	8300.00	8300.00	4.00	1.00	4.82	-0.14
13	W34 M3	2	2	2	0.73	7300.00	3650.00	1.00	0.50	2.74	1.43
14	27 6m	1	3	1	1.25	12500.00	12500.00	3.00	1.00	2.40	0.14
15	14 2m	1.7	3	2	0.84	8400.00	4941.18	1.50	0.50	3.57	0.84
16	W28 4m	1.88	3	2	0.63	6300.00	3351.06	1.50	0.50	4.76	1.02
17	W28 1m	2	2	3	0.39	3900.00	1950.00	0.67	0.33	5.13	1.43
18	W28 5m	1.72	3	2	0.64	6400.00	3720.93	1.50	0.50	4.69	0.86
19	7m2	2	10	4	0.73	7300.00	3650.00	2.50	0.25	13.70	-0.86
20	25m1	2	2	2	0.59	5900.00	2950.00	1.00	0.50	3.39	1.43
21	8m3	1	2	4	0.48	4800.00	4800.00	0.50	0.25	4.17	0.43
22	12.2m	2	9	3	0.59	5900.00	2950.00	3.00	0.33	15.25	-0.57
23	24. m4	1	4	1	0.8	8000.00	8000.00	4.00	1.00	5.00	-0.14

24	6. m1	1	4	1	0.77	7700.00	7700.00	4.00	1.00	5.19	-0.14
25	W3. 4m6	1	1	1	0.39	3900.00	3900.00	1.00	1.00	2.56	0.71
26	27. 10m	1	1	1	0.89	8900.00	8900.00	1.00	1.00	1.12	0.71
27	14. 1m	2	6	2	0.65	6500.00	3250.00	3.00	0.50	9.23	0.29
28	27. 8m	1	3	1	0.53	5300.00	5300.00	3.00	1.00	5.66	0.14
29	14. 9m	1	1	1	1.04	10400.00	10400.00	1.00	1.00	0.96	0.71
30	21. 4m	1	2	2	0.81	8100.00	8100.00	1.00	0.50	2.47	0.43
31	W31-m4	1	2	1	0.81	8100.00	8100.00	2.00	1.00	2.47	0.43
32	13-m4	1	1	2	0.94	9400.00	9400.00	0.50	0.50	1.06	0.71
33	29-m6	1	6	1	0.85	8500.00	8500.00	6.00	1.00	7.06	-0.71
34	W33. M3	5	3	1	0.51	5100.00	1020.00	3.00	1.00	5.88	4.14
35	W31.m1	1	4	1	0.41	4100.00	4100.00	4.00	1.00	9.76	-0.14
36	28_m7	1	2	1	0.21	2100.00	2100.00	2.00	1.00	9.52	0.43
37	15-2m	1	3	2	0.53	5300.00	5300.00	1.50	0.50	5.66	0.14
38	15-1m	4	5	2	0.48	4800.00	1200.00	2.50	0.50	10.42	2.57
39	15-3m	1	4	1	0.64	6400.00	6400.00	4.00	1.00	6.25	-0.14
40	23-m7	1	10	1	1.2	12000.00	12000.00	10.00	1.00	8.33	-1.86
41	33-m1	2	6	2	1.23	12300.00	6150.00	3.00	0.50	4.88	0.29
42	W30-4m	1	2	1	0.83	8300.00	8300.00	2.00	1.00	2.41	0.43
43	13-m6	1	3	1	1.3	13000.00	13000.00	3.00	1.00	2.31	0.14
44	29-m2	1	5	1	0.7	7000.00	7000.00	5.00	1.00	7.14	-0.43
45	20-5m	1	4	1	0.92	9200.00	9200.00	4.00	1.00	4.35	-0.14
46	23-m6	1	7	1	0.47	4700.00	4700.00	7.00	1.00	14.89	-1.00
47	20-3m	2	3	2	0.1	1000.00	500.00	1.50	0.50	30.00	1.14
48	W30-1m	1	2	2	1.11	11100.00	11100.00	1.00	0.50	1.80	0.43
49	28-m2	1	4	2	0.19	1900.00	1900.00	2.00	0.50	21.05	-0.14
50	23-m2	2	2	2	0.34	3400.00	1700.00	1.00	0.50	5.88	1.43

51	28-m10	1	2	2	0.41	4100.00	4100.00	1.00	0.50	4.88	0.43
52	Brulus 41	2	5	3	1.28	12800.00	6400.00	1.67	0.33	3.91	0.57
53	Brulus 42	2	3	3	1.89	18900.00	9450.00	1.00	0.33	1.59	1.14
54	Brulus 43	2	2	2	0.83	8300.00	4150.00	1.00	0.50	2.41	1.43
55	Brulus 44	5	6	6	2.55	25500.00	5100.00	1.00	0.17	2.35	3.29
56	Brulus 45	1	2	2	0.57	5700.00	5700.00	1.00	0.50	3.51	0.43
57	Brulus 46	1	2	3	0.45	4500.00	4500.00	0.67	0.33	4.44	0.43
58	Brulus 47	2	5	4	1.14	11400.00	5700.00	1.25	0.25	4.39	0.57
59	Brulus 48	2	1	3	0.69	6900.00	3450.00	0.33	0.33	1.45	1.71
60	Brulus 49	3	4	3	0.78	7800.00	2600.00	1.33	0.33	5.13	1.86
62	Brulus 50	2	7	2	1.6	16000.00	8000.00	3.50	0.50	4.38	0.00
Min.		1.00	1.00	1.00	0.10	1000.00	1000.00	500.00	0.17	0.96	-1.86
Max.		5.00	10.00	6.00	2.55	25500.00	25500.00	13000.00	1.00	30.00	4.14
Av.		1.63	3.66	1.92	0.76	7776.19	7776.19	5509.52	0.66	6.05	0.58
Arenaceous sediments (IAEA, 1979 and Boyl, 1982)		1	3	-	1.4	-	9475	-	-	-	-

7.4 Radium equivalent activity

It is calculated according to the equation ($Ra_{eq} = A_U + 10/7A_{Th} + 10/130A_K$) proposed by Tufail et al. (1992), where A_U, A_{Th}, and AK are the specific activities of Ra, Th, and K, respectively, in Bq kg^{-1}. The mean Ra_{eq} values for the studied sediments (63.1 Bq kg^{-1}), which is much lower than the criterion limit of 370 Bq kg^{-1}.

7.5 External and internal hazard index

According to Saito et al. (1998), the annual external gamma ray dose can be calculated by the equation: $H_{ex} = A_U/370 + A_{Th}/259 + A_K/4810$. However, the internal exposure to 222Rn and its radioactive progeny is controlled by the internal hazard index (Hin = AU/ 185 + ATh/259 + AK/4810), as given by Nada (2003); El-Feky et al. (2011); El Kammar et al. (2017) and Harpy et al. (2019). These indices must be less than unity in order to keep the radiation hazard insignificant (Lakehal et al. 2010; Baykara et al. 2011). The radiation exposure due to radioactivity in construction materials must be limited to 1.5 mSv year−1, (Table 2). The values of the external hazard (Hex) and internal hazard (Hin) for the studied sediments is less than unity, which agrees with the recommended values.

7.6 Activity concentration index (I_γ)

Another radiation hazard index called the representative level index, I_γ, is defined by Nuclear Energy Agency (1979) and El-Farrash et al. (2020) as follows ($I_\gamma = A_U/150 + A_{Th}/100 + A_K/1500$), where A_U, A_{Th}, and A_K are the activity concentrations of ^{238}U, ^{232}Th, and ^{40}K, respectively, in Bq kg^{-1}. The safety value for this index is ≤ 1, whereas the obtained Iγ averages for the studied sediments is 0.46, which show values lesser than the permissible level.

Table (2): Results of radionuclide activity concentrations, the dose rate (*D*), the annual effective dose equivalent (*AEDE*), radium equivalent activity (Ra_{eq}), external (H_{ex}), internal (H_{in}) hazard indices and gamma index (I_γ) for the studied sediments.

Serial	Sample	eU	eTh	Ra	K	Abs. Dose ^{238}U(nGy/h)	Eff. Dose (mSv)	Ra_{eq}	H_{ex}	H_{in}	I_γ
No.	No.	Bq/kg	Bq/kg	Bq/kg	Bq/kg						
1	6 m	18.60	8.08	11.10	237.88	19.93	0.02	40.94	0.11	0.14	0.31
2	8 m	24.80	12.12	11.10	147.11	18.58	0.02	39.73	0.11	0.14	0.29
3	7 m	12.40	12.12	11.10	147.11	18.58	0.02	39.73	0.11	0.14	0.29
4	5 m	24.80	4.04	33.30	200.32	26.18	0.03	54.48	0.15	0.24	0.40
5	-1 m	12.40	24.24	33.30	222.23	39.29	0.05	85.02	0.23	0.32	0.61
6	4 m	24.80	12.12	33.30	147.11	28.84	0.04	61.93	0.17	0.26	0.44
7	2 m	37.20	32.32	22.20	200.32	38.13	0.05	83.78	0.23	0.29	0.60
8	3 m	24.80	24.24	22.20	284.83	36.77	0.05	78.74	0.21	0.27	0.58
9	M8	12.40	8.08	11.10	181.54	17.58	0.02	36.61	0.10	0.13	0.28
10	sm	18.60	12.12	22.20	275.44	29.06	0.04	60.70	0.16	0.22	0.45
11	W28.3m	24.80	16.16	11.10	84.51	18.41	0.02	40.69	0.11	0.14	0.29
12	2 m	12.40	16.16	11.10	259.79	25.72	0.03	54.17	0.15	0.18	0.41
13	W34 M3	24.80	8.08	22.20	228.49	24.66	0.03	51.32	0.14	0.20	0.38
14	27 6m	12.40	12.12	11.10	391.25	28.76	0.04	58.51	0.16	0.19	0.46
15	14 2m	21.08	12.12	22.20	262.92	28.54	0.04	59.74	0.16	0.22	0.44
16	W28 4m	23.31	12.12	22.20	197.19	25.80	0.03	54.68	0.15	0.21	0.40
17	W28 1m	24.80	8.08	33.30	122.07	61.28	0.08	133.61	0.36	0.69	0.89
18	W28 5m	21.33	12.12	22.20	200.32	25.93	0.03	54.92	0.15	0.21	0.40
19	7m2	24.80	40.40	44.40	228.49	54.44	0.07	119.69	0.32	0.44	0.85
20	25m1	24.80	8.08	22.20	184.67	22.84	0.03	47.95	0.13	0.19	0.35

21	8m3	12.40	8.08	44.40	150.24	31.66	0.04	67.50	0.18	0.30	0.48
22	12.2m	24.80	36.36	33.30	184.67	45.05	0.06	99.45	0.27	0.36	0.71
23	24. m4	12.40	16.16	11.10	250.40	25.33	0.03	53.45	0.14	0.17	0.40
24	6. m1	12.40	16.16	11.10	241.01	24.94	0.03	52.72	0.14	0.17	0.40
25	W3. 4m6	12.40	4.04	11.10	122.07	12.66	0.02	26.26	0.07	0.10	0.20
26	27. 10m	12.40	4.04	11.10	278.57	19.18	0.02	38.30	0.10	0.13	0.30
27	14. 1m	24.80	24.24	22.20	203.45	33.38	0.04	72.48	0.20	0.26	0.53
28	27. 8m	12.40	12.12	11.10	165.89	19.37	0.02	41.18	0.11	0.14	0.31
29	14. 9m	12.40	4.04	11.10	325.52	21.14	0.03	41.91	0.11	0.14	0.33
30	21. 4m	12.40	8.08	22.20	253.53	25.71	0.03	53.25	0.14	0.20	0.40
31	W31-m4	12.40	8.08	11.10	253.53	20.58	0.03	42.15	0.11	0.14	0.32
32	13-m4	12.40	4.04	22.20	294.22	24.97	0.03	50.60	0.14	0.20	0.38
33	29-m6	12.40	24.24	11.10	266.05	30.86	0.04	66.19	0.18	0.21	0.49
34	W33. M3	62.00	12.12	11.10	159.63	19.11	0.02	40.69	0.11	0.14	0.30
35	W31.m1	12.40	16.16	11.10	128.33	20.24	0.02	44.06	0.12	0.15	0.32
36	28_m7	12.40	8.08	11.10	65.73	12.75	0.02	27.70	0.07	0.10	0.20
37	15-2m	12.40	12.12	22.20	165.89	24.49	0.03	52.28	0.14	0.20	0.38
38	15-1m	49.60	20.20	22.20	150.24	28.72	0.04	62.61	0.17	0.23	0.45
39	15-3m	12.40	16.16	11.10	200.32	23.24	0.03	49.59	0.13	0.16	0.37
40	23-m7	12.40	40.40	11.10	375.60	45.19	0.06	97.71	0.26	0.29	0.73
41	33-m1	24.80	24.24	22.20	384.99	40.95	0.05	86.44	0.23	0.29	0.65
42	W30-4m	12.40	8.08	11.10	259.79	20.84	0.03	42.63	0.12	0.15	0.33
43	13-m6	12.40	12.12	11.10	406.90	29.42	0.04	59.71	0.16	0.19	0.47
44	29-m2	12.40	20.20	11.10	219.10	26.47	0.03	56.81	0.15	0.18	0.42
45	20-5m	12.40	16.16	11.10	287.96	26.90	0.03	56.34	0.15	0.18	0.43
46	23-m6	12.40	28.28	11.10	147.11	28.34	0.03	62.82	0.17	0.20	0.45
47	20-3m	24.80	12.12	22.20	31.30	18.88	0.02	41.92	0.11	0.17	0.29
48	W30-1m	12.40	8.08	22.20	347.43	29.62	0.04	60.47	0.16	0.22	0.46
49	28-m2	12.40	16.16	22.20	59.47	22.50	0.03	49.86	0.13	0.19	0.35
50	23-m2	24.80	8.08	22.20	106.42	19.57	0.02	41.93	0.11	0.17	0.30
51	28-m10	12.40	8.08	22.20	128.33	20.49	0.03	43.61	0.12	0.18	0.31
52	Brulus 41	24.80	20.20	33.30	400.64	44.29	0.05	92.98	0.25	0.34	0.69
53	Brulus 42	24.80	12.12	33.30	591.57	47.37	0.06	96.12	0.26	0.35	0.74
54	Brulus 43	24.80	8.08	22.20	259.79	25.97	0.03	53.73	0.15	0.21	0.40
55	Brulus 44	62.00	24.24	66.60	798.15	78.69	0.10	162.62	0.44	0.62	1.22
56	Brulus 45	12.40	8.08	22.20	178.41	22.58	0.03	47.47	0.13	0.19	0.35
57	Brulus 46	12.40	8.08	33.30	140.85	26.14	0.03	55.68	0.15	0.24	0.40
58	Brulus 47	24.80	20.20	44.40	356.82	47.59	0.06	100.70	0.27	0.39	0.74
59	Brulus 48	24.80	4.04	33.30	215.97	26.83	0.03	55.68	0.15	0.24	0.41
60	Brulus 49	37.20	16.16	33.30	244.14	35.33	0.04	75.17	0.20	0.29	0.55
62	Brulus 50	24.80	28.28	22.20	500.80	48.22	0.06	101.12	0.27	0.33	0.76
Min		12.40	4.04	11.10	31.30	12.66	0.02	26.26	0.07	0.10	0.20
Max		62.00	40.40	66.60	798.15	78.69	0.10	162.62	0.44	0.69	1.22
Av.		20.73	15.01	21.85	243.39	29.78	0.04	63.01	0.17	0.23	0.46
international average		33	45	32	412	57	0.48	370	1	1	0.5

8. Conclusions

Baltim-El Burullus coastal sediments lie between Longitudes 30° 10′ and 34° 15′E, and Latitudes 31° 10′and 31° 35′N. Laboraty measurements indicate that the studied area contains low contents of radioelements. The studied stream sediments could be intimated to different sources, due to the variability in sourced and different contribution of the radioelements, thus the activity concenteration of 40K > 232Th > 226Ra> 238U. eU/RaeU ratios average show the disequilibrium in the studied coastal sediments, suggesting disequilibrium with uranium migration out. The inter-element relations show the presence of U and Th in discrete minerals or their separate migration. Nearly, all the studied samples are within the recommended permissible levels for most radiological parameters, with the exception of some samples lie out of the limits. The samples have significantly high activity and so, lie out of range with respect to the recommended levels may be related to the presence radioelement-bearing minerals such as zircon and monazite.

9. References

[1] Abdel-Razek, Y.A. and Bakhit, A.F. (2009): Comparison between the Measurements of Radon Gas Concentration and γ-ray Intensities in Exploring the Sand at El Burullus Beach", Arab Journal of Nuclear Sciences and Applications, 42(4), 393-404.

[2] Abu Elatta, S.A., Hassan, S. F., El-Farrash, A. H., El-Feky, M. G. and Refaat, M. (2014): Recent Uranium Mobilization and Radioactivity of Metamorphosed Sandstones at Sikait Area, Southeastern Desert, Egypt. J. Geo, Min, Res., 6, 2, 28-38.

[3] Adams, J. A. S. and Weaver, C.E.(1958): Thorium to uranium rations as indicator for sedimentary processes; An example of geochemical facies. Am. Assoc. Petrol. Geol., 42 pp. 397-430.

[4] Adewumi, A.: Determination of Excess Lifetime Cancer Risk Due to Gamma Radiation in Some Agricultural Products Obtained From Two Oil Fields in the Niger Delta Region of Nigeria.IJSR.2016;5(10): 656-60.

[5] Ball, J. (1942): Egypt in the classical geographers. Surv. Min. Dept., Cairo.

[6] Boyle R.W. (1982): Geochemical Prospecting for Thorium and Uranium Deposits [M]. pp.498, Elsevier, Amsterdam and New York.

[7] De Putter, T. h., Charlet, J.-M., Quinif, Y. (1999): REE, Y and U concentration at the fluid-iron oxide interface in Late Cenozoic cryptodolines from Southern Belgium. Chem. Geol. 153, 139–150.

[8] De Putter, T. H., Bernard, A., Perruchot, A., Yans, J., Verbrugghe, F. R., Dupuis, C. h. R. (2002): Neoformed mineral paragenesis in acid weathering systems: sedimentary vs volcanic environments. Proceedings of the 12th International Clay Conference, Elsevier Oxford.

[9] El-Afandy, A.H., M. G. El-Feky, S. Taha, S.M. El Minyawi, H.A. Sallam (2016): Distribution of Radioelements and Evaluation of Radiological Hazard effects on Stream Sediments and Cataclastic Rocks of Wadi Abu Rusheid, Southeastern Desert, Egypt. Greener Journal of Geology and Earth Sciences, 4, (3), 056-069.

[10] El-Farrash, A. H., El-Feky, M. G., Hassan, S. F., Abu Elatta, S. A., and M. Refaat (2020): Radiological and environmental studies on the metamorphosed sandstones at Wadi Sikait area, southeastern desert, Egypt. IOP Conf. Series: Materials Science and Engineering 975 (2020) 012017 IOP Publishing doi:10.1088/1757-899X/975/1/012017.

[11] Mohamed Galal El-Fekya, Hesham Samir Mohammed, Ahmed Mohammed El-Shabasy, Mohammed Ramadan Ahmed, Yasser Kamal Abdel-Monem & Hamed Ibrahim Mira (2021): Mobilisation of radionuclides during uranium and gold processing of granitic rock at El-Missikat area, central Eastern Desert, Egypt. International Journal of Environmental Analytical Chemistry ISSN: (Print) (Online) Journal homepage: https://www.tandfonline.com/loi/geac20

[12] El Galy, M. M., Nada, A., El Feky, M.G. and Roz, M. (2007): The radioactive disequilibrium state in U-series at G-II of U-mineralization, Gabal Gattar area, North Eastern Desert, Egypt". Arab Jour. Nuc. Sci. Appl., 40, (1), 17-26.

[13] EL-Kammar, A. M., EL-Feky, M. G., Abu zied, H. T. and Othman, D. A. (2018): RAdioactivity and environmental impacts of some Carboniferous and Cretaceous kaolin deposits in southwestern-Sinai, Egypt. Nuclear Sciences Scientific Journal, 7, 87- 100.

[14] El-Kammar, A., Abu-Zied, H. T., El-Feky, M.G. Osman, D. (2017): Composition, radioactivity, and possible applications of kaolin deposits of Sinai, Egypt. Arab J Geosci, 10:463

[15] El Feky M.G., El Mowafy A. A., and Abdel Warith A (2011): Mineralogy, geochemistry, radioactivity and environmental impacts of Gabal Marwa granites, southeastern Sinai, Egypt. Chin.J.Geochem., 30, 175–186.

[16] Harpy, N. M., El Dabour, S. E., Sallam, A. M., Nada, A. A., El Aassy, A. E. & El Feky, M. G. (2019): Radiometric and environmental impacts of mill tailings at experimental plant processing unit, Allouga, Egypt. ISSN: 1527-5922 (Print) 1527-5930 (Online) Journal homepage: https://www.tandfonline.com/loi/uenf20

[17] IAEA (1979): International Atomic Energy Agency. "Gamma-Ray Survey in Uranium Exploration". IAEA Technical Report Series, No. 186, Vienna, Austria, pp. 90.

[18] Kuenan, (1962): Nile sediments in marine geology: J. Willey

[19] Malczewski, D., Taper, L., Dorda, J.(2004): Assessment of natural and anthropogenic radioactivity levels in rocks and soils in the environs of Swieradow Zdroj in Sudetes, Poland by in situ gamma-ray spectrometry. Jour. Environ. Radioact. 73, 233 – 245.

[20] Matolin, M.(1991): Construction and use of spectrometric calibration pads, Laboratory of gamma-ray spectrometry. A report to the government of the Arab Republic of Egypt, 4, 030-03.

[21] McLennan, S.M., and Taylor, S.R.(1980): Rare earth element–thorium correlations in sedimentary rocks and the composition of the continental crust. Geochimica et Cosmochimica Acta, 44, 1833–1839.

[22] McLennan, S.M., and Taylor, S.R.(1991): Sedimentary rocks and crustal evolution: tectonic setting and secular trends. J. Geol., 99, 1–21.

[23] McLennan,S.(2009): Geochemistry: Branches, processes, phenomena. Geophysics and Geochemistry, Encyclopedia of Life Support Systems, 1, 99–118.

[24] Hamed Mira, Hussein Hussein, Sameh Tawfik and Neveen Abed (2020): Stream Sediments Geochemical Exploration in Wadi El Reddah area, Northeastern Desert, Egypt, Mediterranean Journal of Chemistry 10, (8), 809-827.

[25] Nada, A.(2003): Evaluation of natural radionuclides at Um-Greifat area, Eastern Desert of Egypt. Appl. Radiat. Isot. 58, 275–280.

[26] Nuclear Energy Agency(1979): Exposure to Radiation from Natural Radioactivity in Building Materials. Report by NEA Group of Experts. OECD, Paris.

[27] Rudnick, R. L. and Gao, S. (2004): Composition of the Continental Crust. In: Treatise on Geochemistry. Holland, H.D. and Turekian, K.K. (Editors), Elsevier, Amsterdam. 3: 1-64.

[28] Örgün ,Y., Altınsoy N., S¸ Ahin, S.Y., Gu¨ngo¨r, Y., Gu¨ltekin, A.H., Karahan, G., Karacık, Z., (2007): Natural and anthropogenic radionuclides in rocks and beach sands from Ezine region (Canakkale), Western Anatolia, Turkey. Appl Radiat Isot 65:739–747

[29] Saito K, Pattoussi H, Zankl M (1998): Calculation of the effective dose and its variation from environmental gamma ray sources. Health Phys., 74, 698–706.

[30] Shives, R.B.K., Ford, K.L., and Charbonneau, B.W.(1995): Applications of gamma ray spectrometric/magnetic/VLF-EM surveys—Workshop Manual; Geological Survey of Canada, Open File 3061, 82 p.

[31] Shukri, N.M. (1950): The mineralogy of some Nile sediments. Q.J.G.S., London, 105: 511-534.

[32] Tufail, M., Ahmad, N., Mirza, S.M., Mirza, N.M., Khan, H.A.,(1992): Regular paper, natural radioactivity from the building materials used in Islamabad and Rawalpindi, Pakistan. Sci. Total Environ. 121, 283–291.

[33] UNSCEAR,(2000): United Nations Scientific Committee on the effects of atomic radiation, sources and effects of ionizing radiation. Report to General Assembly, with Scientific Annexes United Nations. United Nations, New York.

[34] UNSCEAR (2008): United Nations Scientific Committee on the Effect of Atomic Radiation. Sources and Effects of Ionizing Radiation. Report to General Assembly, with Scientific Annexes, United Nations, New York.

[35] Wignall, P.B., Meyers, K.J., (1988): Interpreting the benthic oxygen levels in mudrocks: a new approach. Geology 16, 452-455.

[36] Yang, Y.; Wu, X.; Jiang, Z.; Wang, W.; Lu, J.; Lin, J.; Wang, L., and Hsia, Y.(2005) Radioactivity concentration in soils of the Xiazhuang granite area, China. Appl. Radiat. Isot., 63, 255–259.

Radionuclide Content and Radiation Dose Rates of Uraniferous and Mineralized Granites at GV Occurrence in Gabal Gattar, Northeastern Desert, Egypt.

F M Abdelwahab[1], N H El Gendy[2], M Kh Barakat[2], M S Mohamed[1], A M El-Tohamy[1], R A El Terb[1] and M G El Feky[1]

[1]Nuclear Materials Authority of Egypt, Kattamiya, Cairo
[2]Faculty of Science, Geology, Geology Department, Tanta University, Egypt

Abstract. G. Gattar is located in the northern part of the Eastern Desert of Egypt, bounded by latitudes 26° 52′ - 27° 08′ N and longitudes 33° 13′ – 33° 26′ E and covers an area of about 178 km2. Eighteen samples, nine from uraniferous and nine from mineralized granites were measured radiometrically for 238U, 232Th, 226RaeU and 40K determination by Na-I detector and the radioactive minerals identified by alpha track method. Based on the petrography, radioactive mineral content and radioactivity, Gattarian granites were classified into uraniferous and mineralized granites. Uranium contents reach up to 229 ppm in the uraniferous granites and to 20300 ppm in the mineralized granites, which is extremely higher than the international limits. eTh/eU ratio of the studied uraniferous and mineralized granites show average ratios of 0.05 and 0.0044, respectively, which is extremely lower than the average of igneous rocks, suggesting uranium enrichment in reducing conditions. The uraniferous and mineralized granites U/Th ratios average are 62.23 and 4234.45, respectively, indicating that the studied two granitic types are considered highly productive rocks. The radioelements interrelations manifest that the distribution of radioelements not only magmatic but also related to hydrothermal redistribution of radionuclides associating with uranium deposition. The radiological hazard parameters, such as the dose rate (D), the annual effective dose equivalent (AEDE), radium equivalent activity (Raeq), external (Hex), internal (Hin) hazard indices and gamma index (Iγ) were estimated for these granites and compared with the international levels. Nearly, all the studied samples are to a great extent higher than the recommended permissible levels for all environmental parameters.

Keywords: Gattarian granite- GV occurrence, Radionuclides, Environmental parameters, Radioactive mineralization.

1. Introduction

The Nuclear Materials Authority of Egypt started a systematic program for detection of nuclear raw materials in the Eastern, Western Desert and Sinai Peninsula. The various exploration activities resulted in the discovery of several radioactive anomalies and/or uranium-thorium mineralization at different localities. Among these localities, Gabal Gattar is the most important one due to hosting uranium mineralization. The geologic, mineralogic and radiometric studies revealed that it includes more than eleven uranium occurrences. These occurrences are named as, according to the date of discovery, G-I,

G-II, G-III...etc. Gattarian granite is considered as uraniferous granite containing (8-21 ppm U) and (11-34 ppm Th), (Moharam, 1997), whereas the normal uranium content in granites (4 ppm), (Rogers and Adams, 1969; Cuney, 1978). Several studies have suggested that uranium especially in uraniferous granites, is readily mobilized (Rosholt et al., 197١; Nkomo et al., 197٨; Stuckless et al., 1981) particularly when it occur in the labile form in minerals such as biotitic, chlorite, epidote, iron and iron-titanium oxides and hydroxides grain boundaries and crystal defects of minerals. There is a critical feature in the formation of the Gattarian granite, it was both highly enriched in uranium and much of uranium was located in labile sites, that fluids could readily mobilize uranium and this confirms that uranium of mineralized zone in all probabilities was derived from the surrounding uranium enriched granites of Gattar area.
The distribution of the naturally occurring radionuclides mainly uranium (^{238}U), thorium (^{232}Th), potassium (^{40}K) and other radioactive elements, depends on the distribution of rocks from which they originate and the processes which concentrate them. Exposure to ionizing radiation from natural sources is a continuous and unavoidable feature of life on earth. The radiation dose to the population is from indoor exposure, since people spend most of their time indoors and almost all building materials are radioactive to some degree. The indoor absorbed dose rate varies depending on the type of dwelling, age and the materials used for construction, and location (UNSCEAR, 2000).
This work is aims to determine the radiological and environmental impacts on the workers and human-beings in the study area.

2. Geologic outlines

G. Gattar represents the northern part of a big granite batholith. It is located in the northern part of the Eastern Desert of Egypt. It is bounded by latitudes 26° 52′ - 27° 08′ N and longitudes 33° 13′ – 33° 26′ E and covers an area of about 178 km^2. The pluton of G. Gattar bounded by diorite and granodiorite from the eastern north corner and Hammamat sediments from the western north corner. W. Belih extends between Hammamat sediments and G. Gattar following a major fault trending N75°– 80°E, more or less parallel to the local reverse fault separating these sediments from the intruded granites to the south (Salman et al., 1986; Shalaby, 1996; Abu Zeid et al. 2021). W. Gattar and its tributaries located in the western side of the pluton extending NNW-SSE (Fig. 1). The country rocks are invaded by intensive dike swarms trending in the ENE-WSW to NE-SW directions and comprise porphyritic granites, granophyres, andesites and basic types. These dikes do not cut the granitic pluton except the basic one (Roz, 1994).
The younger granites are the major rock type in the mapped area, two granitic phases encountered at the northern part of G. Gattar area are identified, i.e., phase II and phase III pink granites. The phase III pink granite is the most abundant type; it forms an elongated body of 35 km long and about 10 km wide containing some xenoliths of older granitoids (Ghabrial and Lotfi, 1967). At the vicinity of faults and fractures these pink granites acquire a reddish brown colour and are considered as hydrothermally altered pink granite (Salman et al., 1990). The pink granite has been dated at 570 Ma with Sr^{87}/Sr^{86} initial ratio (0.7053) (Mousa, 1998; Shalaby et al. 2015).
Near the northern contacts of the pluton with the older Hammamat sediments, the NNE-SSW trending faults and shear zones are sometimes associated with U-Mo mineralization (GI occurrence). More than seven U-mineralized occurrences are distributed along and near the northern contacts of the plutons, (Fig. 1). All these U-occurrences are structurally controlled.
The uranium mineralization usually found associated with gangue minerals, commonly represented by quartz, fluorite and carbonates. Secondary generated quartz and uranium mineralization show an intimate intergrowth in microfractures. In some parts, the amorphous silica coating the mineralized surface, and may protect them from alteration (El Feky, 2011).

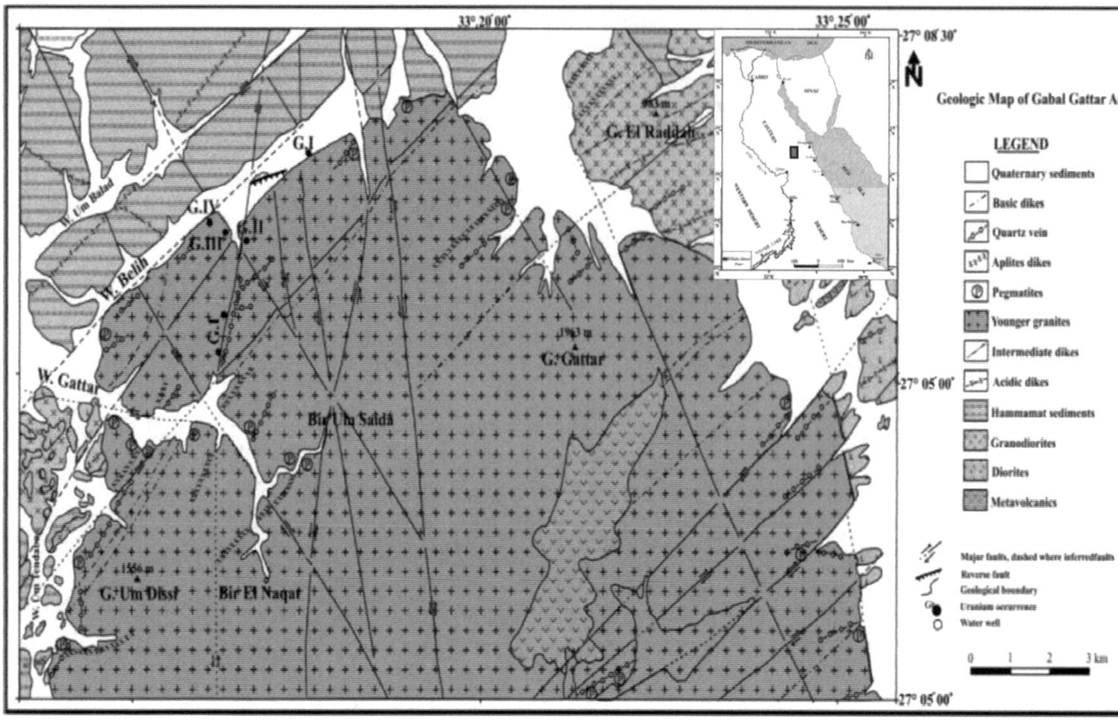

Figure 1: Geological map of Gabal Gattar area (after Roz, 1994).

3. Analytical technique:

A total of (18) representative samples (9 from the uraniferous and 9 from the mineralized granites) were collected from GV occurrence in Gabal Gattar for more detailed radiometric measurements. This was carried out using the high efficiency γ- ray spectrometer (NaI"Tl"-Detector). The main radioelements; ^{238}U, ^{232}Th, ^{226}Ra and ^{40}K are determined through four energy regions of interest (ROIs). Since U and Th are not γ- emitters, they are measured indirectly using their most common γ- ray emitting radionuclides, ^{234}Th and ^{212}Pb, respectively. Ra is determined through the effective γ- line of ^{214}Pb while K is determined directly through the line of ^{40}K (Matolin, 1991). An autoradiographic investigation was carried out for studying the forms of the radioactive mineralization in the studied samples. Alpha-sensitive cellulose-nitrate film was sandwiched between the thin-polished section and a glass slide. Exposure time varied from 21 to 30 days.

4. Results

Microscopically, the radioactive gattarian granites can be categorized to uraniferous and mineralized granites according to their radioactive mineral contents. They are mainly composed of potash feldspars (perthites), quartz, plagioclase and brotite. Zircon, fluorite opaques monazite are the main minerals with or without uranophane in the uraniferous granite. The mineralized granites contain kasolite, uranophane, uraninite and pitchblende, in addition to these minerals. Chlorite, sericite kaolinite and saussurite are the essential secondary products. The hematitized granites are characterized by staining and coating of most of the rock forming minerals by iron oxides. Sometimes the plagioclase and feldspars of the hematized granites are sericitized, kaolinized and albitized (Fig. 2a & b). Hematite is usually coating or staining with a thin film mostly of the secondary uranium mineral, with partially dissolved plagioclase and, participating fuzzy quartz (silicified) ((Fig. 2c & d). The kaolintized granites are characterized by the presence of kaolinite associated with sericite, hematite and/or carbonates. The silicified granites are characterized by presence of porphyroclastic texture (due fracturing and brecciation effects) and occurrence of secondary quartz (Fig. 2c & d).

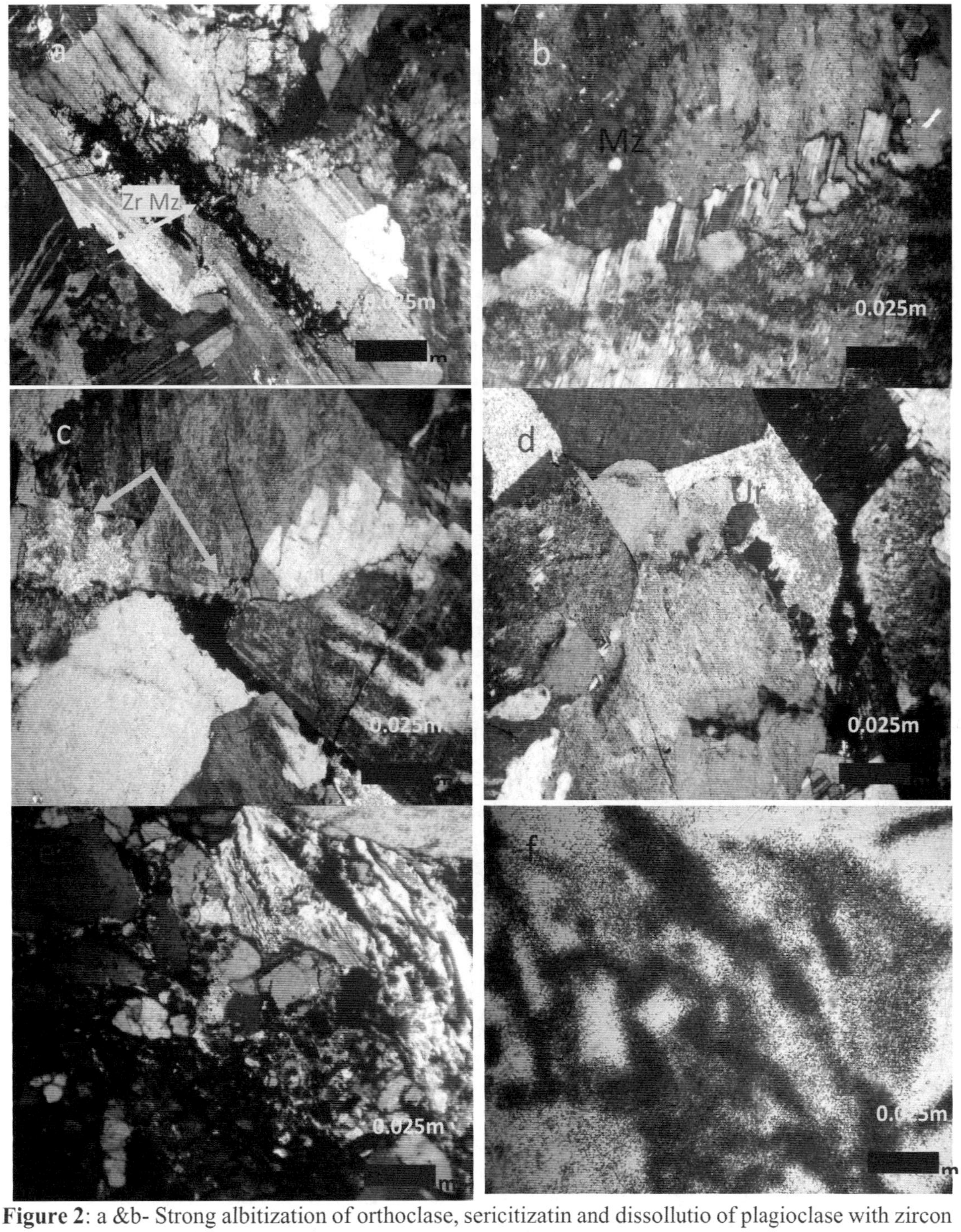

Figure 2: a &b- Strong albitization of orthoclase, sericitizatin and dissollutio of plagioclase with zircon and monazite, c & d- Fussy form of quartz crystals with uranophane stained on iron oxides. e & f- uranium mineralization and its corresponding alpha tracks.

Alpha-track's studies also indicate that uranium mineralization is mainly found in primary, secondary uranium minerals and along the grain boundaries, in addition to being contained in refractory minerals (e.g. allanite, zircon and monazite), (El Feky, 2011). Uranium mineralization and their corresponding alpha tracks are recognized in the mineralized granites (Figs. 2e & f).

5. Radiometric investigation

Based on the petrography, radioactive minerals content and radioactivity, Gattarian granites are classified into uraniferous granites, which contain more than twice of Clark value (4ppm), (Clark et al. 1966) and as characterized by Darnly, (1982), Assaf et al., (1997), Raslan and El Feky (201٢), El Mezayen et al. (2019) and El Feky et al., (2021). In addition, their mineral contents are composed mainly of accessory minerals such as zircon and monazite and don't exceed 229 ppm in their concentrations. On the other hand, the mineralized granites contain beside these minerals upon uranophane, kasolite, uraninite and pitchblende and their uranium contents reach up to 20300 ppm.

Eighteen samples, nine from uraniferous and nine form mineralized granites were measured radiometrically for ^{238}U, ^{232}Th, ^{226}RaeU and ^{40}K determination by Na-I detector. ^{238}U, ^{232}Th, ^{226}RaeU and ^{40}K are varied between (31 and 229 ppm), (1.45 and 6 ppm), (44 and 167 ppm) and (1.62 and 3.26 wt. %), with averages 118.18, 2.94, 87.73 and 2.2, respectively. However, the mineralized granites contain^{238}U, ^{232}Th, ^{226}RaeU and ^{40}K content ranges between (1830 and 20300 ppm), (1.50 and 40 ppm), (600 and 11100 ppm) and (1.45 and 3.33 wt. %) with averages of 8066.36, 9.49, 4450 and 2, 46, respectively. The average of uranium in uraniferous granites is much higher than the average of silicic intrusive rocks (Adams et al., 195٨), granites of Saudi Arabia (Stuckless et al. 1984) and the values of UNSCEAR (2008) but has lower thorium. In comparison, the mineralized granites have extremely higher radioelements contents relative to the international averages (Tables 1& 2).

The geochemistry of U and Th didn't change over magmatic crystallization and so, the Th/U ratio could be stable during magmatic differentiation (Raslan and El Feky, 201٢). The Th/U ratio in granites (3.5 to 4), as indicated by Rogers and Adams (1969), which either increase (Whitfield et al., 1959; Rogers and Ragland 1961) or decrease (Larsen and Gottfried, 1960, El Feky et al., 2011, El Mezayen et al., 2020, El Feky et al. 2021) depending on the redox conditions, the volatile content, or alterations by endogen or supergene solutions (Falkum and Rose-Hansen, 1978). eTh/eU ratios in igneous rocks are 3 or 4:1 (El Feky et al., 2011), however eTh/eU ratio of the studied uraniferous and mineralized granites have average ratios of 0.05 and 0.0044, respectively, which is extremely lower than the average of igneous rocks, suggesting uranium enrichment in reducing conditions.

The eU/eTh ratio is an important radiometric technique to identify the sites of U-mineralization. It is used also to detect the redox conditions at which U mobilized (Naumov, 1959; Anjos et al. 2005; El Nahas et al. 2011 and El Mezayen et al. 2020). The productive uraniferous rocks have U/Th ratio > 1 (Darnely and Ford, 1989; Anjos et al. 2005). The U/Th ratios average of uraniferous and mineralized granites are 62.23 and 4234.45, respectively, indicating that the two studied granitic types are highly productive rocks (Tables 1 & 2).

6. Geochemistry of U and Th

eU-eTh variation diagram of the studied uraniferous and mineralized Gattarian granites clarifies strong positive and moderate correlation in the two types, respectively, suggesting that the magmatic processes played an important role in the concentration of radioelements (Stuckless et al. 1984; Raslan and El Feky, 2012), (Figs. 3a & 4a). eTh/eU ratio is moderately positive correlated with U and weakly to moderate correlated with thorium in the uraniferous granites, weakly positive with U and strongly positive with thorium in the mineralized granites, respectively, manifests that the distribution of radioelements not only magmatic but also related to hydrothermal redistribution of radionuclides due to uranium deposition (Figs. 3b &c, 4b & c). K with U and Th correlation, indicate the presence of ill-defined to weakly negative correlation in the uraniferous granite to ill-defined correlation in the mineralized granites, indicating the absence of the role of alteration of K-feldspars in redistribution of radioelements (Figs. 3d & e, 4 d & e). eTh/eU-eTh/K intercept shows that most samples in uraniferous and mineralized granite samples lie in the ultra-fixed-U zone (Figs. 3f, 4f).

Table 1: Radioelements content and their ratios in the studied uraniferous granites.

Ser. No	Sample Name	eU (ppm)	eTh (ppm)	eRa (eU)	K%	eTh/eU	eU/eTh	eU/Ra	Th/K	K (ppm)	K/U	eU-eTh/3.5
1	GV.4	229	1.45	167	1.62	0.01	157.93	1.37	0.9	16200	70.74	228.59
2	GV.4.1	154	6	80	1.78	0.04	25.67	1.93	3.37	17800	115.58	152.29
3	GV.5	128	2	94	1.94	0.02	64	1.36	1.03	19400	151.56	127.43
4	GV.6	62	2	52	2.03	0.03	31	1.19	0.99	20300	327.42	61.43
5	GV.6.1	56	2	61	3.26	0.04	28	0.92	0.61	32600	582.14	55.43
6	GV.7	148	2	91	1.9	0.01	74	1.63	1.05	19000	128.38	147.43
7	GV.7.1	191	1.66	120	2	0.01	115.06	1.59	0.83	20000	104.71	190.53
8	GV.8	31	1.8	44	2.27	0.06	17.22	0.7	0.79	22700	732.26	30.49
9	GV.8.1	41	6	45	2.55	0.15	6.83	0.91	2.35	25500	621.95	39.29
Minimum		31	1.45	44	1.62	0.01	6.83	0.7	0.61	16200	70.74	30.49
maximum		229	6	167	3.26	0.15	157.93	1.93	3.37	32600	732.26	228.59
average		**118.18**	**2.94**	**87.73**	**2.2**	**0.05**	**62.23**	**1.29**	**1.45**	**22027.27**	**330.70**	**117.45**
Stuckless et al. (1984)		**4.8**	**14.2**	-	-	**2.96**	**0.39**	-	-	-	-	-
UNSCEAR (2008)		**2.7**	**8.63**	**3.06**	**1.32**	**3.2**	**0.31**	**0.88**	**6.54**	**13200**	**4888.89**	**0.23**

Table 2: Radioelements content and their ratios in the studied mineralized granites.

Ser. No	Sample Name	eU(ppm)	eTh(ppm)	eRa (eU)	K%	eTh/eU	eU/eTh	eU/Ra	Th/K	K (ppm)	K/U (ppm)	eU-eTh/3.5
1	GV.1	1830	1.50	1420	3.20	0.0008	1220	1.29	0.47	32000	17.49	1829.57
2	GV.1.1	3420	2.00	2160	2.33	0.0006	1710	1.58	0.86	23300	6.81	3419.43
3	GV.2	2640	1.70	1500	1.55	0.0006	1552	1.76	1.10	15500	5.87	2639.51
4	GV.2.1	2000	10.00	1200	3.33	0.0050	200	1.67	3.00	33300	16.65	1997.14
5	GV.3	20300	2.00	9000	2.36	0.0001	10150	2.26	0.85	23600	1.16	20299.43
6	GV.3.1	17900	1.55	11100	3.10	0.0001	11548	1.61	0.50	31000	1.73	17899.56
7	GV.5.1	2010	40.00	660	1.45	0.0199	50.25	3.05	27.59	14500	7.21	1998.57
8	GV.9	3500	2.30	2150.00	2.98	0.0007	1521.74	1.63	0.77	29800	8.51	3499.34
9	GV.9.1	13000	1.85	8000.00	1.99	0.0001	7027.03	1.63	0.93	19900	1.53	12999.47
Minimum		1830	1.50	660.00	1.45	0.0001	50.25	1.29	0.47	14500.00	1.16	1829.57
Maximum		20300	40.00	11100.00	3.33	0.0199	11548.39	3.05	27.59	33300.00	17.49	20299.43
Average		**8066.36**	**9.49**	**4450.00**	**2.46**	0.0044	**4234.45**	**1.89**	**5.83**	**24609.09**	**7.78**	**8064.64**
Stuckless et al. (1984)		**4.8**	**14.2**	-	-	**2.96**	**0.39**	-	-	-	-	-
UNSCEAR (2008)		**2.7**	**8.63**	**3.06**	**1.32**	**3.2**	**0.31**	**0.88**	**6.54**	**13200**	**4888.89**	**0.23**

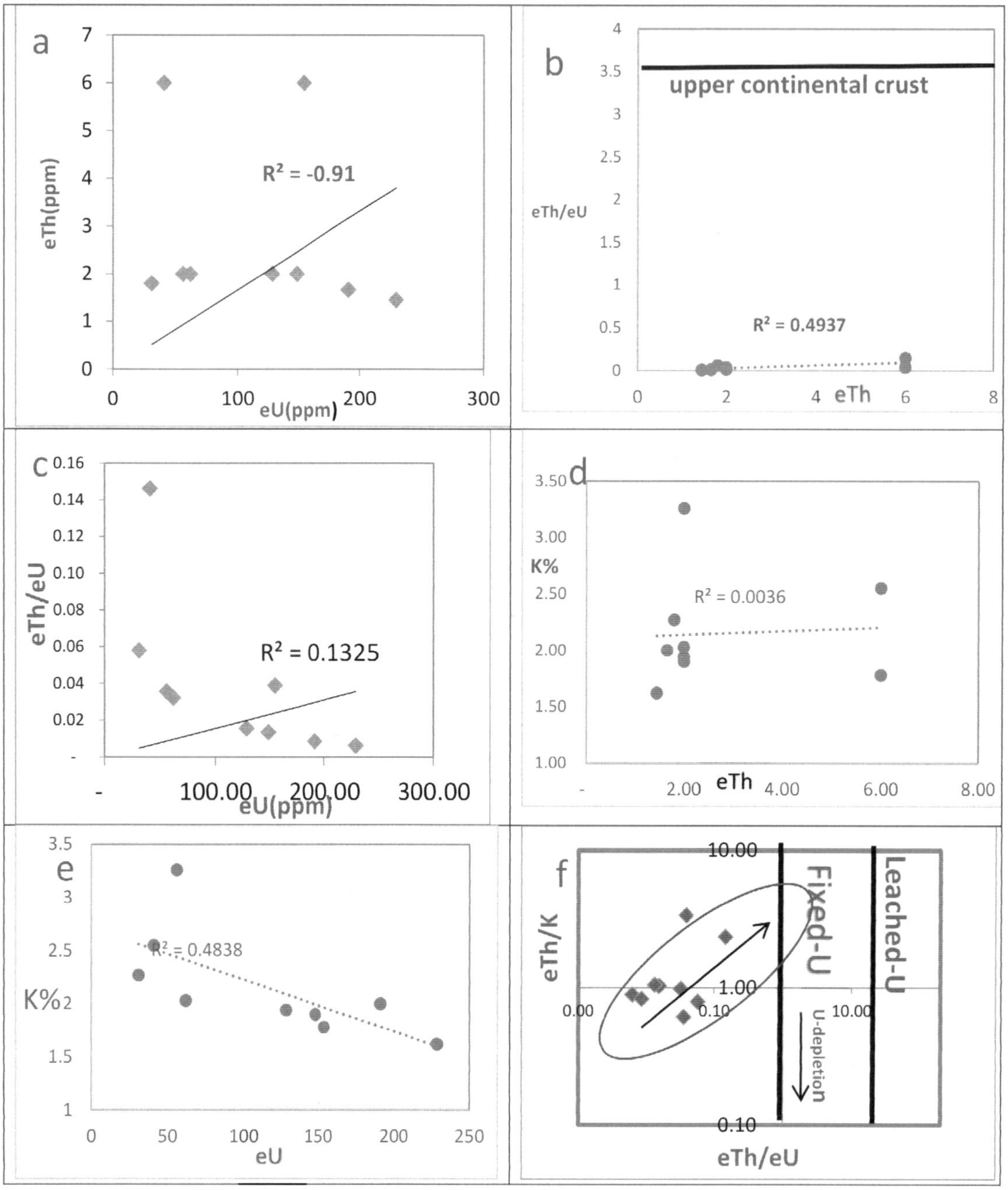
a
eTh(ppm)
eU(ppm)
R² = -0.91
b
eTh/eU
upper continental crust
R² = 0.4937
eTh
c
eTh/eU
R² = 0.1325
eU(ppm)
d
K%
R² = 0.0036
eTh
e
K%
R² = 0.4838
eU
f
eTh/K
Fixed-U
Leached-U
U-depletion
eTh/eU

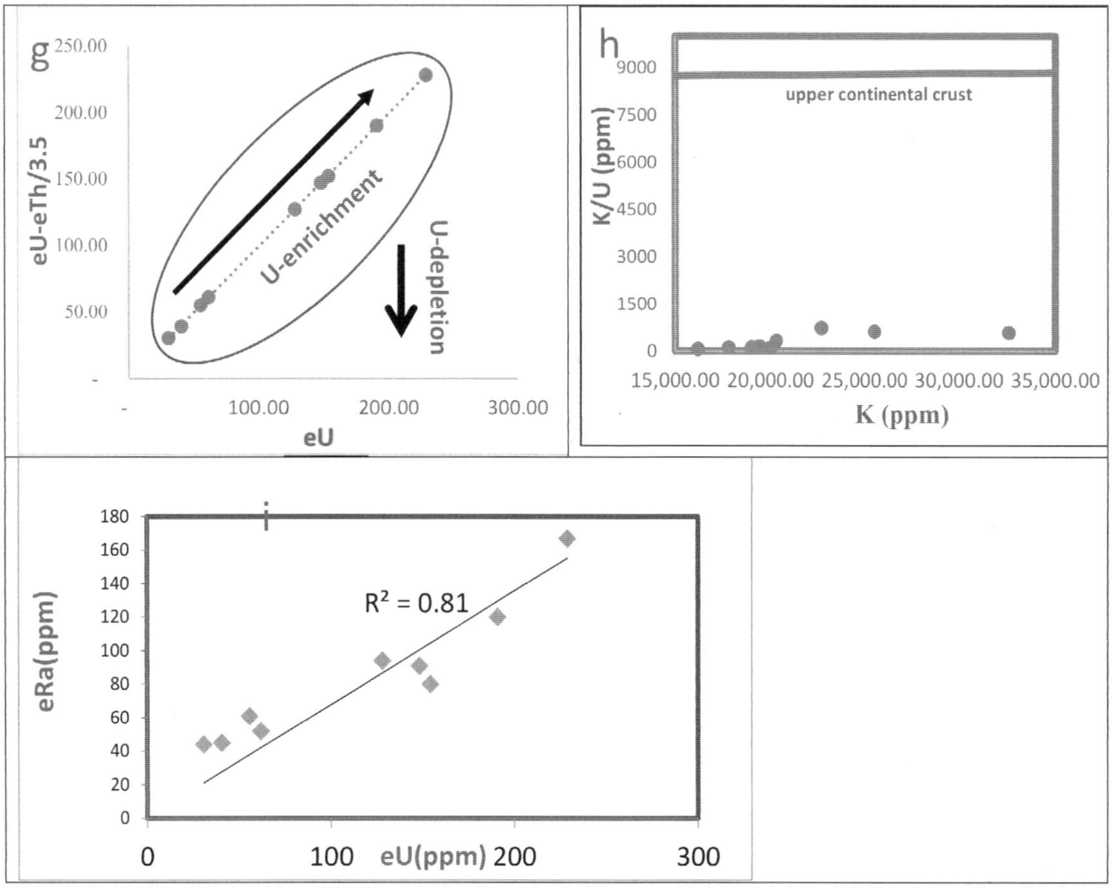

Figure 3: a-eU-eTh, b-eTh-eTh/eU, c-eU-eTh/eU, d-eTh-K, e-eU-K, f- eTh/eU-eTh/K g- eU-eU-eTh/3 h- K-K/U and i -eU-RaeU of the studied uraniferous granites.

6.1. Authigenic U

Enrichment of such authigenic U can be calculated by the formula of Wignall and Meyers (1988): (Ua) = (U) – (Th)/3, where the square brackets represent concentration expressed in ppm. All the resulted uranium values in the uraniferous and mineralized granites have positive values, suggesting that uranium in the studied granites is authigenic (Figs. 3g, 4g and tables 1 & 2.). K-K/U cross plot indicate that K/U values in the granitic rock types lie under the corresponding values in the continental crust, which is mainly related to uranium enrichment in these granites (Figs. 3h, 4h and tables 1 & 2).

6.2. Radioactive equilibrium/disequilibrium

The study of radioactive equilibrium/ disequilibrium explains the isotopic variation of some radioactive nuclides and consequently can be used as a tool for U-exploration processes. The equilibrium state means a closed system (for physical and chemical exchanges of the radionuclides) with the surrounding environments whereas disequilibrium indicates an open system, i.e., the radioactive nuclides may migrate out from their parents. The liberated radionuclides can be accumulated through several chemical and geological processes, including; chemical weathering, precipitation of minerals from aqueous solutions by biological and inorganic processes, adsorption on suspended clay minerals, etc.

The equilibrium factor (P) is given as P=eU/RaeU **(El Galy et al., 2008: El Feky et al. 2021)** by the measurement of both the equivalent uranium (eU) and radium (RaeU) concentrations (in ppm) radiometrically. The equilibrium is achieved if the activity ratio between the parent/daughter or between daughter/daughter is equal to unity. In the present study, the radioactive equilibrium/ disequilibrium states of the studied granites are discussed through the equilibrium factor (P), i.e. P= eU/RaeU. The obtained averages are 1.29, 1.89 in the uraniferous and mineralized granites, respectively, which is higher than unity and indicate the positive disequilibrium in the studied granites with uranium additions (Figs. 3i, 4i and tables 1 & 2).

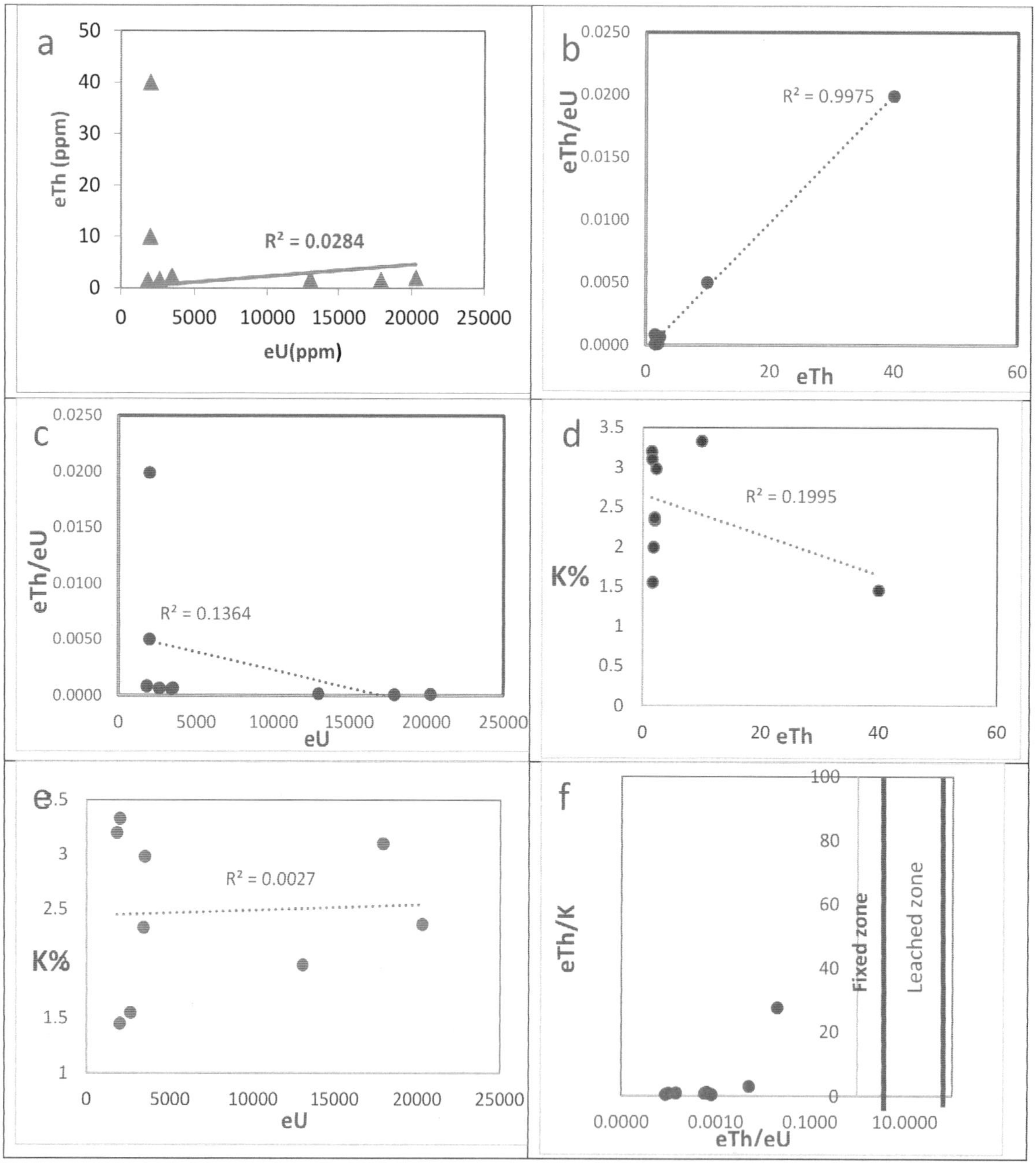

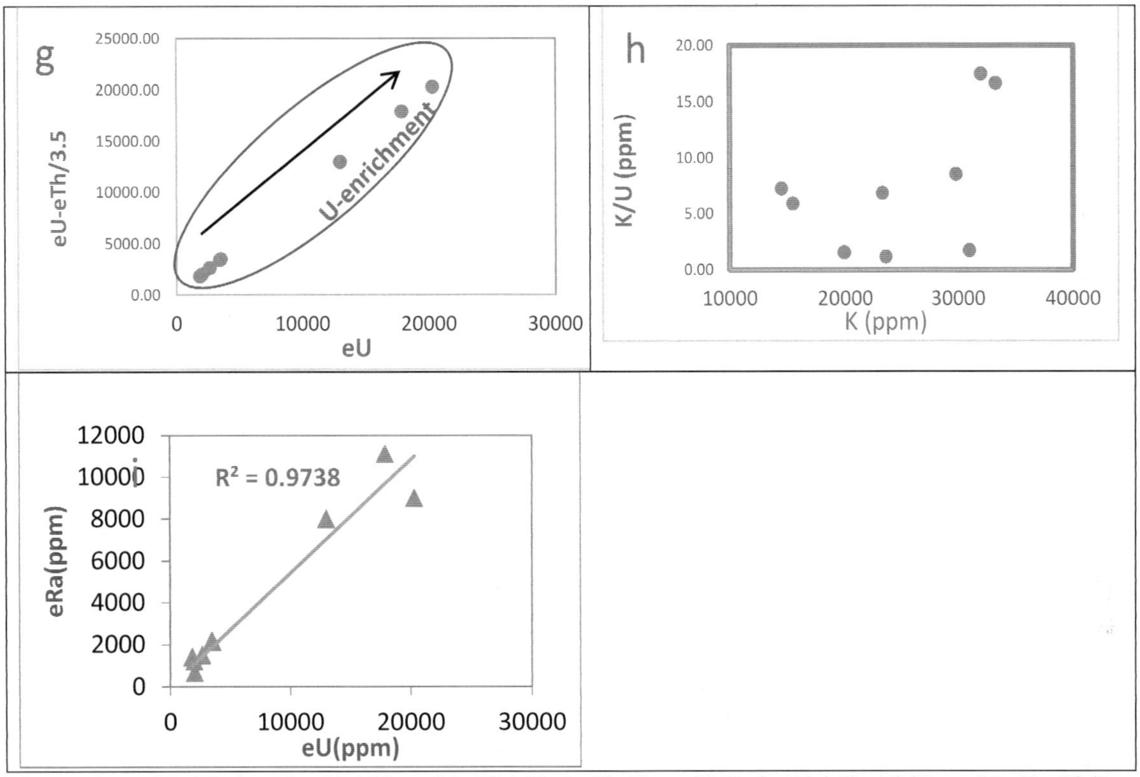

Figure 4: a-eU-eTh, b-eTh-eTh/eU, c-eU-eTh/eU, d-eTh-K, e-eU-K, f- eTh/eU-eTh/K g- eU-eU-eTh/3 h- K-K/U and i -eU-RaeU of the studied mineralized granites.

7. Environmental Impacts

7.1. Activity concenterations in the studied granitic rocks

Activity concentrations values of eU and eTh in ppm, as well as K, in %, were converted to activity concentration, Bq kg^{-1}, using the conversion factors given by the Polish Central Laboratory for Radiological Protection (Malczewski et al. 2004), where A_U, A_{Th}, and A_K are the average specific activities of ^{238}U, ^{232}Th, and ^{40}K in Bq kg^{-1}, respectively.

The radioactive Gattarian granites are catigorized according to their mineral contents into uraniferous and mineralized granites. The ^{238}U activity concentration of the uraniferous granites range between 384.40 and 2839.60, averaging 1465.46 Bqkg^{-1}, ^{232}Th activity concentration varies between 5.86 and 24.24 Bq kg^{-1} with 11.88 Bq kg^{-1} as an average. ^{226}Ra varies between 488.48 and 1853.70 with 973.77 Bq kg^{-1} as an average. ^{40}K ranges between 507.06 and 1020.38 Bq kg^{-1} with average 689.45 Bq kg^{-1}. The studied uraniferous samples have much higher activity than the world concentration limits for ^{238}U, ^{226}Ra and ^{40}K but lower for ^{232}Th. The permissible levels are 33, 45, 34 and 412 Bq.Kg-1for ^{238}U, ^{232}Th and ^{40}K, respectively, (UNSCEAR, 2008).

Activity concentrations of 238U, 232Th, ^{226}Ra and 40K in the mineralized granites range from (22692 to 251720), (6.06 to 161.60), (7326 to 123210) and (453.85 to 1042.29) with averages (100022.91, 38.34, 49395 and 770.26), respectively. All the mineralized granites samples have extremely higher activity concentrations than the permissible levels, (Tables 3 & 4).

7.2. Absorbed gamma dose rates

According to Örgün et al. (2007), the absorbed gamma dose rates in air (D), at 1 m above the ground surface, for the uniform distribution of radionuclides (^{238}U, ^{232}Th, and ^{40}K), were estimated on the basis of guidelines provided by UNSCEAR (2000) by the following equation: D (nGy h^{-1}) = $0.462A_U + 0.604A_{Th} + 0.0417A_K$. Table clarifies the estimated indoor gamma dose rate values for the studied granitic rock samples. The indoor D values for the uraniferous granite samples range from 211.61 to 1336.58 nGy h^{-1} with a mean of 710.86 nGy h^{-1}. For mineralized granite samples, they ranges from 3501.14 to 3501.14 nGy h^{-1} with a mean of 22875.51 nGy h^{-1}. The mean indoor D values for all the studied granitic rocks exceed to a great extent the worldwide average value (55 nGy h^{-1}, Yang et al. 2005), (Tables 3& 4).

Table 3: Results of radionuclide activity concentrations, the dose rate (*D*), the annual effective dose equivalent (*AEDE*), radium equivalent activity (Ra_{eq}), external (H_{ex}), internal (H_{in}) hazard indices and gamma index (I_γ) for the uraniferous granite.

Ser. No	Sample Name	eU	eTh	Ra	K	Abs. Dose 238U(nGy/h)	Eff. Dose (mSv)	Raeq	Hex	Hin	Iγ
		Bq/kg	Bq/kg	Bq/kg	Bq/kg						
1	GV.4	2839.60	5.86	1853.70	507.06	1336.58	1.64	2886.97	7.80	15.48	19.33
2	GV.4.1	1909.60	24.24	888.00	557.14	920.11	1.13	1987.09	5.37	10.53	13.34
3	GV.5	1587.20	8.08	1043.40	607.22	763.49	0.94	1645.45	4.45	8.74	11.07
4	GV.6	768.80	8.08	577.20	635.39	386.56	0.47	829.22	2.24	4.32	5.63
5	GV.6.1	694.40	8.08	677.10	1020.38	368.24	0.45	784.43	2.12	4.00	5.39
6	GV.7	1835.20	8.08	1010.10	594.70	877.54	1.08	1892.49	5.11	10.07	12.71
7	GV.7.1	2368.40	6.71	1332.00	626.00	1124.36	1.38	2426.13	6.56	12.96	16.27
8	GV.8	384.40	7.27	488.40	710.51	211.61	0.26	449.44	1.21	2.25	3.11
9	GV.8.1	508.40	24.24	499.50	798.15	282.80	0.35	604.43	1.63	3.01	4.16
Minimum		384.40	5.86	488.40	507.06	211.61	0.26	449.44	1.21	2.25	3.11
maximum		2839.60	24.24	1853.70	1020.38	1336.58	1.64	2886.97	7.80	15.48	19.33
average		1465.46	11.88	973.77	689.45	710.86	0.87	1531.10	4.14	8.10	10.31
international average		33	45	32	412	57	0.48	370	1	1	0.5

Table 4: Results of radionuclide activity concentrations, the dose rate (*D*), the annual effective dose equivalent (*AEDE*), radium equivalent activity (Ra_{eq}), external (H_{ex}), internal (H_{in}) hazard indices and gamma index (I_γ) for the mineralized granites.

Ser. No	Sample Name	eU	eTh	Ra	K	Abs. Dose 226Ra (nGy/h)	Abs.Dose 238U (nGy/h)	Eff. Dose (mSv)	Raeq	Hex	Hin	Iγ
		Bq/kg	Bq/kg	Bq/kg	Bq/kg							
1	GV.1	22692	6.06	15762	1001.60	7327.47	10529.13	12.91	22777.70	61.56	122.89	152.01
2	GV.1.1	42408	8.08	23976	729.29	11112.20	19627.79	24.07	42475.64	114.80	229.42	283.29
3	GV.2	32736	6.87	16650	485.15	7716.68	15148.41	18.58	32783.13	88.60	177.08	218.63
4	GV.2.1	24800	40.40	13320	1042.29	6221.71	11525.47	14.13	24937.89	67.40	134.43	166.43
5	GV.3	251720	8.08	99900	738.68	46189.48	116330.32	142.67	251788.36	680.51	1360.83	1678.71
6	GV.3.1	221960	6.26	123210	970.30	56967.26	102589.76	125.82	222043.58	600.12	1200.01	1480.44
7	GV.5.1	24924	161.60	7326	453.85	3501.14	11631.42	14.26	25189.77	68.08	135.44	168.08
8	GV.9	43400	9.29	23865	932.74	11070.14	20095.31	24.64	43485.02	117.53	234.82	290.05
9	GV.9.1	161200	7.47	88800	622.87	41056.09	74504.89	91.37	161258.59	435.83	871.51	1075.16
Minimum		22692	6.06	7326	453.85	3501.14	10529.13	12.91	22777.70	61.56	122.89	152.01
Maximum		251720	161.60	123210	1042.29	56967.26	116330.32	142.67	251788.36	680.51	1360.83	1678.71
Average		100022 .91	38.34	49395	770.26	22875.51	46258.36	56.73	100118.71	270.59	540.92	667.59
international average		33	45	32	412	57	57	0.48	370	1	1	0.5

7.3. Annual effective dose equivalent
It was calculated from the absorbed dose by applying the dose conversion factor of 0.7 Sv/Gy and the outdoor occupancy factor of 0.2 (Örgün et al. 2007). The mean values of the uraniferous and mineralized granites are 0.87 and 46258.26 mSvy−1, which are so much higher than the recommended worldwide average of the annual effective dose (0.48 $mSvy^{-1}$), as suggested by UNSCEAR (2000), (Tables 3 & 4).

7.4. Radium equivalent activity
It is calculated according to the equation ($Ra_{eq} = A_U + 10/7A_{Th} + 10/130A_K$) proposed by Tufail et al. (2007), where A_U, A_{Th}, and AK are the specific activities of Ra, Th, and K, respectively, in Bq kg^{-1}. The mean Ra_{eq} values for the uraniferous and mineralized granites are 1531.10 and 100118.71 Bq kg^{-1}, respectively, which much higher values than the criterion limit of 370 Bq kg^{-1}, as recommended by UNSCEAR (2000).

7.5. External and internal hazard index
According to Saito et al. (1998), the annual external gamma ray dose can be calculated by the equation: $H_{ex} = A_U/370 + A_{Th}/259 + A_K/4810$. However, the internal exposure to 222Rn and its radioactive progeny is controlled by the internal hazard index ($H_{in} = A_U/185 + A_{Th}/259 + A_K/4810$), as given by Nada (2003), El-Feky et al. (2011), El Kammar et al. (2017) and Harpy et al. (2019). These indices must be less than unity in order to keep the radiation hazard insignificant (Lakehal et al. 2010; Baykara et al. 2011). The radiation exposure due to radioactivity in construction materials must be limited to 1.5 mSv $year^{-1}$, (Tables 3 & 4).
The values of the external hazard (H_{ex}) and internal hazard (H_{in}) for the uraniferous and mineralized granites are to a great extent higher than international standards, suggesting that these samples cannot be used as a building or decorative material of dwelling (Tables 3 & 4).

7.6. Activity concentration index (I_γ)
Another radiation hazard index called the representative level index, I_γ, is defined by Nuclear Energy Agency (1979) as follows ($I_\gamma = A_U/150 + A_{Th}/100 + A_K/1500$), where A_U, A_{Th}, and A_K are the activity concentrations of ^{238}U, ^{232}Th, and ^{40}K, respectively, in Bq kg^{-1}. The safety value for this index is ≤ 1, whereas the obtained Iγ averages for the uraniferous and mineralized granites are 10.31 and 540.92, respectively, which have to a great extent values higher than the recommended safety value.

8. Conclusions
Based on the petrography, radioactive minerals content and radioactivity, Gattarian granites were classified into uraniferous and mineralized granites. Uranium contents are extremely higher than the international limits. eTh/eU ratio of the studied uraniferous and mineralized is extremely lower than the average of igneous rocks, suggesting uranium enrichment in reducing conditions. The uraniferous and mineralized granites U/Th ratios average indicate that the two granitic types are considered highly productive rocks. The radioelements interrelations manifest that the distribution of radioelements not only magmatic but also related to hydrothermal redistribution of radionuclides associating with uranium deposition. Nearly, all the studied samples environmental parameters are to a great extent higher than the recommended permissible levels.

References
[1] Abu Zeid, E 2021 *Radioactivity, petrogensis and rare earth elements tetrad effect of the mineralized granite of Gabal Gattar, North Eastern Desert, Egypt.* Nuclear Science Scientific Journal, 9A, 63-80
[2] Adams, J and Weaver, C 1958 *Thorium to uranium rations as indicator for sedimentary processes; an example of geochemical facies*. Am. Assoc. Petrol. Geol., 42, 397-430

[3] Anjos, A, Veiga, R, Soares, T, Santos, A, Aguiar, J, Frascá, M, Brage, J, Uzêda, D, Mangia, L, Facure, A, Mosquera, B, Carvalho, C and Gomes, P 2005 *Natural radionuclide distribution in Brazilian commercial granites*. Radiation measurements, vol.39, pp245-253

[4] Assaf, H, Mahdy, M, El-Afandy, A, 1997 *Egyptian Younger Granites; an approach to define parameters favoring formation of uranium deposits.* 3rd Int. Conf. Geochem., Alex. Univ. Egypt. P.409-420

[5] Baykara, O, Karatepe and S, Do˘gru, M 2011 *Assessments of natural radioactivity and radiological hazards in construction materials used in Elazig, Turkey*. Technical Report. Radiation Measurements, 46, 153-158.

[6] Clark, S, Peterman, J and Heier, K 1966 *Abundance of uranium, thorium and potassium. In: S. P. Clark, Jr. (ed.), Handbook of physical contents*, Geol. Soc. Am. Mem. 97, section 24, p. 521-541.

[7] Cuney, M 1978 *Geologic environment, mineralogy, and fluid inclusions of the bois Noirs-Limouzat uranium vein, Forez, France*. Econ. Geol., V. 73, 1567-1610.

[8] Darnely, A and Ford, K 1989 *Regional airborne gamma –ray surveys Areviem .Proceedings of the 3th intern. conf. Geophys . Geochem .Expl. mineral and groundwater*. Ontario, Geol. Surv., Spec., V. 3, 229-240.

[9] Darnely, A 1982 *"Hot granite"; Some general remarks. In: Uranium in granites (ed), Y. Mauric*, Geological Survey of Canada, 81-23, 1-10.

[10] El Feky M, Mohammed, H, El-Shabasy, A., Ahmed, M, Abdel-Monem, Y and Mira, H 2021 *Mobilisation of radionuclides during uranium and gold processing of granitic rock at El-Missikat area, central Eastern Desert, Egypt.* International Journal of Environmental Analytical Chemistry ISSN: (Print) (Online) Journal homepage: https://www.tandfonline.com/loi/geac20

[11] El Feky, M, El Mowafy, A and Abdel Warith, A 2011 *Mineralogy, geochemistry, radioactivity and environmental impacts of Gabal Marwa granites, southeastern Sinai, Egypt.* China J. Geochem., 30, 175–186.

[12] El-Feky, M 2011 *Mineralogical, REE-geochemical and fluid inclusion studies on some uranium occurrences, Gabal Gattar, Northeastern Desert, Egypt*, China J. Geochem., 30, 430–443.

[13] El Galy, M, El Mezayn, A, Said, A, El Mowafy, A and Mohamed, M 2008 *Distribution and environmental impacts of some radionuclides in sedimentary rocks at Wadi area, southwest Sinai, Egypt*. J. Environ. Radioact. 99, 1075–1082.

[14] El-Kammar, A, Abu-Zied, H, El-Feky, M and Osman, D 2017 *Composition, radioactivity, and possible applications of kaolin deposits of Sinai, Egypt*. Arab J. Geosci, 10:463

[15] El Mezayen, A, Heikal, M, El-Feky, M, Shahin, H, Abu Zeid, I and Lasheen, S 2019 *Petrology, geochemistry, radioactivity, and M–W type rare earth element tetrads of El Sela altered granites, south eastern desert, Egypt.* Acta Geochimica, Vol. 38, pp. 95–119.

[16] El Mezayen, A, Ibrahim, E, El-Feky, M, Omar, S, El-Shabasy, A and Taala, S 2020 *Physico-chemical conditions controlling the radionuclides mobilisation in various granitic environments,* International J. of Environmental analytical chemistry.

[17] El Nahas H, El Feky M, and Mira H 2011 *Mineralogy, M-type tetrad effect and radioactivity of altered granites at the G. Abu Garadi shear zone, central Eastern Desert, Egypt*, China J. Geochem., 30, 153–164.

[18] Falkum, T and Rose-Hansen J 1978 *The application of radioelement studies in solving petrological problems of the Precambrian intrusive Homme granite in the Flekkefjord area, South Norway*, J. Chemical Geology. 23, 73–86.

[19] Ghobrial, M and Lotfi, M 1967 *The geology of Gabal Gattar and Gabal Dokhan areas*. Gen. Egypt Organ. Geol. Res. Min., Cairo, Egypt, Paper no. 40,26p.

[20] Harpy, N, El Dabour, S, Sallam, A, Nada, A, El Aassy, A and El Feky, M 2019 *Radiometric and environmental impacts of mill tailings at experimental plant processing unit, Allouga, Egypt.* ISSN: 1527-5922 (Print) 1527-5930 (Online) Journal homepage: https://www.tandfonline.com/loi/uenf20

[21] Lakehal, Ch, Ramdhane, M, and Boucenna, A 2010 *Natural radionuclide concentrations in two phosphate ores of East Algeria*. Jour. Environ. Radioact. 101, 377–379.
[22] Larsen, E and Gottfried, D 1960 *Uranium and thorium in selected suites of igneous rocks*, J. Am. J. Sci. 258A, 151–169.
[23] Malczewski, S, Teper L, Dorda, J. 2004 *Assessment of natural and anthropogenic radioactivity levels in rocks and soils in the environs of ŚwieradówZdrój in Sudetes, Poland, by in situ gamma-ray spectrometry*, Journal of Environmental Radioactivity, 73, 233–245.
[24] Matolin, M, 1991 *Construction and use of spectrometric calibration pads, Egy./ 4/030-03,Laboratory Gamma-Ray Spectrometry*. A report to the government of the Arab Republic of Egypt.
[25] Raslan, M and El-Feky, M 2012 *Radioactivity and mineralogy of the altered granites of the Wadi Ghadir shear zone, South Eastern Desert, Egypt.* China J. Geochem., 31, 030–040
[26] Moharem, A 1997 *Geological and geochemical studies on the uranium bearing granites in the northern part of Gabal Gattar, North Eastern Desert, Egypt*, M. Sc. Thesis, Faculty of Science, Ain Shams University, 195 p.
[27] Mousa, M 1998 *Geochronological studies of some granitoids: Application to geochemical evolution and tectonic history of the Northern Eastern Desert, Egypt.* Ph. D. Thesis, Ain Shams University, Cairo, Egypt, 284p.
[28] Nada, A 2003 *Evaluation of natural radionuclides at Um-Greifat area, Eastern Desert of Egypt.* Appl. Radiat. Isot. 58, 275–280.
[29] Naumov, G 1959 *Transportation of uranium in hydrothermal solution as carbonate*, Geochemistry, V.1, pp. 5-20
[30] Nkomo, I, Stuckless, J, Thaden, R and Rosholt, J 1978 *Petrology and uranium mobility of a granite of early Precambrian age from the Owl creek Mountains,Wyoming*, in Wyoming Geol. Assoc. Guide book Ann. Field Conf., 30th, 335-348.
[31] Nuclear Energy Agency (NEA-OECD) 1979 *Exposure to Radiation from Natural Radioactivity in Building Materials.* Report by NEA Group of Experts, OECD, Paris.
[32] Örgün,Y, Altınsoy, N, Ahin, S, Gu¨ngo¨r, Y Gu¨ltekin, A, Karahan, G and Karacık, Z 2007 *Natural and anthropogenic radionuclides in rocks and beach sands from Ezine region (Canakkale), Western Anatolia, Turkey*. App. lRadiat. Isot. 65:739–747
[33] Raslan, M and El Feky, M 2012 *Radioactivity and mineralogy of the altered granites of the Wadi Ghadir shear zone, South Eastern Desert, Egypt*, China J. Geochem.,31,030–040.
[34] Rogers, J and Ragland, P 1961 *Variation of thorium and uranium in selected granitic rocks*, J. Geochim. et Cosmochim. Acta. 25, 99–109.
[35] Rogers, J and Adams, J 1969 *Uranium and thorium. In: Wedepohl, K. H. (ed.), Handbook of geochemistry*, VII-3, 92-B-1 to 92-0-8 and 90-B-1 to 90-0-5, Springer Verlag, Berlin.
[36] Rosholt, J, Prijana, and Noble, D 1971 *Mobility of uranium and thorium in glassy and crystallized silicic volcanic rocks*, Earth and planetry sci . Letters, v. 7: 1061-1069.
[37] Roz, M 1994 *Geology and uranium mineralization of Gabal Gattar, north Eastern Desert, Egypt.* M. Sc. Thesis, Faculty of Science, Cairo University, pp175.
[38] Saito, K, Petoussi, H and Zanki, M 1998 *Calculation of the effective dose and its variation from environmental gamma ray sources*. Health Phys. 74, 698–706.
[39] Salman, A, El Aassy, I and Shalaby, M 1986 *New occurrence of uranium mineralization in Gabal Gattar, north Eastern Desert, Egypt.* Annals of Geological Survey of Egypt XVI pp31–34.
[40] Salman, A, El-Aassy, I and Shalaby, M 1990 *New occurrence of uranium mineralization in Gabal Gattar, Northetn Eastern Desert, Egypt*, Annal. Geol. Surv. Egypt, v. XVI: 31-34.
[41] Shalaby, M 1996 *Structural controls of uranium mineralization at Gabal Gattar, North Eastern Desert, Egypt.* Egyptian Academy of Sciences, vol. 46, p521–536.
[42] Shalaby, M, Abu Zeid, E and Mahdy, N 2015 *On the petrogenesis and evolution of U-rich granite: insights from mineral chemistry studies of Gattar granite, North Eastern Desert, Egypt.* Arab J. Geosci., vol. 8, pp3565–3585.

[43] Stuckless, J, Nkomo, I, Wenmer, D, and Van Trump, G. 1984 *Geochemistry and uranium favourability of the postorogenic granites of the north-eastern Arabian Shield, Kingdom of Saudi Arabia*, J. Bull Fac. Earth Sci. King Abdel Aziz Univ. 6, 195–209.

[44] Stuckless, J and Miesch, A 1981 *Petrogenetic modeling of a potential uranium source rock, Granite Mountains, Wyoming.* U.S. Geol. Surv. Prof. Paper 1225, 34p.

[45] UNSCEAR 2000 *United Nations Scientific Committee on the effects of atomic radiation, sources and effects of ionizing radiation.* Report to General Assembly, with Scientific Annexes United Nations. United Nations, New York

[46] UNSCEAR 2008 *United Nations Scientific Committee on the Effect of Atomic Radiation. Sources and Effects of Ionizing Radiation.* Report to General Assembly, with Scientific Annexes, United Nations, New York.

[47] Whitfield, J, Rogers, J and Adams, J 1959 *The relationship between the petrology and uranium contents of some granitic rocks.* Geochem. Cosmochem. Acta, 17: 248-271.

[48] Wignall, P and Meyers, K 1988 *Interpreting the benthic oxygen levels in mudrocks: a new approach.* Geology 16, 452-455.

[49] Yang, Y, Wu, X, Jiang, Z, Wang, W, Lu, J, Lin, J, Wang, L and Hsia, Y 2005 *Radioactivity concentration in soils of the Xiazhuang granite area. China.* Appl. Radiat. Isot. 63, 255–259.

Isotopic Ratio of Radionuclides in Natural Materials

Fawzia Mubarak[1] and Abdallah Ali[2]
Nuclear and Radiation Safety Research Center[1], Nuclear Research Center[2],
Egyptian Atomic Energy Authority, Cairo, Egypt
fawziamubarak@yahoo.com, Abdallah.ph999@yahoo.com
Corresponding author : Fawzia Mubarak
Email : fawziamubarak@yahoo.com

Abstract

In this work, the $^{235}U/^{238}U$ ratio and ^{235}U concentration were discussed. $^{235,238}U$ isotopes were measured and calculated for uranium mines samples and reference material.The difference and deviation in experimental and calculated values for ^{235}U were discussed. It can be concluded that the $^{235}U/^{238}U$ ratio was varied naturally with a range of 0.0442 to 0.0478. The measured ratio was deviated upon the reference value (0.0462) with a range of -0.02% to 1.5%, as we can conclude that the measured values were very accurate. While the values of the calculated ratio were varied from -0.089% up to 17%, which represents low accuracy. Isotopic uranium ratio changes over time depending on spontaneous fission half-life, radioactive decay rate and gamma-ray intensity for $^{235,238}U$. These fractions in uranium isotopic were discussed and here we assume a tool for correction and normalization. It can be concluded that the $^{235}U/^{238}U$ ratio and ^{235}Uconcentration can be varied naturally with a slight deviation. The difference and reliability of $^{235,238}U$ isotope on experimental results of uranium mines depend on several reasons such as decay rate of natural uranium, Spontaneous fission half-life and Gamma-ray intensity for $^{235,238}U$. Therefore, the experimental results for ^{235}U are very accurate than the calculated results to improve the isotopes of U and $^{235}U/^{238}U$ ratio in Uranium mines.

Keywords: $^{238}U/^{235}U$ isotopic ratio, Spontaneous fission half-life, $^{235,238}U$concentration, Gamma spectroscopy and Reference material.

1. INTRODUCTION

Sinai Peninsula is the Asiatic part of Egypt, but it is of major economic importance in northern Africa. It has a triangular shape, covering an area of 61,000 km2, and is separated geographically from the Eastern Desert by the Gulf of Suez. It has a special interest in natural resources prospecting after the discovery of oil and many other mineral deposits, such as manganese, iron, copper, zinc, lead, cobalt, nickel, silver and uranium, which could be of great value for development purpose. Special attention has been given to the study of the natural radionuclides, especially in southwestern part of Sinai. Within natural radionuclides, uranium has special economic importance. From many isotopes of uranium, only three isotopes ^{238}U, ^{235}U and ^{234}U occur with a significant abundance of about 99.274%, 0.7204%, and 0.00548% respectively [1-4]. Fractionation in uranium isotopes must be present in invariant values but due to chemical reactions such as oxidation and reduction processes, this value can be varied [5-8]. ^{238}U and ^{235}U have a very long half-life and non-overlapping decay chains; hence they are important tools for geochronology [9]. ^{235}U was significantly more abundant in the past than ^{238}U, but due to its shorter half-life, it becomes much less abundant.$^{235}U/^{238}U$ ratio (0.0072) may slightly be varied in nature but it can be considered as a good indicator of the origin and activities associated with the naturally occurring samples [10,14]. An artificial neural network model was used for the determination of uranium isotopic ratio in gamma spectroscopy measurements [15]. This study aims to investigate the limit of natural variation of $^{235}U/^{238}U$ ratio and to test the accuracy of ^{235}U concentration in natural material.We will examine the uranium isotopes ratio in uranium mines samples and to detect any signature in $^{235,238}U$ isotopic ratio according to several reasons effected on uranium isotopes concentration measurements such as the alteration processes affected this variation in different times. This study aim to compare the experimental measurements of $^{238}U/^{235}U$ isotopic ratio with the calculated methods.

2. MATERIALS AND METHODS

2.1. Samples Collection and Preparation

Different samples from uranium mines were collected from different locations in southwestern part of Sina (Abu Hamata, Adadia, El Hashash, and Wadi al Hammam). Samples were powdered, dried in an oven at 110°C until the dry weight of each sample became constant. 500 grams of each sample was packed and well-sealed in a plastic container. To prevent the escaping of uranium and thorium daughters especially radon isotopes, the container was tied with adhesive tape. For allowing the occurring of secular equilibrium between^{226}Ra and its daughter products samples stored for one month. Activity concentration was measured using the nondestructive technique, gamma spectroscopy.$^{235,238}U$ stander reference materials with specific activities (79.99×10^3Bq/g, 12.44×10^3Bq/g) respectively were used [16].

2.2. Gamma Spectroscopy

A high efficiency and energy resolution P-type coaxial HPGe detector was used which shielded by 10 cm Pb thickness followed by1 mm Cd and 1 mm Cu for reducing cosmic rays and cutting off x-ray from lead and Cd [17]. The detector has FWHM of 1.9 keV at 1.33 MeV, relative efficiency of 50% connected to a multi-channel analyzer (MCA) coupled with software program Genie 2000. The advantage of this detector is the measurement of low-level activity samples and the ability to estimate the detection limits and minimum detectable activity. Efficiency calibration of the system was carried out by using standard sources (RGU-1, 400 μg/g of U-238) and (RGTh-1, 800 μg/g of Th-232) produced by IAEA.

2.3. Activity Measurements

^{238}U is an alpha emitter, so it can be indirectly characterized through the decay chain product ^{234m}Pa with 1001 keV energy line. ^{235}U can be characterized through 143.8, 163.4, 185.7 and 205.3 keV energy lines. The most precise method to determine the 235U content with gamma spectroscopy is termed the enrichment meter method; accuracies of better than ±0.015 wt% can be achieved [18]. In this method, the absolute peak area of the 185.7 keV gamma peak from 235U is measured, and compared to well-defined reference standards to determine the amount of 235U present. Although the ^{235}U energy line 185.7 keV has a relatively higher branching ratio (57%), but it is very close to the ^{226}Ra energy line of 186.2 keV and overlapping occur. So, U-235 concentration was calculated from this energy line and the result was compared with that measured through the other lines. Ra-226 was measured directly through the 186.1keV (after subtracting the 185.7 keV line of ^{235}U) and indirectly by measuring the high-intensity lines of ^{214}Pb (351.9 keV) and ^{214}Bi (609.3, 1120.2 and 1764.5 keV) gamma lines. Long measurement time improves the precision and accuracy of the uranium composition measurements. Samples and background were measured for about 6 hours for each [19-23].

2.4. Theoretical Aspects:

2.4.1 Activity Concentration:

Activity concentration (A) of each radionuclide can be calculated as [10]:

- $$A\ (Bq/kg) = \frac{net\ C/s}{Eff.\times I.m} \quad (1)$$

Where; net C/s is the net count /sec., Eff. Is the specific efficiency for the specified energy line, I is the branching ratio and m is the mass of the sample (kg).

2.4.2 ^{235}U Calculation:

^{235}U was measured through 143.8, 163.4, and 205.3 keV energy lines, but we also calculate it from 186.2 keV energy line according to the following equations [10]:

- $$c/s_{Ra} = 0.583\ c/s_T \quad (2)$$
- $$c/s_{U5} = 0.417\ c/s_T \quad (3)$$
- $$A_{U5} = \frac{c/s_{U5}}{eff._{186.2} \times I_{U5}} \quad (4)$$

Where; c/s $_{Ra}$, c/s $_{U5}$ and c/s$_T$ are the net c/s for ^{226}Ra, ^{235}U and total 186.2 photo peak respectively, and A_{U5} is the activity of ^{235}U

2.4.3. ^{235}U% Concentration:

The ^{235}U percentage concentration can be calculated as following [4]:

- $$235_U\% = \frac{m(235_U)}{m(235_U)+m(238_U)} \times 100 \quad (5)$$

- $$m(235_U) = \frac{A(235_U)x235xT_{\frac{1}{2}}(235)}{N_A \times ln2} \quad (6)$$

- $$m(238_U) = \frac{A(238_U)x238xT_{\frac{1}{2}}(238)}{N_A \times ln2} \quad (7)$$

Where; m(^{235}U) and m(^{235}U) are the mass of ^{235}U and ^{238}U (in kg) respectively, A(^{235}U) and A(^{238}U) are the activity concentration (Bq/kg) of ^{235}U and ^{238}U respectively and NA is the Avogadro's no.

3 Result and discussion

In our study, we will discuss the difference and reliability of ratio and rate between two members of $^{235,238}U$ isotope on experimental results of uranium mines in Egypt.

In the future, the atomic percentage of ^{235}U will become smaller than 0.72% depending on the fact of the half-life of ^{235}U smaller than the half-life of ^{238}U and the spontaneous fission for ^{238}U is very low rate than ^{235}U, especially the spontaneous fission half-life of ^{238}U nucleus is about 10^{15} years. Also depending on the fact of absorption fission of ^{235}U nucleus undergoes with very low neutron energy (thermal neutron), on the other hand, ^{238}U nucleus undergoes with neutron energy more than 0.6 MeV (fast neutron).

Table1 shows activity concentrations of ^{226}Ra, ^{235}U and ^{238}U in different samples. $^{235}U/^{238}U$ ratio was measured for different samples and the % concentration of ^{235}U was calculated. There was variation between the calculated concentration of ^{235}U and the measured value.These difference in activity concentration are due to some reasons. Firstly, the disequilibrium conditions between a parent and its daughters which may occur as a result to the presence of some very long-lived daughters. Secondly, the difference in the emission probability ~ 3.29 % of the 186.1 keV peak from ^{226}Ra and ~ 57 % of the 185.7

Journal of Physics: Conference Series **2305** (2022) 012007 doi:10.1088/1742-6596/2305/1/012007

keV peak from ^{235}U. Lastly, due to the difference in the error in the calibration curve due to the energy range of interest.

Results show that activity concentrations of different samples were ranged between 925 -3060 Bq/kg for ^{226}Ra- series, 830– 2836Bq/kg for ^{238}Uand 39 -135Bq/kg for ^{235}U. ^{235}U/^{238}U ratio was varied from 0.0442to 0.0478with bias up to 10% (The theoretical value is 0.0462). ^{235}U concentration was varied naturally from 0.69% up to 0.78% (The theoretical value is 0.7257%). Validations were carried out by measuring a reference material (RGU-1) and calculate the ^{235}U/^{238}U ratio and ^{235}Uconcentration and results show that their values were in good agreement with the certified values.

3.1 The radioactive decay rate of natural uranium

Stander references of natural ^{235,238}U in secular equilibrium with specific activity (79.99× 10^{3}Bq/g) and (12.44× 10^{3}Bq/g) respectively were used. Figures 1,2 show that the radioactive decay rate of natural uranium in secular equilibrium with its decay products ^{235,238}U with a variety of time.

Table (1): Activity concentrations and isotopic ratio of $^{235,\,238}$U in uranium mines samples.

Sample	^{226}Ra	^{238}U	^{235}U measured	^{235}U calculated	Ratio meas. (^{235}U/^{238}U)	Ratio cal. (^{235}U/^{238}U)	Ratio ^{235}U (cal. /meas.)
S1	3060	2836	135	67.7	0.0476	0.0239	0.5014
S2	3020	2569	122	79.8	0.0475	0.0311	0.6541
S3	2982	2146	101	82.7	0.0470	0.0385	0.8188
S4	1847	1526	71	50.8	0.0465	0.0333	0.7155
S5	1538	1300	59	36.5	0.0454	0.0281	0.6186
S6	1462	1205	56	44.5	0.0464	0.0369	0.7946
S7	1375	1348	64	41.7	0.0475	0.0309	0.6515
S8	1358	1159	54	38.6	0.0457	0.0333	0.7148
S9	1332	1381	66	41.6	0.0478	0.0301	0.6303
S10	1305	1378	64	43.8	0.0464	0.0318	0.6843
S11	1230	1063	47	33.2	0.0478	0.0312	0.7063
S12	1217	1133	53	28.3	0.0468	0.0250	0.5339
S13	1051	945	43	32.8	0.0455	0.0347	0.7628
S14	1034	1368	65	36.7	0.0475	0.0268	0.5646
S15	1007	943	45	36.2	0.0477	0.0384	0.8044
S16	972	830	39	19.2	0.0470	0.0231	0.4923
S17	925	1014	48	21.6	0.0473	0.0213	0.4500
AV	1571	1420	66.6	43.2	0.0468	0.0305	0.6529
SD	729	569	27.4	18.0	0.0010	0.0052	0.1140
Min	925	830	39	19.2	0.0442	0.0213	0.4500
Max	3060	2836	135	82.7	0.0478	0.0385	0.8188

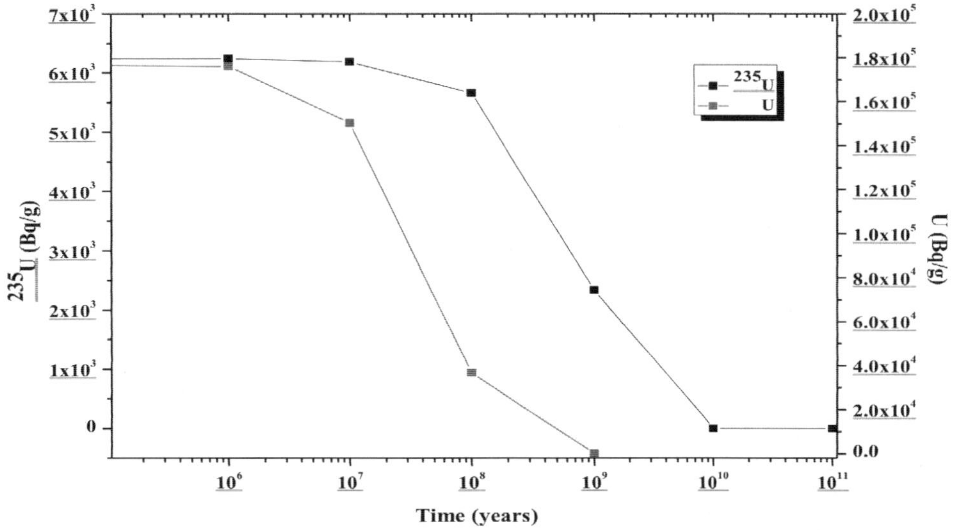

Figure 1. The radioactive decay rate of natural uranium (1gram) in secular quilibrium with its decay products ^{235}U with a variety of time (years) in log time scale.

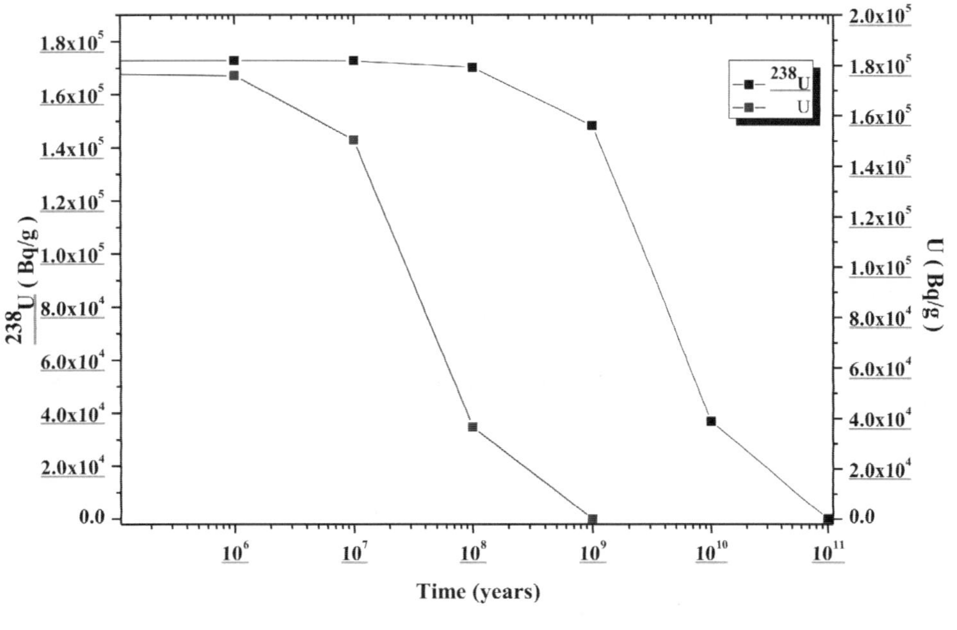

Figure 2. The radioactive decay rate of natural uranium (1gram) in secular equilibrium with its decay products ^{238}U with a variety of time (years) in log time scale.

3.2 Gamma-ray intensity for ^{235,238}U decay series

For accurate measurement, we used the strongest and pure gamma line energies for ^{235,228}U. The selection of high gamma-ray intensity without interference with another gamma- lines.Therefore, we measured the energy of a gamma-ray that corresponds to uranium nuclides and its isotope, while the accurate activity corresponds to the intensity of the gamma-ray in uranium radioactive in samples measured as well as some considerations like interferences, detector efficiency and geometry.^{235,238}U nuclides were investigated in natural isotopic composition for nuclear material from uranium mines. Fig.3 shows the^{235,238}U decay series in secular equilibrium for nuclear material from uranium mines in Egypt.

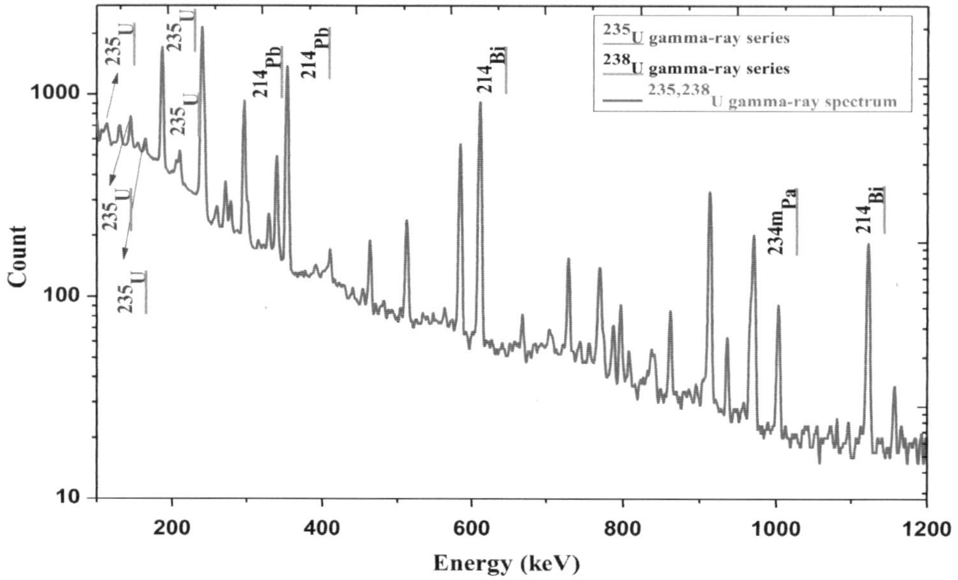

Figure 3. Gamma-ray spectrum for ^{235,238}U decay series in secular equilibrium for nuclear material from uranium mines in Egypt.

3.3 Spontaneous fission half-life

The spontaneous fission half-life ($T_{1/2}$ SF) for ^{238}U is very low rate than ^{235}U, especially the spontaneous fission half-life of ^{238}U nucleus is about (8.4×10^{15}) years and (3.5×10^{17}) years for ^{235}U [24]. (Z^2/A) represent the ratio between the square atomic number (Z^2) and mass number (A). Where Z is the atomic number and A is the mass number, ($T_{1/2}$SF), Spontaneous fission half-life. A correlation equation exists between spontaneous fission half-life of ^{235,238}U was $(T_{1/2}\,SF) = \frac{Z^2/A - 35.59}{1.17} \times 10^{18}$ as shown in Figure 4.

3.4 The isotopic ratio for ^{235}U/^{238}U

^{235}U/^{238}U ratio for uranium mines sample as a function between ^{235,238}U with average ratio value (0.04809± 0.00047). A correlation exists between Activity of^{235}U and ^{238}U in uranium samples (R^2 = 0.95811) with fitting equation A (^{235}U) = 0.048×A (^{238}U) - 1.71) as shown in Figure 5.

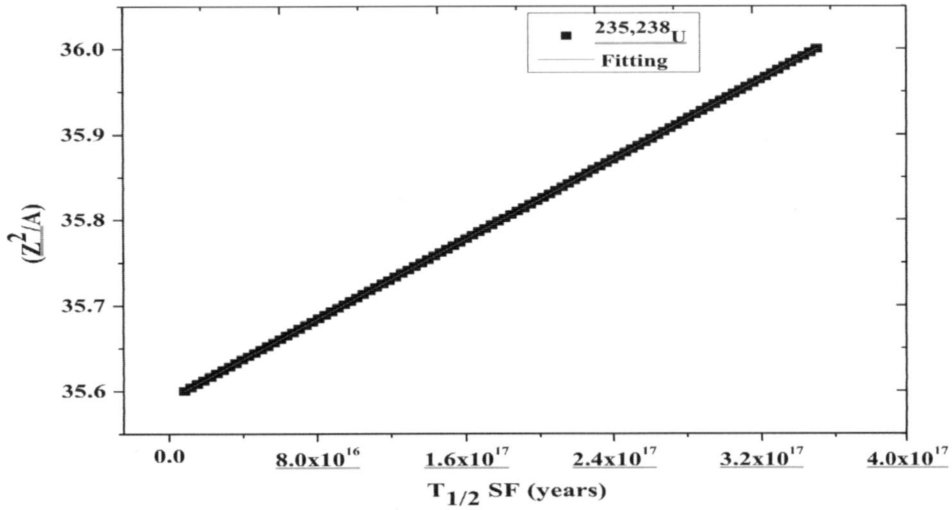

Figure 4. Spontaneous fission half-life for $^{235,238}U$ nuclides depending on their Z^2/A ratio.

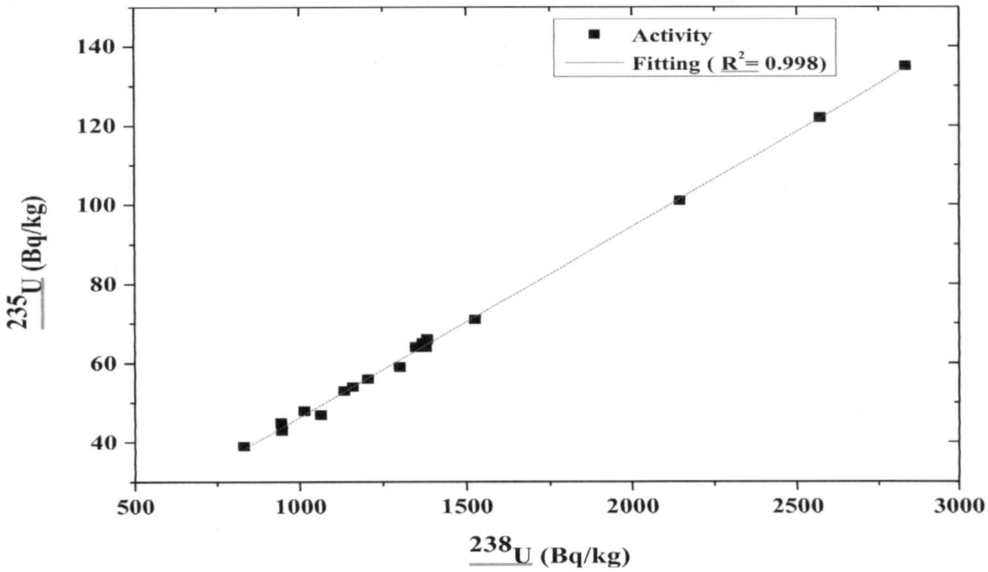

Figure 5. A correlation between activity for $^{235,\,238}U$ isotopic ratio of uranium mines samples.

4. Conclusion

Uranium mining is of great economic importance. The quality of the product depends on an accurate estimate of the ratio of isotopes in uranium to each other. From this study, it can be concluded that the $^{235}U/^{238}U$ ratio and ^{235}Uconcentration can be varied naturally with a slight deviation. The difference and reliability of $^{235,238}U$ isotope on experimental results of uranium mines depend on several reasons such as decay rate of natural uranium, Spontaneous fission half-life and Gamma-ray intensity for $^{235,238}U$. Therefore, the experimental results for ^{235}U are very accurate than the calculated results to improve the isotopes of U and $^{235}U/^{238}U$ ratio in Uranium mines.

References

1- Mishra S., S. Sahoo, P. Chaudhury, K. Pradeepkumar 2017 Measurement and validation of uranium isotope ratio in uranium ore for isotopic fingerprinting, *Radiat. Prot. Environ.*. *https://doi.org/10.4103/rpe.rpe_36_16*.
2- Weyer S., A.D. Anbar, A. Gerdes, G.W. Gordon, T.J. Algeo, E.A. Boyle 2008 Natural fractionation of 238U/235U, *Geochim. Cosmochim. Acta.* https://doi.org/10.1016/j.gca.2007.11.012.
3- Varga Z., M. Krachler, A. Nicholl, M. Ernstberger, T. Wiss, M. Wallenius, K. Mayer 2018 Accurate measurement of uranium isotope ratios in solid samples by laser ablation multi-collector inductively coupled plasma mass spectrometry, *J. Anal. At. Spectrom.* https://doi.org/10.1039/c8ja00006a.
4- Luca A. 2008 Experimental determination of the uranium enrichment ratio, Rom. Reports Phys.
5- Schauble E.A. 2006 Equilibrium uranium isotope fractionation by nuclear volume and mass-dependent processes, *AGU Fall Meet.*
6- Schauble E.A. 2007 Role of nuclear volume in driving equilibrium stable isotope fractionation of mercury, thallium, and other very heavy elements, *Geochim. Cosmochim. Acta.* https://doi.org/10.1016/j.gca.2007.02.004.
7- Stirling C.H., M.B. Andersen, E.K. Potter, A.N. Halliday 2007 Low-temperature isotopic fractionation of uranium, *Earth Planet. Sci. Lett.* https://doi.org/10.1016/j.epsl.2007.09.019.
8- Condon D.J., N. McLean, S.R. Noble, S.A. Bowring 2010 Isotopic composition (238U/235U) of some commonly used uranium reference materials, *Geochim. Cosmochim. Acta.* https://doi.org/10.1016/j.gca.2010.09.019.
9- Brennecka G.A., L.E. Borg, I.D. Hutcheon, M.A. Sharp, A.D. Anbar 2010 Natural variations in uranium isotope ratios of uranium ore concentrates: Understanding the 238U/235U fractionation mechanism, *Earth Planet. Sci. Lett.* https://doi.org/10.1016/j.epsl.2010.01.023.
10- Ebaid Y.Y. 2010Use of gamma-ray spectrometry for uranium isotopic analysis in environmental samples, *Rom. Reports Phys.*
11- Xiao G., R.L. Jones, D. Saunders, K.L. Caldwell 2014 Determination of 234U/238U, 235U/238U and 236U/238U isotope ratios in urine using sector field inductively coupled plasma mass spectrometry, *Radiat. Prot. Dosimetry*. https://doi.org/10.1093/rpd/ncu023.
12- Liu C., B. Hu, J. Shi, J. Li, X. Zhang, H. Chen 2011 Determination of uranium isotopic ratio (235U/ 238U) using extractive electrospray ionization tandem mass spectrometry, *J. Anal. At. Spectrom.* https://doi.org/10.1039/c1ja10054h.
13- James D. Wells, 2018 U-235 and U-238 isotopic fractions of natural Uranium throughout geological time, LCTP, Physics Dept, University of Michigan, Ann Arbor, MI 48109, USA.
14- Fujikawa Y., K. Shizuma, E. Ikeda, T. Baba, T. Yoshimoto, M. Fukui, M. Saito, H. Yamanishi, T. Uda, 2002 Uranium and plutonium isotope ratio measurement as a tool for environmental monitoring experiences in Osaka, Gifu and Hiroshima, Japan, J. *Nucl. Sci. Technol.* https://doi.org/10.1080/00223131.2002.10875531.
15- E.K. Elmaghraby, M. Tohamy, M.N.H. Comsan 2019 Determination of isotopes activity ratio using gamma ray spectroscopy based on neural network model, *Appl. Radiat. Isot.* https://doi.org/10.1016/j.apradiso.2019.03.014.
16- Sampsom TE. 2003. "Verification and validation of the FRAM Isotopic Analysis Software." LA-14057, *Los Alamos National Laboratory.*

17- Senthilkumar R.D., R. Narayanaswamy 2016 Assessment of radiological hazards in the industrial effluent disposed soil with statistical analyses, *J. Radiat. Res. Appl. Sci.* https://doi.org/10.1016/j.jrras.2016.07.002.
18- Yucel H, E Yeltepe, AO Yuksel, and H Dikmen. 2015. "U-235 isotopic characterization of natural and enriched uranium materials by using multigroup analysis (MGA) method at a defined geometry using different absorbers and collimators." *Nukleonika 60(3):615–620.*
19- Entesar H. El-Araby, Khaled A. Salman and Fawzia Mubarak 2021 Human Risk due to Radon and Heavy Metals in soil. *Iran J Public Health, Vol. 50, No.8, pp.1624-1634*
20- Kaspar TC, CA Lavender, MW Dibert 2017 Evaluation of Uranium-235 Measurement Techniques, United State Dept. of Energy, USA.
21- Mubarak F., M. Fayez-Hassan, N.A. Mansour, T. Salah Ahmed, A. Ali, 2017 Radiological Investigation of High Background Radiation Areas, *Sci. Rep.* https://doi.org/10.1038/s41598-017-15201-2.
22- Fawzia Mubarak and H. M. Saleh 2020 Estimation of Occupational Exposure to Workers during Ceramic Manufacture" *Iran J Med Phys; 17: 175-182*
23- Mathew K.J., R.M. Essex, C. Gradle, U. Narayanan 2015 Uncertainties achievable for uranium isotope-amount ratios: estimates based on the precision and accuracy of recent characterization measurements, *J. Radioanal. Nucl. Chem.* https://doi.org/10.1007/s10967-014-3828-0.
24- . Shultis J.K, R.E. Faw, 2007 Fundamentals of nuclear science and engineering, second edition, https://doi.org/10.1201/b12824.

Process simulation and optimization of dodecyl benzene sulfonic acid production plant

Ahmed Fathy Mohamed[1], Waleed Abd El Azim Ibrahim El Sayed[1], Nabil Abdel Moneim[2] and Faisal Hassan Sorour[3]

[1] Egyptian Armed Forces, Cairo, Egypt.
[2] Chemical Engineering Department, Cairo University, Cairo, Egypt.
[3] Canal High Institute for Engineering and Technology, Suez, Egypt.

* sorourfh@outlook.com

Abstract. The aim of work is to design sulfonating plant on the basis of a hydrophilic sulfonate head-group and a hydrophobic alkyl benzene tail-group reaction mechanism, study of kinetics and thermodynamics followed by a steady-state simulation for the process was done by ASPEN HYSYS (V8.8) through using Peng-Robinson fluid package. Then heat integration of the process ,done by using Aspen Energy Analyzer and studies of the influence parameters on the simulated process of dodecyl benzene sulfonic acid. Alkyl benzene sulphonate is produced due to the sulphonation reaction of an alkylbenzene (LAB) plus a sulphonating agent in a falling film sulphonator to produce sulfonic acid (LABSA).The modeling of dodecyl benzene sulfonic acid plant depends on : Air drying, Sulphur melting and proportioning, Sulphur combustion and SO_3 production, Film sulfonation. Dodecyl benzene sulfonic acid production rate is influenced by *changing* of the following factors: the sulfur molar flow rate, molar flow rate of air and pressure of dodecyl benzene, the results of optimization shows that : The optimum operating Molar flow rate of sulfur is lies in the range of [3:3.5] Kg.mol/h ,The optimum operating Molar flow rate of Air is 25 Kg.mol/h, The optimum operating Molar flow rate of Dodecyl Benzene is 3.1 Kg.mol/h and The optimum operating Pressure of Dodecyl Benzene is 70 KPa. For heat integration, a proposed heat exchanger network is generated which made the process functional without external heating. Generation of the top 10 heat exchanger network (HEN) leads to: the best case scenario from the top 10 scenarios from the point of cost and also from the point of cooling and heating required was design (9). After heat integration , the total amount of cooling demand decreased from 912620.50 KJ/h to 890261.56 KJ/h and the total amount of heating demand decreased from 22358.94 KJ/h to 0KJ/h.

Keywords: Dodecyl Benzene Sulfonic acid, simulation, modelling, aspen hysys.

1. Introduction

The aim of this work is to develop and implement an effective knowledge on Simulation of plant in Aspen-Hyses. ASPEN HYSYS, PLUS, CHEMCAD etc, these simulators are widely used in different ways such as chemical, oil, petrochemical and industries of gas, pulp, mining, pharmaceutical, paper, biotechnology and power plants. Today, the main process modelling tool is Aspen HYSYS where Aspen HYSYS is used for performance monitoring for oil & gas production, petroleum refining, optimization of process, conceptual design, processing of gas, and industries of air separation. Finally, Aspen Hysys has giant importance for chemical engineers to simulate a process Aspen-HYSYS modelling used the best technologies and has the ability to provide the design of the process to improve the efficiency and profitability of the business.

Nowadays Aspen HYSYS software is the master process modeling solution that gives large economic benefits throughout the process engineering lifecycle. Chemical engineers used Aspen HYSYS to design new plants and also optimize production for greater margins. [1] The steady-state simulation design of the process of dodecyl benzene sulfonic acid production plant is done by ASPEN HYSYS software (v 8.8) and heat integration for the process of dodecyl benzene sulfonic acid production plant is done by Aspen Energy Analyzer and studies of the factors influence on production rate of dodecyl benzene sulfonic acid.

LABSA is intermediate of LAB and LAS, produced on LAB sulfonation, HLAS on neutralization results in to LABS salt (LAS). Alkyl benzene sulfonate is made from the sulfonation reaction of alkyl benzene (LAB) with a sulfonating agent (a dilute solution of SO3 in air -typically 10% SO3) in a falling film sulfonator to give the sulfonic acid (LABSA).

It represents commercially the most important supply form. The dodecyl benzene sulfonic acid is liquid of a low viscosity, even at manufacturing concentrations of 96% and can be stored for extended periods, due to the stability of the carbon–Sulphur bond which allows the use of the concentrated sulfonic acid a highly cost-effective way for production of the surfactant.

Linear alkyl benzene sulfonic acid (LABSA) has foaming property and good decontamination, and is stable in acidic, and some alkalic solution. The biodegradation percent is more than 90%. (LABSA) is used widely in production of detergents including laundry powders (washing powder), laundry liquids, dishwashing liquids and other household cleaners. In addition to its usage in numerous industrial applications like textile cleaning agent, dyeing auxiliary (assistant); electroplating industry, leather making industry and degreasing agent (degreaser of plating); made Paper deinking agent industry.

1.1. Overview of Sulfonic acids

Sulfonic acids are defined as a group of organic acids which contain one or more sulfonic (SO3H) groups. This section is an overview of organic sulfonic acids which contains the formula RSO3H, where the R-group may be derived from many different sources. The R-group contains a large variety of functionalities such as carboxylic acid, amine, amide, ester, ether, ketone, phenol, nitrile, etc. Sulfonic acids also find have application as catalysts in both alkylation (qv) processes and general organic reactions. Sulfonic acid derivatives have found more applications in the areas of custom organic syntheses and biological research. The derivatives of sulfonic acids, which contain the R-group is derived from an inorganic source such as a halide, oxygen (i.e., sulfate), or amine (i.e., sulfamic acid).[2]

Sulfonic acid is used in production of synthetic detergents, rubber and synthetic fibers, emulsifying and wetting agent, also cationic exchangers used with LAS for chromatographic analysis (HPLC) [3] steam boilers used LAS also as anti-scaling agent [4].

1.2. Physical Properties of Sulfonic acid

Sulfonic acids physical properties are depending on the nature of the R-group. Sulfonic acids are foundd in both the solid and liquid phases at room temperature and have the same acidity characteristics of sulfuric acid. Sulfonic acids are subject to thermal decomposition, i.e., desulfonation, at elevated temperatures.

Table 1. indicates that Arene-based sulfonic acids are thermally unstable and therefore must be distilled under extreme vacuum conditions using a minimal amount of heating to avoid thermal decomposition. [5-6].

Table 1. Physical properties of different types of Sulfonic acid

Acid	CAS Registry Number	Mp / °C	Bp / °C	Density g/cm3
dodecyl benzene sulfonic acid	[27176-87-0]	6.3	189	1.06
Methane sulfonic acid	[15-15-2]	20	122	1.4
Ethane sulfonic acid	[59445-6]	-17	123	1.33
Propane sulfonic acid	[28553-80-2]	-37	159	1.19
Butane sulfonic acid	[30734-86-2]	-15	149	1.19
Pentane sulfonic acid	[35452-30-3]	-16	163	1.12
Hexane sulfonic acid	[13595-13-8]	16	174	1.10
Benzene sulfonic acid	[98-11-3]	44	172	
p- toluene sulfonic acid	[104-154]	106	182	
Trifluoro methane sulfonic acid	[1493-13-6]	------	161	1.70

The sulfonic acids are stronger than the corresponding carboxylic acid a million times. As example, methane sulfonic acid and p-Toluene sulfonic acid have pKa (Acid dissociation constant) values of −2.8 and −1.9, while those of benzoic acid and acetic acid are 4.20 and 4.76, respectively. Nevertheless, their consequence of strong acidity and pKa values cannot be measured directly, and values commonly quoted should be regarded as indirect estimates with significant uncertainties. As example, different sources have reported the pKa of methane sulfonic acid to reach a high of −0.6 [7] or low of − 6.5. [8] Sulfonic acids react with solid sodium chloride (salt) to form the sodium sulfonate and hydrogen chloride.[9]

This property indicates an acidity within two or three orders of magnitude of that of HCl (g), whose pKa was recently accurately determined (pKa aq = −5.9).

Due to the polarity of sulfonic acids, it is tended to be crystalline solids or viscous, high-boiling liquids. The sulfonic acids are colourless and non-oxidizing, [10] and this makes them suitable for usage as acid catalysts in the organic reactions. This acid is soluble with water, non-oxidizing and colourless. It is highly used in the chemical industries, due to its high acidity, standard measures must be observed in terms.

2. Methodology:

The modelling of dodecyl benzene sulfonic acid process plant depends on these steps: sulfur oxidation step to produce sulfur dioxide, conversion of sulfur dioxide to sulfur trioxide, production of dodecyl benzene sulfonic by reaction of sulfur trioxide with dodecyl benzene. In this paper, Dodecyl Benzene Sulfonic Acid Production Plant was simulated by using ASPEN HYSYS (V8.8) with steady state mode, which considered as one of the best chemical engineering simulators.

2.1 Process Description:

The film sulfonation plant is based on SULFUREX continuous process, where the sulfonation/sulfation reaction takes place in a multi tube film reactor, using SO3 as sulfonation gas in variable quantity and concentration. The SO3 gas is produced by sulfur combustion.

2.1.1. Plant configuration consists of the following units:

- Air drying
- Sulphur melting and proportioning
- Sulphur combustion and SO_3 production
- Film sulfonation.

2.1.1.1. Air Drying

The air required for Sulphur combustion and SO3 production is compressed by a low-pressure compressor. The air is cooled by cooling water, in order to remove the heat of compression and for moisture condensation.

2.1.1.2 Sulphur melting and proportioning

The solid sulfur is melted and filtered to avoid the pump valves clogging and then fed to the sulfur burner by a proportioning pump. The molten sulfur is fed under mass control, by means of a mass flow meter. The pump and its relevant piping are heated by steam in order to keep the temperature constant and, consequently, to minimize the molten Sulphur viscosity and relevant pipe pressure drop.

2.1.1.3. Sulphur combustion and SO3 production

The molten Sulphur is fed to the burner, counter currently with the combustion air. This easy system allows the complete combustion of Sulphur without any spraying nozzles, which might often clog due to the Sulphur impurities.

The SO2 gas leaves the burner at around 700°C at a SO2 concentration of approx. 7% by volume and it is cooled to about 420°C before reaching the SO2/SO3 conversion tower. The SO3 gas is finally cooled to about 55°C, by means of heat exchangers (shell and tube type).

$$S + O_2 \longrightarrow SO_2$$

$$SO_2 + 1/2O_2 \xrightarrow{V_2O_5} SO_3$$

2.1.1.4. Film Sulphonation

The special design of the reactor distribution heads indicates excellent advantage of the film sulfonation plant which give the possibility to dose accurately and to keep the exact mole ratio between the raw material and the sulfonation gas in each single reaction tube, and to self-equilibrate it, thus eliminating any risk of over-sulfonation.

The exact mole ratio between SO3 and organic feedstock is essential to obtain the maximum conversion degree with optimum color .In film reactors of different types the equilibrium is achieved only by the utmost accuracy in the mechanical construction; it is rather difficult to obtain such accuracy in large size plants and to keep it unchanged during the life of the equipment (owing to assembly and disassembly, during maintenance, caused by charring of the distribution area, which has a negative influence on the homogeneity of the gas/liquid distribution), while the reactor is self-equilibrating as described before. The heat of reaction is removed in the shell side by cooling water.

Another advantage of sulfonation plant consists in the emergency system, which, in case of a sudden lack of raw material due for example to power failure, enters automatically in operation, cutting off the flow of SO3 and feeding fresh organic feedstock through the reactor, flushing organic feeding line with compressed air, thereby avoiding any over sulfonation of the product under processing and any need of cleaning the reactor for charring due to lack of liquid; the product partially sulfonated is recycled to the reactor when plant running is resumed. The procedure for product change-over is very simple, as the product residence time in the sulfonation system is very low.

$$C_{18}H_{30} + SO_3 \longrightarrow C_{18}H_{30}O_3S$$

2.2. Software:

The software used in the simulation of dodecyl benzene sulfonic acid plant design is Aspen Hysys (v.8.8) with steady state mode.

2.3. Process simulation:

The main steps for dodecyl benzene sulfonic acid process simulation by using Aspen Hysys are the following:

1) Selection of component list.
2) Selection of fluid package.
3) Defining reactions and formation of reaction sets.
4) Installing the feed streams.
5) Drawing flow sheet.

Fig (2.8) shows the Process Flow Diagram (PFD) for the production of dodecyl benzene sulfonic acid process, generated by Aspen Hysys.

2.3.1. Components list selection:

In this simulation, the reactant component list contains (Oxygen, Nitrogen, Water, Sulfur, Sulfur dioxide, Sulfur trioxide, Dodecyl benzene, Dodecyl benzene sulfonic acid).

Figure (2.1) demonstrates the used component list in aspen Hysys programming.

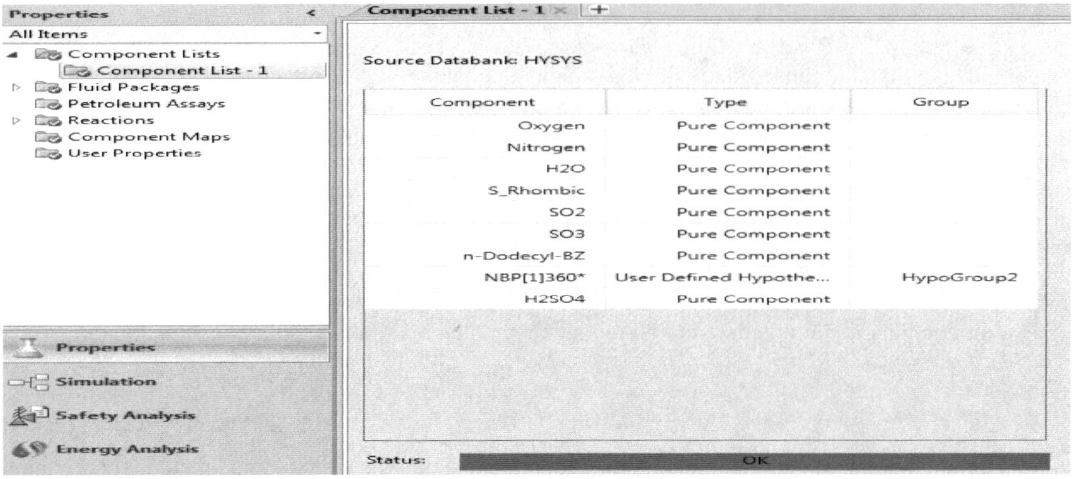

Figure (2.1) component list in aspen Hysys software.

2.3.2. Fluid package selection:
In this simulation, the used fluid package is Peng- Robinson (PR), which is the most improved model in Aspen HYSYS.

2.3.3. Defining reactions and formation of reaction sets:
In this simulation, the procedure of Dodecyl benzene sulfonic acid production involves sets of reactions; Sulfur dioxide production, Sulfur trioxide production, Dodecyl benzene sulfonic acid production. Figures from (02) to (04) give input information to the making of different reactions sets.

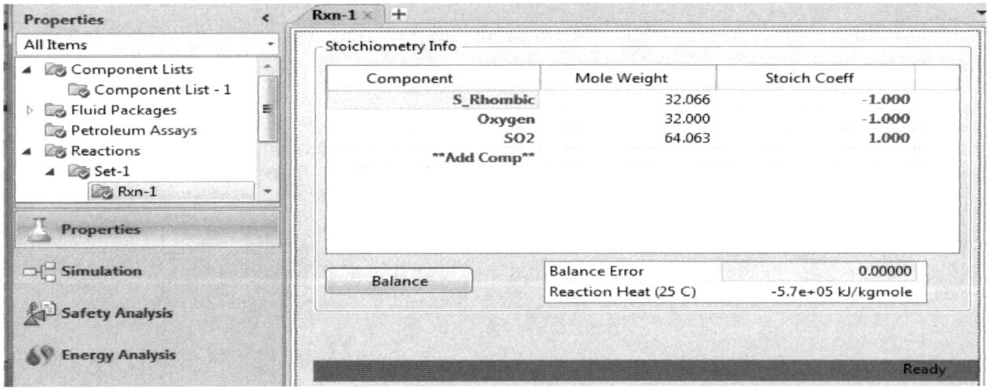

Figure (2.2) reaction set of Sulfur dioxide production.

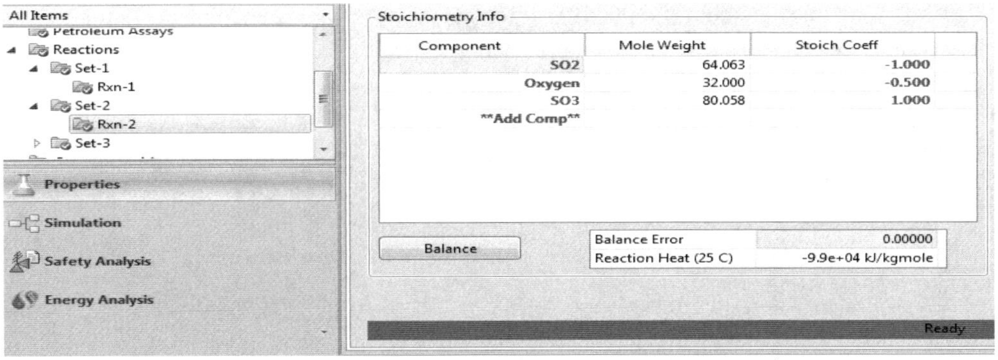

Figure (2.3) reaction set of Sulfur trioxide production.

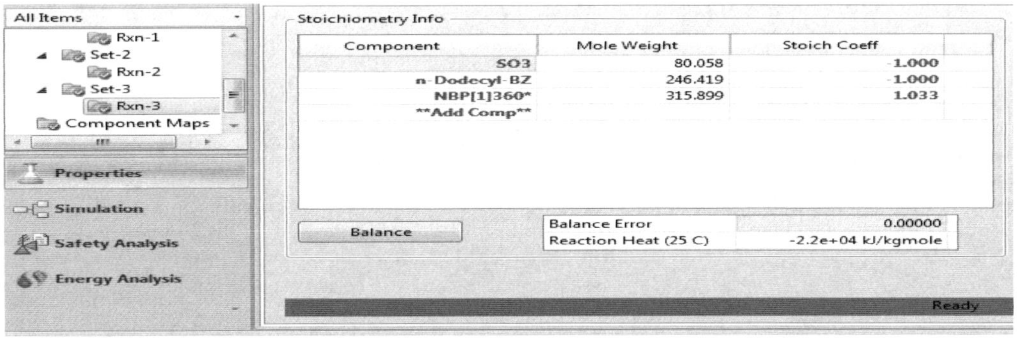

Figure (2.4) reaction set of Dodecyl benzene sulfonic acid production.

2.3.3.1. 1st Conversion reactor (Sulphur combustion)

Input and output streams condition.

2.3.3.2. 2nd Conversion reactor (SO3 production)

Input and output streams condition.

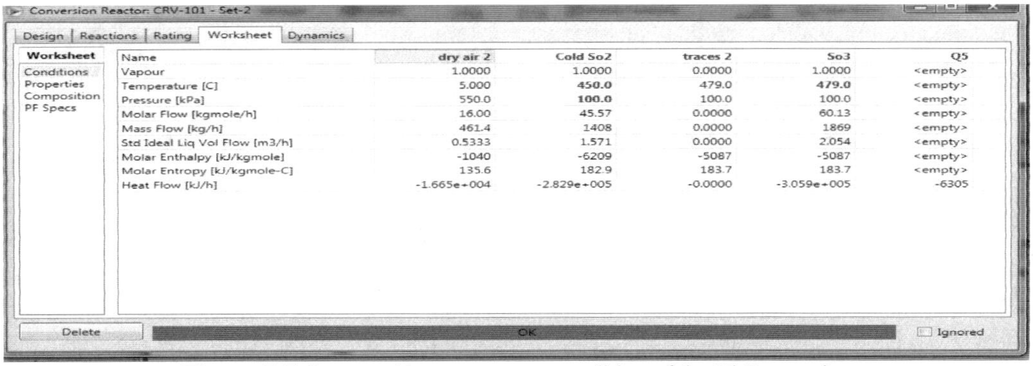

Name	dry air 2	Cold So2	traces 2	So3	Q5
Vapour	1.0000	1.0000	0.0000	1.0000	<empty>
Temperature [C]	5.000	450.0	479.0	479.0	<empty>
Pressure [kPa]	550.0	100.0	100.0	100.0	<empty>
Molar Flow [kgmole/h]	16.00	45.57	0.0000	60.13	<empty>
Mass Flow [kg/h]	461.4	1408	0.0000	1869	<empty>
Std Ideal Liq Vol Flow [m3/h]	0.5333	1.571	0.0000	2.054	<empty>
Molar Enthalpy [kJ/kgmole]	-1040	-6209	-5087	-5087	<empty>
Molar Entropy [kJ/kgmole-C]	135.6	182.9	183.7	183.7	<empty>
Heat Flow [kJ/h]	-1.665e+004	-2.829e+005	-0.0000	-3.059e+005	-6305

Figure (2.5) Input and output streams condition of the 1st Conversion reactor.

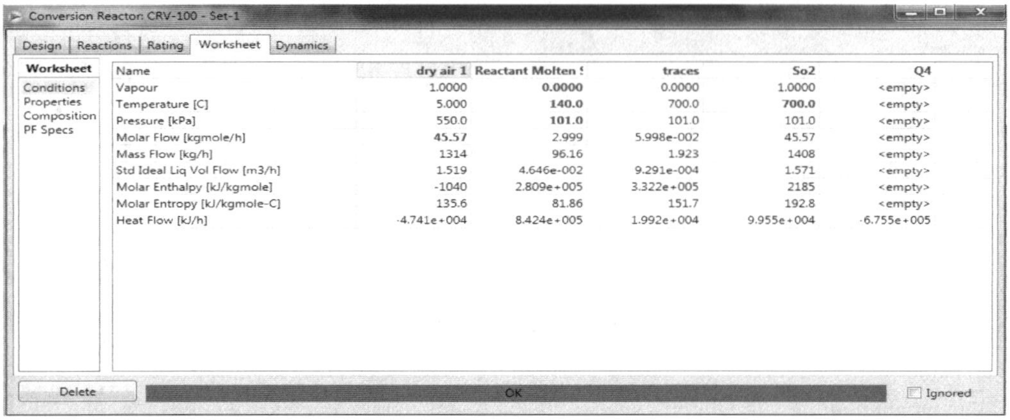

Name	dry air 1	Reactant Molten !	traces	So2	Q4
Vapour	1.0000	0.0000	0.0000	1.0000	<empty>
Temperature [C]	5.000	140.0	700.0	700.0	<empty>
Pressure [kPa]	550.0	101.0	101.0	101.0	<empty>
Molar Flow [kgmole/h]	45.57	2.999	5.998e-002	45.57	<empty>
Mass Flow [kg/h]	1314	96.16	1.923	1408	<empty>
Std Ideal Liq Vol Flow [m3/h]	1.519	4.646e-002	9.291e-004	1.571	<empty>
Molar Enthalpy [kJ/kgmole]	-1040	2.809e+005	3.322e+005	2185	<empty>
Molar Entropy [kJ/kgmole-C]	135.6	81.86	151.7	192.8	<empty>
Heat Flow [kJ/h]	-4.741e+004	8.424e+005	1.992e+004	9.955e+004	-6.755e+005

Figure (2.6) Input and output streams condition of the 2nd Conversion reactor.

2.3.3.3 3rd Conversion reactor: (dodecyl benzene sulfonic acid production)

Input and output streams condition.

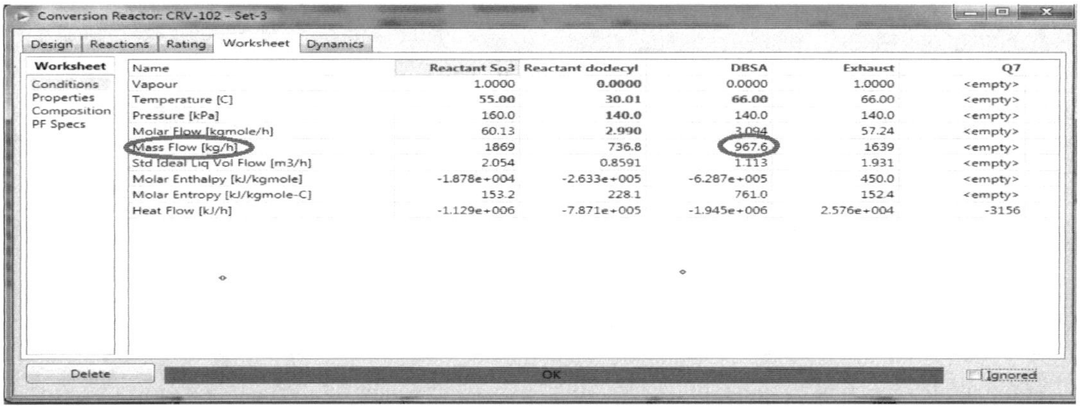

Name	Reactant So3	Reactant dodecyl	DBSA	Exhaust	Q7
Vapour	1.0000	0.0000	0.0000	1.0000	<empty>
Temperature [C]	55.00	30.01	66.00	66.00	<empty>
Pressure [kPa]	160.0	140.0	140.0	140.0	<empty>
Molar Flow [kgmole/h]	60.13	2.990	3.094	57.24	<empty>
Mass Flow [kg/h]	1869	736.8	967.6	1639	<empty>
Std Ideal Liq Vol Flow [m3/h]	2.054	0.8591	1.113	1.931	<empty>
Molar Enthalpy [kJ/kgmole]	-1.878e+004	-2.633e+005	-6.287e+005	450.0	<empty>
Molar Entropy [kJ/kgmole-C]	153.2	228.1	761.0	152.4	<empty>
Heat Flow [kJ/h]	-1.129e+006	-7.871e+005	-1.945e+006	2.576e+004	-3156

Figure (2.7) Input and output streams condition of the 3rd Conversion reactor.

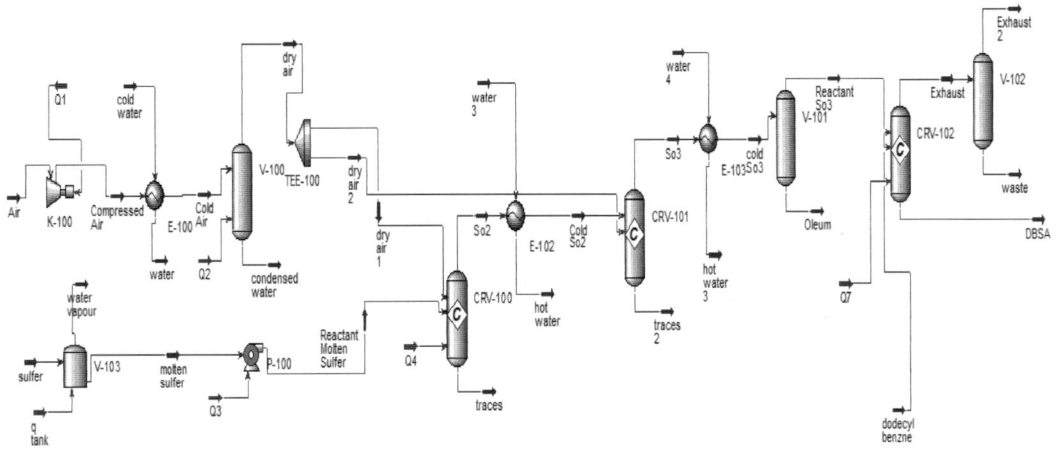

Figure (2.8) Hysys simulation of dodecyl benzene sulfonic acid process plant.

3. Heat Integration

Heat integration is a method used for the heat generated through adiabatic compression or chemical reactions to be conserved. The recovered heat can be used to heat cool streams by other means than using external heat in the form of utility streams.

For optimal heat –recovery in a system, a developed heat exchanger network (HEN) for the system is improved by using pinch analysis method where in pinch analysis method, the most important factor is the potential difference of temperature between the cold and hot streams where this temperature is the minimum temperature difference between two streams which are matched in the HEN; ΔT min can be less than 10^{0}C. [11]

The streams requiring external heating or cooling are given in table (3.1)

The streams requiring external heating or cooling are given in table (3.1)

Table (3.1) streams requiring external cooling or heating:

streams	**T_{in} (^{0}C)**	**T_{out} (^{0}C)**
Cold streams		
Compressed Air to Cold Air	**277.8**	**20**
SO_2 to Cold SO_2	**700**	**450**
SO_3 to Cold SO_3	**479**	**55**
Hot stream		
Sulfur to Molten Sulfur	**30**	**140**

In the heat integration network, the temperature of hot and cold streams plotted as a function of enthalpy which generates a composite curve as indicated in Figure (3.1).

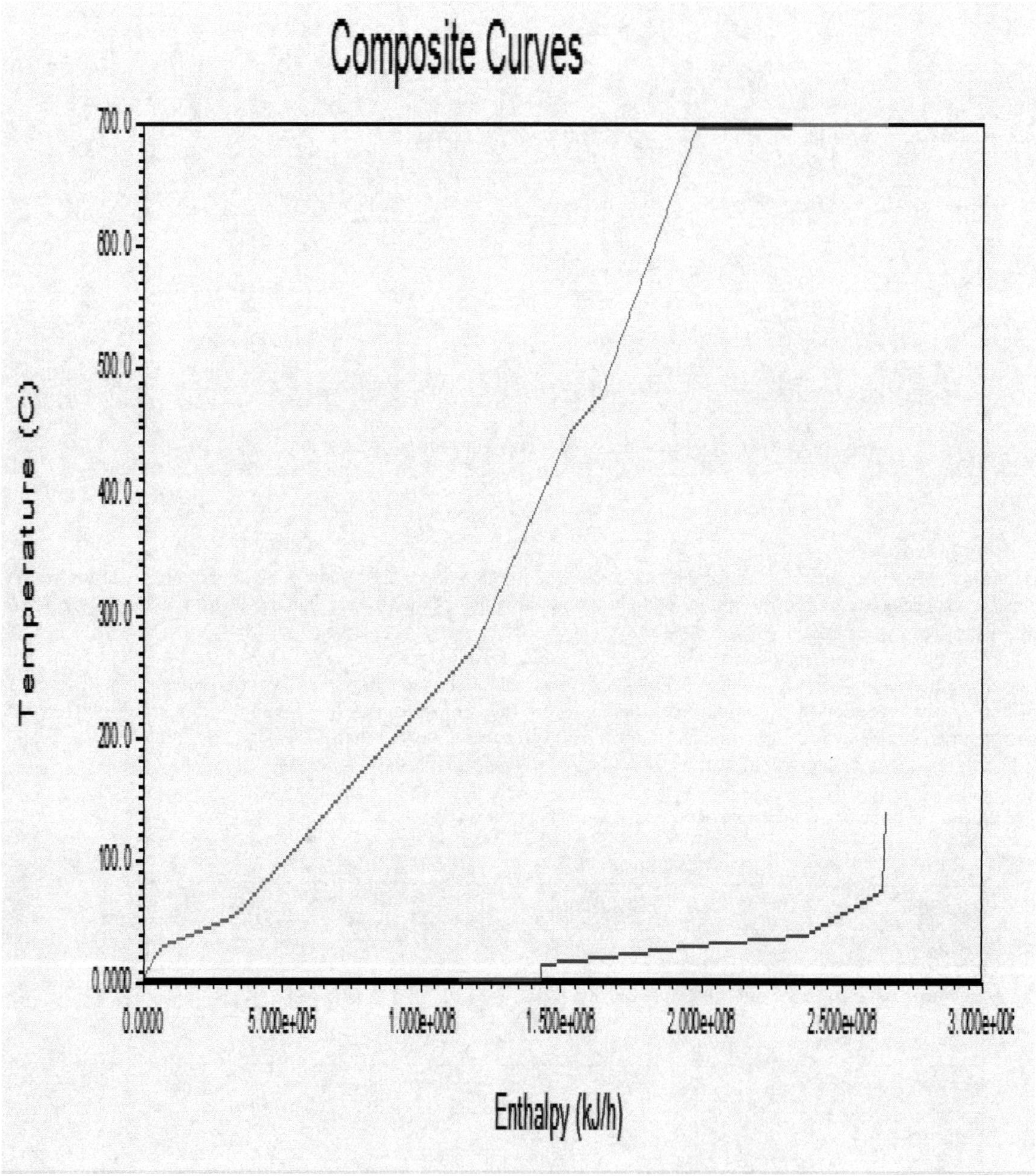

Figure (3.1) Temperature of hot and cold streams vs. enthalpy.

For the system heat integration (HI), a proposed heat exchanger network (HEN) was proposed using Aspen Energy Analyzer. Aspen Energy Analyzer generates the best design of heat exchanger network is generated in Figure (3.2).

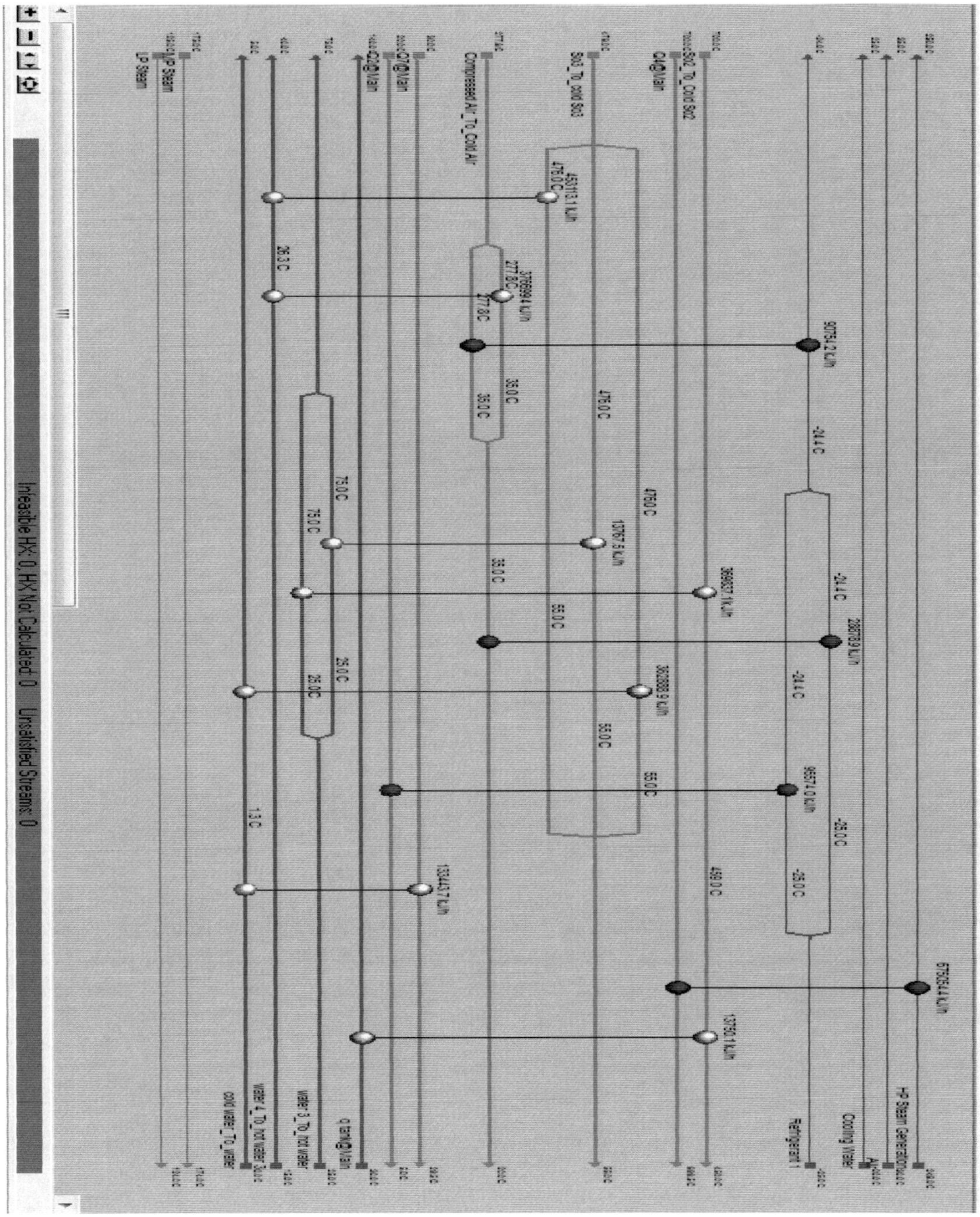

Figure (3.2) The best design of heat exchanger network.

4. Results and discussion

4.1. The effect of different processing parameter on dodecyl benzene sulfonic acid production rate:

4.1.1. Molar flow rate of molten sulfur:
The influence of the **molar flow rate** of the molten sulfur on dodecyl benzene sulfonic acid production rate can be easily realized from figure (4.1) the figure indicates that as the molten sulfur flow rate increase in the intervals (1: 3) Kg.mol/hr., the production rate of dodecyl benzene sulfonic acid increase and as the molten sulfur flow rate increase in the intervals (3.5: 5) Kg.mol/hr., the production rate of dodecyl benzene.

Sulfonic acid slightly increases (i.e. the reaction reached the state of equilibrium).

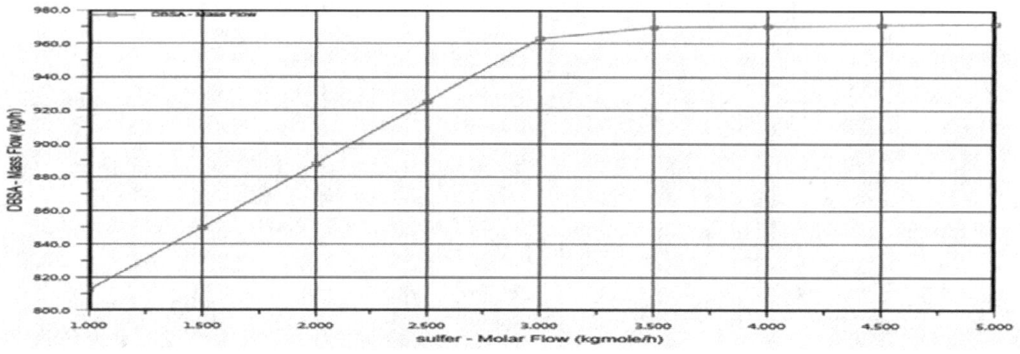

Figure (4.1) the molar flow rate of the molten sulfur vs. dodecyl benzene sulfonic acid production rate.

4.1.2. Temperature of molten sulfur

The influence of the temperature of the molten sulfur on dodecyl benzene sulfonic acid production rate can be easily realized from figure (4.2). This figure, indicates that there is no influence of the temperature of the molten sulfur on dodecyl benzene sulfonic acid production rate where an increase of the temperature of molten sulfur the production rate still constant.

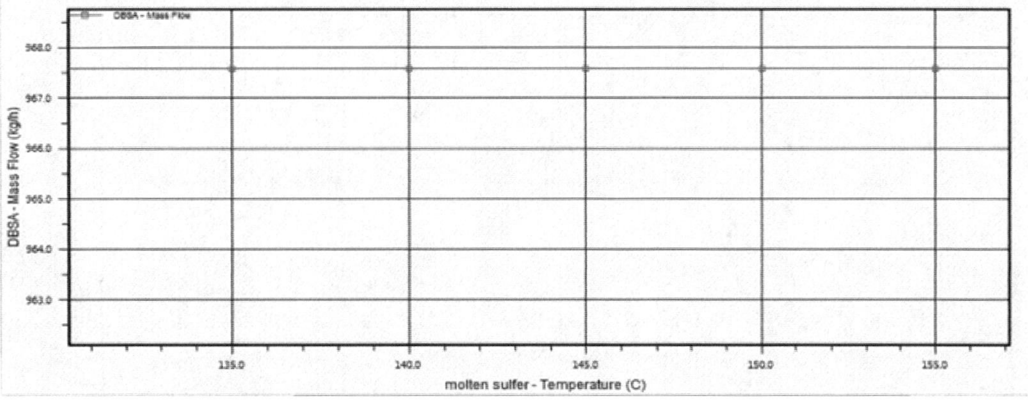

Figure (4.2) Temperature of molten sulfur vs. Mass flow rate of dodecyl benzene sulfonic acid product.

4.1.3. Molar flow rate of dodecyl benzene:

The influence of the **molar flow rate** of the **dodecyl benzene** on dodecyl benzene sulfonic acid production rate can be easily realized from figure (4.3). This figure, shows that as the **dodecyl benzene** flow rate increase the production rate of dodecyl benzene sulfonic acid increase.

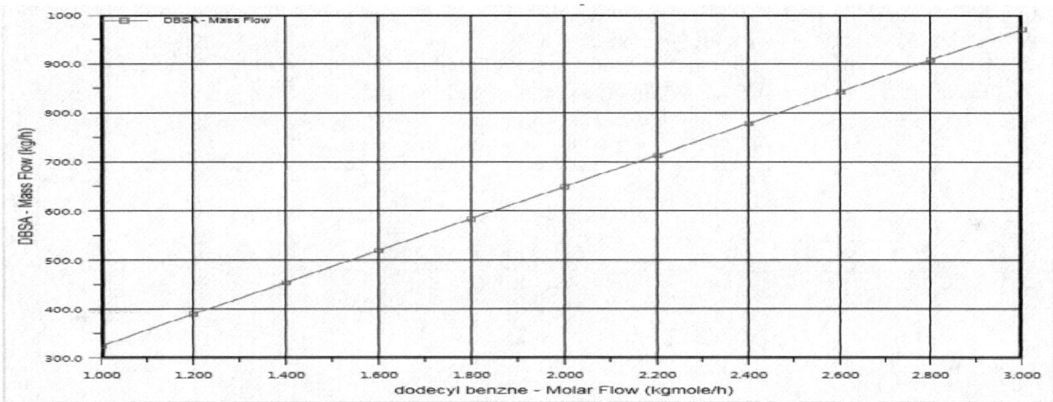

Figure (4.3) Molar flow rate of the **dodecyl benzene** vs. dodecyl benzene sulfonic acid production rate.

4.1.4. Pressure of dodecyl benzene

The influence of the pressure of the dodecyl benzene on dodecyl benzene sulfonic acid production rate can be easily realized figure (4.4). This figure, indicates that as the dodecyl benzene pressure increase the production rate of dodecyl benzene sulfonic acid increase.

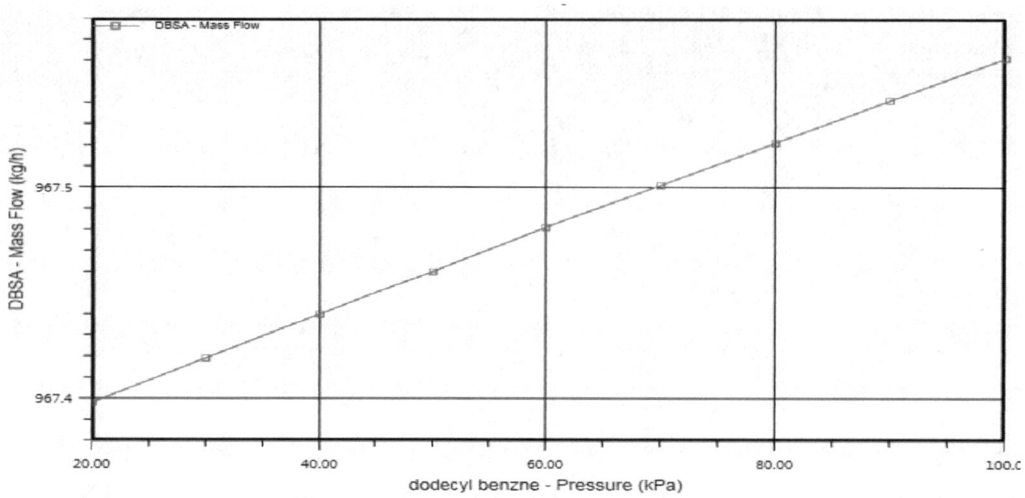

Figure (4.4) The **pressure** of the **dodecyl benzene** vs. dodecyl benzene sulfonic acid production rate.

4.1.5. Molar flow rate of air

The influence of the **Air molar flow rate** on dodecyl benzene sulfonic acid production rate can be easily realized from figure (4.5). From the figure, it is clear that as the **Air** flow rate increase in the intervals (20: 25) Kg.mol/hr., the production rate of dodecyl benzene sulfonic acid increase.

The maximum dodecyl benzene sulfonic acid production rate production rate is obtained at 25 Kg.mol/hr. Air flow and in the interval higher than 25 Kg.mol/hr., the production rate slightly decreases.

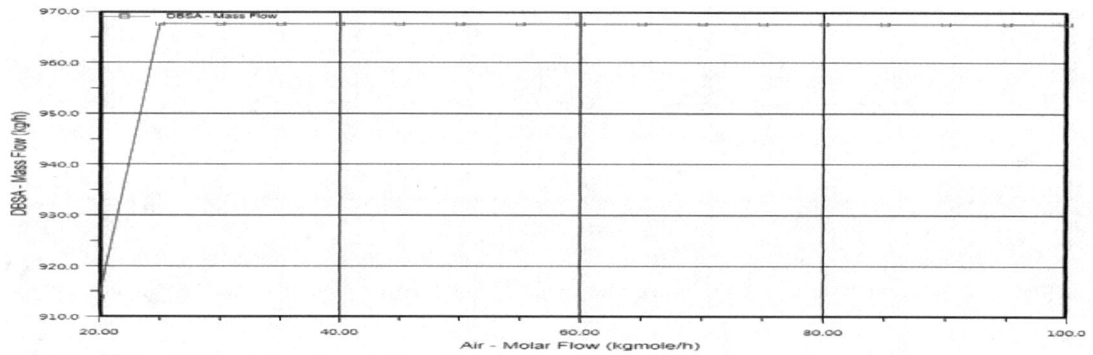

Figure (4.5) The Molar flow rate of Air vs. dodecyl benzene sulfonic acid production rate.

5. Conclusion

The followings are the conclusions drawn from this research work:

1. Computer helped modelling and simulation in many fields of engineering and especially in the field of chemical engineering also used in the industries of oil, gas and petrochemical. Nowadays, for engineers to make fast and pinpoint calculations, many simulators are developed or updated in order to execute these calculations. Today, simulators for chemical engineering helped them to simulate design of equipment of chemical plants and also operation of these plants, which saves time and money.
2. Simulators for chemical engineering also enable them to design large plants by using defined thermodynamic or using equations of state, kinetic data, mathematical and practical models and also equations, and input data of the user. Without using these simulators for modelling and design of large plants, this would be more expensive and also consuming of time.
3. In this thesis, dodecyl benzene sulfonic acid production plant was designed & simulated by using one of simulators for chemical engineering, called ASPEN HYSIS (V8.8) with steady state mode using Peng-Robinson fluid package. In this thesis modelling of dodecyl benzene sulfonic acid plant depends on these steps:

- Air drying
- Sulphur melting and proportioning
- Sulphur combustion and SO_3 production
- Film sulfonation.

4. After performing the simulation, dodecyl benzene sulfonic acid production rate is influenced by molar flow rate of sulfur, molar flow rate of air, molar flow rate of and pressure of dodecyl benzene.

- The optimum operating Molar flow rate of sulfur is in the rage [3:3.5] Kg.mol/h
- The optimum operating Molar flow rate of Air is 25 Kg.mol/h.
- The optimum operating Molar flow rate of Dodecyl Benzene is 3.1 Kg.mol/h.
- The optimum operating Pressure of Dodecyl Benzene is 70 KPa.

5. For heat integration, a proposed heat exchanger network is generated which made the process functional without external heating.
6. Generation of the top 10 heat exchanger network (HEN).
7. The best-case scenario from the top 10 scenarios from the point of cost and also from the point of cooling and heating required was design (9).
8. After heat integration for the system the total amount of cooling demand decreased from 912620.50 KJ/h to 890261.56 KJ/h and the total amount of heating demand decreased from 22358.94 KJ/h to 0KJ/h.

6. Recommendations

The following recommendations are for future work in order to either continue or improve the present work:

- Chemical integration for this process in order to use side products in the process as recycles for environmental consideration.
- Equipment integration for this process in order to perform more than one function in one piece of equipment.
- Perform the process in dynamic mode instead of steady state mode.
- It is highly important to perform economic evaluation and feasibility studies for the process.

References

[1] Banks J Introduction to simulation *2000 Winter Simulation Conference Proceedings (Cat. No.00CH37165)* **1** 9–16
[2] M.M.Chakrabraty " Chemistry of Oils and Fats ", Allied Publication Ltd., New Delhi, 2003
[3] www.ijsr.net : impact factor (2015) (internet).
[4] Robert L.Pecsok, L.Donald Shields, Thomas C.Arins," Modern Methods of -15.
[5] R. Weast and J. Grasselli, Handbook of Data on Organic Compounds, 2nd ed., CRC Press, Boca Raton, Fla., 1985, pp. 930, 2955
[6] S. Budavari, ed., The Merck Index, 11th ed., Merck & Co., Inc., Rahway, N.J., 1989, pp. 167, 938, 1009, 1501.
[7] Bordwell F G 2002 Equilibrium acidities in dimethyl sulfoxide solution *Accounts of Chemical Research* **21** 456–63
[8] Smith M B and March J 2006 March's Advanced Organic Chemistry *March's Advanced Organic Chemistry*
[9] Clayden J, Geeves N and Warren S 2012 Organic Chemistry *Ed. Oxford University Press.* 1392
[10] Gernon M D, Wu M, Buszta T and Janney P 1999 Environmental benefits of methanesulfonic acid. Comparative properties and advantages *Green Chemistry* **1** 127–40
[11] Song C 2006 Global challenges and strategies for control, conversion and utilization of CO2 for sustainable development involving energy, catalysis, adsorption and chemical processing *Catalysis Today* **115** 2–32.

Acknowledgements

The authors are grateful and would like to express their appreciation to Military Technical College, Cairo, Egypt; for its technical support.

Declarations

Conflicts of interest/Competing interests

The authors declare that they have no known competing financial interests or personal relationships that could have appeared to influence the work reported in this paper.

The authors declare the following financial interests/personal relationships which may be considered as potential competing interests: Non

Hybrid Hydro-pyrometallurgical Process for Zn Recovery from Electric Arc Furnace Dust

R.A. Bayoumi[1*], N. M. Abdelmonem[2], M. A. Soliman[1], I.M. Ismail[2], A. A. Refaat[2,3]

1 Chemical Engineering Department, Faculty of Engineering, The British University in Egypt (BUE) El-Sherouk 11837, Cairo, Egypt.
2 Chemical Engineering Department, Faculty of Engineering, Cairo University, Cairo, Egypt.
3 Faculty of Engineering, King Salman International University, Sinai Egypt.

Corresponding author's e-mail address: rana.adel@bue.edu.eg

Abstract. Electric arc furnace dust (EAFD) which is considered hazardous waste, contains significant amounts of Zinc (Zn), iron (Fe), and some dangerous elements. Zn exists in the dust as Zinc Oxide (ZnO) and Zinc Ferrite ($ZnFe_2O_3$) which is difficult to be leached with traditional methods. The main objective of the suggested process is to recover Zn from EAFD by using a hybrid pyro/hydrometallurgical method. This method includes mixing cement dust and EAFD and heating the mixture in a pyrometallurgical treatment step to convert $ZnFe_2O_3$ to ZnO. This step is followed by leaching the treated dust with sodium hydroxide (NaOH) solution using different Solid/Liquid (S/L) ratios through different time intervals in a hydrometallurgical treatment step. Finally, Zn-metal can be recovered from the leached solution in an electrochemical step (electro-winning). About 82 % of the Zn originally presented in the EAFD sample was successfully extracted by using a treatment mixture of 2:1 cement dust: EAFD at 1000 °C for 2 hrs., then leaching with 2M NaOH solution with an S/L ratio of 1/20 for 6 hrs. at a stirring rate of 250 rpm.
Keywords: EAFD; Cement Dust; Leaching; Pyro/hydrometallurgical Process; Zn Recovery

1. Introduction

Electric arc furnace dust (EAFD) is a by-product produced in the off-gas cleaning system during steel industries. Around 15-20 kg of dust (EAFD) is produced for each ton of steel. EAFD contains mainly Zn, Fe, Pb, and a significant amounts of toxic and hazardous metals, such as Cd, As, Cr, in addition to Pb. So, it is considered as one of the most dangerous wastes faced by steel production industries. EAFD is deemed to be a hazardous material by the EPA (the United States Environmental Protection Agency), as it is unsafe for water sources if not well treated or stored [1-5]. Recycling this dust back to the furnace is preferred since it contains some metals like iron (Fe) and manganese (Mn), but the presence of considerable amount of Zn in the dust makes it difficult to be recycled back to the electric arc furnace as it hurts the furnace refractory [6, 7].

Despite of the dangerous nature of EAFD, there are three main reasons account for the necessity of its processing. The first one is to recover iron concentrates that can be used as raw material in steel or pig production. Secondly, zinc (Zn) which is the second main constituent of the dust (about 8-33% of the dust) can be recovered to be used for marketable products. Finally, dust processing decreases the number of harmful materials. So, EAFD processing for metals recovery is necessary for the conservation of resources and to avoid the environmental pollution [4, 6].

The main objective of the present study is to investigate the possibility of the hybrid hydro-pyrometallurgical process for Zn recovery from Electric Arc Furnace Dust using cement dust. Processing of the dust can be carried out through two main processes, pyrometallurgical process, and hydrometallurgical process, or hybridization of both processes [4, 6]. The hybrid pyro/hydrometallurgical process is mainly applied to maximize Zn-extraction. A pyrometallurgical treatment step is firstly applied to convert zinc ferrite to easily leached ZnO. Then, a hydrometallurgical step is applied for Zn leaching from the treated dust. In the literature, different leaching agents were used to recover Zn in the oxide form [7], some of them are acidic [8-13], others are alkaline [14-18]. Recovery of zinc in the metallic form occurs through an electrochemical step [19]. A variety of hybrid process were studied using tetrabromobisphenol (TBBPA) [20, 21], poly(vinyl chloride) (PVC) [22, 23] , carbon-containing material [24], and Calcium Oxide (CaO) [25, 26]. In addition to getting rid of halides, the hydrometallurgical process has low capital & operating costs. But it has some drawbacks like acidic leaching is not suitable for selective Zn recovery [6]. Alkaline leaching agents were found to be more effective in selective Zn leaching, while Fe remains in the dust. Hydrometallurgical processes are not used on an industrial scale till now, which may be due to that the number of researches in this area is not significant. Treatment of EAFD through Pyrometallurgical processes depends mainly on the reduction of Zn by a suitable reducing agent (mainly carbon monoxide (CO)) under suitable thermodynamic conditions[20, 21]. The most common pyrometallurgical process is the Waelz process. In environmental terms, despite its commercial success, the Waelz process produces a large quantity of waste, and it is an energy-intensive process [2, 22].

Zinc, as an oxide (ZnO) and/or as a ferrite ($ZnFe_2O_4$,) can be found in EAFD. The optimum result of hydrometallurgical processing is selective zinc extraction into a solution with iron remaining as a solid residue. Alkaline leaching can achieve this selective zinc extraction since metals like Zn, Pb, and Cd flow into the solution while iron remains in the solid residue. Furthermore, because zinc in the ferritic form is not leached out during alkaline leaching, the pyrometallurgical reduction is required as an intermediate step. Alkaline leaching has various drawbacks, such as complicated solution recycling and high reagent costs. [8]. This introduces the idea for the hybrid pyro/hydrometallurgical treatment process used in this research.

2. Experimental work

2.1. Raw materials

The raw materials that were used in the experiment are as follow:

a. Electric Arc Furnace Dust (EAFD) obtained from Egyptian Steel factory.
b. Cement bypass dust obtained from Egyptian Cement Factory.
c. Sodium Hydroxide 96 % Obtained from Al Ahram Chemical Company.

All reagents are of analytical grade and were used without further purification. The water used for sample preparation was double distilled water.

2.2. Assessment of Raw Materials

For elemental composition, X-ray fluorescence (XRF) is utilized to determine the concentrations of specific oxides. The phases present in the substance were determined using X-ray diffraction (XRD). X-ray diffraction patterns for the dust were carried out by XRD-6100 X-ray diffractometer, with CuKα ($\lambda = 1.5406$ Å) radiation in the 2θ range from 10 to 90°. The scanning mode is continuous with scan speed 2 deg./min, the sampling pitch 0.02 deg., and the pre-set time 0.6 sec. Energy Dispersive X-ray (EDX) test was performed to determine the composition of the dust. Field emission Scanning electron microscope (FESEM, Quattro S, Thermo Scientific) was used for this purpose.

2.3. Zn Recovery from EAFD

As mentioned above, a hybrid pyro/hydrometallurgical process is suggested to recover Zn from EAFD. First, EAFD was treated with cement dust (a rich calcium oxide-CaO stream). Through this step, Zinc ferrite is converted into zinc oxide and calcium ferrite, this facilitate the extraction of Zn from the dust.

During this step, a uniform mixture of EAFD and cement by-pass dust was prepared at a ratio of 2:1 (cement by-pass dust: EAFD). This solid mixture is placed in an alumina crucible in a muffle furnace at 1000 °C for 2 hours. After firing, the fired solid mixture is subjected to the leaching stage using sodium hydroxide solution at a Solid/Liquid (S/L) ratio of 1/300 and a stirring rate of 250 rpm for 2 hours which are considered the optimum conditions as mentioned in the literature [26], the setup is shown in figure 1. Since the used S/L ratio was too large, different S/L ratios 1/20, 1/10, and 1/5 for different time intervals 2, 4, and 6 were tested. During the leaching process, sodium hydroxide reacts with zinc oxide to form sodium zincate according to Eq. (1). Finally, liquid solutions formed during the leaching that contain sodium zincate are subjected to electro-winning step to precipitate Zn metal by using an electroplating cell shown in figure 2. The precipitation of Zn is according to Eq. (2).

Romchat, study 2012 [26] suggested performing the pyrometallurgical reduction using pure calcium carbonate ($CaCO_3$), but our proposed work aims to study the possibility of mixing the EAFD waste with another waste stream (cement by-pass dust) which is more economic and beneficial, then the results were compared together.

$$ZnO + 2NaOH \rightarrow Na_2ZnO_2 + H_2O \quad (1)$$

$$Na_2ZnO_2 + 2H_2O \rightarrow 2NaOH_{aq} + Zn_s + H_{2_g} \uparrow \quad (2)$$

Sodium Zincate | Sodium Hydroxide | Zinc | Hydrogen

Figure 1 Experimental set-up for leaching experiment

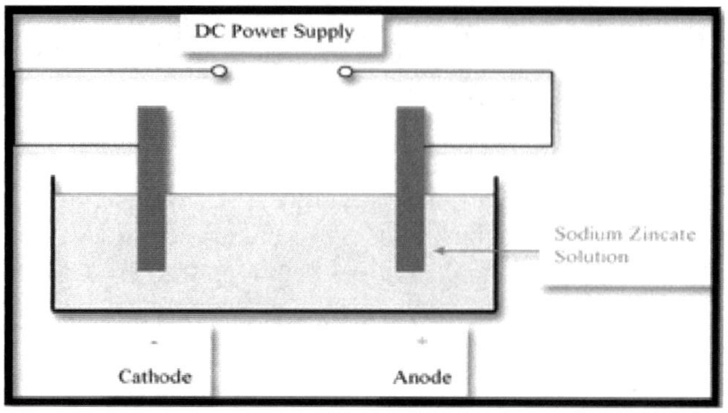

Figure 2 Electro-winning Experiment

2.4. Assessment of Resulted Solid Powder after each Experimental Step

XRD analysis is performed after each thermal treatment step to determine the percentage of ZnO formed. The amount of Zn-metal recovered during the leaching step can be determined by performing EDX analysis for the resulting solid mixture.

3. Results and Discussion

3.1 Chemical Analysis Results (Raw materials)

3.1.1 XRF Analysis of EAFD. Table 1 shows the results of X-ray fluorescence (XRF) analysis for EAFD. The solid waste consists mainly of Fe_2O_3, ZnO, CaO, and SiO_2 oxides with other oxides.

Table 1 Chemical analysis of EAFD

Oxides	Wt%	Oxides	wt%
Fe_2O_3	32.2	Na_2O	2.3
ZnO	23.1	K_2O	3.42
Cao	19.4	SO_3	2.5
SiO_2	5.9	TiO_2	0.2
Al_2O_3	0.6	P_2O_5	0.02
MnO	1.1	Cl	2.6
MgO	0.9	L.O.I.	5.8

3.1.2 XRF for Cement Dust. Table 2 shows the results of X-ray fluorescence (XRF) analysis for cement dust. The solid waste consists mainly of CaO oxide with other oxides. The loss on ignition has a high value because of the composition of CaCO3 to CaO and CO_2.

Table 2 Chemical analysis of cement dust

Oxides	wt%	Oxides	wt%
Fe_2O_3	2.76	K_2O	3.93
Cao	46.5	SO_3	1.6
SiO_2	8.92	TiO_2	0.31
Al_2O_3	1.45	P_2O_5	0.04
MnO	0.07	Cl	5.4
MgO	1.03	L.O.I.	26.8
Na_2O	0.8		

3.2 XRD Analysis Results

3.2.1 XRD Analysis for Raw Materials. The mineralogical analysis of cement dust is shown in figure 3. It was found that this waste is mainly composed of the following phases: Calcite ($CaCO_3$), Lime (CaO), Silica (SiO_2), and Calcium aluminum silicate hydrate ($CaAl_2Si_2O_8 . 4H_2O$). The mineralogical analysis of EDAF dust is shown in figure 4. It was found that this waste is mainly composed of the following phases: Zincite (ZnO), Zinc Iron Oxide ($ZnFe_2O_4$), Potassium Manganese Sulfide ($K_2Mn_3S_4$), and Sodium Manganese Silicate Hydroxide ($NaMn_2Si_3O_8(OH)$).

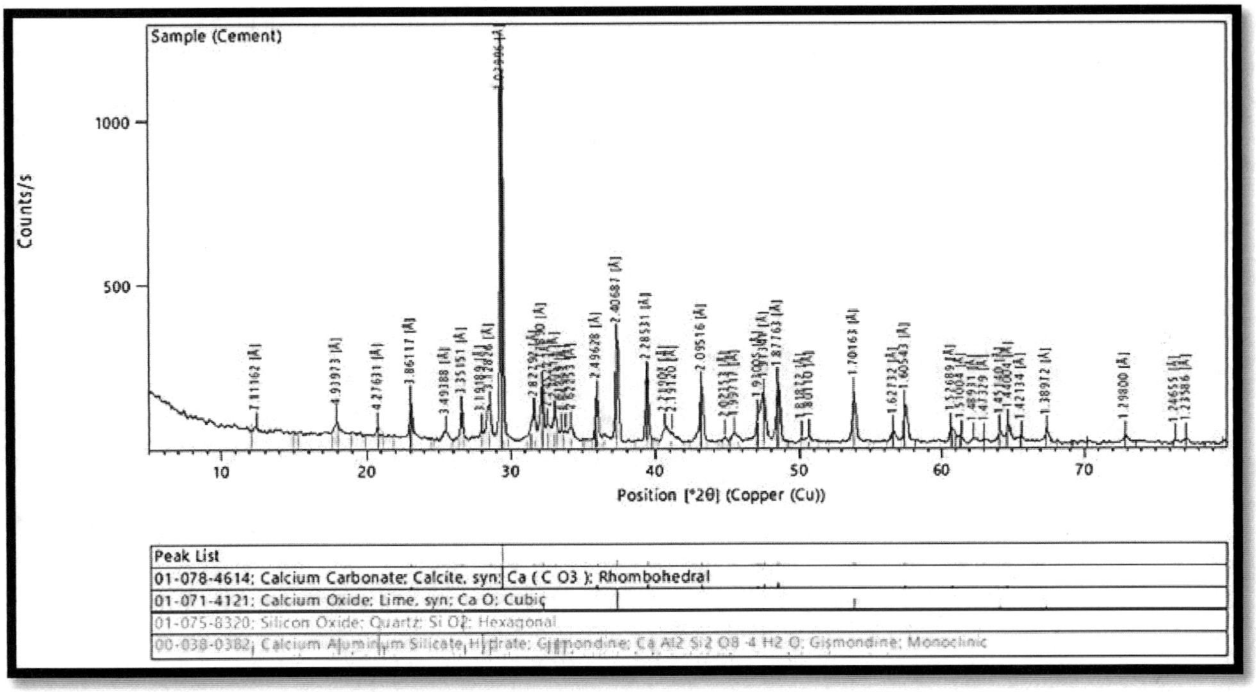

Figure 3 XRD Analysis for Cement Dust

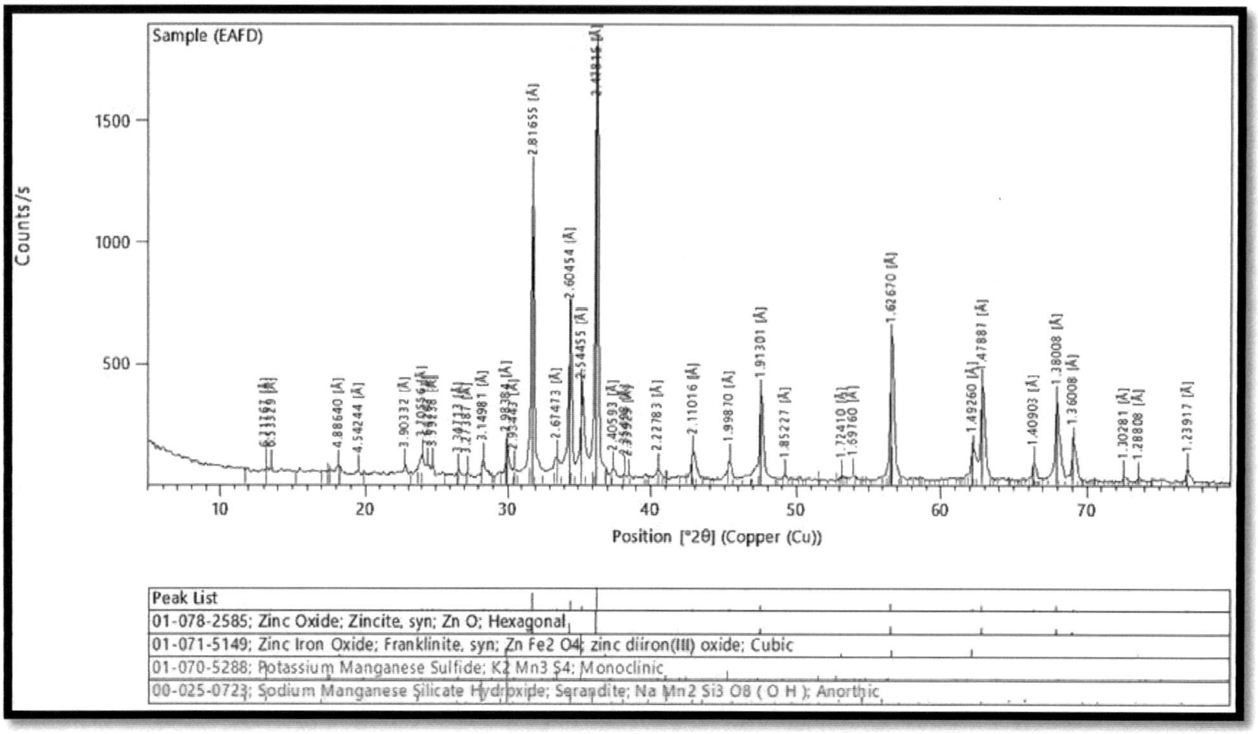

Figure 4 XRD Analysis for EAFD

3.2.2 XRD Analysis for Mixture before Treatment. The mineralogical analysis of a mixture of cement dust and EAFD with a ratio of (2:1) is shown in figure 5. It was found that this waste is mainly composed of the following phases: Zincite (ZnO), Zinc Iron Oxide ($ZnFe_2O_4$), Magnesium oxide (MnO), Potassium chloride KCl, Sodium Chloride (NaCl), Ferric Oxide (Fe_2O_3), Calcite ($CaCO_3$), Silica (SiO_2) and Calcium aluminum silicate hydrate ($CaAl_2Si_2O_8$. $4H_2O$).

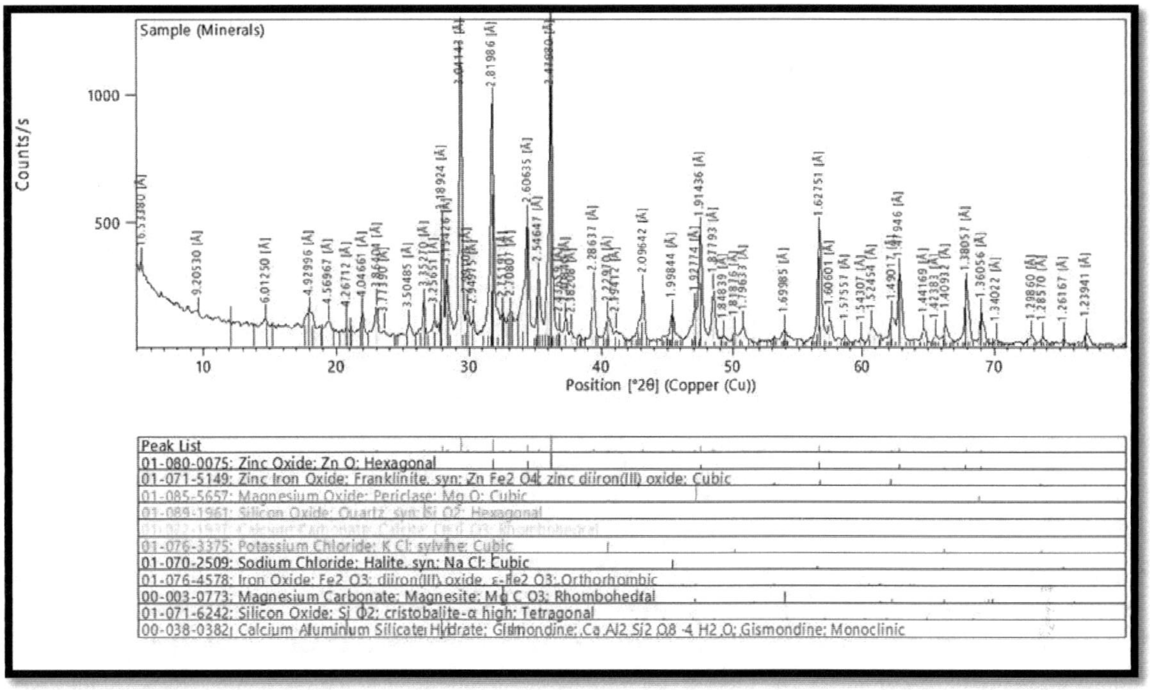

Figure 5 XRD Analysis for Mixture of 2 (cement dust):1(EAFD)

3.2.3 XRD Analysis for Mixture after Combustion and before Leaching. The mineralogical analysis of the mixture is shown in figure 6. It was found that this waste is mainly composed of the following phases: Zincite (ZnO), Lime (CaO), Zinc Iron Oxide ($ZnFe_2O_4$), Magnesium oxide (MnO), Calcium Silicate (Ca_2SiO_4), Calcium Iron Oxide ($Ca_2Fe_2O_5$), and Silica (SiO_2). It was found that the peaks for $ZnFe_2O_4$ decreased and that for ZnO increased, also new peaks for $Ca_2Fe_2O_5$ appeared, which means that the treatment process was successful.

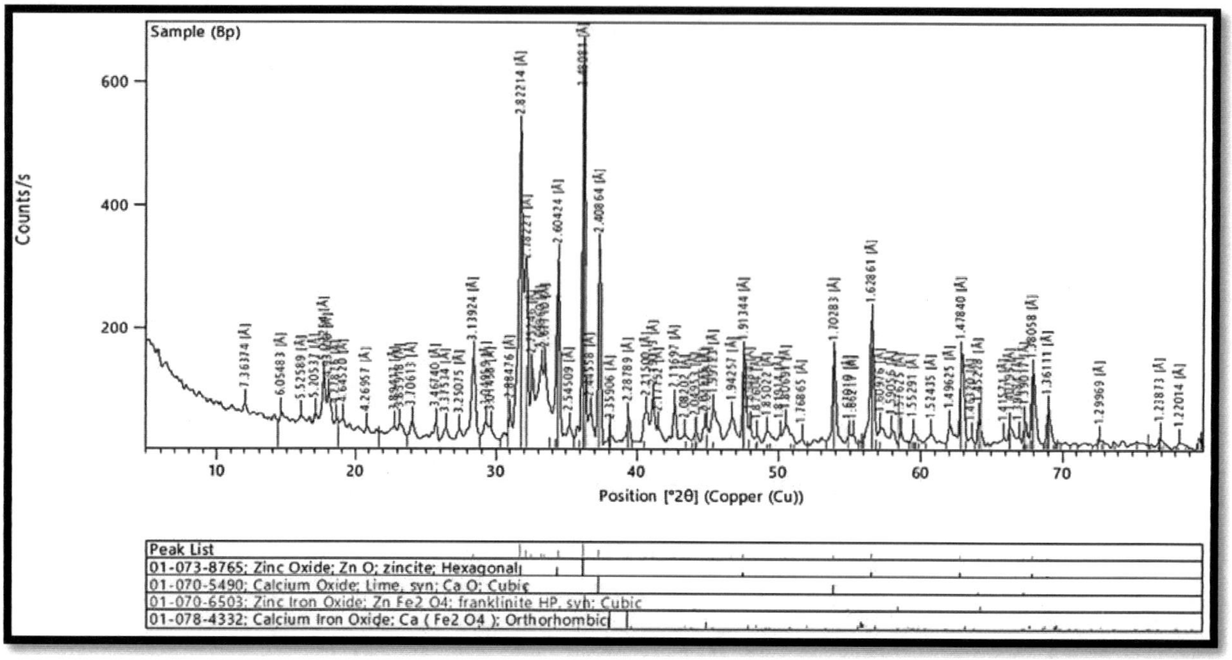

Figure 6 XRD Analysis for Mixture of 2 (cement dust):1(EAFD) after combustion and before leaching

3.2.4 XRD Analysis for Mixture after Leaching. The mineralogical analysis of the mixture is shown in figure 7. It was found that this waste is mainly composed of the following phases: Calcium Zinc hydroxide ($Ca(Zn_2(OH)_6\ (H_2O)_2$), Calcium hydroxide ($Ca(OH)_2$), Zincite (ZnO), Calcium Carbonate ($CaCO_3$), Sodium Calcium Carbonate Hydrate ($Na_2Ca(CO_3)2.2H_2O$), Calcium Silicate (Ca_2SiO_4), Calcium Manganese Carbonate ($Ca\ Mn(CO_3)_2$) and Silica (SiO_2). It was found that the peaks for ZnO decreased which means that most of the zinc oxides were leached in the solution.

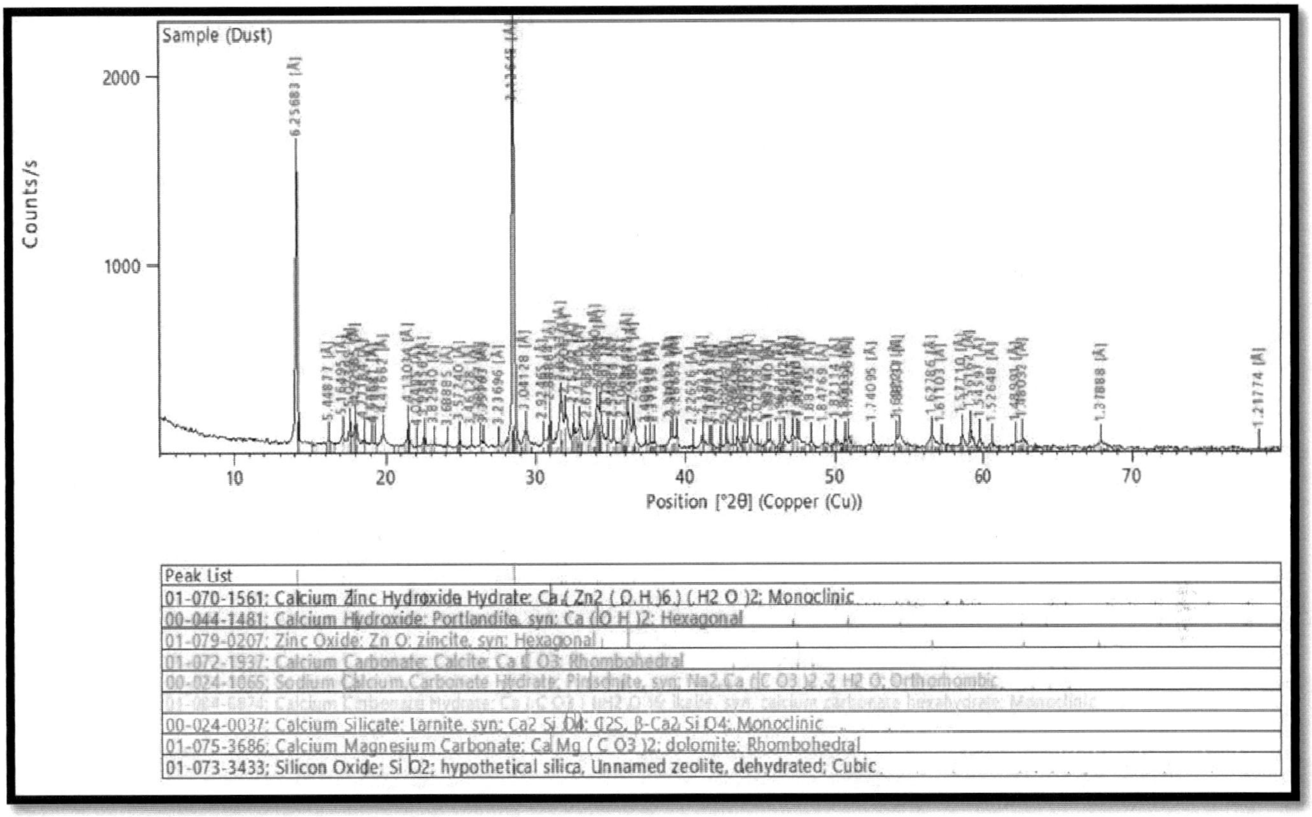

Figure 7 XRD Analysis for Mixture of 2 (cement dust):1(EAFD) after leaching

4. EDX Analysis Results

4.1 EDX Analysis for Raw Materials. The results of EDX analysis of cement dust and EAFD mixture with ratio (2:1) are shown in table 3. It was found that this mixture is mainly composed of the following elements: Zinc (Zn), Calcium (Ca), Oxygen (O), and other elements with a small percentage.

Table 3 EDX Analysis for Mixture of 2 (cement dust):1(EAFD)

Element	Wt%	Atomic%
O K	28.86	56.11
Mg K	0.77	0.98
Al K	0.76	0.87
Si K	1.58	1.75
Mo L	2.22	0.72
Cl K	7.35	6.45
K K	2.2	1.75
Ca K	14.16	10.99
Fe K	4.24	2.36
Zn K	37.86	18.01

4.2 EDX Analysis for Mixture before Leaching and after Leaching. EDX analysis is used to determine the Zn removal due to the leaching process using different Solid to NaOH solutions and also different leaching times. Starting with Solid to NaOH solution ratio equals to 1:300 and leaching time of 2 hrs, the analysis of samples before and after the leaching process is shown in table 4. It was found that the percentage removal of Zn is 93.3%, while for the study of Rochat, 2012 [26] the % Zn removal was about 90%. The results for both studies are almost the same, but for this work, a waste material (cement dust) was used instead of raw $CaCO_3$, which is beneficial in terms of environmental and economic issues. The obtained extraction efficiency is accepted but according to the used S/L ratio (1:300), the volume of the reactor is large which will increase the cost of the process. Accordingly, different S/L ratios for different time intervals were studied to overcome this problem.

Table 4 EDX Analysis for Mixture of 2 (cement dust):1(EAFD) before and after leaching using solid to solution ratio equals 1:300

Before leaching			After leaching		
Element	Wt.%	Atomic%	Element	Wt.%	Atomic%
O K	25.41	48.34	O K	44.31	10.19
MgK	0.73	0.91	MgK	1.46	9.64
AlK	1.3	1.46	AlK	0.68	9.49
SiK	5	5.42	SiK	3.52	5.36
K K	10.5	8.7	K K	0.53	16.88
CaK	27.85	21.15	CaK	43.47	1.53
FeK	5.2	2.83	FeK	4.42	4.24
ZnK	24.06	11.2	ZnK	1.61	10.94

Using solid to NaOH solution ratio equals to 1:10 gives Zn removal percentage equals 69.64 % for 2 hrs leaching, 72.4% for 4 hrs leaching, and 78.93 % for 6 hrs as shown in figure 8. It is obvious that, the removal efficiency increases with increasing the leaching time which is consistent with the mass transfer and separation concepts.

Figure 9 illustrates the relation between the percentage Zn-recovery and the solid to NaOH solution ratio. Solid to NaOH solution ratios equal to 1:20, 1:10, 1:5, and 1:3 are studied at constant leaching time of 6 hours. It was found that as the amount of NaOH solution increases or as the ratio decreases, the percentage Zn recovered increases which is consistent with mass transfer concepts, since increasing the amount of solvent used increases the separation efficiency.

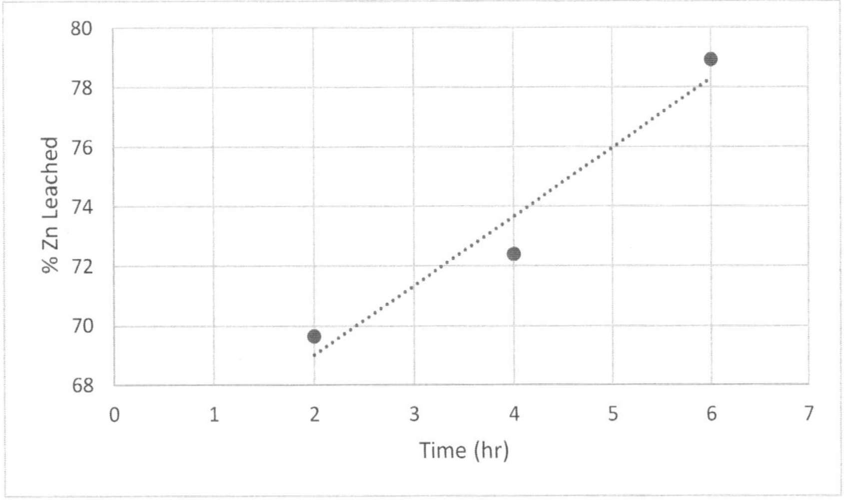

Figure 8 Leaching at S/L ratio of 1:10

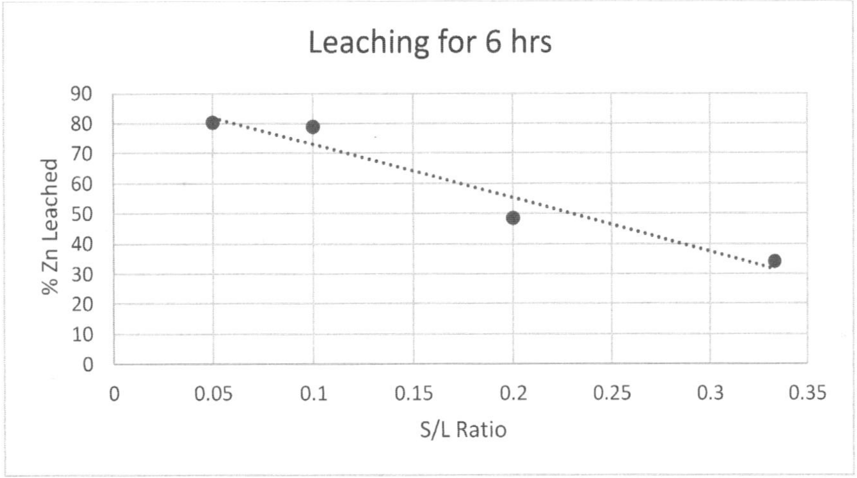

Figure 9 Leaching for 6 hours at different S/L ratios

4.3 Comparison with Previous Studies

Table 5 shows a comparison between the results obtained from the current work and those obtained from other previous work. It was noticed that, the current work has the following advantages compared with the other previous work:

- Using cement dust waste stream during the thermal treatment or pyrometallurgical step of the EAFD instead of using pure $CaCO_3$ results in an environmental pollution reduction in addition to it is more economic.
- The possibility of using a small solid/liquid ratio (S/L) during the leaching step, allows using a small reactor volume which makes our process more economic.
- The current process is more selective towards Zn-metal recovery unlike other processes that recover lead with zinc and lead is a toxic metal.

Table 5 Comparison between the results obtained in this paper and the results obtained from recent papers

Hydrolysis step	Pyromettalurgical process		Hydrometallurgical Process			% Zn and Pb (Lead) Recovery	Reference
	Reagent	Temperature	Reagent	Temperature	S/L ratio		
Yes	NaOH pellets	350 ◦C	NaOH	-	-	80 % Zn and 66.67 % Pb	[30]
No	NaOH	450 ◦C	Citric Acid	40 ◦C	100g/L	100 % Zn and ≥ 80 % Pb	[4]
No	CaO	1100 ◦C	NaOH	70 ◦C	1g/300 ml	100 % Zn with no Pb	[25]
No	TBBPA	350 ◦C	Water	50 ◦C	1g/300 ml	90.5 % Zn and 100 % Pb	[31]
No	TBBPA	Microwave heating ≥ 350 ◦C	Water	-	-	52 % Zn and 31 % Pb	[32]
No	PVC	Microwave heating ≥ 350 ◦C	Water	-	-	97 % Zn and 59 % Pb	[33]
No	Carbon Containing material	1200 ◦C	NaOH	80 ◦C	1g/100 ml	70 % Zn	[24]
No	Cement Dust	1100 ◦C	NaOH	100 ◦C	1g/ 20 ml	82 % Zn with no Pb	Present Work

5. Conclusion

Electric arc furnace dust (EAFD) consists mainly of zinc and iron which are considered to be hazardous solid waste. The main objective of the current work is to recover Zn in the metallic form from EAFD by using a hybrid pyro/hydrometallurgical process. This is done by mixing EAFD with a waste stream of a cement-by pass dust. The resulting solid mixture is first subjected to heat treatment in a pyrometallurgical step to covert the Zn ferrite to zinc oxide in order to enhance the leaching process. The resulted solid mixture is then subjected to a leaching step using sodium hydroxide (NaOH) solution with different S/L ratios and time intervals to extract ZnO from the solid mixture. Finally, Zn in the metallic form will be recovered electrochemically in an electro-winning process. It was found that, the percentage Zn-recovery increases with the leaching time since increasing the contact time is preferred to increase the separation efficiency. Also, increasing the amount of NaOH leaching solution enhances the efficiency of recovery. Actually, 82% of the Zn originally presented in the EAFD sample was successfully extracted by using a treatment mixture of 2:1 cement dust: EAFD at 1000 °C for 2 hrs, then leaching with 2 M NaOH solution with an S/L ratio of 1/20 for 6 hrs at a stirring rate of 250 rpm.

References

[1] S. Mauthoor, R. Mohee, and P. Kowlesser, "An assessment on the recycling opportunities of wastes emanating from scrap metal processing in Mauritius," Waste Manag., vol. 34, no. 10, pp. 1800–1805, 2014, doi: 10.1016/j.wasman.2013.12.014.

[2] S. Małecki et al., "High-performance method of recovery of metals from eaf dust—processing without solid waste," Materials (Basel)., vol. 14, no. 20, 2021, doi: 10.3390/ma14206061.

[3] T. Suetens, B. Klaasen, K. Van Acker, and B. Blanpain, "Comparison of electric arc furnace dust treatment technologies using exergy efficiency," J. Clean. Prod., vol. 65, pp. 152–167, 2014, doi: 10.1016/j.jclepro.2013.09.053.

[4] P. Halli, J. Hamuyuni, M. Leikola, and M. Lundström, "Developing a sustainable solution for recycling electric arc furnace dust via organic acid leaching," Miner. Eng., vol. 124, no. May, pp. 1–9, 2018, doi: 10.1016/j.mineng.2018.05.011.

[5] Y. Chen et al., "Efficient extraction and separation of zinc and iron from electric arc furnace dust

by roasting with $FeSO_4 \cdot 7H_2O$ followed by water leaching," Sep. Purif. Technol., vol. 281, no. October 2021, p. 119936, 2022, doi: 10.1016/j.seppur.2021.119936.
[6] M.Sc. Hatim Abdalla Sasi Issa, "MECHANO-CHEMICAL AND THERMAL TREATMENT OF IRON BEARING WASTE MATERIALS: ECOLOGICAL BENEFITS AND SYNERGETIC EFFECTS," 2015.
[7] P. Halli, J. Hamuyuni, H. Revitzer, and M. Lundström, "Selection of leaching media for metal dissolution from electric arc furnace dust," J. Clean. Prod., vol. 164, pp. 265–276, 2017, doi: 10.1016/j.jclepro.2017.06.212.
[8] F. Kukurugya, T. Vindt, and T. Havlík, "Behavior of zinc, iron and calcium from electric arc furnace (EAF) dust in hydrometallurgical processing in sulfuric acid solutions: Thermodynamic and kinetic aspects," Hydrometallurgy, vol. 154, pp. 20–32, 2015, doi: 10.1016/j.hydromet.2015.03.008.
[9] P. Oustadakis, P. E. Tsakiridis, A. Katsiapi, and S. Agatzini-Leonardou, "Hydrometallurgical process for zinc recovery from electric arc furnace dust (EAFD). Part I: Characterization and leaching by diluted sulphuric acid," J. Hazard. Mater., vol. 179, no. 1–3, pp. 1–7, 2010, doi: 10.1016/j.jhazmat.2010.01.059.
[10] T. Havlík, B. V. e Souza, A. M. Bernardes, I. A. H. Schneider, and A. Miškufová, "Hydrometallurgical processing of carbon steel EAF dust," J. Hazard. Mater., vol. 135, no. 1–3, pp. 311–318, 2006, doi: 10.1016/j.jhazmat.2005.11.067.
[11] M. Kul, K. O. Oskay, M. Şimşir, H. Sübütay, and H. Kirgezen, "Optimization of selective leaching of Zn from electric arc furnace steelmaking dust using response surface methodology," Trans. Nonferrous Met. Soc. China (English Ed., vol. 25, no. 8, pp. 2753–2762, 2015, doi: 10.1016/S1003-6326(15)63900-0.
[12] T. Havlik, M. Turzakova, S. Stopic, and B. Friedrich, "Atmospheric leaching of EAF dust with diluted sulphuric acid," Hydrometallurgy, vol. 77, no. 1–2, pp. 41–50, 2005, doi: 10.1016/j.hydromet.2004.10.008.
[13] R. A. Shawabkeh, "Hydrometallurgical extraction of zinc from Jordanian electric arc furnace dust," Hydrometallurgy, vol. 104, no. 1, pp. 61–65, 2010, doi: 10.1016/j.hydromet.2010.04.014.
[14] G. Orhan, "Leaching and cementation of heavy metals from electric arc furnace dust in alkaline medium," Hydrometallurgy, vol. 78, no. 3–4, pp. 236–245, 2005, doi: 10.1016/j.hydromet.2005.03.002.
[15] A. Kekki, J. Aromaa, and O. Forsen, "Leaching characteristics of eaf and AOD stainless steel production dusts," Physicochem. Probl. Miner. Process., vol. 48, no. 2, pp. 599–606, 2012, doi: 10.5277/ppmp120223.
[16] Q. Li, Y. Zhao, J. Jiang, and C. Zhang, "Optimized Hydrometallurgical Route to Produce Ultrafine Zinc Powder from Industrial Wastes in Alkaline Medium," Procedia Environ. Sci., vol. 16, pp. 674–682, 2012, doi: 10.1016/j.proenv.2012.10.093.
[17] B. S. Yu, Y. R. Wang, and T. C. Chang, "Hydrothermal treatment of electric arc furnace dust," J. Hazard. Mater., vol. 190, no. 1–3, pp. 397–402, 2011, doi: 10.1016/j.jhazmat.2011.03.056.
[18] P. Palimąka, S. Pietrzyk, M. Stępień, K. Ciećko, and I. Nejman, "Zinc Recovery from Steelmaking Dust by Hydrometallurgical Methods," Metals (Basel)., vol. 8, no. 7, p. 547, 2018, doi: 10.3390/met8070547.
[19] C. Lanzerstorfer, "Electric arc furnace (EAF) dust: Application of air classification for improved zinc enrichment in in-plant recycling," J. Clean. Prod., vol. 174, pp. 1–6, 2018, doi: 10.1016/j.jclepro.2017.10.312.
[20] M. Al-Harahsheh, M. Aljarrah, A. Al-Otoom, M. Altarawneh, and S. Kingman, "Pyrolysis kinetics of tetrabromobisphenol a (TBBPA) and electric arc furnace dust mixtures," Thermochim. Acta, vol. 660, no. August 2017, pp. 61–69, 2018, doi: 10.1016/j.tca.2017.12.022.
[21] M. Grabda, S. Oleszek, E. Shibata, and T. Nakamura, "Study on simultaneous recycling of EAF dust and plastic waste containing TBBPA," J. Hazard. Mater., vol. 278, pp. 25–33, 2014, doi: 10.1016/j.jhazmat.2014.05.084.

[22] M. Al-Harahsheh et al., "Pyrolysis of poly(vinyl chloride) and-electric arc furnacedust mixtures," J. Hazard. Mater., vol. 299, pp. 425–436, 2015, doi: 10.1016/j.jhazmat.2015.06.041.
[23] G. S. Lee and Y. J. Song, "Recycling EAF dust by heat treatment with PVC," Miner. Eng., vol. 20, no. 8, pp. 739–746, 2007, doi: 10.1016/j.mineng.2007.03.001.
[24] M. C. Siame, J. Kaoma, N. Hlabangana, and G. Danha, "An attainable region approach for the recovery of iron and zinc from electric arc furnace dust," South African J. Chem. Eng., vol. 27, no. December 2018, pp. 35–42, 2019, doi: 10.1016/j.sajce.2018.12.002.
[25] R. Chairaksa-Fujimoto, K. Maruyama, T. Miki, and T. Nagasaka, "The selective alkaline leaching of zinc oxide from Electric Arc Furnace dust pre-treated with calcium oxide," Hydrometallurgy, vol. 159, pp. 120–125, 2016, doi: 10.1016/j.hydromet.2015.11.009.
[26] ROMCHAT CHAIRAKSA, "Development of New EAF Dust Treatment Process With CaO Addition Method."
[27] C. C. Wu, F. C. Chang, W. S. Chen, M. S. Tsai, and Y. N. Wang, "Reduction behavior of zinc ferrite in EAF-dust recycling with CO gas as a reducing agent," J. Environ. Manage., vol. 143, pp. 208–213, 2014, doi: 10.1016/j.jenvman.2014.04.005.
[28] C. A. Pickles, "Thermodynamic analysis of the selective carbothermic reduction of electric arc furnace dust," J. Hazard. Mater., vol. 150, no. 2, pp. 265–278, 2008, doi: 10.1016/j.jhazmat.2007.04.097.
[29] P. J. W. K. De Buzin, N. C. Heck, and A. C. F. Vilela, "EAF dust: An overview on the influences of physical, chemical and mineral features in its recycling and waste incorporation routes," J. Mater. Res. Technol., vol. 6, no. 2, pp. 194–202, 2017, doi: 10.1016/j.jmrt.2016.10.002.
[30] Z. Youcai and R. Stanforth, "Integrated hydrometallurgical process for production of zinc from electric arc furnace dust in alkaline medium," J. Hazard. Mater., vol. 80, no. 1–3, pp. 223–240, 2000, doi: 10.1016/S0304-3894(00)00305-8.
[31] M. Al-Harahsheh, M. Aljarrah, F. Rummanah, K. Abdel-Latif, and S. Kingman, "Leaching of valuable metals from electric arc furnace dust—Tetrabromobisphenol A pyrolysis residues," J. Anal. Appl. Pyrolysis, vol. 125, no. May, pp. 50–60, 2017, doi: 10.1016/j.jaap.2017.04.019.
[32] M. Al-harahsheh, S. Kingman, and I. Hamilton, "Microwave treatment of electric arc furnace dust with Tetrabromobisphenol A: Dielectric characterization and pyrolysis-leaching," J. Anal. Appl. Pyrolysis, vol. 128, no. June, pp. 168–175, 2017, doi: 10.1016/j.jaap.2017.10.013.
[33] M. Al-harahsheh, S. Kingman, L. Al-Makhadmah, and I. E. Hamilton, "Microwave treatment of electric arc furnace dust with PVC: Dielectric characterization and pyrolysis-leaching," J. Hazard. Mater., vol. 274, pp. 87–97, 2014, doi: 10.1016/j.jhazmat.2014.03.019.

Kinetic Decomposition Models for Ammonium Percholorate Hybrid Catalyzed with Nanothermite Particles

Abdelaziz Hamed [1,*], Mohamed Gobara [1], Mohamed Mokhtar[1] and Sherif Elbasuney [1,2]

[1] School of Chemical Engineering, Military Technical College (MTC), Cairo, Egypt.
[2] Head of Nanotechnology Research Center, Military Technical College (MTC), Cairo, Egypt.

* a.m.mostafa@mtc.edu.eg

Abstract. Ferric oxide is a universal catalyst. Reactive metal fuel can act as a high energy dense material such as aluminum which is marked by the very high gravimetric and volumetric heat output. This manuscript reports on the fabrication of colloidal ferric oxide nanoparticles of 5 nm. Colloidal ferric oxide/aluminum nanothermite mixture was integrated into ammonium percholorate (AP) via the coprecipitation technique. The shape and particles size of the prepared nano ferric oxide were investigated by using TEM instrument. Uniform dispersion of Al/Fe2O3 in the ammonium perchlorate matrix was verified using EDAX instrument. Nanothermite particles offered enhanced AP decomposition enthalpy by 120 % using DSC. Nanothermite colloid offered a decrease in AP activation energy by 51 and 40 % using Friedman and Ozawa models respectively. AP decomposition mechanism was reported to go through three consequent mechanisms including the first order mechanism, two-dimensional diffusion reactions, and one- dimensional diffusion mechanism according to extent of reacted fraction (α) of 0-0.25, 0.3-0.6, and 0.6-0.9 respectively. The results show that the catalyzing ability of the nanothermites was confirmed and has shown a superior effect on the AP energetic system.

Keywords: Ammonium perchlorate; Catalyst; Thermal behavior; Energetic systems; high temperature decomposition (HTD), low temperature decomposition (LTD)

1. Introduction

Ammonium perchlorate (AP) is the most common oxidizer used in various energetic systems [1]. It's widely accepted that thermal decomposition kinetics play a vital role in its combustion behavior [2]. The thermal decomposition of AP is a complex process that is affected by different conditions such as shape, crystalline phase, and crystal purity [3-7]. AP thermal decomposition depends on physical and chemical interaction among decomposition products [8, 9]. AP decomposition could be altered via controlled morphological properties, as well as the integration of advanced catalyst particles such as ferric oxide, and copper oxide [10-14]. Ferric oxide is the common catalyst to boost AP decomposition with positive impact on its energetic power [15]. On the other hand, thermite is a pyrotechnic heterogeneous mixture of metal oxide/metal powders that can produce an exothermic oxidation–reduction (redox) reaction [16-20]. The reaction rate of thermite particles is quite low as it is limited to the inter-particle distance [21-28]. The contact surface area between reactive components (fuel and oxidizers) must be enhanced, increase the reaction rate and diminish the ignition delay [29]. The distinctive feature of nanothermites, to conventional (micron-scale) thermites, is their extremely high ignition sensitivity, high reaction rates [18, 30, 31]. Certain kinds of energetic nanothermite reactions can evolve high gaseous products [32]. Therefore nanothermites have been suggested for use as igniters and lead-free primers [33]. Nanothermite could have a potential role in catalytic decomposition of AP particles with the increase in its decomposition enthalpy. Performance of the catalyzed energetic systems such as AP can be accessed via investigation of decomposition reaction kinetics [9]. Isoconversional (model free) and model fitting are the two main models for decomposition kinetic study [34]. It is widely accepted that AP decomposes at two kinetic steps Low-temperature decomposition (LTD) and high-temperature decomposition (HTD) [9, 35]. LTD is the first decomposition step which occurs at nearly 300 °C. The second step is HTD decomposition which arises above 450 ^{0}C. Decomposition is the initial step of AP combustion; nanocatalyst could be integrated to enhance decomposition and combustion of AP [36]. Introducing nanothermite such as ferric oxide/ aluminum binary mixture could enhance the catalytic effect and boost decomposition enthalpy. The current study reports on the facile synthesis of ferric oxide catalyst of 5 nm particle size. Aluminum nanoplates of 100 nm particle size were employed. Colloidal Fe_2O_3/Al nanothermite mixture was developed and integrated into AP matrix via co-precipitation. AP nanocomposite with uniform particle size of 5 μm was developed. Enhanced particle dispersion into energetic matrix was verified via elemental mapping. Nanothermite particles offered enhanced AP decomposition enthalpy by 125 % using DSC. The maximum decomposition temperatures for the two exothermic decomposition peaks were decreased by 14 ^{0}C and 59 ^{0}C respectively. Nanothermite particles offered decrease in AP activation energy by 51 and 40 % using Friedman model and Ozawa models respectively. AP nancomposite demonstrated complex decomposition reaction with three decomposition stages. AP decomposition includes first order mechanism, two dimensional diffusion reaction, and dimensional diffusion mechanism according to different extent of reacted fraction.

2. Experimental work

2.1 Synthesis of Fe_2O_3 NPs

Ferric oxide NPs were fabricated by using a hydrothermal water system. Superheated water at 300 ^{0}C and 225 bars (20 ml/min). An aqueous solution of 0.05 M (ferric nitrate) was employed as metal salt precursors at 20 ^{0}C, 225 bars (10 ml/min.). Fe2O3 particles were fabricated in a continuous manner at the interface of the flow inside the reactor. Morphology of developed ferric oxide and aluminum NPs were investigated using TEM.

2.2 Integration of thermite NPs into AP

Nanothermite particles were integrated into the AP matrix via co-precipitation technique. Nanothermite particles were re-dispersed in acetone using ultrasonic probe homogenizer. AP was dissolved in acetone colloid. Finally, anti-solvent was introduced for effective coating of thermite NPs with AP matrix. The final nanocomposite was filtered and dried 'figure 1'. Then the co-precipitated integrated AP composite filtrated and dried for further characterization.

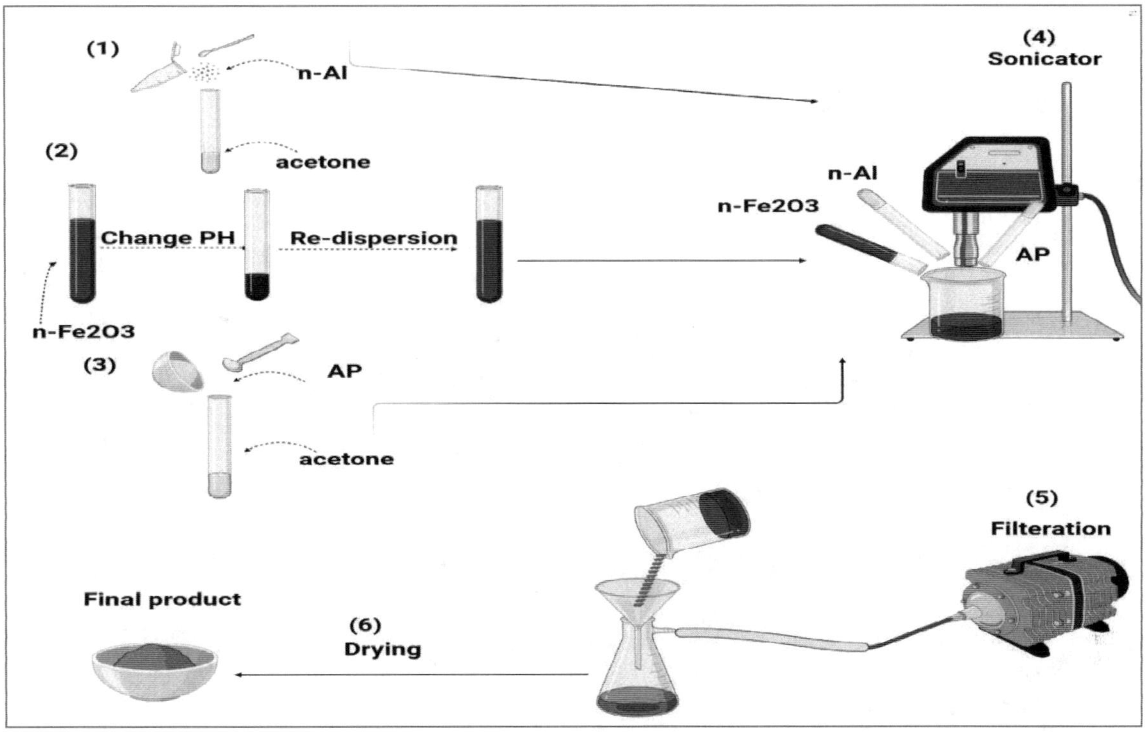

Figure 1. Development of AP nanocomposite via co-precipitation technique.

2.3 Characterization of AP nanocomposite

Morphology (shape and size) of developed AP nanocomposite was investigated using SEM ZEISS SEM EVO 10 MA integrated with EDAX for elemental analysis. Interpretation of nanothermite particle dispersion into AP matrix was assessed using EDAX detector.

2.4 Thermal behavior of AP nanocomposite

Thermal behavior of pure AP crystals and AP nanocomposite were assessed by DSC Q20 by TA, USA. The tested sample was heated at 10 ^{0}C min^{-1} up to 500 ^{0}C, under N_2 gas flow of 50 ml min^{-1}.

2.5 Decomposition Kinetics of AP nanocomposite

Kinetics of energetic materials and their thermal behavior is vital for sensitivity, stability, and lifetime evaluation. Isoconversional and model fitting are the two main methods for predicting decomposition kinetic parameters. In this manuscript, the kinetic decomposition parameters were evaluated by isoconversional methods. The first model was the differential isoconversional method of Friedman equation (1) [37].

$$\ln\left(\frac{d\alpha}{dt}\right)\alpha i = \ln[f(\alpha)A\alpha] - \frac{E\alpha}{RT\alpha{,}i} \quad (1)$$

Where $\left(\frac{d\alpha}{dt}\right)$ is the rate of change of extent of conversion (rate of reaction), $f(\alpha)$ is the solid state reaction model, A is the pre-exponential factor, $E\alpha$ is the apparent activation energy, R is the universal gas constant, and T is the temperature at certain extent of conversion. The second model is the integral isoconversional method of Ozawa, Flyn and Wall (OFW) equation (2) [38, 39].

$$\ln(\beta i) = \text{Const} - 1.052\left(\frac{E\alpha}{RT\alpha}\right) \quad (2)$$

Where βi is the heating rate. Decomposition kinetics were evaluated via DSC; the tested sample was heated at controlled heating rate of 3, 5, and 7 ^{0}C. min^{-1}, under N2 flow of 50 ml/min.

3. Results and Discusssions

3.1 Characterization of AP nanocomposite

TEM microggraphs demonstrated mono-dispersed ferricoxide particles of 5 nm averagep article size 'figure 2 a.b'. Aluminum nanoplates of 100 nm particle size were reported from TEM micrograhs 'figure 2 c.d'.

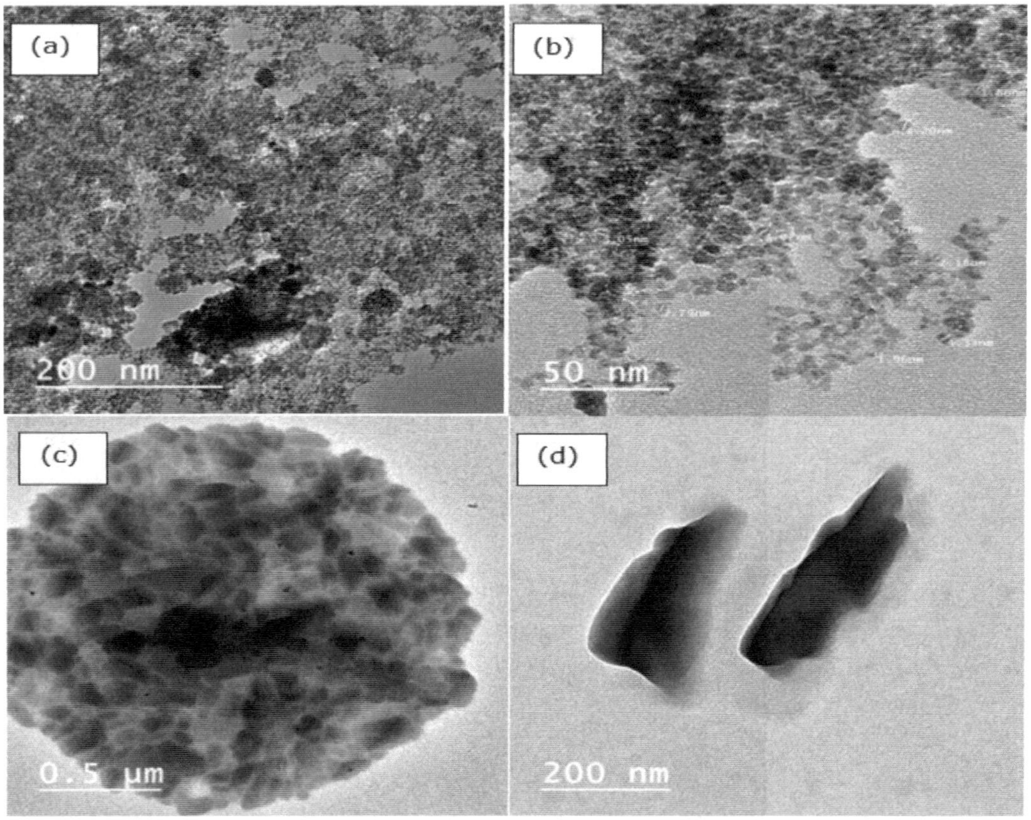

Figure 2. TEM micrographs of ferric oxide nanoparticles (a, b), aluminum nanoplates (c, d).

SEM was employed to investigate morphology (shape and particle size) of AP nanocomposite to pure AP. While starting AP demonstrated orthorhombic crystal of 150 μm size 'figure 3 a'; AP nanocomposite revealed ultrafine particles of 5 μm size 'figure 3 b'.

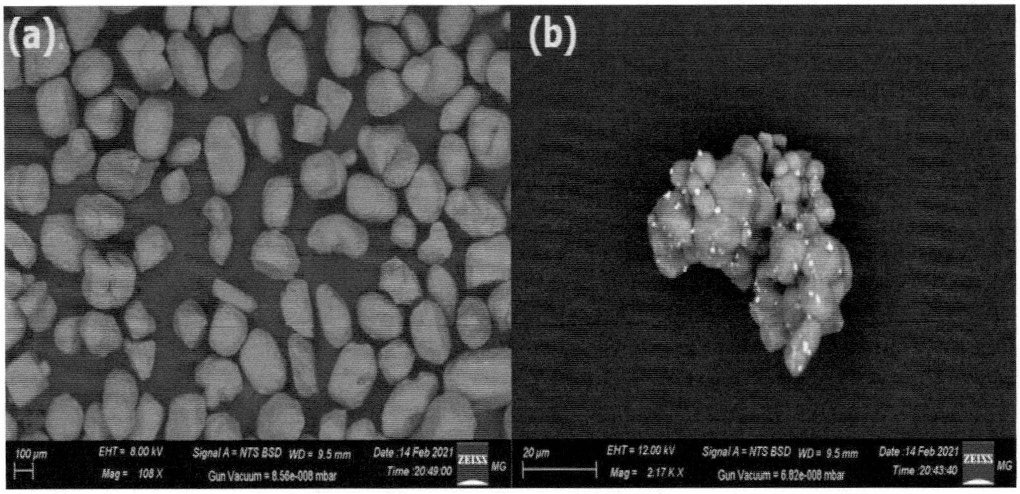

Figure 3. SEM micrographs of pure AP (a), and AP nanocomposite (b).

The difference in crystalline size was due to the crystallization technique that has been employed. It is obvious that uniform dispersion of nanothermite particles within the energetic matrix was accomplished. Therefore, it can be deduced that the integration of thermite NPs into the APC matrix offered uniform dispersion. Elemental mapping was conducted for pure AP and AP nanocpmposite. Elemental analysis of pure AP using EDAX detector confirmed the existence of the main AP elements. AP elemental composition was 12.2 % for nitrogen, 56.6 % for oxygen, 30.7 % for chlorine, and 0.5 % for carbon was due to the sample holder used in the image processing 'figure 4'.

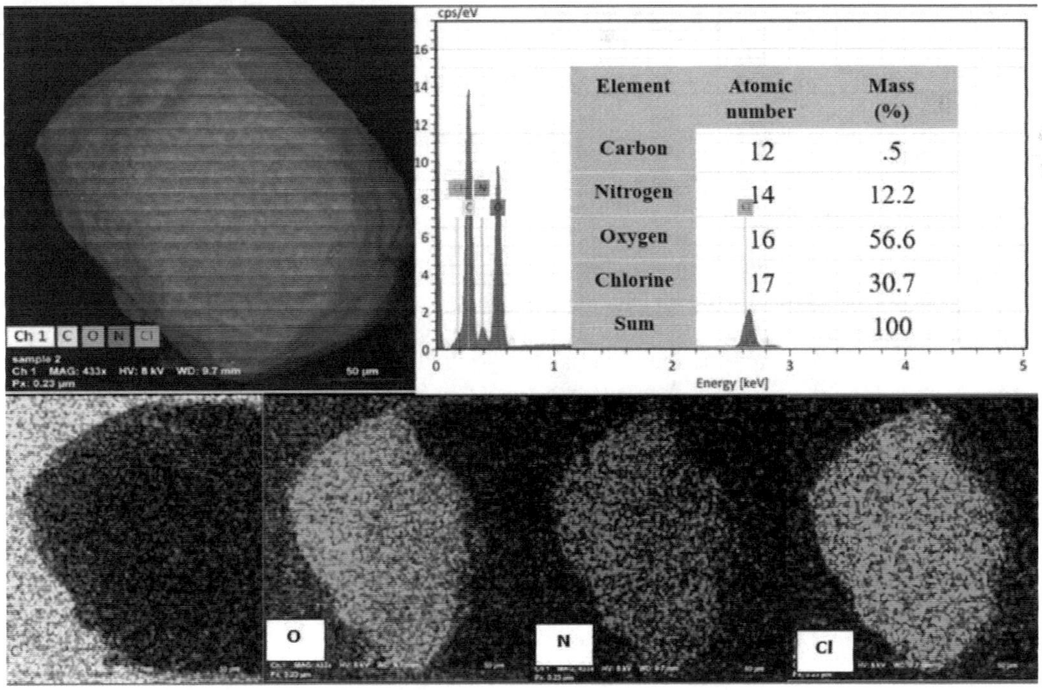

Figure 4. Elemental mapping of pure AP using EDAX detector.

In addition, elemental mapping for AP nanocomposite was conducted to verify the uniform nanoparticle dispersion. Enhanced nanothermite dispersion was accomplished. The elemental composition demonstrated proper percentage of each element. The mass percentage was 11.4 % for nitrogen, 53.8 % for oxygen, 29.1 % for chlorine, 0.9 % for iron, 4.3 % for aluminum 'figure 5'. Therefore, superior performance could be accomplished due to the enhanced particles distribution.

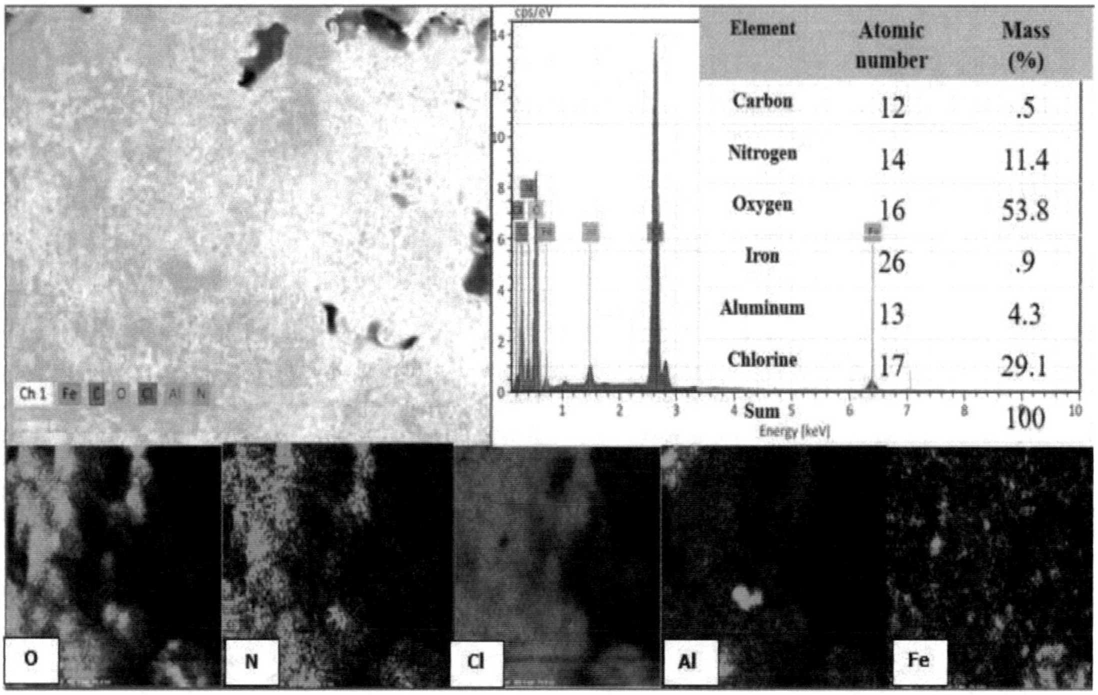

Figure 5. Elemental Mapping of AP nanocomposite using EDX detector.

3.2 Thermal behaviour of AP nanocomposite

Thermal behavior of AP nanocomposite was assessed using DSC. The tested sample was heated at controlled heating rate of 10 0C/min. The obtained DSC outcomes were compared to pure AP particles. Nanothermite particles offered an increase in AP total heat release by 125 %. The maximum decomposition temperature for the two exothermic decomposition peaks was reduced. LTD and HTD were decreased by 14 ^{0}C and 59 ^{0}C respectively 'figure 6'.

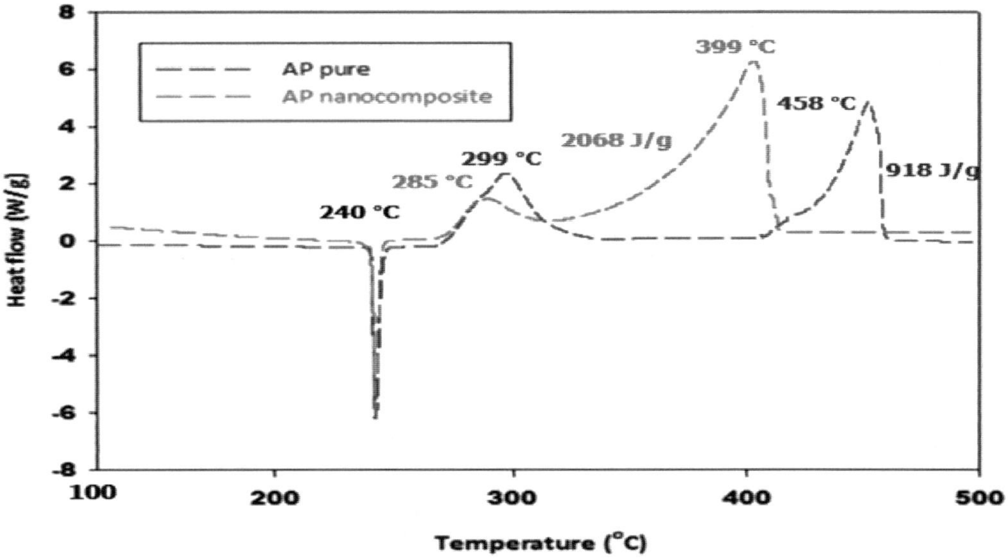

Figure 6. DSC thermograms of pure AP and AP nanocomposite.

The decrease in the main two decomposition peak temperature (LTD, and HTD) could be ascribed to the catalytic effect of ferric oxide particles. Furthermore, the enhanced decomposition enthalpy could be ascribed to the aluminum particles as high energy dense materials.

3.3 Decomposition kinetics of AP nanocomposite

Kinetic decomposition parameters and activation energy were evaluated using Friedman and OFW models respectively. AP nanocomposite was heated at different heating rates of 3, 5, 7 ^{0}C min^{-1} using DSC 'figure 7'.

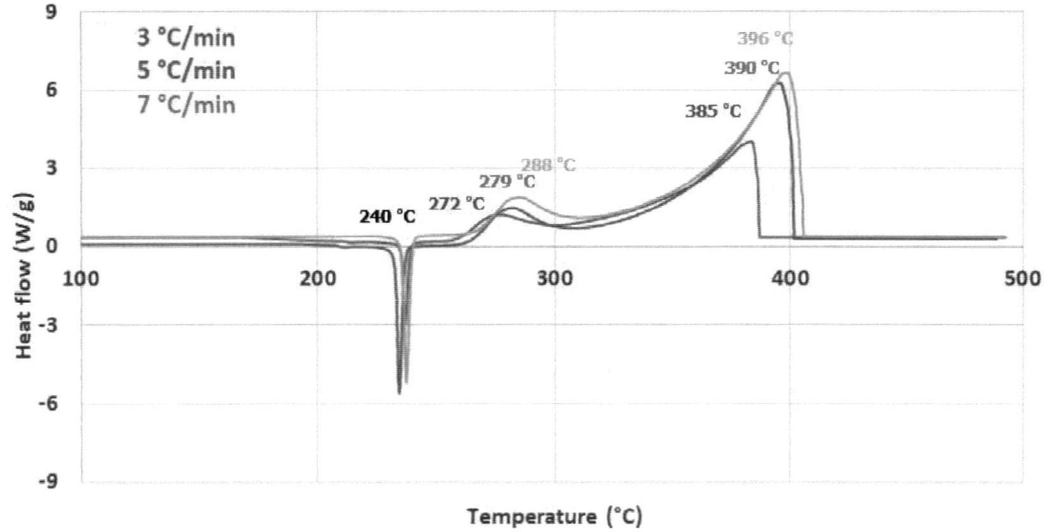

Figure 7. DSC thermograms of different heating rates for AP nanocomposite

From DSC thermograms it is obvious that the LTD and HTD shifted to higher value with increase in heating rate [40]. In addition, the reacted fractions with temperature were plotted 'figure 8'.

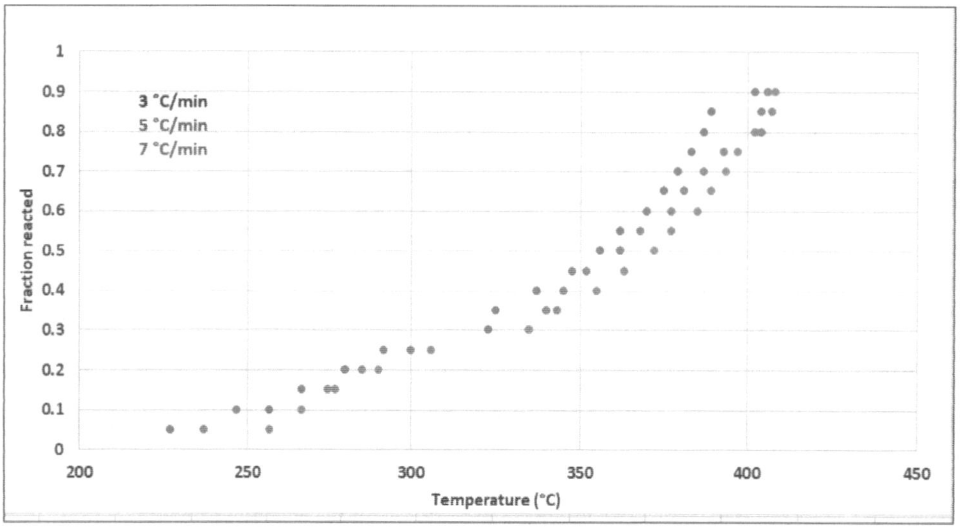

Figure 8. Fraction reacted with temperature at different heating rates.

Solid state decomposition mechanism of AP nanocomposite could be predicted from 'figure 8' by monitoring the degree of conversion at different temperature.

3.3.1 Kinetic study using Friedman model

Kinetic parameters obtained from Friedman differential isoconversional model were tabulated in table 1.

Table 1. Kinetic parameters of AP nanocomposite using Friedman model.

α	E_a (kJ mol^{-1})	Log A (s^{-1})	R_2
0.05	064.88	02.50	0.881
0.10	060.90	03.20	0.956
0.15	035.00	03.82	0.857
0.20	030.00	03.93	0.893
0.25	025.00	03.10	0.911
0.30	054.20	04.00	0.961
0.35	060.00	04.50	0.933
0.40	066.50	05.60	0.922

0.45	075.00	06.60	0.917
0.50	079.80	06.10	0.929
0.55	088.10	07.78	0.937
0.60	095.60	08.57	0.961
0.65	099.20	09.30	0.924
0.70	105.40	08.90	0.974
0.75	110.00	09.50	0.928
0.80	113.00	10.90	0.991
0.85	115.10	10.20	0.927
0.90	111.5	10.50	0.899
Mean	**077.00**	**06.61**	

From the previous table, the apparent activation energy was found to be 77 KJ mol^{-1} for AP nanocomposite compared with 157.9 KJ mol^{-1} for pure AP [41]. It can be concluded that nanothermite particles offered a decrease in AP activation energy by 51 %. This result confirmed the enhanced catalytic activity of thermite NPs on AP decomposition kinetics. Activation energy at different reacted fractions is represented in 'figure 9'.

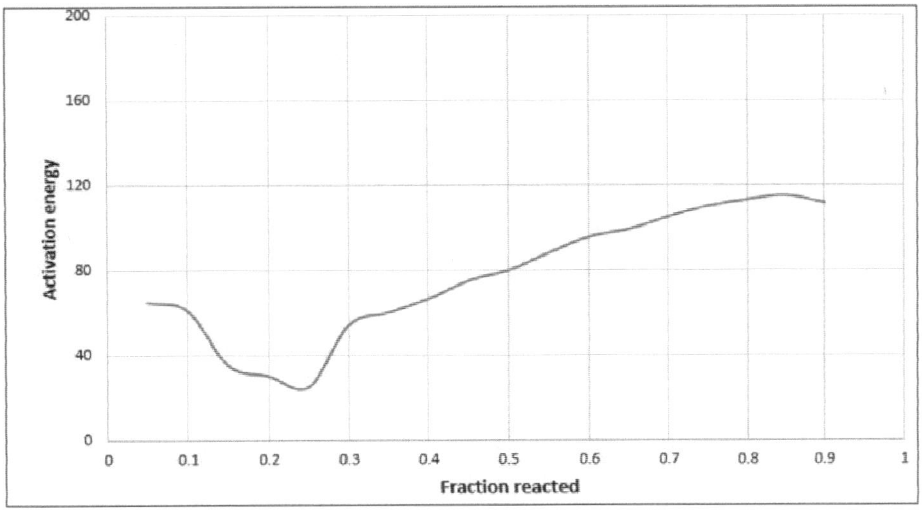

Figure 9. Activation energy of AP nanocomposite at different fraction reacted using Friedman model

The fluctuated values of the activation energy at different extent of conversion indicated the complexity of decomposition mechanism. Extent of conversion below 0.25 (α=0 to 0.25) demonstrated decrease in activation energy. Extent of conversion higher than 0.3.3 (α= 0.3 to 0.9) demonstrated an increasing values.

3.3.2 Kinetic study using Ozawa model

Kinetic parameters obtained from the isoconversional method of Ozawa model at different extent of conversion were tabulated in table 2.

Table 2. Kinetic parameters of AP nanocomposite using Ozawa model

α	E_a (kJ mol^{-1})	Log A (s^{-1})	R_2
0.05	055.00	02.30	0.881
0.10	045.00	03.12	0.956
0.15	027.00	03.77	0.857
0.20	025.00	03.82	0.893
0.25	030.00	03.26	0.911
0.30	035.00	04.25	0.961
0.35	037.00	04.33	0.933
0.40	049.80	05.56	0.922
0.45	058.20	06.78	0.917
0.50	074.60	06.02	0.929
0.55	077.40	07.89	0.937
0.60	102.70	08.62	0.961
0.65	103.00	09.45	0.924
0.70	103.50	08.82	0.974
0.75	120.00	09.14	0.928
0.80	125.00	10.45	0.991
0.85	127.00	10.66	0.927
0.90	115.00	10.83	0.899
Mean	**072.78**	**06.62**	

The results obtained from Ozawa model of analysis was in a consistence with Friedman model outcomes. While the apparent activation energy of pure AP was 121.5 KJ mol^{-1} [42]; AP nanocomposite demonstrated apparent activation energy of 72.78 KJ mol^{-1}. Nanothermite particles experienced decrease in AP activation energy by 40 % [42]. The activation energy value at different fraction reacted is demonstrated in 'figure 10'.

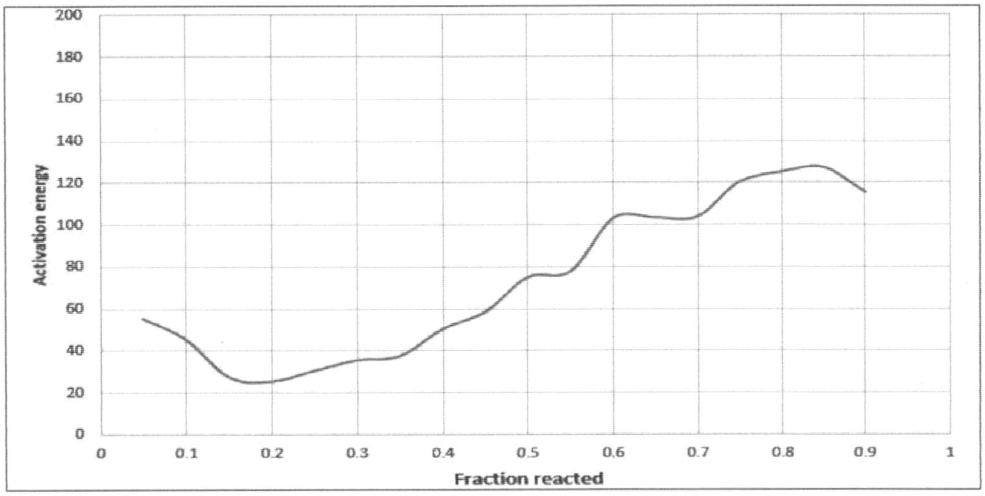

Figure 10. Activation energy of AP nanocomposite at different fraction reacted using OFW model

'Figure 10'outcomes confirmed results obtained from 'figure 9'. It can be assured that thermal decomposition of AP is a complex process that occurred in consecutive steps.

3.4 Decomposition model prediction of AP nanocomposite

Variables of the decomposition mechanism can express the decomposition reactions [43]. The solid-state mechanism of decomposition has been predicted via the relations of different decomposition parameters such as the extent of conversion, rate of the extent of conversion change, and temperature. 'Figure 11'represents the relation of fraction reacted with temperature.

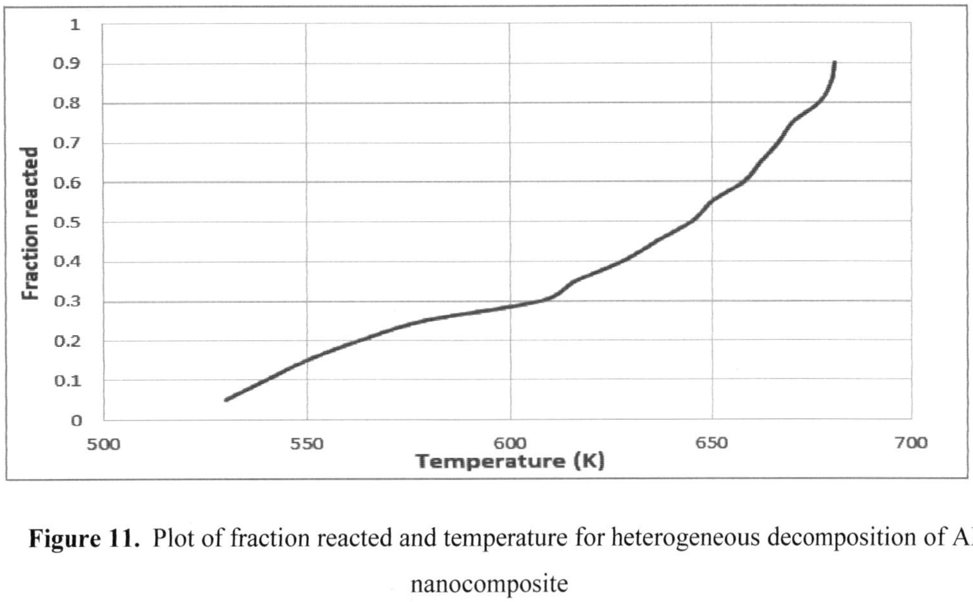

Figure 11. Plot of fraction reacted and temperature for heterogeneous decomposition of AP nanocomposite

AP nanocomposite experienced decomposition in more than one step. This result demonstrated the degree of complexity of AP decomposition. It can be concluded that the mechanism of decomposition started with first-order decomposition and ended with diffusion decomposition reaction. Moreover, the relation between the extent of conversion and the rate of changing the degree of conversion is represented in 'figure 12'.

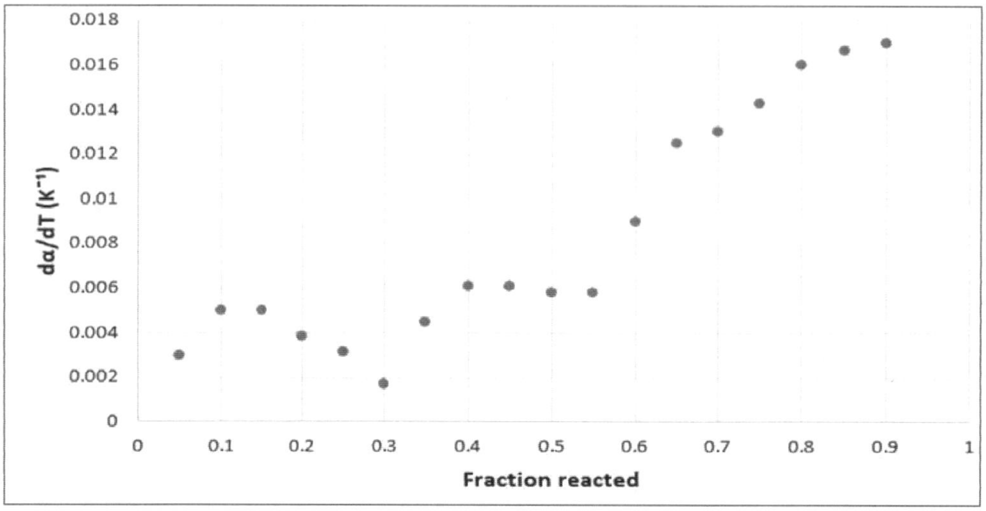

Figure 12. Plot for α and dα/dT for heterogeneous decomposition reaction of AP nanocomposite

It is obvious that the mechanism of decomposition is running in three steps as follow.

- Decomposition reaction obeys first order mechanism with Ea=35 kJ mol^{-1} (α = 0-0.25).
- Decomposition reaction obeys two dimensional diffusion mechanism with Ea=75 kJ mol^{-1} (α = 0.25-0.6).
- Decomposition reaction obeys one dimensional diffusion mechanism with Ea=113 kJ mol^{-1} (α=0.6-0.9).

From the previous results the mechanism of decomposition of AP nanothermite was found to be complex as it undergoes through three different decomposition steps.

4. Conclusion

Fe2O3 nanoparticles were successfully fabricated by using hydrothermal synthesis. Colloidal nanothermite particles Fe2O3/Al were integrated into AP matrix via co-precipitation technique. Nanothermite particles revealed uniform dispersion of nanothermites into AP matrix. While Fe2O3 NPs could act as a catalyst aluminum NPs could act as a high energy dense material. Nanothermite particles offered enhanced AP decomposition enthalpy by 120 %. The maximum decomposition temperatures for the two exothermic decomposition peaks were decreased by 14 ^{0}C and 59 ^{0}C respectively. Nanothermite particles offered decrease in AP activation energy by 51 and 40 % using Friedman model and Ozawa modesl respectively. AP decomposition mechanism of is running in three steps including first order mechanism, two dimensional diffusion reaction, and dimensional diffusion mechanism for different extent of fraction reacted. This confirms the complexity of AP decomposition.

Reference

[1] Rajić M. and M. Sućeska 2000 Study of thermal decomposition kinetics of low-temperature reaction of ammonium perchlorate by isothermal TG. *Journal of thermal analysis and calorimetry* **63** (2) p. 375-386.

[2] Raha K, S. Ramamurthy and D. Patil 1989 The catalytic effect of rare earth oxides on the thermal decomposition of ammonium perchlorate. *Journal of Thermal Analysis and Calorimetry* **35** (4) p. 1205-1212.

[3] Elbasuney S. and M. Yehia 2019 Thermal decomposition of ammonium perchlorate catalyzed with CuO nanoparticles. *Defence Technology* **15** (6) p. 868-874.

[4] Elbasuney, S. and M. Yehia 2019 Ammonium Perchlorate Encapsulated with TiO2 Nanocomposite for Catalyzed Combustion Reactions. *Journal of Inorganic and Organometallic Polymers and Materials* **29**(4) p. 1349-1357.

[5] Elbasuney, S. and M. Yehia 2020 Ferric Oxide Colloid: A Novel Nano-catalyst for Solid Propellants *Journal of Inorganic and Organometallic Polymers and Materials* **30**(3): p. 706-713.

[6] Elbasuney S, et al 2021 Ferric oxide colloid: novel nanocatalyst for heterocyclic nitramines . *Journal of Materials Science: Materials in Electronics* **40**(4).

[7] Maycock J, V.P. Verneker and L. Rouch Jr 1968 Influence of growth parameters on the reactivity of ammonium perchlorate. *Inorganic and Nuclear Chemistry Letters* **4** (2) p. 119-123.

[8] Longuet B and P. Gillard 2009 Experimental investigation on the heterogeneous kinetic process of the low thermal decomposition of ammonium perchlorate particles. *Propellants, Explosives, Pyrotechnics: An International Journal Dealing with Scientific and Technological Aspects of Energetic Materials* **34**(1) p. 59-71.

[9] Vyazovkin, S. and C.A. Wight 1999 Kinetics of thermal decomposition of cubic ammonium perchlorate. *Chemistry of materials* **11**(11) p. 3386-3393.

[10] Elbasuney S, et al 2018 Chemical stability, thermal behavior, and shelf life assessment of extruded modified double-base propellants. *Defence Technology* **14**(1) p. 70-76.

[11] Elbasuney S, et al 2017 Stabilized super-thermite colloids: A new generation of advanced highly energetic materials. *Applied Surface Science* **419** p. 328-336.

[12] Elbasuney S, M. Gobara and M. Yehia 2019 Ferrite Nanoparticles: Synthesis, Characterization, and Catalytic Activity Evaluation for Solid Rocket Propulsion Systems. *Journal of Inorganic and Organometallic Polymers and Materials* **29**(3): p. 721-729.

[13] Elbasuney S , et al 2020 Synthesis of CuO-distributed carbon nanofiber: Alternative hybrid for solid propellants. *Journal of Materials Science: Materials in Electronics.* **31**(11): p. 8212-8219.

[14] Elbasuney S , et al 2019 The significant role of stabilized colloidal ZrO2 nanoparticles for corrosion protection of AA2024. *Environmental Nanotechnology, Monitoring & Management* **12**: p. 100242.

[15] Yan Q.-L, et al, 2016 Catalytic effects of nano additives on decomposition and combustion of RDX-, HMX-, and AP-based energetic compositions. *Progress in Energy and Combustion Science.* **57**: p. 75-136.

[16] Elbasuney S, et al 2018 Super-Thermite (Al/Fe2O3) Fluorocarbon Nanocomposite with Stimulated Infrared Thermal Signature via Extended Primary Combustion Zones for Effective Countermeasures of Infrared Seekers. *Journal of Inorganic and Organometallic Polymers and Materials.* **28** p. 77-158.

[17] Elbasuney S and G.S. El-Sayyad 2020 The potentials of TiO2 nanocatalyst on HMX thermolysis . *Journal of Materials Science: Materials in Electronics,.* **31**(17): p. 14930-14940.

[18] Elbasuney S, et al 2020 Multi-component nanocomposite infrared flare with superior infrared signature via synergism of nanothermite and reduced graphene oxide. *Journal of Materials Science: Materials in Electronics* **31**(14) p. 11520-11526.

[19] Elbasuney S et al 2020 Facile synthesis of RGO-Fe2O3 nanocomposite: A novel catalyzing agent for composite propellants. *Journal of Materials Science: Materials in Electronics* **31**(23): p. 20805-20815.

[20] ElbasuneyS, et al 2020 Novel (MnO2/Al) thermite colloid: an opportunity for energetic systems with enhanced performance. *Journal of Materials Science: Materials in Electronics.* **31**(23): p. 21399-21407.

[21] Elbasuney S 2014 Dispersion characteristics of dry and colloidal nano-titania into epoxy resin. *Powder Technology.* **268**(0): p. 158-164.

[22] Elbasuney S 2015 Continuous hydrothermal synthesis of AlO(OH) nanorods as a clean flame retardant agent. *Particuology*. **22**: p. 66-71.

[23] Elbasuney S 2015 Surface engineering of layered double hydroxide (LDH) nanoparticles for polymer flame retardancy. *Powder Technology.* **277**: p. 63-73.

[24] ElbasuneyS 2017 Sustainable steric stabilization of colloidal titania nanoparticles. *Applied Surface Science.* **409**: p. 438-447.

[25] Elbasuney S 2017 Novel multi-component flame retardant system based on nanoscopic aluminium-trihydroxide (ATH). *Powder Technology.* **305**: p. 538-545.

[26] Elbasuney, S 2018 Novel colloidal molybdenum hydrogen bronze (MHB) for instant detection and neutralization of hazardous peroxides. *TrAC Trends in Analytical Chemistry.* **102**: p. 272-279.

[27] Elbasuney S 2018 Novel Colloidal Nanothermite Particles (MnO2/Al) for Advanced Highly Energetic Systems. *Journal of Inorganic and Organometallic Polymers and Materials.* **28**(5): p. 1793-1800.

[28] Elbasuney S 2019 Steric Stabilization of Colloidal Aluminium Particles for Advanced Metalized-Liquid Rocket Propulsion Systems. *Combustion, Explosion, and Shock Waves.* **55**(3): p. 353-360.

[29] Yang Y, D. Xu and K. Zhang 2012 Effect of nanostructures on the exothermic reaction and ignition of Al/CuOx based energetic materials. *Journal of Materials Science.* **47**(3): p. 1296-1305.

[30] Elbasuney S et al 2020 Novel nanocomposite decoy flare based on super-thermite and graphite particles. *Journal of Materials Science: Materials in Electronics.* **31**(8): p. 6130-6139.

[31] Elbasuney S et al 2020 Novel high energy density material based on metastable intermolecular nanocomposite. *Journal of Inorganic and Organometallic Polymers and Materials.* **30**(10): p. 3980-3988.

[32] Rogachev A and A. Mukasyan 2010 Combustion of heterogeneous nanostructural systems. *Combustion, Explosion, and Shock Waves.* **46**(3): p. 243-266.

[33] Higa K T 2007 Energetic nanocomposite lead-free electric primers. *Journal of propulsion and power.* **23**(4): p. 722-727.

[34] Vyazovkin S et al 2011 ICTAC Kinetics Committee recommendations for performing kinetic computations on thermal analysis data. *Thermochimica acta.* **520**(1-2): p. 1-19.

[35] Zhu Y L et al 2014 Kinetics of thermal decomposition of ammonium perchlorate by TG/DSC-MS-FTIR. *Journal of Energetic Materials.***32**(1): p. 16-26.

[36] Rossi C 2014 Two Decades of Research on Nano-Energetic Materials. *Wiley Online Library.* Chapter 4 pp 258–290

[37] Friedman H L1964 Kinetics of thermal degradation of char-forming plastics from thermogravimetry. Application to a phenolic plastic. *In Journal of polymer science* part C: polymer symposia.

[38] Ozawa T, A 1965 new method of analysing thermogravimetric data. *Bulletin of the chemical society of Japan.* **38**(11): p. 1881-1886.

[39] Flynn J.H and L.A. Wall 1966 General treatment of the thermogravimetry of polymers. *Journal of research of the National Bureau of Standards.* Section A, Physics and chemistry. **70**(6): p. 487.

[40] Yan Q.-L, et al 2013 Thermal behavior and decomposition kinetics of Formex-bonded explosives containing different cyclic nitramines. *Journal of thermal analysis and calorimetry.* **111**(2): p. 1419-1430.

[41] Li X R Yang and X. Li 2004 The functional additveis CNTs. *Conference paper in Propellants Explos.* Xi'an Modern Chemistry Research Institute p. 228-231.

[42] Joshi S, S P.R. Patil and V. Krishnamurthy 2008 Thermal Decomposition of Ammonium Perchlorate in the Presence of Nanosized Ferric Oxide. *Defence Science Journal.* **58**(6) p. 317-360.

[43] Khawam, A. and D.R. Flanagan 2006 Solid-state kinetic models: basics and mathematical fundamentals. *The journal of physical chemistry B.* **110**(35): p. 17315-17328.

Utilization of Nano thin films for solar cell self-cleaning.

Mariam Abu Elezz(*), Nabil Abdelmonem , Ibrahim Ismail, Amr Abdelghany.

Chemical Engineering Department, Faculty of Engineering, Cairo University, Giza, Egypt. (*)Email of the corresponding author: mariam.eng8899@gmail.com.

Abstract: The present work for preparing titanium dioxide as thin film to improve the PV system and put alternatives to manual cleaning with no loss in the efficiency. Using self-cleaning coatings instead of cleaning chemicals on PV panels is a very suitable process in the application and cost. A set of experiments were conducted on a laboratory scale by using the coating layer. The results showed that using pre-coated with titanium dioxide on PV panels is more effective than no coating layer. As increasing photo catalytic activity and the stability of the coated layer are more effective.

Keywords: Nano-titanium dioxide, solar cell, thin-film, self-cleaning, dip coating.

1. Introduction

The available energy production today on earth is 89 Pet watts (PW), making solar energy the most abundant and promising source of power. More than a million solar cell systems were installed in the US with efficiency from 20 to 23 % in 2018. Unfortunately, efficiency drops quickly with solar panel contamination since any dirt or obstruction on the way between the sun and the solar cells block sunlight and reduce collected energy [1] [2].

When dust accumulates on the PV panel's surface, it affects the electricity production as the amount of energy production is directly proportional to the amount of light reaching the panel surface. When there is dust on the panel, the amount of electricity gain reduced, so we must get rid of any dust to obtain a high amount of electricity and energy [3]. Many other factors affected energy production: solar irradiation intensity, wind velocity, air temperature, and incidence angle. These factors can be optimized by choosing a suitable place for the panels and proper fitting with the environmentSelf -cleaning coatings reduce the maintenance costs, amount of water, and human resources, so it is a very effective process. Different types of coating depending on the application, their behavior to water as hydrophobic or hydrophilic, and their decomposition in organic compounds contain photocatalytic particles, as Nanocrystalline titanium dioxide in the form of Anatase phase [4]. Some of the panels are located in the sunny desert environment and this dry environment, and the presence of wind makes dust on PV panel's surface. Other panels are located near some industrial areas where there is dust everywhere. By 2022 India plans to have 40 GB of solar energy, the environmental dust residue on the surface. It reduces energy efficiency from solar panels because of decreasing the amount of light on the solar cell surface. Many cleaning techniques can be used as failing rains, the wind, mechanically by vibration, blowing and ultrasonic driving, self-cleaning Nano-films or electrostatic dust removal is significant to obtain high efficiency of energy [5]. The self-cleaning phenomenon has a relationship with the contact angle obtained at the three phase solid, liquid, and vapor. Generally, when the contact angle is less than 90° the solid surface is hydrophilic, and when it is more than 90° it is a hydrophobic surface. When it is 0° it is called ultra (super) hydrophilic, and when it is more than 150° is usually named as ultra (super) hydrophilic [6].In Middle East isa bad condition for solar panels because of the amount of dust in the air and the sandstorms that occurs frequency; This makes acclimated Thick layer of dust in less than

1hour on the surface of solar cell. So that, we can observe the loss of efficiency for the PV by up to 80% And that's means we should clean the surface of PV regularly at least daily [7].

2. Experimental Methods

2.1. Substrates preparation

The Substrates used in the experimental work were: Glass microscope(2cmx5cm)for SEM, EDX, Contact Angle and XRD characterization ; Ultrasonic bath was used to cleaned the glass substrates by distilled water, HCL, and Acetone in the end it's dried at 100°C to be sure there is no dust affect in the analyse of SEM or XRD .;and Glass slides to cover solar cell (15 cmx18 cm) to cover solar panels during Daly testing and the Commercialized solar panels (characteristics are presented in the table 1);It illustrates a graphical flowchart summarized in reactions and coating processing used in this study before comparing the solar cells self-cleaning efficiency.

Table 1: Solar panel characteristics are presented in the table

Solarpanelcharacteristics	
Technology	a-Si(amorphoussi
Voltage	2 Volt
Amber	700mA
Rangeof temperature	-40°/+80°C
dimensions (L*W*H)	185x135x22mm
Weight	0.25kg
Brand	PHYWE

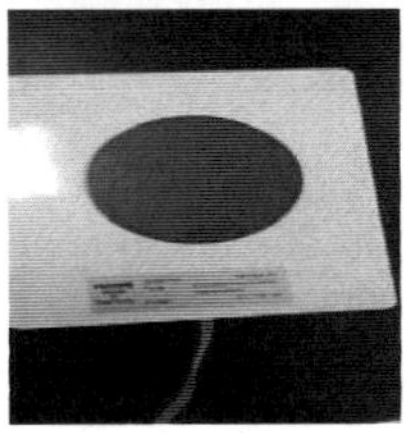

2.2. Preparation of (titanium dioxide) thin film

Raw Material and Chemicals for titanium dioxide liquidcoating:

Titanium tetra isopropoxide (TTIP) 97% (Sigma-Aldrich made in Germany) was used as a precursor in preparing TiO_2 sol, isopropanol (i-PrOH) as a solvent, deionized water, Hydraulic acid HCl, and Methylene blue. Solutions to be composed of the molar ratio of 1:25:0.3:1 of TTIP to (i-PrOH) to HClto deionized water and were prepared as below: by dissolving TTIP in 15 Mi-PrOH, with stirring at room temperature for one hour, after that a mixture of 10Mi-PrOH with deionized water was added dropwise to the overhead solution with the continuous stirring for another hour, then adding acid in natural atmospheric at room temperature and subsequently, until getting the value of a pH 4. These reactions must be carried out in the fume hood because of HCl gases' exhaustion during mixing. Physical-chemistry properties of the precursor solutions as pH, density, and viscosity were controlled to allow the tailoring of particle size and morphology in the thin films. The heat treatment consists of three-step (drying), which can improve adhesion between the substrate and thin film and release the stresses through solvent evaporation slowlyFigure 1 illustrates a graphical flowchart of the processing used in this study[8].

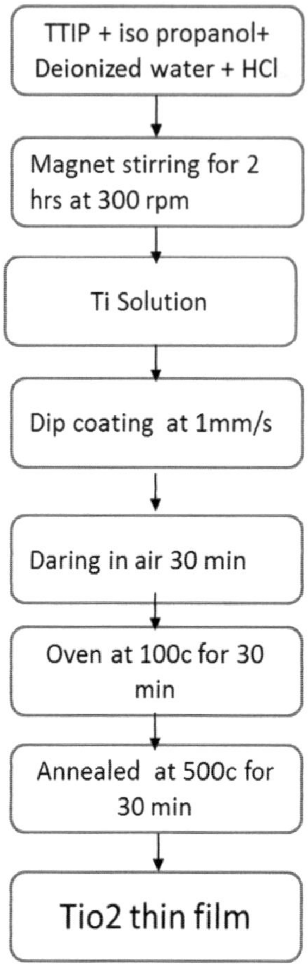

Figure1: A graphical flowchart summarizes reactions and coating processing.

2.3. Dip coating process

Pre-cleaned process for glass substrates in an ultrasonic bath then dried at 100°C then on clean glass substrates, a thin film was deposited through dip-coating method at room temperature and drawing speed of about 10 mm/s. Since a condition used in dip-coating for the deposition thin films, this step of the process should be carried out at a controlled withdrawal speed and free from vibration or external interference to allow the deposition of a homogeneous film with controlled thickness[5].

2.4. Drying: in three steps as the following

In Air: The thin films become dry at room temperature after 24 h to allow the solvent to evaporate slowly during the condensation process because of the fast sol-gel reaction of titanium precursor. Then in Oven: heating temperature increases at the rate of 10°C /min and then stops at 100°C. Temperature is kept for 30 min for the adhesion of the samples and to free the lagging stresses, finally in Muffle furnace: the furnace was gradually increased at heat rate of 10°C/min until the temperature is (500°C). Temperature is kept here for 30 min to achieve the crystallization formation of the thin film. In the end, the films were cooled and the heat in the furnace was lost to room temperature [9].

2.5. Photocatalytic activity:

By using Ultra-violet (UV) champers with mercury lamp to measure change in hydrophilicity. The cleaning process has been more effective as The hydrophilicity makes the water diffusion above the surface that makes the dirt combine faster and dry the surface more quickly, UV Change the contact angle from 22 to zero,these glass slides were submitted to irradiation by ultraviolet for 2 hours using a fluorescent mercury lamp, power 26 W and λ = 254 nm[10].

The photocatalytic affectivity was investigated by investigating the dye colour degradation of Methylene blue with the time change of UV irradiation. First, the methylene blue (MB) aqueous solution prepared with the concentration of 25ppM, the MB calibration curve by measuring MB's absorbance of MB in different concentrations.

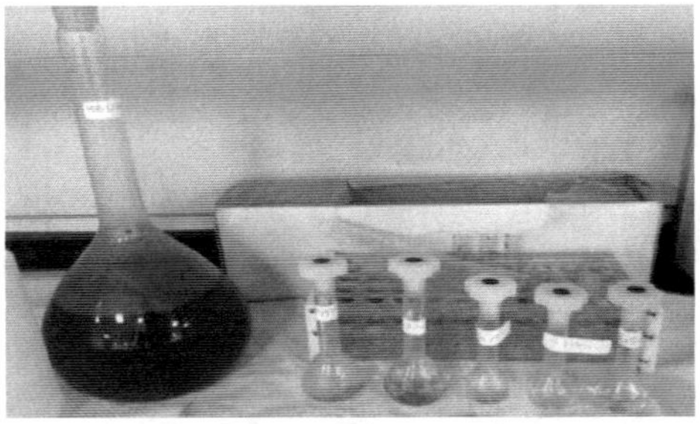

Figure 2: preparation of different concentration MB to calculate the calibration curve

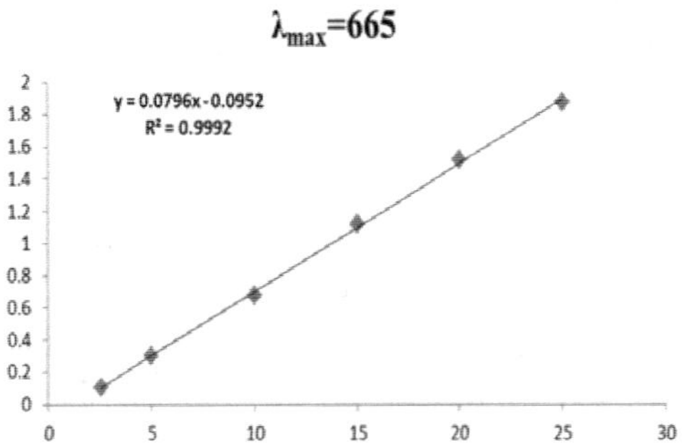

Figure 3: calibration curve for MB at λmax=664

The thin film's photocatalytic activity studied the change ofmethyleneblue MB degradation solution under ultra-violet (UV) light irradiation. While 25 mL, the prepared MB as a sol stock with concentrated 25ppm has been taken in a small beaker from the stock. The TiO_2 coated glass substrates (3x 7cm) have been dipped into the MB solution horizontally. Before exposing the samples and MB to irradiation mercury lamp, the beaker has been kept in the dark with stirring for 1 hour to be sure activation equilibrium has been done between sample and MB. Under the UV mercury lamp, the beaker exposed to this lamp irradiated for 2 hours the distance from the lamp to the sample and beaker determined by 5 cm as figure 4. The change of concentration of MB solution has been determined every 15 min using a UV– spectrophotometer at a maximum wavelength of λmax=664 nm after drawing the calibration curve to get the concentration function in time equation [11].

Figure 4: setup of self-cleaning testMercury lamp with 36 watts and magnet stirrer inside the handmade UV chamber

2.6. *Photovoltaic (PV) set up in natural weather and observes the change before and after self-cleaning coating.*

- *Steering solar panels*

In which direction are the fixed cells routed to obtain the most significant productivity. In the northern hemisphere, the cells are directed towards the south, while in the southern half, they are directed north. In the northern hemisphere (this is the case in all Arab countries) [12].

- *Calculate the best angles of cells in the summer*

In the case of the grid-independent system, the angles of the solar panels shall be calculated in order to obtain the highest productivity in the summer according to the following equation.

If the site is located between the equator and latitude of less than 25, the following equation is used: Best grade to guide cells = (latitude * 0.05) +1, If the location is between latitude 25 and 50, the following equation is used: Best grade to guide panels = (latitude * 0.93) -21, and If the site is located on a latitude more significant than 50, the following equation is used: Best grade to guide the panels = Latitude - 24.5 [13].

Figure 5: PV set up Real PV in natural weather

3. Characterization and Analytical, Testing and Examination Methods

3.1. *Characterization of Nano particles size and morphology*

In solution phase before coating process Measurement of pH the pH was measured using Cyber scan pH 510, Eutech instruments. It can measure the pH, vary from 0.00 to 14.00. The addition of sodium hydroxide pellets adjusted the pH of the bath. The pH electrode was calibrated using stock solutions of pH 4, 7, and 10. These solutions were prepared by dissolving buffer tablets from Fischer Scientific in deionized water. The pH electrode was left in each solution for a time until the reading is stabilized. The Zeta potential measurements by dynamic light scattering (DLS) it is gives the overestimate in average hydrodynamic diameter of the partials in solutions. Measurements were conducted with a Malvern Zetasizer NanoZS90 5001. For all sizes, disposable cuvettes were used to take at least five concordant measures for mean hydrodynamic diameter at 20°C.

For the thin film after coating use the X-ray diffraction system (XRD): is a required method closely target to examine the microstructure, phase identifies, and competent analysis of materials. A diffraction pattern is a result of the interaction of X-rays with crystal maters. A diffraction pattern has been achieved for all crystal maters; on the other hand, the amorphous materials can't be identified with x-ray technique. The goniometer is the XRD machine's mechanical unit, which consists of three parts; a sample holder, a detector arm, and an associated gearing. Scanning Electron Microscopy (SEM): The scanning electron microscope (FE-\SEM; QUATA FEG250) and the elemental analyses of the sample have been identified by using Energy-dispersive X-ray spectroscopy (EDAX). It was used to obtain micrographs of higher resolution. The SEM can also be used to obtain -dimensional images that help get a better morphological characterization. The area to be examined is irradiated with a finely focused electron beam. In the case of the glass substrate, it was covered with a thin layer of gold. The voltage applied to the beam was 15 or 20 kV. The low vacuum range is from 6 Pa to 270 Pa. topographic images can be determined by the high sensitive four segments BSE detector. It is also equipped with EDX for elemental and crystallographic information mapping.

3.2. Characterization of the coatings self-cleaning properties

- *Measuring the (static water contact angle)*

Water contact angle (Θ) above the flattened surface TiO_2 thin film has been used to investigate the hydrophilicity property with a contact angle meter (OCA 15EC) angle model produced by the company of Data Physics Instrument Gmbh (in National Research Centre). The Contact angle has been determined using the pendant dropped shape process with a GBX. The images were captured with a camera with an image resolution of 768 ×576 pixels (check images from email). The contact angles presented in this study were measured before and after the coating reaction was over and also with and without addition of surfactants. Analysed System has been calculated the right and left contact of the drop chap with an accurate information ±0.2°. All measurements were made at a controlled ambient temperature of 20° C and relative humidity (79%). Placed at four different positions for the samples, and the average measurement has been estimated to be the final contact angle [14].

- *Spectrophotometer measurements*

To know if the coating is with photocatalytic activation, the photocatalytic activity for thin films has been measured with methylene blue (MB) as the organic dye photo degradation using spectrophotometer (Evaluations 600). At maximum wavelength of maxi=664 nm and the range of the wavelength between 400 and 700.

3.3. Daily testing for PV panels

Finally, the comparison of solar cells efficiency by measurement of volt and amber to get the work for each PV coated and uncoated every day and observing the change using voltmeter.

4. Results and discussion

4.1. Particle size in solution

Nanoparticles are obtained using a dynamic light scattering device. The table 2 shows the size distribution of (TiO_2) solution nanoparticles for the different conditions of preparation solutions experiment before coating process the suitable average size is (72 nm).

Table 2: size of (TiO_2) nanoparticles in different conditions before the coating process.

EXP. No	TTIP:Iso-Pro: H_2O / ml	Amount of HCl /ml	Stirring time /hrs.	Gelation time/ hrs.	Dip coating speed mm/s	Drying			pH	observation	size /nm
						Air	oven	furnace			
1	01:25:01	0.3	2	24	1	30	100	500	2	big pikes + no stability	124
2	1.8:45:.54	1.83	2	24	1	30	100	500	1	2 big pikes	611 - 9
3	0.4:23:.9	0.09	2	24	1	30	100	500	4	very heavy no gelation	14
4	03: 19 : .18	0.09	2	24	1	30	100	500	4	quick gelation	50
5	1:19:.18	0.09	2	24	1	30	100	500	4	medial	23
6	1.5:19: .18	0.09	2	24	1	30	100	500	4	best Nano size	72

This test can give additional evidence about the (TiO_2) prepared nanoparticles grain size in the solution step and there's different parameter was studied their effects on the particle size as the following:

- *Effect of pH in partial size of TiO2 solution*

Nanoparticle size could be changed and controlled with a change in pH value. Nanoparticles with the best size range were synthesized at a pH of 4 with a good uniform size distribution, as in figure 6. In general, nanoparticles tend to be significant with irregular size ranges at acidic pH values compared with different pH values, which gave the relatively uniform size as in figure 7. At a pH of 2, particle size is provided by 124 nm, which is enormous and has no stability as the curve appears. That means there will be a different size in the particle surface after coating.

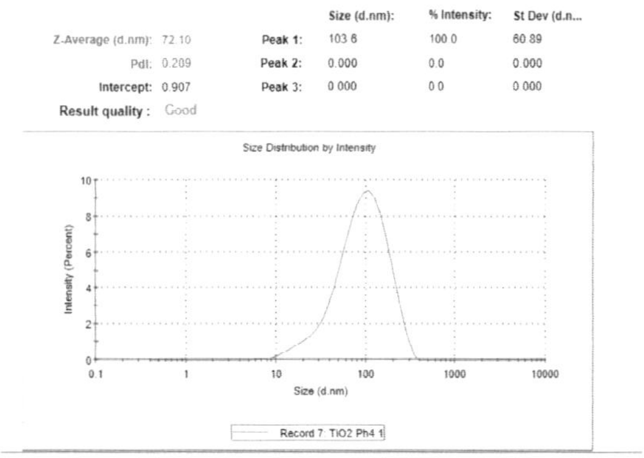

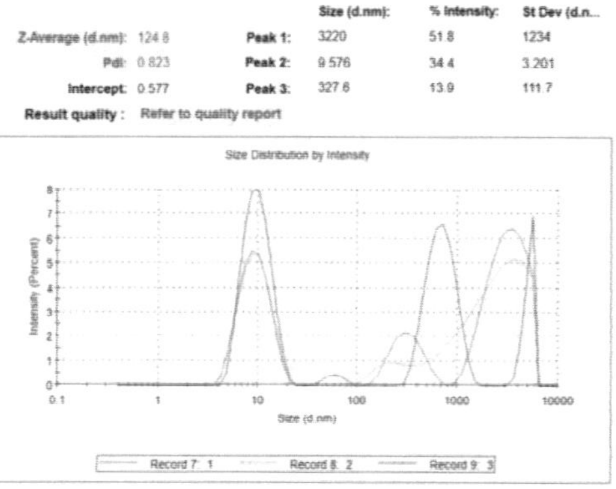

Figure 6: measurement TiO2 Nanoparticles at atpH=4

Figure 7:measurement TiO2 Nanoparticles size at pH=2

In table 2, the change of the practical size of TiO_2 in solution changes with the amount of HCl acid (a source of catalyst). In figure 6, when the pH= 2 in the TiO_2 solution the particle size is 184 nm which is not stable with low gelation, so it is not favorable for the coating and the pick is not symmetric that's means the solution have different size and it's not suitable for coating process. On the other hand, in figure 7, at pH=4, the particle size is 72 nm which is not heavy and has a promising gelation nature to make coating with suitable size and pick is symmetric which mean all the Nanoparticle in solution have the same size and the same properties so that its suitable for coating process.

- Effect of precursor's ratios in partial size of TiO_2 solution

As shown in table 2, it is noticed that the practical size of TiO_2 changed with the molar ratio of titanium source (titanium tetra isopropoxide (TTIP) in the solution. In figure 8, when 1 mole of TTIP is added, the particle size is 23nm which has no prospect of gelation to make the coating. Also In figure 9, when adding 0.4 moles of TTIP, the particle size is 14 nm which is small with no potential of gelation to make coatings and the not symmetric pick that mean the solution have not the same size of nanoparticles which not suitable for coating process.

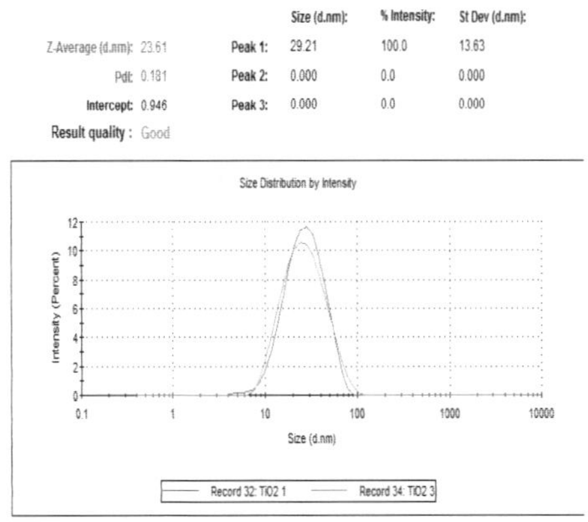

Figure 8: Nanoparticles size for TiO_2 of solution is 23 atmolar ratio of TTIP=1

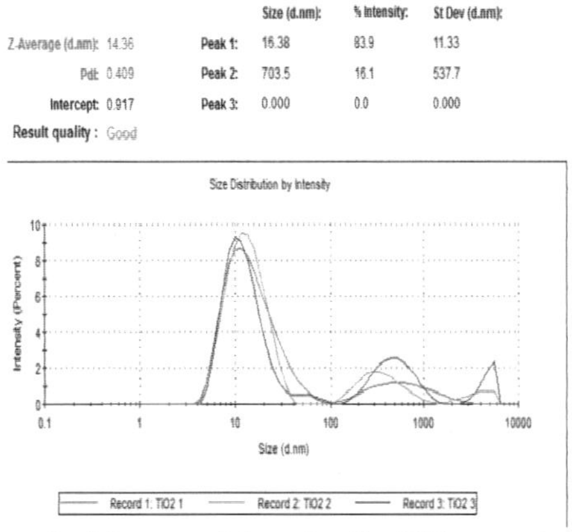

Figure 9: Nanoparticles size for TiO_2at molar ratio solution14 nm at molar ratio of TTIP=0.4

4.2. Structure Characterizations consist of: X-ray diffraction (XRD) and Scanning Electron Microscopy (SEM). These tests were achieved for the optimized samples of TiO_2 thin film.

- *XRD pattern for the sample (TiO2) thin film*

The XRD patterns for the selected coating samples on a microscope glass slide prepared at pH 4 and temperature at 500°C were analyzed. The XRD patterns of TiO_2 thin film, as shown in figure 10, show that the film was polycrystalline, having an Anatase phase. It is observed that the film exhibited characteristic peaks of Anatase crystal planes (011), (020), and (220). The nearly sharp peak detected at (26) can be related to the Anatase phase with crystal plane (011), while the intensity of the other peaks is fragile, rutile or brookite phases were not found. Because of its large Band gap=3.7, it is the most able for the photocatalytically active, super hydrophilicity, and self-cleaning process.

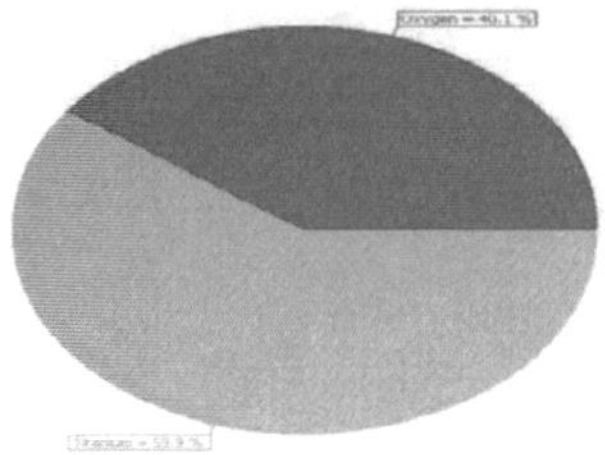

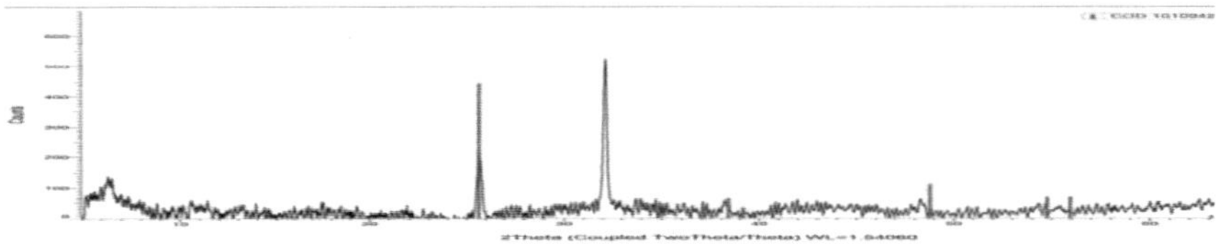

Pattern List #1

Show	Color	Name	Pattern #	Compound Name	Formula	Quality	Y-Scale
Yes		COD 1010942	COD 1010942	Anatase	O2 Ti	Quality Unknown	56.58 %

System	Space Group
Tetragonal	I 41/a m d (141)

a	b
3.73000 Å	

c	alpha	beta	gamma	Z	Volume	Density	Cell Tuned	F (N)
9.37000 Å				4	130.36 Å³		No	

Figure 10: XRD pattern list of TiO_2 thin film

The average crystallite size of the sample (TiO_2) is calculated using equation (Scherer equation). It is found that the average crystallite size utilized from the sharp peak located at (26°) is (27 nm).

- *Morphological analysis*

The surface morphology of the TiO_2 films was examined using Scanning electron microscope. (SEM) images of TiO_2 film, the surface being very smooth. One can observe from figure 11 that the coating is crack-free, without any porosity, transparent, and uniform. Without any irregularities and in figure 12 EDX analyses for TiO_2 thin film appearing the thin film's campsites [15].

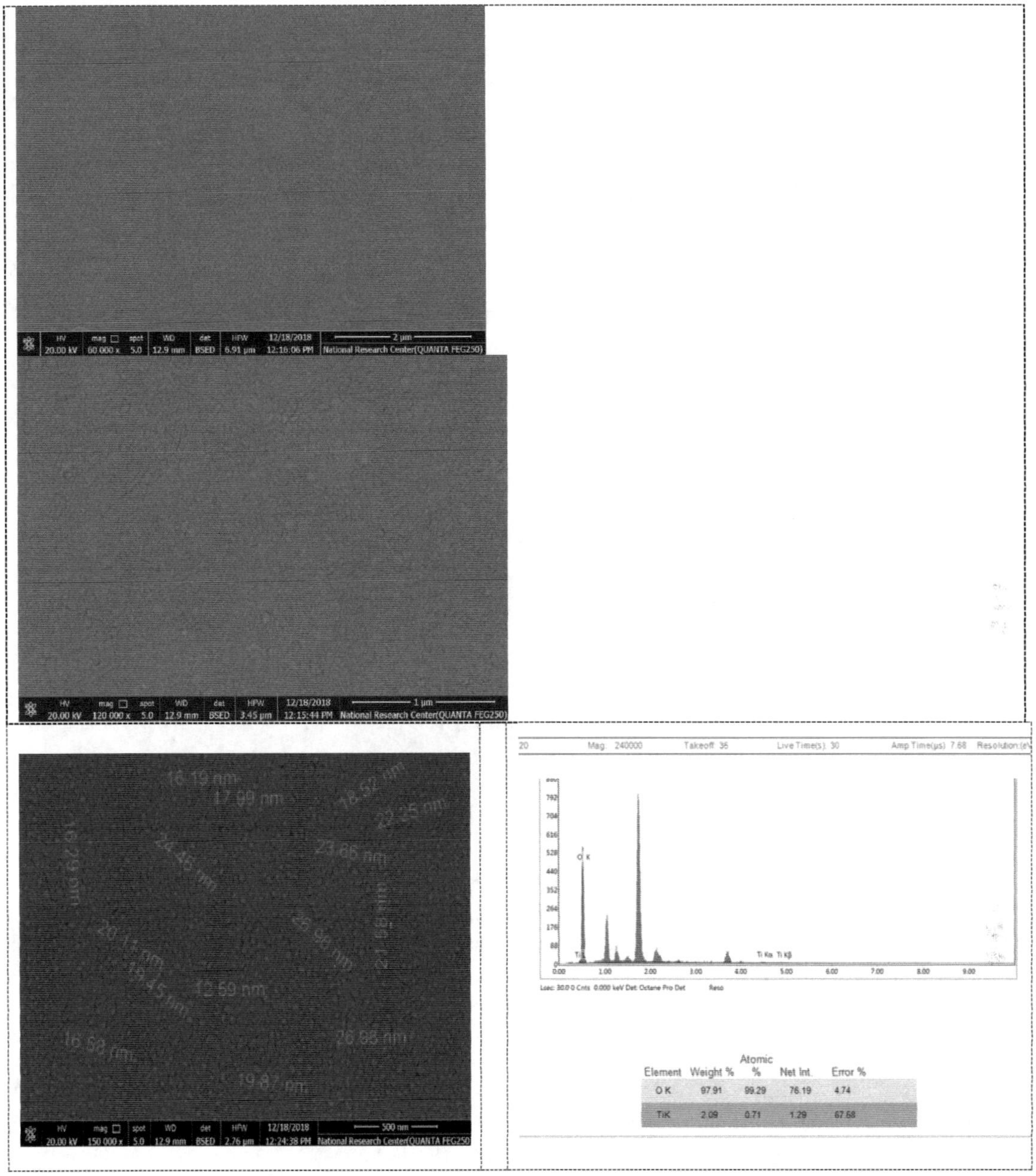

Figure 11:SEM And EDX analyses of the TiO_2 coating.The TiO_2 coating is crack-free, with porosity free, transparent and uniform.

4.3. Self-Cleaning Properties

Self-cleaning properties can be achieved through hydrophilicity and photocatalytic activity of the samples. The surface hydrophilicity of the films was quantified from water droplet contact angle measurements. The photocatalytic activity test was achieved by using photo degradation of Methylene Blue (MB) dye as a pollutant model.

- *Photocatalytic activity*

Photocatalytic activity of TiO_2 films: the photocatalytic thin layers' performance was studied using the methylene blue aqueous solution degradation under UV irradiation. That illustrates the TiO_2 thin layer photocatalytic performance as a function of irradiation time. The MB solution was stable in the case of no thin films and irradiated with UV light. In figure 13 we plot on (C_0/C) against time; when time increases, the concentration of MB decreases, and the photocatalytic activity of the thin layer rises also. While comparing the slope result of the two layers, the SiO2/TiO2 thin film rate was better than that of the TiO_2 layer, which means high photocatalytic activity.

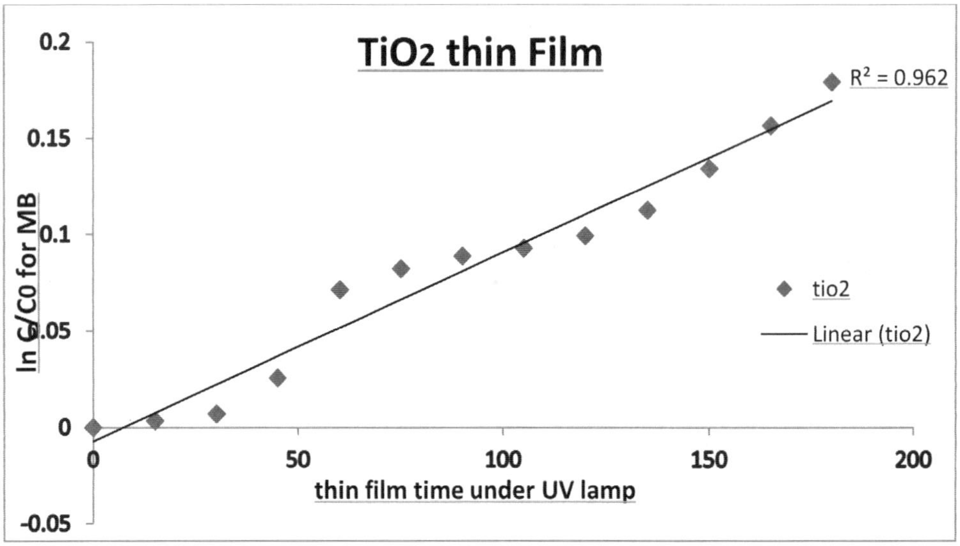

Figure 12: Photocatalytic activity of TiO_2 films

- Surface wettability of thin films (super-hydrophilicity)

The prepared thin films were irradiated under (36W), handmade UV sources champers lamp for different time intervals. After each time interval, the substrate was transferred to a contact angle system. Contact angle measurements were made under air setting using the sessile drop method. Contact angle measurements were taken for each prepared sample in different 3 points. Table 3 shows WCA of all TiO_2 samples before and after UV irradiation every 30 min for 120 min. WCA Measurements were performed with two weeks aged samples, and the results of super hydrophilic state were similar. At dark environments, a TiO_2 film tends to reestablish the hydrophobicity quickly. After UV irradiation, all samples presented super hydrophilic behavior (angle < 1°), except uncoated glass, showing how coating applications efficiently gave superhydrophilic property to the glass. Superhydrophilic surface is promising for photovoltaic module due to its self-cleaning effect.

Table 3: degradation of contact angle for TiO2 thin film under UV lamp with time

Exp. No	Time Under UV	Contact Angle
1	0	48
2	30	35
3	60	24
4	90	14
5	120	0

- *Contact angle for glass without any coating*

With estimating the property of the Photo-induced hydrophilic for the thin layer by adjusting water contact angle before and after UV irradiation, and the angle of glass substrate without coating was 33 as shown in figure 14.

Figure 13: The water contact angle of the glass substrate without coating

- *Contact angle for TiO2 thin film under UV lamp*

The TiO_2 layer one thin films, the contact angle of water for TiO_2 layer obtained was 48, 35, 24, and 116 and after UV irradiation it has been decreased below 0 within 120 min of UV irradiation as shown in table 3. Figure 15 Images of water droplets on the sample (SiO_2/TiO_2): a- without UV irradiation(48), b- contact angle 35º) after 30 min. of UV irradiation, c- contact angle (24º) after 60 min of UV irradiation and d- contact angle (0º)after 120 min. of UV irradiation [16].

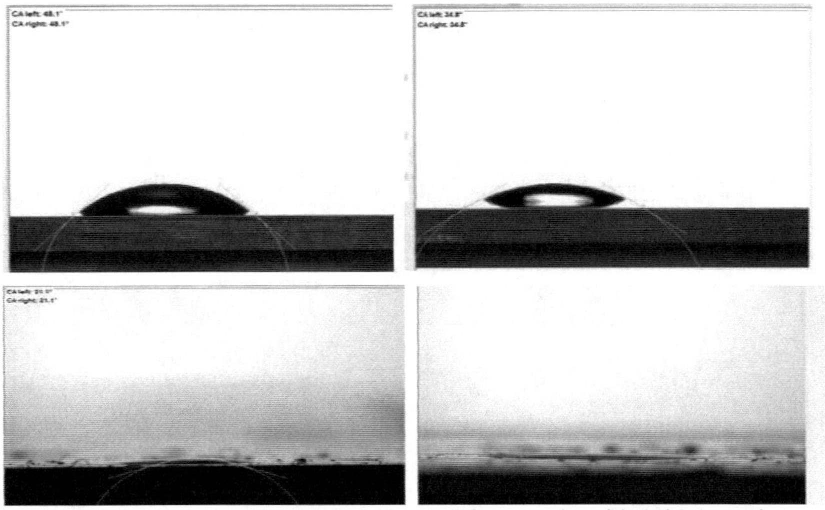

Figure 14: Images of water droplets on the sample with (TiO_2) coating.

4.4. Comparison of solar cells power without coating, With the TiO2 coating

After all the preparations were completed, the testing was ready to begin. Tested data is recording for tow the solarpanel in a natural atmosphere every day at the same time. Each panel was photographed before and after the test time. For photographing the accumulated dust, a 40x zooming lens was used on a mobile phone camera. Due to the instability of a hand-held camera, the photos may not have always been precisely from the same spot, but they were sufficiently close for determining the effect of the cleaning. Figure 16 shows the photographs during the test process of the tested panel

Figure 15: PV with glass coated with TiO2 thin film

Outdoor testing shows that the system daily energy output is nearly identical over 28 days. There was a significant difference between the cumulative power collected by the uncoated and coated PV modules. This means that the coating reduces the dust accumulation as difference between uncoated PV modules surface and otherone coated with TiO2 shown in figure 17. The graph is drawn by considering the readings taken daily. The time is taken in x-axis and power generated by the solar panel s is taken in y-axis. The red curve shown in the graph is the power generated by the tested panel with Tio2 coating, the blue one is belong to power generated by the panel without coating.

Figure 16: difference between uncoated PV modules surface and otherone coated with TiO2

The test of self-cleaning coatings occur by set up 2 PV in the month of July start from the Date 1 to 28 in year 2019, by measuring the current ant voltage for each PV every day the optimum power is obtained from 12pm to 2pm. Due to improved weather conditions. The generation of power is increased in reading of coating TiO2 all the days as compared to without coating as shown in figure 18 which mean it's had ability to self-clean dust with calculation is between 2:2.5%.

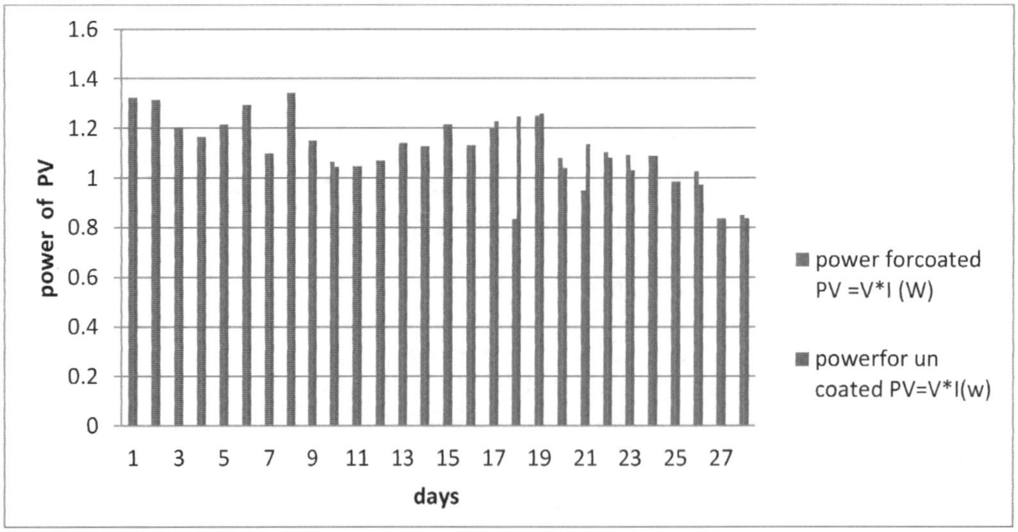

Figure17: Comparison of solar cells power without coating, With the TiO2 coating

5. Conclusion

The main goal of the presented research was to produce nanostructured TiO2 by sol-gel method using a titanium isopropoxide precursor in simplified conditions. According to the exposed results and discussions, we could draw the following conclusions: In the present study, TiO2 films the glass substrate which was pre-coated was prepared by sol-gel dip-coating process using different solutions ratios. A comparison of differences in film morphologies, photocatalytic activities, hydrophilicity, and optical properties with nanoparticle size between 15-72 nm was conducted. After the heating treatment for the TiO2 films at 500°C, it becomes transparent, uniform in color and thickness, and they were in Anatase phase, orientated across the (101) plane as EDX results showed .Scanning Electron Microscopic data showed that the thin layer films made an excellent distribution with TiO2 particles, and no agglomeration may affect the properties of the coating and weaken it. The methylene blue photocatalytic degradation was enhanced due to the presence of an intermediate TiO2 layer Also notice that increase in the effect of the self-cleaning and super hydrophilicity. It was recognized that the tendency of electrical losses of PV modules and dust accumulation mechanism depends on the environment of the test site and type of dust, and according to results of outdoor exposure tests in Cairo, Egypt. The PV with double layer TiO2 was found to save output loss power by 2.5% from the solar panel.

Nomenclature:

θ	Water contact angle.
PV	Photovoltaic.
WCA	water Contact angle.
SEM	Scanning electron microscopy.
i-prop	Isopropanol.
TiO_2	Titanium dioxide or Titania.
XRD	X-ray diffraction.
V	Voltage.
P	power.
DLS	Dynamic light scattering.

6. Reference:

[1] Boris Kobrin. 2018. Self-cleaning Technologies for Solar Panels - N-Tech. N-tech Res. Publ. May (2018), pp15.18

[2] A Pan · 2019 · Cited by 35 — The super-hydrophobic coating with micro-nano structures (glass sample D) achieves the highest spectral transmittance and the lowest PV power efficiency .

[3] A.O. Mohamed and A. Hasan. 2012. Effect of dust accumulation on performance of photovoltaic solar modules in Sahara environment. J. Basic Appl. Sci. Res. 2, 11 (2012),pp 11030–11036. .

[4] Annalisa Andaloro and Francesco Frontini. 2017. Application of Titanium Dioxide Self-Cleaning Coatings on Photovoltaic Modules for Soiling Related Losses Reduction. Am. J. Eng. Appl. Sci. 10, 1 (2017), 78–93. DOI:https://doi.org/10.3844/ajeassp.2017.pp 78.93.

[5] Tarun; Dodhia, Devendrasinh; Darbar, Saravana; Sara, and Karthikeyan. A K. 2016. Experimental Study of Air Pollutants Deposition Impact on the Performance of Solar Photovoltaic Panel. Retrieved from http://sajms.com/volume-3-issue-2/.pp1233

[6] Quesada-González, M. *et al.* (2017)Interstitial boron-doped anatase TiO_2 thin-films on optical fibres: atmospheric pressure-plasma enhanced chemical vapour deposition as the key for functional oxide coatings on temperature-sensitive substrates. *J. Mater. Chem. A* **5**, 10836–10842.

[7] K Abushgair · 2021 · Cited by 11 studied the development in different self-cleaning methods for PV panels. They mainly focused on super-hydrophilic and superhydrophobic pp[10] ...

[8] A Shokuhfar, M Alzamani, E Eghdam, M Karimi, and S Mastali. 2012 . SiO_2-TiO_2 Nanostructure Films on Windshields Prepared by Sol-Gel Dip-Coating Technique for Self-Cleaning and Photocatalytic Applications. pp, 16–21. DOI:https://doi.org/10.5923.

[9] Ibrahim Mustapha Alibe, Khamirul Amin Matori, Hj Abdul, Aziz Sidek, Yazid Yaakob, Umer Rashid Id, Ali Mustapha Alibe, Mohd Hafiz, and Mohd Zaid (2016). Effects of Calcination Holding Time on Properties of Wide Band Gap Willemite Semiconductor Nanoparticles.

[10] Francesco Frontini Dye-sensitized Tio. 2017. Visible-Light-Driven, Dye-Sensitized TiO 2 Photo-Catalyst for Self-Cleaning Cotton Fabrics.pp 017). DOI:https://doi.org/10.3390/coatings7110192.

[11] T Weber · Cited by 6 — Thomas Weber, PVMRW 2015, 24.02.2015, Denver ... PI Photovoltaik-Institut Berlin AG, Wrangelstraße 100, 10997 Berlin. Impact ... Cleaning of PV Modules.

[12] E Elshazly · 2021 — Tilt angles 15° and 30° show to be optimal for the installation of the PV solar system, as they produce the highest amount of output power.

[13] Abdallah, R.; Juaidi, A.; Abdel-Fattah, S.; Manzano-Agugliaro, F. **2020**Estimating the Optimum Tilt Angles for South-Facing Surfaces in Palestine. *Energies* **2020**, *Page 13*, 623. https://doi.org/10.3390/en13030623

[14] Annalisa Andaloro and Francesco Frontini. 2017. Application of Titanium Dioxide Self-Cleaning Coatings on Photovoltaic Modules for Soiling Related Losses Reduction. Am. J. Eng. Appl. Sci. 10, 1 (2017), pp 78–93. DOI:https://doi.org/10.3844/ajeassp.2017.78.93.

[15] W Zhao · 2021 — Dust deposition on solar photovoltaic (PV) cell surface will significantly decrease the PV power efficiency, as the transmittance of the solar cells would pp16. —.

[16] Boris Kobrin. 2018. Self-cleaning Technologies for Solar Panels - N-Tech. N-tech Res. Publ. May (2018), pp15.

Smart, Self-cleaning coating Based Ultra-Hydrophobic Copolymer Nano-composites

S. Ibrahim[1,2]* **and A. Labeeb**[3,4]

[1] Packaging Materials Department, National Research Centre, Elbuhoth Street 33, Dokki, Cairo, 12622, Egypt
[2] Nanomaterials Investigation Laboratory, central Laboratories Network, National Research Centre, Elbuhoth Street 33, Dokki, Cairo, 12622, Egypt
[3] Microwaves Physics and Dielectrics Department, National Research Centre, Elbuhoth Street 33, Dokki, Cairo, 12622, Egypt
[4] Liquid crystal Institute, chemical Physics Interdisciplinary Program, Kent State University, Kent, Ohio, 44242

* *sa.ibrahim@nrc.sci.eg*

Abstract Ultra-hydrophobic (UH) surfaces have gained a great attention due to the high potential for tremendous applications. Simplified preparation technique was used with low-cost value to produce UH nano-composites coating. Copolymer nano-composites have numerous advantages as a flexible coating through the huge covering volume/area. The molecular weight of prepared copolymer was defined with gel permeation chromatography. UH coating films were investigated with differential scanning calorimetry, dynamic laser scattering, transmission electron microscopy, atomic force microscope, and dielectrically study. The contact angle of UH coating film and surface area were measured. Copolymer coating with 9% loaded ratio of nanocomposite given 132° contact angel with excellent nano particle size, and surface area. The zeta potential results indicated to perfect stability of ecofriendly emulsified coating. The main aim of research was verified the ability of prepared copolymer nano-composites as a UH self-cleaning coating film on different applications especially solar panel and packaging substrates.

Keywords: ultrahydrophobic; smart coating; self-cleaning; nanocomposites

1. Introduction

An important character such as wettability for hard surfaces is mostly determined by different factors, i.e. the chemical composition and surface topography [1]. Ultrahydrophobic surfaces, having a larger contact angle (CA) than 120° which have fascinated attention as a result of their useful uses [2–4].

Ultrahydrophobic is based on minimizing the surface wetting processes. The surface treatment increases the surfaces self-cleaning ability – allowing rain or dew to wash away the dust and dirt that could accumulate on applied coating film. Because of its ability to make water bead up and roll off, the surface is classified as superhydrophobic [5]. Self-cleaning surfaces, water-proof fabrics, non-stick containers and microfluidics are some of the available applications [5]. However, advances in nanotechnology and promising experimental results have allowed the area to rapidly develop in recent years. There is an extensive range of ultrahydrophobic surface preparation methods already available, but these are expensive and even require specialized equipment. The current challenge is to find the best approach to produce these useful products in a cost-effective way.

Ultrahydrophobic surfaces can be succeeded usually by chemical modification and/or physical engineering treatment of surface topography [6]. A hierarchical formation in nanoscale / micro dual forms, alongside using chemical modification for lowering surface energy materials are used as hydrophobic treatments [7]. Two models, the Wenzel model and the Cassie-Baxter model, are most frequently used to interpretation of experimental results[8]. The Wenzel model defines wetting behavior which a liquid droplet on a rough surface is in close contact with surface brusqueness. Whereas, the Cassie model defines states where the droplet settled on a solid–air composite surface [9]. The discrepancy between the two models is about the contact angle hysteresis. Therefore, such a transition is dependent on the contact mode and the deposition of the liquid drops on the rough surfaces, such as, dragging on rough surface or being impacted [10,11]. Thus, many studies were carried out in the droplet impacting action on superhydrophobic surfaces by several groups. Lei *et. al.* established the impacting battle between aqueous droplets and the solid surface which are depend on the geometric factors of the surface [11]. Pontus *et al.* has improved the criterion of the transition from the Cassie model to the Wenzel model. Theses transition was defined by geometric parameters (length, width and height of patterns, pitch between patterns), and impact velocity. The former researches are aimed on the impact of geometric factors of surfaces on wettability mode [12]. Low fractions surface composition would not have an effect on the surface wettability, but it would affect the surface wetting transitions. Moreover, the contact angle hysteresis could change due to the droplet–surface interaction manipulations. The dynamic water wetting behavior of a forced droplet on approachable macromolecules brushes anodized metal with low concentration of initiator was investigated [13]. The changes of CAs are superficial and providing the surface dual-faced wetting physical characteristics.

On the other hand, Sol-gel polymerization technique offers the possibility for such optimization, but many parameters have to be considered. The challenges are certainly the optimization of the surface properties regarding charge [14]. colloidal particles are particularly attractive for large-scale fabrication of ultrahydrophobic and switchable surfaces at low cost [4]. In fact, colloidal particles can simultaneously provide two properties required for the design of ultrahydrophobic materials: roughness and intrinsic hydrophobicity [15]. The successful design of ultrahydrophobic surfaces requires, however, accurate control of these two parameters [16]. For example, it was found that ultrahydrophobicity cannot be achieved by using layers of densely packed, smooth hydrophobic particles, and fractal structures must be used. Thus, it will be very important to adapt the best physical characterization methods for the new polymer particles and to include application tests [17].

The success of this research is strongly relied on the potential to optimize easy and cheap self-cleaning coating through well controlling in particle size, particle distribution, surface charge, and stability of the particles, and the contact angle as well as the thermal properties of the coating films. The topography and surface morphology were investigated as roughness parameters. Highly effective auto-cleaning coated substrates are examined as ultrahydrophobic surface with most advanced techniques to verify its characters to applied as multi-coating purpose. Moreover, the dielectric properties of the coating films are investigated to determine the electrical changes based on the chemical compositions.

The handled study is dealing with preparation of nano-composites copolymer coating through ecofriendly easy and cheap method. The silica was applied with different ratios to optimize the hydrophobic character of applied coating. It is not only cutting-edge research, but also of great industrial interest. One step nano-composites preparation was successfully created ultra-hydrophobic coating that is usable for different outdoor applications.

2. Materials and methods

2.1. *Materials*

1,1,1,3,3,3-hexafluoroisopropyl methacrylate (HF) and 2-hydroxylethylmethacrylate (HMA), Sigma, were refined before using. Hexadecane (HD), potassium peroxide sulfate, sodium bicarbonate, ammonium chloride, ethanol, hydroquinone (HQ), sodium silicate, sodium dodecyl sulfate (SDS) and chlorotrimethylsilane (CTMS) were bought from Sigma.

2.2. *Methods*

2.2.1 Synthesis of hydrophobic silicon precursor nanoparticles.

The hydrolysis reaction was performed for sodium silicate (26.1 g) in preparation of Silica (SiO_2) nanoparticles was dispersed in 450 ml of Millipore water and adding of 150 ml of 0.36 M NH_4Cl solution. The solution was stirred for 3 hr. at 80°C. The product was separated and rinsed with Millipore water and ethanol to remove Cl^-. Consequently, silicon dioxide nanoparticles were modified by hydrophobic CTMS. The e product plus 0.075 moles of CTMS were dispersed in 300 mL of ethanol and heated to 80°C for 3 hr. under stirring. After the reaction completion, the product was dried under vacuum condition at for 24 h at ambient temperature to gain CTMS/SiO_2.

2.2.2 Preparation of polymer hydrophobic coating.

A round bottom flask with three-necks, a monomer (HF/HMA with equal mole ratio) and 0.049 moles of HD were added to 0.043 moles SDS in 109.47 g of Millipore water, within concentration of forming lyotropic liquid crystalline phases after CMC critical point. Degassing process was carried out for the mixture (nitrogen/vacuum cycles, magnetic stirring at 300 rpm for 30 min under nitrogen) followed by further stirring at 800 rpm for 60 min. Then, a miniemulsion was formed by applied ultrasonic energy for 14 min (80% amplitude, Q-Sonic, 750W). A stream of N_2 was pressed over the emulsion within cooling. Various quantities of modified SiO2/CTMS were distributed in 30 ml of Millipore water under sonication for 4.5 min at 30% amplitude followed by addition to the reaction flask. an aqueous solution 540 mg of KPS in 10.8 g of Millipore water was added to reaction at 72◦C with stirring at 800 rpm for 4 hr and let the flask to reach 25°C. The reaction flask was instead in a cooling bath to ambient temperature. The emulsion was dropped in 450 mL of methanol to precipitate latex. HF/HMA was dried in a air oven over night.

2.2.3. Dynamic leaser scattering (DLS)

The particle size distribution and over all charges (zeta potential) of polymer nano-composites and silica particles were analyzed using NICOMP 380 ZLS, PSS, Santa Barbara, CA, USA).

2.2.4. Surface area investigation

N_2 ads./des. were measured by NovaTouchLX4 Quantachrome, USA to calculate surface areas of copolymer nano-composites.

2.2.5. Thermal behavior

DSC131 evo, SETARAM Inc., France was applied to measure the differential thermal behavior for nanocomposites samples. The test was computerized with heating range from 25°C to 400°C with a rising temperature rate 10 °C / min. All measured samples were processed using (CALISTO Data processing software v.149).

2.2.6. contact angle

Theata Optical Tensiometer, Dataphysics, Model OAC 13Ec, GmbH, Germany was used to determine CA. The water drop image was taken by the camera and analyzed by Dataphysics software find the CAs.

2.2.7. Transmission electron microscopy (TEM) investigation.

A JEOL 120/JEOL 200 TEM (carl Zeiss NTS) electron microscope (S160-3 Plano GmbH) was used to investigate the polymer and nano-composite materials. The size detection of the polymer composites particles was scaled by Scandium software.

2.2.8. Gel permeation chromatography

GPC was used to determine Number average molecular weights (Mn), weight average molecular weights (Mw) and polydispersity index (PDI). Mn and Mw were computed through the Polymer Lab caliber® software.

2.2.9. Atomic force microscope (AFM)

AFM imaging was presented with a multimode AFM Nanoscope IV in the tapping mode. Topography and phase scans of tested surfaces was obtained.

2.2.10. Broadband dielectric spectroscopy (BDS).

The BDS measurements were carried out by broadband dielectric network analyzer equipped system. This system is mainly consisted of the Solartron 1260 impedance/gain-phase network analyzer (Schlumberger Ltd. Co., UK) using frequency range 0.1Hz-10MHz. The low and high frequencies were connected by a relay connector to gathering all frequencies dependent data together. Monitoring the signals was performed by the oscilloscope for both in and out signals from the measuring cell test fixture. Good electromagnetic shielding was fulfilled on the sample holder which is maintained in and implemented to diminish noise problems at low frequencies specifically. The data acquisition throughout the interface GPIB cable and IEE488 was automatically obtained using software Lab VIEW 7.1. Additionally, the calibration was performed before each measurement, to eliminate the effect of stray capacitance. The error percentage in the permittivity ε′ and dielectric loss ε″ is about 5%. To provide a dry environment with low humidity.

Samples were placed between two stainless steel plates as a disc shaped for dielectric properties measurements. Then, the capacitance and dielectric loss tangent in accordance with the frequency range 0.1Hz- 1MHz were obtained directly from the network analyzer. Permittivity was calculated using the formula;

$$\varepsilon' = Cd/\varepsilon_o A \quad \ldots\ldots\ldots\ldots\ldots\ (1),$$

where, d is the thickness of the sample, c is the capacitance and A the area of the sample, ε' is the permittivity of the sample. The ac-conductivity σ_{ac} of these samples was evaluated using the relation;

$$\sigma_{ac} = 2\pi f \varepsilon_o \varepsilon' \tan\delta \quad \ldots\ldots\ (2),$$

where, f is the frequency of the applied field and $tan\delta$ the loss factor. The measured values of ε' and $tan\delta$ were used for the calculation of the ac-conductivity.

3. RESULTS AND DISCUSSION

3.1. *Surface contact Measurements*

The wettability of treated surface with prepared PS, HF/HMA and nano-composites were investigated in related to contact angle measurement as shown in Fig. 1. The HF/HMA copolymer was exhibited high contact angle in comparison to polystyrene as a reference coating [5]. The static contact angles were shifted from 90° for PS to 115° according to halogenated hexa-substituted polymer with HMA copolymer.

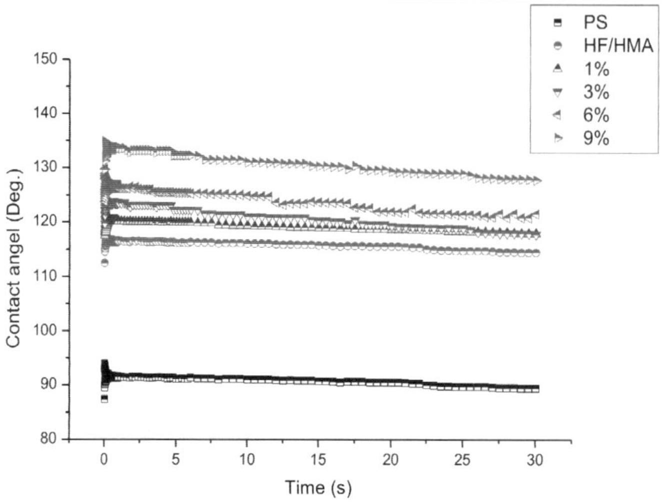

Fig. 1: contact angles of PS, HF/HMA and nano-composites copolymers with different ratios.

HF/HMA copolymer modified surfaces as waterproof surfaces with different ratios of SiO_2/CTMS exhibited hydrophobicity higher than unmodified one. The contact angle was gradually increased with of SiO_2/CTMS from 1% to 9%. All copolymer nanocomposites were presented grater contact angle than 120° which considered as ultra-hydrophobic surfaces character. The first two loaded ratios 1% and 3% are demonstrated final static contact angle at ≈ 122°. This may be attributed to low difference loaded ratios with complete encapsulation in copolymer. Additionally, the copolymer nano-composites with 6% and 9% are given 126.9° and 132.6°, respectively. Overall, the hydrophobicity of loaded copolymer with SiO_2/CTMS increase the waterproof surface character with loaded ratios as self-cleaning coating (high contact angles).

3.2. *Surface Area*

The surface area of PS, HF/HMA and nano-composites copolymers with different ratios in adsorption isotherm branch are tabulated in Table 1. The prepared copolymer and its nano-composites with expected nanoparticles behaviors were investigated to confirm the area/ volume relationship.

The surface area was decreased with increased of SiO_2/CTMS loaded ratios as shown in Table 1. This is due to the packaging effect of loaded nanoparticles of SiO_2/CTMS which encouragement the aggregation effect [18]. Moreover, the correlation coefficient is nearly ≈1 as first order relationship. Although, the decreasing in surface area through loading percentage, the results are still high values in agreement with nano-composites structures. Overall, the surface areas measurements were presented high degree of agreement with particle size investigation as shown in Table 1 and 2.

Table 1: Surface area of PS, HF/HMA copolymer and nano-composites copolymers with different ratios in adsorption isotherm branch.

Sample	Isotherm branch	Slope	Intercept	correlation coeff. (R)	Surface Area (m^2/g)
PS	Absorption	18.196	0.312	0.9998	188.2
HF/HMA		8.873	0.119	0.9996	387.3
1%		9.023	0.233	0.9912	396.2
3%		11.284	0.194	0.9998	303.4
6%		12.747	0.179	0.9999	269.4
9%		18.035	0.169	0.9999	191.3

3.3. *Thermal investigation*

DSC was used as an effective and high precision heat gain instrument to investigate the thermal behaviors of HF/HMA copolymer and nano-composites loaded with different ratios of SiO_2/CTMS. Fig. 2 exhibited the thermograms in range from 0°C up to 140°C of the prepared copolymers in presence and absence of nanosilica.

HF and HMA were presented smooth thermographs with two characteristic peaks at 56°C and 80°c, respectively [19,20]. According to DSC thermograms measurements, the glass transition temperature (T_g) of HF in range 56°C to 59°C, where HMA gives T_g from 81°c toward 70°c. copolymer thermograph the T_g are highly compatible with literature survey results. In addition, the T_g peaks of HF and HMA were closed together with increase the loaded ratio of SiO_2/CTMS. The maximum two peak differences were 25, 21, 17, 13and 11°c for HF/HMA, 1%, 3%, 6% and 9%, respectively.

The T_g shifts can be handled according to two main assumptions. First, the ability of HMA and HF to form hydrogen-bonds between the hydroxyl groups in a chain terminal. This is suppressed the chain mobility in the copolymer matrix lower the T_g, although the cleavage of the OH···OH encourages the association of the C=O···HO type of hydrogen-bond [21]. This phenomenon due to the main chain mobility affected through the diversity in hydrogen-bonds assembly at T_g by shifting in corresponding to nanocomposite loading ratio. The second assumption, the addition of SiO_2/CTMS was enhanced the phase interaction and improved the compatibility of polymers/inorganic nano-composites phases which reflected on the reduction of T_g temperatures differences between HF and HMA.

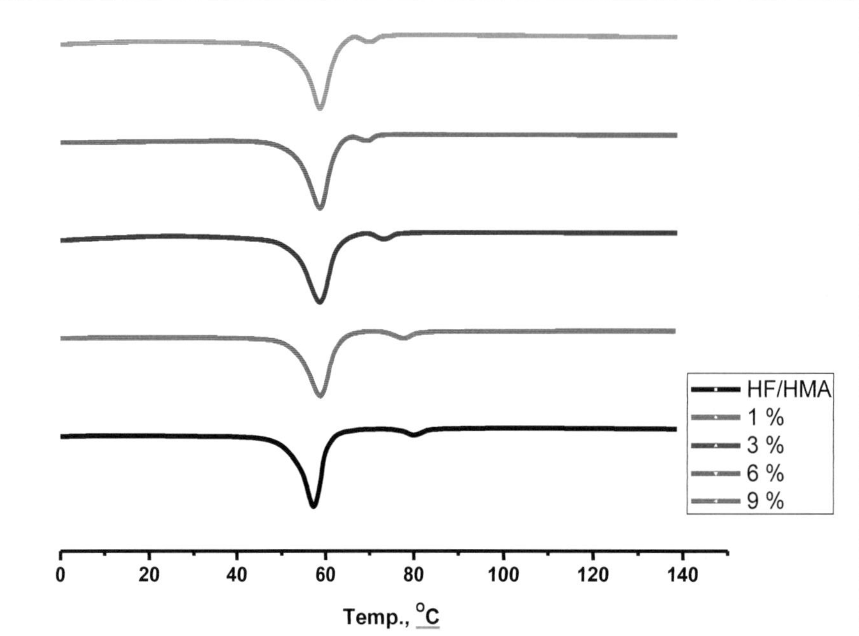

Fig. 2. DSC thermogram of HF/HMA and copolymer nanocomposites with different ratios.

The absence of crystalline peak in DSC thermographs can be attributed to the amorphous nature of copolymer nano-composites and/or the slow cooling in preparation process of coating film [22]. Wherein, the quenching cooling process are mostly encouragement to ordering of polymer chain to exhibit crystalline form with detectable peak in DSC thermograph [23].

3.4. *Particle Size and Zeta Potential*

3.4.1. Mean Particle size

The particle size Gaussian distribution as function of intensity-weight of PS, HF/HMA (HF) and nano-composites copolymers with diverse SiO_2/CTMS loaded ratios were shown in Fig. 3. Uni-modal homogeneous distributions of particle size were exhibited for all polymer, copolymer and nano-composites. During, the loading ratios from 1% to 9% demonstrated narrow distribution with 1% and gradually increase in 3%, 6% and 9%.

The mean diameters of these dispersed copolymer nano-composites were tabulated in Table 2. The particle size was decreased in case of 1% lower than related HF/HMA copolymer. Additionally, the mean diameter was increase from SiO_2/CTMS loading ratio 3% through 6% and 9%. These results are in agreement with surface area measurement based on volume/area ratio [24].

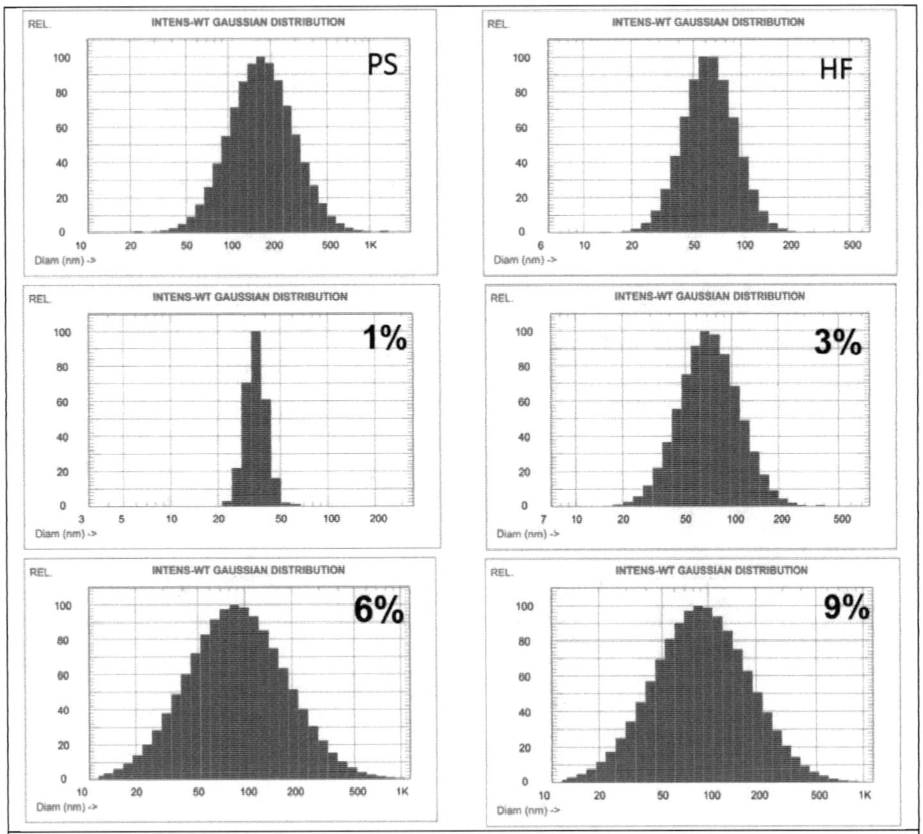

Fig. 3: Shows the particle size distribution of PS, HF/HMA copolymer in presence of silicon nanoparticles with different concentration

Table 2. The polydispersities and hydrodynamic diameters of PS, HF/HMA and polymer nano-composites.

Sample	Average diameter (nm)	Stnd. Dev. (nm)	PDI	Chi-Square
PS	196.6	104	0.28	34.80
HF/HMA	67.2	25.8	0.15	13.58
1%	35.1	5.5	0.02	2.49
3%	78.5	34.4	0.19	41.41
6%	114.4	84.3	0.54	57.93
9%	115.1	82.0	0.51	3.45

The polydispersity of all samples is located at 0.02 to 0.54 which revealed to excellent narrow size distribution [25,26]. The diversity in particle size is minimum in sample HF/HMA with SiO_2/CTMS (1%) as perfect encapsulated particles. The degree of diversity was slightly increase in case of HF/HMA with SiO_2/CTMS (3%) and nearly duplicated with 6% and 9%.

In conclusion, the hydrodynamic particle size distributions presented a nice concept of nano-composites dispersed in an aqueous medium as homogenized ultra-hydrophobic coating.

3.4.2. Zeta measurement

The potential of zeta measurements as an overall charged were used as indicator to the stability of dispersed copolymer nano-composites solutions [27]. All samples were prepared with constant concentration under sonication condition. The zeta potential was measured as an average value to multi-measurements under low voltage. The average phase shift and mobility of charged particles in dispersed solution were pointed to a low packing effect by charge/size mobility particles [28]. The refractive index applied current, and viscosity of aqueous dispersed medium are constant for all samples.

The average zeta potential measurement of PS, HF/HMA copolymer and nano-composites system in presence of HF/HMA with SiO_2/CTMS (1%, 3%, 6% and 9%) are collected in Table 3. The zeta potential of copolymer is 38 mV which indicated to moderated emulsion stability. Moreover, the zeta potential of nano-composites was in range between 42 and 48 mV over the loading ratio from 1% to 9% SiO_2/CTMS. According to previous studies, emulsion solution with average zeta potential (±40 - ±60) is good stable homogenized solution [29]. Therefore, copolymer nano-composites emulsions are an excellent stable solution.

Table 3. Average zeta potential and relative indices of PS, HF/HMA and copolymer Nano-composites.

Sample	cell current, mA	Avg. Phase Shift, rad/sec.	Avg. Mobility, M. U.	Avg. Zeta Potential, mV
PS	0.23	10.21	0.41	35.46
HF/HMA	0.27	15.79	0.60	38.03
1%	0.29	4.63	0.18	42.42
3%	0.31	26.90	1.04	43.93
6%	0.36	34.70	1.28	47.23
9%	0.35	38.26	1.39	48.61

3.5. *Gel-Permeation Chromatography*

The detectable Mns bands of copolymer nano-composites are illustrated in Fig. 4 with nice bell shape distribution. The molecular weight distribution of copolymers in presence and absence of modified nanosilica with different ratios were exhibited homogenous polymer chain fractionation through separating column.

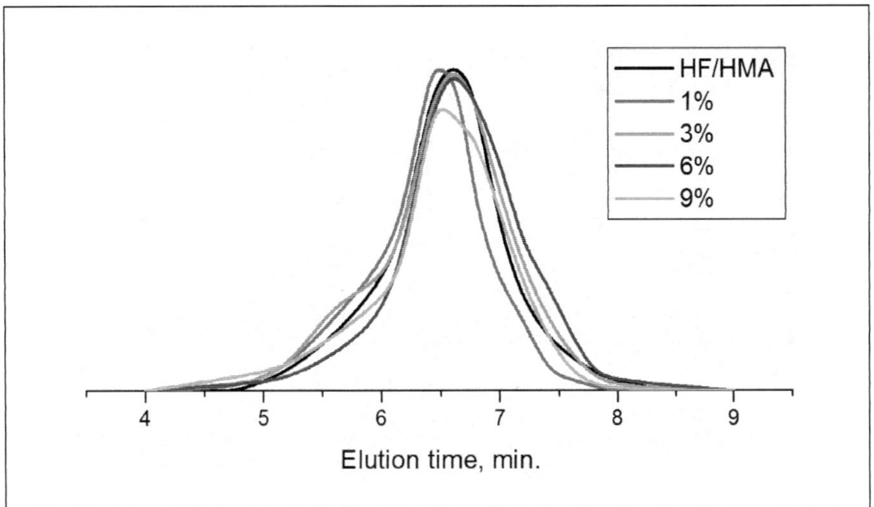

Fig. 4. GPC chromatographs for the polymerization of HF/HMA/SiO_2 nanoparticles

The molar mass distribution for copolymer nano-composites (HF/HMA/SiO_2, 1%, 3% and 6%) show smooth bands while HF/HMA/SiO_2 9% presence a little shoulder in the top of band. Narrow molecular weight distributions are indicated to well-defined polymerization process through nano-encapsulation for modified silica. The conversion and molar mass ratios are tabulated in Table 4. The molecular weights are varied from 44800 to 46300 g/mol with low polydispersity indices. The PDIs of prepared polymer exhibited low value (1.3) of HF/HMA copolymer.

The polydispersities of prepared copolymer is less than 1.3 with quite high conversion of HF/HMA. Presence of silica nanoparticles have a slight deficient (3% - 5%) in degree of polymerization with a little higher value of PDI than copolymer.

Table 4. Molar mass ratios of HF/HMA and copolymer nano-composites.

Sample	conversion, %	Mn, g/mol	Mw, g/mol	PDI
HF/HMA	89	46300	56900	1.23
1%	87	45100	57700	1.28
3%	85	46000	60700	1.32
6%	86	46200	58600	1.27
9%	84	44800	55800	1.24

3.6. *Atomic Force Microscope*

Inorganic/organic nano-composites for distinguish phases are detected with AFM in microphase structure of polymer SiO_2 NPs. copolymers can be used as polymer matrices stabilizer for inorganic nanomaterials as nano-composites [30]. Phase imaging of coated surface and 3D building image were shown in Fig. 5. The contact mode was direct attach of cantilever with coated film depending on difference in hardness between SiO_2 as hard material and copolymer as semi-soft segments. The contrast in phase image is related to the actual fact in compositional glassy and/or semi-rubber like state. The brighter zones in the phase images are corresponded to higher areas or hard form, while deep color is indicated to bottom areas or low stiffness copolymer. The created 3D AFM topographic image was reflecting the roughness of the coating film which play a vital role in water/UH film phase interaction.

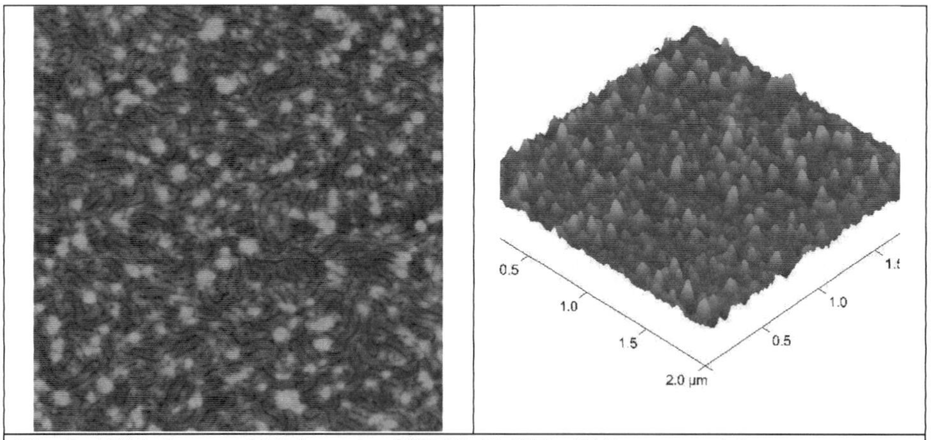

Fig. 5. AFM phase and 3D images of HF/HMA/SiO_2 (9%) nanoparticles

3.7. TEM Investigations

The transmission electron micrographs of HF/HMA copolymer in presence and absence of SiO_2-CTMS were shown in Fig. 6. The copolymer micrographs (A&B) are describing the morphological form of polymerized di-monomers through a controlled miniemulsion technique.

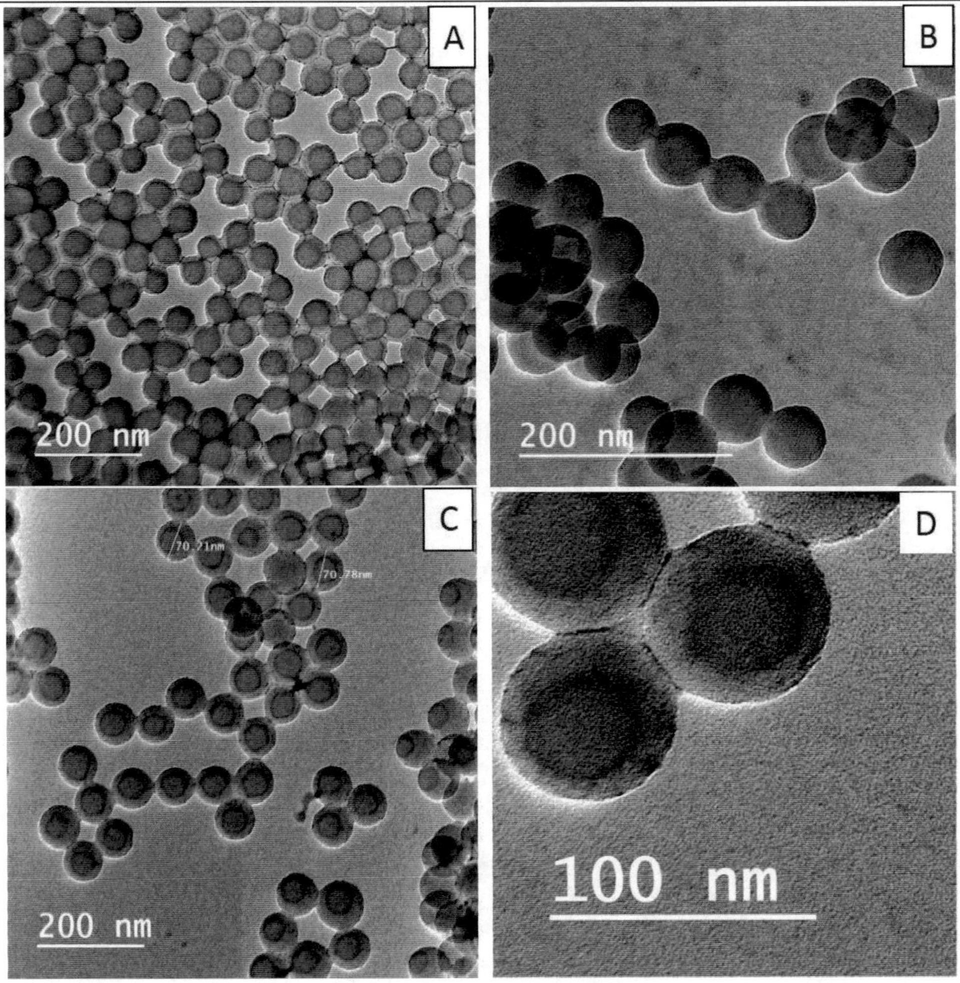

Fig. 6: TEM micrographs of HF/HMA copolymer (A) & (B) polymer nanoparticles and (C) & (D) copolymer nanocomposite (9%)

The perfect copolymer images expressed excellent smooth and round particles with different magnifications. Additionally, the particle size appeared in semi one size in range from 60-80 nm. The micrographs of copolymer nano-composites with modified silicon nanoparticles exhibited perfect encapsulated particles with zoom out and zoom in two magnifications. The images are clearly presented the SiO_2 core and copolymer outer shell. Smooth particles with round ball forms were detected in

micrographs (C&D). The particle size was pointed to a uni-size in range 70 nm as presence with scale bar measurements.
In comparison of size by TEM and DLS measurements of copolymer nano-composites (9%) indicated to 78 nm and 70 nm, respectively. Excellent agreement was verified with only 11 % division between the two sizes by 8 nm. In brief, the results of TEM, DLS and surface area measurements have confirmed each other and indicated to perfect nano-composites behaviors in size, morphology, and surface area.

3.8. *Dielectric Properties*

The dielectric measurements were fulfilled for both polyhydroxyethylmethacrylate (HMA) and polyhexafluoroisopropyl methacrylate (HF). The coated silica within polymer was improved the hydrophobicity of the polymers. The blend series were consisted of four silica ratios 1%, 3%, 6% and 9% alongside the pure HF/HMA.

The dielectric parameters as real and imaginary parts of complex dielectric permittivity are namely real permittivity $\varepsilon'$and dielectric loss ε''. The equation roles the relation between ε' and ε'' is[31]:

$\varepsilon^* = \varepsilon' - i\varepsilon''$...................... (3)

Both real permittivity $\varepsilon'$and dielectric loss ε'' were gotten from the following equations

$\varepsilon' = Cd/\varepsilon_0 A$, ... (4)
$\varepsilon'' = \varepsilon' \tan\delta$ .... (5)
where, *c* is a capacitance of the material as a dielectric between two parallel plates which have active area *A* and separated with *d* thickness.

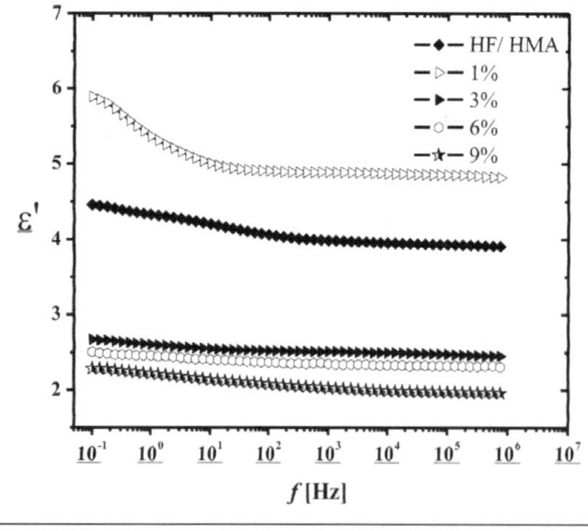

Fig. 7: The Permittivity dielectric permittivity ε' of HF/HMA and the four-silicate series were measured against the frequency from 0.1 Hz to 1MHz.

The permittivity $\varepsilon_0$of air or vacuum is considered which its value (8.85×10^{-12} F/m). Moreover, both $\varepsilon'$and ε'' are unitless physical quantities because they are representing ratios between two capacitances with and without dielectric materials $\varepsilon' = C/C_o$, where c_o is a capacitance without dielectric material. The imaginary part ε'' of equation (1) represents the loss in the energy or the amplitude of an applied signal as a dissipated factor $(\tan\delta)$. The real permittivity ε' and loss tangent $\tan\delta$ were attained from high-resolution BDS [32]. The computations was carried out by LabVIEW 7.1. The relative errors for both ε' and ε'' is up to 1% and 3% respectively. The measurements were done in the room temperature.

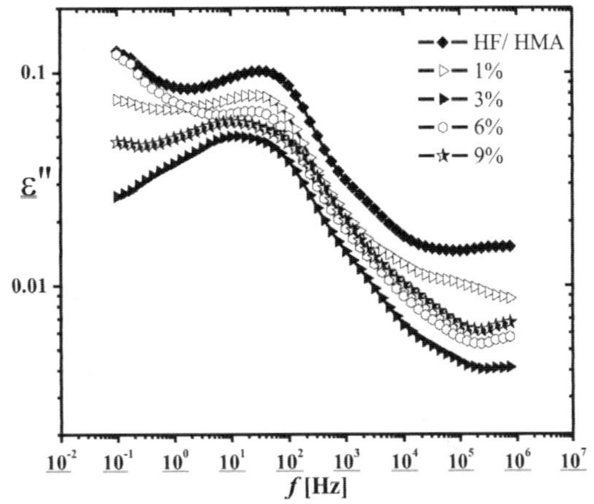

Fig. 8: The dielectric loss of HF/HMA blended with different ratios of silica is plotted against frequency.

In the figure 7, the dielectric permittivity ε' of HF/HMA and the four-silicate series were measured against the frequency from 0.1 Hz to 1MHz at room temperature. The HF/HMA shows ε' around 4 which increased to 5 at 100 Hz. The increase in permittivity after adding 1% of SiO_2/CTMS could be explained as consequence of slightly presence of more oxalated silicon groups in the polymer network which are acting as a platform for silica groups. The raise of real permittivity in range 0.1-10 Hz is regarding to the electrode polarization [33]. It seems that the electrode polarization decreased with increase the silica additions. Moreover, the values of permittivity decreased with increased the ratio of silica which is proving the OH groups are occupied or substituted by Silica. More silica percentage addition, more decrease in the permittivity ε'.

The dielectric loss ε'' refers to the dissipated energy by internal dynamic of the polymer. Thus, a part of energy is used for doing rotational motion which can be performed in the radio and microwave frequency regions. On other hand, the contribution of polymers with their compatibility with each other can be revealed from the Maxwell-Wagner-Sillars interfacial polarization. Thus, a part of energy is used for doing which can be performed in the radio and microwave frequency regions. The rotational motions could be happened to the backbone of polymer structure and often namely α-relaxation which existed at lower frequencies. If the rotation motion occurs in the branched or attached groups or atoms to the polymer backbone, then a second process called β-relaxation will be existed at higher frequencies. In our case, there are two peaks which overlapped certainly, and obviously their amplitude decreased for HF/HMA after adding 1% of SiO_2/CTMS. The reason for that, the free volume around HF/HMA is decreased due to copolymerization processes in slightly presence of SiO_2/CTMS. After that, there were a dramatic decrease in the copolymer HF/HMA when 3% SiO_2/CTMS is added. The reason for this dramatic decrease could be the substitution of silicate group to hydroxyl group. Thus, the backbone of the copolymer become more rigid rather than before. Moreover, the conductivity which could be deduced from the plot it is looking too poor and would have a very small value.[34] The contribution of the Maxwell-Wagner-Sillers effect cannot be seen only in HF/HMA with 9%wt of silica, while the electrode effect is clearly exhibited in 1% SiO_2/CTMS. Thus, the explanation for homogeneity between contents such as HF/HMA plus 1wt% Silica and other series can be recognized from the contribution of both interfacial Maxwell-Wagner polarization and the electrode polarization [35]. Then with increasing the SiO_2/CTMS weight ratio as in 6% and 9%, the loss values start increase, and the relaxation peak slightly shifts to higher frequencies as indication to increase of the polarity alongside the increase of dynamic rotational motion of silicate group via silica aggregations. The silica series shows an increase in the conductivity with increasing SiO_2/CTMS groups due to electrode polarization.

4. CONCLUSION

The ultrahydrophobic coating films are successfully prepared with different silicon nanoparticles ratios. The contact angles exhibit ultra-surface energy polymer nano-composites over 132° in the highest weight percentile of SiO_2/CTMS nano-composites with 9% SiO_2 nanoparticles. Effective coating film with high efficiency self-cleaning character as cheap and easy colloidal coating based on *in-situ* nanoparticle encapsulation. The glass transition T_g temperatures were shifted toward each other for both HF and HFM copolymer with increasing SiO_2/CTMS amounts. Both TEM and AFM results confirm with DLS a well unified nanoparticle size (60-80 nm) and dispersion. Tremendous demands are recommended for external applications in stadium chairs, solar panel, building faces, cars furniture, glass and metals decoration parts, submarines, historical monuments …. etc. The prepared colloidal with high zeta potential around ±42mV - ±48mV for all SiO_2/CTMS ratios give a chance to apply the ultrahydrophobic coatings by different methods like immersing, brushing, or spraying techniques.

AKNOWLEDGMENT

Authors have deep grateful to Nanomaterials Investigation Lab., Central Network Laboratories., National Research Centre (NRC), Egypt for complete testing of prepared polymer nanocomposites.

5. REFERENCES

[1] Dahlström M, Jonsson H, Jonsson P R and Elwing H 2004 Surface wettability as a determinant in the settlement of the barnacle Balanus Improvisus (DARWIN) *J. Exp. Mar. Bio. Ecol.* **305** 223–32

[2] Nguyen-Tri P, Tran H N, Plamondon C O, Tuduri L, Vo D-V N, Nanda S, Mishra A, Chao H-P and Bajpai A K 2019 Recent progress in the preparation, properties and applications of superhydrophobic nano-based coatings and surfaces: A review *Prog. Org. Coatings* **132** 235–56

[3] Vazirinasab E, Jafari R and Momen G 2018 Application of superhydrophobic coatings as a corrosion barrier: A review *Surf. Coatings Technol.* **341** 40–56

[4] Ibrahim S and Sultan M 2019 Superhydrophobic Coating Polymer/Silica Nanocomposites: Part I Synthesis and Characterization as Eco-Friendly Coating *Silicon*

[5] Ibrahim S, Labeeb A, Mabied A F, Soliman O, Ward A, Abd-El-Messieh S L and Abdelhakim A A 2017 Synthesis of super-hydrophobic polymer nanocomposites as a smart self-cleaning coating films *Polym. Compos.* **38** E147–56

[6] Ibrahim S A and Sultan M Z 2019 Superhydrophobic coating polymer/silica nanocomposites: Enhancing Water Vapor Barrier Properties of Packaging Paper with Ultra-Thin PS/silica Nanocomposite Polymer Coating *Egypt. J. Chem.* **62**

[7] Zhao X, Li Y, Li B, Hu T, Yang Y, Li L and Zhang J 2019 Environmentally benign and durable superhydrophobic coatings based on SiO2 nanoparticles and silanes *J. Colloid Interface Sci.* **542** 8–14

[8] Huré M, Olivier P and Garcia J 2022 Effect of Cassie-Baxter versus Wenzel states on ice adhesion: A fracture toughness approach *Cold Reg. Sci. Technol.* **194** 103440

[9] Yonemoto Y, Tomimitsu I, Shimizu K and Kunugi T 2021 Wettability model for water-ethanol binary mixture droplet on roughened low-surface-energy solids *Int. J. Multiph. Flow* **137** 103569

[10] Guo Z, Liu W and Su B-L 2011 Superhydrophobic surfaces: From natural to biomimetic to functional *J. Colloid Interface Sci.* **353** 335–55

[11] Yao C W, Garvin T P, Alvarado J L, Jacobi A M, Jones B G and Marsh C P 2012 Droplet contact angle behavior on a hybrid surface with hydrophobic and hydrophilic properties *Appl. Phys. Lett.* **101** 111605

[12] Forsberg P, Nikolajeff F and Karlsson M 2011 Cassie–Wenzel and Wenzel–Cassie transitions on immersed superhydrophobic surfaces under hydrostatic pressure *Soft Matter* **7** 104–9

[13] Wang H, Zhang Z, Zheng J, Zhao J, Liang Y, Li X and Ren L 2021 Multifunctional superhydrophobic surface with dynamically controllable micro/nanostructures for droplet manipulation and friction control *Chem. Eng. J.* **417** 127944

[14] Meawad A and Ibrahim S 2019 Novel bifunctional dispersing agents from waste PET packaging materials and interaction with cement *Waste Manag.* **85** 563–73

[15] Ferrari M, Piccardo P, Vernet J and Cirisano F 2021 High Transmittance Superhydrophobic Coatings with Durable Self-Cleaning Properties *Coatings* **11**

[16] Rossi E M, Phani P S, Guillemet R, Cholet J, Jussey D, Oliver W C and Sebastiani M 2021 A novel nanoindentation protocol to characterize surface free energy of superhydrophobic nanopatterned materials *J. Mater. Res.* **36** 2357–70

[17] Camalan M and İhsan Arol A 2022 Preliminary assessment of spray coating, solution-immersion and dip coating to render minerals superhydrophobic *Miner. Eng.* **176** 107357

[18] Follain N, Belbekhouche S, Bras J, Siqueira G, Chappey C, Marais S and Dufresne A 2018

Tunable gas barrier properties of filled-PCL film by forming percolating cellulose network *Colloids Surfaces A Physicochem. Eng. Asp.* **545** 26–30

[19] Lee H B, Jhon M S and Andrade J D 1975 Nature of water in synthetic hydrogels. I. Dilatometry, specific conductivity, and differential scanning calorimetry of polyhydroxyethyl methacrylate *J. Colloid Interface Sci.* **51** 225–31

[20] Zhou D, Koike Y and Okamoto Y 2008 Synthesis of poly[1,1,1,3,3,3-hexafluoro-2-(pentafluorophenyl)propan-2-yl methacrylate]: Thermal and optical properties of the copolymers with methyl methacrylate *J. Fluor. Chem.* **129** 248–52

[21] Morita S 2014 Hydrogen-bonds structure in poly(2-hydroxyethyl methacrylate) studied by temperature-dependent infrared spectroscopy *Front. Chem.* **2** 1–5

[22] Kasmi N, Terzopoulou Z, Chebbi Y, Dieden R, Habibi Y and Bikiaris D N 2022 Tuning thermal properties and biodegradability of poly(isosorbide azelate) by compositional control through copolymerization with 2,5-furandicarboxylic acid *Polym. Degrad. Stab.* **195** 109804

[23] Kotresh T M, Ramani R, Jana N, Minu S, Shekar R I and Ramachandran R 2021 Supermolecular Structure, Free Volume, and Glass Transition of Needleless Electrospun Polymer Nanofibers *ACS Appl. Polym. Mater.* **3** 3989–4007

[24] Dai L, Sun C, Wei Y, Mao L and Gao Y 2018 Characterization of Pickering emulsion gels stabilized by zein/gum arabic complex colloidal nanoparticles *Food Hydrocoll.* **74** 239–48

[25] Ibrahim S and El-Khawas K M 2019 Development of eco-environmental nano-emulsified active coated packaging material *J. King Saud Univ. - Sci.* **31** 1485–90

[26] Ibrahim S, El Saied H and Hasanin M 2018 Active paper packaging material based on antimicrobial conjugated nano-polymer/amino acid as edible coating *J. King Saud Univ. - Sci.*

[27] Lu Z, Hou D, Hanif A, Hao W, Li Z and Sun G 2018 Comparative evaluation on the dispersion and stability of graphene oxide in water and cement pore solution by incorporating silica fume *Cem. Concr. Compos.* **94** 33–42

[28] Ferrari L, Kaufmann J, Winnefeld F and Plank J 2014 Impact of particle size on interaction forces between ettringite and dispersing comb-polymers in various electrolyte solutions *J. Colloid Interface Sci.* **419** 17–24

[29] Lu G W and Gao P 2010 Emulsions and Microemulsions for Topical and Transdermal Drug Delivery *Handb. Non-Invasive Drug Deliv. Syst.* 59–94

[30] Ibrahim S and Voit B 2009 Synthesis and characterization of well-defined block copolymers by combing controlled radical and cationic polymerization *Macromol. Symp.* **275–276** 59–66

[31] Sutton L E 1956 Dielectric Behavior and Structure. Dielectric Constant and Loss, Dipole Moment and Molecular Structure. *J. Am. Chem. Soc.* **78** 507–8

[32] Rozik N N, Khalaf A I and Ward A A 2015 Studies the behaviors of polyaniline on the properties of PS/PMMA blends *Proc. Inst. Mech. Eng. Part L J. Mater. Des. Appl.* **230** 526–36

[33] Samet M, Levchenko V, Boiteux G, Seytre G, Kallel A and Serghei A 2015 Electrode

polarization vs. Maxwell-Wagner-Sillars interfacial polarization in dielectric spectra of materials: Characteristic frequencies and scaling laws *J. Chem. Phys.* **142** 194703

[34] Serghei A and Kremer F 2008 Broadband dielectric studies on the interfacial dynamics enabled by use of nanostructured electrodes *Rev. Sci. Instrum.* **79** 026101

[35] Reislöhner U and Ronning C 2012 Maxwell-Wagner polarization in Cu(In,Ga)(S,Se) $_2$ *Appl. Phys. Lett.* **100** 252111

Facile Synthesis and Catalytic Activity Assessment of Copper Chromite Nanoparticles for Ammonium Perchlorate Decomposition

Shukri Ismael [1,*], A Deif [2], Ahmed Maraden [1], Sherif Elbasuney [1,3]

[1] School of Chemical Engineering, Military Technical College (MTC), Cairo, Egypt.
[2] Space technology Centre, Cairo, Egypt.
[3] Head of Nanotechnology Research Center, Military Technical College (MTC), Cairo, Egypt.
[*]Corresponding authors: a.m.mostafa@mtc.edu.eg.

Abstract. The universal burning rate modifier used in composite solid propellants is copper chromite. The fabrication of copper chromite nanoparticles (CCNs) is highly appreciated; as superior performance could be accomplished. This research looks at how copper chromite nanoparticles can be made in a sustainable way using hydrothermal processing. Mono-dispersed particles with a medium particle size of 12 nm were seen in TEM micrographs. The XRD diffractogram revealed a crystalline structure. The co-precipitation approach was used to incorporate CCNs into ammonium perchlorate (AP). Differential scanning calorimetry (DSC) and thermal gravimetric analysis were used to analyze the activity of catalysis in CCNs on AP breakdown (TGA). The enthalpy of AP breakdown was enhanced from 742 J/g to 1391 J/g. CCNs caused the Peak of high-temperature decomposition shift toward lower temperatures, resulting in the overlapping of the two peaks into a single peak. At 244 ^{0}C, CCNs showed a decrease in AP endothermic phase transition. These characteristics could ensure high-burning-rate catalyzed decomposition processes.

Keywords: Copper chromite; Encapsulation; Ammonium perchlorate; Hydrothermal synthesis; Thermal behavior.

1. Introduction

Ammonium perchlorate (AP) is an efficient oxidizer mostly used in (composite solid propellants (CSPs) as a result of the high oxygen content[1]. It typically makes up 60-90 wt% of propellant, and its thermal breakdown has an impact on CSP combustion efficiency[2]. Lowering the temperature at which AP decomposes increases the rate of burning and decreases the delay ignition time[3]. additionally, the specific impulse of CSPs can be increased by increasing the heat released during AP breakdown[4]. As a result, boosting AP's thermal breakdown can improve CSPs' thermal properties[5]. Thermal decomposition can be aided by decreasing the size of AP's particles [6], but decreasing the size of AP's particles is a hazardous process [7]. Fe_2O_3, CuO, Co_3O_4, and Cr_2O_3 nanocatalysts can be added to CSPs to lower the breakdown temperature and increase the heat release of AP[8]. In the AP breakdown process, metal oxide acts as a catalyst that is heterogeneous, Increasing the rate of ammonia/perchloric acid breakdown on the catalyst surface [9, 10]. The active sites of a heterogeneous catalyst are the most important factor[11], particle size, and surface area [12-14]. to increase the action of a catalyst, it must have a big surface area and a small particle size. For the combustion of CSPs, copper chromite is used as a rate modifier[15]. The degradation of AP is accelerated by copper chromite, which increases the rate of combustion and the amount of heat released[16]. Copper chromite is produced utilizing a hydrothermal method to produce copper chromite with high purity and uniformity at the nanoscale. The regularity of the metal oxides' solid-solid interaction with AP, as well as the dispersion of metal oxides in AP, are critical factors in determining the catalytic efficacy of nanocatalysts for full combustion. Because of their nanosize, metal oxides can easily collect, decreasing the performance of nanocatalysts in AP breakdown. There is poor contact between the oxide NPs and the AP due to nano-metal oxide aggregation, Heat flow from the burning surface to the condensed AP could be difficult. To optimize their catalytic activity and AP thermal degradation, homogeneous nanocatalyst dispersion in the AP matrix is necessary. To achieve high performance and reduce aggregation, the encapsulation approach can achieve a good dispersion of nanocatalyst in AP particles. The anti-solvent to solvent ratio affects nanocatalyst encapsulation [17-19]. Anti-solvent concentrations rise, resulting in higher nucleation rates and smaller crystals. AP encapsulation with catalyst NPs can ensure a higher burning rate due to the high homogeneity, dispersion, and interfacial surface area. This work describes the easy synthesis of copper chromite nanoparticles via hydrothermal processing. Copper nitrate trihydrate and chromium nitrate nonahydrate were used as precursors in the fabrication of CCNs. The activity of catalysis in copper chromite on AP was determined using TGA and DSC techniques.

2. Experimental work

2.1 Synthesis of copper chromite (CCNs)

1mmole of $Cu(NO_3)_2.3H_2O$, and 2mmol of $Cr(NO_3)_3.9H_2O$ were dissolved in 200 ml of DI water and magnetic stirrer for 1hr for complete dissolving and formation of a homogeneous solution. The mixture was transferred into the hydrothermal batch reactor at 200 ^{0}C for 8 hours for reaction completion.

2.2 Integration of CCNs into AP

Using an ultrasonic probe, Colloidal CCNs were re-dispersed in acetone after being recovered from their manufacturing medium. After then, the AP particles subsequently dissolved with acetone. The antisolvent approach was used to incorporate colloidal CCNs into the AP matrix. (Figure 1).

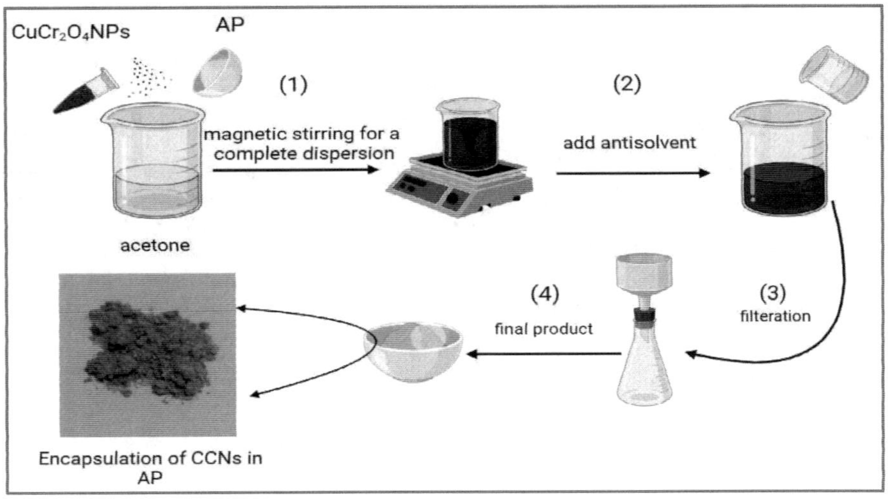

Figure 1. Schematic diagram for encapsulation of AP with CCNs

2.3 Nano-catalyst characterization

The shape (size and appearance) of copper chromite nanoparticles (CCNs) was studied using transmission electron microscopy (TEM; JEM-2100F by Jeol Corporation) and scanning electron microscopy (SEM) (SEM; ZEISS SEM EVO 10 MA supported with EDAX detector). Using an EDS detector, the distribution of CCNs into the AP matrix was examined. With a Hiltonbrooks X-ray diffractometer (XRD), the crystallinity of copper chromite was investigated over a range of 2θ from 5° to 90°. Increase your progress by (Burker Corporation, Japan).

2.4 Catalytic activity assessment

The influence of CCNs on the thermal degradation of AP was evaluated using TA's DSC Q20. Under the inert gas flow of N_2 at 50 ml min^{-1}, the investigated specimen was burned to 500 °C at 10 °C min^{-1}. TGA 55 by TA was utilised to track the AP nanocomposite's thermal response. The studied specimen was non-isothermally heated to 500 °C at a rate of 10 0C min^{-1}.

3. Results and discussions

3.1 Characterization of the synthesized CCNs

As revealed in the TEM micrographs, scattered CCNs with a medium particle size of 12 nm were found (figure 2). CCNs are stable Nanospheres formed up of a high number of nanoparticles that are connected.

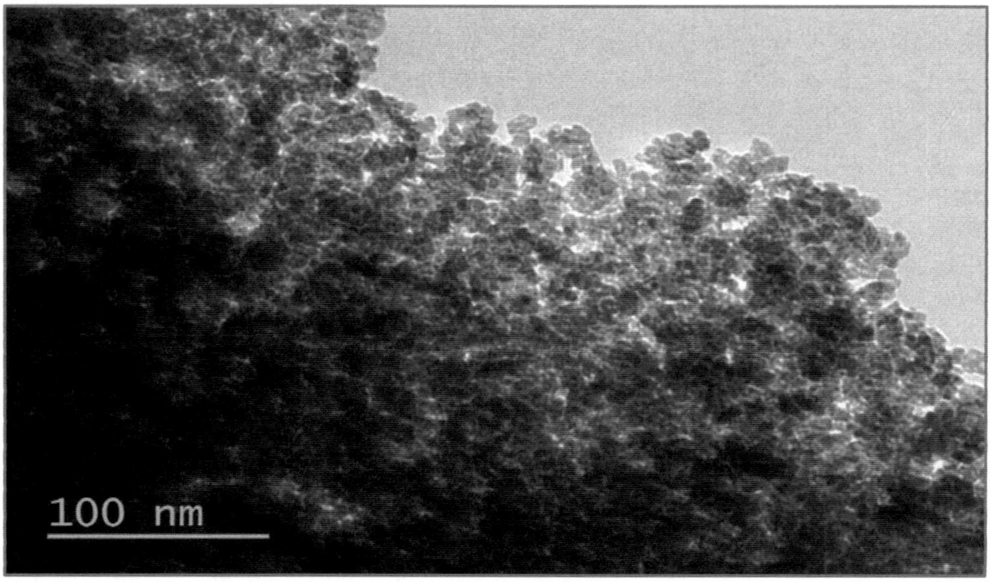

Figure 2. TEM of nano synthesized copper chromite

XRD diffractograms of high-quality crystalline copper chromite revealed a sphere structure with distinctive sharp peaks. with 2 Θ=16.06, 30.56, 35.74, 36.3 degrees respectively in agreement with JCPDA card no. 01-089-0540 (Figure 3). Pure nanocatalysts with no interfering components were revealed by XRD diffractograms.

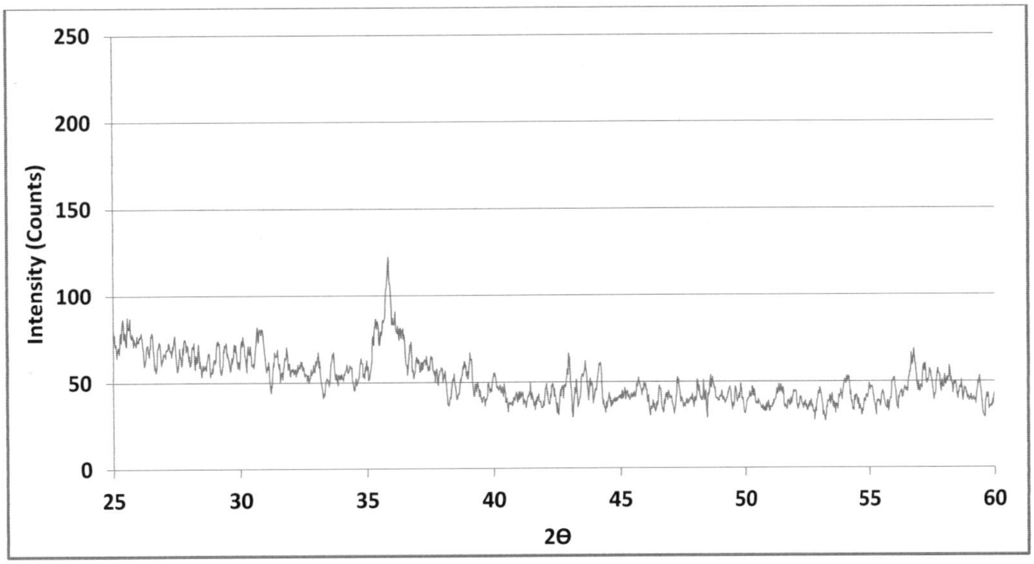

Figure 3. XRD diffractogram of developed copper chromite nanoparticles (CCNs).

SEM micrographs of the CCNs produced show that they are mono-dispersed and spherical in shape, with a size distribution of 12 nm, confirming the TEM analysis as shown in (figure 4)

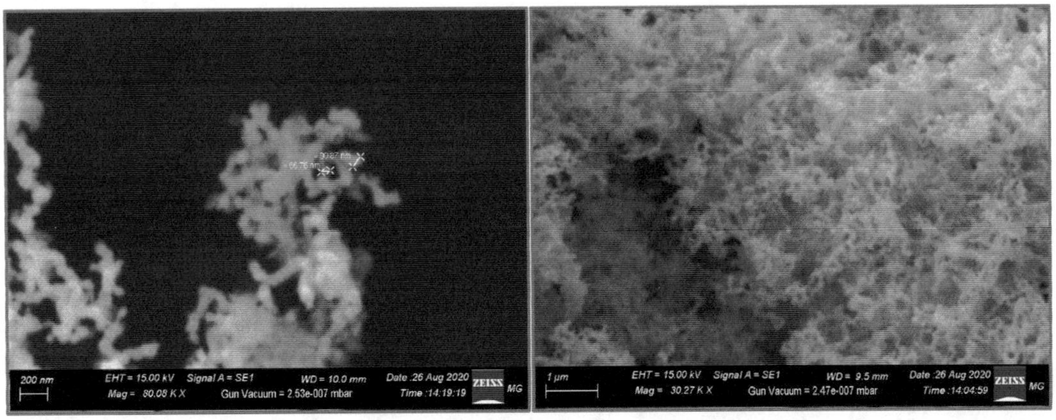

Figure 4. SEM micrographs of developed copper chromite nanoparticles (CCNs).

The use of an EDX detector to map the elements of synthesized CCNs revealed consistent dispersion of copper and chromium within the material structure. The chemical makeup of the elements was measured and found to be balanced and free of interfering substances. It was discovered that the Cr: Cu ratio is 2:1. (Figure 5).

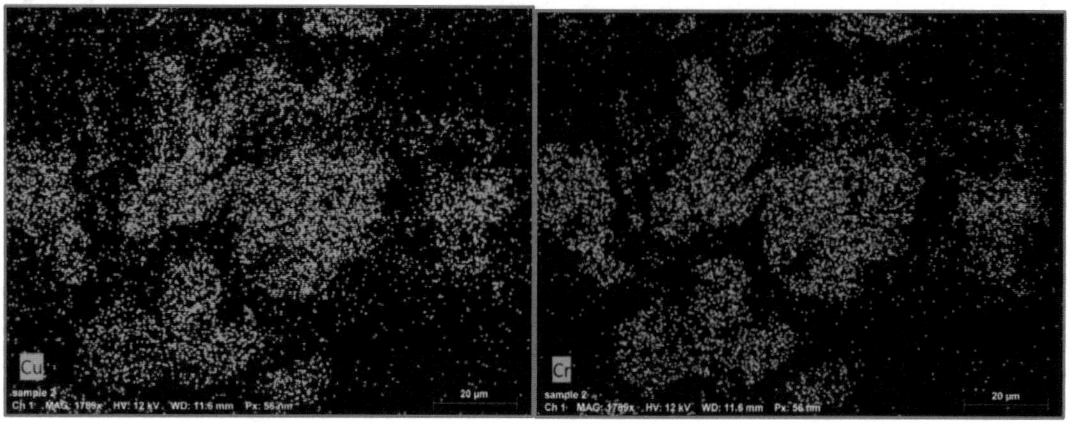

Element	At. No.	Line s.	Netto	Mass [%]	Mass Norm. [%]	Atom [%]	abs. error [%] (1 sigma)
Oxygen	8	K-Serie	65291	11.78	42.92	72.38	1.39
Chromium	24	K-Serie	34630	9.85	35.90	18.63	0.31
Copper	29	K-Serie	6047	5.81	21.18	8.99	0.23
			Sum	27.45	100.00	100.00	

Figure 5. Elemental mapping and elemental composition of copper chromite.

3.2 Characterization of AP nanocomposite

To begin AP, SEM was used to analyze the morphology of an AP nanocomposite. AP initially indicated 150 µ sphere particles (Figure 6-a); however, AP nanocomposite revealed ultra-small particles of 5 µ. (Figure 6-b). The fast crash antisolvent technique utilized to integrate colloidal particles into AP resulted in uniform particles with uniform dispersion in AP, which can be deduced. SEM was used to confirm the homogenous dispersion of nanocatalyst particles in AP. The increased amount of copper chromite particle dispersed into the AP nanocomposite was visible in SEM micrographs of AP nanocomposite.

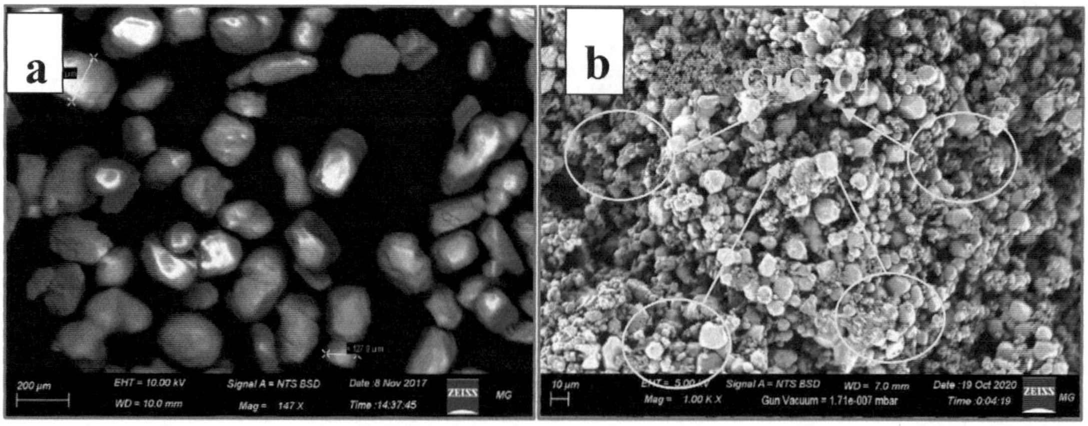

Figure 6. SEM micrographs of starting AP (a), AP nanocomposite (b).

To ensure catalyst particle distribution in the AP matrix, element mapping was used. Particle dispersion into the AP matrix was confirmed by elemental mapping of the major catalyst components (Figure 7).

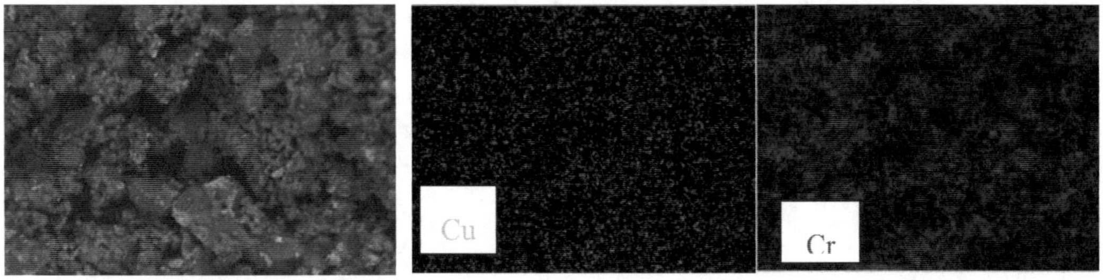

Figure 7. Elemental Mapping of CCNs-AP nanocomposite using EDX detector

3.3 Thermal behaviour of AP nanocomposite

At a temperature of 242°C, AP showed an endothermic peak with a crystallographic phase shift from orthorhombic to cubic with the energy required 102 J/g and two exothermic peaks as shown in (Figure 8)[20, 21].

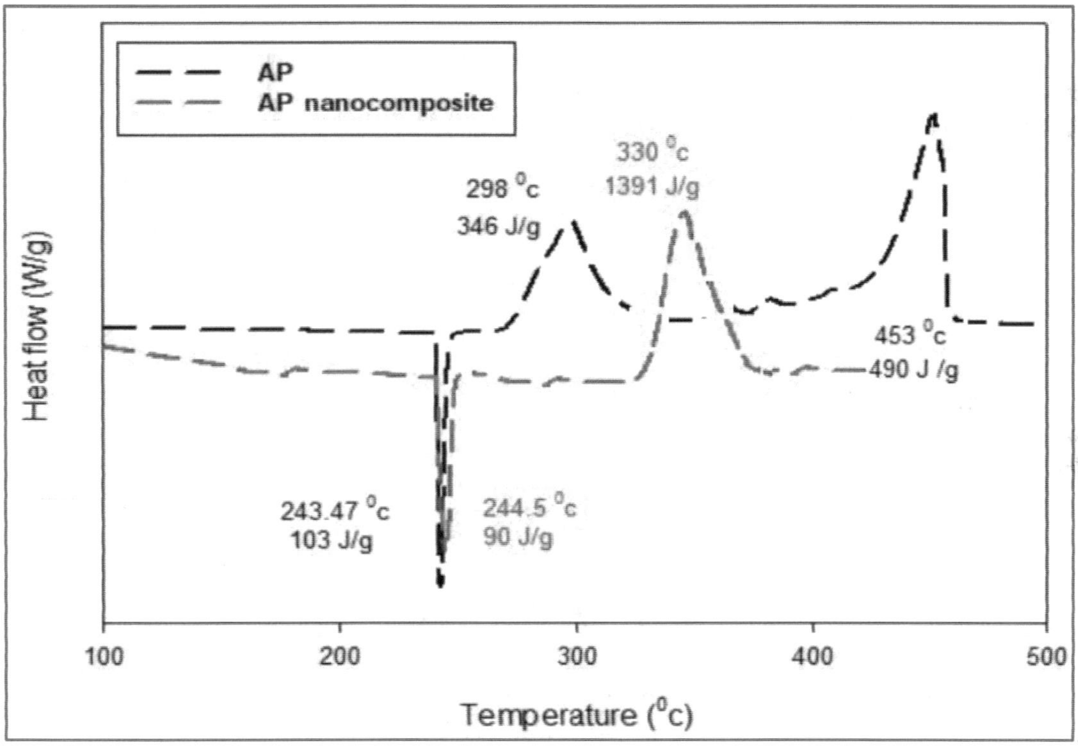

Figure 8. DSC thermograms of AP (a), and AP nanocomposite (b)

The low-temperature decomposition (LTD) peak is related to partial decomposition with the formation of some intermediate products at 298 °C and energy released 346 J/g. AP decomposed to NH_3(g) and $HClO_4$(g) according to (Eqn. (1)). NH_3(g) then substantially adsorbed on the AP's morphology and exhibit an inhibition effect on AP and cause the reaction to cease. $HClO_4$(g) can escape easily from the topography of the AP. increasing the temperature to the high-temperature decomposition (HTD) leads to desorption of NH_3(g) which may be oxidized in the oxidized phase according to Eqns (2,3,4,5).

$$NH_4ClO_4 \longrightarrow NH_4^+ + ClO_4^- \longrightarrow NH_3 + HClO4 \quad (1)$$

$$HClO_4 \longrightarrow Cl_2 + O_2 + ClO_2 + H_2O \quad (2)$$

$$Cl_2 + H_2O \longrightarrow HCl + O_2 \quad (3)$$

$$NH_3 + ClO_2 \longrightarrow N_2O + Cl_2 + H_2O \quad (4)$$

$$N_2O + O_2 \longrightarrow NO + NO_2 \quad (5)$$

CCNs have a large surface area, which allows them to absorb NH_3(g) produced by AP breakdown and prevent NH_3(g) inhibition on their surface, allowing AP to decompose further according to the Le Chatelier principle. As a result, the (HTD) will be moved and overlapped with an (LTD) to generate a single peak at 330oC, with 1391 J/g of heat emitted. TGA was utilized to look at the influence of catalyst CCNs on the thermal properties of the AP (Figure 9). Traditional electron-transfer theory predicts the occurrence of a partially full 3d orbit in CCNs, aids in an electro-transfer process that can absorb electrons from the AP ion and its intermediate products, enhancing the thermal degradation of AP. The activity of catalysis in several types of catalysts on the thermal degradation of AP is shown in Table 1, revealing that CCNs have a higher heat evolution and a similar decomposition temperature than other catalysts.

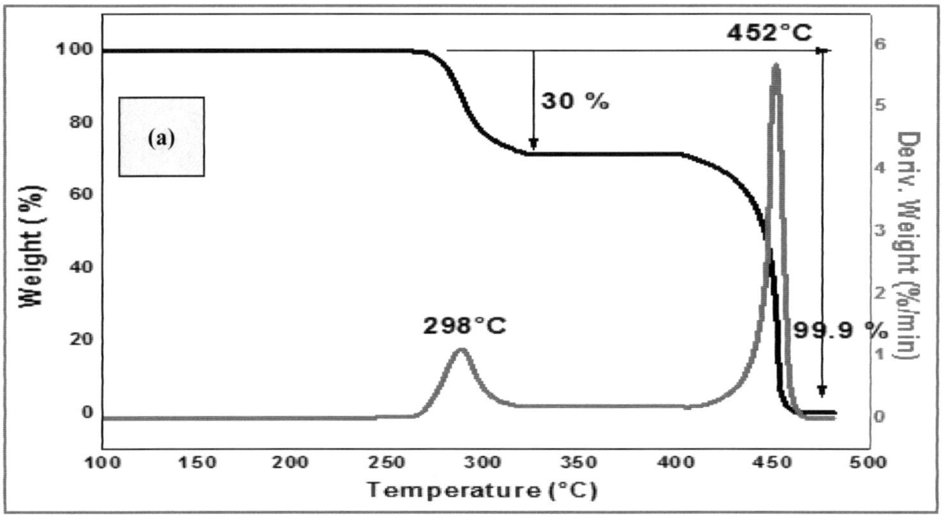

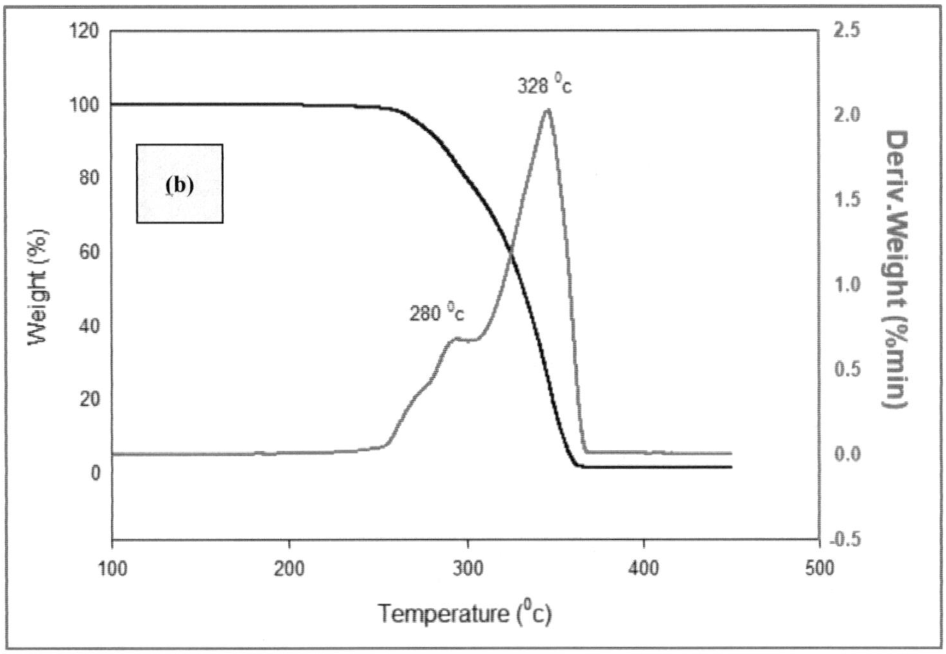

Figure 9. TGA thermograms of AP (a), and AP nanocomposite (b).

The main stage in the high-temperature decomposition process is the conversion of O2 to O2-, which can subsequently mix with NH_3 to form N_2O, NO_2, and H_2O. As a result, the catalyst's high electron transfer capacity and big zones of specific surface have a significant impact on AP thermal degradation. $CuCr_2O_4$ as a p-type semiconductor material has effective sites (positive holes on the surface of the catalyst) for accepting emitted electrons from perchlorate, as well as the extraction of atomic oxygen from the perchlorate ion, according to the electron transfer process..[22] as shown in (Figure 10)

$$e^-_{oxide} + ClO_4^- \longrightarrow O_{oxide} + ClO_3^- \longrightarrow 0.5\, O_2 + e^-_{oxide} + ClO_3^- \quad (6)$$

Table 1. The catalytic activity of several types of catalysts on the thermal degradation of AP

Catalyst system	Mass %	AP decomposition temperature (°c)	Heat evolved (J/g)	Ref.
Pure AP	------	۲۹۸, 453	836	This work
$CuCr_2O_4$ CCNs	1	330	1391	This work
Fe_2O_3 NPs	1	390.0	1230	[23]
$NdCrO_3$	2	334.0	1074	[24]
CuO	1	350.0	1268	[25]
Co_3O_4	5	305.5	1068	[26]
Ferrocene derivative	5	344.8	1291	[27]

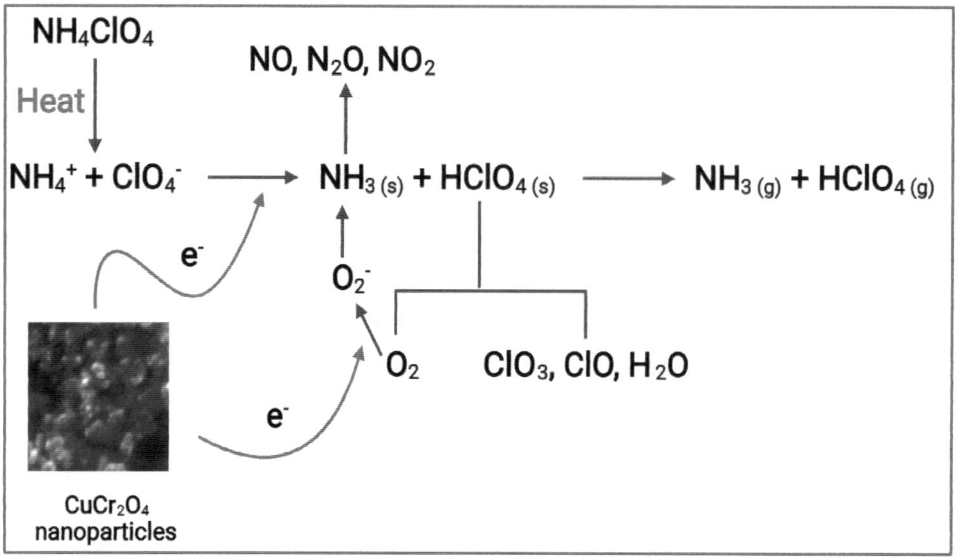

Figure 10. Schematic illustration of the CCNs catalyst to AP decomposition

Where e^{-}_{oxide} refers to a positive hole in the valence band of the oxide, and O_{oxide} refers to an atom removed from the oxide. Because the super oxide ion's presence O_2^- on the architecture of the catalysts enhances catalytic action, the $CuCr_2O_4$ catalyst may aid in the breakdown of ClO_4^- molecules into ClO_3 and O_2 more quickly. Furthermore, the ion of a transition metal Cr^{3+} in the architecture of $CuCr_2O_4$ exhibits hole conductivity because its outermost d orbitals are not filled by electrons. It readily accepts the ClO_4^- released electron to make Cr^{2+}, which was then moved to the catalyst's surface and interacted with NH_4^+ to produce ammonia and hydrogen atoms[28].

4. Conclusion

The researchers created copper chromite nanoparticles (CCNs) with a particle size on average of 12 nm. Before being properly incorporated into the AP, the CCNs were re-dispersed in an organic solvent. the most common oxidizer in composite solid propellant systems. According to SEM, the distribution of CCNs in AP was uniform. DSC and TGA were used to determine the catalytic efficiency of CCNs. At 242 °C, there was an endothermic crystallographic phase shift followed by two exothermic decompositions. CCNs' endothermic phase shift was decreased, showing that they have significant catalytic activity despite having low activation energy. CCNs increased the amount of heat released from AP to 1391 J/g, previously AP only released 742 J/g. A single peak was created by combining the two big exothermic decomposition peaks. CCNs were identified as a one-of-a-kind emergent nanocatalyst with exceptional performance. Using sophisticated kinetic models on AP kinetic decomposition, the effect of CCNs on kinetic decomposition will be explored.

Reference

[1]. Sherif, E., et al., Thermochemical Aspects of Activated Ammonium Perchlorates with Superior Thermal Stability, Decomposition Enthalpy, Propagation Index, and Decomposition Kinetic Parameters. *Journal of Thermal Science*, 2021. **30**(6): p. 2196-2201.

[2]. Elbasuney, S., S. Ismael, and M. Yehia, Ammonium Perchlorate/HMX Co-crystal: Bespoke Energetic Materials with Tailored Decomposition Kinetics via Dual Catalytic Effect. *Journal of Energetic Materials*, 2021: p. 1-20.

[3]. Dolgoborodov, A.Y., et al., Thermal decomposition of mechanoactivated ammonium perchlorate. *Thermochimica Acta*, 2018. **669**: p. 60-65.

[4]. Boldyrev, V., Thermal decomposition of ammonium perchlorate. *Thermochimica acta*, 2006. **443**(1): p. 1-36.

[5]. Keenan, A.G. and R.F. Siegmund, Thermal decomposition of ammonium perchlorate. Quarterly Reviews, *Chemical Society*, 1969. **23**(3): p. 430-452.

[6]. Kumari, A., et al., Nano-ammonium perchlorate: preparation, characterization, and evaluation in composite propellant formulation. *Journal of Energetic Materials*, 2013. **31**(3): p. 192-202.

[7]. Li, G.P., et al. The preparation and properties of AP-based nano-limit growth energetic materials. in *Advanced Materials Research*. 2014. Trans Tech Publ.

[8]. Chaturvedi, S. and P.N. Dave, Nano-metal oxide: potential catalyst on thermal decomposition of ammonium perchlorate. *Journal of Experimental Nanoscience*, 2012. **7**(2): p. 205-231.

[9]. Alizadeh-Gheshlaghi, E., et al., Investigation of the catalytic activity of nano-sized CuO, Co3O4 and CuCo2O4 powders on thermal decomposition of ammonium perchlorate. *Powder technology*, 2012. **217**: p. 330-339.

[10]. Aijun, H., et al., Preparation of nano-MnFe2O4 and its catalytic performance of thermal decomposition of ammonium perchlorate. *Chinese Journal of Chemical Engineering*, 2011. **19**(6): p. 1047-1051.

[11]. Zhao, S., D. Ma, and W. Jin, Preparation of Co Fe 2 O 4 Nanocrystallites by Solvothermal Process and Its Catalytic Activity on the Thermal Decomposition of Ammonium Perchlorate. *Journal of Nanomaterials*, 2010. **2010**: p. 28.

[12]. Elbasuney, S., et al., Synthesis of CuO-distributed carbon nanofiber: Alternative hybrid for solid propellants. *Journal of Materials Science: Materials in Electronics*, 2020. **31**(11): p. 8212-8219.

[13]. Said, A. and R. Al-Qasmi, The role of copper cobaltite spinel, CuxCo3− xO4 during the thermal decomposition of ammonium perchlorate. *Thermochimica acta*, 1996. **275**(1): p. 83-91.

[14]. Elbasuney, S. and M. Yehia, Thermal decomposition of ammonium perchlorate catalyzed with CuO nanoparticles. *Defence Technology*, 2019. **15**(6): p. 868-874.

[15]. Patil, P.R., V.e.N. Krishnamurthy, and S.S. Joshi, Effect of nano-copper oxide and copper chromite on the thermal decomposition of ammonium perchlorate. Propellants, Explosives, Pyrotechnics: *An International Journal Dealing with Scientific and Technological Aspects of Energetic Materials*, 2008. **33**(4): p. 266-270.

[16]. Inami, S., et al., Solid propellant kinetics. I: The ammonium perchlorate-copper chromite-fuel system. *Combustion and Flame*, 1971. **17**(2): p. 189-196.

[17]. Elbasuney, S. and M. Yehia, Ammonium perchlorate encapsulated with TiO 2 nanocomposite for catalyzed combustion reactions. *Journal of Inorganic and Organometallic Polymers and Materials*, 2019. **29**(4): p. 1349-1357.

[18]. Wu, C., et al., Encapsulation of perchlorate salts within metal oxides for application as nanoenergetic oxidizers. *Advanced Functional Materials*, 2012. **22**(1): p. 78-85.

[19]. Hu, Y., et al., Ammonium perchlorate encapsulating nanothermites as high energetic composites: Preparation, thermal decomposition and combustion properties. *Chemical Engineering Science*, 2019. **207**: p. 334-343.
[20]. Bircumshaw, L. and B.H. Newman, The thermal decomposition of ammonium perchlorate-I. Introduction, experimental, analysis of gaseous products, and thermal decomposition experiments. *Proceedings of the Royal Society of London. Series A. Mathematical and Physical Sciences*, 1954. **227**(1168): p. 115-132.
[21]. Jacobs, P.W. and A. Russell-Jones, Sublimation of ammonium perchlorate. *The Journal of Physical Chemistry*, 1968. **72**(1): p. 202-207.
[22]. Kumar, H., et al., The effect of reduced graphene oxide on the catalytic activity of Cu–Cr–O–TiO 2 to enhance the thermal decomposition rate of ammonium perchlorate: an efficient fuel oxidizer for solid rocket motors and missiles. *RSC advances*, 2017. **7**(58): p. 36594-36604.
[23]. Joshi, S.S., P.R. Patil, and V. Krishnamurthy, Thermal Decomposition of Ammonium Perchlorate in the Presence of Nanosized Ferric Oxide. *Defence Science Journal*, 2008. **58**(6).
[24]. Sharma, J., et al., Review on the catalytic effect of nanoparticles on the thermal decomposition of ammonium perchlorate. *Energy and Environment Focus*, 2014. **3**(2): p. 121-130.
[25]. Chen, L., L. Li, and G. Li, Synthesis of CuO nanorods and their catalytic activity in the thermal decomposition of ammonium perchlorate. *Journal of Alloys and Compounds*, 2008. **464**(1-2): p. 532-536.
[26]. Yu, C., et al., Shape-controlled syntheses of Co3O4 nanowires arrays with excellent catalytic performances upon ammonium perchlorate decomposition. *Materials Research Bulletin*, 2018. **97**: p. 483-489.
[27]. Zhao, H., et al., Ferrocene and [3] ferrocenophane-based β-diketonato copper (ii) and zinc (ii) complexes: synthesis, crystal structure, electrochemistry and catalytic effect on thermal decomposition of ammonium perchlorate. *RSC advances*, 2016. **6**(41): p. 34476-34483.
[28]. Wang, Y., et al., Catalytic activity of nanometer-sized CuO/Fe2O3 on thermal decompositon of AP and combustion of AP-based propellant. *Combustion Science and Technology*, 2010. **183**(2): p. 154-162.

Removal of Co(II) from Water Using Graphene Oxide Functionalized with TOPO

Waleed F. Khalil, Ahmed Agha. R. A. Rayan.,

Nuclear and Radiological Safety Research Center (NRSRC). Egyptian Atomic Energy Authority (EAEA).
Corresponding author's e-mail (Waleed F. Khalil): waleed_fekry12878@yahoo.com

Abstract. One of the most immensely adsorbents for removing radioactive waste from water is graphene oxide nanosheet (GO). It has a large specific surface area and a variety functional groups, including epoxide, carbonyl, carboxyl, and hydroxyl. This work has prepared GO and functionalized with TOPO (Tri-octyl phosphine oxide). Raman spectroscopy, XRD, High-Resolution Transmission Electron Microscopy (HRTEM), and Fourier transform infrared spectroscopy utilized the unfunctionalized and functionalized GO samples. The results indicate that the produced GO-TOPO adsorbents are very effective Co (II) adsorbents in the acidic medium. The maximum adsorption of Co (II) adsorbed on GO-TOPO was approximately 134 mg/g. The ideal conditions for eliminating Co (II) from water are pH of 6 and temperature 40 °C. Langmuir and Freundlich equations have been used to describe the isothermal data. According to R square, the Langmuir model was better than the Freundlich model.

Keywords: Cobalt (II); Graphene oxide; TOPO; Adsorption; water treatment

1. Introduction

Cobalt one of three metals that have natural magnetic sources, together with iron and nickel. Cobalt-based alloys are used to make turbine blades for aircraft, jet engines, orthopedic implants, and superalloys like stellite, the world's largest user of cobalt metal. Lithium-ion batteries, electroplating, and oxidation catalysts are all made from cobalt. Furthermore, cobalt is an essential component of Cobalamin (vitamin B12) and plays a role in animal metabolism [1].

Nuclear technology has advanced rapidly in recent years in response to rising energy demand, resulting in the production of radioactive waste that is damaging to both the human body and the environment. Cobalt 60 (60Co) radioisotope frequently utilised in cancer therapy, food preservation, and medical equipment sterilization. In hospitals and research facilities, these use generate radioactive waste[2].

The radionuclides 60Co and 58Co found in the liquid wastes emitted by pressurised water nuclear power reactors as fission products [3]. During the spent fuel treatment process, large amounts of cobalt radionuclides are inevitably discharged into the environment; these radionuclides are hazardous to human health. Cobalt 60 is one of the most dangerous radioactive elements due to its ability to degrade and to accumulate via the food chain [4].

Nanomaterials could provide effective adsorbents for efficiently removing radioactive pollutants from wastewater compared to traditional adsorbents, which have a limited adsorption capacity. Nanomaterials have a greater surface-to-volume ratio, making them more reactive. Experts are intrigued by these compounds and are using them to remove radioactive and heavy metals from wastewater [5] . Due to its high specific surface area and the presence of functional groups such as hydroxyl, carboxyl, and epoxy groups on the sheet edges, GO shows superior metal ion adsorption capabilities [6, 7]. Because of its large theoretical specific surface area, GO has the potential to be changed and used in the future as a better adsorbent for the removal of various pollutants from water. These oxygen groups that protrude from its layers may be utilized as anchoring sites to attach ions electrostatically and/or coordinately, especially for metals with different valences [8-11].

Sato (1972), studied the mechanism of extraction of divalent ions such as Co, Cu, Zn, and Cd with tri-octyl phosphine oxide. Also found that the order of the extraction efficiency of TOPO for divalent metals at low aqueous acidity [12]. One of the most common procedures for separating metal ions from aqueous solutions is solvent extraction at acid solubility. Solvent extraction has been widely investigated for separating cobalt and nickel from solution for industrial processes by TOPO [13]. The GO-TOPO is concluded to be a suitable adsorbent for heavy metal elimination via aqueous acidity adsorption. The functionalization of GO surface with TOPO by covalently bonded improves the removal of metal ions [14].

In this study, GO was synthesized and functionalized using TOPO. The crystal structure and morphology of GO were investigated using XRD and TEM. The functionalization of GO and GO-TOPO were compared using FTIR. The adsorbent GO-TOPO was evaluated the capacity for adsorb Co (II) from water. The contact period, metal concentration, pH, and temperature affected the adsorption process.

2. Experimental

2.1. Materials

The following resources have been used: Merck supplied very pure graphite fine powder (Germany). Sigma (Sigma Aldrich, UK) provided 98 percent sulphuric acid (H2SO4), 95 percent phosphoric acid (H3PO4), absolute ethanol,36.5 percent hydrochloric acid (HCl), and TOPO. The following materials were purchased: EDTA, NaOH, and KMnO4 (WINLAB, UK). 30% of the H2O2 was obtained by Alfa Company, India. All trials in this paper utilized deionized water.

2.2. Graphene oxide preparation

Graphene oxide was synthesized using the improved Hammer method with slight changes, natural graphite powder, H3PO4, KMnO4, and H2SO4 [15] . Purification of the oxidation product was washed with 10 % HCl solution and repeatedly washed with deionized water by a centrifuge (6000 rpm) to get exfoliated GO with a few layers redispersed in deionized water with sonication for 4 h. The yielded GO was dried in a vacuum oven at 70°C for 6 hours the desired GO[5] .

2.3. Preparation of GO-TOPO

According to Mishra's 2016 study, GO was functionalized with TOPO. 0.8 g of prepared graphene oxide was dissolved in 400 mL of methanol and sonicated for 120 minutes in a flask with a capacity of 1000 mL. After mixing the graphene oxide suspension with 5.0 g of TOPO, it was refluxed for 16 hours at 60 °C. 200 mL of methanol was added to remove any unreacted molecules. After one hour of centrifugation at 10,000 rpm, the result was separated with several washes of purified water and methanol. For 12 hours, the prepared sample was dried in a vacuum oven set to 60 °C [16].

2.4. Characterization

The KBr pellets technique was used to investigate the FTIR spectra of GO, TOPO, and GO-TOPO using an FTIR spectrometer (SHIMADZU 8400, Japan). To better understand the interaction between graphite and graphene oxide, it was chosen to investigate each spectrum using an X-ray diffractometer

(PANalytical Empyrean, Netherlands) and CuK radiation (wavelength = 1.54045). (GO). On the other hand, transmission electron microscopy was applied (JEOL-JEM 2100, Japan).

2.5. Batch adsorption experiments

To conduct adsorption experiments on batches of 120 mL plastic caps, the following approach was used: All GO or GO-TOPO (adsorbent) experiments were suspended in deionized water in a stock solution of 0.5 g for all adsorption experiments. The adsorbate stock solution was dissolved in 1.0 L of deionized water to produce the adsorbate. Diluting the solution produced standard cobalt solutions with concentrations of 25, 50, 75, and 125 ppm. HCL or NaOH were both used to adjust the pH.

The adsorption suspension was shaken at 200 rpm for four hours and then centrifuged at 5000 rpm for thirty minutes. According to H. Amer (2018), cobalt concentrations were evaluated through a colorimetric method utilizing SPADNS[17]. We determined the amount of cobalt adsorbed on the GO or GO-TOPO by lowering the starting concentration (Co) from the equilibrium concentration (Ce). The adsorption capacity was determined by utilizing equations 1 and 2 to calculate adsorption percentage (percent) and adsorption capacity:

$$\text{Adsorption } \% = \frac{C_o - C_e}{C_o} \times 100\% \quad (1)$$

$$\text{Adsorption capacity} = \frac{C_o - C_e}{C_e} \times \frac{V}{m} \quad (2)$$

The starting and equilibrium concentrations of cobalt are given by Co (mg/L) and Ce (mg/L). Volume V (ml) and adsorbent m (g) are the measurements of the suspension (g). All of the data is the average of at least three different experiments.

3. Results and discussion

3.1. GO and GO-TOPO characterization

XRD measurements of graphite powder before and after oxidation were investigated using graphite oxidation techniques. Figure 1a shows that graphite powder with d-spacing = 3.34 has a high peak at 2 theta = 26.1695o. A new graphene oxide peak occurred at 2-theta = 10.93° when the graphite sheets were oxidized, displacing the graphite peak. In addition, the d-spacing value was increased to 8.0427 by introducing function groups between the graphite layers. This was attributed to the extensive oxidation of graphite used in the synthesis of graphene oxide. These results correspond with earlier studies [18-20]. A HRTEM imaging of graphene oxide is shown in Figure 2b. The graphene oxide layered structure's surface appears to be smooth and transparent, with slight folds and wrinkles.

The FT-IR spectra of GO and TOPO modified GO shown in Figure 1c. Several oxygen-containing functional groups may be seen in the GO FT-IR spectra. These correspond to stretching vibrations of the C-O, C-OH, and COOH bonds and the O–H and COOH bonds in C–C and C–C molecules. Immediately after loading TOPO on the graphene oxide surface, the O-H stretching was seen at 3173 cm1. A new peak was identified at 1692 cm1, connected to the stretching and vibrations in the TOPO spectrum. From the result of FTIR, we conclude that the graphene oxide surface was successfully modified with TOPO [14, 21]. Raman spectroscopy is an essential tool for the characterization of GO. The Raman spectral data is shown in Fig. 1d. The obtained Raman spectroscopy of GO revealed two significant bands: the D-band at 1345 cm-1 and the G-band at 1597 cm-1. The D and G bands had an intensity ratio (ID/IG) of 0.842. These findings agree with previous studies [22].

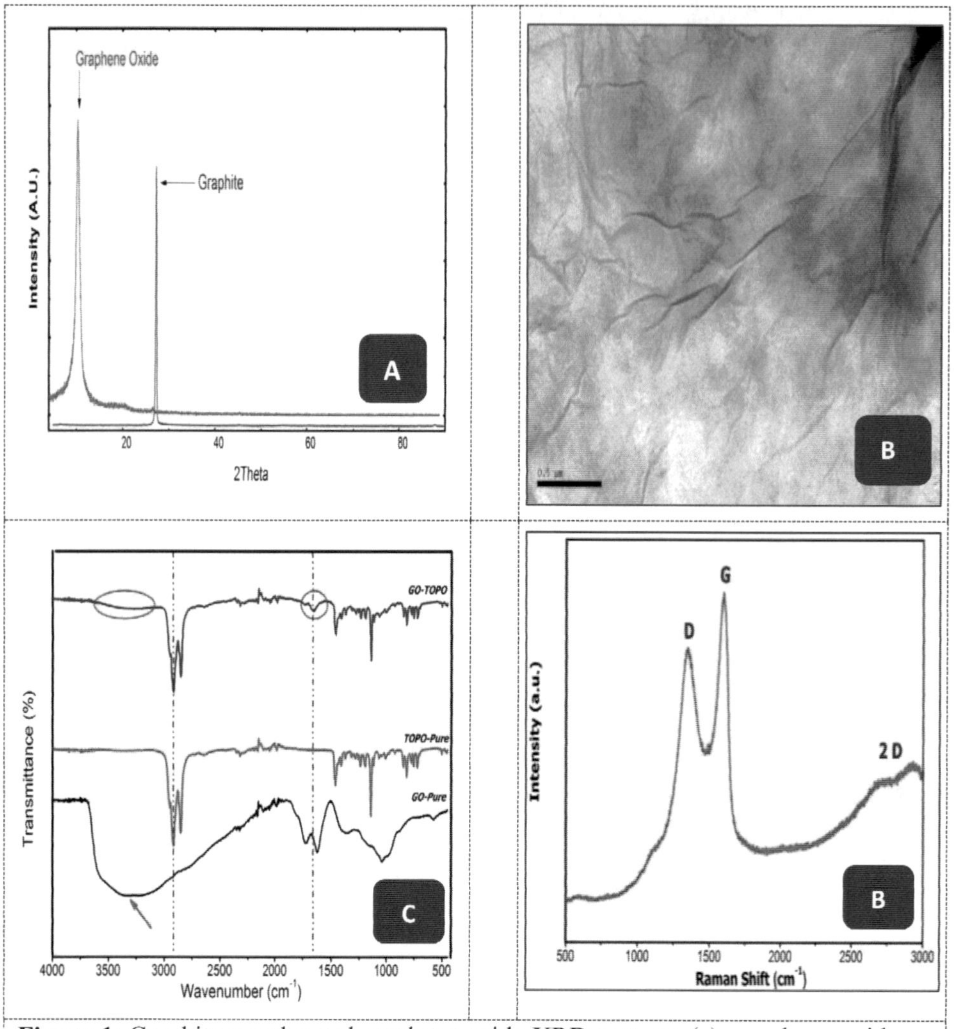

Figure 1: Graphite powder and graphene oxide XRD patterns (a), graphene oxide TEM image (b),graphene oxide FT-IR and TOPO modified GO FT-IR spectra (c), and graphene oxide Raman spectra (d)

3.2. Adsorption studies

3.2.1 The effect of adsorbent concentration

The behaviour of cobalt adsorption on GO-TOPO was investigated at three different adsorbent concentrations (0.25, 0.5, and 0.75 g/L). To increase the adsorption capacity for Co from 90 to 134 mg/g at the same co concentration, pH, and temperature, increase the adsorbent concentration from 0.25 to 0.50 g/L (Fig. 1a). The increased in the active adsorption sites and the metal cation capture through electrostatic interaction between negatively charged graphene oxide surfaces are responsible for this effect. The adsorption capacity was reduced to 133 mg/g when the adsorbent was increased to 0.75 g/L. Adsorption capacity decreases when adsorbent concentration increases due to adsorption site overlap or aggregation [23]. It is because the high capacity of Co (II) adsorption, GO-TOPO was tested at a concentration of 0.5 g/L in this study.

3.2.2 Effect of ion concentration

Cobalt concentrations of 25, 50, 75, and 125 ppm were used to evaluate the adsorption capacities of a 0.5 g/L adsorbent at a constant pH of 6 at temperature 25 °C. Figure 3a shows that increasing the adsorbate concentration from 25 ppm to 125 ppm increasing the adsorption capacity of GO-TOPO to cobalt from 43, 80, and 108 mg/g to 134 mg/g. It is due to the high number of metal ions competing for active sites in GO and GO-TOPO, adsorption capacity improves as cobalt concentration increases. These metal cations increase the concentration gradient's pulling power as their concentrations increase. The adsorption surface is initially high, but as adsorption occurs and equilibrium is reached, it progressively reduces. This is due to the GO surface has a significant number of empty active sites available for adsorption during the first stage, but the remaining unfilled sites become more difficult to fill after a period of time [5].

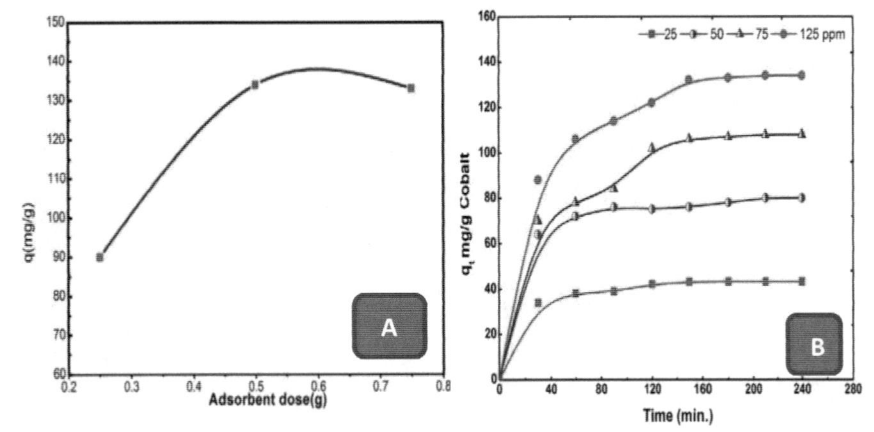

Figure 2. a) Effect of adsorbent dose GO-TOPO concentration b) Effect of Cobalt concentrations adsorbed on GO-TOPO.

3.2.3 Effect of temperature

To completely optimize the optimal conditions for cobalt removal from solution, the adsorption operation was performed at several temperatures (25, 40, and 50 °C) under the specified conditions (0.5g/L adsorbent, 125 ppm metal ion, and pH 6) showed in Figure 3 . A temperature increase from 25 to 40 degrees Celsius increased the cobalt adsorption capacity on GO-TOPO from 134 to 152 mg/g. The GO-TOPO capacity is reduced to 129 mg/g when heated to 50 degrees Celsius. An increase in adsorption capacity at 40°C showed that cobalt was responding endothermically on the GO surface. An increase in the number of active sites produced has also been associated with bond rapture [24]. On the other side, the temperature is 50°C. If cobalt removal is declining, it is almost definitely because it is more mobile at higher temperatures. Temperatures more than 40 degrees Celsius may reduce surface complexation and electrostatic interactions. [25].

3.2.4 Effect of initial pH.

Solution pH is an important controlling parameter that has a significant impact on adsorbent performance. Figure 3b shows the influence of pH on the adsorption of cobalt on GO and GO-TOPO (0.5 g/L adsorbent and 125 ppm cobalt). The pure GO adsorption capacity for cobalt was found to be 96, 125, and 83 mg/g at pH 4, 6, and 9, respectively. The adsorption capacities of GO-TOPO were 132,

134, and 112 mg/g. When the pH was reduced from 6 to 4, the adsorption capacity were reduced by 19%; however, raising the pH from 6 to 9 reduced the adsorption capacity by 33.6 percent in the case of pure GO. On the other hand, the adsorption capacity for GO-TOPO was slightly reduced by 1.5 % when the pH was reduced from 6 to 4. In the case of pure GO, however, raising the pH from 6 to 9 reduced the adsorption capacities by 8.9%. Both the pure GO and the GO-TOPO show weak adsorption of cobalt at low pH. This poor removal performance can be attributed to the intense competition for protonated adsorption sites between the numerous H with Co(II) ions. The removal efficiency of Co(II) rapidly increased to its maximum level as the pH values gradually increased from acidic to neutral. The interactions between the progressively deprotonated functional groups and the target cobalt ions are credited with the high adsorption ability. At high pH values, soluble OH- Co(II) complexes predominate in solution, which are difficult to adsorb on GO and GO-TOPO. From the results, it can be concluded that the modification of GO with TOPO is related to the stability of removal cobalt in acidic solutions with a comparison with pure GO. These results are in complete agreement with those of the previous study [13, 26].

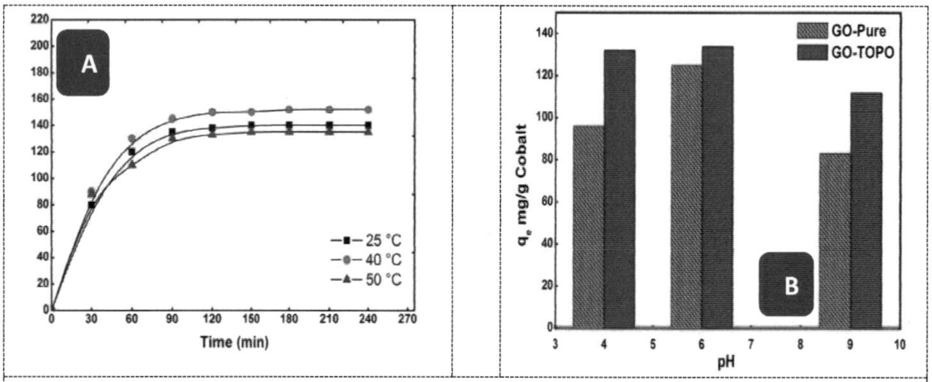

Figure 3: A) Effect of temperature on adsorption capacity of Cobalt GO-TOPO at pH 6. B) Effect of pH on adsorption capacity at room temperature.

4. Adsorption isotherms

Equilibrium isotherm tests were conducted to determine the maximum adsorption capacity of GO-TOPO for Co (II), with initial Co (II) concentrations ranging from 25 to 125 ppm and other parameters remaining constant (adsorbent concentration = 0.5 g/L and pH 6.0). In Fig. 5, the adsorption capacity of GO-TOPO is plotted against the equilibrium concentration of cobalt in aqueous water, and the data was simulated using the widely used Langmuir and Freundlich isotherm models to better understand the adsorption mechanism.

$$\frac{C_e}{q_e} = \frac{1}{q_{max} K_L} + \frac{1}{q_{max}} \quad (3)$$

$$ln\, q_e = K_F + \frac{1}{n} ln\, q_e \quad (4)$$

Eqs. (3) and (4) describe the Langmuir and Freundlich isotherm models, respectively, where Q max is the maximum adsorption capacity for Co (II) and KL is the equilibrium constant. The Freundlich constants KF and n are related to adsorption capacity and adsorption intensity, respectively. Table 1 shows the Langmuir and Freundlich isotherm parameters determined via fitting techniques. With a correlation coefficient (R2) of 0.9998, the Langmuir equation fits the experimental data better than the

Freundlich model. According to the Langmuir model, cobalt is adsorbed by specific GO-TOPO sites and forms a monolayer.

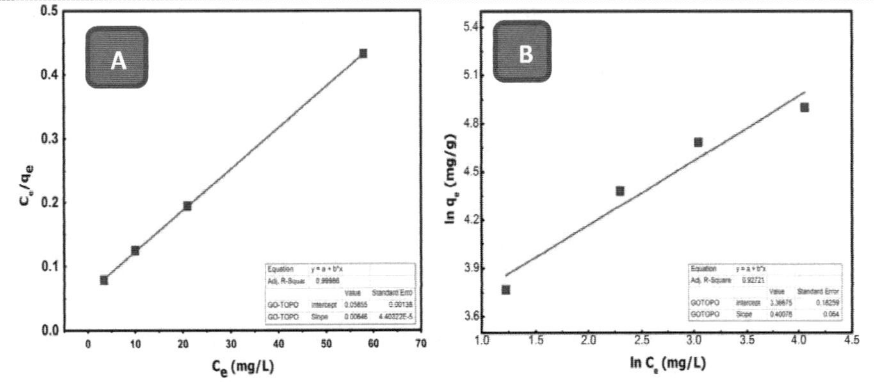

Figure 4. Langmuir a) and Freundlich b) adsorption isotherms model of cobalt on GO-TOPO.

Table 1. Langmuir and Freundlich adsorption constants associated to the adsorption isotherms of cobalt on GO-TOPO.

Langmuir		**Freundlich**	
q_{max} (mg/g)	154.7	K_F (mg/g (mg/L)^n)	3.3668
K_L (mg/L)	0.1103	*n*	2.495
R^2	0.9998	R^2	0.981

5. Conclusion

The graphene oxide was synthesized using the improved hammer process. It was characterized using HRTEM, FTIR, XRD, and Raman Spectroscopy. The TOPO was successfully used to functionalize the GO. The GO-TOPO was used for the removal of cobalt from water. It was revealed that functionalizing GO with TOPO enhanced Co (II) removal in acidic media. The effects of contact time, pH, and temperature on the adsorption process were investigated. When compared to GO and GO-TOPO, the results indicate that has an effect on the stability of cobalt removal in acidic solutions. pH and temperature values of 6.0 and 40 °C were found to be ideal for removing Co (II) from water respectively. Langmuir and Freundlich equations have been used to describe the isothermal data. According to R square, the Langmuir model was better than the Freundlich model.

Reference

[1] A. Mohanty, N. Devi, L.B. Sukla, N. Swain, Materials Today: Proceedings 30 (2020) 262-266.
[2] K. Rao, Current Science 81 (2001) 1534-1546.
[3] L. Chen, S. Lu, S. Wu, J. Zhou, X. Wang, Journal of Molecular Liquids 209 (2015) 397-403.
[4] M. Liu, C. Chen, J. Hu, X. Wu, X. Wang, The Journal of Physical Chemistry C 115 (2011) 25234-25240.
[5] W.M. El Rouby, A.A. Farghali, M. Sadek, W.F. Khalil, Journal of Inorganic and Organometallic Polymers and Materials 28 (2018) 2336-2349.
[6] J. Zhao, W. Ren, H.-M. Cheng, Journal of Materials Chemistry 22 (2012) 20197-20202.
[7] G. Zhao, J. Li, X. Ren, C. Chen, X. Wang, Environmental science & technology 45 (2011) 10454-10462.
[8] C. Bielawski, D. Dreyer, S. Park, R. Ruoff, Chem. Soc. Rev 39 (2010) 228-240.
[9] T. Sreeprasad, S.M. Maliyekkal, K. Lisha, T. Pradeep, Journal of hazardous materials 186 (2011) 921-931.
[10] G. Zhao, X. Ren, X. Gao, X. Tan, J. Li, C. Chen, Y. Huang, X. Wang, Dalton Transactions 40 (2011) 10945-10952.
[11] G. Ramesha, A.V. Kumara, H. Muralidhara, S. Sampath, Journal of colloid and interface science 361 (2011) 270-277.
[12] T. Sato, M. Yamatake, Zeitschrift für anorganische und allgemeine Chemie 391 (1972) 174-182.
[13] R.A. Kumbasar, Separation and purification technology 68 (2009) 208-215.
[14] C.J. Madadrang, H.Y. Kim, G. Gao, N. Wang, J. Zhu, H. Feng, M. Gorring, M.L. Kasner, S. Hou, ACS applied materials & interfaces 4 (2012) 1186-1193.
[15] D.C. Marcano, D.V. Kosynkin, J.M. Berlin, A. Sinitskii, Z. Sun, A. Slesarev, L.B. Alemany, W. Lu, J.M. Tour, ACS nano 4 (2010) 4806-4814.
[16] S. Mishra, J. Dwivedi, A. Kumar, N. Sankararamakrishnan, New Journal of Chemistry 40 (2016) 1213-1221.
[17] H. Amer, W.M. Moustafa, A.A. Farghali, W.M. El Rouby, W.F. Khalil, Zeitschrift für anorganische und allgemeine Chemie 643 (2017) 1776-1784.
[18] X. Wang, Z. Chen, S. Yang, Journal of Molecular Liquids 211 (2015) 957-964.
[19] X. Fan, W. Peng, Y. Li, X. Li, S. Wang, G. Zhang, F. Zhang, Advanced Materials 20 (2008) 4490-4493.
[20] X. Meng, D. Geng, J. Liu, M.N. Banis, Y. Zhang, R. Li, X. Sun, The Journal of Physical Chemistry C 114 (2010) 18330-18337.
[21] J. Park, S.-J. Park, S. Kim, Research on Chemical Intermediates 40 (2014) 2509-2517.
[22] H. Chen, D. Shao, J. Li, X. Wang, Chemical Engineering Journal 254 (2014) 623-634.
[23] P. Nuengmatcha, R. Mahachai, S. Chanthai, Oriental journal of chemistry 30 (2014) 1463-1474.
[24] A. Fakhri, Journal of Saudi Chemical Society (2013).
[25] C. Wang, H. Luo, Z. Zhang, Y. Wu, J. Zhang, S. Chen, Journal of Hazardous materials 268 (2014) 124-131.
[26] Y. Cai, X. Wang, J. Feng, M. Zhu, A. Alsaedi, T. Hayat, X. Tan, Environmental Pollution 249 (2019) 434-442.

Gamma Radiation Synthesis and Characterization of Poly (Polyaniline /Glycidyl methacrylate) Containing TiO_2 Nanoparticles and Studying Their Physicochemical Properties.

M Salah[1], M B El-Arnaouty[1], M Eid[1], El-Sayed Soliman[2] and El-Sayed A Hegazy[1]
[1]Polymer chemistry department, National Centre for Radiation Research and Technology, Egyptian Atomic Energy Authority, Cairo, Egypt.
[2]Factually of Science, Ain Shams University, Cairo, Egypt.
Corresponding author's e-mail (M Salah): mohamed.soliman.ncrrt@gmail.com

Abstract. Gamma radiation was successfully utilized to formulate conducting polyaniline/glycidyl methacrylate containing TiO_2 [(PANI/GMA)-TiO_2]. The pattern of X-ray diffraction of the produced polymers indicated the existence of TiO_2 inside the composite, with peaks connected to the TiO_2 centred at $2\Theta = 25.3°$, 27.5°, 37°, 38°, 41.37°, 48.1°, 54.26°, and 55.2°. The interaction between TiO_2 and PANI/GMA caused the distinctive FTIR peaks of PANI to shift to a higher wavenumber in [(PANI/ GMA)-TiO_2] nanocomposites. Scanning electron microscopy was utilized to evaluate these composites' morphological and structural aspects. The thermal stability tests utilizing thermal gravimetric analysis demonstrated that the inclusion of TiO_2 leads to enhancing thermal stability. The electrical conductivity of the material as determined by the LCR Meter was around 10^{-3} (S/cm), indicating that it might be used as an electro-conductive substance.

Keywords: TiO_2, PANI, X-ray diffraction, FTIR, electrical conductivity.

1. Introduction

Conductive polymers, also called intrinsically conducting polymers, are organic polymers capable of conducting electricity [1]. Conducting polymers have been used in various applications, including batteries, electrical devices, sensors, and anti-corrosive inorganic coatings [2].

Polyaniline (PANI) is an electrically conducting polymer that has been widely studied due to its ease of processing, environmental stability, and prospective applications in batteries, biosensors, catalysis, and electronic technologies [3]. Additionally, it is amenable to doping with inorganic and organic acids. Furthermore, PANI has emerged as a viable candidate for practical applications such as electromagnetic radiation shielding, metal corrosion prevention, transparent electrodes, light-emitting diodes, and humidity sensing battery applications. The PANI polymer chain contains reactive –NH– groups bordered on each side by phenylene rings, providing the system with chemical flexibility and significantly increasing its potential [4].

Epoxy resin is a substantial thermosetting polymer that is widely used in high-performance composite materials, electronics, coatings, adhesives, and aerospace industries due to its superior chemical and mechanical characteristics, which include good chemical resistance, a high heat distortion temperature, and high tensile and compressive strength [5]. Incorporating inorganic fillers in order to boost the qualities of neat epoxy nanocomposites has been done for a long time. They are classified as specialty products due to inorganic fillers scattered on the nanoscale and at least one dimension of the filler phase less than 100 nm.

The use of inorganic TiO_2 nanocrystals with a big energy gap and a high refractive index leads to a modified polymer with novel optical characteristics that can be tweaked by modifying the concentration of the embedded inorganic part, size, and composition, as well as increased thermo-mechanical stability [6].

In this research paper, Gamma radiation was used to create polyaniline/glycidyl methacrylate (PANI/GMA) and polyaniline/glycidyl methacrylate incorporating TiO_2 [(PANI/GMA)-TiO_2]

Journal of Physics: Conference Series 2305 (2022) 012015 doi:10.1088/1742-6596/2305/1/012015

nanocomposites. Next, the products were evaluated by X-ray diffraction (XRD), Fourier-transform infrared spectroscopy (FT-IR), Transmission electron microscope (TEM), Thermogravimetric analysis (TGA), electrical conductivity (EC), and scanning electron microscope (SEM). The purpose of producing PANI/GMA and [(PANI/GMA)-TiO_2] is to use them as charge storage materials or solar cell materials due to their exceptional photovoltaic characteristics.

2. Materials and Methods

2.1. Materials
Titanium dioxide (size =100 nm) and Glycidyl methacrylate (2, 3 epoxy propyl methacrylate) were procured from Sigma Aldrich (USA). Aniline was supplied by the Oxford laboratory (India). Methanol (99%) and Hydrochloric acid were obtained from Beta Chem. for laboratory chemicals (Egypt).

2.2. Preparation of polyaniline (PANI)
20 ml aniline was mixed with HCl (40 ml) and then dissolved in methanol (40 ml) to get 20% aniline. The mixture was agitated at 25 °C for 2 h before being irradiated with a 60Co gamma-ray irradiator at a dosage of 10 kGy.

2.3. Preparation of PANI/GMA
As a stock solution, Glycidyl methacrylate (GMA) was dissolved in methanol in a 50:50 volume ratio at a temperature of 25 °C. The produced PANI was added to the GMA solution in an 80:20 volume ratio and stirred for 2 h to prepare PANI/GMA, then placed into test tubes and irradiated with gamma rays at a 10 kGy irradiation dosage.

2.4. Synthesis of [PANI/GMA)-TiO_2] nanocomposites
0.3 % nano-sized titanium dioxide (TiO_2) was added to the prepared PANI/GMA under stirring (2h and 25 °C) and sonicated for 12 min. Then, the blends were put in glass dishes and irradiated with gamma rays at a dose of 10 kGy. The resulting polymers were dried in an oven at 70°C for 24 h.

3. Characterization

3.1. FT-IR spectroscopy
FTIR examination of materials was carried out by ASTM E 1252- (02) "Standard Practice for General Techniques for Obtaining Infrared Spectra for Qualitative Analysis," via FTIR Perkin Elmer-Spectrum (400-4000 cm-1), which was found in Egyptian Petroleum Research Institute (EPRI).

3.2. X-ray diffraction
The samples were analyzed with an XRD spectrophotometer that was found in Egyptian Petroleum Research Institute (EPRI), [Model: X'pert Pro, Manufacture: PAN analytical, equipped with X-ray tube (40 KV, 30 mA, Cu target). The X-ray survey was conducted between 4 and 80 (degrees) 2Θ at a scanning speed of 8 (deg/min) and in continuous scanning mode.

3.3. Thermogravimetric analysis
TGA was carried out by TGA-DSC (Model: SDTQ 600, USA), found in Egyptian Petroleum Research Institute (EPRI). The temperature was raised from ambient to 600° C at a rate of ten degrees Celsius per minute. Using primary TGA thermograms, we investigated the rate of thermal degradation and thermal stability of several nanocomposites.

3.4. Scan Electron Microscope
A scanning electron microscope (JOEL JSM 5300, Japan), found in Egyptian Petroleum Research Institute (EPRI), was used to investigate the morphology of various nanocomposites using an intense electron beam at high resolution and magnification.

3.5. EC measurements
EC measurements were accomplished via LCR Meter (TEGAM, Model: 3550), found in the faculty of science, Ain Shams University.

4. Result and discussion

4.1. FT-IR measurements
Figure 1 (a) displays the FT-IR spectra of PANI/GMA, indicating the presence of C=C stretching vibration bands associated with the benzenoid and quinonoid units at 1496 and1602 cm^{-1}, respectively. Further, the stretch vibration of the N–H bonds was seen at 3353 cm^{-1} [7]. Moreover, a stretching band ascribed to C-N was identified at 1320cm^{-1} [8]. The ester C=O group (due to the GMA core) and the C-O stretching vibration were assigned to the absorption bands at 1719 and 1161cm^{-1}, respectively. The C-O-C stretching (1257cm^{-1}) with symmetric stretching vibration, asymmetric stretching vibration (943 cm^{-1}), and a strong band (748 cm^{-1}) may be ascribed to hydrogen atoms nearby in the epoxide ring. A strong band attributed to (=C-H) out-of-plane bending of the aromatic ring was seen at 692 cm^{-1} [9]. PANI/GMA characteristic peaks indicate that the PANI shell has been successfully coated.

In Figure 1 (b), the spectrum of [(PANI/GMA)-TiO_2] exhibits the same characteristic peaks, as well as a peak at 670 cm^{-1}, which correlates to Ti–O stretching [10]. However, there was peak displacement when TiO_2 nanoparticles were introduced to PANI/GMA. 3353 cm^{-1} to 3391 cm^{-1} and 1719 cm^{-1} to 1724 cm^{-1} are the displacements. The associated peaks were moved to a higher wavenumber, and their intensities were altered after adding TiO_2 nanoparticles. Furthermore, methylene's vibration band was shifted from 2964 cm^{-1} to 2940 cm^{-1}. These considerable changes indicate that PANI and its inorganic analog, TiO_2 nanoparticles, exhibit intermolecular bonding (interfacial interaction). It is widely established that titanium in Titania forms coordination complexes with nitrogen atoms in PANI structures [11].

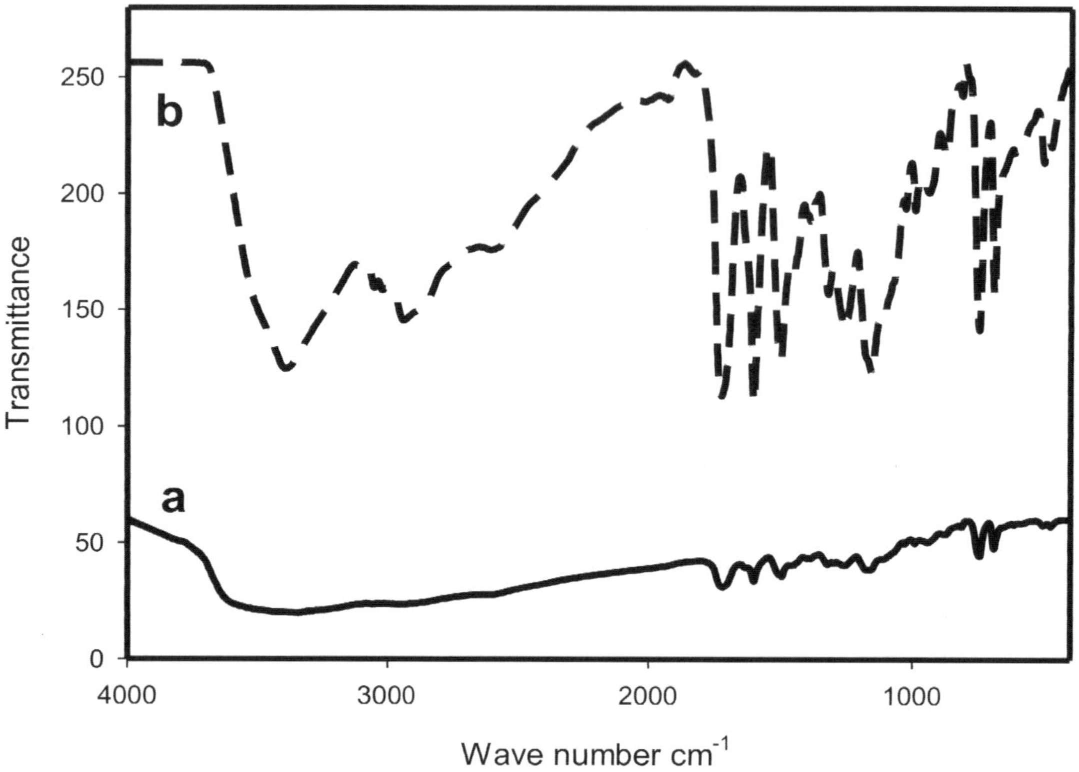

Figure 1: The FTIR spectra of (a) PANI/GMA, (b) 0.3% of [(PANI/GMA)-TiO_2].

4.2. Scan Electron Microscopy analysis

Scanning electron microscopy (SEM) is used for determining the porosity and shape of polymer matrices. The SEM picture of PANI/GMA displays an amorphous surface with a spongy look, as seen in Figure 2(a) [12]. The SEM image of the [(PANI/GMA)-TiO_2] nanocomposite in Figure 2(b) reveals a granular structure resembling coral reef with an interlocking pattern. This clearly demonstrates that most TiO_2 are PANI-coated [13]. TiO_2 is distributed and packed more uniformly in the PANI/GMA matrix, and no big lumps [7].

Figure 2: SEM image of (a) PANI/GMA, (b) 0.3% [(PANI/GMA)-TiO_2] at a 10 kGy irradiation dosage.

4.3. X-ray diffraction measurements

XRD analysis was used to determine the structure of synthesized PANI/GMA and [(PANI/GMA)-TiO_2] and the effect of TiO_2 on the structure PANI/GMA. The XRD patterns of PANI/GMA and [(PANI/GMA)-TiO_2] are shown in Figure 3(a & b). The XRD pattern of PANI/GMA has several crystalline peaks at $2\Theta = 21°$, 22°, 26.7°, and 35.6°, as well as a single massive amorphous peak at $2\Theta = 16°$, As exposed in Figure 3(a). The peaks at 21°, 22°, 26.7°, and 35.6°, respectively, correspond to periodicity parallel and perpendicular to the conjugation chains of PANI. The crystalline peak indicates that the PANI doped with HCl is not entirely amorphous and that the π-π interchain stacking has been reinforced [14], [15].

The XRD patterns of [(PANI/GMA)-TiO_2] exhibit typical peaks for both PANI and TiO_2 nanoparticles, demonstrating the presence of TiO_2 nanoparticles inside the composite, as shown in Figure 3(b). In the case of [(PANI/GMA)-TiO_2] nanocomposite materials with TiO_2 concentrations of 0.3 percent, an amorphous peak at $2\Theta=20°$ was seen, while the peaks associated with TiO_2 nanoparticles were located at $2\Theta=25.3°$, 27.5°, 37°, 38°, 41.37°, 48.1°, 54.26°, and 55.2°.

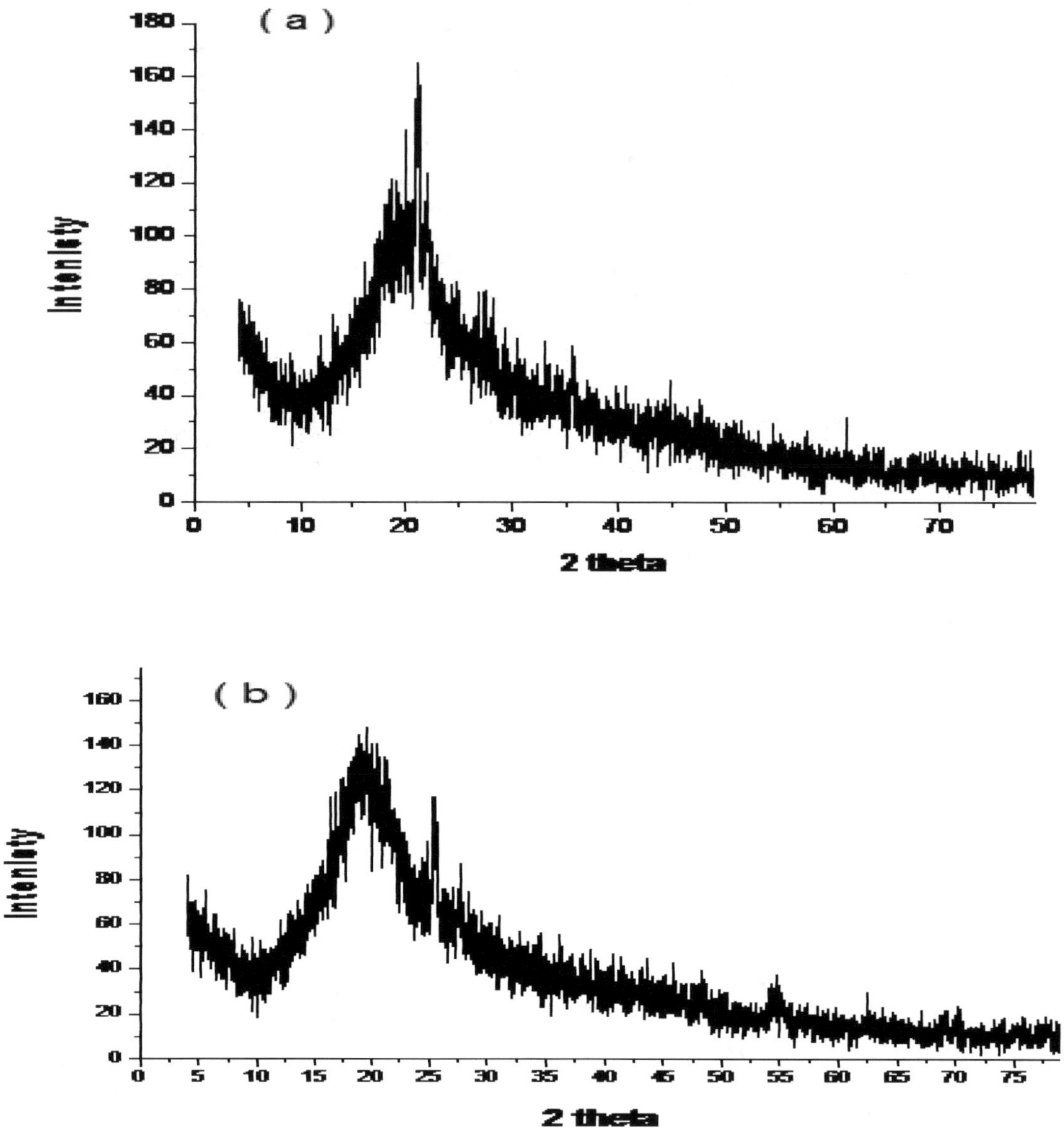

Figure 3: XRD of (a) PANI/GMA, (b) 0.3% of [(PANI/GMA)-TiO_2] at a 10 kGy irradiation dosage.

4.4. Thermogravimetric analysis

The Thermogravimetric analysis is usually regarded as the most feasible and extensively used approach for demonstrating polymer thermal stability across a broad temperature range. The multistage thermal breakdown of the PANI/ GMA is shown in Figure 4(a). The first stage reaches temperatures of up to 172°C because of moisture evolution, with a thermal breakdown rate of 5.3 percent. The second stage, up to 303°C, may be caused by the thermal breakdown of functional groups in the polymer's side chain, with a thermal decomposition percentage of 27.4 percent. Significant breakdown temperatures were 346°C, 413°C, and 600°C, respectively, with thermal decomposition rates of 47.6, 76.72, and 93.2 percent, which might be attributable to anhydride formation and backbone destruction [16], [17].

The thermal decomposition of the 0.3% of [(PANI/ GMA)-TiO_2] composite is depicted in Figure 4(b). The first stage, which reaches 107°C, may be caused by residual water in the matrix with a deteriorated weight percent of 5.3 percent. The second and third phases, which reach temperatures of up to 201 and 302 °C, respectively, may be caused by the breakdown of the polymer's function groups and desorption of TiO_2 nanoparticles in the side chains at degraded weight percentages of 17.4 and 37.1 percent, respectively. Significant decomposition temperatures were between 347 and 394°C, with degraded weight percent of 61.1 and 79.4 percent ascribed to main chain breakage, respectively. At 600°C, the remaining weight percent was determined to be 7.2 percent, indicating that carbonization prevailed [18].

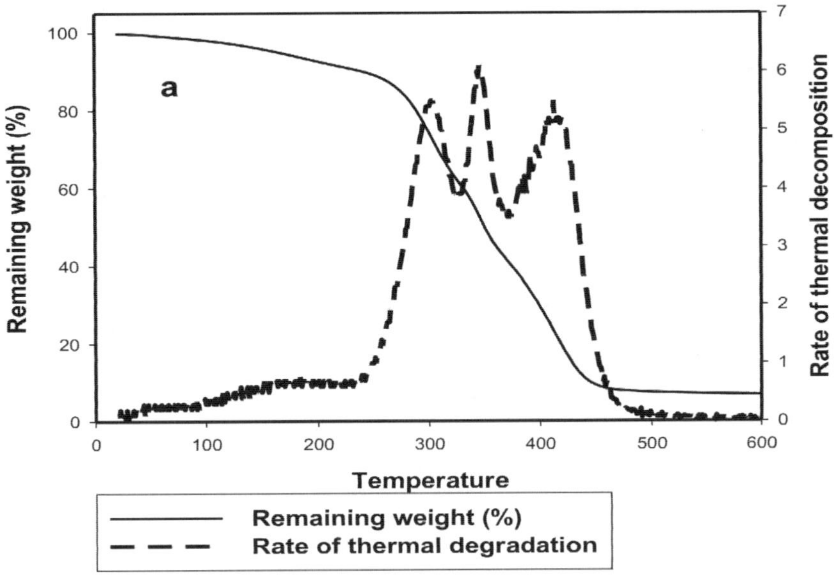

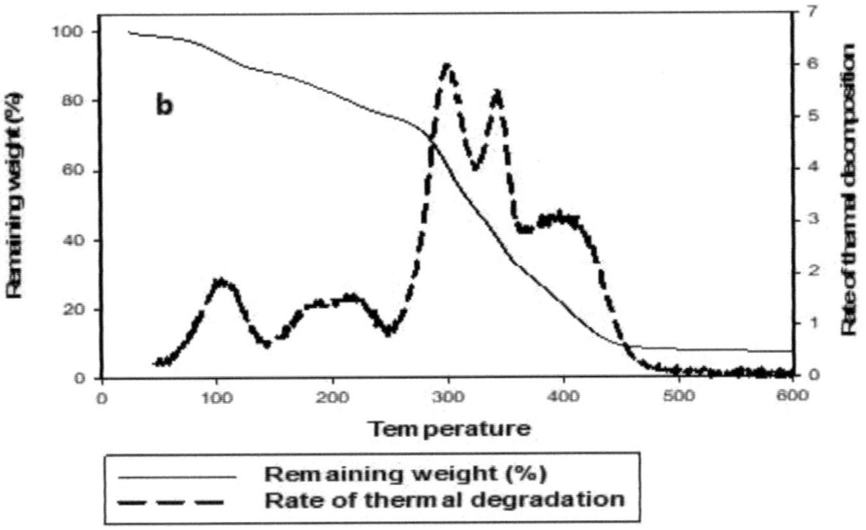

Figure 4: The Thermogravimetric analysis of (a) PANI/GMA, (b) 0.3% of [(PANI/GMA)-TiO_2] at a 10 kGy irradiation dosage.

The thermal breakdown rate (dw/dt) of [(PANI/GMA)-TiO_2] is shown in Table 1 for a variety of TiO2 concentrations, maximum decomposition temperatures (Tm), and decomposition start temperatures (Td). Tm and Td values were reduced when TiO_2 was added to PANI/ GMA. Additionally, it can be shown that the incorporation of TiO_2 reduces the maximum thermal breakdown rate, implying an improvement in thermal stability. Based on the remaining weight percent at 600°C, it can be deduced that the integration of TiO_2 enhances the remaining weight percent for various TiO_2 concentrations, indicating an increase in thermal stability [12].

Table 1: The thermal decomposition rate of [(PANI/GMA)-TiO_2] with varying TiO_2 concentrations, maximum decomposition temperatures (Tm), and decomposition start temperature (Td).

Sample description	Temperature (°C)	Rate of thermal decomposition	T_m	Remaining weight (%)	T_d	Remaining weight (%) at 600°C
P(ANI/ GMA)	172	0.6	172	94.7	104	6.8
	303	5.56	303	72.6	243	
	346	6.05	346	52.4	326	
	413	5.52	413	23.2	375	
P(ANI/ GMA)- TiO_2 (0.3%)	107	1.85	107	94.7	60	7.2
	201	1.57	201	82.6	142	
	302	5.9	302	62.9	249	
	347	5.3	347	38.9	323	
	394	3.02	394	20.55	367	

4.5. EC measurements
The conductivities of PANI/GMA and [(PANI/GMA)-TiO_2] nanocomposites are summarized in Table 2. The EC of PANI varies between 10^{-10} and 10^3 S/cm depending on the acid dopant and fillers used [10]. EC measurements on PANI/GMA and [(PANI/GMA)-TiO_2] nanocomposites show that when TiO_2 is added to PANI/GMA, the EC of PANI/GMA drops. The formation of bonds in the structure results from the –NH of PANI adsorbing on the surface of TiO_2 nanoparticles. The conductivity of a polymer is regulated by the dopant composition and the concentration of inorganic components, affecting the conductivity of nanocomposites [12].

Table 2: The EC of PANI/GMA and [(PANI/GMA)-TiO_2] nanocomposites with varying TiO_2 contents of 0.3 percent after 10 kGy irradiation.

Formulation	Electrical conductivities (S/cm) x 10^{-3}
PANI/GMA	1.343
[(PANI/GMA)- TiO_2] (0.3%)	1.137

5. Conclusion
Using gamma radiation, a fast and easy approach for the preparation of P (ANI/GMA) and P (ANI/GMA)- TiO_2 nanocomposite has been demonstrated. According to the FTIR data, a chemical interaction between P(ANI/GMA) and TiO_2 has been developed. According to XRD analysis, the P (ANI/GMA) includes crystalline peaks in addition to the amorphous one. The addition of TiO_2 increases

the thermal stability of different preparations, according to research on thermal stability. Because TiO_2 nanoparticles are packed in the P(ANI/GMA) matrices with uniform distribution, SEM revealed that TiO_2 nanoparticles had a substantial effect on the morphology. The observed electrical conductivity was roughly around10^{-3} S/cm. Finally, P(ANI/GMA)-TiO_2 nanocomposites have the potential to be used in a variety of industrial applications, such as solar cells and charge storage.

References

[1] Inzelt G 2008 Chapter 1: Introduction. In Scholz F Conducting Polymers: *A New Era in Electrochemistry Monographs in Electrochemistry* Springer pp 1.
[2] Akbarinezhad E, Ebrahimi M , and Faridi H R 2009 Corrosion inhibition of steel insodium chloride solution by undoped polyaniline epoxy blend coating *Prog.Org. Coat.* **64** 361.
[3] Karim M R, Lee C J, Park Y -T, and Lee M S 2005 SWNTs coated by conducting polyaniline: Synthesis and modified properties *Synth. Met.* **151** 131.
[4] Pron A and Rannou P 2002 Processible conjugated polymers: from organic semiconductors to organic metals and superconductors *Prog. Polym. Sci.* **135** 90.
[5] May C A 1988 *Epoxy Resins Chemistry and Technology, second ed.*, Marcel Dekker Inc.
[6] Convertino A, Leo G, Tamborra M, Sciancolepore C, Striccoli M, Curri M, and Agostiano A 2007 TiO2 colloidal nanocrystals functionalization of PMMA: a tailoring of optical properties and chemical adsorption *Sens. Actuators B* **126** 138.
[7] Strawhecker K E and Manias E 2000 Structure and Properties of Poly(vinyl alcohol)/Na+ Montmorillonite Nanocomposites *Chem Mater* **12** 2943.
[8] Li X W, Chen W, Bian C Q, He J B, Xu N, and Xue G 2003 Surface modification of TiO2 nanoparticles by polyaniline *Appl. Surf. Sci.* **217** 16.
[9] Sharma B K, Khare N, Dhawan S K, and Gupta H C 2009 Dielectric properties of nano ZnO-polyaniline composite in the microwave frequency range *Journal of Alloys and Compounds* **477** 370.
[10] Gurunathan K, Amalnerkar D P, and Trivedi D C 2003 Synthesis and characterization of conducting polymer composite (PAn/TiO_2) for cathode material in rechargeable battery *Mater. Lett.* **57** 1642.
[11] Zhang Wen L, Piao Shang H and Choi H 2013 Facile and fast synthesis of polyaniline-coated poly(glycidyl methacrylate) core–shell microspheres and their electro-responsive characteristics *Journal of Colloid and Interface Science* **402** 100.
[12] Mostafaei A and Nasirpouri F 2014 Epoxy/polyaniline–ZnO nanorods hybrid nanocomposite coatings :Synthesis, characterization and corrosion protection performance of conducting paints *Progress in Organic Coatings* **77** 146.
[13] Sathiyanarayanan S, Syed Azim S, and Venkatachari G 2007 A new corrosion protection coating with polyaniline–TiO2 composite for steel *Electrochimica Acta* **52** 2068.
[14] Shi L, Wang X, Lu L, Yang X, and Wu X 2009 Preparation of TiO2 / polyaniline nanocomposite from alyotropic liquid crystalline solution *Synthetic Metals* **159** 2525.
[15] Long Y, Chen Z, Wang N, Li J, and Wan M 2004 Electronic transport in PANI-CSA/PANI-DBSA polyblends *Physica B: Condensed Matter* **344** 82.
[16] Li F, Xu X, Li Q, Li Y, Zhang H, Yu J and Cao A 2006 Thermal degradation and their kinetics of biodegradable poly(butylene succinate- co-butylene terephthate)s under nitrogen and air atmospheres *Polymer Degradation and Stability* **91** 1685.
[17] Kwanghee L, Shinuk C, Sung Heum P, Heeger A J, Chan-Woo L, and Suck-Hyun L 2006 Metallic transport in polyaniline *Nat. Lett.* **441** 65-8.
[18] El-Arnaouty M B and Eid M 2010 Synthesis of Grafted Hydrogels as Mono-Divalent Cation Exchange for Drug Delivery *Polymer-Plastics Technology and Engineering* **49** 182.

Preparation and application of titanium dioxide nanoparticles in offset ink

Sayeda Sadek Ali[a,b,*], Yasser Kamal Abdel-moneam[c], Mohammad Hegazy[c,*]

a Egyptian Organization for Standardization and Quality (EOS), Egypt
b Department of Chemistry, Faculty of Science, Mansoura University, Egypt
c Department of Chemistry, Faculty of Science, Menoufia University, Shebin El-Kom, Egypt

E-mail: sosadek@yahoo.com; dr_hegazy2000@yahoo.com

Abstract. TiO_2 nano pigments in pure crystallographic forms (anatase) were prepared via a hydrothermal approach. The produced pigments were investigated physiochemically by several techniques, then they were utilized in paper-coating mixtures and their effect on the properties of coated papers was investigated. XRD revealed that these pigments produced a pure anatase phase pigment. TEM showed that the crystallite size for anatase is 50 nm. Whether optical or physical properties for the treatment before and after application were tracked. Although TiO_2 is applied as a white dye, it provided good results on the black ink that is used in dyeing, whether ordinary TiO_2 or that in the range of nano, in small quantities. The titanium oxide nanoparticles gave better results because of the increase in the surface area and optical or electromagnetic properties obtained when converting it to the range of nanoscale. Most interestingly, the addition of TiO_2 nanoparticles provided an improved ink that can carry out heat, humidity, oxygen, sunlight, and vibration for a long time, as indicated by the accelerated aging test. Thus, the prepared nano pigments of the TiO_2 phase could be applied in jobbing offset ink and commercial paper sheets as a promising and novel pathway.

Keywords: offset ink, TiO_2 nanoparticles, dispersion, optical density, improved rheological properties

1. Introduction

TiO_2 is quintessential photo-catalyst (i.e. a material which permits a certain reaction to occur in the existence of light, that otherwise does not happen). Among the different TiO_2 polymorphs, pure anatase is considered the most promising polymorph for photo catalytic applications as it offers higher electron mobility, lower dielectric constant, and lower density compared to rutile and brookite [1,2]. TiO_2 also has the added benefit of offering superior brightness, which also increases the quality of a finished product considering its appearance. The abrasive nature of this compound does offer a problem to printers, especially gravure printers, because of its transitivity, which wears down the engraved imaged cylinder and ink metering doctor blade [3].

TiO_2 nanostructures (in various forms) have many applications thanks to the size-related features. For nanometer-scaled TiO_2, the energy band structure gets separated because of its unique surface,

photochemical as well as photophysical properties. Therefore, lots of works have reported the synthesis of nanocrystallineTiO_2 featured with high surface areas [4].
TiO_2 nanostructures have gained much attention in resolving many pollution or environmental problems. Hence, using TiO_2 for H_2 production and photo-assisted water-splitting devices introduces a good way for clean and low-cost production of hydrogen through the mean of solar energy [5].
In this paper, TiO_2 nanostructures were successfully synthesized as reported in the literature, tracking its effects on comparison with commercial ones, and measuring the optical and physical properties of the treatment before and after application using an offset ink (jobbing). The aim of this research is to make an improved ink with nanoparticles used in writing important printed paper documents.

2. Experimental

2.1. Materials

Titanium tetrachloride ($TiCl_4$, 98%) was purchased from Organochem Company, India. Ammonium hydroxide (NH_4OH, 35%) and ethanol (C_2H_5OH, 99%) were purchased from Adwic Pharmaceutical and Chemicals Company, Egypt. Titanium dioxide (TiO_2, 98%) was purchased from Alpha Chemika, India. Jobbing inks were purchased from Pachin-Elobour Paints and Chemical Industry, Egypt.

2.2. Preparation of titanium dioxide nanoparticles

By the hydrothermal method of titanium tetra chloride, a 4ml solution was dropped into a 400 ml mixture of EtOH & distilled H_2O (ratio of 4:1). The mixture was heated to around 80° C with a continuous stirring. A white suspension of TiO_2 nanoparticles was formed after heating for 2 h at the mentioned temperature. The experiment was continued for another 2 h to get rid of chloride ions as hydrogen chloride gas. TiO_2 nanoparticles were settled down through centrifugation at 5400 rpm for 30 min, then the precipitate was filtrated and washed in water several times until it became free from impurities, then dried in a drying oven at 50 °C. After calcination for about 4 h at 400 °C, the aimed TiO_2 nanoparticles were obtained as a white powder.

2.3. Preparation of the modified inks

Modified inks mixtures were prepared by mixing the prepared nano titanium or commercial Degussa TiO_2 with offset inks (jobbing) with weight ratios (0.1, 0.15, 1, 3) % jobbing offset ink (J) JNx, and JDx. Where N refers to nanoTiO_2 modification, D is commercial Degussa modification, and x is the ratio of modified TiO_2 either nano or Degussa, where x = (0.1, 0.15, 1, 3).

2.4. Characterization of the prepared pigments and offset inks

The crystal structures, phase identifications, crystallinities, purities, as well as crystallite sizes for the as-prepared pigments were performed through XRD (Bruker axs D8; Germany) via Cu-Kα (where, k = 1.5406 A°) radiation, and secondary monochromatorin a range of 2h, from 10 to 80. Crystallite sizes are computed from XRD information. The morphological structures for the as-prepared pigments were investigated via high-resolution TEM (Jeol, JEM-2010, Japan) that operates at maxima of 200 kV. X-ray Photoelectron Spectroscopy studies were performed using a KRATOS XSAM-800, which ranged from 0 to 1300 eV. The properties of ink offset including the prepared nano pigments were assessed via certain standard tests for physical and/or optical properties.

Preparation of materials and mixing them with the offset ink (jobbing), the physical (morphological and rheological) properties were measured to analyze whether the modified offset set ink make suffice on its role or not. Viscosity and yield value were measured using Laray viscometer (France). Tack property was measured using an electronic digital inkometer (USA). The ink flow property was measured using a glass flowmeter, while the printed density was measured by a densitometer.

3. Results and discussion

3.1. Characterization of the prepared titanium dioxide nanopigments

The XRD patterns of nano and Degussa TiO_2 are shown in figure 1 and figure 2, respectively. The peaks of these materials were recognized to the following crystal planes: (101); (004); (200); (105); (211); (204); (116); (220); (215); (303). Furthermore, all the peaks were defined and could be successively corresponding to the anatase TiO_2 structure (JCPDS-21-1272). As known, the anatase titanium dioxide nanoparticles are photoactive and practically applicable for water purification. In addition, the XRD pattern had no characteristic peaks to the other crystalline forms, indicating a pure product of the anatase phase.

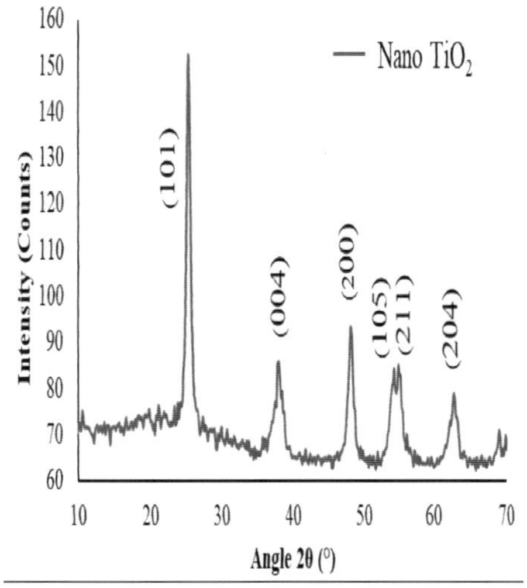

Figure 1. XRD pattern of nano TiO_2.

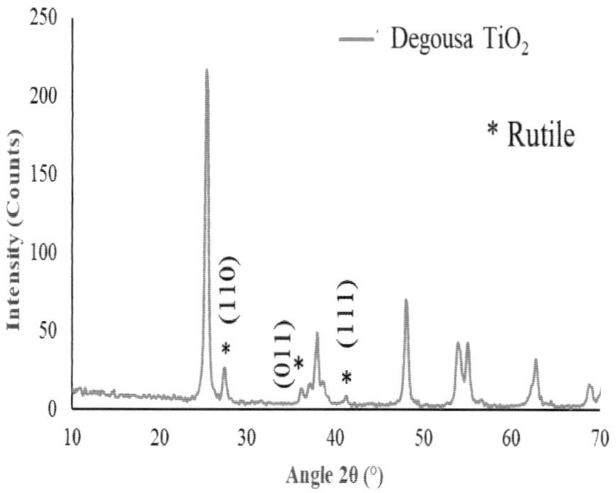

Figure 2. XRD pattern of DegussaTiO_2.

X-ray photoelectron spectroscopy (XPS) was used to make surface elemental analysis qualitatively, and quantitatively, besides studying the different oxidation states of the components' elements.

As shown in figure 3 and Table 1, the surface elemental analysis for investigated nano and DegussaTiO_2 showed the presence of Ti, O elements of metal oxide composition, besides a significant appearance for C, due to cell-instrumentation technique, which confirms the precise formation and preparation of pure TiO_2.

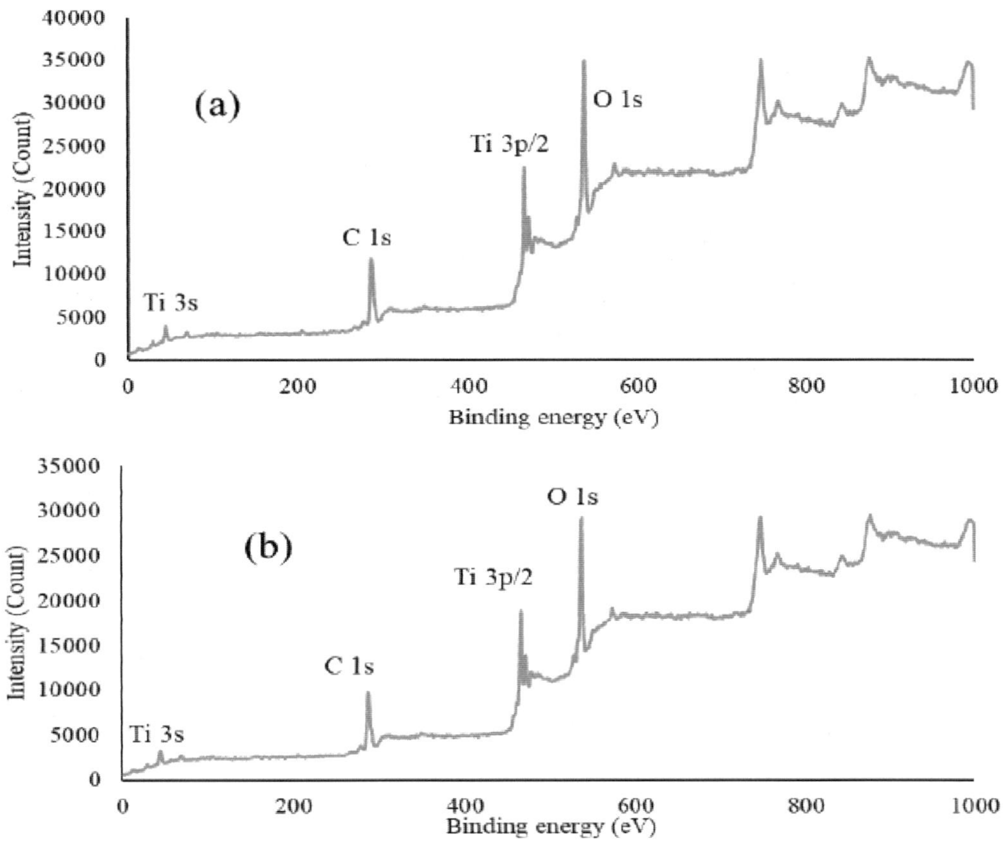

Figure 3. XPS charts of (a) Degussa TiO_2 and (b) nano TiO_2 samples.

Table 1. Elemental analysis of TiO_2 samples by XPS technique.

TiO_2 Samples	**Elements ratio (%)**		
	Ti	**O**	**C**
Nano	**35.0**	**31.6**	**33.4**
Degussa	**35.2**	**32.1**	**33.7**

The TEM micrographs of the nano and Degussa TiO_2 samples are represented in figure 4. It showed that nanoparticles had an almost spherical shape. The mean size-resolved from the width of arbitrarily chosen nanoparticles is similar to the size of the nanoparticles determined from the XRD results.

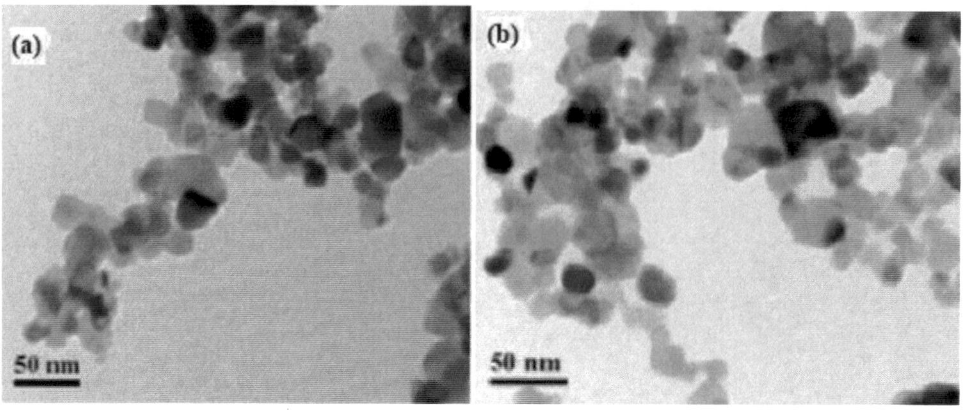

Figure 4. TEM of (a) Degussa TiO_2 and (b) nano TiO_2 samples.

3.2. Study the physical properties of modified ink

3.2.1. Rheological properties

In this section, we have studied the rheological properties of the modified ink with its different ratios to assure that the modification with TiO_2 doesn't demolish the rheological properties of the ink. On the contrary, it enhances the properties.

3.2.1.1. Viscosity

The main ingredients of any ink are pigments, additives, as well as solvents. A small changes in the related composition results in different chemical and/or physical characteristics of the ink which may affect the printing route negatively [6].

Both density and viscosity fora definite ink are of great interest for the producers of that ink because these physical characteristics can provide significant data on the utility as well as the quality, whether for the raw materials or the final products. Hence, the addition of nano and Degussa TiO_2 to this offset ink (jobbing ink), with 4 different ratios (0.1, 0.15, 1, 3) %, affects the offset ink, and the comparison to standard data of the pure offset ink are recorded in Table 2.
It's noticed that the addition of nano TiO_2 led to increasing the viscosity, while JD viscosity decreased with Degussa TiO_2. This is due to the hydrophilicity of Degussa TiO_2 besides the hydrogen bond between TiO_2, and oxygenated groups in jobbing offset ink, which allows Degussa to form a good mixture with jobbing offset ink and thus decreases the viscosity [5,6].

Table 2. Viscosity values of pure and modified jobbing ink in different ratios.

TiO_2 ratio (%)	Ink viscosity η (Pa.s)		
	St. value	JN	JD
0.1	107	106	110
0.15		108	109
1		110	105
3		115	102

3.2.1.2. Yield value

The "yield value" is the required stress for beginning a movement of a certain fluid. This stress is because of the electrical charges located on the studied surfaces of a particle. The Bingham's yield value is detected through a direct-indicating Viscom [7].
Hence, the addition of nano and Degussa TiO_2 to the offset ink (jobbing), with 4 different ratios (0.1, 0.15, 1, 3) %, impacts such ink. The comparison according to yield value to standard data of the pure offset ink are listed in Table 3.
It's observed that the addition of nano TiO_2 leads to a decrease in the yield value, that's because of the high density of solid TiO_2 in comparison with jobbing ink, besides the hydrogen bond between TiO_2, and oxygenated groups in jobbing offset ink. This allows Degussa to create a good mixture with jobbing offset ink and decreases the yield value [8].

Table 3. Yield value values of pure and modified jobbing ink in different ratios.

TiO_2 ratio (%)	Yield value Y.V (Pa)		
	St. value	JN	JD
0.1	1655	2150	2100
0.15		1695	1670
1		1655	1600
3		1610	1580

3.2.1.3. Tack & flow

It is a property of a printing ink that defines the cohesion existed between particles of an ink film. The ink with a high extent of tack (tacky ink) needs more forces for splitting than that with a less tacky film. However, tacky inks, cannot print the solids well. In multicolor printing technique, it is necessary for the first printed ink to have a greater tack than the printed ink on its top, otherwise the latter ink will not trap successively on the first ink. The tack of a given ink is easily measured by a Tackoscope or an Inkometer [9].
The effect of TiO_2 modifications to the offset inks as well as the comparison according to yield value to standard data of the pure offset inks are listed in Table 4.
As seen, the tack value is highly decreased by increasing nanoTiO_2 with small ratios in JN 0.1 and JN 0.15 due to high dispersion between nanoTiO_2. The slightly decrease at high ratios in JN 1 and JN 3 is due to the high density of solid TiO_2. JD increases with increasing TiO_2 ratio, due to the hydrogen bond between TiO_2 and oxygenated groups in jobbing offset ink, which allows Degussa to constitute a good mixture with jobbing offset ink, and increase tack value, due to viscosity decrement [10-11].

Table 4. Tack values of pure and modified jobbing ink in different ratios.

TiO_2 ratio (%)	Tack (N)		
	St. value	JN	JD
0.1	8.6	9.3	8
0.15		8.4	8.2
1		8.2	8.4
3		8	8.7

3.2.2. Optical properties

In this section, we discuss the ink optical properties as they are important factors for the inkjet print color reproduction. The applied ink was sized with different modified ratios that affect ink optical characteristics, but with a minor effect on the other characteristics.

3.2.2.1. Print density

It refers to a measurement of the light reflected off of a substrate. From its definition, it is clear that the materials that occur in a direct contact to a printing surface will play an essential role for the quality of the print density. Nevertheless, many materials can be applied for metering the ink to a substrate. Herein, we will highlight on some foam metering applications [12].
The influence of TiO_2 modifications to the offset ink together with the comparison to standard data of the pure offset inks are listed in Table 5.
The modified jobbing ink with Degussa TiO_2 (JD) or with nanoTiO_2 (JN) confirmed that the addition of TiO_2 (either nano or Degussa) leads to make a sinusoidal relation between average print density (Avg. ρ) and TiO_2 ratio. The highest Avg. ρ was acquired in JN 0.15 and JD 0.15. The highest statistical standard deviation of print density (st. ρ) was acquired in JN 0.1 and JD 0.1, due to hydrophilicity of TiO_2, besides the hydrogen bond between TiO_2 and oxygenated groups in the jobbing offset ink [13].

Table 5. Average and standard deviation of print density for modified jobbing ink.

TiO_2 ratio (%)	Average		St. deviation	
	JN	JD	JN	JD
0.1	1.484	1.414	0.078344	0.061805
0.15	1.495	1.499	0.025927	0.03414
1	1.44	1.453	0.031869	0.03093
3	1.451	1.418	0.042544	0.051164

3.2.2.2. Accelerated aging

It refers to testing the agitated conditions of humidity, heat, sunlight, vibration, oxygen, etc., in order to accelerate the normal aging routes of studied items. For instance, cycles of cold and heat can simulate the consequence of night and day for a few minutes or hours. Thus, physical or chemical tests are carried out via subjecting the studied product to representative stages or rapid changes in stress at long periods. The unusually high levels of such stress are directed to accelerate the influences of a natural aging or the influences that cause failures intentionally, in which the mechanical parts must run at more elevated speed than that they could receive in normal usages.
Polymers for example are usually kept at high temperatures for speedup the chemical breakdowns by using environmental chambers for the process. Lots of accelerated aging techniques applied in this case are depended on the assumption of first-, zero-, or pseudo-first-order kinetics, according to the following function for Arrhenius reaction rate (eq. (1) [14]).

$$r = Ae^{-\phi/kT} \quad \text{eq. (1)}$$

Where r represents the rate of the related reaction; A is the constant for the material; represents the apparent activation energy; k represents Boltzmann's constant (0.8617 x 10^{-4} eV/K); T° is the absolute temperature.
Accelerated aging is tested by taking a quantity of ink that has been greened using the 4 different ratios % of nanoTiO_2 and Degussa, printing it on paper, then placing it in a dark room "UV chamber" designated for this test. The process continued with exposing it to a UV lamp for three days, and this means exposing this paper to sunlight for several years. After that, we compare the color of the paper

before exposure to UV rays and after exposure to UV rays because accelerated aging measures the extent to which the ink carries out sunlight and moisture over time.

Both figure 5a & figure 5b explain the comparison between the printed paper with 4 different ratios (0.1, 0.15, 1, 3) % of TiO_2 nano, and the same ratios of Degussa, before exposing to a UV lamb for three days in a "UV chamber". After the process, we reported a comparison according to the resulting accelerated aging to standard data of the pure offset ink to know the effect of the sunlight as well as the moisture on the quality of the modified ink with time.
It's worthy noticed that the aging rate has a changeable relation of rate which increased and decreased as has a sinusoidal wave, with nanoTiO_2 ratio, due to photoactivity of TiO_2, and the high dispersion between nanoTiO_2. The aging rate is slightly decreased with increasing nanoTiO_2 to high ratios JD3, which is increased the aging rate due to high dispersion between Degussa TiO_2 and photoactivity of rutile-anatase of Degussa TiO_2 [15-18], as shown in figure 5a & figure 5b.

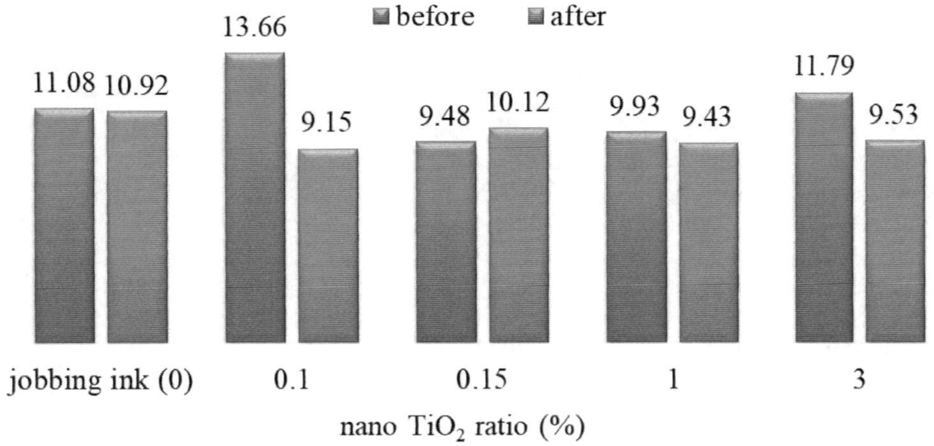

Figure 5a. Accelerated aging for pure and modified ink with nano TiO_2, before and after exposing to a UV lamb for three days.

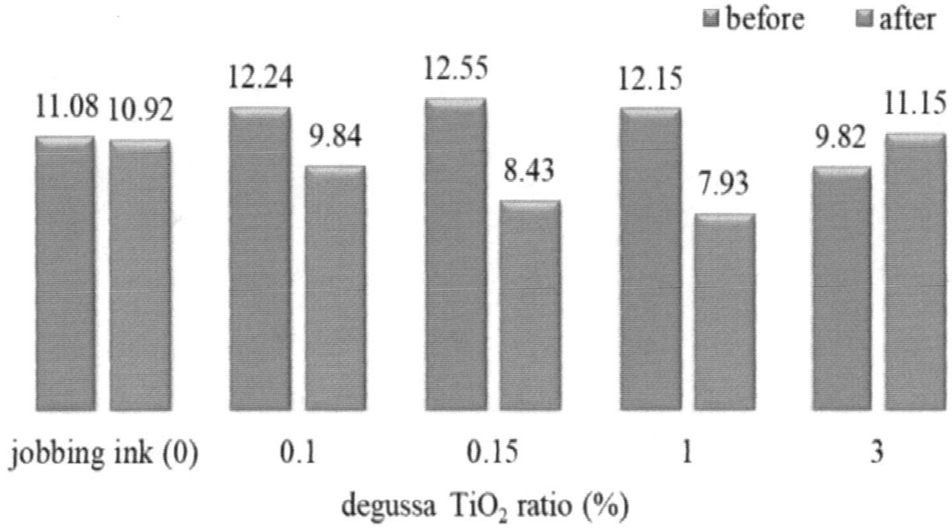

Figure 5b. Accelerated aging for pure and modified ink with Degussa TiO_2, before and after exposing to a UV lamb for three days.

4. Conclusions

Herein, TiO_2 was prepared with the hydrothermal method (although a low yield was obtained via such method, the nanomaterials can be formed within a broad range of temperatures varying from room temperature to even very high temperatures beside the morphology control of the prepared materials). According to XRD, the prepared TiO_2 was nanostructured with anatase allotropy, while Degussa had 85% anatase and 15% rutile. Also, XRD, XPS, TEM, proved that the prepared TiO_2 was pure without any impurities. The addition of titanium dioxide nanoparticles didn't alter the morphology of the used offset ink, according to TEM. The rheological properties of the offset ink (viscosity, yield value, flow, tack) were improved with the addition of TiO_2 (either nano or Degussa). In addition, the optical properties of the offset ink (print density and accelerated aging) were altered by the addition of TiO_2 (either nano or Degussa). The optical properties for modified ink were better with the addition of a small ratio of nanoTiO_2, due to the high photo-activity of anatase phase structure.

5. References

[1] J. Li, T.Ishigaki, and X. Sun, "Anatase, Brookite, and Rutile Nanocrystals via Redox Reactions under Mild Hydrothermal Conditions: Phase-Selective Synthesis and Physicochemical Properties", J. Phys. Chem. C, 111 (2007) 4969–4976.

[2] R. Kaplan, B. Erjavec, G. Dražić, J. Grdadolnik, and A. Pintar, "Simple synthesis of anatase/rutile/brookite TiO2 nanocomposite with superior mineralization potential for photocatalytic degradation of water pollutants", Appl. Catal. B: Envir., 181, (2016) 465-474.

[3] A. M. Barreiro, G. K. Pinheiro, B. N. Wesling, D. Müller, L. T. Scarabelot, L. V. de Souza, D. Hotza, and C. R. Rambo, "Aerogel-Based TiO2 Stable Inks for Direct Inkjet Printing of Nanostructured Layers", Adv. Mater. Sci. Eng., 4273097 (2020) 1-9.

[4] N. Rahoui, P. Zhou, N. Taloub, M. Hegazy and Y. D. Huang. "Synthesis and Evaluation of Titanium Oxide Modified Mussel Inspired Nanoparticles for Synergetic Photothermal and Photodynamic Effect", IOP Conf. Ser. Mater. Sci. Eng., 389 (2018) 012009.

[5] A. J. Haider, Z. N. Jameel, and I. H. M. Al-Hussaini, "Review on: Titanium Dioxide Applications", Ener. Proc., 157 (2019) 17–29.

[6] F. Loffredo, I. A. Grimaldi, A. D. Del Mauro, F. Villani, V. Bizzarro, G. Nenna, R. D'Amato, and C. Minarini, "Polyethylenimine/N-doped titanium dioxide nanoparticle based inks for ink-jet printing applications", J. Appl. Polym. Sci., 122 (2011) 3630–3636.

[7] J. Drelich, and D. Chibowska, "Spreading kinetics of water drops on self-assembled monolayers of thiols: Significance of inertial effects". Langmuir, 21 (2005) 7733-7738.

[8] V. Bonfrate, D. Manno, A. Buccolieri, S. K. Padmanabhan, A. Licciulli, A. Serra, E. Braione, L. Calcagnile, and G. Giancane, "Design and synthesis ofiron-doped nanostructured TiO2 and its potential use in the photodegration of hazardous materials present in personal care products", Chem. Select., 2 (2017) 5095–5099.

[9] S. Rousu, J. Preston, J. Gustafsson, , and P. Heard, "Interactions between UV curing, hybrid-UV and sheetfed offset inks and coated paper – Part 2 Commercial print trials". TAGA Journal, 2 (2006) 174-189.

[10] B. Fouchet, P. Noirot, D. Desjumaux, and R. Van Gilder, "Diffusion dynamics of offset ink components in very thin latex films of different structure and composition." International Printing and Graphic Arts Conference, Vancouver, British Columbia, October Montreal: Pulp and Paper Technical Association of Canada (2004).

[11] M. Gros A. Gandini, A. Blayo, Y. Xiang, and D. Bousfield "The influence of ink pigment concentration and filter cake resistance coefficient on ink setting rates." 11th International Printing and Graphic Arts Conference, Bordeaux, France, October. Paris, France (2002) 8pp.

[12] J. Adams, and P. Dolin "Printing Technology". Albany, NY 5th ed. (2002).

[13] A. Solomon, "Synthesis, Modification, Applications and Challenges of Titanium Dioxide Nanoparticles", Res. J. Nanosci. Eng., 3 (2019) 10-22.

[14] H. Kamal, G. Ström, K. Karlström, J. Schoelkopf, and P. Gane, "Effect of excess dispersant on surface properties and liquid interactions on calcium carbonate containing coatings." Nordic Pulp and Paper Research Journal, 25 (2010) 82-92.

[15] W. Hoa, J. C. Yu, and S. Lee, "Low-temperature hydrothermal synthesis of S-doped TiO2 with visible light photocatalytic activity", J. Solid State Chem., 179 (2006) 1171–1176.
[16] F. A. Morsy, S. El-Sherbiny, M. Samir, and O. A. Fouad, "Application of nanostructured titanium dioxide pigments in paper coating: a comparison between prepared and commercially available ones", J. Coat. Tech. Res., 13 (2016) 307–316.
[17] A. Lewis, H. Taha, A. Strinkovski, A. Manevitch, A. Khatchatouriants, R. Dekhter, and E. Ammann, "Near-field optics: From subwavelength illumination to nanometric shadowing", Nat. Biotechnol., 21 (2003) 1378-1386.
[18] M. Kawashima, and D. Bousfield. "The influence of blanket type on printing: Part II: Effect on ink tack". Journal of Pulp and Paper Science, 35 (2009) 70-73

A Mini-Review on Recent Developments in Metal-Organic-Frameworks (MOFs) Derived Carbon as A Microwave Absorption Materials (MAMs)

Mohamed E. Elmowafy[1], Osama Abuzalat[1], Hesham Tantawy[1], Mohamed A. Elsayed[1*]

[1] Chemical Engineering Department, Military Technical College, Kobry Elkobbah, Cairo, Egypt.

* m.aboelfotoh@mtc.edu.eg

Abstract: As a result of its ability to reduce electromagnetic waves (EMW), microwave absorption materials (MAMs) have gained a lot of interest recently. Thermal or other sources of energy can be used to disperse EMW. Ideal MAMs have many loss processes and must be light, thin, absorb a wide range of frequencies, and have high absorption capabilities. Because of the high concentration of electromagnetic radiation in our surroundings that are detrimental to people, MAMs (microwave absorption materials) are a hot issue. Radar stealth is further aided by MAMs, which are used to hide weaponry. Nanomaterials have been tested as MAMs, however, not all of them have worked out. In this study, we hope to shed light on the design and manufacturing of MOF-derived Carbon-based MAMs, along with their basics and applications.

Keywords Microwave absorption materials; Metal–organic frameworks; Porous carbon; Effective absorption bandwidth; Reflection loss

1. Introduction

MA materials have a wide range of uses, including military stealth, electronic stability, medical diagnostics, microwave interference, and pollutant abatement (among others). Recently carbon-based MOFs as MA materials demonstrate a lot of interest due to their low observed density, remarkable good conductivity, and outstanding stability. Limitations on their overall performance, such as characteristic impedance mismatch and restricted absorption bandwidth, were imposed by a single electrical loss. Integrating nanoparticles into a carbon matrix has become a viable strategy for increasing MA capabilities because of their high efficiency[1].

Metal-organic frameworks (MOFs) have good properties like adjustable chemical structures, their huge specific surface area, and good distributions of their pores which make them good materials for use in different fields. When MOF materials expose to the pyrolysis process at a given temperature, MOF can change into porous carbon (PC) decorated with metal species. Metal-organic frameworks (MOFs) can be used as a composite with other loss materials to enhance the MA efficiency. In this paper, we will discuss many types of porous carbon-based MOFs which have a dielectric or magnetic loss and acts as MAMs. The numerous MA techniques and processes are explored in detail. This article concludes with a discussion of MOF-derived PC-based MAMs drawback/problem/future potential[2, 3].

Metal-organic frameworks (MOFs) can be composed by coordination between organic ligands and metal ions or clusters. MOFs come in a wide range of compositions and topologies due to the wide range of organic ligands that may be utilized. As far as we know, over 20,000 MOFs have been discovered. Due to their vast range of characteristics, MOFs have sparked a lot of scientific curiosity, which might lead to extensive applications in domains including electrochemical energy storage, purification, and sensing [4, 5].

Metal and carbon-based compounds from MOFs able to be produced by using the thermal decomposition process under a vacuum. Fortunately, following pyrolysis, MOF morphologies are mostly retained. In addition to that, the pyrolysis process makes MOF-derived PC nano-composites have good features like MOFs, such as a good distribution of pores, large surface area, tunable appearance shape, and stable chemical structure which enable them to be good candidates for MAMs. Although MOF-derived PC as nano-composites materials have a relatively simple attenuation effect in MAMs, they can be used with other materials which have specific loos mechanisms to enhance the microwave radiation attenuation process[6, 7].

MOF-derived PC as nano-composites materials, including Ni, Co, and Fe incorporated to MOF-derived PC as nano-composites materials which used as MAMs are precised in that study. Besides pure MOFs, multi-metal MOFs and adjustable chemical compositions have also been used to generate MAMs.

2. Microwave Absorption Theories

When EMW incident on the MAMs surface, three scenarios may be happened as in Figure 1. EMW is reflected on the MAMs surface and absorbed by the MAMs, but some of it is also injected into the MAMs interior (transmitted EMW). During the synthesis process of desired MAMs, the researchers expect that the incident EMW able to be dissipated inside the MAMs as abundant as possible and reduce the reflectance and transmittance through it. Consequently, the high performance of MAMs is achieved by two factors: high attenuation to EMW and also good impedance to EMW. Incident EMW must penetrate the MAMs as deeply as possible to accomplish efficient impedance matching[7].

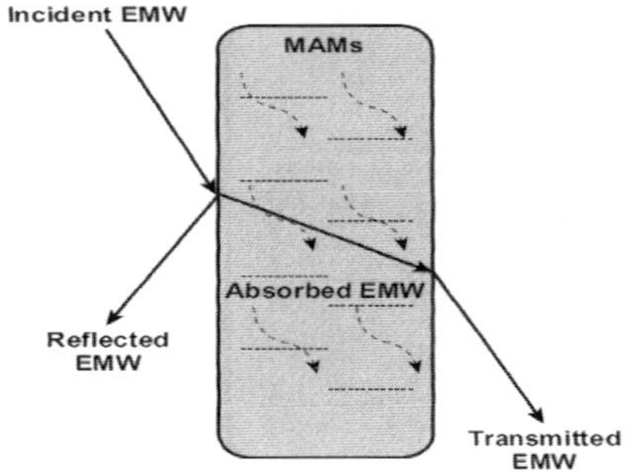

Figure 1. The interaction between MAMs with microwaves.

Common MAMs classifications demonstrated in Table 1 are separated into three kinds in accordance with loss mechanisms as follows: dielectric loss, multiple loss, and magnetic loss materials. Representative materials for dielectric loss mechanism are carbon-based materials, non-magnetic metal oxides, polymers, and so on. These materials have excellent properties like lightweight, high thermal stability, high electrical conductivity, and high mechanical strength whereas their MA performance and effective absorption bandwidth (EAB) maybe not be adequate. Representative materials to the Magnetic loss mechanism are ferrite, magnetite, magnetic compounds and metals, and others. Conversely, the higher observed density and relatively lower stability restricted their experimental applications[8].

Table 1 Common MAMs classifications[9].

MAMs types	Representative materials	Mechanisms of loss
Dielectric loss	Nanocarbon materials, metal oxides as non-magnetic, polymer, etc	Polarization relaxation loss and Electrical conductivity loss
Magnetic loss	Magnetic compounds and metals, ferrite, magnetite, etc	Eddy current loss, residual loss, and Hysteresis loss
Multiple loss	Mixture of the directly above	Multiple loss

3. MAMs are made of magnetic single metal MOF-derived PC nano-composites.

Fe, Co, and Ni are three of the most common magnetic metals on the market. In MOF synthesis, they are commonly used as inorganic metal centers. When MAMs are quickly pyrolyzed, the loss process might cause an impedance mismatch, resulting in poor MA performance. Impedance matching may be improved by using them with dielectric loss material [10]. Microstructure design can create a variety of loss processes.

It is the most widely studied, PC-based nano-composites generated from MOFs as MAMs. Self-assembly utilizing Co^{2+} and 2-methylimidazole is used to make ZIF-67 (zeolitic imidazolate framework). To make porous Co/C composites at varied pyrolysis temperatures, Kuang uses an inert environment to make the thermal decomposition of Co-based MOF (ZIF-67)[11]. Before and after sintering, the material's morphology was almost unchanged; only there are some wrinkles on the surface, as seen in Figure 2. Increased MA performance can be attributed to the sample pyrolysis at 500°C. RL (reflection loss) of -35.3 dB at 5.8 GHz and EAB of 5.80 GHz (8.40 – 14.20 GHz) may be accomplished at a thickness reach to 2.5 mm of the obtained Co/C-500, which has a maximum thickness of 4 mm. Although the MA performance is outstanding, RL and EAB are not, due to the high dielectric loss of carbon, the porous structure, and the magnetic loss of Co[12].

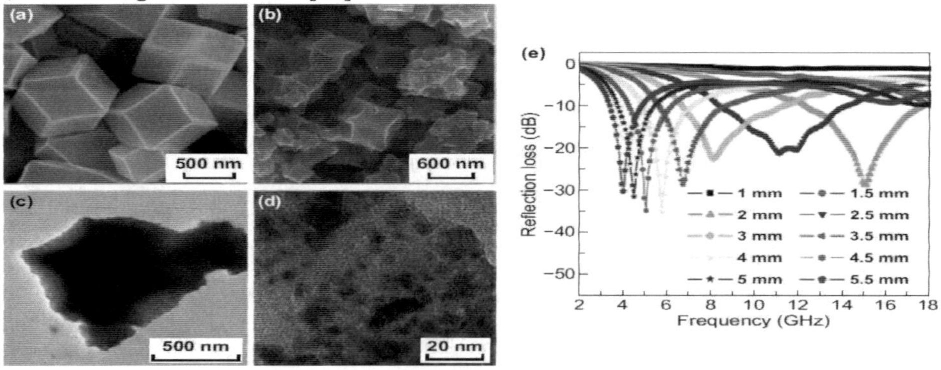

Figure 2. (a) Demonstrate SEM images of obtained Co/C before carrying out the calcination process. (b) Demonstrate SEM images of obtained Co/C after carrying out calcination process at 500 °C for 5 hours under using Ar gas. (c) Demonstrate TEM images of obtained Co/C-500 at low magnification (d) Demonstrate TEM images of obtained Co/C-500 at high magnification. (e) Demonstrate calculated reflection loss against frequency for obtained Co/C-500 with various suggests thicknesses. Is reprinted with approval from Ref.[11]

Dong created MOF/RGO (reduced graphene oxide) hybrids using an in situ formation of Co-based MOF on GO (graphene oxide) nano-sheets and a carefully regulated calcination method (see Figure 3). The sample which had 4.1 mm thickness achieved -52 dB of maximum RL at 9.6 GHz. Even though the MOF/RGO combination is just 3.2 mm thick, the EAB of the mixture may reach 7.72 GHz, which is higher than other MOF and RGO-based MAMs[13]. This high MA performance can be attributed to the improved match between the dielectric loss mechanism and magnetic loss mechanism in MOF/RGO composite[14].

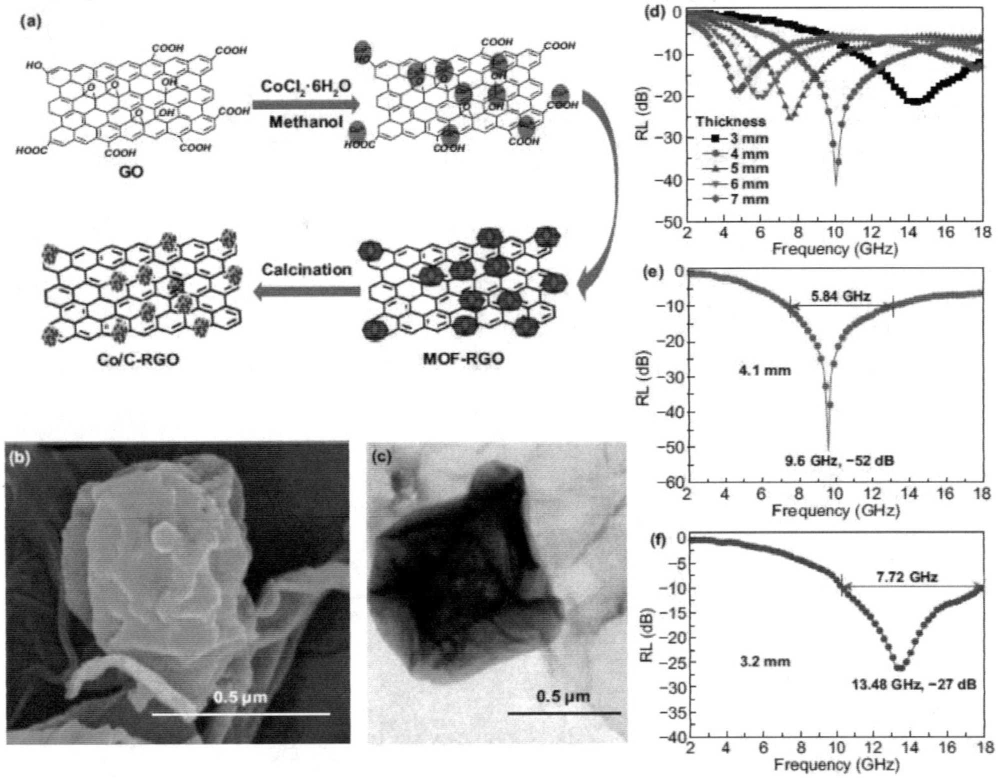

Figure 3. (a) Procedure for the synthesis of MOF/RGO composite. (b), (c) Demonstrates SEM and TEM images of MOF/GO composite which was heated for 6 h at 500 °C. (d–f) Demonstrates RL curves of MOF/GO-500 composite and undergo 4.1 and 3.2 mm thickness at 6 wt% as a filler packing percentage in wax. Is reprinted with approval from Ref.[13]

Ni-based MOF is another compound which extensively used as MAM. To make Ni@C composites, Zou and his colleagues used the thermal breakdown of pure Ni-MOF. At calcinated temperature (800°C), the sample had a maximum RL of -55.7 dB, whereas the EAB at 1.85 mm thickness was 6.0 GHz[15]. The high performance of MAMs is related to the synergistic effect between magnetic nickel and carbon nanoparticles in addition to their hollow structure[16].

The hollow spheres of Ni-based MOF which have different morphology of the surface were made by Yang's team using a hydrothermal technique. The morphologies of the surface are controlled by the reaction time of the hydrothermal process[17]. The morphology of the surface has different appearances such as smooth, rod-resemble, and hair- resemble matching to the time of reaction that is 6, 10, and 8 hr respectively.

The RLmax (maximum reflection loss) in reaction time (10 hr) for a sample with 1.5mm thickness was -58 dB at EAB reached to 6 GHz however with a suitable thickness of 4.6 mm, the obtained EAB was in value equal 6.2 GHz as shown in Figure 4.[18] .

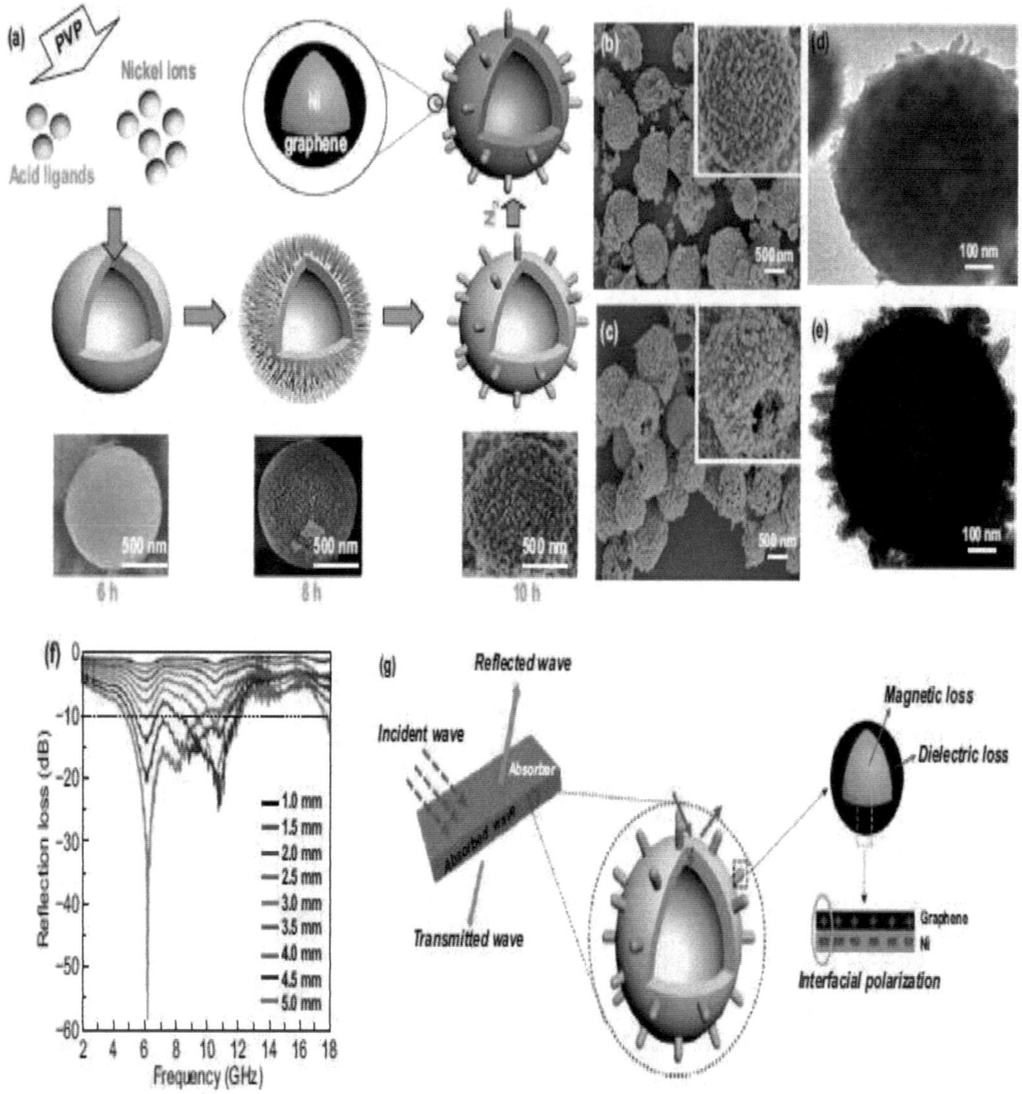

Figure 4. (a) Demonstrate the construction of Ni-MOF in the shape of a hollow sphere with adjustable surface constructions. (b), (c) Demonstrate the SEM images of the samples of Ni-MOFs in a duration time equal to 10 h before and after introducing annealing temperature (600 °C). (d), (e) Demonstrate TEM images of the samples of Ni-MOFs in duration time equal 10 h before and after introducing annealing temperature at (600 °C). (f) Represent the reflection losses value of obtained Ni- MOF sample at various times of reaction. (g) Represent the mechanism of EMW absorption by ready samples. This is Reprinted with approval from Ref.[17]

The in situ synthesis of Fe/C nanocubes by thermal decomposition of Prussian blue MOF at suitable temperature had been reported by Xu et al. Figure 5 shows that when the thickness equals 5 mm, the RLmax reach (-22.6 dB) with 4 GHz at a temperature equal to 650 °C. However, at a thickness equal to 2mm, the EAB reach 7.2 GHz. The synergetic effect of magnetic loss and dielectric loss, provide Fe/C nanocubes with high MA performance[19].

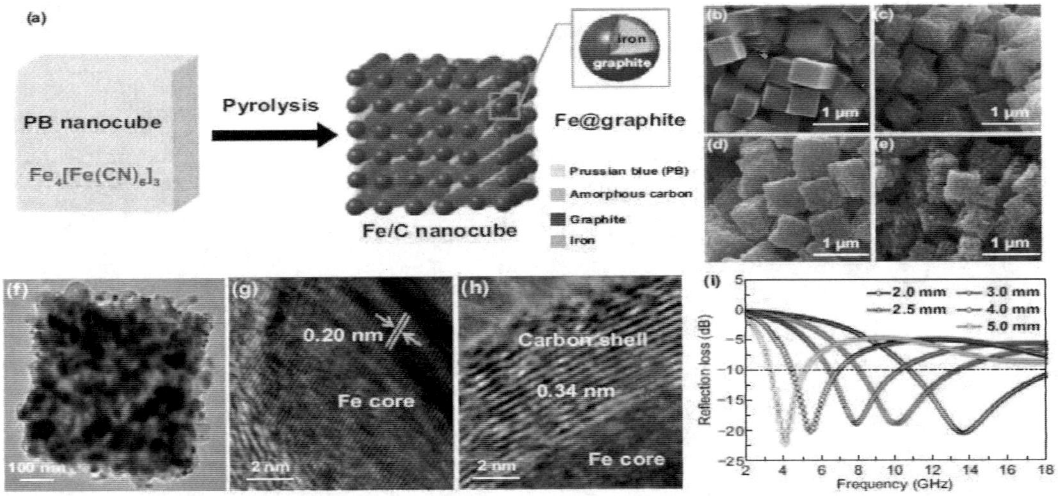

Figure 5. (a) Demonstration diagram of the pyrolysis process of Prussian blue MOF nanocube (PB) to obtain Fe/C nanocubes. (b) SEM image of Prussian blue (PB) nanocubes. (c – e) SEM image of Fe/C nanocubes which were acquired at various temperatures of pyrolysis as follows: 600, 650, and 700 °C. (f) Demonstrate a TEM image of the obtained sample at 650 °C. (g) Demonstrate HRTEM image related to Fe core and also (h) HRTEM related to graphitic carbon shell. (I) Demonstrate Reflection loss of obtained sample produced at temperature 650 ^{O}C with different thickness as MAM. Reprinted with permission from Ref.[19]

4. MAMs are made of Multi-metal MOF-Derived Porous Carbon-Based Nanocomposites.

Lu's group generated multilayered NiCo nanoparticles/nanoporous carbon (NiCo/NPC) composites using in situ pyrolysis of the bimetallic NiCo-MOF. Impedance matching optimization and EMW attenuation augmentation were aided by the synergistic interactions of dielectric loss and magnetic loss among NiCo NPs(nanoparticles), graphitized carbon layer, and NPC[20]. Dong had also prepared hollow and porous CoNi@C microspheres which derived from CoNi MOFs[21]. With a thickness of 1.6 – 4.0 mm, the RLmax can achieve (-44.8) dB at a frequency equal to 10.7 GHz, and the obtained EAB is 13.3 GHz, as shown in Figure 6.

To make the NiCo/NPC composites, Lu's team used bimetallic NiCo-MOF in situ pyrolysis to melt the metals in place. The synergistic interactions of dielectric loss and magnetic loss among NiCo NPs, graphitized carbon layer, and NPC improve the EMW attenuation. Dong had previously used CoNi-MOFs to make hollow and porous CoNi@C microspheres. At thickness (3.2 mm), the RLmax reaches to (-44.8) dB at a frequency 6.8 GHz whereas the obtained EAB is 4 GHz at 2.4 thickness as illustrated in Figure 6.

Also disclosed were MAMs built from PC-based nanocomposites produced from multi-magnetic metal MOFs. Multi-magnetic metal MOFs, such as FeCo@C@CNGs(carbon nanocages)[22], NiCo alloy/carbon nanorod@CNT[23] , Fe–Co/NC/rGO[24], FeNi@CNT/CNRs(carbon nanorods)[25], and CoFe@C@MnO2[26], often have dielectric loss added to improve impedance match. Each of these samples has outstanding electromagnetic wave impedance and attenuation [27].

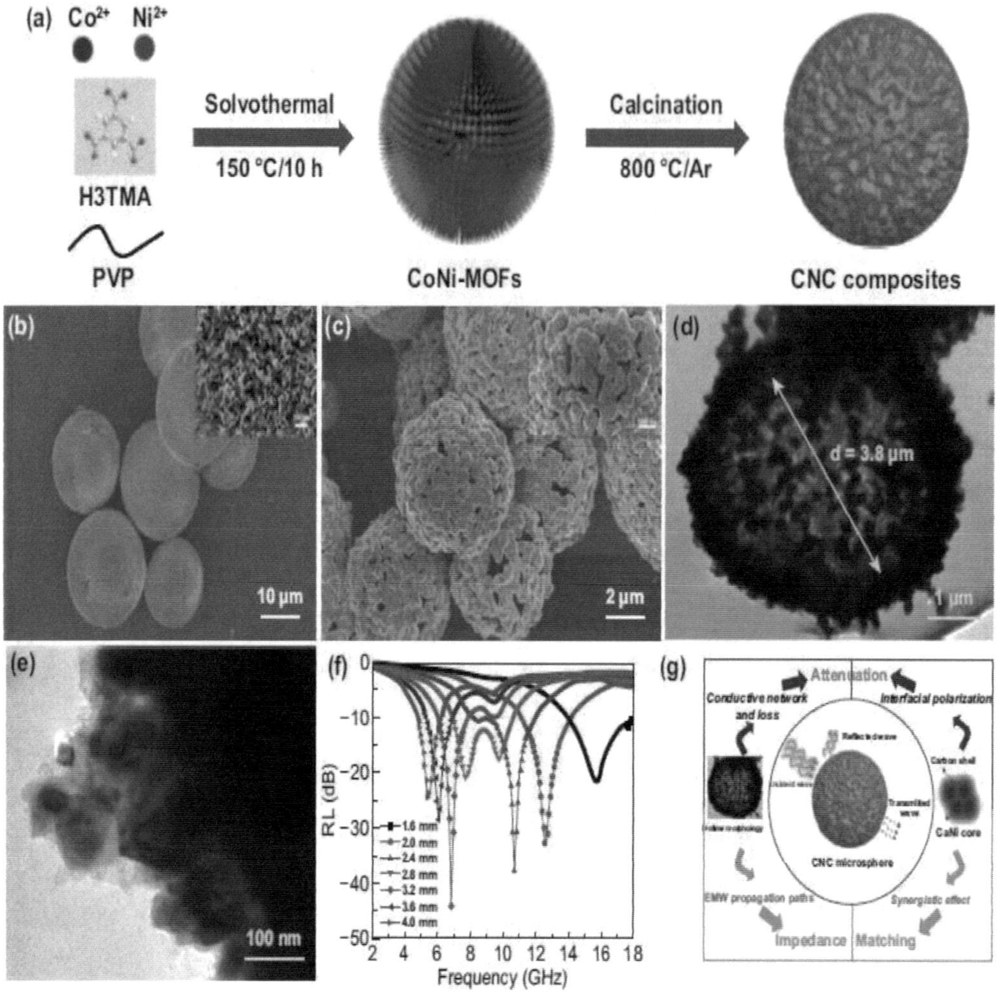

Figure 6. (a) Diagram for the artificial procedure of hollow CNC microspheres composites. (b), (c) Demonstrate SEM pictures, (d), (e) Demonstrate TEM pictures, and (f) Demonstrate RL curves of obtained CoNi@C samples. g) Diagram which demonstrates the MA mechanisms for CNC microspheres composites. Reprinted with approval from Ref.[21]

5. Conclusion

MOF-derived PC-based nanocomposites which are used as MAMs are the focus of this review study. A lightweight, highly effective MAM may be developed by pyrolytically forming PC-based MAMs from MOFs, according to these findings. After carrying out the pyrolysis process, the PC-based MAMs which produced from the MOFs demonstrate good properties like high electrical conductivity, low density, high porosity, and good dielectric loss. It is also possible that the inorganic metalcore will cause magnetic or dielectric loss. MOF-derived PC-based nanocomposites are frequently incorporated with other materials which have specific loss mechanisms to enhance the MA performance. A well-designed nano-composites with numerous benefits will exhibit good EMW attenuation performance due to the various loss mechanisms and also the synergistic impact of the multi-components MOF-derived PC-based nanocomposites in conjunction with other loss materials which represent a viable development path for MAMs in the near future.

References

[1] Z. Zhang *et al.*, "A review on metal–organic framework-derived porous carbon-based novel microwave absorption materials," vol. 13, no. 1, pp. 1-29, 2021.

[2] H. Zhao *et al.*, "Biomass-derived porous carbon-based nanostructures for microwave absorption," vol. 11, no. 1, pp. 1-17, 2019.

[3] D. Liu *et al.*, "Waxberry-like hierarchical Ni@ C microspheres with high-performance microwave absorption," vol. 7, no. 17, pp. 5037-5046, 2019.

[4] F. Chen *et al.*, "1D magnetic nitrogen doped carbon-based fibers derived from NiFe Prussian blue analogues embedded polyacrylonitrile via electrospinning with tunable microwave absorption," vol. 224, p. 109161, 2021.

[5] H. Zhao, Y. Cheng, H. Lv, G. Ji, and Y. J. C. Du, "A novel hierarchically porous magnetic carbon derived from biomass for strong lightweight microwave absorption," vol. 142, pp. 245-253, 2019.

[6] H.-B. Zhao, Z.-B. Fu, H.-B. Chen, M.-L. Zhong, C.-Y. J. A. a. m. Wang, and interfaces, "Excellent electromagnetic absorption capability of Ni/carbon based conductive and magnetic foams synthesized via a green one pot route," vol. 8, no. 2, pp. 1468-1477, 2016.

[7] Y. Du *et al.*, "The electromagnetic properties and microwave absorption of mesoporous carbon," vol. 135, no. 2-3, pp. 884-891, 2012.

[8] P. Zhou *et al.*, "Coral-like carbon-based composite derived from layered structure Co-MOF-71 with outstanding impedance matching and tunable microwave absorption performance," vol. 108, pp. 10-17, 2022.

[9] S. Gao, G. S. Wang, L. Guo, and S. H. J. S. Yu, "Tunable and ultraefficient microwave absorption properties of trace N-doped two-dimensional carbon-based nanocomposites loaded with multi-rare earth oxides," vol. 16, no. 19, p. 1906668, 2020.

[10] C. Zhou *et al.*, "Lightweight hollow carbon nanospheres with tunable sizes towards enhancement in microwave absorption," vol. 108, pp. 234-241, 2016.

[11] Y. Lü *et al.*, "MOF-derived porous Co/C nanocomposites with excellent electromagnetic wave absorption properties," vol. 7, no. 24, pp. 13604-13611, 2015.

[12] X. Liang, G. Wang, W. Gu, and G. J. C. Ji, "Prussian blue analogue derived carbon-based composites toward lightweight microwave absorption," vol. 177, pp. 97-106, 2021.

[13] K. Zhang *et al.*, "Electromagnetic dissipation on the surface of metal organic framework (MOF)/reduced graphene oxide (RGO) hybrids," vol. 199, pp. 340-347, 2017.

[14] N. Li, G. Huang, H. Xiao, Q. Feng, and S. J. C. Fu, "Investigations on structure-dependent microwave absorption performance of nano-Fe3O4 coated carbon-based absorbers," vol. 144, pp. 216-227, 2019.

[15] Z. Zhang *et al.*, "Ni@ C composites derived from Ni-based metal organic frameworks with a lightweight, ultrathin, broadband and highly efficient microwave absorbing properties," vol. 12, no. 1, p. 011001, 2018.

[16] W. Li *et al.*, "Co nanoparticles supported on cotton-based carbon fibers: a novel broadband microwave absorbent," vol. 772, pp. 760-769, 2019.

[17] Z. Yang *et al.*, "Surface architecture of Ni-based metal organic framework hollow spheres for adjustable microwave absorption," vol. 2, no. 12, pp. 7888-7897, 2019.

[18] L. Yan *et al.*, "Core-shell, wire-in-tube and nanotube structures: Carbon-based materials by molecular layer deposition for efficient microwave absorption," vol. 173, pp. 145-153, 2021.

[19] Liu Y, Chen Z, Xie W, Qiu F, Zhang Y, Song S, Xiong C, Dong L J J o A and Compounds 2019 Enhanced microwave absorption performance of porous and hollow CoNi@ C microspheres with controlled component and morphology **809** 151837

[20] Wang F, Wang N, Han X, Liu D, Wang Y, Cui L, Xu P and Du Y J C 2019 Core-shell FeCo@ carbon nanoparticles encapsulated in polydopamine-derived carbon nanocages for efficient microwave absorption **145** 701-11
[21] Wang L, Wen B, Bai X, Liu C and Yang H J A A N M 2019 NiCo alloy/carbon nanorods decorated with carbon nanotubes for microwave absorption **2** 7827-38
[22] Wang S, Xu Y, Fu R, Zhu H, Jiao Q, Feng T, Feng C, Shi D, Li H and Zhao Y J N-m l 2019 Rational construction of hierarchically porous Fe–Co/N-doped carbon/rGO composites for broadband microwave absorption **11** 1-16
[23] Xu X, Ran F, Lai H, Cheng Z, Lv T, Shao L, Liu Y J A a m and interfaces 2019 In situ confined bimetallic metal–organic framework derived nanostructure within 3D interconnected bamboo-like carbon nanotube networks for boosting electromagnetic wave absorbing performances **11** 35999-6009
[24] Zhang Y, Yang Z, Li M, Yang L, Liu J, Ha Y and Wu R J C E J 2020 Heterostructured CoFe@ C@ MnO2 nanocubes for efficient microwave absorption **382** 123039
[25] Wang J, Wu F, Cui Y, Zhang A, Zhang Q and Zhang B J C 2021 Efficient synthesis of N-doped porous carbon nanoribbon composites with selective microwave absorption performance in common wavebands **175** 164-75

Sensitivity and Performance of Dinitroanisole (DNAN) in comparison with TNT

Ahmed K. Hussein[1], Ahmed Elbeih[1], Mahmoud Abdelhafiz[1], Hesham S. Ahmed[2], Ahmed S. A. Shalaby[2], Ahmed Azazy[2], Ahmed Saleh[2]

[1] Military Technical College, Kobry Elkobbah, Cairo, Egypt
[2]Science and Technology Center of Excellence (STCE), Ministry of Military Production, Cairo, Egypt.

Abstract: The constant search for protection the soldier during training and participation in hostilities led us to aspire to develop types of energetic materials of a special nature that qualify them to reach the maximum levels of safety during handling, transportation and uses. In this work, we focus on one of these compounds, which is the main component of the preparation of low sensitive compositions. DNAN is an explosive with low sensitivity. Preparation of DNAN in laboratory scale was performed; explosive characterization was presented. Impact and friction sensitivities of DNAN, heat of combustion and detonation velocity were specified. TGA and DSC were used to investigate the DNAN thermal behavior under specific conditions. It was concluded that sensitivity of DNAN is lower than TNT and the chosen cyclic nitramines. The detonation properties of DNAN are slightly lower than TNT however DNAN is candidate individually or with other explosives to replace TNT in low sensitive compositions to full fill the safety and security manipulation of ammunitions.

Keywords: DNAN; TNT; Sensitivity; Performance; Detonation parameters.

1. Introduction

Dinitroanisole, which is commonly known as DNAN, has been used in various applications like the synthesis of explosives. Low sensitivity helps DNAN to be investigated to take the place of TNT in PAX-21 composition [1]. The performance of several interesting advanced high explosives has been published but the high sensitivity is a main problem in their applications [2-4]. Also, several compositions were studied based on low sensitive explosives [5-7] and other based on energetic binders [8-11]. DNAN has been utilized instead of conventional explosives in the formation of IMX-101 composition in the 155 mm M795 artillery projectile and in the formation of IMX104 composition in mortar systems [12-14]. The kinetics of decomposition and thermal stability of new advanced energy materials were studied in literature [15-17]. CL-20 and BCHMX were studied as plastic explosives [18, 19] and as melt cast compositions with high performance [20] but there sensitivity were high in comparison with DNAN. The performance of several traditional and advanced explosives have been studied in different shaped charges [17-20] while the DNAN is not published in literature yet.

In this paper, preparation of DNAN was taking our interest to fulfill the new security requirements in manipulation, storage and transportation of the military ammunitions. DNAN was prepared in first time in our laboratory. The characterization of the prepared DNAN was presented and compared with the standard TNT explosives. Thermochemical code (EXPLO5) was used to calculate the detonation characteristic. Thermal behavior was determined using Thermogravimetry (TGA) and differential Scanning colometry (DSC).

2. Experimental

2.1. Materials and preparation method

DNAN (Fig. 1) was successfully synthesized via the reaction of dinitrochlorobenzene (DNCB), sodium hydroxide, and methanol at room temperature [21] (Fig. 2). Purification and crystallization of the crude DNAN was performed in boiling methanol. TNT was a product of Military factory 81 military grade A with melting point 80.6 °C. Structural formula for TNT presented in (Fig. 1). The preparation of DNAN charge was achieved by melting DNAN at 97-100°C with continuous stirring for 30 min, followed by the addition of PVC and left to cool. The same procedure was performed for TNT at 90 °C.

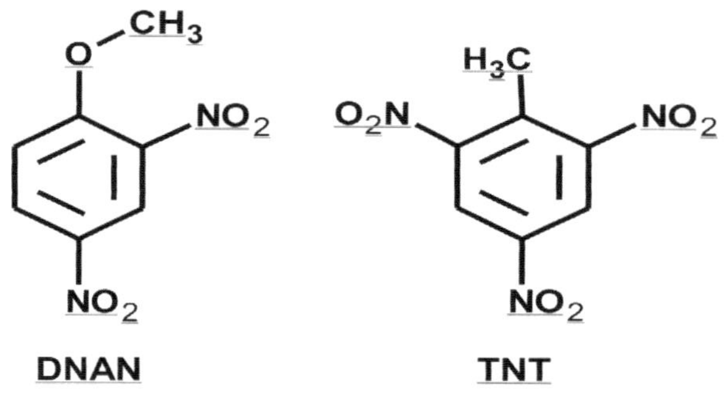

Figure 1: Structural formulas of DNAN and TNT

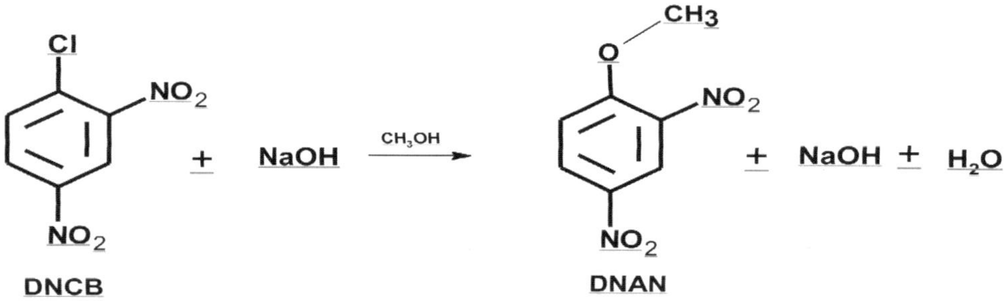

Figure 2: Scheme of DNAN preparation from DNCB.

2.2. Characterization Procedure

The determination of DNAN combustion heat was achieved via BCA 500 Bomb calorimeter, model, OZM Company, Czech Republic [22]. The sensitivity of DNAN against impact and friction stimuli was measured using BAM impact sensitivity instrument [22] via dropping 2 and 5 kg weight hammers and BAM friction test apparatus using the Probit analysis [23]. Detonation velocity was investigated using the EXPLOMET-FO-2000 (created by KONTINITRO AG). EXPLO5 code, version 5.04, was utilized to calculate the theoretical detonation characteristics of the individual explosives, while, the thermogravimetric analysis (TGA55, TA Instruments, USA) was used to track the thermal behavior of DNAN samples. Heat flow properties were resolute by DSC, Netzsch 200 F3 instrument.

3. Results and Discussion

Relationship between experimental product ρD^2 and calculated detonation pressure is presented in Fig. 3. This relationship shows that the data obtained for all the used samples relatively correspond to a linear relationship; an order of them in Fig. 3 is according to the more powerful explosives to the less one. It is clear that the performance of the DNAN is less than TNT. However ε-Cl20 is the highest detonation pressure among all the studied explosives. It is also important to say that the DNAN and TNT are nitro polyarenes and their molecular structure different than the other studied explosives but it is correlate well with each other.

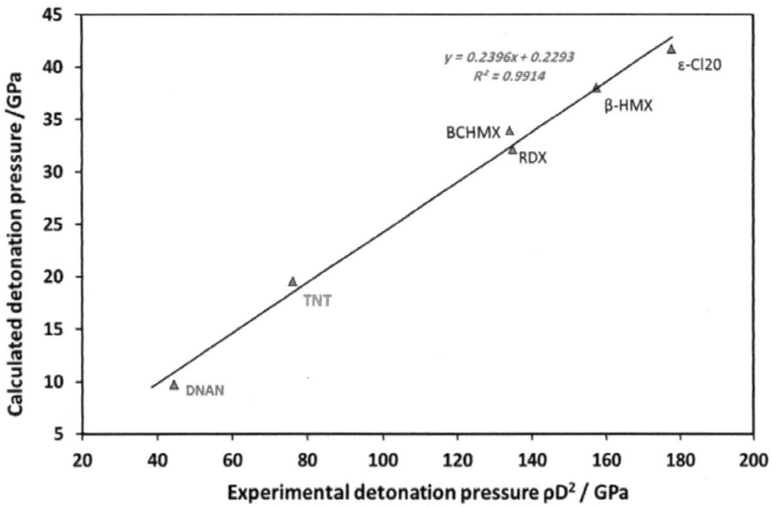

Figure 3. The relationship between the calculated detonation pressure and representative of the detonation pressure ρD^2

It is commonly known in the field of explosives that the explosive performance is directly proportional with the explosive sensitivity against external stimuli. This confirms the validity of the later presumption that increasing the explosive performance should be accompanied by an increase in the sensitivity values, as illustrated in Fig. 4.

As it can be seen from Fig. 4 that the DNAN will provide higher safety in manipulation as the impact sensitivity of the DNAN is the lowest. The heat of detonation of DNAN was the lowest of all the studied samples. This result is compatible with the general rule that decreasing the sensitivity leads to lower detonation characteristics and the vise verse is correct. CL-20 has the highest heat of explosion and the lowest impact sensitivity.

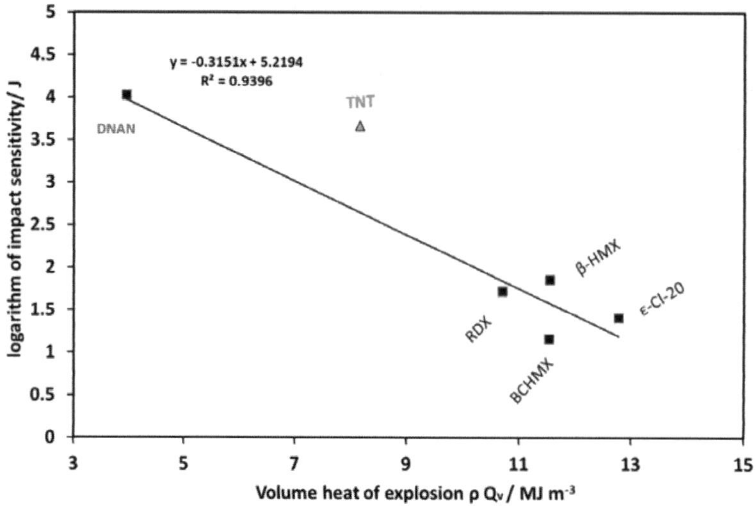

Figure 4. Semi-logarithmic relationship between impact sensitivity and volume heat of explosion

As far as friction sensitivity is concerned, a comparison between the friction and impact sensitivities is presented in Fig. 5. It is obvious here that the studied nitramines are near to each other for friction sensitivity and BCHMX is considered the most sensitive explosives among all the studied explosives In contrary, DNAN is considered the most insensitive explosive among all the studied explosives as its friction sensitivity reaches to the same level as TNT which is the maximum value of the BAM friction apparatus.

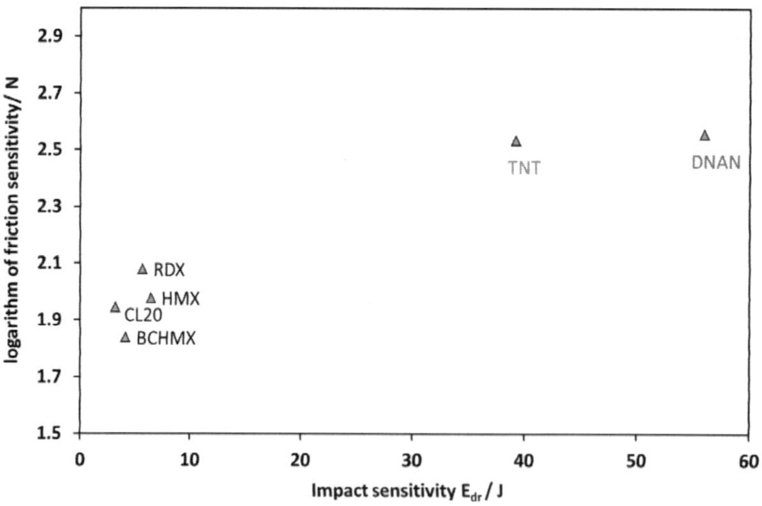

Figure 5. A comparison between the logarithm of friction sensitivity and the impact sensitivity of the studied explosives

The non-isothermal DSC of pure DNAN were measured under 5 K min-1 heating rate (see Fig. 6 a). Recorded curves are presented in Fig. 6, while the data of For TGA/DTG and DSC curves are listed in table 1.

Table 1: DSC, TGA/DTG data for pure DNAN Explosives

Sample	Heating rate	First peaks (DSC)				Second Peaks			
	β/ K min $^{-1}$	T_{ot} (°C)	T_p (°C)	T_{ed} (°C)	ΔH_1 (J g^{-1})	T (°C)	T_p (°C)	T_e (°C)	ΔH_2 (J g^{-1})
DNAN	5	85.8	87.0	88.3	-89.7	276.5	273.6	276.9	-362.0
		TGA				DTG			
		T_o/° C		Mass loss / %		T_p / °C		T_e/ °C	
		192.5		-99.8		216.1		222.7	

It is obvious from the Fig. 6 a that the DNAN shows two endothermic peaks at 87.0°C and 273.6°C in which the complete decomposition is appear however in TGA/DTG curve Fig.6b show a complete decomposition at 216.1°C. Actually, it is different than DSC as the thermal measurement techniques is different; DSC closed pan and the TGA is open crucible which provide evaporation of DNAN. This observation was clear with the complete mass loss of the DNAN sample and the same as our previously investigated paper about TNT mixtures with nitramine.

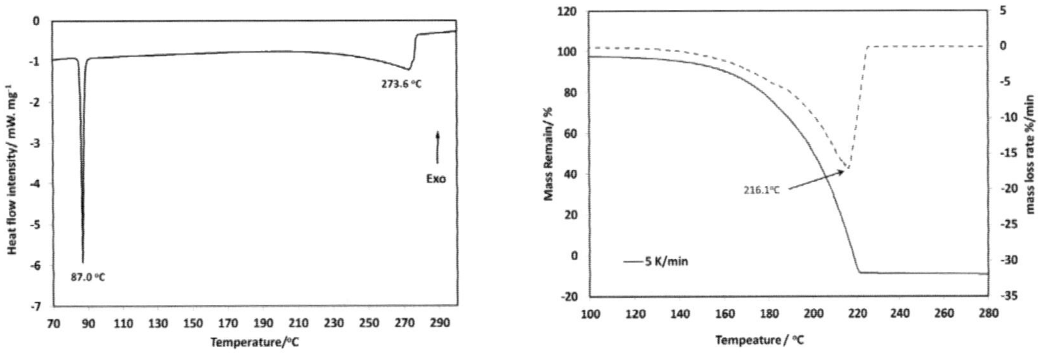

a. b.

Fig. 6 Thermal behavior of Pure DNAN at heating rate 5 K min^{-1}: a DSC, b. TGA/DTG

4. Conclusion

2,4-Dinitroanisole (DNAN) successfully prepared in our laboratory for the first time. The performance of the insensitive DNAN considered modest compared with the studied nitramines but slightly lower than TNT. However, its sensitivity toward impact is very interesting. Regarding to the friction sensitivity, the DNAN shows the same level as TNT however it reached to the maximum limit of the BAM friction apparatus. The DSC thermal behavior for the DNAN shows to Endothermic peaks at 87.0 oC but the decomposition peaks is at 273.6oC the complete mass loss on weight (decomposition) for DNAN was observed at 216.0 oC and it is different than DSC which connected with the different techniques and the test conditions of DSC and TGA apparatus (closed pan and Open Crucible). Last.

The detonation properties of DNAN which are slightly lower than TNT but easily to melt out candidate to replace TNT in low sensitive ammunitions.

5. References

[1] Walsh MR, Walsh ME, Taylor S, Ramsey CA, Ringelberg DB, Zufelt JE, Thiboutot S, Ampleman G, Diaz E, 2013, *Propellants, Explosives, Pyrotechnics* **38(3)**, 399.
[2] Elbeih A. and Zeman S. 2014, *Central European Journal of Energetic Materials* **11**, 501.
[3] Hussein A.K., Zaki M.G. and Elbeih, A. 2022. Influence of a novel nano-thermite colloid based on CuO coated CNTs on the thermo-analytical characteristics of 1, 3, 5-trinitro-1, 3, 5-triazinane. *Combustion Science and Technology*, in press.
[4] Elbeih A., Abdelhafiz M. and Hussein A.K. 2021. *Defence Science Journal*, **71**(2), 177.
[5] Hussein A.K., Elbeih A. and Zeman S. 2017, *Thermochimica Acta* **655**, 292.
[6] Hussein A.K., Zeman S. and Elbeih, A. 2021. Manifestations of replacing 2, 4, 6-trinitrotoluene by 2, 4-dinitroanisole (DNAN) in compositions based on several interesting nitramines, *Journal of Energetic Materials*, in press.
[7] Elshenawy T., Li Q.M. and Elbeih A. 2020. Experimental and numerical investigation of zirconium jet performance with different liner shapes design. *Defence Technology*, in press.
[8] Abd-Elghany M., Elbeih A. and Klapötke T.M. 2018, *Journal of Analytical and Applied Pyrolysis* **133**, 30.
[9] Hussein A.K., Elbeih A. and Zeman S. 2018, *RSC advances* **8**, 17272.
[10] Feldman D. 2021, *Biointerface Research in Applied Chemistry* **11**, 8179
[11] Hussein A.K., Zeman S. and Elbeih A. 2018, *Thermochimica Acta* **660**, 110.
[12] Trzciński W., Cudziło S., Dyjak S. and Nita M. 2014, *Central European Journal of Energetic Materials,* **11**, 443.
[13] Sun XL, Qi Z, Li QQ, Zhang XR, Shao XL, Jin Y, Zhang JJ and Liu YY. 2020, *Biointerface Research in Applied Chemistry* **10**, 5815.
[14] Faisal N. 2020, *Biointerface Research in Applied Chemistry* **10**, 7223.
[15] Abd-Elghany M., Klapötke T.M. and Elbeih A. 2017, *Journal of Analytical and Applied Pyrolysis* **128**, 397.
[16] Elbeih A., Abd-Elghany M. and Klapötke T.M. 2017, *Propellants, Explosives, Pyrotechnics* **42(5)**, 468.
[17] Abd-Elghany M., Klapötke T.M. and Elbeih A. 2017, *Propellants, Explosives, Pyrotechnics* **42(12)**, 1373.
[18] Zeman S., Elbeih A. and Akštein Z. 2011, *Hanneng Cailiao/Chinese Journal of Energetic Materials*, **19(1)**, 8.
[19] Pham TT. 2020, *Biointerface Research in Applied Chemistry* **10**, 4853.
[20] Zeman S., Hussein A.K., Jungova M. and Elbeih A. 2019, *Defence Technology*, **15(4)**, 488.
[21] Elshenawy T., Elbeih A. and Li Q.M. 2016, *Central European Journal of Energetic Materials* **13**, 927.
[22] Elbeih A., Elshenawy T., Zeman S. and Akstein Z. 2018 *Central European Journal of Energetic Materials* **15**, 3.
[23] Abd-Elghany M, Klapötke TM, Elbeih A, Hassanein S and Elshenawy T. 2017, *Huozhayao Xuebao/Chinese Journal of Explosives and Propellants* **40**, 24.
[24] Elshenawy T., Elbeih A. and Klapötke T.M. 2018, *Journal of Energetic Materials* **36**, 127.
[25] Sabatini J.J. and Oyler K.D. 2016, *Crystals*, **6(1)**, 5.
[26] Suceska M, 1995, *Test methods of explosives*, Springer, Heidelberg.
[27] Šelešovský J. and Pachmáň J. 2010, *Central European Journal of energetic materials*, **7**, 269.

High Performance Sheet of PBX based on NTO

Ahmed Elbeih[1], Mohamed Elnogomy[2], Tamer Elshenawy[2], Ahmed Saleh[3*], Ahmed Azazy[3], Ahmed k. Hussein[1], M. G. Zaky[1], Hesham S. Ahmed[3]

[1] Military Technical College, Kobry Elkobbah, Cairo, Egypt
[2] Technical Research Center, Cairo, Egypt
[3] Science and Technology Center of Excellence (STCE), Ministry of Military Production, Cairo, Egypt.

Abstract: 3-nitro-1,2,4-triazol-5-one (NTO) is a low sensitive high explosive which has been used in the formation of low sensitive ammunitions. In this study, the preparation method of NTO explosive was discussed. A sheet of PBX on the basis of NTO bonded by HTPB binder matrix was prepared (NTO-PBX). The formulation is investigated for explosive properties to understand the influence of NTO on the performance of the PBX sheet. Impact and friction sensitivities of NTO-PBX, heat of Combustion and Detonation velocity were specified. EXPLO5 code, version 5.04, was utilized to calculate the theoretical detonation characteristics of NTO-PBX. For comparison, Data-sheetC, EPX-1, Semtex 1A and Formex P1, were also studied. By comparing the obtained results, confirmation of some relationships was investigated. The detonation velocity of the NTO sheet explosive was the highest of all the studied samples. The experimental and calculated results of NTO-PBX and the other compositions are in a good agreement.

Keywords: sheet explosive, NTO, sensitivity, detonation parameters.

1. Introduction

The well-known sheet explosive consists of a matrix of polymer including high energy material such as PETN [1]. The high energy material is the main source of energy produced from the system. The elastic polymeric behavior is used to decrease the sensitivity of the explosive and perform flexibility of the prepared composition to enhance the adhesive characteristics of the sheet explosive and give the possibility of rolling the formed sheet [2-5]. The recent researches are looking for low sensitive explosives and oxidizers as a replacement for the traditional components in the high energy compositions [6-10]. In addition, the selection of polymer matrix to form a flexible sheet explosive is a target for most of the scientists working in the energetic materials field [11-13]. The destructive effect of the sheet explosive is depends mainly on the power and the detonation properties of the high energy filler [14-17]. Several commercial plastic bonded explosives are available with known characteristics for the applications in different fields [18]. Also there are several advanced high energy materials which had been studied as PBXs to obtain high performance PBXs with high loading density but there is a problem with the mechanical characteristics and the flexibility of these compositions [19-21]. The low sensitive high energy material named 3-nitro-1,2,4-triazol-5-one (NTO) is considered a promising candidate to be used in the formation of high performance sheet explosive. NTO has been approved to be a main

component in the formation of insensitive compositions such as IMX-101 (replacement of TNT in warheads).
The performance properties of NTO seem to be similar to the traditional RDX but it has very low sensitivity in comparison with RDX [22]. The study of NTO as a high energy material with matrix of polymer to prepare high performance explosive sheet with high percentage of explosive filler and low sensitivity is our target in this study. HTPB polyurethane matrix has been selected to be the flexible matrix for collecting the NTO particles and form NTO-sheet explosive. The prepared NTO-sheet explosive (NTO-PBX) has 88 mass percentage of NTO and 12 mass percentage of HTPB polymer matrix. The details of the HTPB matrix are included in the experimental part. Several other traditional PBXs are studied in this work for comparison. Formex plastic explosive [23], EPX1 [24], Semtex 1A [25] and sheet explosive based on PETN have been studied. Mutual comparison between the explosive properties of these compositions with the prepared NTO-sheet explosive are discussed and presented to clarify the influence of NTO on the formation of advanced high performance sheet explosive. Thermochemical code (EXPLO5) was used to calculate the detonation characteristic.

2. Experimental

2.1. Preparation of Explosive Sheets

NTO based composition (NTO-PBX) has been prepared using vertical mixer under vacuum. NTO was prepared in explosives lab., MTC, Cairo. Recrystallization process was used to obtain NTO with particle size of 200-280 μm. The source of the polymeric matrix ingredients and the steps of NTO-PBX preparation was the same as reported in ref. [26]. The lacquer ingredients (HTPB, DOZ and MAPO) was poured and mixed for 15min. under vacuum in the mixer. NTO crystals were divided to three portions and added to the lacquer, each portion was added at 55 °C and left under mixing for 30 min. in vacuum. Lastly, the curing agent HMDI was added to the composition at 40 °C and mixed for 25 min. in vacuum. The obtained NTO-PBX was casted and cured in the form of blocks, then cutting nife was used to form a sheets with thickness 6mm. The obtained NTO-PBX has 88 wt% NTO and 12 wt% HTPB matrix. A curing ratio of 0.78 NCO/OH was applied.

Table 1. The mass ratio of the NTO-PBX ingredients

Abbreviations	NTO (wt. %)	HTPB (wt. %)	HMDI (wt. %)	DOZ (wt. %)	MAPO (wt. %)
NTO-PBX	88	9.14	0.54	1.82	0.5

2.2. Characterization Procedure

The determination of NTO-PBX combustion heat was achieved via BCA 500 Bomb calorimeter, model, OZM Company, Czech Republic [27]. The sensitivity of NTO-PBX against impact and friction stimuli was measured using BAM impact sensitivity instrument [27] via dropping 2 and 5 kg weight hammers and BAM friction test apparatus using the Probit analysis. Detonation velocity was investigated using the EXPLOMET-FO-2000 (created by KONTINITRO AG). EXPLO5 code, version 5.04, was utilized to calculate the theoretical detonation characteristics of the prepared NTO-PBX and the other studied PBXs,

Table 2. The measured parameters of the studied samples

No.	Composition	Impact sensitivity (J)	Friction Sensitivity (N)	Combustion Heat (J·g^{-1})	Density (g·cm^{-3})	Detonation velocity Measured (m/s)
1	NTO-PBX	34.2	>360	9234	1.59	7728
2	Semtex 1A	21.1	187	14003	1.47	7264
3	Formex P1	13.5	194	12943	1.53	7544
4	EPX-1	13.9	180	11528	1.55	7636
5	Datasheet C	31.5	>360	14176	1.45	6660

Table 3. The calculated detonation characteristics of the tested explosives

No.	Explosive type	Density (g·cm^{-3})	EXPLO5		
			detonation velocity (m·s^{-1})	detonation pressure (GPa)	detonation Heat (J·g^{-1})
1	NTO-PBX	1.59	7514	21.88	3840
2	Semtex 1A	1.47	7014	17.52	5099
3	Formex P1	1.53	7346	20.03	5411
4	EPX-1	1.55	7398	21.17	5742
5	Datasheet C	1.45	6458	12.86	4564

3. Results and Discussion

The sensitivities of NTO-PBX in comparison with the selected explosives are clarified in Fig. 1. The impact sensitivity of NTO-PBX is the lowest; it has impact sensitivity of 34.2 J which is lower than Data sheetC (contains 63 wt% PETN) also it is lower than the other plastic explosives. On the other side, the friction sensitivity of NTO-PBX is in the same level of Data-sheetC. Semtex, Formex and EPX-1 have friction sensitivity close to each other (83-87 wt% of PETN in these compositions). Semtex 1A has lower impact sensitivity than the two plastic explosives due to the lower PETN content in Semtex explosive. Even though, NTO-PBX has 88 wt% of NTO explosive, it has the lowest impact sensivity. This result proved the efficiency of NTO to have very low impact sensitivity when coated by the polyurethane matrix. With respect to the sensitivities to different stimuli, NTO-PBX is a low sensitive composition with respect to the traditional plastic explosives.

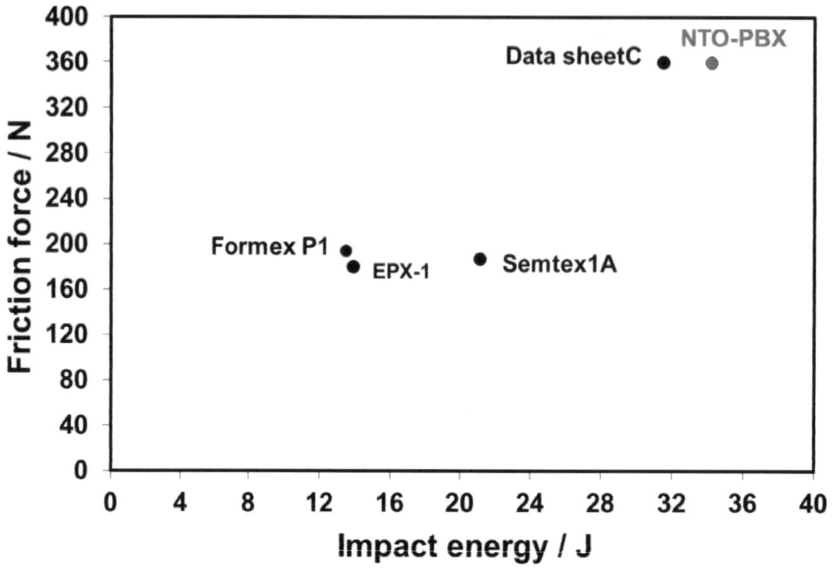

Figure 1. Sensitivity measurements

The reported linear relationship between the density and the detonation velocity of the samples is presented in Fig. 2. Increasing the bulk densities of the explosives lead to increase of their detonation characteristics. Here, this relation was confirmed even with the presence of NTO explosive. Only the data-sheetC has lower detonation characteristics lead to remove its result from the linear relationship. Data-sheetC has only 63 wt% of PETN in addition to 8 wt% of nitrocellulose (used in the double base propellants), this composition with 29 wt% plasticizer caused decreasing of the detonation velocity and keep the performance lower than predicted. The NTO-PBX has the highest detonation velocity and density of all the compositions. The measured detonation velocity is 7728 m/s close to the detonation velocity of PE-4 but has much lower sensitivity. The polyurethane matrix decreased the density of NTO but still the detonation velocity is acceptable.

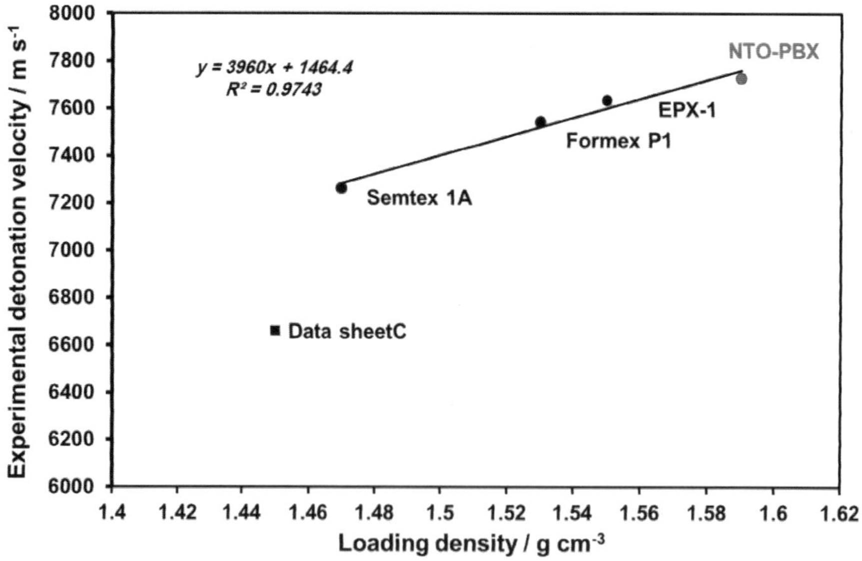

Figure 2. The dependence of detonation velocity on the loading density

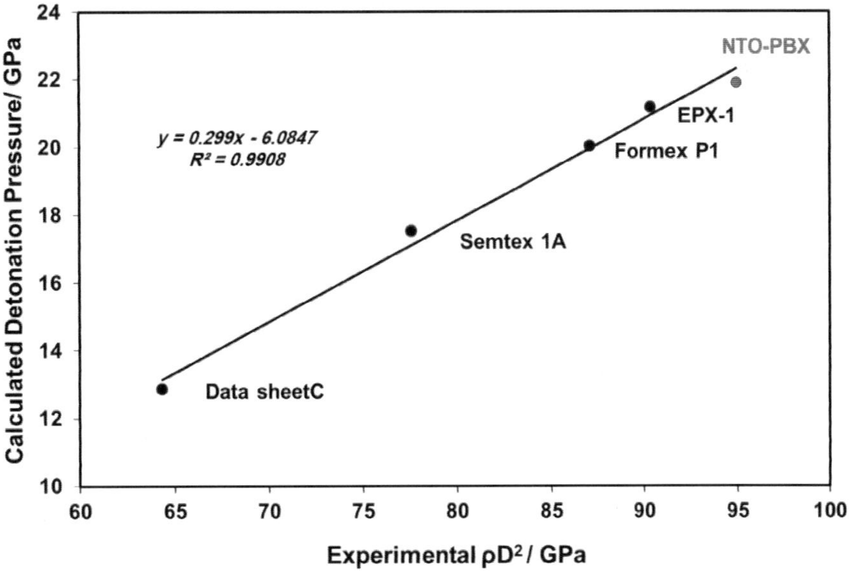

Figure 3. The theoretically obtained detonation pressure in comparison with the experimental ρD^2

The results of EXPLO5 code to determine the detonation parameters has been determined as reported in Table 3. The theoretical results have been checked in comparison with the experimental results based on the known relationship reported by the theories of detonation. The calculated results of the detonation pressure are compared with the experimental ρD^2 (ρ-loading density, D- experimental detonation velocity). In this case, it is obvious that the detonation properties of NTO-PBX are higher than all the other compositions. The high mass percentage of NTO on the composition increased the performance of the prepared PBX. Also EPX-1 has high detonation properties in comparison with the traditional plastic explosives. As predicted, Data-sheetC has the lowest detonation parameters due to the low mass percentage of PETN in this composition as discussed in the previous section. A good agreement between the theoretical and experimental measurements was observed in Fig. 3 with accuracy $R^2 = 0.99$. It means that theoretical results represent the actual measurements of the samples.

On the other side, the heats of combustion have been measured and compared with the heat of detonation calculated theoretical as shown in Fig. 4. The influence of the plasticizers (represent as a fuel in the composition) caused increase of the heat of combustion for the traditional plastic explosives. Data-sheetC has the highest heat of combustion while the NTO-PBX has the lowest value. Unexpected result was observed for the new PBX based on NTO. The heat of combustion and the heat of detonation of NTO-PBX are lower than all the studied samples. This result is mainly based on the low heat of detonation of the NTO explosive. It has been reported in literature that the heat of detonation of NTO is low in comparison with the other energetic materials. In this case, the point of NTO-PBX is out of the relation as shown in fig. 4. Also the types of the polymeric matrix in each composition affect the combustion characteristics of the samples. The results of the heat of detonation seem to be logic. Samples with higher mass percentage of PETN have higher heat of detonation.

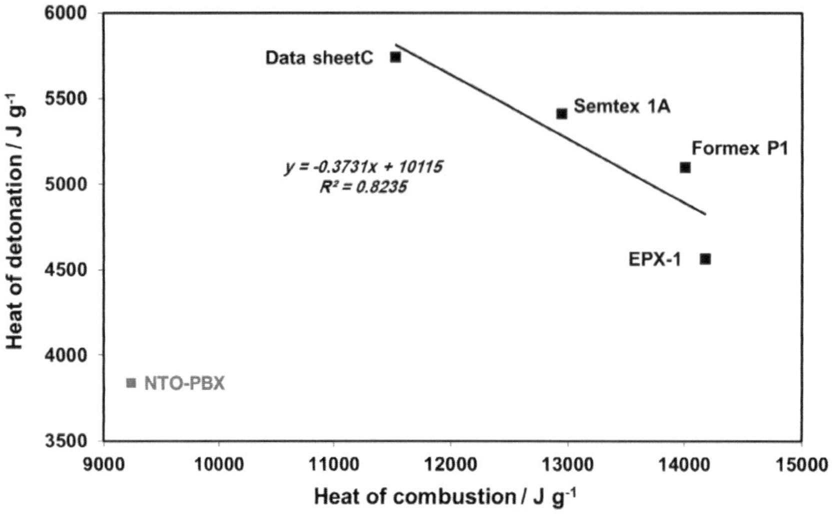

Figure 4. Heat of combustion in comparison with heat of detonation

4. Conclusion

NTO Sheet explosive has been prepared successfully with high mass percentage of NTO as a explosive filler. The impact sensitivity of NTO is the lowest of the studied samples and the friction sensitivity is in the same range of data-sheetC (containing 63 wt% of PETN). The detonation velocity of NTO-PBX is 7728 m/s which is the highest value of the studied samples. The theoritically obtained detonation parameters of NTO-PBX are the highest except for the heat of detonation. It is concluded that the NTO-PBX has low heat of detonation and this might be due to the low detonation heat of the pure explosive filler (NTO). The experimental and calculated results of NTO-PBX and the other compositions are in good agreement. NTO-PBX has very low sensitivities and its performance is high and it is perdicted to have applications in the future as high performance low sensitive sheet explosive.

5. References

[1] Klapötke T M 2015, *Chemistry of high-energy materials*, Walter de Gruyter GmbH & Co KG.
[2] Hussein A.K., Zeman S. and Elbeih A. 2018, *Thermochimica Acta* **660**, 110.
[3] Hussein A.K., Elbeih A. and Zeman S. 2018, *RSC advances* **8**, 17272.
[4] Elbeih A., Abd-Elghany M. and Klapötke T.M. 2017, *Propellants, Explosives, Pyrotechnics* **42(5)**, 468.
[5] Elbeih A. and Zeman S. 2014, *Central European Journal of Energetic Materials* **11**, 501.
[6] Abd-Elghany M., Elbeih A. and Klapötke T.M. 2018, *Journal of Analytical and Applied Pyrolysis* **133**, 30.
[7] Abd-Elghany M., Klapötke T.M. and Elbeih A. 2017, *Journal of Analytical and Applied Pyrolysis* **128**, 397.
[8] Hussein A.K., Elbeih A. and Zeman S. 2017, *Thermochimica Acta* **655**, 292.
[9] Abd-Elghany M, Klapötke TM, Elbeih A, Hassanein S and Elshenawy T. 2017, *Huozhayao Xuebao/Chinese Journal of Explosives and Propellants* **40**, 24.
[10] Abd-Elghany M., Klapötke T.M. and Elbeih A. 2017, *Propellants, Explosives, Pyrotechnics* **42(12)**, 1373.
[11] Elbeih A., Elshenawy T., Zeman S. and Akstein Z. 2018 *Central European Journal of Energetic Materials* **15**, 3.
[12] Feldman D. 2021, *Biointerface Research in Applied Chemistry* **11**, 8179.
[13] Pham TT. 2020, *Biointerface Research in Applied Chemistry* **10**, 4853.

[14] Elshenawy T., Elbeih A. and Li Q.M. 2016, *Central European Journal of Energetic Materials* **13**, 927.
[15] Elshenawy T., Elbeih A. and Li Q.M. 2018, *International Journal of Mechanical Sciences*. **136** 234.
[16] Elshenawy T., Elbeih A. and Klapötke T.M. 2018, *Journal of Energetic Materials* **36**, 127.
[17] Elbeih A., Abdelhafiz M. and Hussein A.K. 2021. *Defence Science Journal*, **71**(2), 177.
[18] Zeman S., Elbeih A. and Akštein Z. 2011, *Hanneng Cailiao/Chinese Journal of Energetic Materials*, **19(1)**, 8.
[19] Licht H-H 2000 *Propellants Explosives, Pyrotechnics* **25** 126.
[20] Sun XL, Qi Z, Li QQ, Zhang XR, Shao XL, Jin Y, Zhang JJ and Liu YY. 2020, *Biointerface Research in Applied Chemistry* **10**, 5815.
[21] Faisal N. 2020, *Biointerface Research in Applied Chemistry* **10**, 7223.
[22] Mukundan T., Purandare G.N., Nair J.K., Pansare S.M., Sinha R.K. and Singh H. 2002 *Defence Science Journal* **52** 127.
[23] Zeman S., Hussein A.K., Jungova M. and Elbeih A. 2019, *Defence Technology*, **15(4)**, 488.
[24] Elbeih A. 2015 Characteristics of a new plastic explosive named EPX-1. *Journal of Chemistry* **2015,** ID861756
[25] Elbeih A., Hussein A.K., Elshenawy T., Zeman S., Hammad S.M., Baraka A., Elsayed M.A., Gobara M. and Tantawy H. 2020, *Defence Technology,* **16,** 487.
[26] Elbeih A., Wafy T.Z. and Elshenawy T. 2017 *Central European Journal of Energetic Materials* **14** 77.
[27] Cooper P.W 1996 Explosives engineering, Wiley-VCH.

Recovery of valuable heavy minerals via gravity and magnetic separation operations from Diit Quaternary stream sediments, southern coast of the Red Sea, Egypt

M M Fawzy[1*], M S Abu El Ghar[2], I M Gaafar[1], A M El shafey[1], M Diab[1] and A W Hussein[2]

[1] Nuclear Materials Authority, P. O. Box 530 Maadi, Cairo, Egypt

[2] Geology Department, Faculty of Science, Fayoum University, Fayoum, Egypt

*Corresponding author: mm1_fawzy@yahoo.com

Abstract. This research aims to concentrate and recover valuable heavy minerals (VHMs) contained in Diit area, southern coast of the Red Sea, Egypt. In this study, three technological samples were collected representing Lens, Delta and Wadi Diit; mineralogical investigation followed by mineral processing was performed in order to recover VHM content. VHMs such as zircon, ilmenite, rutile, sphene, apatite, garnet, and magnetite have been recorded in associations with lesser amounts of uranothorite, monazite, xenotime, fergusonite, khatyrkite, and gold. Initially, two stage wet gravity concentration processes, rougher and scavenger, were carried out via a shaking table to produce a clean concentrate of VHMs, which was used as feed material for the magnetic separation process. Magnetic separation was performed in order to separate ferromagnetic minerals from paramagnetic minerals as well as diamagnetic minerals to obtain clean concentration of these fractions. Assay and Material balance of the concentration steps proved that the THM content of Diit lens increased to 89.70% with a recovery of 91.68% in a weight of 61.08% out of the original sample. While the percentage of THM content in Wadi and Delta Diit samples increased to 48.68 and 47.67% respectively, with recovery equivalent to 71.67 and 82.06% respectively in a weight of 16.51, 18.7% respectively out of the original sample. Finally, a process flowsheet was created according to the optimum conditions for concentration and separation processes.

Keywords: Diit Area, VHMs, valuable heavy minerals, wet gravity concentration, Magnetic separation, flowsheet

1. Introduction and geological context

Heavy minerals (HMs) are minerals with higher specific gravity values than quartz (sp.gr. 2.65), as it is the most common rock-forming mineral. The valuable heavy minerals (VHMs) are heavy minerals of economic importance due to their properties and their prevalence. [1,2] The most valuable and commercially exploitable heavy minerals are; zircon, ilmenite, monazite, rutile, xenotime, kyanite, andalusite, silimanite, staurolite, chromite, garnet, magnetite, columbite-tantalite, cassiterite, wolframite and scheelite as well as gold and platinum. Beneficiation and upgrading of the VHMs in beach sand or in stream sediments are performed firstly using gravity concentration technique; followed by the combination of magnetic and electrostatic separation.[3-6] Flotation can also be used as a later or complementary stage to achieve a cleaner product.[7-9] Diit area is located on the southern coast of the Red Sea of Egypt. It is bounded by latitudes of 22° 18' N and 22° 41' N and longitudes of 35° 55' E and 36° 18' E (figure 1) and covered an area about 760 km^2. Heavy minerals accumulations are observed along the coast of the Red Sea, which were formed not only by transportation processes related to the Red Sea offshore currents, but also by the drainage networks locating in the Eastern Desert of Egypt.[10-12] The rock units of Diit area have a variety of age ranging from Late Proterozoic to Cenozoic rocks. The detailed geological, structural, and petrographical characterization of the study area was previously discussed.[10] The alluvial deposits in the study area cover the area between the Basement Mountains and

Journal of Physics: Conference Series 2305 (2022) 012020 doi:10.1088/1742-6596/2305/1/012020

the Red Sea coast forming fan shape delta called as "Delta Diit". The Delta Diit has a triangular shape, with its apex trending nearly in the southwest direction at the mouth of Wadi Diit (runs about 85 Km in southeastern desert of Egypt, while its most length about 300 Km extends in Sudan), while its base represents the Red Sea coast. It has a considerable concentration of black sands in the form of lens with about 5km length and width of about 30m extending along the shoreline (figure 1). Diit Quaternary deposits are represented by wadi deposits, sand dunes, sand sheets, and sabkha deposits along the southern coast of the Red Sea (figure 1).

Detailed mineralogical characterization of Diit stream sediments was previously discussed by Fawzy et al., (2022).[11] Fawzy (op. cit.) collected and studied ninety four samples that covered the area under investigation (seventy six spot samples in Delta Diit, ten samples in Wadi Diit and eight samples in Diit lens). The mineralogical investigation of this work concluded that the VHMs that were recorded are magnetite, ilmenite, garnet, zircon, sphene, apatite, rutile, pyroxene, amphibole, uranothorite, xenotime, monazite, fergusonite, khatyrkite, and gold.

This paper examines the ability for upgrading and concentrating the VHMs from Diit Quaternary stream sediments by using the simplest and cheapest physical methods and proposes a flowsheet used as a guide for exploiting the economic heavy minerals presented in the area under investigation and similar ones.

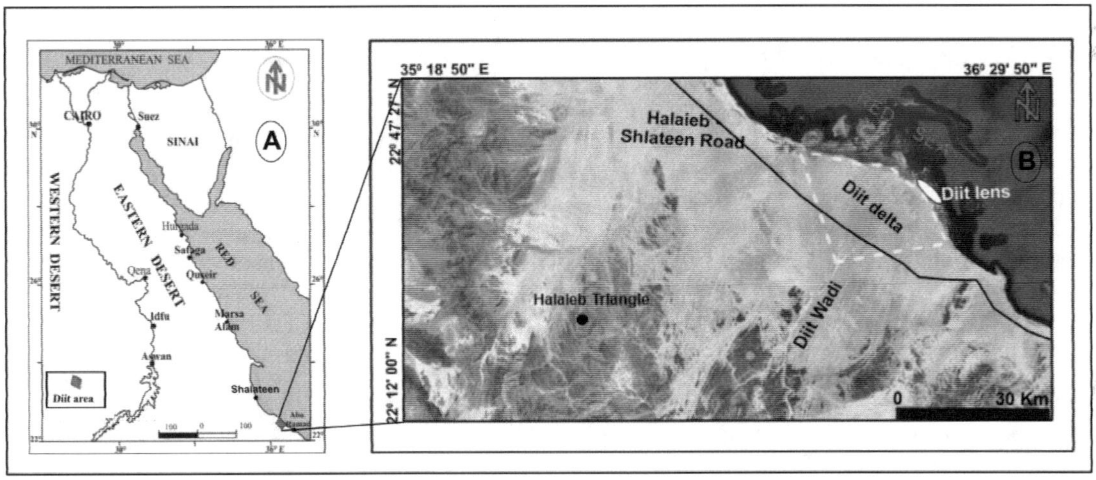

Figure 1. (A) General location map of the study area. (B)Satellite image showing Diit lens, Wadi, and Delta area

2. Sampling and Methodology

Three representative technology samples were used in this work from Wadi, Delta, and Diit lens, each weighing about 20 to 25 kg. Initially, grain size distribution analyses were performed for each of the three samples using a representative sample weighing about 250 g. The samples were sieved using 2, 1, 0.5, 0.25, 0.125 and 0.063 mm (ASTM codes[13]) sieve series representing gravel, very coarse sand (V.C. sand), coarse sand (C. sand), medium sand (M. sand), fine sand (F. sand), very fine sand (V.F. sand) and mud sizes. Heavy-liquid separation tests were carried out on another representative sample weighing about 100 g to determine the THM content using bromoform with a specific gravity of 2.89. Grain size distribution analyzes of heavy minerals obtained from bromoform separation were done to determine the size distribution of these target minerals.

The initial mineralogical identification for the heavy bromoform fractions was carried out with the aid of binuclear microscope, and then subjected to scanning electron microscope (SEM) attached with a Philips XL 30 energy-dispersive spectrometer unit (EDS).

The upgradability tests for VHMs were performed via gravity concentration using Wilfley shaking table No. 13 [14] for get rid of the light gangue minerals as much as possible. After completion of each concentration stage (rougher and scavenger), a representative sample weighting about 100 g of each of the gravity concentration products (concentrate and tail) is subjected to heavy-liquid separation test using bromoform for determination of the THM assay. The obtained heavy mineral fractions were magnetically separated using Carpco dry high intensity magnetic separator (DHIMS) Model MLH (13) III-5" [15] to fractionate and separate the ferromagnetic from paramagnetic and diamagnetic minerals and obtain a clean concentrate from theses fractions. The magnetic separation processes were achieved at pre-optimized factors of a medium air gap of 1.5 cm, magnetic field current at 1 and 3 amperes, magnetic roll speed 30 rpm and optimum feed rate 39.2 g/min.

X-ray fluorescence analyses were performed for feed samples as well as the heavy mineral concentrates for measuring major and some trace element contents. XRF analyses were carried out via Axios advanced, Sequential WD_XRF Spectrometer.[16] To prepare the samples for XRF measurements, representative samples were ground to a powder (<75 μm) inside an agate mortar and then heated for 2 hours at 1000°C to calculate the total loss of ignition (LOI). For the purpose of mineralogical identification and confirmation, X-ray diffraction (XRD) analyzes was performed on magnetic products using a Bruker D8 Discover X-ray Diffractrometer. [17]

3. Results and Discussions

3.1. Mineralogical characterization results

As a results of grain size distribution analyses for the studied samples (lens, Delta and Wadi Diit), the percentages of distribution of gravel, very coarse sand, coarse sand, medium sand, fine sand, very fine sand and mud are presented in (figure 2). The samples show a high degree of homogeneity: indeed the sandy fraction, with about 97.98, 90.85 and 97.32% of total percentages in weight of Diit lens, Delta and Wadi samples respectively on average, was most present, while the others, gravelly and silty-clayey fractions, showed lower percentages in weight. Cumulative percent passing curves for the representative samples of Diit lens, Delta and Wadi were presented in (figure 3). From these curves, the effective diameter values of D_{80} are 0.32, 0.6, and 0.36 for Diit lens, Delta and Wadi respectively.

For calculating the VHM content for each sample of Diit area, about 100 representative grams were separated using bromoform ($CHBr_3$, sp. gr. of 2.89) and the ratio of heavy economic mineral to gangue light silicate minerals were calculated. The light fractions of Wadi and Delta Diit samples represent higher values (80.87% mass for Delta Diit sample and 79.49% mass for Wadi Diit sample) while the lower value represents the lens sample (39.15%) which is relative to the presence of high values heavy mineral content. Microscopic investigation of the light gangue fractions proved that they contain mainly of quartz, feldspars and micas

The content of the heavy fraction in the Diit lens sample (59% mass) is greater than those of Wadi and Delta Diit sample (11.45% and 12.48% mass respectively). The microscopic examinations of heavy fractions with the help of SEM analyses proved that ilmenite, magnetite, zircon, sphene, apatite, garnet, rutile and heavy green silicates (pyroxene, epidot, and amphibole) are the constituent minerals of all Diit area samples, but in varying proportions between Wadi, Delta and lens. Also present, in minor or trace amounts, are uranothorite, xenotime, monazite, fergusonite, and khatyrkite (figures 4 and 5). As shown in (figure 2), grain size distribution analysis of the heavy fractions indicates that most of the heavy minerals are present in the sandy fraction (99% Diit lens, 93% Delta Diit and 90.5% Wadi Diit). The remaining percentage of the heavy minerals is present in the mud fraction, while the gravel fraction in all samples does not contain heavy minerals.

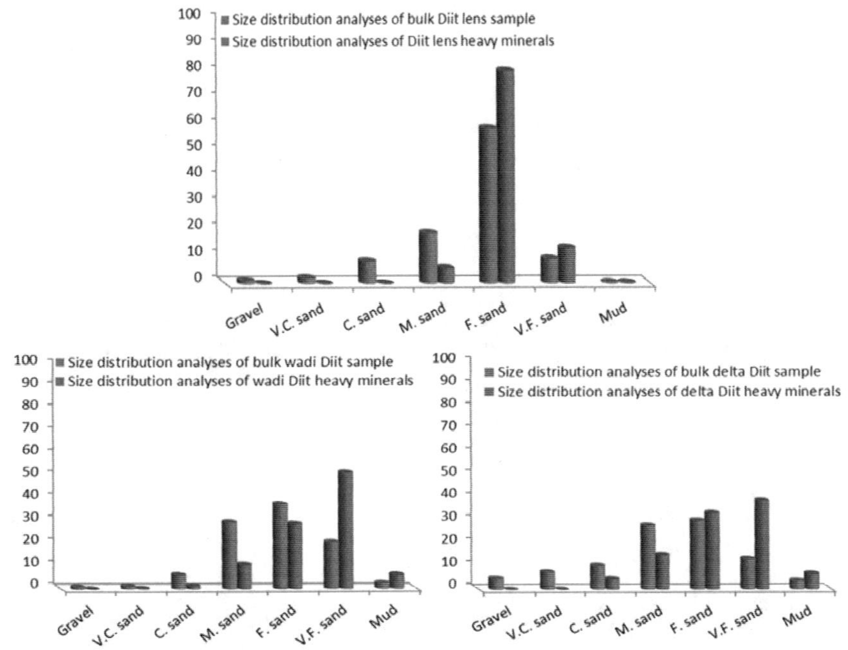

Figure 2. Histogram showing the grain size distribution analyses for Diit samples (lens, Wadi, and Delta) in comparison with their heavy mineral distribution analyses.

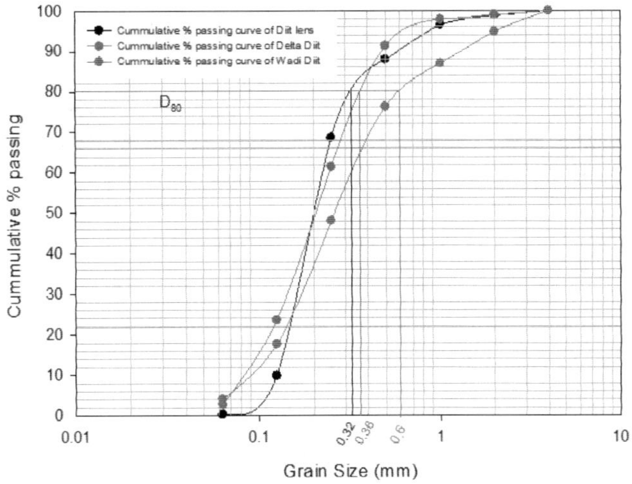

Figure 3. Grain size distribution analyses of Diit lens, Wadi and Delta samples.

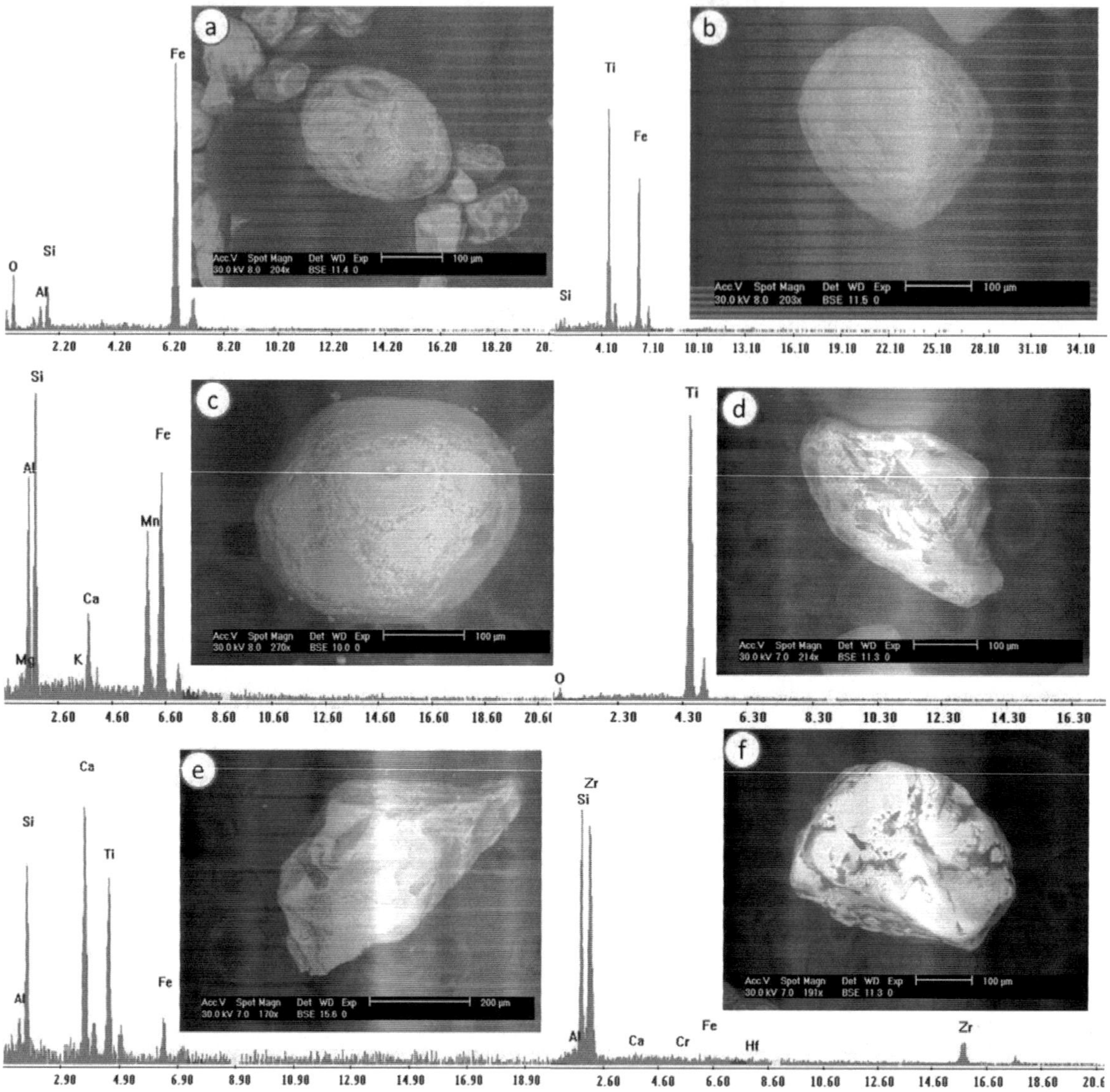

Figure 4. Back-scattered electron (BSE) images and corresponding EDS spectra for magnetite (a), ilmenite (b), garnet (c), rutile (d), sphene (e), and zircon (f)

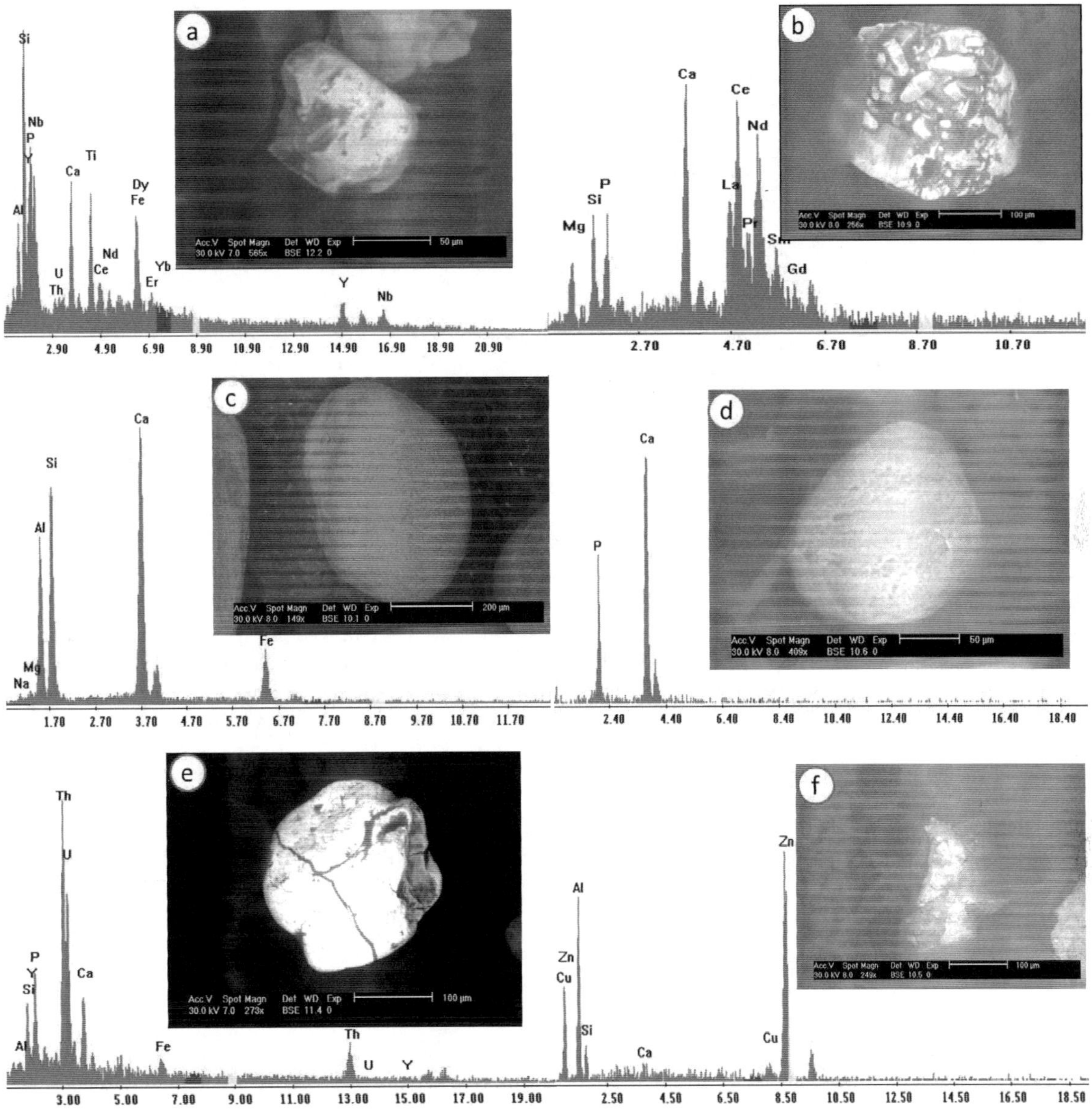

Figure 5. Back-scattered electron (BSE) images and corresponding EDS spectra for fergusonite (a), xenotime (b), green silicates (c), apatite (d), uranothorite (e), and khatyrkite (f)

3.2. Feasibility of physical upgrading

Usually, the physical upgrading processes of the economic heavy minerals found in black sand or stream sediments start with gravity concentration processes in order to eliminate the low density gangue silicate minerals such as quartz and feldspar, which are present in large proportions and produce a heavy mineral concentrate (HMC). Then followed by the stage of magnetic separation of the obtained HMC, in order to separate ferromagnetic from paramagnetic and from diamagnetic heavy minerals.[18-20]

It is necessary to evaluate the gravity concentration process in terms of whether it is possible or not, so the concentration criterion (CC) is commonly used for this purpose. A concentration criterion (CC) can be defined as;

$$Concentration\ Criterion\ (CC) = \frac{Sp.gr.of\ heavy\ mineral - sp.gr.of\ fluid\ medium}{Sp.gr.of\ light\ mineral - sp.gr.of\ fluid\ medium}$$

The separation efficiency decreases with decreasing the criterion value, with values > 2.5, possible separation occurs.[21, 22] Concentration criterion values for the heavy minerals in the Diit sample relative to the main gangue minerals (quartz and feldspar) are tabulated in Table 1. The data shows that gravity separation of some VHMs (magnetite, ilmenite, zircon, rutile, uranothorite, monazite, xenotime, fergusonite, khatyrkite, and gold) from the silicates gangue minerals will be possible, whereas separation of other heavy minerals as apatite, sphene, garnet, amphibole, epidote and pyroxene, from the gangue silicate minerals will be relatively difficult. So, several rounds of scavenging concentration steps were carried out after the rougher step to recover the target minerals that were not recovered during the initial roughing stage. The scavenging concentration stages were achieved by changing the concentration conditions (water flow, feed rate, deck tilt, and amplitude) to make them more rigorous than the initial roughing.

3.2.1. Shaking table concentration. In order to eliminate the high percentages of light gangue minerals (quartz and feldspar) in Diit area samples (80.87%, 79.49%, and 39.15% mass for Delta, Wadi, and lens of Diit sample respectively), wilfely shaking table no. 13 was used to obtain a clean concentrate of the VHMs. To obtain higher recovery values, a rougher concentration step was initially performed, and then the obtained tail was subjected to scavenging concentration step to recover the remaining heavy minerals that were not recovered during the initial roughing stage. Two rounds of the scavenging concentration stages were carried out on the Wadi and Delta Diit samples, while the Diit lens sample required only one round for scavenging operations.

The operation conditions for the roughing and scavenging concentration stages using shaking table as feed rate, water flow rate, stroke length and inclination angle were optimized and listed at (figures 7 and 8). It is quite clear that the scavenging stage has greater values of the operation conditions than the roughing stage, and this comes as a result of the increase in the proportion of light gangue minerals and the decrease in the heavy minerals during the scavenging stages compared to the roughing stages.

The final concentrates obtained after several wet gravity concentration operations for the Wadi, Delta and Diit lens samples were collected, dried, weighed, and analyzed elementally using XRF spectrometry and compared with XRF analyses of the feed samples and presented in (table 2) in addition to the calculated elemental improving ratio (EI). Improving ratio of elementals is calculated by dividing the grade of concentrate by the grade of the feed (c/f) and its value refer to how many times the concentrate has element concentration relative to the feed.

From the results of (table 2), the values of the improving ratio for the major elements SiO_2, Al_2O_3, K_2O, and SO_3 indicate that their grades were reduced in the concentrate than in the feed, and this indicates a reduction in the percentages of the gangue silicate minerals that have low specific gravity (quartz and feldspar) as a result of wet gravity separation processes via shaking table. On the contrary, the improving ratio values for the elements Ti, Ni, Fe, Zn, Mn, Pb, Y, Zr, Cr, and Nb indicate a doubling of their grade in the concentrate than in the feed sample, and this related to a doubling of the ratios of their minerals such as ilmenite, magnetite, rutile, sphene, zircon, and garnet in the concentrate as a result of the high value of its specific gravity during wet gravity separation processes.

Table 1. Concentration criteria for separation of heavy minerals from gangue minerals.

Heavy Minerals		Gangue Minerals	
Mineral	Sp. gr.	Quartz (sp. gr. 2.62)	Feldspar (sp. gr. 2.56)
Magnetite	5.19	2.59	2.69
Ilmenite	4.79	2.34	2.43
Zircon	4.6	2.22	2.31
Apatite	3.2	1.36	1.41
Sphene	3.5	1.54	1.60
Rutile	4.20	1.98	2.05
Garnet	3.50	1.54	1.60
Pyroxene	3.50	1.54	1.60
Amphibole	3.2	1.36	1.41
Epidote	3.4	1.48	1.54
Uranothorite	5.5	2.78	2.88
Monazite	5.26	2.63	2.73
Xenotime	4.75	2.31	2.40
Fergusonite	5.7	2.90	3.01
Khatyrkite	4.35	2.07	2.15
Gold	17	9.88	10.26

Assay and material balance of the wet gravity concentration processes for Wadi and Delta Diit samples after roughing and two rounds of scavenging operations were presented in Figure 7.The results revealed that the THM content was raised from 11.45 to 48.68 % for Wadi Diit sample and from 12.48% to 47.67% for Delta Diit sample in 16.51 and 18. 7% by mass yield for Wadi and Delta Diit samples respectively. It means that 71.67 and 82.06% of the THM content was saved in the final concentrate for Wadi and Delta Diit samples respectively and 28.33 and 17.94% of the THM content was lost at the tailing of the Wadi and Delta Diit samples respectively. The improving ratio of the Wadi and Delta Diit samples are 4.25 and 3.92% respectively, this means that the Wadi and Delta Diit concentrate have 4.25 and 3.92 times respectively the THM concentration of the feed.

Figure 8 displays the material balance of the tabling concentration processes for Diit lens sample after roughing and one round of scavenging operation. The material balance results revealed that the THM content was raised from 61.89 to 89.70% in 61.08 % by mass yield; it means that 91.68% of the THM content was saved in the final concentrate and 8.32% of the THM was lost at the tailing. The value of the Diit lens improving ratio means that the concentrate has 1.5 times the THM concentration of the feed.

Table 2. XRF elemental analyses demonstrating the feed grade, concentrate grade and improving ratio (ER) for Diit samples (lens, Wadi and Delta).

Sample	Diit Lens			Delta Diit			Wadi Diit		
	Feed grade (f)	Conc. grade (c)	IR (c/f)	Feed grade (f)	Conc. grade (c)	IR (c/f)	Feed grade (f)	Conc. grade (c)	IR (c/f)
Major elemental oxide in wt.%									
SiO_2	29.60	23.61	0.80	59.84	51.39	0.86	59.88	45.95	0.77
TiO_2	9.04	20.11	2.22	0.94	2.34	2.49	0.89	1.91	2.15
Al_2O_3	7.41	7.72	1.04	14.37	13.23	0.92	14.81	13.30	0.90
$Fe_2O_3^{Total}$	23.78	33.15	1.39	8.26	14.14	1.71	8.71	11.53	1.32
MnO	0.904	1.40	1.54	0.163	0.325	1.99	0.164	0.260	1.59
MgO	3.85	3.99	1.04	3.14	5.09	1.62	3.00	4.66	1.55
CaO	18.06	6.88	0.38	5.75	7.36	1.28	5.14	6.91	1.34
Na_2O	0.87	1.03	1.18	2.47	2.93	1.19	2.26	3.17	1.40
SO_3	0.12	0.11	0.92	0.13	0.03	0.23	0.08	0.02	0.25
K_2O	0.36	0.23	0.64	2.05	0.69	0.34	2.03	0.74	0.36
P_2O_5	0.25	0.30	1.20	0.30	0.32	1.07	0.33	0.32	0.97
L.O.I.	4.57	0.26		2.31	1.76		2.43	1.86	
Trace elements in ppm									
Ni	----	88.8	88.8	133.58	239.9	1.80	110.01	195.82	1.78
Zn	208.88	549.9	2.63	104.44	494.7	4.74	128.54	583.57	4.45
Cu	87.88	25.7	0.29	95.86	48.4	0.50	119.83	65.91	0.55
Pb	-----	15.7	15.7	46.42	477.0	10.28	------	11.3	11.3
Sr	1818.04	841.3	0.46	752.58	1748.4	2.32	693.39	1698.81	2.45
Rb	54.87	9.2	0.17	100.59	51.2	0.51	91.44	41.15	0.45
Zr	3109.26	6982.9	2.25	555.23	3878.2	6.98	421.97	2958.01	7.01
Cr	2360.66	7605.3	3.22	403.71	2136.0	5.29	451.61	2262.57	5.01
Ga	37.20	29.6	0.80	22.32	99.2	4.44	44.64	188.83	4.23
Y	133.86	247.9	1.85	47.25	254.1	5.38	31.50	173.57	5.51
Co	432.58	67.5	0.16	------	114.9	114.9	--------	100.9	100.90
Nb	167.77	309.0	1.84	------	120.4	120.4	--------	105.44	105.44

3.2.2. Magnetic separation results. Magnetic separation processes via DHIMS produced four magnetic fractions, which are as follows; Ferromagnetic, strongly paramagnetic, weakly paramagnetic and finally the diamagnetic mineral fraction. Ferromagnetic fraction percentages are 1.56%, 2.19% and 3.16% mass for Diit lens, Wadi and Delta samples respectively while the strongly paramagnetic fraction at 1 ampere reaches 38.05% for the lens, and reaches 25.81 and 7.62% for the Wadi and Delta Diit samples respectively. As for the results of separating the weakly paramagnetic fraction at 3 amperes, they are

20.33, 28, and 44% for the lens, Wadi and Delta samples respectively. Finally, the diamagnetic fraction has percentage of 40.07 for lens sample and 44, 44.69 for Wadi and Delta Diit sample respectively.

X-ray diffraction analyses are used to identify and confirm the mineral contents of the different magnetic fractionations and presented in (figure 6). The ferromagnetic fraction diffraction lines are in accordance with COD card no. 9005812 for magnetite, COD card no. 9015065 for hematite, and COD card no. 9002332 for magnetite-h (figure. 6a). Pure concentrate of ilmenite occurred and confirmed at the paramagnetic fraction that has been separated at 1ampere field current and has diffraction lines in accordance with COD card no. 1011033 for ilmenite (figure. 6b). The paramagnetic fraction that has been separated at magnetic field current 3 amperes has diffraction lines in accordance with large group of paramagnetic minerals as follows; COD card no. 9000299 for epidote, COD card no. 9001647 for monazite, COD card no. 9001409 for garnet, COD card no. 1008932 for xenotime, COD card no. 9011839 for fergusonite, COD card no. 9000597 for pyroxene, and COD card no. 9001548 for amphibole (figure 6c). Whereas the non-magnetic fraction that has been separated at 3 amperes magnetic fraction has diffraction lines in accordance with COD card no. 1011263 for zircon, COD card no. 9010050 for apatite, COD card no. 9001681 for rutile, COD card no. 9000820 for uranophane, COD card no. 9008463 for gold, and COD card no. 9012196 for khatyrkite (figure. 6d).

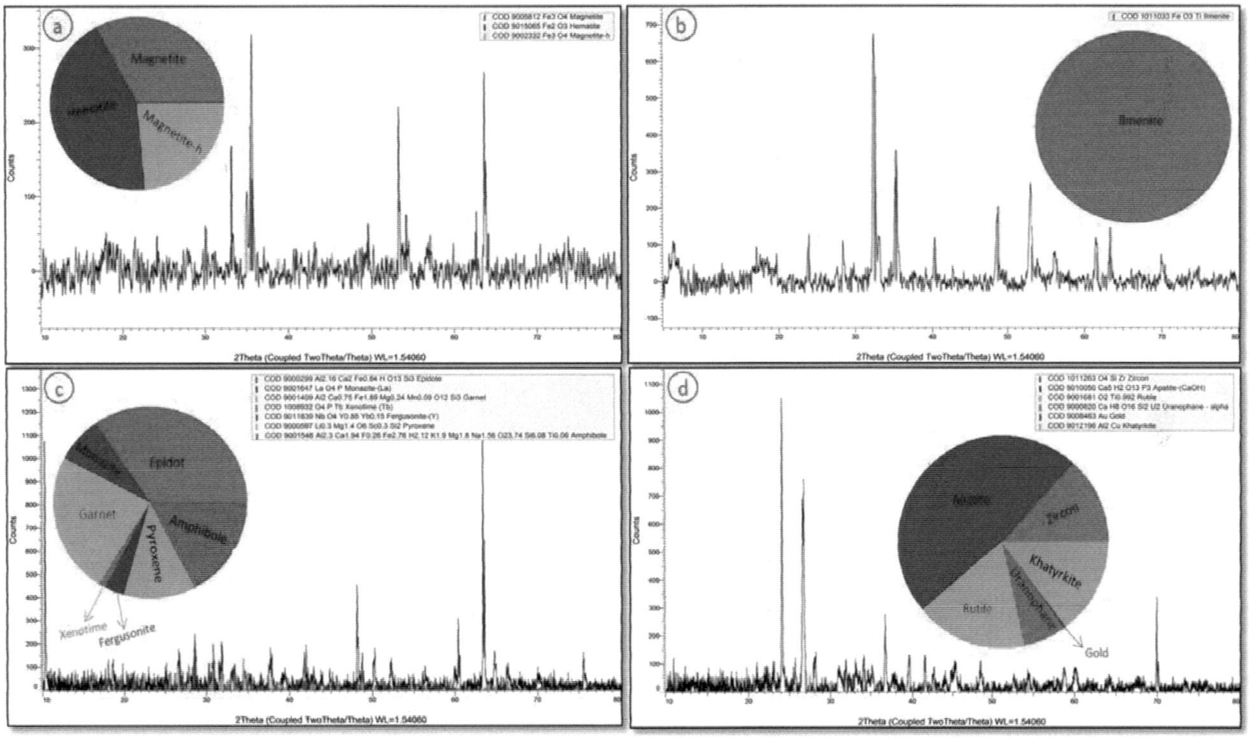

Figure 6. XRD diffractograms showing: a. Mineral content of the ferromagnetic fraction, b. Mineral content of the paramagnetic fraction at 1 ampere, c. Mineral content of the paramagnetic fraction at 3 amperes, and d. Mineral content of the diamagnetic fraction at 3 amperes.

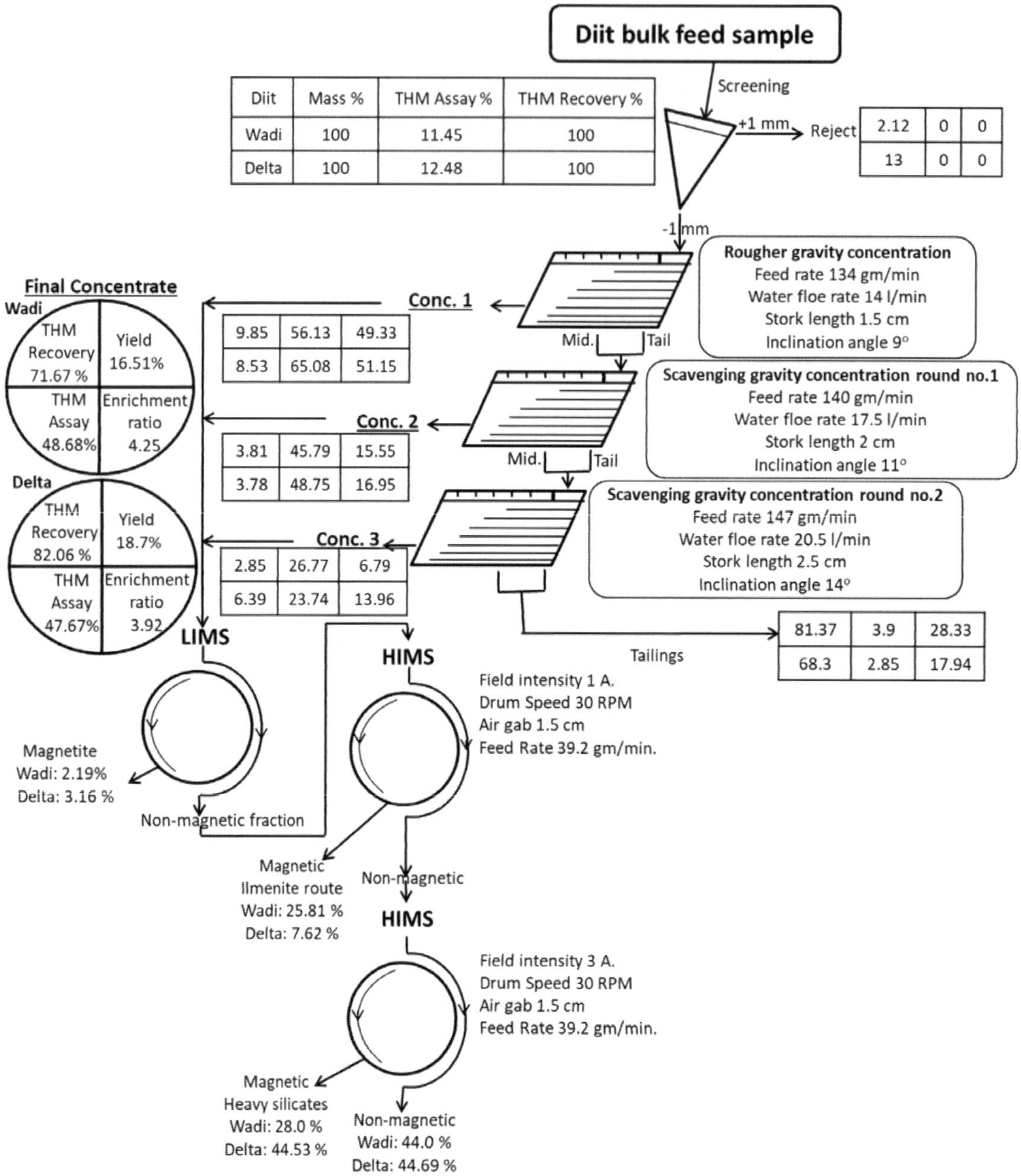

Figure 7. Flowsheet with material balance for recovery of THM from Wadi and Delta Diit feed sample under optimum conditions.

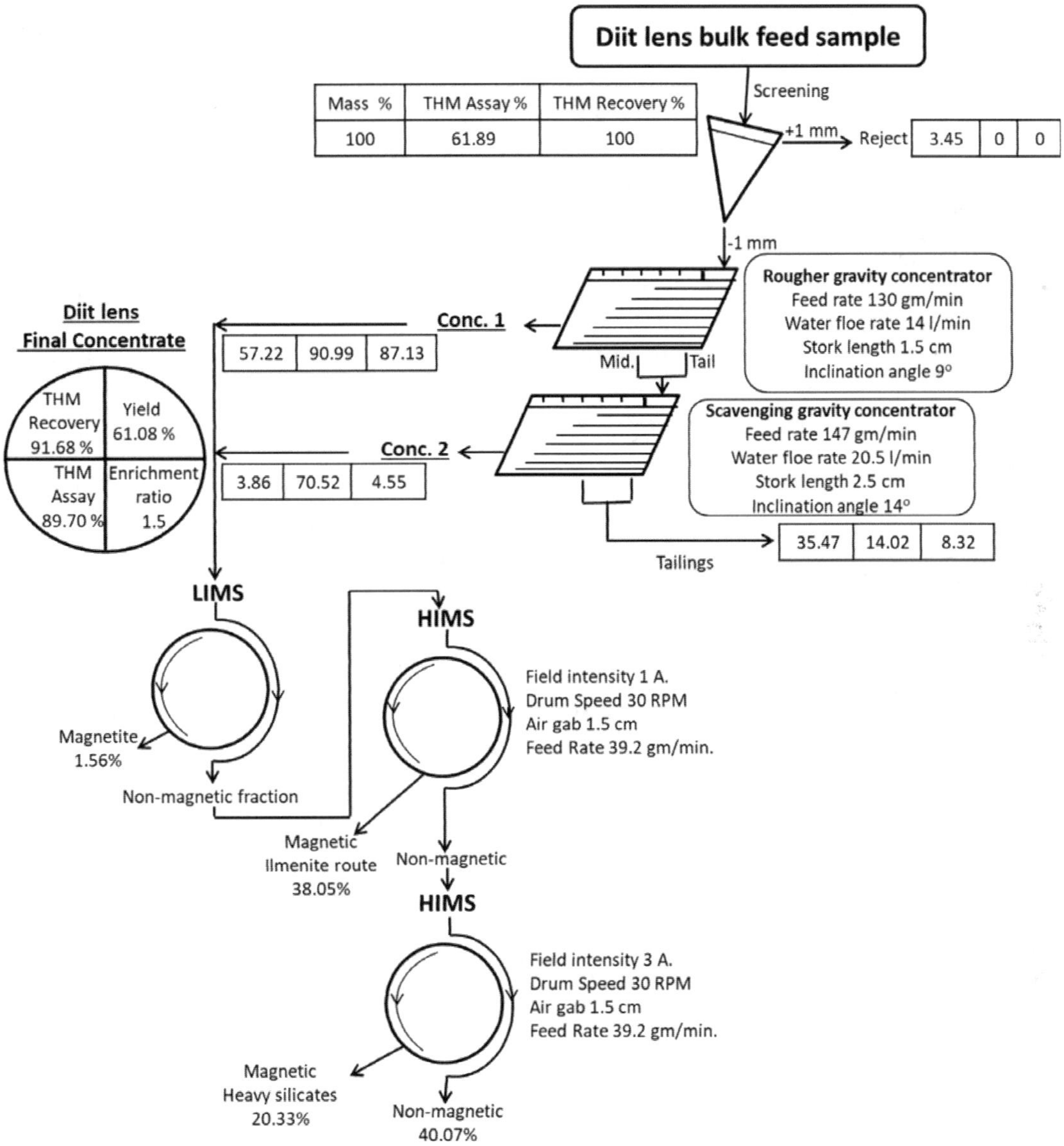

Figure 8. Flowsheet with material balance for recovery of THM from Diit lens feed sample under optimum conditions.

4. Conclusions

Recovery of VHMs from Diit stream sediments was successfully performed in this study using wet gravity concentration tool in conjunction with magnetic separation tool (Carpco DHIMS). Gravity concentration operation tests resulted that the content of SiO_2, Al_2O_3, K_2O, and SO_3 is significantly degraded in the concentrate relative to the feed samples while on the contrary, the contents of Ti, Fe, Mn, Ni, Zn, Pb, Zr, Cr, Y, and Nb were upgraded several times in the concentrate as compared to the feed sample. This means that the contents of the VHMs of ilmenite, magnetite, rutile, sphene, zircon, and garnet were increased in the concentrate while the contents of gangue silicate minerals (quartz and feldspar) decreased in the concentrate and increased in the tail.

Assay and material balance of the scavenging after roughing step of the Diit lens sample revealed that the THM assay was raised from 61.89 to 89.70% in 61.08 % by mass yield with of recovery is 91.68%. The THM assay in the Wadi and Delta Diit samples increased from 11.45 to 48.68 and from 12.48 to 47.67% respectively after two rounds of the scavenging concentration step following the roughing step. The THM recovery for the Wadi and Delta Diit samples was 71.67 and 82.06% respectively in a weight of 16.51, 18.7% respectively out of the feed sample with THM loss 28.33 and 17.94 % respectively for Wadi and Delta Diit at the end of the whole process.

DHIMS succeeded in separating VHMs into four fractions; the first fraction is the magnetite concentrate, the second fraction is the ilmenite concentrate, and the third fraction is a concentrate of the green heavy paramagnetic silicates minerals as pyroxene, amphibole, garnet and epidot as well as trace amounts of monazite, xenotime and fergusonite. Finally, the fourth fraction is a concentrate for the diamagnetic minerals as zircon, rutile, apatite, sphene, gold, and khatyrkyite.

Declaration

Conflict Of Interest

The authors declare that they have no conflict of interest.

References

[1] Gonçalves C C and Braga F A 2019 *Minerals* **9** 176

[2] Moore P 2011 *Int. Min.* **1** 7

[3] Sampaio J A, Luz A B, Alcantera R M and Araújo L S L 2001 *Center for Mineral Technology: Rio de Janeiro, Brazil* 233

[4] Schnellrath J, Monte M B M, Veras A, Júnior H R and Figueiredo CMV 2001 *Center for Mineral Technology: Rio de Janeiro, Brazil* 189

[5] Rosental S, Raras T and Rochas E 2005 *Center for Mineral Technology: Rio de Janeiro, Brazil* 727

[6] Laxmi T, Srikant S S, Rao D S and Rao R B 2013 *J. Min. Sci. Technol.* **23** 725

[7] Routray S and Rao R B 2011 J. *Min. Mater. Charact. Eng.* **10** 1409

[8] Routray S, Laxmi T and Rao R B 2013 *J. Mater. Mech. Eng.* **3** (2) 80

[9] Tranvik E, Becker M, Palsson B I, Franzidis J P and Bradshaw D 2017 *Miner. Eng.* **101** 30

[10] Ramadan T M 2014 Geological and geochemical studies on some basement rocks at Wadi Hodein area South Eastern Desert, Egypt. *Ph.D. Thesis, Al-Azhar University* 192

[11] Fawzy M M, Abu El Ghar M, Gaafar I M, Elshafey A M, Diab M and Hussein A W 2022 Mining, Metallurgy & Exploration, Published online; 21 January https://doi.org/10.1007/s42461-022-00543-x

[12] Diab M, Abu El Ghar M, Gaafar I M, Elshafey A M, Hussein A W, and Fawzy M M 2022 *Journal of Mining and Environment (JME) Published online 31 January DOI:10.22044/jme.2022.11479.2136*

[13] Fawzy, M., Mahdy, N., Mabrouk, S. 2020 *Arab. J. Geosci.* **13** 413 https://doi.org/10.1007/s12517-020-05381-z

[14] ASTM, American Society for Testing and Materials, ASTM C33 Standard Specification for Concrete Aggregates ASTM, Philadelphia, Pennsylvania, 2003

[15] Raslan M, Kharbishb S, Fawzy M, El Dabea M and Fathy M 2021 *Journal of Mining Science* **57** (2) 316

[16] Rydberg J 2014 *J Paleolimnol* **52** 265 doi. org/ 10. 1007/ s10933- 014- 9792-4

[17] Fawzy M M, Kamar M S and Saleh G M 2021 *International Review of Applied Sciences and Engineering* **12** (2) 134 DOI:10.1556/1848.2021.00200

[18] Rahman A, Pownceby M, Tardio J, Sparrow G, Haque N and Hasan F 2021 Kurigram District, Bangladesh. *Minerals* **11**(7) 786 doi.org/10.3390/ min11070786

[19] Moscoso-Pinto F and Kim H 2021 *Minerals* **11**(1) 49 doi.org/10.3390/ min11010049

[20] Raslan M and Fawzy M 2018 *Tabbin Inst. Metallurgical Stud. (TIMS Bulletin)* **107** 52

[21] Wills B A, Napier-Munn T J and Wills' 2006 *Mineral Processing technology: seventh ed., Elsevier Science & Technology Books*

[22] Das A and Sarkar B 2018 *Mineral Processing and Extractive Metallurgy Review* **39** (1) 1

A Novel Approach for Synthesis of Low Sensitive FOX-7 with a high % yield

Mahmoud Abdelhafiz[*1], **Hosam E. Mostafa**[1]

[1] Chemical Engineering Department, Military Technical College (MTC), Egypt

*Corresponding Author: Hafiz_theone@hotmail.com

Abstract: Attaining a high performance is one of the most seeking demands in the energetic materials field. However, achieving that goal while keeping a high level of safety became a serious puzzle. 1,1-diamino-2,2-dinitroethene (FOX-7) could be a promising candidate to solve that issue regarding its favorable explosive characteristics together with its low vulnerability. However, its high cost and low yield obtained via the FOI method limit its use in the field of insensitive munition (IM). In this work, we propose a novel approach to synthesis the FOX-7 crystals with high purity, a relatively high yield, and an extremely low impact, friction, ignition sensitivities. The obtained crystals were characterized using FTIR Spectroscopy, Proton NMR Spectroscopy, IKA impact test apparatus, BAM friction test apparatus, Chilworth deflagration test apparatus, and the adiabatic calorific bomb. Results confirmed the successful preparation of pure FOX-7 crystals with a promising yield of 64%, in comparison to the yield of the FOI method of 55%, and extremely low sensitivities using our proposed approach.

Keywords: Energetic Materials, FOX-7, Insensitive Munitions (IMs), Low-vulnerable, PBXs.

1. Introduction

Many research projects are focused on improving the performance characteristics of energetic materials while maintaining low sensitivity. The progressive deployment of Insensitive Munitions (IM) by numerous defense forces throughout the world is compatible with these objectives. These systems dependably meet their performance, readiness, and operational requirements on-demand, but their violent response to unexpectedly dangerous stimuli is kept to a manageable level [1]. Being dropped [2,3],[4],[5], shell impact [6,7], and detonation by influence are some of the risks that IM is meant to address.

IM-compliant explosive compositions can be created using one of two methods. For example, a Polymer Bonded Explosive (PBX) can be created by combining novel binders like GAP with energetic components already found in flexible polymeric matrices known as binders (glycidyl azide polymer). TATB, NTO, ADN, HNIW, and CL20 are all examples of inherently low-sensitive energetic compounds that can be used to make explosives [9–19]. TATB, NTO, ADN, HNIW, and CL20 are all examples of intrinsically low-sensitive energetic components that can be used to make explosives.

For the time being, the commercial availability of FOX-7 is limited because of its expensive cost. Although it performs in the same ballpark as RDX, the FOX-7 is far less sensitive. When compared to cyclotrimethylene trinitramine (RDX), FOX-7's first findings from the Swedish inventors revealed that it was much less responsive to diverse forms of danger stimuli and somewhat higher in performance. Its processing capabilities and compatibility with other materials are as yet unknown. Future research will undoubtedly focus on these traits. Table 1 compares some of the FOX-7's characteristics to those of RDX. FOX-7's theoretically expected great performance has yet to be shown empirically. FOX-7 has the potential to make a significant contribution to the advancement of non-lethal weapons. This combination might be a new hope in the realm of explosives. It was launched by FOI colleagues in the fall of 1998.

An important part of this study is examining how FOX-7 is synthesized and characterized in contemporary low-sensitive explosive formulations and how it functions as an energetic material in those formulations. DSTO's Explosives Group [20] is reporting on current and prospective research that employs thermochemical calculations to analyze several PBXs based on FOX-7, which might be used to replace the standard Comp B. [20].

Table 1. Properties of FOX-7 in comparison with RDX.

Property	FOX-7	RDX
Density, (ρ), (g/cm^3)	1.885	1.816
Detonation velocity, (D), (m/sec^2)	8870	8750
Detonation pressure, (P), (GPa)	34	34.6
Oxygen balance, (O_b)	-21.6	-21.6
Activation energy, (kcal/mole)	58	40
Heat of formation, (ΔH), (kcal/mole)	-32	16
Impact sensitivity, (Nm)	15-25	7.5
Friction sensitivity, (N)	>340	120
Ignition temperature, (°C)	226	223

2. Experimental Work

2.1 Chemicals
Methanol (HPLC grade) was purchased from Sigma-Aldrich (Steinheim, Germany) and then was exploited as a solvent without any further purification. Sodium Methoxide, Acetamidine Hydrochloride, and Diethyl Oxalate were acquired from Al-Gomhoria company for medicines and medical supplies (Cairo, Egypt). Hydrochloric Acid (37%), Sulphuric Acid (95%), and Nitric Acid were obtained from Harshit Export (Mumbai, India).

2.2 Preparation of 1,1-Diamino-2,2-dinitroethene (FOX-7) according to the modified (FOI) method
The heterocyclic starting material 2-methoxy-2-methylimidazoledinedione was synthesized by condensing diethyl oxalate with acetamidine hydrochloride at high dilution in methanolic sodium methoxide and then recrystallizing it from the methanol. First, 2-methoxy-2-methyl-5-imidazolidinedione was nitrated, resulting in 2,2-dinitomethylene-5-imidazolidinone, which was hydrolyzed to yield FOX-7.

2.2.1 Preparation of 2-methoxy-2-methyl-4,5-imidazolidinedione. The magnetic stirrer bar, drying tube, and dropping funnel were all included in the 2L round bottom flask. At an initial concentration of 0.9455g/cm3, the flask was filled to its full capacity with methanol (430ml) and sodium methoxide (30% in methanol, 112ml). At room temperature, 18.24 g of acetamidine hydrochloride was added to the agitated solution to produce a suspension. Thereafter, the mixture was stirred for an additional hour with the addition of 27.94 grams of diethyl ethyl oxalate in methanol (200 milliliters).
Using strong hydrochloric acid at a temperature below 30 °C, the reaction was brought down to a pH of around 4. Filtration through filter aid (APS 1739) removed the insoluble salts, and the filtrate was evaporated to dryness at a temperature of 30 °C to yield a white solid. Filtration of 160 ml boiling methanol was done to remove the insoluble salts from this material. The filtrate was then diluted to a

final volume of 160 ml. In the morning, the crystallized white substance was collected and dried after cooling overnight in the refrigerator (17.75 g, 64 percent).

2.2.2 Nitration of 2-methoxy-2-methyl-4,5-imidazolidinedione to prepare 2,2-Dinitromethylene-4,5-imidazolidinedione. In an ice bath, 198 ml of concentrated sulphuric acid was added to a 500 ml round bottom flask equipped with a magnetic stirrer, thermometer, dropping funnel, and drying tube. The cooled acid solution was gently diluted with 2-methoxy-2-methyl-4,5-imidazolidinedione (35.4 g), resulting in a bright yellow solution. It took 60 minutes for the color of the combination to change from yellow to deep red, and finally to an orange suspension with the addition of 70 percent nitric acid (43 ml). For 30 minutes, the suspension was swirled at room temperature. It was then air-dried and filtered before being utilized in the following process.

2.2.3 Hydrolysis of 2,2-dinitomethylene-4,5-imidazolidinedione to produce FOX-7. This reaction was carried out in a 500 ml conical flask, with the addition of the crude 2,2-dinitromethylene-4,5-imidazolidinedione to the water (120 ml). In order to keep the temperature stable between 20 and 30 degrees Celsius, 30 percent ammonia in water was supplied to the suspension at a rate that kept the temperature between 20 and 30 degrees Celsius. After 2 hours of stirring at room temperature, the resulting suspension was filtered and the solid product was recovered using filtering and water washing (4 times). The brilliant yellow crystalline material was identified as 1,1-diamino-2,2- dinitroethene after air and vacuum drying at 60 °C and 7 mmHg (19.66 g of yield percentage about 54.1 percent).

2.2.4 Recrystallization of obtained FOX-7 crystals. Recrystallization of FOX-7 crystals was performed utilizing a DMSO ionic solvent as a co-solvent system (20:80 w/w). The crystallization process took place at 60 °C, where the remaining crystals were dissolved in methanol then collected via filtration. A non-agitated cooling was used to accelerate the FOX-7 crystal's growth. The resulting crystals were subjected to different characterization processes to confirm their identity, purity, shape, and quality of the final percentage yield of FOX-7 crystals.

3. Characterization

3.1 Melting Point Determination
A simple technique was used in measuring the melting point of 2-methoxy-2-methyl-4,5-imidazolidinedione. A small amount of a sample was inserted in a capillary tube which was placed into a special metallic block with a glass window and calibrated thermometer. The metallic block was heated over a hot plate. The melting temperature of the sample was carefully observed via the glass window. Three consecutive trials for each sample were performed and the average values were determined.

3.2 FTIR Spectroscopy
Jasco FTIR -6200 was used to identify the product through the FTIR spectrum. If the sample is solid, the solid sample and KBr would be ground, mixed, and pressed in the form of a disc having standard dimensions. The wavenumber range used was 500- 4000 cm-1 which is the appropriate range for most organic compounds. Using the FTIR spectrum made it possible to identify the main functional groups of the prepared materials.

3.3 Proton NMR Spectroscopy
Varian NMR 300 MHz spectrometer is the most powerful tool available for organic structure determination was used to study the nuclei of 1H and 13C. The examined sample was dissolved in a suitable solvent (DMSO) then the sample was measured due to the magnetic field generated from the spinning charged nucleus.

3.4 Determination of Explosive Characteristics of FOX-7

3.4.1 Sensitivity Tests

3.4.1.1 Determination of sensitivity to impact. The upper limit of sensitivity to impact [24] was determined using IKA apparatus which consists mainly of a falling steel hammer (2 or 5 kg) moving on vertical rails provided with a calibrated scale. The testing piston part consists of 2 steel cylinders and a casing. Weighed samples of about 0.01 g of the explosive are spread over the central surface of the bottom cylinder, and then the piston is located in its position. The impact of the weight, through the piston, transferred to the sample may or may not result in its initiation according to the sensitivity of the sample; sound, light, or smoke are usually observed. If none of these effects are observed, initiation failure is reported. The test results were reported as a relation between the percentage of initiations and the drop height having constant mass.

The experiment was performed using a weight of 2 or 5 kg and drop height was increased gradually. For each height, six consecutive trials were performed and the corresponding percentage of initiation was recorded. The upper sensitivity limit, H100 was used to identify the minimum height at which 100% initiation was achieved. Similarly, the lower sensitivity limit, Ho, was used to identify the maximum drop height at which no initiation was achieved. The impact energy for dropping weight at each height is calculated according to the equation:

$$\mathbf{E_i = m\ H\ g}$$

Where:

E_i	Impact energy (J).
m	mass of the falling hammer (kg).
H	drop height (m)
g	acceleration due to gravity = 9.8 m/sec^2

3.4.1.2 Determination of sensitivity to friction. Chilworth BAM friction test apparatus, (UK) was used to determine the sensitivity to friction [24]. The apparatus consists of a hard unglazed porcelain plate of dimensions 25x25x5 mm, fixed in a movable table sliding at constant speed between two marks by an electric motor. A cylinder is fixed on a clamping holder. An apparatus lever can be loaded with different loads at ten different positions. The value of the force, measured in Newton, for a certain load and certain position was obtained from the calibration table. The table is shifted to the starting position after each trial by turning the rotor of the electric motor; sensitivity to friction is determined by spreading about 0.01g of the dry explosive on the surface of the porcelain plate in the form of a thin layer. Different loads are used to change the normal force between the porcelain pistil and the plate; force varies from nearly 5 to 320 N. The sample initiation may be observed through sound effects, smoke appearance, or by the characteristic smell of the decomposition products. Six consecutive trials were performed for each load and lever position.

3.4.1.3 Determination of ignition temperature. The ignition temperature can be experimentally determined by heating at a constant rate, an explosive sample of a given mass until the ignition occurs [24]. The ignition temperature test was performed by using the Chilworth deflagration test apparatus, made in the UK. It consists of an aluminum block comprising three vertical holes to accommodate test tubes, a programmable heating controller in a separated control box allowing remotely controlled temperature, a heat sink, and an electric fan to accelerate the cooling down of the heating block.

Three samples of 0.2 g each were dried up and ground to suitable particle size then inserted into 3 test tubes which were placed vertically into the heating block and the temperature was uniformly increased at the rate of (5°C/min) until deflagration of the sample occurs. The ignition temperature was monitored digitally on the control unit and the average temperature for the three samples was calculated.

3.4.1.4 Determination of ignition temperature at constant delay period of ignition (5 sec). Delay of ignition of an explosive at a certain temperature was experimentally determined by placing an explosive sample of a fixed mass at that temperature which should be higher than the previously determined ignition temperature of the explosive. The test is done using the same apparatus as before. The temperature was adjusted and fixed until the ignition took place and delay time was recorded. The samples of 0.05 g each were inserted into three test tubes already heated in the aluminum block to the investigated temperature and a stopwatch was used to measure the time elapsed till ignition. The test was repeated at different temperatures till the obtained delay period was exactly 5±0.1 sec.

3.4.1.5 Determination of heat of the explosion. The measurement of the explosion heat is analogous to the determination of the heat exchange in various chemical or physical processes. The experiment is performed in an adiabatic calorimeter assembly [24] and is based on the observation of the temperature increase in the calorimetric vessel during the combustion process. The heat capacity of the calorimeter assembly is determined by a calibration procedure during which a substance of known explosion heat is used. Consequently, the explosion heat of the sample under test is determined by a similar run, using the relation:

$$\mathbf{Q = K_w * \Delta T}$$

Where:

K_w	Heat capacity of the calorimeter (units)
ΔT	Temperature difference of the vessel (units)

The adiabatic calorimeter assembly consists of a calorimetric bomb, a calorimetric vessel filled with water, a stirrer, and a thermometer.

4. Results and discussion

Crystals obtained after applying the modified synthesis approach followed by a recrystallization process were found to have an interestingly high yield of ~ 64%. The melting point of 2-methoxy-2-methyl-4,5-imidazolidinedione (step I) was found at 158.2 °C, which coincides with that of FOI [25]. However, the melting point of FOX-7 could not be measured due to the near value of deflagration point (218 °C) from the point of FOX-7 decomposition. FTIR spectrum, as shown in Figures 1 and 2 and listed in Table 2, confirmed the successful preparation of FOX-7 through this technique. Table 2 describes how much the Characteristic peaks coincide with the standard FOI results. The NMR spectrum of 2-methoxy-2-methyl-4,5 imidazolidinedione according to experimental results is the same as the FOI results but with some shifts of peaks due to the presence of impurities in main product; Figure 3 for ^{1}H and Fgure 4 for ^{13}C. The NMR spectrum of 1,1-diamino-2,2- dinitroethene (FOX-7) according to experimental results is the same as the FOI results but with some shifts in peaks due to the traces of impurities found in the main product.

The results developed for the sensitivity values, as given in Table 4, against the impact and friction stimuli were found to be (12-25) J and (>320) N respectively. The ignition temperature of the obtained FOX-7 was 223 °C. These results coincide with the standard sensitivity values given by FOI [26][27]. The activation energy of the obtained FOX-7 was found to be 55 kcal/mol, which is very close to that of FOI [28]. The heat of explosion of the obtained FOX-7 crystals was 4091kJ/kg 4079 kJ/kg which is acceptable in comparison to the standard FOI value, 4091 kJ/kg [26]. These results confirm the successful synthesis of pure FOX-7 crystals through the proposed novel technique.

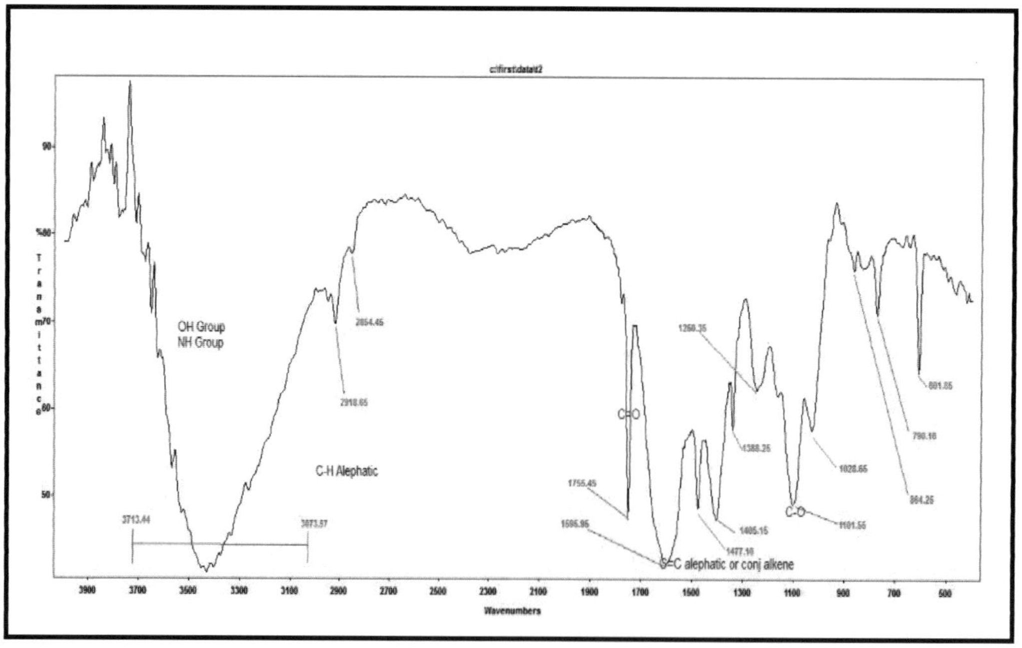

Figure 1. The FTIR characterization of 2-methoxy-2-methyl-4,5 imidazolidinediones (step 1).

Table 2. The FTIR characterization of 2 methoxy-2-methyl-4,5 imidazolidinediones and 1,1-diamino-2,2- dinitroethene (FOX-7)

FTIR spectroscopy		
2 methoxy-2-methyl-4,5 imidazolidinediones (step I)		
Function group	**FOI Results**	**Experimental Results**
	Wave number (cm^{-1})	**Wave number (cm^{-1})**
NH	3249	3270
C=O	1755	1755
	1477	1477
	1427	1405
	1388	1388
	1179	1260
	1124	1101
	1047	1028
	790	790
	729	700
	674	680
	597	601
1,1-diamino-2,2- dinitroethene (FOX-7)		
Function group	**FOI Results**	**Experimental Results**
	Wave number (cm^{-1})	**Wave number (cm^{-1})**
NH_2	3408	3400
	3330	3260
	1636	1654
NO_2	1520	1532
	1472	1440

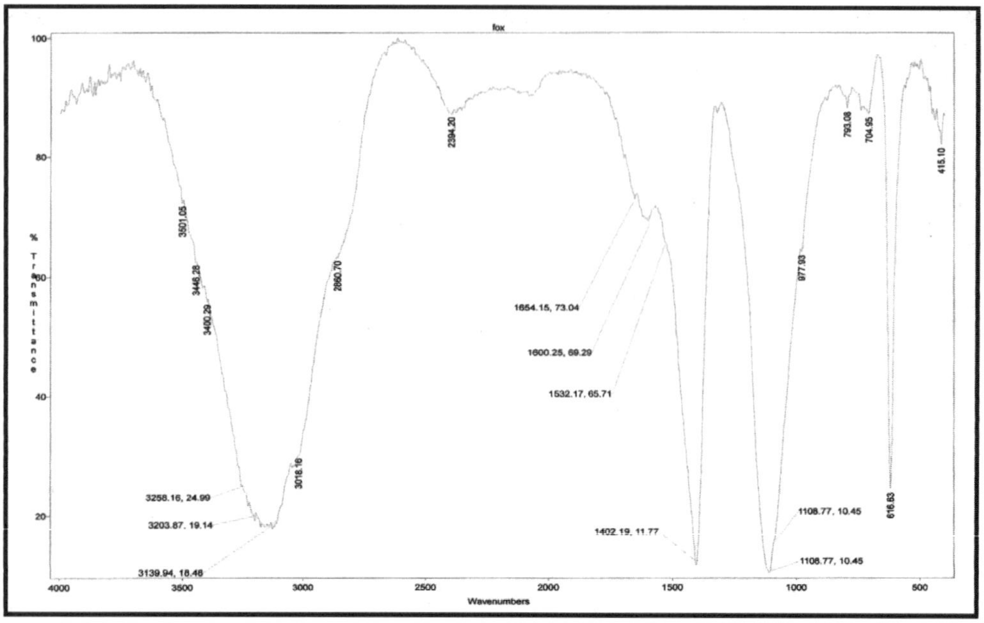

Figure 2. The FTIR characterization of 1,1-Diamino-2,2-dinitroethene [FOX-7] (step 3).

Table 3. NMR characterization of 2 methoxy-2-methyl-4,5 imidazolidinedione and 1,1-diamino-2,2-dinitroethene (FOX-7)

NMR spectroscopy		
2 methoxy-2-methyl-4,5 imidazolidinediones		
Type of atom	**FOI Results**	**Experimental Results**
^{1}**H**	**chemical shift (δ)**	**chemical shift (δ)**
^{3}**H , CH$_3$**	1.58	1.571
^{3}**H , OCH$_3$**	2.98	2.972
^{2}**H, NH**	9.92	9.96
^{13}**C**	**chemical shift (δ)**	**chemical shift (δ)**
^{13}C	26.86	27.8
	47.86	47.88
	91.59	100.7
	159.84	168.96
1,1-diamino-2,2- dinitroethene (FOX-7)		
Type of atom	**FOI Results**	**Experimental Results**
^{1}**H**	**chemical shift (δ)**	**chemical shift (δ)**
^{4}**H, NH**	8.8	8.13
^{13}**C**	**chemical shift (δ)**	**chemical shift (δ)**
^{13}C	128.3	126.76
	158.3	158.39

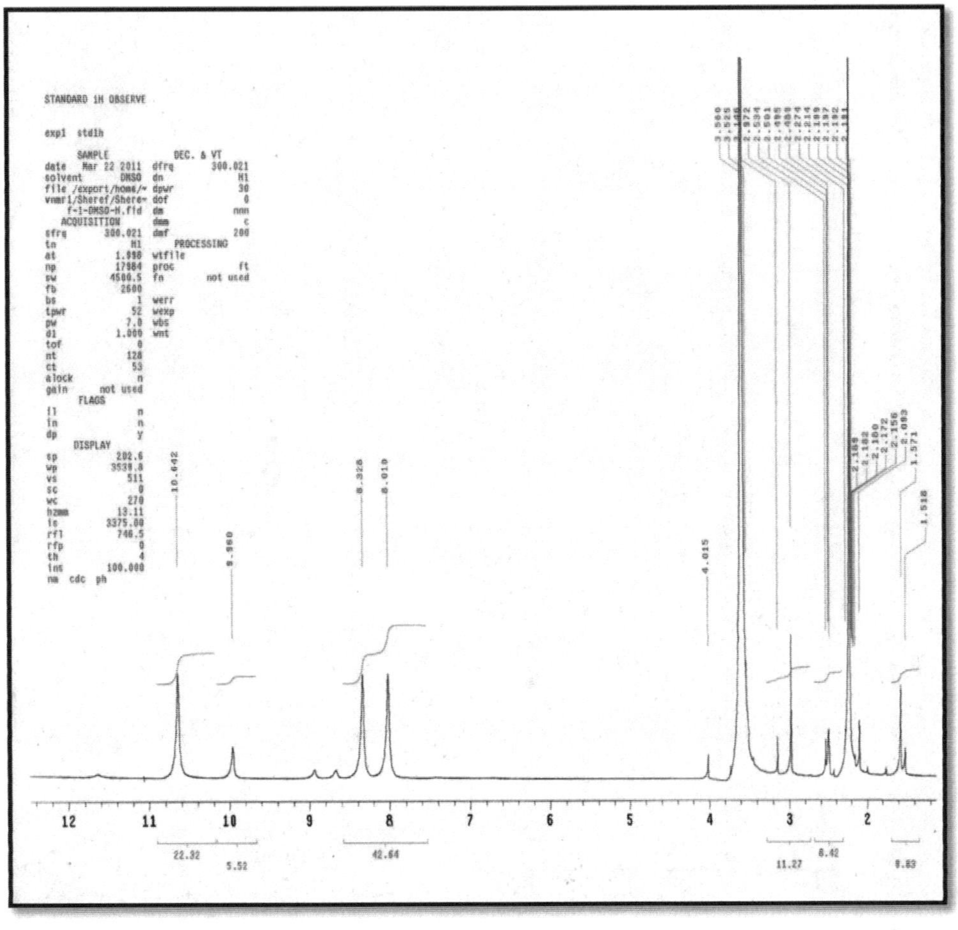

Figure 3. The ^{1}H NMR characterization of 2-methoxy-2-methyl-4,5 imidazolidinedione (step 1)

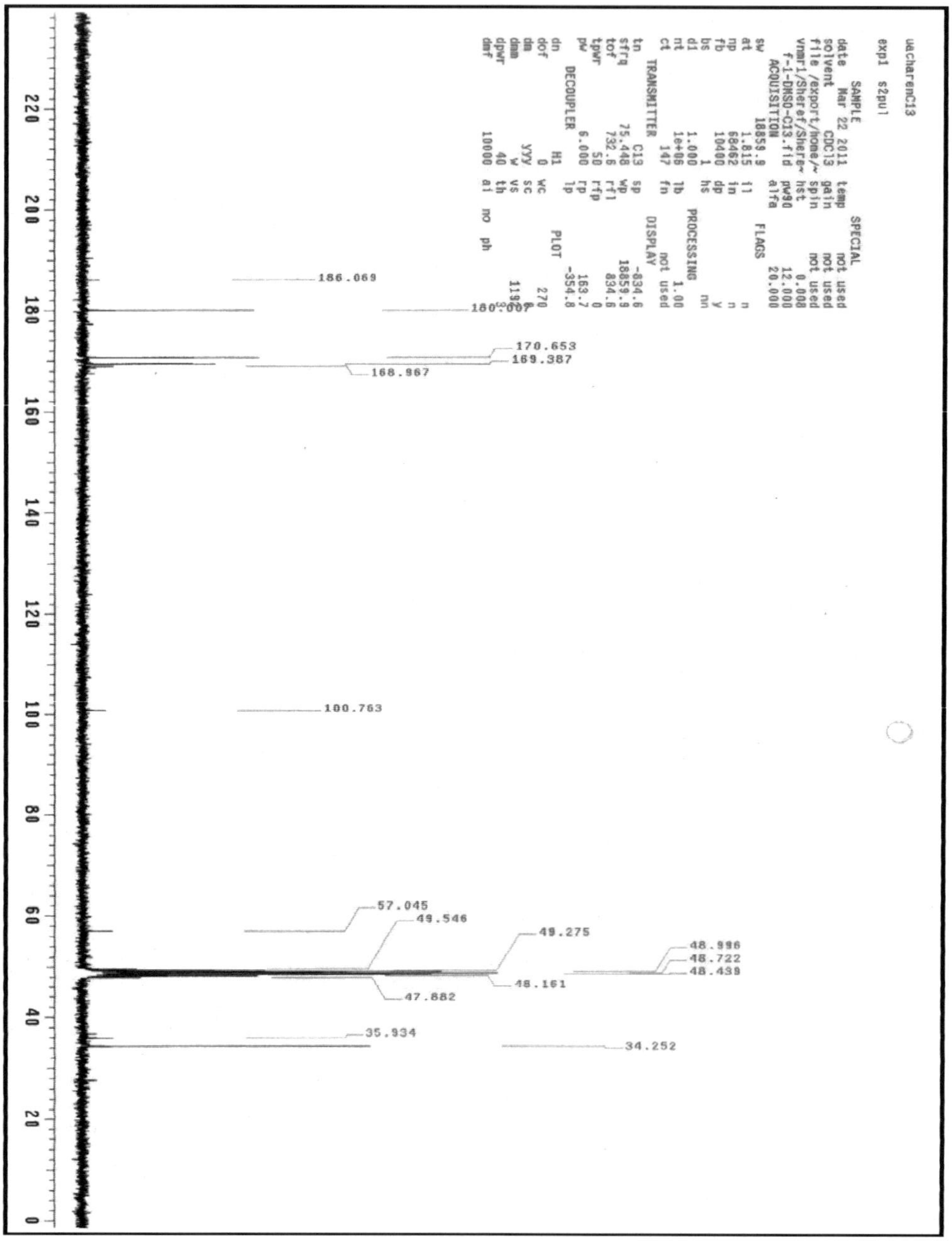

Figure 4. The ^{13}C NMR of 2 methoxy-2-methyl-4,5 imidazolidinediones (step 1).

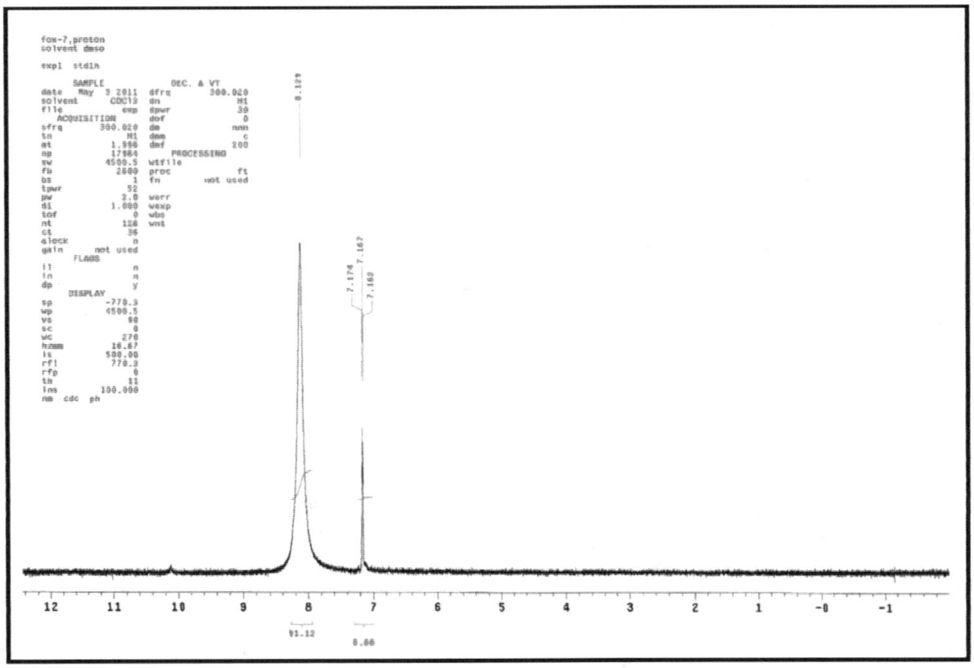

Figure 5. The ^{1}H NMR of 1,1-Diamino-2,2-dinitroethene (step 3)

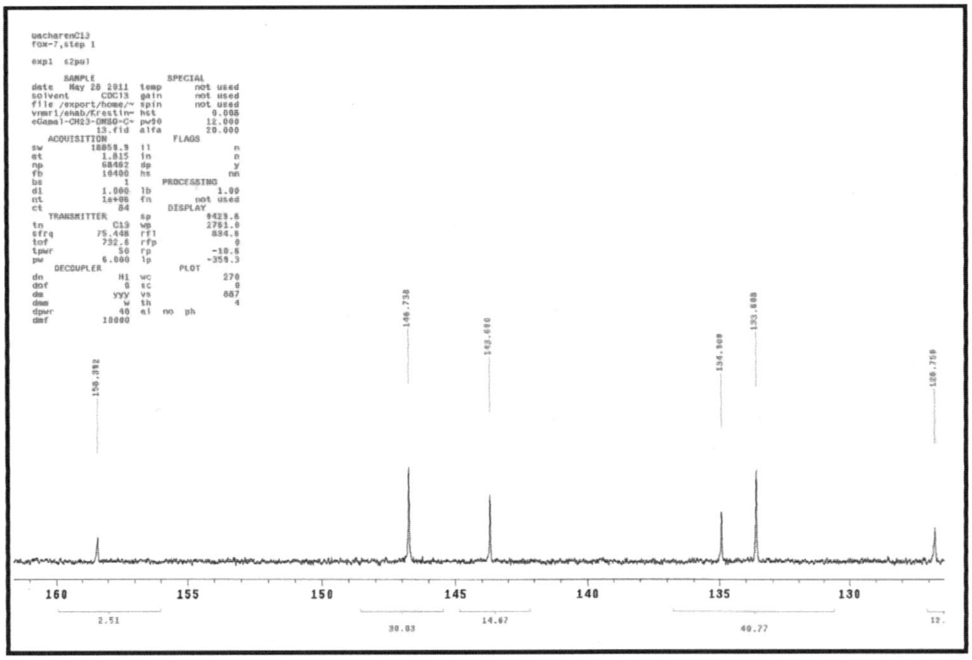

Figure 6. The ^{13}C NMR of 1,1-Diamino-2,2-dinitroethene (step 3).

Table 4. Explosive sensitivity test results of FOX-7.

Explosive properties	FOI	Experimental
Impact sensitivity, (J)	15-25	12-25
Friction sensitivity, (N)	>340	>320
Ignition temperature, (°C)	226	223
Activation energy, (kcal/mole)	58	55

5. Conclusions

1,1-Diamino-2,2-dinitroethene (FOX-7) has been successfully prepared with lower sensitivity and higher performance in comparison to RDX. Interestingly, the % yield of the obtained FOX-7 crystals via our modified approach [64%], was higher than that of the standard method of the Swedish Defense Research Agency (FOI) [58%]. The proposed novel technique involved several modifications; adjusting the reaction stoichiometry at which the exact amount instead of excess amount of sodium methoxide (17.207 g) was used according to the stoichiometric reaction equation, adjusting of the pH value to approximately 9 instead of 4 by the addition of concentrated hydrochloric acid to improve the precipitation and separation of white crystals of NaCl, starting the reaction procedure by reducing the reaction temperature lower than 10 oC instead of 30 oC which reduce the formation of byproducts, and finally, the filtrate was evaporated to dryness by using rotary evaporator instead of vacuum oven to achieve a complete drying at < 30oC to avoid thermal decomposition of step I product through long drying time of vacuum oven. The FTIR spectrum and the NMR spectrum of 2-methoxy-2-methyl-4,5 imidazolidinedione (step I) and 1,1-diamino-2,2- dinitroethene (FOX-7) were the same as the FOI standard results. The sensitivity results were near to that of FOI results and lower than that of RDX. The heat of explosion of FOX-7 was measured as 4079 kJ/kg while it was calculated as 4091 kJ/kg according to FOI results.

References

[1] Anniyappan M, Talawar MB, and Sinha RK, 2020 *Combust. Explos. Shock Waves* **56(5)** 495
[2] Moy L, Yang E, Yauch S, and Bober R, 2018 *ARMY ARMAMENT RESEARCH DEVELOPMENT AND ENGINEERING CENTER PICATINNY ARSENAL NJ PICATINNY ARSENAL, USA*
[3] Brousseau P, and Thiboutot S, 2018 *Ministry of National Defence, Canada*
[4] Blazek B, and Webb J, 2020 *Naval Air Warfare Center, USA*
[5] Graswald M, and Gutser R, 2018 *Insensitive Munitions Energ. Mater. Technol. Symp*
[6] Chang T, Wen Q, and Wang Y, 2021 *J. Therm. Sci. Eng. Appl.* **13(4)**
[7] Xiao Y, Xiao X, Xiong Y, and Wang Z, 2021 *J. Mech. Sci. Technol.* **35(6)**
[8] Fuchs B, Baker E, and Tomasello MK, 2018 *Review and update of STANAG 4526 Shaped Charge Jet, Munitions Test Procedure*
[9] Wu B, Zhou J, Guo Y, Zhu R, and Wang D, 2021 *Def. Technol.*
[10] Xu H, Duan X, and Li H, 2015 *RSC Adv.* **5** 95764
[11] Sirach RR, and Dave PN, 2021 *Chem. Heterocycl. Compd.* **57(7/8)** 720
[12] Du L, Jin S, Shu Q, and Li L, 2021 *Def. Sci. J.*
[13] Lang J, 2021 *Propellants, Explos, Pyrotech* **46** 233
[14] Xie H, and Gou R, 2021 *Cryst. Res. Technol.* **56** 2000203
[15] Bellamy AJ, 2003 *Propellants, Explos, Pyrotech* **28(3)** 145
[16] Zhen Y, and Yijiang X, 2016 *Acta Chim. Sin.* **74(7)** 612
[17] Zhao X, Chen L, Wang J, Ruan J, and Jiang Z, 2021 *Journal of Energetic Materials* 1

[18] Li X, Yang Y, Song C, Sun Y, Han Y, and Wnag J, 2021 *Materials* **14(5)** 1093
[19] Michalchuk L, Rudić S, Pulham CR, and Morrison CA, 2021 *Chemical Communications* **57(85)** 11213
[20] Latypov NV, Bergman J, and Langlet A, 1998 *Tetrahedron* **54(38)** 11525
[21] U Bemm, and Ostmark H, 1998 *Acta Crystallogr. Sect. C* **54** 1997
[22] Latypov NV, Bergman J, and Langlet A, 2001 US Patent 6,312,538 B1
[23] Langlet A, and Wingborg N, 1996 *Proc. Int. Symp. Spec. Top. Chem. Propuls.* 616
[24] Suceska M, 1995 Test methods for Explosives, Springer
[25] Latypov NV, Bergman J, Langlet A, and U. Wellmar, 1998 *Tetrahedron* **54(38)** 11525
[26] Rudolf Meyer AH, and Kohler J, 2007 *Explosives* P **91**, 6th Edition, WILEY-VCH
[27] Östmark H, Langlet A, and Bergman H, 1998 *The 11th International Detonation Symposium* **30**
[28] Waldemar AT, and Amel B, 2016 *Central European Journal of Energetic Materials* **13(2)** 527

Thermal behavior and explosive properties of TNT/RGO

Ahmed Talaat*[1], Ahmed Elbeih[1], Hesham Tantawy[1], Ahmed K. Hussein[1]

1Military Technical College, Kobry El-Kobbah, Cairo, Egypt
*corresponding author; saberbellah0@gmail.com

Abstract: 2,4,6-trinitrotoluene (TNT), is the most widely accessible explosive. Thanks to the melt cast characteristics (melting temperature of TNT is 80-82 °C,), lead to contribute in many applications as mortars, grenades, artillery rounds, warheads, and antipersonnel mines. TNT has high chemical stability, and moderate sensitivity. In this paper, a novel addition of nano particles of reduced graphene oxide (RGO) was added to enhance the TNT characteristics. A composite TNT/RGO 98:2 wt. % was prepared. The morphology of the composite mixture was captured using scanning electron microscopy (SEM). Impact sensitivity and friction sensitivity were evaluated, thermal behavior was investigated using diffeential scanning calorimetry (DSC), the characterization of the prepared composite and pure TNT was examined using an X-Ray diffractometer. Explosive characteristics were determined by detonation velocity and brisance by Kast . The results showed improvement in the sensitivity with reserving the high explosive characteristics of TNT. Interestingly, a decrease in the melting point of TNT due to the presence of RGO in the composite was observed. This addition might be an effective parameter in the melting procedure of TNT.

Keywords: TNT, Reduced graphene oxide, Scanning electron microscopy, explosive properties.

1. Introduction

2,4,6-trinitrotoluene (TNT), is the world's most extensively used explosive. It has a variety of uses, including the manufacture of melt cast explosives for mortars, grenades, artillery rounds, warheads, and antipersonnel mines [1]. TNT has a strong chemical stability and moderate sensitivity, thanks to its melting point (80-82 °C). Also it is resistant to abrasion, leading it to contribute to artillery shells and mines with the addition of other explosives [2]. One of the most interesting composition is (compositionB), which consist of 60% RDX (1,3,5-trinitro-1,3,5-triazinane) and 40 % TNT. Various TNT-based melt cast formulas are used in military and commercial applications [3-5]. Many researchers presented promising high performance energy materials but the expense of their productions are the main problem [6-8]. TNT and other advanced explosives have been studied as the destructive materials in shaped charges [9-11]. The researches proved that the melting characteristics of TNT are the key for its applications [12]. The energetic polymers have also been studied to increase the performance of the explosives but their sensitivities are high in comparison with inert binders [13-15].

Recently, an innovative type of carbon nanomaterial, reduced graphene oxide (RGO), takes place in preserving the high energy when used in energetic materials. It can be used in a very tiny amount but its effect is considered attractive. RGO has a two-dimensional (planar) material that can be extracted by oxidizing and reducing the graphite powder [16]. RGO has already been used extensively in materials research because of its theoretically huge specific surface area, which might affect the characteristics of the materials. Preparation of graphene sheet divided into two-part method, the first one is up to down

(chemical vapor deposition) (CVD) [17] and the second one is divided into two different processes (Mechanical – Chemical); the mechanical by (Cleavage of Graphite crystals) [sonication of exfoliation graphite] and [double sided adhesive][18] and the chemical by (oxidation process) and (graphite intercalation compounds). Oxidation process can be done by 4 different methods; 1-[Brodie ($KClO_3$-HNO_3)], 2-[Staudenmaier use ($KClO_3$,$NaClO_3$,H2SO4)], 3- [Hummer method by using ($NaNO_3$,$KMNO_4$,H2SO4)], 4-[Modified Hummer method by using ($NaNO_3$,$KMNO_4$,H2SO4)].
The impact of TNT on the production of advanced explosive is examined and demonstrated by comparing the explosive characteristics of our prepared composition with the produced TNT/RGO explosive. The detonation characteristic was measured by detonation velocity and brisance. Carbon materials [19] in particular, have been the focus of extensive research because of their unique feature and excellence. Manning et al. found that graphite not only improved insensitivity but also enhanced performance; it was shown that adding 1.0 wt. percent to a recipe increased the yield [20]. Chi et al. found that fullerene reduced the friction and impact sensitivity of raw TNT from 100% to 70% and 60%, respectively [21-23]. CNTs have also been demonstrated to be an excellent desensitizer for TNT and used with thermite mixture [24-26]. The results demonstrated that a composite of molecules in graphene with TNT had a noticeable favorable effect on the impact sensitivity of raw TNT, which lowered from 100% to 27%, and the friction sensitivity, which decreased from 100% to 36%. RGO sheets were found to be compatible with TNT based on DSC data.

2. Experimental

2.1. Materials:

Hydrochloric acid 5% (HCl), Phosphoric acid (H_3PO_4) and Sulfuric acid (H_2SO_4) were purchased from Aldrich. Graphite and Potassium permanganate ($KMnO_4$) were acquired from WINLAB, UK. Hydrogen Peroxide (H_2O_2) was acquired from Merck (Germany). Acetone was supplied by (Alfa Company, India). TNT is a product of Abu Zaabal company for specialty chemicals. All acquired chemicals were used in the experimental work as received without any additional purification.

2.2. Preparation of Reduced Graphene Oxide (RGO)

RGO was prepared using the modified Hummers method by replacing the sodium nitrate in ordinary Hummers method with phosphoric acid. Our experience with this approach gives a successive and producible high quality graphene. Firstly, 90 ml of H_2SO_4 was added to 60 ml of H_3PO_4 in a 250 ml beaker. Then, 3.0 g of graphite was added to the beaker with continuous stirring in the presence of an ice bath to control the reaction temperature. Once the beaker temperature reaches 0 °C, 20 g of $KMnO_4$ was added. After 30 min, the beaker was removed from the ice bath and left to warm up at room temperature. The mixture was subjected to mechanical stirring for 1 hr. afterward, the beaker was placed on a hotplate to keep its temperature at 60 °C for 24 hr. Then 30 ml H_2O_2 was added which turns the brownish color of the mixture into yellow. The mixture was then sonicated for 1 hr at 40 GHz. Beaker content was washed by HCL 5% and distilled water until its pH reaches 6. Finally, the precipitation was dried for 12 hr using a vacuum oven adjusted at 60 °C.

2.3. Preparation of TNT/RGO Composites

The second experimental step after the RGO preparation was to prepare TNT/RGO composite. 50 ml of acetone was added to 10 mg of RGO prepared in a 250 ml beaker. The mixture was subjected to ultra-sonication for 1 hr. Then, 5 g of TNT has added to150 ml of another acetone portion. The mixture was under continuous sonication for 2 hr at 40°C in a beaker of 1 liter. Then, the suspension was poured into a cold-water beaker which acts as an anti-solvent. Finally, the TNT/RGO composite was successfully separated using a 5000 rpm centrifuge followed by washing with distilled water. The composite was then dried at 50°C under vacuum for 12 hr.

2.4.Characterization

Morphology of the crystals of TNT and the TNT/RGO crystals was studied using a scanning electron microscope (SEM, JOEL JSM-6010LA). TNT and TNT-RGO were examined by XRD, Shimadzu XRD-6000 with Cu radiation =1.54056 A°, using a fully computerized X-ray diffractometer. Throughout the experiments, the X-ray tube was run at 40kV and 30mA anode current. During all important measurements, the specified operation conditions were maintained. The full 2-range was scanned continuously from 4 to 90 degrees with a scan speed of 8 degrees per minute and a preset duration of 0.15 seconds...

2.5. Thermal study.

Heat flow was determined by the differential scanning calorimetry technique (Q20 by TA Instruments) with a closed pan and the sample mass was 2 mg under a heating rate of 10 K min-1and a temperature range of 40-350 °C. The mass of samples was 2.5 mg under a nitrogen atmosphere of 40 ml min^{-1}.

2.6 Explosive characteristics

The sensitivity of TNT against impact and friction stimuli was tested using the BAM impact sensitivity instrument [5] via dropping of 2 and 5 kg weight hammers and BAM friction test apparatus using the Probit analysis. Detonation velocity was investigated using the EXPLOMET-FO-2000 (created by KONTINITRO AG). Brisance (tested by the relative performance of the prepared mixtures) was determined using the Kast technique developed by the OZM Company. Deformity of the conventional crusher for copper (10.5 mm height and 7 mm diameter) due to the influence of the sample detonation was measured. Each tested sample has a mass of 2 g. The result is calculated as TNT equivalent and the relative brisance is reported in Table 1.

3. Results and Discussion

3.1. SEM Image

The morphology of the crystals of TNT-RGO and the pure TNT is presented in Figure 1. The composite TNT/RGO crystals have a nature that looks like the individual TNT. This result gives another confirmation about the neutral effect of RGO on the surface of the individual TNT. The RGO appears in the SEM images which might have an improvement of the explosive characteristic of the TNT. It is clear from the SEM image that the TNT was edged with RGO which affects the sensitivity by decreasing the friction between the crystal and other external effect. It was concluded that the RGO around the surface of TNT decreased the sharp edges of TNT crystals without influence on the crystal sizes. So it is predicted to decrease the sensitivity of the individual TNT.

a. b.

Figure 1. Micrographs of SEM; a. Pure TNT b.TNT-RGO2%

3.2 XRD Analysis

The XRD pattern of TNT/RGO composites is shown in Figure 2, along with the TNT diffraction pattern for comparison. From the TNT/RGO composites, as illustrated in Figure 2, it was noticed a decrease of the intensity of count about 200 count. The obtained diffraction peaks agree with CCDC number (1582920) [27]. It is clear that there is not great change in the XRD; the peak observed at 15.9 in case of TNT-RGO sample caused by the monoclinic crystal form was compared to the two peaks at 15.8 and 16.4 originated by the orthorhombic form of TNT [28]. The decrease of the crystallinity might be a reason to reduce the sensitvity of TNT.

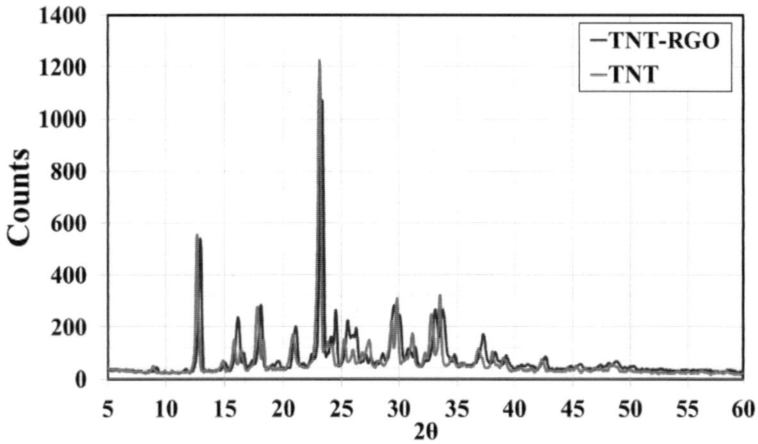

Figure 2. Micrographs of XRD of TNT and TNT-RGO2%

3.3 DSC study

The thermograms obtained by DSC for the thermal action of TNT / RGO 2% composite and pure TNT under a heating rate of 10°C min^{-1} are presented in (Figure 3 a) and (Figure 3b) respectively. While the DSC parameters for the studied TNT composite and the pure TNT are listed in Table 1. The DSC curve of pure TNT shows endothermic decomposition at a maximum peak temperature of 81^0C and a well-formed decomposition peak at 306^0C. In the case of the composite TNT with RGO, The endothermic decomposition peak has been decreased due to the presence of RGO. The high heat conductivity of RGO resulted in the high accumulation of the heat and speed up the melting process of crystals of TNT. Also, it was observed that the presence of oxygenic functional groups on the RGO structure resulted in absorption of the heat to trigger the deoxygenation process and caused the sharp decomposition of TNT/RGO.

Table 1. DSC data of TNT and TNT-RGO 2% at 10° C/min heating rate.

Sample	Heating rate °C/min	Endothermic Peak			Exothermic Peak		
		T_o /°C	T_P /°C	ΔH Endo / Jg^{-1}	T_o/°C	T_P/°C	ΔH Exo / Jg^{-1}
TNT-RGO2%	10	76.96	78.09	88	285.89	295.47	502.69
TNT		79.34	81.03	93.63	300.66	306.21	126.5

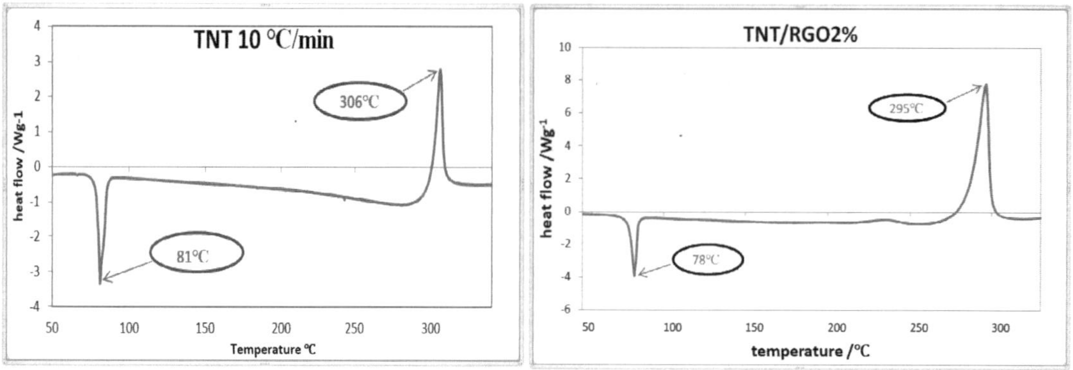

a. b.

Figures 3. Thermogram of DSC for the studied explosive samples .

3.4 Sensitivity measurment.

The sensitivity of TNT/RGO compared with pure TNT is represented in Table 2. the results showed an improvement in impact sensitivity by 40% compaed with pure TNT. The infinitesmal adddition of RGO dramatically enhance the friction sensitivity to be more than the maximum limit of the friction.

Table 2. Sensitivity measurements

No.	Composition	Impact sensitivity (J)	Friction Sensitivity (N)
1	TNT	30	345
2	TNT/RGO	50	>360

3.5 Performance measurement.

The explosive characteristic measured for the TNT composite and pure TNT was represented in Table 3. Despite, the observed change in sensitivity of composite TNT/RGO considered, here the detonation velocity of TNT/RGO composition was reduced by 150 m/s when compared with the highly pressed TNT. The composite TNT/RGO was pressed to lower loading density than the pure TNT, so the detonation velocity was lower. On the other hand, the brisance of TNT/RGO was only reduced by 1 % in comparison with TNT.

Table 3. Performance measurements

No.	Composition	Density (g/cm^3)	Detonation velocity (m/s)	Brisance (equivelant TNT%)
1	TNT	1.60	6950	100
2	TNT/RGO	1.52	6800	99

4. Conclusion

By the addition of 2 wt% of RGO, a composite explosive based on TNT/RGO was successfully prepared. TNT-RGO crystals have the same shape and size as the pure crystals of TNT. For TNT-RGO, although few wrinkles were found, it could decrease the TNT's sensitivity to mechanical stimuli. the decomposition peak of TNT-RGO is lower than pure TNT. Besides, a decrease in the melting temperature of TNT/RGO was observed due to the large heat accumulation in the presence of RGO which caused a speed up of the melting process of the TNT crystals. Also, it was observed that the presence of oxygenic functional groups on the RGO structure resulted in the absorption of the heat to trigger the deoxygenation process and caused the sharp decomposition of TNT/RGO. On the other hand, still the addition of RGO to TNT by 2 % reserving the performance in comparison with raw TNT. As a result, TNT/ RGO composite is a promising composition and needs more investigation.

5. References

[1] Klapötke TM. Energetic Materials Encyclopedia. de Gruyter; 2018.
[2] Elbeih A. and Zeman S. 2014, *Central European Journal of Energetic Materials* **11**, 501.
[3] Hussein A.K., Elbeih A. and Zeman S. 2018, *Thermochimica Acta* **666**, 91.
[4] Hussein A.K., Elbeih A. and Zeman S. 2017, *Thermochimica Acta* **655**, 292
[5] Hussein A.K., Zeman S. and Elbeih A. 2021, Manifestations of replacing 2, 4, 6-trinitrotoluene by 2, 4-dinitroanisole (DNAN) in compositions based on several interesting nitramines *Journal of Energetic Materials*, in press.
[6] Abd-Elghany M., Elbeih A. and Klapötke T.M. 2018, *Journal of Analytical and Applied Pyrolysis* **133**, 30.
[7] Abd-Elghany M., Klapötke T.M. and Elbeih A. 2017, *Journal of Analytical and Applied Pyrolysis* **128**, 397.
[8] Elbeih A., Abd-Elghany M. and Klapötke T.M. 2017, *Propellants, Explosives, Pyrotechnics* **42(5)**, 468.
[9] Elshenawy T., Elbeih A. and Li Q.M. 2016, *Central European Journal of Energetic Materials* **13**, 927.
[10] Elshenawy T., Elbeih A. and Klapötke T.M. 2018, *Journal of Energetic Materials* **36**, 127.
[11] Elbeih A., Elshenawy T., Zeman S. and Akstein Z. 2018, *Central European Journal of Energetic Materials* **15**, 3.
[12] Shehzad R.A., Muhammad S., Iqbal J., Al-Sehemi A.G., Yaseen M., Aloui Z. and Khalid M. 2021, *Journal of Molecular Modeling*, **27(1),** 1.
[13] Hussein A.K., Elbeih A. and Zeman S. 2018, *RSC advances* **8**, 17272.
[14] Abd-Elghany M., Klapötke T.M. and Elbeih A. 2017, *Propellants, Explosives, Pyrotechnics* **42(12)**, 1373.
[15] Hussein A.K., Zeman S. and Elbeih A. 2018, *Thermochimica Acta* **660**, 110.
[16] Lightcap I.V., Kosel T.H. and Kamat P.V. 2010, *Nano letters,* **10**, 577.
[17] Blocher JM, 1974, *Journal of Vacuum Science and Technology,* **11**, 680.
[18] Yi M. and Shen Z. 2015, *Journal of Materials Chemistry A,* **3**, 11700.
[19] Zaky M.G., Elbeih A. and Elshenawy T. 2021. *Central European Journal of Energetic Materials* **18**(1), 63.
[20] Manning et al, 2003, *U.S. patent no. 6,524,706*, Washington, DC.
[21] Chi et al, 2005, *in International Autumn Seminar on Propellants. Explosives and Pyrotechnics* Beijing, 319.
[22] Feldman D. 2021, *Biointerface Research in Applied Chemistry* **11**, 8179.
[23] Pham TT. 2020, *Biointerface Research in Applied Chemistry* **10**, 4853.
[24] Hussein A.K., Zaki M.G. and Elbeih, A. 2022, Influence of a novel nano-thermite colloid based on CuO coated CNTs on the thermo-analytical characteristics of 1, 3, 5-trinitro-1, 3, 5-triazinane. *Combustion Science and Technology*, in press.
[25] Sun XL, Qi Z, Li QQ, Zhang XR, Shao XL, Jin Y, Zhang JJ and Liu YY. 2020, *Biointerface Research in Applied Chemistry* **10**, 5815.
[26] Faisal N. 2020, *Biointerface Research in Applied Chemistry* **10**, 7223.
[27] Fondren N., Fondren Z., Unruh D. and Weeks B., 2019, *Crystal Growth & Design,* **20**, 568.
[28] Lopez-Lopez M., Garcia-Ruiz C., 2014, *TrAC Trends in Analytical Chemistry*, **54**, 36.

Aziridine type bonding agents for composite solid rocket propellant

Mohamed F. Eissa[1*], Maraden, Ahmed[1] M. Elhedery, Tarek [1] M. Abd elall, Ali [1] E. Mostafa, Hosam[1]

[1] Military Technical College

*m7mdmtc@gmail.com

Abstract. Bonding agents, as a paramount component in the composite solid rocket propellant, effectively enhance the interactions between polybutadiene matrix and oxidizer. It mainly affects the processing ability, mechanical behavior, and aging. Bonding agent plays the role of an anonymous soldier to defeat stresses produced by loading conditions, changes in the environmental condition, and transportation of composite propellant. The role and the mechanism of aziridine type bonding agents are discussed focusing on a novel MST aziridine bonding agent (Tris-(2-methyl-1-aziridinyl) phosphine oxide (MAPO) plus Stearic acid plus Tartaric acid) (poly (isopropylamine N(2-methyl-1-aziridinyl) phosphine oxide octadecanoic)-co-(isopropylamine tartrate)). Preparation of composite solid propellant samples containing the same weight percentage of different bonding agents is conducted. Mechanical properties and aging behavior for the prepared samples were monitored to extract the impact of bonding agent type. Mechanical properties of the bonding agents were recorded at three different temperatures in order to simulate various environmental conditions. Results of stress-strain responses during uniaxial tensile tests showed that MST bonding agent has the strongest interaction with solid particles reaches a value of 50.4 % strain and 8.052 Kg/cm^2 stress at normal conditions. Compared to MT-4 (produced from the reaction of 2 moles MAPO plus 0.7-mole Adipic acid and 0.3-mole Tartaric acid) and MAPO, MST improved the aging behavior.

Keywords: composite solid propellant, aziridine type bonding agents, mechanical properties, aging, MST, MT-4, and MAPO.

1. Introduction

Composite propellant's mechanical characteristics have a significant effect on the ballistic performance requirements to satisfy the mission objectives of the rocket motor. These are mainly dependent on the chemical bond between metal fuel, oxidizer, and binder. However, due to their nature, they can also cause

cracks or de-bonding of cured composite propellant if the polymeric matrix is not held together efficiently. [1–2].
Dewetting occurred due to the reduction of strong bond interactions between polymeric matrix and ammonium perchlorate (AP). This is solved by coating the surface with a coating agent, which forms physical and chemical bonds with the polymeric matrix. [3–4].

Although the bonding agent is a low mass ratio component (0.3–0.5%) of the composite rocket propellant formulation, nevertheless, it extremely affects the aging behavior and mechanical properties of composite rocket propellant [5].

Bonding agents are surface-active agents that enhance the solid-liquid bonding interactions and enhance the adhesion properties of composite rocket propellant creating an "elastic bridge" between the matrix and oxidizer filler particles, which produce a more elastic and tougher propellant [6].

The bonding agent consists of functional groups such as hydroxyl, amino, etc. That fortifies the mechanical properties of the composite rocket propellant by creating the chemical and physical bond between the binder and the solid filler [7–8]. Just a little amount of the bonding agent (0.01-1%) is satisfactory to form a single molecular layer on the AP surface. The functions of bonding agents are reaction with a curing agent to construct a tough shell around the oxidizer and interact with the binder by function groups [9].

Bonding agents for ammonium perchlorate (AP) based composite propellant are vinyl-ether bonding agents, amine-bonding agents, and aziridine-bonding agents [10]. The di- and tri-functional aziridinyl phosphine oxide (MAPO) was synthesized to enhance the oxidizer-binder interaction. The molecule includes a polar bond (P=O) surrounded by functional molecules, alkyl chains that carry one or more hydroxyl groups. These compounds have the general structure shown in Scheme 1 [11].

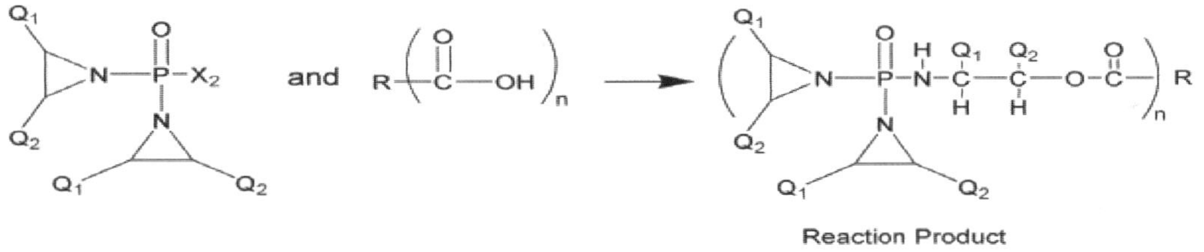

Scheme 1. General structure of MAPO based reaction product, X2 may be the same as the other two aziridine structures or methyl, ethyl, etc. R: alkyl group with hydrogen [11].

At normal conditions, the strain enhances by 4% and the stress enhances by 44% of the reaction product of (MAPO and tartaric acid), however, the strain enhances by 23% and the stress enhances by 182% of the reaction product of (MAPO, tartaric acid, and Adipic acid) in 0.15% Weight of composite propellant [11].

RX bonding agent (MAPO and 12-hydroxystearic acid) is used with ammonium perchlorate. AP opens up the aziridine rings to create a cross-linking network and constructs a hard shell on the surface of AP [5]. Reaction product carry one or more hydrogen that forms the hydrogen bond between molecules of MAPO, AP, and binder matrix chains (OH terminal groups of HTPB), that prevent settling of solid fillers. Examples of the bonding agents are MAPO and MT-4 [12].

MST bonding agent is a novel type of bonding agent that is synthesized from the reaction of tartaric acid, stearic acid, and MAPO (poly (isopropylamine N (2-methyl-1-aziridinyl) phosphine oxide octadecanoic)-co- (isopropylamine tartrate)) [5]. This paper is seeking an enhancement of mechanical properties and aging behavior originated from an effective bonding agent.

2. Experimental

2.1 Materials:

The materials used in this paper are; ammonium perchlorate (AP) with particle sizes 400, 200, and 7-11 (μm) (Abo Zaabal factory, Egypt). Aluminum (Al) with particle sizes 40 and 20 (μm) (Aldrich Chemical Co., England). Hydroxyl-terminated polybutadiene (HTPB) as pre-polymer from (Iverise Co., Brazil) (with viscosity 50.0-70.0 * 10^{-1} (Pa.s), density 0.91 (g/cm^3), average number molecular weight M_n=2000.0-3000.0 and OH value=0.80-0.90).

Dioctylazelate (DOZ) as plasticizer (Aldrich Chemical Co., England). MAPO, MT-4, and MST are bonding agents (Abo Zaabal factory, Egypt), and Hexamethylene diisocyanate (HMDI) as the curing agent (Aldrich Chemical Co., England)

2.2 Preparation of the formulations composite rocket propellant

The casting technique operated for the preparation of the composite rocket propellant formulations was prepared in a vertical kneader having a 3.7L volume [5]. In this job, three composite propellant formulations were formulated. The pre-polymer, and plasticizer (2.633Wt. %) mixed at 50-60 °C in the vertical kneader, then followed by the addition of Aluminum (Al) and mixing for 15 min. then the oxidizer (three stages of Ammonium perchlorate (AP) was added for 15 min). The HMDI curing agent was added to the mixture to attain the NCO/OH ratio of 0.7. The cast propellant slurry was set for 7-10 days in an electric oven at 55 °C. The main ingredients of the propellant formulations are listed in Table 1.

Table 1. Compositions of the composite propellant formulations Wt. percentage.

	AP particle size (μm)			Al (μm)		MAPO %	MT-4 %	MST %	Binder %
	400	200	7-11	10-20	20-40				
MIX1	39	15.5	14.5	6	11	0.3	-	-	14
MIX2	39	15.5	14.5	6	11	-	0.3	-	14
MIX3	39	15.5	14.5	6	11	-	-	0.3	14

2.3 Measurements of the mechanical properties and hardness

The strain-stress behavior, the young's modulus of elasticity, and the hardness were measured for each prepared sample by using the ZWICK test machine. The tensile test was performed at 25, +76, and -40°C for 5 samples and analyzed according to the acceptable mechanical properties for conventional composite propellant [13]. The test specimens were created by using a certain cut press according to (JANNAF) Committee standard.

2.4 Evaluating the Aging Behavior

Exposing propellant samples at elevated temperature ranges will accelerate their chemical change rates according to the Arrhenius equation [14]. Samples were put in aluminum bags in the oven at 90°C for 5 days then cut into standard dumbbell shape (JANNAF). Samples rested at 25°C and mechanical testing was applied to control the aging of the composite propellant samples.

3. Results and discussion:

3.1 Mechanical properties of composite propellant samples

Figures 1, 2, and 3 show the resultant stress-strain curves obtained for the composite propellant samples at temperatures 25, +76, and -40°C respectively.

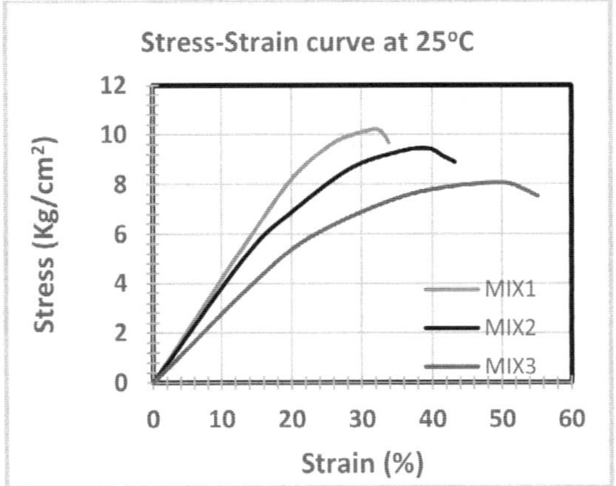

Figure 1. The stress-strain chart at 25 °C of composite propellant formulations

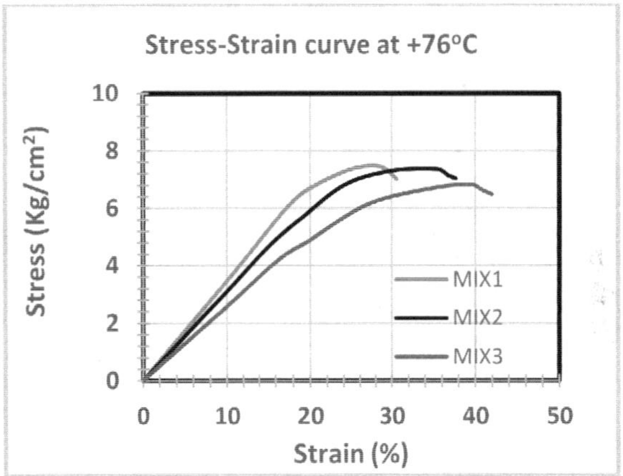

Figure 2. The stress-strain chart at +76°C of composite propellant formulations

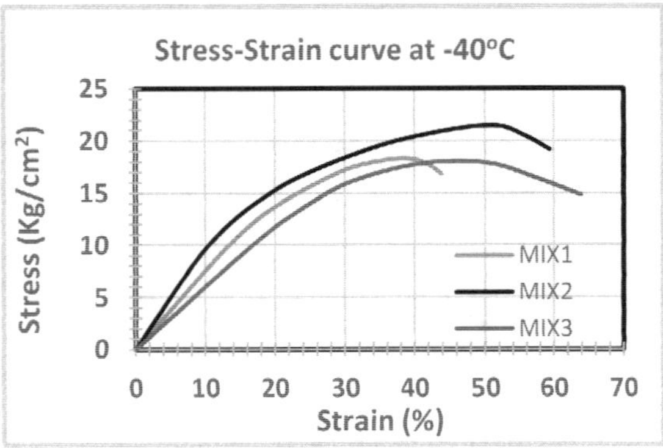

Figure 3. The stress-strain chart at -40°C of composite propellant formulations

According to the results, illustrated in Table 2 and Figure 1, the mechanical properties, which measured at 25 °C, showed that: MIX3 is the greatest value of maximum strain (€) 50.4% at maximum stress (**σ**)

8.05 Kg/cm^2 compared with the other propellant formulations MIX1, and MIX2. In addition, MIX3 has the lower value of Young's modulus (**E₀**) 27.05 Kg/cm^2 and hardness of 58.

In addition, in Figure 2 mechanical properties at +76°C of propellant formulation MIX3 are the highest values in maximum strain (**€**) 39.5% and maximum stress (**σ**) 6.82Kg/cm^2 compared with the MIX1 and MIX2. In Figure 3 at -40°C, the mechanical properties of propellant formulation MIX2 are the greatest value in maximum strain (**€**) 51.1 % but the maximum stress 9.44Kg/cm^2, But MIX3 gives a high value of Max. strain 50% with maximum stress (**σ**) 17.92Kg/cm^2 and modulus (**E₀**) 94.13Kg/cm^2, compared with MIX1 and MIX2.

Table 2. The stress-strain properties of different propellant formulations were measured at different temperatures.

Formulation	MIX1			MIX2			MIX3		
Temperature (°C)	+76	+25	-40	+76	+25	-40	+76	+25	-40
Max. Stress σ (Kg/cm^2)	7.46	10.20	18.30	7.36	9.44	21.47	6.82	8.05	17.92
Max. Strain € (%)	28.40	32.30	39.40	35.40	39.40	51.10	39.50	50.40	50.00
Young's Modulus E_0 (Kg/cm^2)	34.83	41.90	74.97	30.36	37.46	99.71	26.08	27.05	94.13
Hardness Shore A_0	-	67.00	-	-	60.00	-	-	58.00	-

3.2 Mechanical Properties of Aged Propellant Samples

Figure 4 shows the resultant stress-strain curve obtained for the aged composite propellant samples at 25°C. (Aged for 5 days at 90°C).

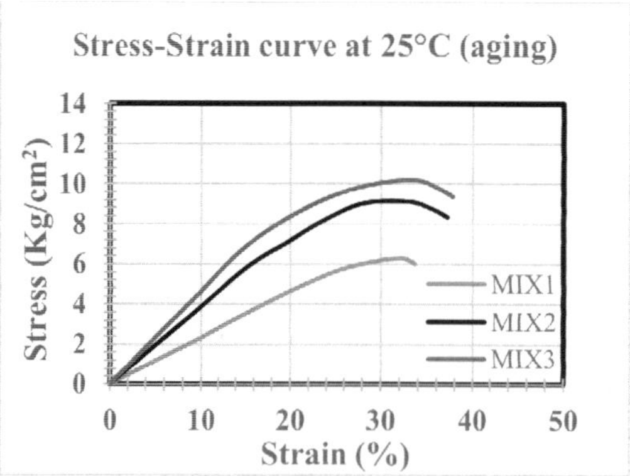

Figure 4. The stress-strain chart at 90°C of aged propellant formulations for 5 days.

According to the program of accelerated aging that performed for 5 days at 90°C (that equals 5 years of natural aging) [16]. the results illustrated in Table 3 *and* Figure 4 showed that the aging of the propellant samples mirrored the stress-strain curves of the non-aged samples, at 25°C MIX3 is the greatest value of maximum strain (**€**) 33.9% at maximum stress (**σ**) 10.18Kg/cm2 compared with MIX1 and MIX2. Because MST bonding agent is suitable for increasing the strain of composite propellant and that required for the aging of composite propellant.

Table 3. The stress-strain properties of different propellant formulations at 25 °C for aging at 90 °C for 5 days.

Formulation	MIX1	MIX2	MIX3
Max. Stress σ *(Kg/cm²)*	6.29	9.12	10.18
Max. Strain *€ (%)*	32.30	33.00	33.90
Young's Modulus E_0 *(Kg/cm²)*	23.67	38.34	46.10
Hardness *Shore* A_0	53.00	58.00	63.00

4. Conclusion

The present study executed the role of bonding agents in the mechanical behavior of composite propellants. The use of bonding agents in a very low percentage is necessary for attaining the wanted mechanical properties, and aging of the composite propellant. In addition, it shows that the choice of the suitable bonding agent gives better results of the composite solid propellant mechanical properties. As the maximum strain, corresponding to maximum stress that depends on the used bonding agent, MST comes higher than MAPO and MT-4 as shown in the results. The results showed that MST bonding agent has the strongest interaction with solid particles giving a high value of strain with 50.4 % and a high value of stress with 8.052 Kg/cm^2 at normal conditions compared with MAPO and MT-4. MST improves aging more than MAPO and MT-4 with a strain of 33.9% because the strain is the mechanical property that affected strongly on the aging of composite solid rocket propellants. MST is a novel bonding agent candidate to offer acceptable and demanded mechanical properties.

5.References

[1] G. S. Tussiwand, V. E. Saouma, R. Terzenbach, L. T. D. Luca, "Fracture Mechanics of Composite Solid Rocket Propellant Grains" Material Testing, (2009).
[2] Yadav, Anil & Pant, Chandra & Das, Sushanta, "Research Advances in Bonding Agents for Composite Propellants" Propellants Explos. Pyrotechnics, (2020).
[3] Y. Y. Yao, J. Q. Zhang, "Research Progress of Bonding Agents for Nitramine Composite Solid Propellants" China Binders, (2009).
[4] C. S. Kim, C. H. Youn, P. N. Noble, A. Gao, "Development of Neutral Polymeric Bonding Agents for Propellants with Polar Composites Filled with Organic Nitramine Crystals" Propellants Explos. Pyrotechnics, (1992).

[5] Ali elall, & Lin, Guo. "Effect of Long-Chain Bonding Agent on the Combustion of Composite Propellant and Modification of Combustion Performance Using Nano Additives" Combustion Science and Technology, (2019).
[6] J. P. Agrawal, "High Energy Materials" Propellants, Explos. Pyrotechnics, (2010).
[7] F. Q. Zhao, W. G. Shan, S. W. Li, "Review on Silane Coupling Agents Used in Solid Rocket Motor Charges and Their Action Mechanism" Energ. Mater, (1998).
[8] B. S. Min, Y. C. Park, J. C. Yoo, "A Study on the Triazole Crosslinked Polymeric Binder Based on Glycidyl Azide Polymer and Dipolarophile Curing Agents" Propellants Explos. Pyrotechnics, (2012).
[9] C. S. Kim, C. H. Youn, P. N. Noble, A. Gao, "Development of Neutral Polymeric Bonding Agents for Propellants with Polar Composites Filled with Organic Nitramine Crystals" Propellants Explos. Pyrotechnics, 1992.
[10] Gan J, Zhang X, Zhang W, Hang R, Xie W, Liu Y, Luo W, Chen Y. "Research Progress of Bonding Agents and Their Performance Evaluation Methods" Molecules, (2022).
[11] H. C. Allen, "Composite Solid Propellant with Additive to Improve the Mechanical Properties thereof" US patent 3,745,074, The United States of America as represented by the Secretary of the Army, Washington, D. C., USA, (1973).
[12] Ali elall, & Lin, Guo. "Synthesis of Long-Chain Bonding Agent of CSRP and Its Effect on Propellant Mechanical Properties" Propellants, Explos. Pyrotechnics, (2018).
[13] G. P. Sutton, O. Biblarz, "Rocket Propulsion Elements", 8th ed., Wiley, New Jersey, (2010).
[14] Cerri, Sara & Manfred, Bohn & Menke, Klaus & Galfetti, Luciano. "Aging of HTPB/Al/AP rocket propellant formulations investigated by DMA measurements" Propellants, Explos. Pyrotechnics, (2012).

Process Engineering Design of Tobacco wastes Incinerator with Utilization of Heat Energy from Combustion Gases

Asmaa Elmansy[1], N. Abdelmonem[1], Ahmed Shaaban[2], Amr Abdelghany[1]*

[1] Chemical Engineering Department, Cairo University, Giza, Egypt
[2] Department of Chemical Engineering and Pilot Plant, National Research Centre (NRC), Egypt.

asmaa_elmansy@hotmail.com, nabil_abdelmonem@yahoo.com, drahmedshaaban@yahoo.com, amr.abdelghany@eng1.cu.edu.eg.

*Corresponding author: amr.abdelghany@eng1.cu.edu.eg

Abstract. An integrated incineration unit was developed to handle tobacco waste within a processing plant in the Eastern Company, Egypt. In addition to the unit, subsequent utilization of the heat content of combustion gases was investigated. The incinerator design was tailored around the current process of solid waste combustion within the tobacco processing plants of the Eastern Company in 6th October City, Egypt. A feeding rate of 1 ton/hr of solid waste consisted of 50% mass of tobacco, 20% paper, and cartons, 20% wooden boxes, and 10% plastics. The volume occupied by the remaining ash after the incineration process would not exceed 5% of the feedstock. The overall material and energy balances were calculated based on a 25490 kg/hr combustion gas discharge with a heat content of 20.09 Gj/hr. Energy from flue gases would be utilized to generate saturated steam or produce hot water. The design included a fired-tube boiler capable of generating 7 ton/hr saturated steam at 185 °C and 10 bar. The temperature of the exhaust effluent combustion gases vented into the atmosphere had to stand at 200 °C to avoid penalties. Moreover, this effluent temperature is considered to be effective and efficient utilization of the heat content in the waste.

Keywords: Tobacco, Solid Waste, Incineration, Fire-tube boiler, Heat Exchanger

1.Introduction

Waste produced at tobacco plants is a significant pollutant. Annual tobacco waste generation ranges from 340 to 680 million kg worldwide (1). After harvesting the tobacco plant and exploiting the tobacco leaves to produce smoking products, the remains of leaves and stalks are considered wastes that need special removal treatment. Tobacco wastes need safe disposal to avoid a harmful effect on the environment as they contain toxic materials such as nicotine and heavy metals (2). Recent attempts have been made to utilize tobacco waste in the extraction of useful chemicals, biomass materials, biodiesel, and production of energy by pyrolysis (3),(4), indoor incineration, and open incineration (5),(6).
Moreover, scrap tires, polypropylene, and polyvinyl chloride were used to improve the pyrolysis of tobacco by reducing char residues and increasing organic gas production (7). Cellulose, Bacterial Cellulose (BC) and Nano-Fiberillated Cellulose (NFC) can be extracted from tobacco stalks (8),(9). Tobacco stalks contain (28 - 40% weight) cellulose as the most predominant polymer (10), followed by Lignin. Through pyrolysis, lignin can be exploited and transformed into bio-products, aromatic compounds, and biofuels (11). Some experiments have been conducted and demonstrated high

polysaccharides content in tobacco waste, including leaves, stalks, and roots. Polysaccharides are important in vital industries, including food and medicine; thus, their extraction is valuable. The polysaccharides were extracted from tobacco waste by response surface methodology (RSM) using hot water (12).

Some experiments have demonstrated the use of tobacco waste in the food industry, dyes, and varnishes due to its citric acid content (more than 10%), making use of its citric acid component and utilizing tobacco waste in the energetic briquettes uses (10). The nitrogen and potassium elements found in tobacco waste may be used to produce agricultural fertilizers. Tobacco waste contains 2.35% nitrogen and 1.95% potassium providing essential nutrients to the soil and plants (13). Furthermore, some experiments extracted activated carbon from tobacco waste (Tobacco-derived activated carbon). The latter is considered a renewable material used in electrical circuits as high-performance capacitor electrodes (14). Recently, the energy scarcity alongside environmental issues mandated the efficient use of wastes to generate energy and decrease the anthropogenic effect of waste. Many studies have been conducted to use agricultural plant residues and primary sources as renewable energy sources, such as bio-oil and biochar (coal). One of these studies is to exploit tobacco residues in energy applications by producing biomass (6),(14). This article exploits tobacco waste as an alternative energy source by burning tobacco waste in incinerators designed for this purpose without hazards affecting the environment. The combustion gases are then exploited to generate steam that will be used to generate electricity.

2.Methodology

Waste materials from tobacco, their chemical composition, and their calorific values were all considered in the design of this study. The materials included tobacco waste, paper, cartons, wooden boxes, and plastic. Equipment used in this study included incinerator and fire-tube boiler; the latter used flue gases emitted from the incinerator to produce saturated steam.

Flue gases may be either utilized for the production of saturated steam or hot water. The results of this study were analyzed and compared with data collected from the incinerator of the Eastern Company in 6th of October City, Egypt. This incinerator handles one ton of solid waste per hour. Waste incinerated in the Eastern Company typically consists of 50% tobacco waste and a combined 50 % paper, carton, aluminum foil, and cellophane paper waste.

3.Materials and Equipment

3.1. Materials used in combustion

3.1.1. Tobacco Waste

Tobacco waste comes from three sources, as follows:

1- Lower-grade tobacco rejected from the production line according to the quality control unit.
2- Fine tobacco particulates collected from the cyclones in the plant.
3- Dicrpencies during production when the plant stops for emergency reasons.

Table (1) presents the chemical composition of the produced tobacco waste from the plant (Burley type). It is composed of 44 % carbon, 7% hydrogen, 43.3% oxygen, 0.9% nitrogen, 0.6% sulfur and 4.2 % ash. The calorific value of the Burley type is 18.3 MJ/kg. (15)

Table 1: The Chemical Composition and Calorific Value of Tobacco

Material	Element content (%)						
	C	H	O	N	S	Ash	Tot.
Tobacco, (Burley type)	44	7	43.3	0.9	0.6	4.2	100%
Gross heat MJ/kg	18.31 Constituent (based on 1 kg)						

3.1.2. *Paper waste*

Paper waste comes from cartons used to transport raw materials to the plant. Table (2) presents the chemical composition of paper waste composed of 44.3% carbon, 6.3% hydrogen, 48.4% oxygen, and 1% ash. (16)

Table 2: The Chemical Composition and Calorific Value of Paper

Material	Element content (%)					
	C	H	O	N	Ash	tot
Sulfite paper [a]	44.3	6.3	48.4	-	1	100%
Gross heat MJ/kg dry basis	17.655 Constituent (based on 1 kg)					

(a) Constituents of sulfite paper: 83% cellulose $C_6H_{10}O_5$; 8% hemi-cellulose $C_6H_{10}O_5$, 6% lignin $C_6H_{10}O_5$, 2% resin $C_6H_{10}O_5$, and 1% ash $C_20H_{30}O_2$.

3.1.3. *Wood waste*

Wood waste primarily comes from the wooden boxes used for the delivery of raw materials to the plant. Table (3) presents the chemical composition of wood waste. the type of wood is spruce wood is composed of 49.8% C, 6.3% H, 43.2% O, 0.13% N, 0.015% S, 0.005% Cl and 0.55% ash content. The calorific value of spruce wood is 18.4 MJ/kg.(17),(18)

Table 3: Chemical Composition of Wood

Type	Element content (%)							
	C	H	O	N	S	Cl	Ash	Tot.
Spruce	49.8	6.3	43.2	0.13	0.015	0.005	0.55	100%
Gross heat MJ/kg dry basis	18.4 Constituent (based on 1 kg)							

3.1.4. *Plastic waste*

Plastic waste comes from the plastic bottles used to store the dye used for production. Plastic consists of polyvinyl chloride (PVC), which is composed chemically of 38.4% C, 4.8% H, 56.8% Cl, and polypropylene (PP), the latter is composed chemically of 85.7% C and 14.3% H. The calorific value of polyvinyl chloride and polypropylene are 19 MJ/Kg and 44 MJ/Kg, respectively (16),(19).

Journal of Physics: Conference Series **2305** (2022) 012024 doi:10.1088/1742-6596/2305/1/012024

3.1.5. Natural Gas

Table (4) details the characteristics of the natural gas from the national grid in Egypt that was used in calculations. The chemical composition and all properties of natural gas are explained in detail at Shaaban *et al.* (16)

3.2. Equipment

3.2.1. Incinerators

The retort multi-chamber was selected for this study of the different types of incinerators. The design of such incinerators guarantees the following:

- Efficient air distribution in the combustion chambers;
- Destruction of harmful materials;
- The "intimate" exposure of the waste fabric to the combustion gases; the simplicity of how combustion gases flow through the individual compartments of the incinerator.
- Efficient heat transfer;
- Ease of ash removal;
- Ease of operation and ease of maintenance.

The essential engineering features that distinguish the retort type of incinerators are the arrangement of the chambers and the gas flow. The incinerator structure consists primarily of the ignition chamber, secondary composition zone, burners, refractory, and external walls. Figure 1 shows the schematic structure of the multi-chamber waste incinerator.

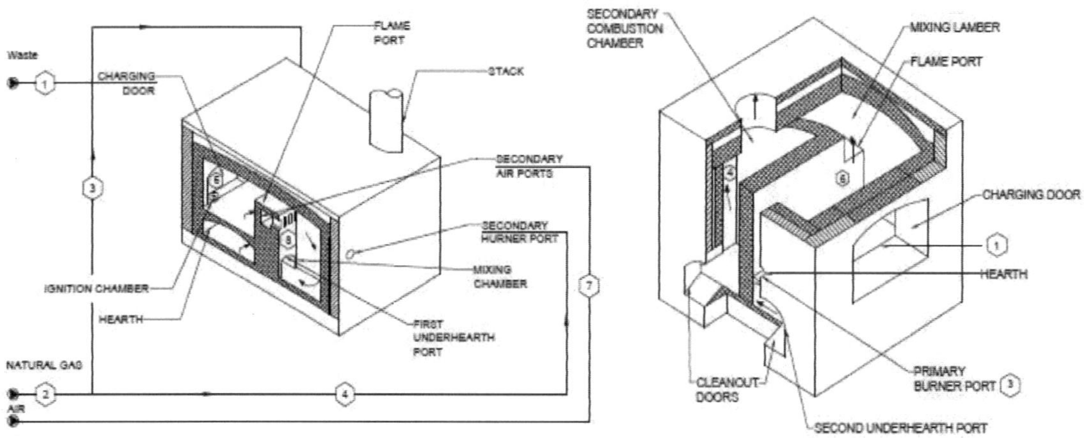

Figure 1: Schematic structure of the multi-chamber waste incinerator (16)

3.2.2. The Fire-Tube Boiler

In many industrial processes, if the output of the process flows at a sufficiently high temperature, the generated heat can be recovered in the form of steam. Many types of steam boilers have been designed that allow the liquid which enters the boilers to be heated and converted into steam. Production of steam depends on many factors, such as the flow rate of liquid, the capacity of steam production, and the exit

pressure. In our case, a fire-tube boiler was chosen, being found to have had a suitable production capacity of steam, the flow rate of water, and the exit pressure of steam. It was thus judged to be more efficient than other choices.
The fire tube boiler has many advantages. It is inexpensive to set up and run, easy to clean, compact in size, easy to maintain (tube replacement), and well suited for space heating and industrial process applications. On the other hand, the disadvantages of the fire tube boilers include not being suitable for high-pressure applications and having limits as far as high capacity steam generation is concerned (22).

4.Incineration Process

Typical waste incineration facilities include the following operations:
- Storage and drying of waste: The upper cranes feed waste to the bouncy of the incinerator, where the drying takes place. Drying is achieved through fuel, radiation, hot waste gases, and, most commonly, air heated. The heated air is pulled through the underside of the ridge surfaces and is allowed to flow over the new waste, leading moisture to evaporate.
- Combustion in a furnace: After the drying process, the combustion process takes place in the middle section of the grate. Dry waste is burned and reduced in size. Since the feeding is fluffy mixture (solid + gas in voids), the excess combustion air is usually between 60-100% mixed gases at temperatures that range between 800-1100 °C.

Reduction of gas temperature and generation of power through overheating steam: At this stage, the burning of residual carbon and other slower-burning elements occurs. The residue is cooled before discharge at the end of the grate. Cooling of the incinerator gas is also necessary before the gas moving into the gas cleaning system. The walls were welded with the pipes of the boiler, then the water passed in pipes for cooling purposes. The heat dissipation process in incinerators produces saturated and superheated steam from water and steam and is then used to generate energy.
Production of residue due to the incineration process; released at the grate discharge point and from the gas cleaning equipment: Combustion residue that remains in the grate is treated by a wet ash extractor and discharged from the bottom of the grate. A top magnetic separator recovers the metal scrap in the combustion residue as it is transferred to the ash hole. The remaining ash on the bottom ash pit is called bottom ash and contains metals, glass, minerals, rocks, ceramics, and non-combustible organic matter.
- The productions of wet or dry flue gas from the burning process: Flue gas is treated by injection of lime powder through an air blower. Adding lime to ash reduces the volume of metals; the release of metals is then quickly filtered under an acidic state. The treated flue gas passes through a dust collection system, such as an electrostatic precipitator, to reduce the dust load to the chimney at a level below 200 mg/m^3. Clean flue gas is discharged into the environment. The dust removed by the separation units is called "fly ash."

The incineration process has advantages over the use of traditional landfills, including the potential to convert water into steam to generate electricity (20),(21), the destruction of combustible toxins and disease-related materials, and the minimization of the volume of solid waste (up to a 90% reduction). (23),(24)

5.Results and Discussions

Some assumptions that were used for the incineration process examined in this study, as described below:

1- The following should be considered constants: (i) the feed rate of waste; (ii) feed rate of fuel (important only at the beginning of the process, after which the incinerator functioned using energy produced from burning exhaust and waste); (iii) gas vaporization rates; and (iv) burning rates.
2- The average temperature of the ignition chamber was typically between 600-850 °C and the exit temperature from the secondary chamber typically stood at 1200 °C.
3- Based on the measurements of the exit temperatures (depending on the results from people was working as this point) from both the ignition and secondary chambers, the required air for

burning waste was found to be 100% above theoretical assumptions. Similarly, the air needed for burning fuel was found to be 20% above theoretical computations. (16)

4- For one ton/hr of waste, 410 m^3 of natural gas/hr and volume flow rate of air 15460 m^3/hr (with 100% excess air) is needed to complete combustion. A total of 20239.6 kg/hr of combustion gases were produced, containing ~20.1×10^6 kJ of heat.
5- Waste combustion reactions were as follows:

$C + O_2 \Longrightarrow CO_2$, $\Delta^o H_{fi} = -393.5$ kJ/mol
$2H_2 + O_2 \Longrightarrow 2H_2O$, $\Delta^o H_{fi} = -242$ kJ/mol
$N2 + 2\,O_2 \Longrightarrow 2NO_2$, $\Delta^o H_{fi} = 33.87$ kJ/mol
$S + O_2 \Longrightarrow SO_2$, $\Delta^o H_{fi} = -297.05$ kJ/mol
$H_2 + Cl_2 \Longrightarrow 2HCl$, $\Delta^o H_{fi} = -92.36$ KJ/mol

Where is $\Delta^o H_{fi}$ the heat of formation of compounds based on standard temperature and pressure which is equal to the heat of reaction for each single reaction mentioned earlier.

Figure 2 shows the block flow diagram of the incineration process, and table (5) summarizes the mass balances of the incinerator. It can be noted that the total mass flow rates of streams 1, 2, 3, 5, and 6 are equal to the total mass flow rate of output stream 9.

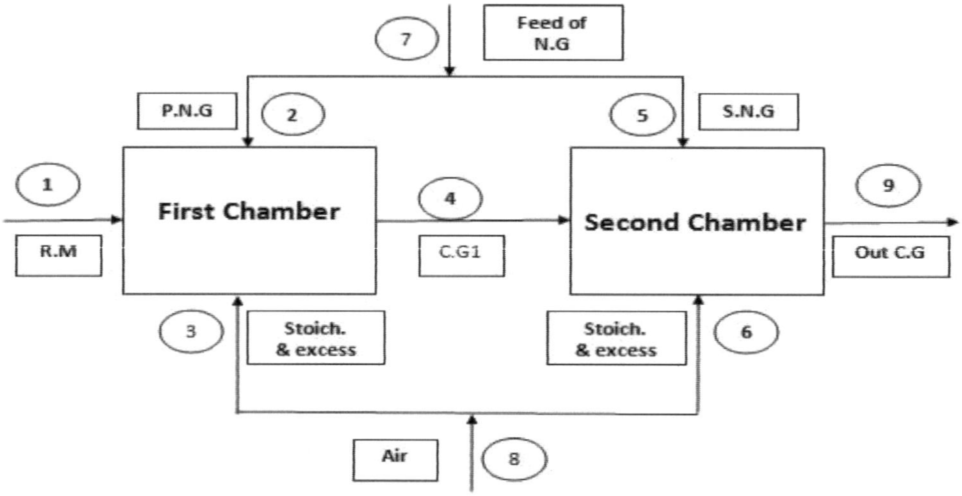

Figure 2: Block flow diagram of the incineration process

R.M : Raw Materials, P.N.G : Primary Natural Gas , C.G1 : Combustion Gases (form first chamber), S.N.G : Secondary Natural Gas , N.G : Natural Gas, C.G : Combustion Gases (form second chamber)

Table 4: mass balances of the system "Incinerator."

Input									
First Chamber	Waste, kg/hr. **(Stream 1)**	1000							Total
	Natural Gas, kg/hr. **(Stream 2)**	198							198
	Air, kg/hr. **(Stream 3)**	13378.9							----
	Components of air, kg/hr.	H_2O: 60.0	O_2: 3103.3	N_2: 10215.6					13378.9
Second Chamber	Natural Gas, kg/hr. **(Stream 5)**	273.8							273.8
	Air, kg/hr. **(Stream 6)**	5388.9							-----
	Composition of air, kg/hr.	H_2O: 24.2	O_2: 1250.0	N_2: 4114.7					5388.9
Output									
		CO_2	H_2O	NO_2	SO_2	HCl	N_2	O_2	
Second Chamber	Combustion gases, kg/hr. **(Stream 9)**	2587.2	1400.6	19.2	5.6	2.9	14494.9	1729.2	20239.6

5.1. Application for Utilization of Combustion Gases Heat Content

5.1.1. Energy Balance to Generate Steam

The heat released from the secondary chamber of the incinerator was used to produce saturation steam at 185 °C and 10 bar by letting the hot flue gases flow into the fire-tube steam boiler.

The total heat that got into the boiler and cooled the hot flue gases from 1200 °C to 200 °C may be represented by the following:

$$Q_{in} = M_{c.g} * Cp_{c.g} * (T_{g.in} - T_{g.out}) \quad (1)$$

where,

$M_{c.g}$ = mass flow rate of combustion gases, kg/hr.

$Cp_{c.g}$= average specific heat of combustion gases, kJ/kg. (1.126 kJ/kg.)

$T_{g.in}$ = temperature of combustion gases which enters the boiler, °C

$T_{g.out}$ =temperature of combustion gases which leaves the boiler, °C

Q_{in} = 20239.6 * 1.126 (1200– 200) = 22789789.6 kJ/hr.

Assuming heat losses of 30%, the actual heat that entered the fire-tube boiler (based on experimental data as near point at this paper to generate steam from combustion gases or burning fuel using using fire tube boiler, taking the assumption of loss of heat approximate 30% of heat entrered at the boiler) may be represented as follows:

$\boldsymbol{Q_{in\cdot act}}$ = 0.7 * 22789789.6 = 15952852.7 kJ/hr

To determine the amount of steam production from cooling hot flue gases at the specific properties of 185 °C and 10 bar, we use the following equation:

$Q_{in} = Q_{out}$

$$M_{c.g} * Cp_{c.g} * (T_{g.in} - T_{g.out}) = M_w * Cp_w * (T_{b.w} - T_{in}) + M_w * \lambda + M_w * Cp_v * (T_s - T_{b.w}) \quad (2)$$

where,

M_w = mass flow rate of water (converted to steam), kg/hr.

Cp_w = specific heat of feed water, kJ/kg. (4.2 kJ/kg.)

T_{in} = temperature of feed water, (25 °C)

$T_{b.w}$ = temperature of boiling water, °C (100 °C)

Cp_v = specific heat of water vapor, °C

T_s = temperature of saturated steam which left the boiler, °C

$$Q_{in\cdot act} = M_w * 4.2 * (100 - 25) + M_w * 2257 + M_w * 2 (185 - 100)$$

M_w = 5818 kg/hr or 1.62 kg/s

The total amount of heat engineered in the fire-tube boiler was 20092347.5 kJ/hr., which allowed for the production of 5818 kg/hr of saturation steam exited at185 °C and 10 bar.

5.1.2. Design of Fire-Tube Steam Boiler

The pressure vessel is the body of a boiler similar to that tested in this study. Water is the fluid used, being converted into steam for process use. The horizontal fire tube boiler has three passes, as shown in Figure 3. The first pass consists of a furnace (hot flue gases), while both the second and third passes consist of tubes. The heat transfer process only operates during the second and third passes.

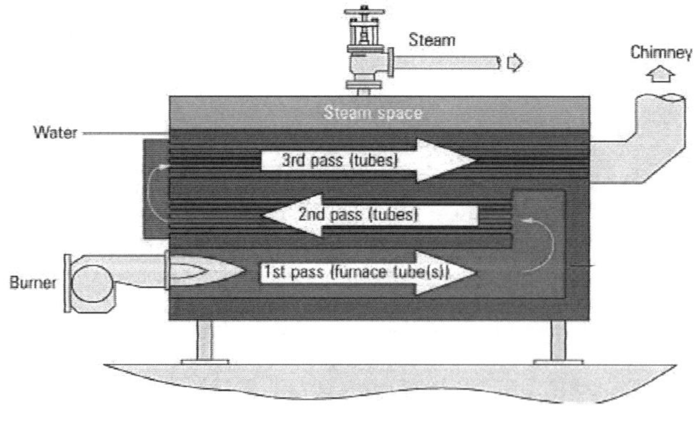

Figure 3: Three Passes Fire Tube Steam Boiler (22)

5.1.3. *Analysis and Calculations of Fire-tube Steam Boiler*

Conduction of heat transfer through piping to the water may be described by the following equation:

$Q/A = (T_{air} - T_w) / (1/h_{air} + t_{tube}/ K_{tube} + 1/h_w)$ (3)

where:

Q = heat transfer rate in kw

T_{air} = temperature of the air in °C

T_w = Temperature of the water in °C

h_{air} = heat transfer coefficient of air in w/m²k, (assume = 75 W/m². °C)

h_w = heat transfer coefficient of water in w/m²k, (assume = 700 W/m². °C)

t_{tube} = thickness of tubes in m, (tube external thickness = 0.003 m.)

K_{tube} = thermal conductivity of carbon steel in w/mk, (assume = 45 W/m. °C)

Engineering Assumptions and Givens:

The hot flue gases had similar properties to that of hot air.

mass flow rate, m· 5818 kg/hr.

operating Steam Boiler Pressure, P = 10 bar

temperature of the furnace = temperature 1st Pass = 1200 °C

temperature of steam = 185 °C

The temperature distribution of water-steam and of the hot flue gases in the fire-tube boiler can be detailed as follows:

T_w = temperature of water (temperature of steam) = 185 °C

$T_{f.g}$ = temperature of flue gases in the 1st pass, (assuming that flue gases to be hot air)

T_3 = temperature of water in the 3rd pass $(T_w + 50\,°C)$

= 185 + 50 = 235 °C

T_2 = temperature of water in the 2nd pass $((T_{f.g}+T_w)/2)$, °C

= (1200 + 235) / 2 = 717.5 °C

$T_{2(air)}$ = temperature of flue gases in the 2nd pass $((T_{f.g}+T_2)/2)$, °C

= (1200 + 717.5) / 2 = 958.75 °C

$T_{3(air)}$ = temperature of flue gases in the 3rdt pass $((T_2+T_3)/2)$, °C

= (717.5 + 235) / 2 = 476.25 °C

Heat transfer over area used for the second and the third passes:

for the second pass:

$Q_2/A_2 = (T_{air2} - T_w) / (1/h_{air} + t_{tube}/ K_{tube} + 1/h_w)$

= (958.75 - 185) / (1/75 + 0.003/ 45 + 1/700)

= 52207.285 w/hr.m²· 52.207 kw/hr.m²

for the third pass:

$Q_3/A_3 = (T_{air3} - T_w) / (1/h_{air} + t_{tube}/ K_{tube} + 1/h_w)$

$= (476.25 - 185) / (1/75 + 0.003/ 51 + 1/700)$

$= 19651.53$ w/hr.m^{2,} 19.651kw/hr.m^2

The total heat transfer over the total area equal:

$Q/A = Q_2/A_2 + Q_3/A_3 = 71.859$ kw/hr.m^2

Thus, the percentage of $Q_2 = 73\%$ from Q_{in} and Q_3 equal 27% from the calculations.

$Q_2 = 0.73 * 4347.25 = 3173.5$ kw/hr.

$Q_3 = 0.27 * 4347.24 = 1173.75$ kw/hr.

Solving for A_2 and A_3using Q_2 and Q_3 values and Q/A values, we get:

$Q_2/A_2 = 52.207$, A2 = 3173.5 /52.207 =60.78 m^2

$Q_3/A_3 = 19.651$, A3 = 1173.75 /19.651 = 59.73 m^2

Tube area and the number of tubes:

The surface area of the tubes was calculated using the assumed tube diameter and length of the boiler taken from guided research conducted on the same case. The following were the results of computations:

diameter of tubes = 0.05 m

length of tubes and boiler = 5m

the diameter of tubes = 2 m

The area of the tube in the boiler:

$A_{tube} = 2\pi * r * L + 2\pi * r^2 = 2*\pi*0.025 * 5 + 2\pi *(0.025)^2 = 0.7889$ m^2. (4)

number of tubes for each pass = A/A_{tube}

number of tubes (pass 2) = $A_2/A_{tube} = 60.78 / 0.7889 = 76.9$

= 77 tubes

number of tubes (pass 3) = $A_3/A_{tube} = 59.73/0.7889 = 75.71$

= 76 tubes

The total number of tubes = 153 tubes.

Table (6) presents a comparison between the performance actual incinerator and boiler at the Eastern Company and designed incinerator and boiler.

Table 5: The comparison between the Performance of Actual Incinerator and Boiler at the Eastern Company and The Designed Incinerator and Fire Tube Boiler.

	Actual Incinerator (Case Study)	Designed incinerator (Theoretical)
Incinerator		
Feed waste, Kg/hr.	1000	1000
Type of waste	50% of Tobacco waste 50 % (paper, carton, aluminum foil, **cellophane** paper)	50% of Tobacco waste 20 % paper 20 % wood 10% plastic
The initial feed of natural gas, m^3/hr.	640	410
Exit temperature from the first chamber, °c	800	850
Exit temperature from the second chamber, °c	Min. 1000	1200
ash from combustion, percent of total feed	5	nearly 4
boiler		
Quantity of production of saturated steam, kg/hr.	7000	5818
The exit temperature of steam, °c	180	185
The operating pressure, bar	8	10

6.Conclusion

Tobacco waste resulting from the manufacturing processes similar to that considered in this study should be regarded as a significant waste compared with other types of waste, such as paper, wood, and plastic waste. Tobacco waste produces a large amount of energy from combustion gases and can be exploited and used in industrial applications.

- One ton of waste similar to that examined in this study consists of tobacco waste, papers or cartons waste, wooden boxes waste, and plastic materials waste at 50%, 20%, 20%, and 10% by weight, respectively. Incineration of such wastes needs about 410 m^3/hr. of natural gas and air volume flow rate at 15460 m3/hr. (with 100% excess air) to complete its combustion.
- Incineration of waste and natural gas produces combustion gases that exist at 1200 °C.
- For every ton/hr of waste, 20239.6 kg/hr of combustion gases are produced, containing 20092348 kJ of heat.
- The utilization of combustion gases can be achieved by designing a fire-tube steam boiler that would be used to produce saturated steam.

  A fire-tube steam boiler would generate saturated steam at 5818 kg/hr flow rate. at 185°C and 10 bar with efficiency approximately 70% of system (taking into consideration the assumption of heat loses of the system depending on the experimental data as near point at this paper.

It is recommended that further larger-scale studies that would examine the characteristics of different types of waste and their possible uses in energy recovery.

References:

[1] WHO. Tobacco and its environmental impact: an overview. 2017. 72 p.
[2] Okunola AA, Babatunde EE, Chinwe D, Pelumi O, Ramatu SG. Mutagenicity of automobile workshop soil leachate and tobacco industry wastewater using the Ames Salmonella fluctuation and the SOS chromotests. Toxicol Ind Health. 2014;32(6):1086–96.
[3] Sun Y, He Z, Tu R, jian Wu Y, chen Jiang E, wei Xu X. The mechanism of wet/dry torrefaction pretreatment on the pyrolysis performance of tobacco stalk. Bioresour Technol. 2019;286:121390.
[4] Yan B, Zhang S, Chen W, Cai Q. Pyrolysis of tobacco wastes for bio-oil with aroma compounds. J Anal Appl Pyrolysis. 2018;248–54.
[5] Onorevoli B, da Silva Maciel GP, Machado ME, Corbelini V, Caramão EB, Jacques RA. Characterization of feedstock and biochar from energetic tobacco seed waste pyrolysis and potential application of biochar as an adsorbent. J Environ Chem Eng. 2018;6(1):1279–87.
[6] Carvalho FS, Fornasier F, Leitão JOM, Moraes JAR, Schneider RCS. Life cycle assessment of biodiesel production from solaris seed tobacco. J Clean Prod. 2019;230:1085–95.
[7] Chen R, Zhang J, Lun L, Li Q, Zhang Y. Comparative study on synergistic effects in co-pyrolysis of tobacco stalk with polymer wastes: Thermal behavior, gas formation, and kinetics. Bioresour Technol. 2019;292:121970.
[8] Ye J, Zheng S, Zhang Z, Yang F, Ma K, Feng Y, et al. Bacterial cellulose production by Acetobacter xylinum ATCC 23767 using tobacco waste extract as culture medium. Bioresour Technol. 2019;274:518–24.
[9] Tuzzin G, Godinho M, Dettmer A, Zattera AJ. Nanofibrillated cellulose from tobacco industry wastes. Carbohydr Polym. 2016;148:69–77.
[10] Pesevski M, Iliev B, Zivkovic D, Jakimovska-Popovska V, Srbinoska M, Filiposki B. Possibilities for utilization of tobacco stems for production of energetic briquettes. J Agric Sci Belgrade. 2010;55(1):45–54.
[11] Guo G, Liu X, Li R, Li Q, Yu HB, Li MJ. Characterization of tobacco stalk lignin using nuclear magnetic resonance spectrometry and its pyrolysis behavior at different temperatures. J Anal Appl Pyrolysis. 2019;142:104665.
[12] Jing Y, Gao Y, Wang W, Cheng Y, Lu P, Ma C, et al. Optimization of the extraction of polysaccharides from tobacco waste and their biological activities. Int J Biol Macromol. 2016;91:188–97.
[13] Shakeel S. Consideration of Tobacco Dust as Organic Amendment for Soil: A Soil & Waste Management Strategy. Earth Sci. 2014;3:117.
[14] Chen H, Guo YC, Wang F, Wang G, Qi PR, Guo XH, et al. An activated carbon derived from tobacco waste for use as a supercapacitor electrode material. Xinxing Tan Cailiao/New Carbon Mater. 2017;32:592–9.
[15] Mijailovic I, Radojicic V, Ecim-Djuric O, Stefanovic G, Kulic G. Energy potential of tobacco stalks in briquettes and pellets production. J Environ Prot Ecol. 2014;15:1034–41.
[16] Shaaban AF. Process engineering design of pathological waste incinerator with an integrated combustion gases treatment unit. J Hazard Mater. 2007;145:195–202.
[17] FAO. Wood Fuels Handbook. Vol. 53, Journal of Chemical Information and Modeling. 2013. 1689–1699 p.
[18] Günther B, Gebauer K, Barkowski R, Rosenthal M, Bues CT. Calorific value of selected wood species and wood products. Eur J Wood Wood Prod. 2012;70:755–7.
[19] Value E, Municipal OF, Landfilled C, The IN, States F. 2014 ENERGY AND ECONOMIC VALUE OF MUNICIPAL SOLID WASTE (MSW), INCLUDING NON-RECYCLED PLASTICS (NRP), Nickolas J . Themelis and Charles Mussche. 2014;1–40.
[20] Batterman S. Findings on an Assessment of Small-scale Incinerators for Health-care Waste. World Heal Organ Prot Hum Environ. 2004;77.
[21] World Bank. Municipal solid waste incineration. Munic waste Combust. 1999;111.
[22] Banoub B, Belgrad R, Gruscinski J. Thermodynamic Design of a Fire - Tube Steam Boiler. 2014;
[23] Yen TF. Combustion and Incineration. Chemical Processes for Environmental Engineering. 2007. 315–358 p.
[24] World Bank. Municipal solid waste incineration - Technical Guidance Report. Munic waste Combust. 1991.

Removal techniques of ammonia from drinking water and wastewater in Egypt and developing countries: a review study

Reham Samy[1], Nabil Abdelmonem[1], Ibrahim Ismail[1], and Amr Abdelghany[1]

[1]Chemical Engineering Department, Faculty of Engineering, Cairo University, Giza, Egypt.

Abstract. The ammonia pollution/contamination reveals at the inlet of drinking water treatment plants, especially on Northern coast of Egypt, due to the drainage of sewage and industrial wastes from factories along the Nile River, resulting in more neutralization that clogs inlet of water treatment plants. Ammonia causes many diseases; some of them lead to coma or death. Excess ammonia also causes the winter dam in North Egypt to force the drinking water treatment stations to stop operation because of the high pollution load. Therefore, many researchers removed ammonia from drinking water as it was limited to 0.5 ppm in the European association and WHO. In this review, a criticizing summary of researches work (1983-2021) is presented where removal of ammonia by physical (i.e. adsorption, air stripping, and membrane distillation), biological, and chemical (oxidation, struvite, photocatalyst, cold plasma) methods were reviewed.

Keywords: Ammonia, Adsorption, Oxidation, Struvite, Removal Efficiency.

1. Introduction

Pollution increase and industrialization negatively affected the quality and availability of surface and groundwater.[1] Groundwater sources contain high concentrations of minerals because of the contact with rocks, and soil, where minerals dissolve into the water. Thus, the high quality of drinking water comes from the groundwater as they were protected from surface contamination of the earth [2].

The winter rainfalls as a water source are distributed in Egypt from Alexandria to El-Sallum: fewer amounts are received on the northern coast with a mean level of 100 mm/year. On the other hand, in El-Arish region rises to 250 mm/year. In addition, some reservoirs were constructed in Sinai to collect heavy rainfalls in the valleys to be used [3].

The annual average amount of rainfall water utilized effectively was 1.3 BCM/year. However, this amount was not considered reliable due to its high dependence on the place and time variability [4].

So, the Nile River, the primary source of drinking water in Egypt, has to be preserved and treated to get drinking water needs. It might be polluted from sewage disposal or water drainage with fertilizers, leading to a high concentration of harmful contaminants, especially ammonia. Waste discharging of domestic and industry to the Nile leaded to increasing of water contamination load ; Ammonia and its ions were most common existing in equilibrium, the quantity of each species is dependent on both pH and temperature. Ammonium ions might cause undesired odors, and many diseases such as loss of balance, headache, insomnia, and coma sometimes may lead to death.

Also, for water disinfection drinking water treatment plants used chlorine; which may react with ammonia, forming chloramine that is toxic and causes many diseases. As a result, many researchers studied ammonia removal techniques from drinking water, using different initial concentrations, pH, temperature, and contact time.

Initially, water should be characterized, before treatment, by its physical and chemical characteristics. Physical characteristics like turbidity due to silt and clay particles, sewage discharges or industrial wastes, or significant microorganisms' numbers. Additionally, the presence of colloidal solids gives the liquid a cloudy appearance that could be harmful [5,6]. Tebbutt [5] stated that even pure water is not colorless. Moreover, it was necessary to distinguish color due to material in solution from color due to suspended matter. So Arcadio et al. [7] defined the measurement unit of color as the platinum (K_2PtCl_6). 1 colour unit is 1 mg/l of Pt in K_2PtCl_6 . Also, the color imparts to water from the decaying materials from vegetation and inorganic matter.

Lawrence et al.[8] and Daniel [9] studied the rapid changes and climatic events that may affect the number of suspended solids (organic, inorganic) in surface water, especially in rivers. He concluded that the muddy color of the water is due to these suspended solids. Tebbutt [5] Besides, Daniel [9] mentioned that electrical conductivity is a relative indicator for TDS.

Chemical characteristics help more in understanding the water sample properties. As pH measurement is an important and frequently used test, every phase of water and wastewater treatment, waste quality management, and many chemical processes, such as metal precipitation, water softening, coagulation, is pH-dependent [7,10].

After physical and chemical characterization of water samples, the highlighted pollutant is Ammonia, a colorless gas with a specific distinct odor, soluble in water and forming ammonium ions. Ammonia may arise from sewage disposal, fertilizer, and Industrial Processes. It causes failure and corrosion in copper alloy pipes and fittings. [11]

Also, in Nepal, Andrew [12] did a thesis about quality assessment for drinking water there. It was founded that nitrates were more in the groundwater during four weeks due to anthropogenic sources. While ammonia is found in aquifers with higher concentrations due to the deposition of lignite and peats, it is the sum of ammonium ions and ammonia gas depending on pH. Many samples were taken and analyzed for both nitrates and ammonia. It was found that 23 % of samples taken from different sites in Nepal were above WHO limits reached 96 mg/l at 300 m depth and 12 mg/l at 200 m depth (i.e., 1.5 mg/l).

Mohamed et al. [13] outlined some pollution sources as sewage disposal into fresh water and the industrial effluents that could cause ammonia pollutants. Moreover, mentioned some examples of disasters (i.e., toxicity of fish, pollution of crops) that occurred in the Nile River in the last few years due to ammonia. Maha et al. explained tap water quality [physico-chemical and biological quality] in Tanta city (North Egypt). They showed that ammonia ranges from 0 to 0.86 mg/l, resulting in nitrite formation in the distribution systems, leading to manganese removal filters' failure. That causes taste and odor problems. [14]

Rabie et al. [15] studied water quality in Menoufia Governorate located in North Egypt as shown in figure (1) at different sites in different villages presented in the following map figure (2) for the governorate sites to guarantee safe drinking water to the population. The parameters of the samples were within accepted levels except for ammonia, iron, and manganese. Therefore regular monitoring was recommended.

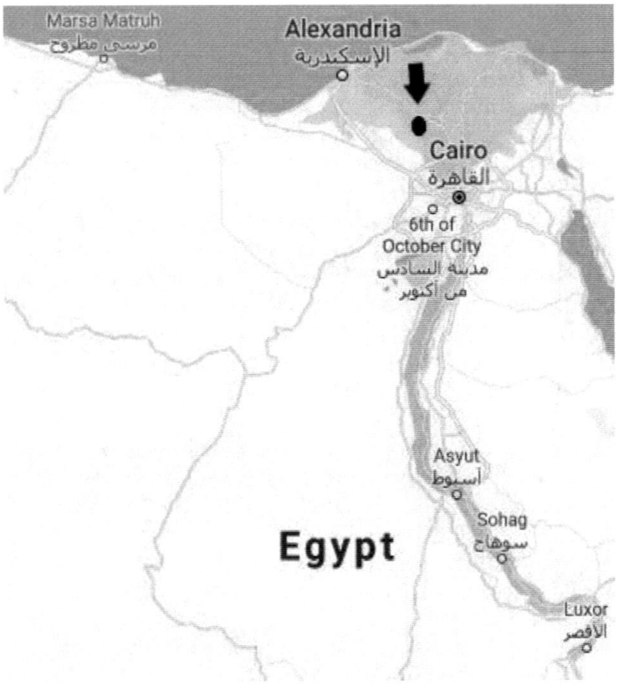

Figure 1. Location of Menoufia Governorate in Egypt.

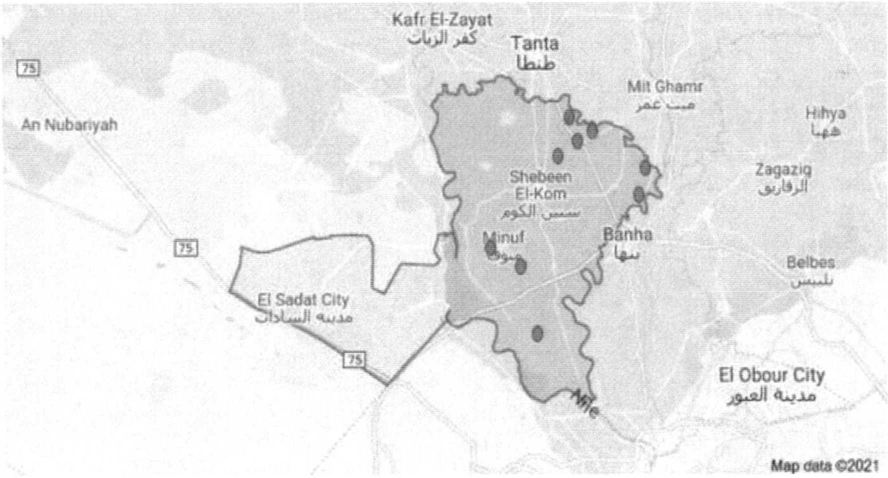

Figure 2. Diff. Sites of sampling in Menoufia Governorate.

*Figures 1 and 2 were edited from google maps

Amaal et al. calculated and investigated 24 sites along the Nile River by measuring water quality index, heavy metal pollution, and contamination indices. They found that ammonia concentrations were higher than recommended limits 0.5 mg/l according to EWQS (2007) and 1.23 mg/l for WHO (2011) in some sites and vary in different seasons. [16]

Dakkak [17] explained the water crisis in Egypt, facing an annual water shortage of around 7 BCM. The factors affecting water security were population explosion, inefficient irrigation, pollution as the Nile River is polluted by municipal and industrial wastes. Furthermore, disturbances because of Ethiopia's high dam would decrease water share. And those are the common sources for water pollution as Kenneth stated in his book. [2]

Researchers used different ammonia determination methods and ways to simulate contaminated water. Suna and Diana [18,19] simulated contaminated water by dissolving anhydrous ammonium chloride salts in deionized water, then analyzing ammonia using ammonium selective electrode; Qi et al. and Ismail determined ammonia by nesslerization. [20,21]

Hassimi, et al.[22] simulated contaminated water by dissolving ammonium sulfate salts. All effluent samples were collected in 1 liter PP bottles to be analyzed with a spectrophotometer with Nessler technique at wavelength 425 μm. In addition to Xingyu et al., Eman and Zhao determined ammonia concentrations by using a UV spectrophotometer. [23–25]

Ying et al. determined ammonia nitrogen by flow injection analysis of the NH_3- o-phthal aldehyde (OPA) -Na_2SO_3 reaction. The reaction was preferred in the alkaline state. This method was considered reliable, simple, and reproducible. [26] and Feng et al. [27] used the indophenol blue method to measure ammonia concentration.

Xiang et al. reviewed different types of wastewater and other ammonia removal techniques as biological, physical, and chemical separation. Including industrial, municipal, poultry wastewater, and landfill leachate. [28]

In Egypt and developing countries, ammonia and its ions were the most known nitrogenous compounds in wastewater, existing in equilibrium; each species' quantity depends on pH and temperature. According to WHO [29] ammonia solubility in water is 421 g/l at 20 °C, and the limit was 1.5 mg/l. Ammonium ions may cause undesired odors, and many diseases such as loss of equilibrium, headache, insomnia, and coma sometimes may lead to death. Therefore, many researchers studied different ammonia removal techniques from drinking water, using other parameters as initial concentration, pH, temperature, and contact time. The following review covers these various techniques as treating water is vital due to building the dam of Ethiopia that shall lower the water share in many countries in the future.

2. Ammonia Removal Techniques:

2.1. Physical removal

2.1.1. Adsorption

Clinoptilolite for ammonia removal was started in 1983 by Hlavay et al., [30] in Hungary drinking water. Three columns filled with sodium clinoptilolite were used in series with an initial concentration of 10 mg/l; when reaching 5 mg/l at the outlet, the column was regenerated by 20 g/l sodium chloride until it went 0.17 mg/l at the outlet of 3rd column. The pH was fixed at 12.2 ±1 for 90 hours. These results are valid for water hardness not exceeding 100 mg/l Ca.

Balci [18] studied ammonia removal using Sepiolite (a silicate clay mineral) ion adsorption and analyzed equilibrium data with several isotherms. The initial ammonium concentration used to describe the adsorption isotherm of NH_4^+ ion sepiolite mixture ranged from 150 to 7000 mg/l. An increase in the adsorption for the particle size of 2.8 mm from 1.82 to 3.7 mmol NH_4^+/l resulted from changing sepiolite concentration ratios in the solution from 1/10 to 1/70.Sepiolite could be used in many applications, as it is characterized by good adsorption capacity, pores structure looks like channels, and high surface area. It is an inexpensive material found in Turkey and a good alternative in liquid phase adsorption applications. [18]

Du et al. [20] used natural Chinese Clinoptilolite [i.e., Clinoptilolite, one of the natural zeolites, to remove ammonia by ion exchange] for ammonia removal from aqueous solution. They performed a batch study and studied contact time, pH, and initial ammonia concentration. Ammonium removal occurred after 15 min. The exchange equilibrium reached after 4 hours. Furthermore, the reduction increased with increasing the concentration of ammonia. Finally, pH affects ammonia removal by affecting ion exchange and adsorbent itself.

Flow velocities variations and breakthrough amounts of Chinese Clinoptilolite for ammonia were estimated in a continuous study. The regeneration of Clinoptilolite was performed by sodium chloride with a concentration of 0.5 mol/l. The conditions for complete renewal were from 15 to 20-bed volume (BV) at pH 11-12.[20]

Zheng et al. investigated the removal of ammonium ions by a composite of hydrogel chitosan grafted polyacrylic acid over rectorite.[31]. Zheng et al. studied different isotherms, and all of them described the results. The equilibrium was achieved in 5 min. The adsorption capacity for ammonium ions was high in a wide range of pH, starting from 4 to 9. No noticeable changes in the adsorption capacity were found over the temperature range studied. Thus, 0.1 mol/l NaOH solution can regenerate this hydrogel adsorbent.

Ismail et al. [32] studied the adsorption behavior of four different ammonia compounds on municipal sludge. The batch reactor used 300 mg/l as the initial concentration at temperature 22°C and reached equilibrium in 4 hours with 89% removal.

Cui et al. [23] removed ammonia in a batch mode using zeolites synthesized from coal fly ash [ZFA-A]. The higher removal efficiency was achieved within a range of pH between 6.5 and 8.5 for the ZFAs. At initial concentration 100 mg/l, pH 7.5, dosage 15 gm/l, and time 30 min., the ammonia removal reached 82.1 %. The experimental results showed that ZFAs is a promising material for ammonia purification from water/wastewater.

Zheng et al. [33] studied statistical Response surface methodology (RSM) based on a three-level analysis, three-variables, Box–Behnken design was used to optimize the preparation parameters of a hydrogel consisted of polyvinyl alcohol, acrylic acid and tourmaline . As a result, the adsorption capacities were 42.74, 37.74 mgN/g at pH range 3-8, initial concentration 100 mg/l, and 0.05 gm of adsorbent dosage. In addition, the prepared hydrogel was observed to be stable and can be reused after addition of strong acid (HCl) or alkali (NaOH) as a regeneration method.

In a batch reaction, Zheng et al. [34] evaluated the Chitosan grafted polyacrylic acid/unexpanded vermiculite (CTS-g-PAA/UVMT) as an adsorbent to remove ammonia. It was observed that a constant adsorption capacity of 21.7 mg/g could be achieved in a wide pH range of 4.0–8.0, initial concentrations range from 25- 1000 mg/l, and 0.05 gm of adsorbent dosage.

Ismail [21] determined the adsorption isotherms, kinetics, and thermodynamics parameters of ammonia removal from water using Clinoptilolite. In a batch study, with ammonia initial concentrations 30, 60, 100, 160, and 250, pH range 4-8, adsorbent dosage 5gm. The adsorption capacities were 14.5 and 13.8 mg/g.

Yu et al. [35] Studied aerobic granules to remove ammonia from synthetic wastewaters. In sequential batch reactors, about 60% of total nitrogen was removed from the water.

Zhao et al. [24] investigated ammonia removal using zeolite synthesized from red mud; five adsorption equilibrium isotherms were studied to find the best fit for the data, the Koble model was the best fit. The maximum capacity of ammonium adsorption obtained was 17.5 mg/g. The studied parameters were 2-11 pH, 5-500 mg/l initial concentration, and 0.5- 25 adsorbent dosage.

Soetardji et al. [36] removed ammonia by modified zeolite mordenite using sodium hydroxide. This modified zeolite was used for the first time in ammonia removal compared to natural zeolite. This study proved that modified zeolite had a higher adsorption capacity than natural zeolites, and 53.91 mg/g adsorption capacity was reached.

Guaya et al. [19] removed ammonium and phosphate using modified natural zeolite by incorporating hydrated aluminum oxide. Three different experiments were conducted; sorption capacity without pH adjustment, sorption capacity at a concentration of 25 mg/l at equilibrium, pH from 2 to 11, and sorption capacity with other ions competing in the wastewater without pH adjustment. The desorption efficiency using 0.1 M sodium bicarbonate was highest for ammonia and phosphate, reaching 92 %, 79 %, respectively.

Feng et al. [27] studied removing ammonia from low concentration ammonia (1-10 mg/l) in natural and secondary water. The adsorbent used was a polymer-supported sorbent on its CuO micro-particles in a batch cation exchange. The results were compared with trials for removal in a fixed bed column, and it indicated that the capacity of adsorption and selectivity was enhanced in the exchanger. Furthermore, the hybrid cation exchanger could be regenerated and reused using 0.25 Molar NaOH, 0.25, and 1 molar of sodium chloride.

Keeratirawee [37] studied coffee residue as a low-cost adsorbent for ammonia in a batch reactor; the coffee residue was un-carbonized and carbonized at 250, 500 °C. It was modified for 24 hours by 5M sulphuric acid; the residue modified and carbonized at 500 °C reached a maximum adsorption percentage of 90% and capacity of 361.85 mg/g.

Ashour et al. [25] prepared activated carbon from activated date pits (ADP); the results showed that ADP's maximum adsorption capacity for removing ammonia was about 37% at 60 °C. The concentration of ammonia was used from 5 – 90 mg/l, 0.5 – 2.5 g of adsorbent.

Ghany et al. [38] studied the adsorption of ammonia on activated carbon from rice straw in batch mode. They compared the results statistically using the linear least square method, trial and error non-linear method of the most common isotherms. Wide range of initial concentrations used in the range of 10 – 350 mg/l, at temperatures 25, 40, 50, and 60 °C; and a fixed amount of adsorbent 0.1 gm, the adsorption capacity reaches 96.4 mg/g after a contact time of 180 min.

Mohamed et al. [39] used natural zeolite (Clinoptilolite) mechanically activated and synthetic zeolite with an initial concentration of 2.5 mg/l and optimum pH 7; the degree of activation affected the adsorption capacity directly; it increased when activation time increased. Capacity reached 71, 80, and 88 mg/g for activation periods 4, 6, and 8 hours.

Bernardi et al. [40] evaluated the efficiency of ammonia removal by using chitins [i.e., shell of fresh shrimps], three types of natural Chitosan, and Chitosan created in the laboratory. The commercial Chitosan removed the minor initial concentration of 0.09 mg/l, while the higher initial concentration of 22.91 mg/l was removed about 42%.

Cruz et al. [41] investigated polymer hydrogel with a sorption capacity of 8.8–32.2 mg NH_4–N/g, which were equivalent to removal efficiencies ranging from 68% to 80% , unaffected by changes in pH, as the sorption capacity remained constant from pH 5.0 – 8.0. Notably, the hydrogels could be regenerated in weak acidic conditions [i.e., pH 4.0] leading to minimal loss in adsorption efficiency following many sorption/desorption cycles.

Huang et al. [42] reviewed the removal of ammonia from water and wastewater by lower-cost adsorbents. The study focused on more than 70 types of adsorbents; as mentioned in the review, adsorption and ion exchange possessed removal efficiencies from 80 to 95 %. Factors determining the cost-effectiveness of adsorbents were: source abundance, low cost, handling process, removal capacity and time for removal, and easiness of regeneration or any other application. Many adsorbents were mentioned as natural and synthetic zeolite, polymers, carbon-based adsorbents, hydrogels, industrial and agricultural wastes, and nanoparticles.

Safie et al. [43] tested various materials to remove ammonia from synthetic solutions; zeolite was the material with a maximum adsorption capacity of 3.16 mg/g, reaching equilibrium after 2 minutes. While Chitosan with adsorption 2.577 mg/g higher than activated carbon from rice husk 2.2 mg/g and bleached fibers 0.018 mg/g. The authors reviewed other studies concerning equilibrium time; their study was the lowest in zeolite, Chitosan, and activated carbon.

Bellahsen et al. [44] investigated a new agriculture waste pomegranate peel in ammonia removal, a low-cost by-product. The optimum parameters for 30 mg/l initial concentration were at pH 4, speed 150 rpm, and 400 mg of peels powder reaching 6.18 mg/g maximum adsorption capacity (i.e., 36.7%). In addition, six more agricultural wastes as wheat husk were tested but with lower removal percentage of

6.7%. Table 1 summarizes the adsorption techniques conditions used in ammonia removal mentioned above.

It was observed that Clinoptilolite or zeolites, in general, need to be chemically modified before use to give higher efficiencies at lower and higher initial concentrations. As of date pits, the agricultural and organic wastes had lower efficiency, around 37 %, but enhanced with lower concentrations and higher reaction time with coffee residue.

Table 1. Adsorption technique Summary.

Added material	Treatment conditions					Removal efficiency %	Regeneration	Reference
	pH	C_o, mg/l	Adsorbent dosage, g	Agitation time, min.	Temp. °C			
Sodium clinoptilolite	12.2 ± 1	10		5400	25	~ 98	20 g/l Nacl	[30]
Sepiolite (synthetic zeolite, silicate clay minerals)	N/A	150-7000	1/10, 1/20, 1/35, 1/70 (g/ml)	360	25	- 85 % for 1/10 dosage - 24.7 % for 1/70 dosage	Not mentioned	[18]
Natural Chinese Clinoptilolite	6	2-94	1	15-30	----	90-100	Sodium chloride pH 11–12	[20]
Hydrogel composite chitosan grafted poly (acrylic acid)/rectorite	4-9	10-1000	0.05	3-60	23, 30, 37, 44, 51	25 % for 10 wt% rectorite	0.1mol/L NaOH solution	[31]
Municipal sludge	4-10	300	1 g/l	2	22	89	Not mentioned	[32]
Zeolites Synthesized from Coal Fly Ash	6.5-8.5	100	5-30	30	25	82.1	Not mentioned	[23]
Hydrogel (PVA, AA, Tm)	3-8	100	0.05	1-60	30-50	75 – 85%	Strong acid or strong base	[33]
Chitosan grafted poly (acrylic acid)/unexpanded vermiculite	4-8	25-1000	0.05	0-60	30	55 %	0.1 mol/L NaOH solution for 10 min.	[34]
Clinoptilolite	4-8	30, 60, 100, 160 and 250	2	100	10, 25, and 40	58 -96 %	Not mentioned	[21]
Aerobic granules	7.5-9	600-2000		720	30	60	Not mentioned	[35]
Zeolite synthesized from red mud	2-11	5-500	0.5-25	30	25	46.6 %	Not mentioned	[24]
Modified zeolite mordenite by sodium hydroxide	6.5	10	0.1-1	60	30, 35, 40	66.9 %	Not mentioned	[36]
Modified natural zeolite by hydrated aluminium oxide	2-11	25	0.1		21	92	1M NaOH	[19]

Added material	Treatment conditions					Removal efficiency %	Regeneration	Reference
	pH	C_o, mg/l	Adsorbent dosage, g	Agitation time, min.	Temp. °C			
Hybrid Resin (Cu-HCIX) loaded with CuO micro-particles	4-8	1-10	0.2-0.29	180	25	77 - 95 %	0.25 molar NaOH, 0.25, and 1 molar NaCl	[27]
Coffee residue		0.1-2	0.05	200	25	90.6 %		[37]
Modified activated carbon derived from date pits		5-90	0.5	240	25	37	Not mentioned	[25]
Activated carbon prepared from rice straw		10-350	0.1	180	400 - 700	55 -70 %	Heat treated at 25, 40, 50, and 60 °C	[38]
Mechanically Activated Clinoptilolite and Synthetic Zeolite-A from Egyptian kaolinite	2-7	0.25-2.5	0.01-0.08	10-120	25	75 – 99 %	Not mentioned	[39]
Chitins and chitosans	7.5	0.09; 0.95; 2.17; 5.62; 7.95; 9.12 and 22.91	1.0; 2.5; 5.0; 7.5 and 10.0% (mass per unit mass mm)	1-300	25	100— 42%	Not mentioned	[40]
Polymer hydrogels	5-8	40-60	2.5	240	25	68-80%	pH 4.0	[41]
Pomegranate peels	4	30	0.4	120		36.7	Not mentioned	[44]

2.1.2. Air Stripping

Deĝermenci et al. [45] removed ammonia by air stripping at high pH in a semi-batch jet loop reactor; reached a removal efficiency of 99% at pH 11. Various parameters were tested, including the initial concentration of 10-500 mg/l, airflow rate 5-50 L/min., and liquid circulation rate of 35-50 L/min.. The used reactors provide high specific areas, mass, and heat transfer without mechanical stirrers. The most effecting parameters on ammonia removal efficiency were the temperature (20-50 °C) and the gas flow rate.

Yuan et al. [46] recovered ammonia using a rotating packed bed; the initial ammonia concentration was 1000 mg/l. They performed experiments at temperatures from 25 to 40 °C, 0.05 L/min. liquid flow rate, gas flow rate 80 L/min., and pH 11. The efficiency of ammonia removal became 81% within 13.3 s at 40 °C.

Zhang et al. [46] used 40 ml/min. pure oxygen or 100 ml/min. compressed air in a chemical catalytic filter to remove ammonia from groundwater. Although ozone was added to the filter to enhance the removal, the filtration rate changed from 8 to 12 m/h. 4.6 mg/l of oxygen needed to remove 1 mg of ammonia.

While Iddya et al. [47] recovered ammonia from wastewater to be used as a fertilizer (Nutrients), recovery of ammonia was by gas stripping membrane electrically conducted.

2.1.3. Membrane Distillation (MD)

There were generally four well-known MD configurations: Direct contact MD, air gap MD, sweep gas MD, and vacuum MD; the following table (2) lists some of their advantages and disadvantages.

Table 2. Advantages & Disadvantages of MD main configurations.

Configuration	Advantages	Disadvantages
Direct contact	Less consuming input energy	High sensible heat loss between condenser & evaporator sides
Sweep gas	The volatile compounds distilled with low surface tension Lower heat loss of conduction Low mass transfer capability Provides a superior permeate flux and evaporation efficiency	When temperature increases, selectivity decrease
Vacuum	Large membrane surface area lowering the boiling point	Need special device for vacuum generation
Air gap	High thermal insulation	The air gap acts as an extra barrier for mass transfer

MD is a thermally driven process, where vapor molecules are transferred through the microporous hydrophobic membrane, with the following main advantages:

- Minimal energy requirement.
- Ability for recycling and reuse of wastewater.

The membrane is used to separate feed solution (hot side) from permeate (cold side). Depending on the configuration, the molecules were either condensed or removed in the vapor phase at the permeate side. MD removed only the volatile ammonia molecules. The amount of ammonia that can be removed depends mainly on the pH and temperature of the solution. Regeneration of the membrane with ethanol to enhance both hydrophobicity and effective surface porosity.

Semmens et al. [48] used hollow fiber membranes to remove ammonia and volatile components rapidly.The air stripping method was modified to be a closed-loop with an air recycled and acid scrubber unit to avoid releasing ammonia into the atmosphere. In their investigation, hollow fibers with an acid stripping solution simulate the closed-loop of the air system. The hollow fibers were hydrophobic polypropylene with 400 μm internal diameter, 30-μm thickness, and 35 % porosity.

EL-Bourawi et al. [49] investigated the usage of vacuum membrane distillation for ammonia removal with inlet temperature of 50 °C, low downstream pressures of 6.3 Kpa instead of 20 Kpa in another study. It was concluded that high initial feed concentrations range from 0.29 to 1.21 M, and high pH levels around 11 enhance ammonia removal efficiency as it may reach more than 90%.

Xie et al. [50] used sweep gas MD to remove ammonia from wastewater; the initial concentration was 100 mg/L, pH 11.5, 70% membrane porosity, and 3 L/min. gas flow at a high temperature around 70 °C; results in 97% removal efficiency.

Duong et al. [51] tried sweep gas and vacuum membrane distillation to remove ammonia from dilute solutions (~ 100 mg/l). Feed and airflow rates, degree of vacuum, and feed temperature were investigated in this study. Industrial wastewater was studied, and 95 % removal efficiency of volatile ammonia was achieved at pH 11.

While; Qu et al. [52] compared direct contact where thermostat heated the feed, permeate was cooled, and there was no receiving solution in the permeate. And modified direct contact MD methods contained a receiving solution of 0.01 mol/l sulfuric acid on the permeate side, resulting in 99.5 % removal efficiency for the modified direct contact MD versus 52% for direct communication, with pH 12.2 as when pH increased, it enhances the removal efficiency. Temperature ranges from 30 to 55 °C and 80 % membrane porosity.

Shirazi et al. [53] reviewed membrane distillation and its applications in water and wastewater treatment processes. According to the configuration, the removal efficiencies were: 99.5 % in the case of modified direct contact 97% in sweep gas flow configuration.

Taqsim et al. [54] studied combining forward osmotic and membrane distillation in a high concentration solution. The removal efficiency reached almost 99 % from the wastewater generated from the anaerobic digestion process. Osmosis could be a continuous process if it were cleaned from fouling every 15 min. during 24 hours, and the membrane could be regenerated, and it was combined with membrane distillation to recover the draw solution of forwarding osmosis to get high osmotic pressure.

Batool et al. [55] summarized membrane distillation removal of the dissolved ammonia. The membrane distillation technique was suitable for high-temperature wastewater treatment, with a low concentration of volatile compounds.

Most membrane distillation researches were at high pH of around 11 to enhance removal efficiency.

2.2. *Biological Treatment*

Biological treatment for wastewater with high contaminants concentration. However, kollar [56] investigated ammonia removal by biological and physical treatment (ion exchange, chlorination, and biological nitrification) as in Hungary many wells exceeds the required limits. A pilot plant was constructed to reduce ammonia concentration and to produce ammonia-free water.

While Aboabboud et al. [57] investigated the removal of ammonia from groundwater as it is the primary source for drinking in Libya. A synthetic solution of around 3 mg/l was used and entered two fluidized reactors containing a sand and granular activated carbon supporting a biofilm. Bioreactors were preferred as there was no need for backwash. It was stated that 1 mole of ammonia requires 1.5 moles of chlorine; and 7.5 parts of chlorine per part of NH_3-N by weight, and these high chlorine dosages could produce harmful ions water. Granular activated carbon oxidized 90 % ammonia to nitrate 20 days earlier (80 days) than sand but more expensive on an industrial scale.

Hasan et al. [22] investigated the removal of ammonia nitrogen, COD, and manganese by the aerated biological filter. The optimum conditions were COD load of 0.90 kg/m^3, aeration rate (AR) of 0.30 L/min, and 7.5 hours hydraulic retention time (HRT); the removal percentages when using these parameters were 95.5%, 93.9%, and 94.8%, respectively. These optimum conditions were used to study the investment and operating cost of a biological aerated filter system for capacity of 100,000 m^3/day.

Takó [58] compared the two methods used mainly in Hungary; breakpoint chlorination and nitrification. Ammonium ions are toxic when it transforms into nitrites ions, and it reacts with chlorine and produces chloramine, which reduces the amount of disinfectant in the water. The first method used in Hungary was to convert all ammonia to nitrogen, but this produced harmful by-products; thus, activated carbon was used to remove them. While; nitrification process oxidizing bacteria converts ammonium first into nitrite then into nitrate and releases energy used in other methods.

Konrad [59] made an overview in Hungary about nitrifying organisms used to remove ammonia from drinking water. Four microorganisms were mentioned that perform ammonia oxidation to nitrogen or nitrate. The main factors affecting the nitrification process were pH, temperature, and growth time of bacteria.

Hassimi et al. [60] applied on/off control using an aerated biological filter for simultaneous ammonia and manganese removal. Their experimental results showed that after 6 hours of aeration, ammonia and manganese reached acceptable limits of 0.7 mg/l and 0.08 mg/l at a dissolved oxygen saturation level of 3 mg/l. Han et al. built a pilot-scale biological aerated filter for simultaneous removal of ammonia and manganese. Using actual samples and lava with particle size 1-3 mm and density 0.922 g/cm^3 as a packed biological bed, investigating the operation factors such as hydraulic loading for ammonia and manganese removal. Reaching 84.1% removal for ammonia and 86.1% for manganese. [61]

Sheela et al. [62] considered the bio-remediation process cost-effective and can easily convert ammonia existing in industrial effluents into nitrogen gas. pH, temperature, dissolved oxygen concentration factors in high salinity were tested on the nitrification process.

Hassimi et al. [63] studied the kinetics of simultaneous removal of ammonia, chemical oxygen demand, and manganese from drinking water in an aerated biological filter. Hassimi et al. investigated the effect of loading,aeration rates, and retention time on ammonia removal and used the Monod model to calculate the kinetics coefficients.

Jin et al. [64] examined solutions with high ammonia concentrations using a biological selector and anaerobic oxidation, reaching 56% ammonia removal. Inlet feed with a concentration of about 2000 mg/l led to nitrite 8.25 kg.N/m^3.d. Finally, ANAMMOX bacteria convert nitrite to nitrogen gas.

Even in sewage, zhang et al. [65] tried to remove the ammonia using an activated sludge system and a bio-film system (binary systems). Results showed that the biofilm system could use high dissolved oxygen to treat low ammonia sewage, while activated sludge could treat ammonia as partial nitrification. DGGE, clone sequencing, and FISH techniques were used to determine functional organisms' distributions in the two systems. Cheng et al. [66] explained the simultaneous ammonia, iron, and manganese from groundwater. Actual groundwater samples were treated in a pilot-scale bio-filter with an ammonia removal efficiency of around 91 % at 0.4 m depth.

Li et al. [67] studied ammonia removal in a vertical flow in an integrated constructed wetland (Constructed wetland is a new technique to grow plants without soil, and the plants grow while removing pollutants). The highest ammonia removal occurred using simultaneous nitration anaerobic ammonia oxidation (ANAMMOX) denitrification processes. Four configurations were constructed in 8 wetlands vegetated and non-vegetated with furnace slag recirculating in a vertical flow with the aim to reach a 98% ammonia removal rate. Arrangements were according to the layers' depth (5 and 20 cm), the dissolved oxygen and removal efficiency varied according to the following Figure.3.

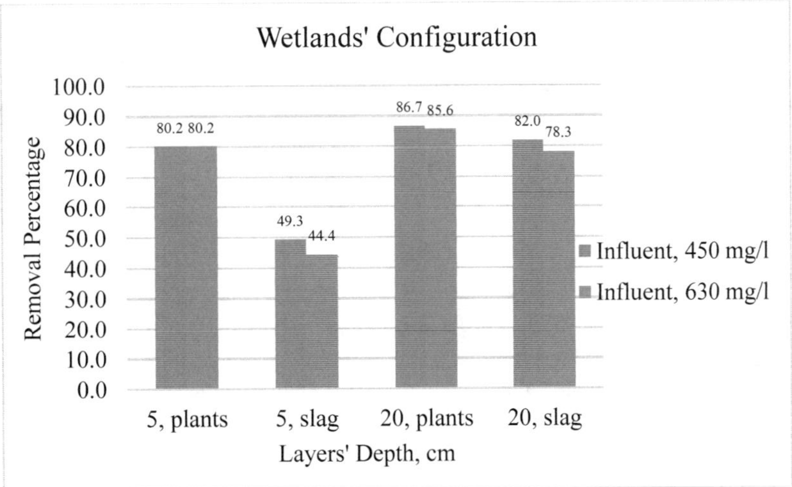

Figure 3. Removal percentage vs. layers' depth of diff. wetlands' configuration.

Liu et al. [68] ensured that aerated biological filters had the highest removal efficiency for ammonia, and it had 92.6% removal in summer compared to 77.5% in winter. In a moving bed batch reactor, Yadu et al. [70] explained the relation between COD/ammonia ratio and ammonia removal in surface water. The studied ratio was 2 to 10, inlet COD concentration range 100 to 2500 mg/l, and ammonia concentration range 50 to 250 mg/l. Maximum ammonia removal efficiency 99.4% was obtained at the lower COD/ammonia ratio of 2. The optimization between variables as retention time, inlet concentrations, and the ratio were performed numerically on Matlab software.

Santos et al. [69] reviewed the removal of ammonia and organic matters by biological filters of activated carbon. Different forms of activated carbon were mentioned and picked the granular form to review by saying the formation way of the film, the bacterial communities, and operating conditions of filters and biofilms.

Hamidi et al., [70] used two biofilters with sand and granular activated carbon, then changed pH and alkalinity as calcium carbonate to observe ammonia and total organic carbon removal. For ammonia at low pH 6-7.5, removal efficiency reaches 53 % even if alkalinity is high (180-220 mg/l) or low (25-50 mg/l). And at high pH 9-10 and high alkalinity, the removal reaches 90 %. The initial concentration of ammonia was 1.99 mg/l at temp. 15 to 23 °C. The opposite occurred with total organic carbon as at low pH, the removal reached 66.7 %, while at higher pH and alkalinity, the removal was 32 %.

Wu et al. [71] biologically treated ammonia in the packed reactor with porous polyurethane carriers over 500 days, with variant retention times from 1.3 to 0.5 hours. At an initial concentration of 10 mg/l, 84% removal was achieved in 0.5 hours, oxidation of ammonia was tested with aeration maximizing removal by 4.5 to 9 times more than oxidation without aeration.

2.3. Chemical removal

2.3.1. Cold plasma

Dung et al. described cold plasma as a green technology that can be used in two different techniques: a) Dielectric barrier discharges [DBD], b) Corona discharges [CD], which are considered as the highest efficiencies methods for surface water treatment.

Sumariyah et al. [72] explained the difference between DBD and CD. The current ion saturation flow in the corona discharge system is more than the flow of dielectric barrier discharges.

Dung et al. tested cold plasma for treating surface water from microorganisms. After 30 minutes of treatment, it was noticed that cold plasma increased nitrate and nitrite concentration and decreased E-coli to 1.0×10^5 CFU/ml (Colony-forming unit/ ml), ammonia decreased from 1.6 mg/l to 0.4 mg/l. [73]

Cold plasma operated at high voltage, weaker acidity, and low initial concentration could be used to remove ammonia from wastewater. Plasma was combined with zeolites to remove ammonia from wastewater. [74] Fan et al. reached around 70 % ammonia removal in 30 minutes with the previous combination, which is better than 53.7% removal with zeolite model alone and 8.4% for cold plasma systems alone.

The oxidation of ammonia was enhanced by the generation of ozone and hydrogen peroxide. The system also compared different gases like argon, nitrogen, and air with oxygen as the working gas. [74]

2.3.2. Electrolysis

Electrochemical oxidation advantages were: the absence of sludge production, the less occupied area by the plants, and relatively low investment costs. The researchers investigated this method and applied two different techniques.

First, Li et al.,[75] used electrochemical oxidation in an undivided cell, studied the most influencing factors (current density, time of electrolysis, and dosing quantity of sodium chloride), and optimized them by response surface method. Ammonia was removed by this method and oxidized into nitrogen gas. The BBD developed mathematical models to tell in advance electrochemical ammonia removal geometry. It was stated that the optimal NaCl dosage, current density, and electrolysis time were 0.31 g/L, 42.75 mA/cm^2, and 100.64 min, respectively. The whole ammonia was removed at optimal conditions and almost 99% removal for nitrite in the treated solution from an initial concentration of 100 mg/l, and these specs for treated water shall meet the standards for drinking water.

In 2013, Eckelberry et al. [76] stated that ammonia removal by electrochemical technique requires chlorine with high concentration, as the electric field in a series of connected reactor tubes changes it into hypo-chlorous acid (i.e., Breakpoint chlorination). The authors investigated different factors like reaction time and the amount of chlorine used.

Shu et al. [77] studied ammonia oxidation by pulse electrolysis; this method simultaneously removed ammonia and recovered manganese. The parameters and technological conditions, including temperature, initial pH, and concentration of addition NaCl, were investigated. The results showed that ammonia nitrogen removal efficiency of 99.8% for 2.5 h PC electrolysis at 30 mA/cm^2, 1000 Hz, and duty cycle 50% were achieved with Ti/SnO_2-IrO_2-RuO_2 electrodes, at initial pH 10.0 and NaCl addition 0.008 mol/L at room temperature.

Aoudj et al. [78] studied removing ammonia from wastewater using electrocoagulation and electro flotation as a post-treatment technique after clarification. The primary oxidation of ammonia occurred at the electro flotation step in chlorine ions, leading to a concentration of 0.22 mg/l lower than desired 0.5 mg/l. The following table (3) summarizes the research conducted using the electrolysis technique.

Table 3. Electrolysis technique conditions Summary.

Device used	C_o, mg/l	NaCl dosage, g/l	Current density, mA/cm^2	Time, min.	Electrodes	Removal efficiency %	References
Box-behnken design	100	0.31	42.75	100.64	- Anode:Ti/IrO_2 - Cathode:Cu-Zn	100	[75]
Pulse current electrolysis	80	0.008 mol/l	30	150	- Anode:Ti/SnO_2-IrO_2-RuO_2 - Cathode: SS Plate	99.87	[77]

2.3.3. Microwave Energy

Rabah et al. [79] performed two experiments using Batch and continuous reactors; Batch studied initial ammonia concentration, initial pH, and radiation time. The optimum removal was obtained in 4 min, with the initial ammonia concentration of 67.5 mg/l at pH 11, ammonia was reduced to 5.0 mg/l (i.e., 92.6% removal efficiency).

Continuous investigated parameters were initial ammonia concentration (60, 80, 100, and 120 mg/l), detention time (5.26, 9, 15, and 22 min.). The results showed that, for all initial concentrations, detention time of 22 min reduced ammonia nitrogen concentration to lower than 5 mg/l. The system's power consumption was relatively high; though, the power consumption can be reduced significantly using efficient heat exchangers, which would lower the required energy for heating. However, many side benefits can be achieved when using MW energy, such as a collection of ammonia gas and heating buildings. [79]

Dong [80] applied microwave radiation followed by aeration to synthetic solutions and landfill leachate, confirming that the successive microwave/aeration process was helpful for ammonia removal. Maximum ammonia removal of 81.7% for the synthetic solution and 70% for leachate by applying microwave energy of 7.8 KJ/L sample followed by 10 minutes' aeration at pH 10.5. Four sets of experiments were carried out for the synthetic solution: water bath (WB), microwave (MW), microwave followed by aeration (MW+A), and water bath followed by aeration (WB+A).

2.3.4. Nano-Technology

Liu et al. [81] investigated an improved catalytic performance of Pt/TiO_2 nanotubes for ammonia oxidation under UV illumination. Sun et al. [82] oxidized ammonia by photocatalytic by palladium-modified nitrogen-doped titanium oxide (TiO_2xNx/PdO) nanoparticles, under visible light source, that was synthesized by sol-gel process.

Experiments were conducted @25°C, 300 W xenon lamp, and glass filters blocked the light with wavelengths below 400 nm and above 700 nm; the initial concentrations of ammonia range from 25 to 150 mg/l, pH 10, the light intensity was 80 MW/cm^2, and the exposure time was two hours. The removal ratio reaches 90%.

Zhang et al. [83]investigated the short (i.e., 3 hours) and long-term (i.e., 60 days) effects of CuO nanoparticles (NPs) on Partial Nitrification (PN) performance. They concluded that CuO NPs with a concentration of 1 mg/L slightly affected the PN process, while that in 3–10 mg/L enhanced ammonia oxidation in short-term exposure. However, long-term exposure within low CuO NPs (5 mg/L) effectively improved the bioactivity.

2.3.5. Oxidation

Khuntia et al. [84] used ozone microbubbles on filtered tap water through an iron removal cartridge at pH 10. Additionally, the microbubbles flow rate generator was 1.7 cm^3/s. As a result, removal efficiency reaches 90%.

Wang et al. [85] studied selective catalytic oxidation of ammonia to nitrogen on $CuO-CeO_2$ mixed oxides. Surfactant template method attached Cu component into ceria CeO_2, at 10% by weight of Cu, temperature 500°C the catalyst had the highest molar ratio, and therefore the highest level of activity was gained. This result appeared after a comparison of catalysts and different preparation methods.

Ichikawa et al. [86] performed oxidation by ozone over metal oxide catalysts, at pH 5.4, with a dosage of 0.1 g catalyst, 1.88 mmol/l as O_3, a total flow rate of 100 cm^3/min. agitated for 360 min. and NH_4^+ was oxidized to N_2 with 88% Selectivity.

Luo et al. [87] performed two-stage ozone oxidation, concentration: 100 mg/l, parameters: ozone flow rate, initial pH, and reaction time. Removal efficiency after the 2nd stage was 85 %. Ammonia was mainly transformed into nitrates, less into nitrites not into N_2.

Jabłońska et al. [88] and Magdalena et al. [89] compared three different copper oxides catalysts for selective ammonia oxidation to N_2 and water vapor. First, copper exchanged zeolites with copper content from 3.7 to 8.7 wt. % reaches selectivity more than 95% at 450 °C. Second, copper-modified clays with 1.4% copper content reach selectivity 90% at 450 °C. Third, Copper oxides are supported on Al_2O_3 or TiO_2 with a copper content of 5 to 15 wt. % and selectivity 90% at 350 °C.

Guo et al. [90] removed around 90% of ammonium and manganese using iron-manganese co oxide filter film acting as a catalyst. Ammonium was oxidized to nitrate by catalytic oxidation, producing hydrogen. The startup time for the film is reduced by potassium permanganate to less than 30 days. The film was regenerated by air recirculation 50 $m^3/(m^2.h)$ for 3 minutes, then air and water for 4 minutes, finally with 36 $m^3/(m^2.h)$ for 3 minutes.

2.3.6. Photocatalytic decomposition

Ou et al. [91] studied the photocatalytic oxidation of ammonia on platinized Nano-tubes. The selectivity of nitrogen was 87.8% after 1 hour. In comparison, Obata et al. [92] used platinum-supported metal doped with titanium dioxide to decompose ammonia into N_2 and H_2 under UV light at room temperature. The photocatalytic reaction took place in a batch reactor, then purged with Argon gas for 15 minutes.

Sun et al. [82] mentioned that photocatalytic oxidation is an efficient removal technique but still needs enhancement in UV requirements, requires high pH, results in increased concentrations of nitrite and nitrates, which are toxic. However, in this study, the sol-gel process was used to produce palladium-modified nitrogen-doped titanium oxide nanoparticles reaching over 90% ammonia removal at pH 8 by visible light only, and the formed product was nitrogen.

Mohammadi et al. [93] achieved 95% ammonia removal by hybrid photocatalyst from titanium dioxide and zinc oxide within 3 hours by UV illumination at pH 11. Photocatalyst regenerated first by distilled water, 3 g/l sodium chloride, then exposed to UV for 2 hours.

Zou et al. [94] studied new graphene-supported $BiFeO_3$ to oxidize ammonia into nitrogen by visible light achieved 91% removal at initial concentration 50 mg/l, at pH 11.

2.3.7. Struvite Crystallization

Hao et al. [95] used recovered struvite as a direct or a valuable indirect fertilizer. Struvite stone ($MgNH_4PO_4{\cdot}6H_2O$) was mainly controlled by concentrations of Mg^{2+}, NH_4^+, PO_4^{3+}, Ca^{2+}, pH, and temperature. For struvite content of more than 90%, the pH may range from 7.5 to 9.

Huang et al. [96] used the product of struvite decomposition in ultrasound stripping to get rid of ammonia from the landfill's leachate. The 90% ammonia removal conditions were at 55 °C for 40 minutes, at pH 9 for many recycles times.

Peng et al. [97] studied struvite formation to simultaneously remove ammonia and heavy metals such as copper. The initial ammonia concentration was 1000 mg/l, and copper ranged from 200 to 800 mg/l, pH 9, reaching 99% removal for ammonia and copper.

Bianchi et al. [98] investigated the kinetics and reaction mechanisms to recover ammonia and phosphorous by precipitation of struvite. The energy required for ammonia decomposition of struvite was determined as 1.87 KJ/g, including water and ammonia evaporation, and decided that ammonia couldn't release before 250 °C. Then struvite can be used as a fertilizer after recovering phosphorus and ammonia, as mentioned in the previous paragraph.

2.3.8. Ultrasound Irradiation

Wang et al. [99] removed ammonia by ultra-sound irradiation while applying sonication power of 200 W for 180 min., pH 11, and initial concentration 680 mg/L, removal efficiency reaches more than 80%. The ammonia-nitrogen removal mechanism by ultrasound irradiation appears to be as if the ammonia molecules were attached to the cavitation bubbles; then, they are transformed into nitrogen and hydrogen molecules via pyrolysis under instant high temperature and pressure of the cavitation bubbles.

Ozturk et al. [100] applied ultrasound irradiation in a short sonication period of 60 to 600 s. The removal efficiency ranged from 8 to 64%, at pH 8.2 to 11; Sonication has been used as an advanced oxidation process (AOP) in wastewater treatment. The mechanism of Sonication was based on sonochemical reactions, which were pyrolytic reactions and cavitation.

2.3.9. Ultraviolet

Zhang et al. [101] investigated a combination of chlorination and ultraviolet radiation for ammonia removal, which led to a higher removal efficiency with lower chlorine demand. In the UV/chlorine process, the oxidation of ammonia was achieved at Cl_2/NH_3 ratios ranging from 0 to 1.6 through the 254 nm wavelength of UV photolysis of chloramines and advanced oxidation reactions. 50% ammonia removal was achieved with Cl_2/NH_3 molar ratio of 0.8 and UV dose 800 mJ/cm^2. Ammonia has a weak absorption wavelengths range greater than 220 nm. The final nitrogenous products formed in the UV/chlorine process were nitrite and nitrate.

The following figure (4) shows the progress of maximum removal efficiency in different techniques according to the last six years' timeline, while figure (5) shows a comparison of the maximum time required for treatment in the different techniques.

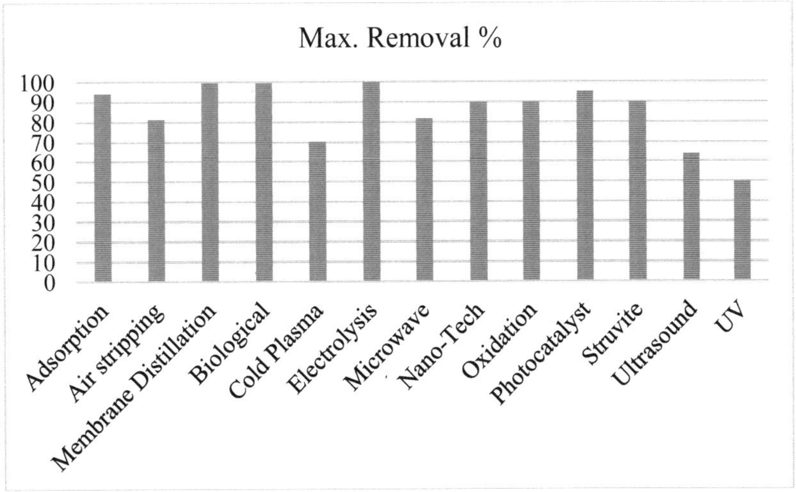

Figure 4. Max. Removal efficiencies in different separation techniques.

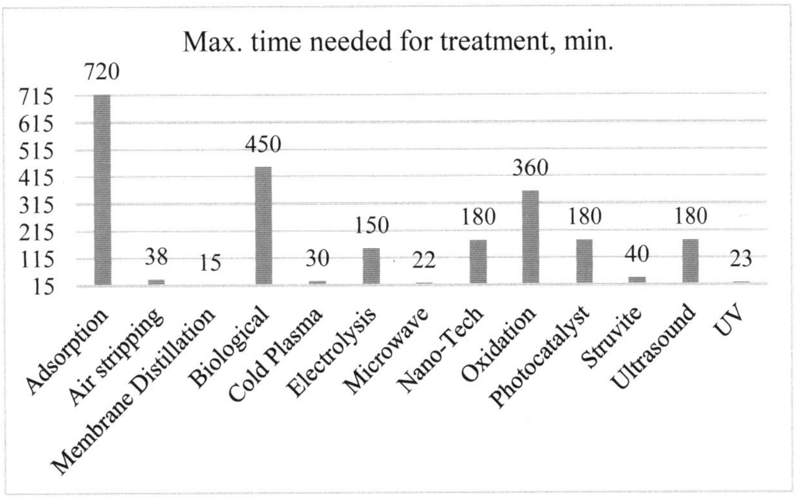

Figure 5. Max. time needed in different separation techniques.

3. Concluding remarks

- As a result of the harmful effect of ammonia on humans, removal methods were studied for its removal. The most effective way was electrolysis as it almost removed 100% of ammonia; it could be combined with solar cells as a renewable source of energy to produce green hydrogen, which is the target of the world to get green H_2 from the environmentally friendly technique.
- More research is needed on Struvite crystallization as it is a technique with low efficiency. Still, it has more potential for a friendly environment as the product can be used as fertilizer to be aligned with the ecosystem (Zero waste technique).
- The most expensive methods were Electrolysis, Microwave, and Nanotechnology
- The cheapest methods were adsorption, and struvite crystallization.

Appendix: A

List of abbreviations	
ADP	Activated Date Pits
AR	Aeration Rate
ANAMMOX	Anaerobic Ammonia Oxidation
AOP	Advanced Oxidation Process
BCM	Billion Cubic Meter
BV	Bed Volume
BBD	Box-Behnken Design
COD	Chemical Oxygen Demand
EWQS	Egyptian Drinking Water Quality Standards
FISH	Fluorescence In Situ Hybridization
HRT	Hydraulic Retention Time
JLRs	Jet Loop Reactors
MD	Membrane Distillation
MW	Microwave
PP	Polypropylene
RSM	Response Surface Methodology
TDS	Total Dissolved Solids
WB	Water Bath
WHO	World Health Organization
ZFA	Zeolite from Fly Ash
ZFA-A	Zeolite A from Fly Ash

References

[1] Tyagi S, Sharma B, Singh P and Dobhal R 2013 Water Quality Assessment in Terms of Water Quality Index *Am. J. Water Resour.* **1** 34–8

[2] Vigil K M 2003 *Clean water: an introduction to water quality and pollution control* **41**

[3] Abdel-Shafy H I and Aly R O 2002 Water Issue in Egypt Resources, Pollution and Protection Endeavors *Cejoem* **8** 3–21

[4] Ministry of Water Resources and Irrigation 2014 Water Scarcity in Egypt: The Urgent Need for Regional Cooperation among the Nile Basin Countries *Minist. Water Resour. Irrig.* 1–5

[5] Tebbutt T H . 1998 *Principles of Water Quality Control*

[6] Wang, Lawrence K.; Hung, Yung Tse; Shammas N K 2005 *Advanced Physicochemical Treatment Processes* (New Jersey: Humana Press, Totowa)

[7] Sincero G and Sincero A 2003 *Physical-Chemical Treatment of Water and Wastewater*

[8] Wang, Lawrence K.; Hung, Yung Tse; Shammas N K 2006 *Advanced Physicochemical Treatment Processes*

[9] Flynn D J and Soediono B 2009 *The nalco water handbook* **53** (Mc Graw Hill)

[10] Nelson M J, Nakhla G, Zhu J, Ammendola P, Chirone R, Spencer Davies P, Bashaar Y A, HACH, Zane Satterfield, P.E. Nescen S, Goel R K, Flora J R V., Chen J P, Eigenberger G, Grady C P L, Daigger G T, Lim H C, Yang J, Trela J, Płaza E, Bioscience Inc. and Mulinix S 2005 Nitrification (Ammonia Oxidation) In Wastewater Treatment Plants *Biosci. Inc.* **3** 236–8

[11] Water Security Agency;Saskatchewan Ministry of Environment 2012 *EPB 431- Ammonia Fact Sheet*

[12] Bittner A 2000 *Nepal Drinking Water Quality Assessment : Nitrates and Ammonia* (University of Michigan)

[13] Mohamed A G, El Safty A M and Siha M S 2013 Current situation of water pollution and its effect on aquatic life in Egypt *Egypt. J. Occup. Med.* **37** 95–119

[14] Maha Abdelfattah Khalil, Zenhom El-Said Salem, Saly Farouk Gheda and Moustafa Mohamed El-Sheekh 2013 Quality Assessment of Drinking Water in Tanta City, Egypt *J. Environ. Sci. Eng. B* **2** 257–75

[15] El Sheikh G, El Bahnasy R, El Shazly H, Al Batanony M and Gabr H 2014 Quality of drinking water in Menoufia Governorate *Menoufia Med. J.* **27** 617–22

[16] Abdel-Satar A M, Ali M H and Goher M E 2017 Indices of water quality and metal pollution of Nile River, Egypt *Egypt. J. Aquat. Res.* **43** 21–9

[17] Dakkak A 2017 Egypt's Water Crisis - Recipe for Disaster *ecomena*

[18] Balci S 2004 Nature of ammonium ion adsorption by sepiolite: Analysis of equilibrium data with several isotherms *Water Res.* **38** 1129–38

[19] Guaya D, Valderrama C, Farran A, Armijos C and Cortina J L 2015 Simultaneous phosphate and ammonium removal from aqueous solution by a hydrated aluminum oxide modified natural zeolite *Chem. Eng. J.* **271** 204–13

[20] Du Q, Liu S, Cao Z and Wang Y 2005 Ammonia removal from aqueous solution using natural Chinese clinoptilolite *Sep. Purif. Technol.* **44** 229–34

[21] Tosun I 2012 Ammonium removal from aqueous solutions by clinoptilolite: Determination of isotherm and thermodynamic parameters and comparison of kinetics by the double exponential model and conventional kinetic models *Int. J. Environ. Res. Public Health* **9** 970–84

[22] Hasan H A, Abdullah S R S, Kamarudin S K and Kofli N T 2011 Response surface methodology for optimization of simultaneous COD, NH_4^+-N and Mn^{2+} removal from drinking water by biological aerated filter *Desalination* **275** 50–61

[23] Cui X, Chen S, Zhang X, Ma J and Li R 2011 Ammonia removal from aqueous solution by zeolites synthesized from coal fly ash *Proceedings - 4th International Conference on Intelligent Computation Technology and Automation, ICICTA 2011* **2** 806–8

[24] Zhao, Ying; Niu, Yongchao; Hu, Xiang; Xi, Beidou; Peng, Xing; Liu, Wenfang; Guan, Weixiong; Wang L 2015 Removal of ammonium ions from aqueous solutions using zeolite synthesized from red mud *Desalin. Water Treat.* **57** 4720–31

[25] Ashour E and Tony M 2016 Adsorption behaviours of ammonia onto modified activated carbon derived from date pits *The International Conference on Chemical and Environmental Engineering* **8** 185–201

[26] Liang Y, Yan C, Guo Q, Xu J and Hu H 2016 Spectrophotometric determination of ammonia nitrogen in water by flow injection analysis based on NH 3 - o-phthalaldehyde -Na $_2$ SO $_3$ reaction *Anal. Chem. Res.* **10** 1–8

[27] Feng Z and Sun T 2015 A novel selective hybrid cation exchanger for low-concentration ammonia nitrogen removal from natural water and secondary wastewater *Chem. Eng. J.* **281** 295–302

[28] Xiang S, Liu Y, Zhang G, Ruan R, Wang Y, Wu X, Zheng H, Zhang Q and Cao L 2020 New progress of ammonia recovery during ammonia nitrogen removal from various wastewaters *World J. Microbiol. Biotechnol.* **36** 1–20

[29] World Health Organization (WHO) 2003 *Ammonia in Drinking-water* **2**

[30] Hlavay J, Vigh G, Olaszi V and Inczédy J 1983 Ammonia and iron removal from drinking water with clinoptilolite tuff *Zeolites* **3** 188–90

[31] Zheng Y and Wang A 2009 Evaluation of ammonium removal using a chitosan-g-poly (acrylic acid)/rectorite hydrogel composite *J. Hazard. Mater.* **171** 671–7

[32] Z.Ismail, Zainab;Tezel U G P 2010 Sorption of quaternary ammonium compounds to municipal sludge *Water Res.* **44** 2303–13

[33] Zheng Y, Liu Y and Wang A 2011 Fast removal of ammonium ion using a hydrogel optimized with response surface methodology *Chem. Eng. J.* **171** 1201–8

[34] Zheng Y, Xie Y and Wang A 2012 Rapid and wide pH-independent ammonium-nitrogen removal using a composite hydrogel with three-dimensional networks *Chem. Eng. J.* **179** 90–8

[35] Yu X, Wan C, Lei Z, Liu X, Zhang Y, Tay J H and Lee D J 2014 Use of aerobic granules for treating synthetic high-strength ammonium wastewaters *Environ. Technol. (United Kingdom)* **35** 1785–90

[36] Soetardji J P, Claudia J C, Ju Y H, Hriljac J A ., Chen T Y, Soetaredjo F E, Santoso S P, Kurniawan A and Ismadji S 2015 Ammonia removal from water using sodium hydroxide modified zeolite mordenite *R. Soc. Chem.* **5** 83689–99

[37] Keeratirawee K 2015 Coffee Residue-Based Adsorbent for Ammonia Removal from Aqueous Solution 61–5

[38] Ghany H A, Ahmed M and Ashour. E 2017 Adsorption of Ammonia Onto Activatedcarbon Prepared From Rice Straw. *Int. J. Adv. Res.* **5** 1180–96

[39] Shaban M, AbuKhadra M R, Nasief F M and Abd El-Salam H M 2017 Removal of Ammonia from Aqueous Solutions, Ground Water, and Wastewater Using Mechanically Activated Clinoptilolite and Synthetic Zeolite-A: Kinetic and Equilibrium Studies *Water. Air. Soil Pollut.* **228** 228–450

[40] Bernardi F, Zadinelo I V, Alves H J, Meurer F and dos Santos L D 2018 Chitins and chitosans for the removal of total ammonia of aquaculture effluents *Aquaculture* **483** 203–12

[41] Cruz H, Luckman P, Seviour T, Verstraete W, Laycock B and Pikaar I 2018 Rapid removal of ammonium from domestic wastewater using polymer hydrogels *Sci. Rep.* **8** 1–6

[42] Huang J, Kankanamge N R, Chow C, Welsh D T, Li T and Teasdale P R 2018 Removing ammonium from water and wastewater using cost-effective adsorbents: A review *J. Environ. Sci. (China)* **63** 174–97

[43] Safie N N ., Zahrim A Y ., Rajin M ., Ismail N M ., Saalah S ., Anisuzzaman S M ., Rahayu A D ., Huslyzam H ., Jennisha R . and Calvin T T H 2019 Adsorption of Ammonium Ion Using Zeolite, Chitosan, Bleached Fibre and Activated Carbon *IOP Conference Series: Materials Science and Engineering* **606**

[44] Bellahsen N ., Varga G ., Halyag N ., Kertész S ., Tombácz E . and Hodúr C 2020 Pomegranate peel as a new low-cost adsorbent for ammonium removal *Int. J. Environ. Sci. Technol.*

[45] Değermenci N, Ata O N and Yildiz E 2012 Ammonia removal by air stripping in a semi-batch jet loop reactor *J. Ind. Eng. Chem.* **18** 399–404

[46] Zhang W, Zhang R, Yang Y, Huang T and Wen G 2019 Removal of high concentrations of ammonium from groundwater in a pilot-scale system through aeration at the bottom layer of a chemical catalytic oxidation filter *Int. J. Environ. Res. Public Health* **16**

[47] Iddya A, Hou D, Khor C M, Ren Z, Tester J, Posmanik R, Gross A and Jassby D 2020 Efficient ammonia recovery from wastewater using electrically conducting gas stripping membranes

Environ. Sci. Nano **7** 1759–71

[48] Semmens M J, Foster D M and Cussler E L 1990 Ammonia removal from water using microporous hollow fibers *J. Memb. Sci.* **51** 127–40

[49] EL-Bourawi M S, Khayet M, Ma R, Ding Z, Li Z and Zhang X 2007 Application of vacuum membrane distillation for ammonia removal *J. Memb. Sci.* **301** 200–9

[50] Xie Z, Duong T, Hoang M, Nguyen C and Bolto B 2009 Ammonia removal by sweep gas membrane distillation *Water Res.* **43** 1693–9

[51] Duong T, Xie Z, Ng D and Hoang M 2013 Ammonia removal from aqueous solution by membrane distillation *Water Environ. J.* **27** 425–34

[52] Qu D, Sun D, Wang H and Yun Y 2013 Experimental study of ammonia removal from water by modified direct contact membrane distillation *Desalination* **326** 135–40

[53] Shirazi M M A . and Kargari A 2015 A review on applications of membrane distillation (MD) process for wastewater treatment *J. Membr. Sci. Res.* **1** 101–12

[54] Husnain T, Mi B and Riffat R 2015 A Combined Forward Osmosis and Membrane Distillation System for Sidestream Treatment *J. Water Resour. Prot.* **07** 1111–20

[55] Shahroie B, Rajabi L and Ashraf Derakhshan A 2016 Short review on membrane distillation techniques for removal of dissolved ammonia *J. Appl. Res. Water Wastewater* **6** 277–80

[56] Kollar, Gy.; Ribari L 1991 Removal of Ammonia With Biological Process in the Treatment of Drinking Water *Period. Polytech. Civ. Eng.* **35** 27–35

[57] Aboabboud M, Ibrahim H and Awad A 2008 Biological ammonia removal from drinking water in fluidized bed reactors *WIT Trans. Ecol. Environ.* **111** 453–63

[58] Takó S 2012 Ammonium removal from drinking water - comparison of the breakpoint chlorination and the biological technology *Conference of Junior Researchers in Civil Engineering* 248–54

[59] Lájer K 2012 Ammonium removal by nitrification in drinking water treatment *J. AM. water Work. Assoc.* **10** 47–53

[60] Abu Hasan H, Sheikh Abdullah S R, Kamarudin S K, Tan Kofli N and Anuar N 2013 Simultaneous NH 4 + -N and Mn2+ removal from drinking water using a biological aerated filter system: Effects of different aeration rates *Sep. Purif. Technol.* **118** 547–56

[61] Han M, Zhao Z wei, Gao W and Cui F yi 2013 Study on the factors affecting simultaneous removal of ammonia and manganese by pilot-scale biological aerated filter (BAF) for drinking water pre-treatment *Bioresour. Technol.* **145** 17–24

[62] B S and Beebi. Sk K 2014 Bioremediation of Ammonia from Polluted Waste Waters- A Review *Am. J. Microbiol. Res.* **2** 201–10

[63] Abu Hasan H, Sheikh Abdullah S R, Kamarudin S K, Tan Kofli N and Anuar N 2014 Kinetic evaluation of simultaneous COD, ammonia and manganese removal from drinking water using a biological aerated filter system *Sep. Purif. Technol.* **130** 56–64

[64] Jin Y ., Wang D . and Zhang W 2015 Treatment of high-strength ammonia containing wastewater using partial nitritation system with biological selector *Glob. Nest J.* **17** 738–43

[65] Zhang X, Liang Y, Ma Y, Du J, Pang L and Zhang H 2016 Ammonia removal and microbial characteristics of partial nitrification in biofilm and activated sludge treating low strength sewage at low temperature *Ecol. Eng.* **93** 104–11

[66] Cheng Q, Nengzi L, Bao L, Huang Y, Liu S, Cheng X, Li B and Zhang J 2017 Distribution and

genetic diversity of microbial populations in the pilot-scale biofilter for simultaneous removal of ammonia, iron and manganese from real groundwater *Chemosphere* **182** 450–7

[67] Li H and Tao W 2017 Efficient ammonia removal in recirculating vertical flow constructed wetlands: Complementary roles of anammox and denitrification in simultaneous nitritation, anammox and denitrification process *Chem. Eng. J.* **317** 972–9

[68] Liu H, Zhu L, Tian X and Yin Y 2017 Seasonal variation of bacterial community in biological aerated filter for ammonia removal in drinking water treatment *Water Res.* **123** 668–77

[69] dos Santos P R and Daniel L A 2020 A review: organic matter and ammonia removal by biological activated carbon filtration for water and wastewater treatment *Int. J. Environ. Sci. Technol.* **17** 591–606

[70] Hamidi H P, Kenari S L D and Basu O D 2020 Simultaneous TOC and Ammonia Removal in Drinking-Water Biofilters: Influence of pH and Alkalinity *J. Environ. Eng.* **146** 040200801–7

[71] Wu Y J, Liu Y W, Cheng H H, Ke C W, Lin T F and Whang L M 2021 Biological pre-treatment system for ammonia removal from slightly contaminated river used as a drinking water source *Process Saf. Environ. Prot.* **147** 385–91

[72] Sumariyah S, Hanafi I, Muchisin Z, Arianto F and Nur M 2018 Comparison between corona and dielectric barrier discharges plasma using of pin to single and dual ring electrodes configuration *IOP Conf. Ser. Mater. Sci. Eng.* **434**

[73] Van Nguyen D, Ho P Q, Van Pham T, Van Nguyen T and Kim L 2019 Treatment of surface water using cold plasma for domestic water supply *Environ. Eng. Res.* **24** 412–7

[74] Fan J, Wu H, Liu R, Meng L, Fang Z, Liu F and Xu Y 2021 Non-thermal plasma combined with zeolites to remove ammonia nitrogen from wastewater *J. Hazard. Mater.* **401** 123627

[75] Li M, Feng C, Zhang Z, Liu X, Ma W, Xue Q and Sugiura N 2011 Optimization of electrochemical ammonia removal using Box-Behnken design *J. Electroanal. Chem.* **657** 66–73

[76] Eckelberry, Nicholas;Sanchez, Jose;Davis A 2013 Removing Ammonia from Water

[77] Shu J, Liu R, Liu Z, Du J and Tao C 2016 Manganese recovery and ammonia nitrogen removal from simulation wastewater by pulse electrolysis *Sep. Purif. Technol.* **168** 107–13

[78] Aoudj S ., Khelifa A . and Drouiche N 2017 Removal of fluoride, SDS, ammonia and turbidity from semiconductor wastewater by combined electrocoagulation–electroflotation *Chemosphere* **180** 379–87

[79] Rabah F K J and Darwish M S 2012 *Characterization of Ammonia Removal from Municipal Wastewater Using Microwave Energy: Batch Experiment* (Islamic University)

[80] Dong S 2015 *Application of Sequential Microwave / Aeration Process for the Removal of Ammonia from Landfill Leachate* (Ottawa)

[81] Liu J, Liu B, Ni Z, Deng Y, Zhong C and Hu W 2014 Improved catalytic performance of Pt/TiO 2 nanotubes electrode for ammonia oxidation under UV-light illumination *Electrochim. Acta* **150** 146–50

[82] Sun D, Sun W, Yang W, Li Q and Shang J K 2015 Efficient photocatalytic removal of aqueous NH4+-NH3 by palladium-modified nitrogen-doped titanium oxide nanoparticles under visible light illumination, even in weak alkaline solutions *Chem. Eng. J.* **264** 728–34

[83] Zhang X, Zhou Y, Yu B, Zhang N, Wang L, Fu H and Zhang J 2017 Effect of copper oxide nanoparticles on the ammonia removal and microbial community of partial nitrification process *Chem. Eng. J.* **328** 152–8

[84] Khuntia S, Majumder S K and Ghosh P 2013 Removal of ammonia from water by ozone microbubbles *Ind. Eng. Chem. Res.* **52** 318–26

[85] Wang Z, Qu Z, Quan X, Li Z, Wang H and Fan R 2013 Selective catalytic oxidation of ammonia to nitrogen over CuO-CeO2 mixed oxides prepared by surfactant-templated method *Appl. Catal. B Environ.* **134–135** 153–66

[86] Ichikawa S I, Mahardiani L and Kamiya Y 2014 Catalytic oxidation of ammonium ion in water with ozone over metal oxide catalysts *Catal. Today* **232** 192–7

[87] Luo X, Yan Q, Wang C, Luo C, Zhou N and Jian C 2015 Treatment of ammonia nitrogen wastewater in low concentration by two-stage ozonization *Int. J. Environ. Res. Public Health* **12** 11975–87

[88] Jabłońska M and Palkovits R 2015 Copper based catalysts for the selective ammonia oxidation into nitrogen and water vapour-Recent trends and open challenges *Appl. Catal. B Environ.* **181** 332–51

[89] Jabłońska M, Beale A M, Nocuń M and Palkovits R 2018 Ag-Cu based catalysts for the selective ammonia oxidation into nitrogen and water vapour *Appl. Catal. B Environ.* **232** 275–87

[90] Guo Y, Huang T, Wen G and Cao X 2017 The simultaneous removal of ammonium and manganese from groundwater by iron-manganese co-oxide filter film: The role of chemical catalytic oxidation for ammonium removal *Chem. Eng. J.* **308** 322–9

[91] Ou H H, Hoffmann M R, Liao C H, Hong J H and Lo S L 2010 Photocatalytic oxidation of aqueous ammonia over platinized microwave-assisted titanate nanotubes *Appl. Catal. B Environ.* **99** 74–80

[92] Obata K, Kishishita K, Okemoto A, Taniya K, Ichihashi Y and Nishiyama S 2014 Photocatalytic decomposition of NH3 over TiO2 catalysts doped with Fe *Appl. Catal. B Environ.* **160–161** 200–3

[93] Mohammadi Z, Sharifnia S and Shavisi Y 2016 Photocatalytic degradation of aqueous ammonia by using TiO2[sbnd]ZnO/LECA hybrid photocatalyst *Mater. Chem. Phys.* **184** 110–7

[94] Zou C Y, Liu S Q, Shen Z, Zhang Y, Jiang N S and Ji W C 2017 Efficient removal of ammonia with a novel graphene-supported BiFeO3 as a reusable photocatalyst under visible light *Cuihua Xuebao/Chinese J. Catal.* **38** 20–8

[95] Hao X D, Wang C C, Lan L and Van Loosdrecht M C M 2008 Struvite formation, analytical methods and effects of pH and Ca 2+ *Water Sci. Technol.* **58** 1687–92

[96] Huang H, Zhang P, Xiao J, Xiao D and Gao F 2017 Repeatedly using the decomposition product of struvite by ultrasound stripping to remove ammonia nitrogen from landfill leachate *Ultrason. Sonochem.* **38** 622–8

[97] Peng C, Chai L, Tang C, Min X, Song Y, Duan C and Yu C 2017 Study on the mechanism of copper–ammonia complex decomposition in struvite formation process and enhanced ammonia and copper removal *J. Environ. Sci. (China)* **51** 222–33

[98] Bianchi L, Kirwan K, Alibardi L, Pidou M and Coles S R 2020 Recovery of ammonia from wastewater through chemical precipitation: Investigating the kinetic mechanism and reactions pathway of struvite decomposition *J. Therm. Anal. Calorim.* **142** 1303–14

[99] Wang S, Wu X, Wang Y, Li Q and Tao M 2008 Removal of organic matter and ammonia nitrogen from landfill leachate by ultrasound *Ultrason. Sonochem.* **15** 933–7

[100] Ozturk E and Bal N 2015 Evaluation of ammonia-nitrogen removal efficiency from aqueous solutions by ultrasonic irradiation in short sonication periods *Ultrason. Sonochem.* **26** 422–7

[101] Zhang X, Li W, Blatchley E R, Wang X and Ren P 2015 UV/chlorine process for ammonia removal and disinfection by-product reduction: Comparison with chlorination *Water Res.* **68** 804–11

Improving the Performance of Biological Tanks in Wastewater Treatment Plants by Modifying the Conventional Ceramic Aerator

Ahmed Sobhy Khalil [1]*, Swellam Sharshir [2]

[1] Department of civil engineering, Faculty of Engineering, Kafrelsheikh University.
[2] Department of Mechanical Engineering, Faculty of Engineering, Kafrelsheikh University.

Abstract. The purpose of the article is to study the process of aeration of wastewater using an aerator with an air-lift effect to ensure a uniform distribution of the concentration of suspended solids in the aeration tank of the model installation. Based on the result of the work carried out, experimental data were obtained and a theoretical description of the parameters of the aeration process, which increase the coefficients of mass transfer and the use of the reactor volume, and the costs of the process. The airlift effect increases the concentration of sludge. The airlift also helps in the spread of suspended solid in the tank, as mass transfer reactions during aeration increase by an average of 11.8%. Also, an increase in dissolved oxygen concentration from 3.02 mg / L to 3.5 mg / L, i.e., by 15%, which indicates an improvement in mass transfer and, accordingly, a decrease in energy consumption for aeration. However, the existing technological potential requires the introduction of variable equipment. An example of an alternative air dispersion system can be aerators with airlift effect.

Keywords: Aerator; Biological tanks; Wastewater; Treatment; Airlift

1. Introduction

Improving the general effect of industrial and domestic wastewater purification is a topic of increasing interest from the scientific community [1-3]. To achieve optimal indicators of maximum permissible concentrations of fishery water bodies, new promising technologies for wastewater treatment are being developed [4]. One of the innovative but insufficiently deeply studied water treatment methods is the technology of using aerators with airlift effect [5]. It is known that aeration of wastewater is the process of oxygenation of the stream of treated water to create aerobic conditions during biological or chemical treatment. This process aims to oxidize organic and mineral contaminants and remove volatile substances contained in wastewater [6]. Recently, the most widespread aeration has been carried out by pneumatic, mechanical, or combined equipment. The air supplied to the liquid to be purified presented in bubbles moving through the water layer and transferring oxygen. The size and number of bubbles determine the efficiency of aeration and energy costs. A pneumatic aeration system is a supply of air (technical oxygen) under a pressure of 10-50 kPa through main and air distribution pipelines to dispersers of various designs. The aeration tank's location and number are determined based on the calculation results.

The pneumatic aeration system is characterized by the fact that, in addition to providing the activated sludge with oxygen, it also performs the function of mixing the entire contents of the aeration tank. An

air bubble, as it floats through the liquid layer, on the one hand, transfers oxygen to the liquid by diffusion, and on the other, causes the fluid to move, causing mechanical stirring.
The mechanical aeration system is carried out when the mixer-aerator rotates, drawing air from the environment into the reactor. Depending on the location of the rotor axis of rotation, it is divided into two types: horizontal and vertical. In contrast to a pneumatic aeration system, a mechanical one is characterized by several advantages such as the absence of clogging and overgrowing of pores, the possibility of timely inspection and repair.
The combined aeration system combines the technological capabilities of pneumatic and mechanical aerators. The main purpose of the system is wastewater treatment of high concentration. Thus, the ubiquitous distribution of aeration equipment for wastewater treatment determines the further search for technical developments to improve the course of water purification processes in the formation of an optimal gas hydrodynamic environment [7].
The research aims to study the possibility of improving the functionality of biological tanks by increasing air and oxygen in the tank and ensuring air distribution over the entire tank sector. In this research, the improvement of air distribution in the traditional identities made of ceramic and porcelain and knowing the effect of the improvement on the efficiency of air distribution in the laboratory.

2. Materials and methods

The analysis of the operation of the aerators was carried out on a Pilot aeration tank with a volume of 0.3 m3. The aeration tank was filled with activated sludge withdrawn from the wastewater treatment plant as illustrated in Fig. 1. The bottom of the aeration tank is provided with a conventional ceramic aerator produced by the company "Bakor".

Figure 1. Model installation with ceramic aerator "Bakor".

The experiment was carried out in 2 stages: I. the study of flows of sludge mixture in aeration tanks on activated sludge of existing treatment facilities. II. Study of flows with suspensions when using aerators as a mixer and / or an oxidizer - with an artificial opacifier of tap water with loam.Stage I. Determination of the uniformity of the distribution of activated sludge in the aeration tank volume by the volume of settled suspensions in 20 - 30 minutes of thawing in samples taken at six points of the model.
Parameters of the test mode for the aerator of the original design Fig. 2-(a): air flow Q = 50 l / min, oxygen content in water O2 = 3.02 mg / l, pressure P = 0.165 MPa. According to visual observations, this aeration mode was characterized by incomplete mixing, the air "torch" of the aerator had a small cross section, the activated sludge was not mixed throughout the entire volume of the model installation, the circulation flows did not reach the bottom, where There were zones of stagnation of sludge Fig. 2-(b).

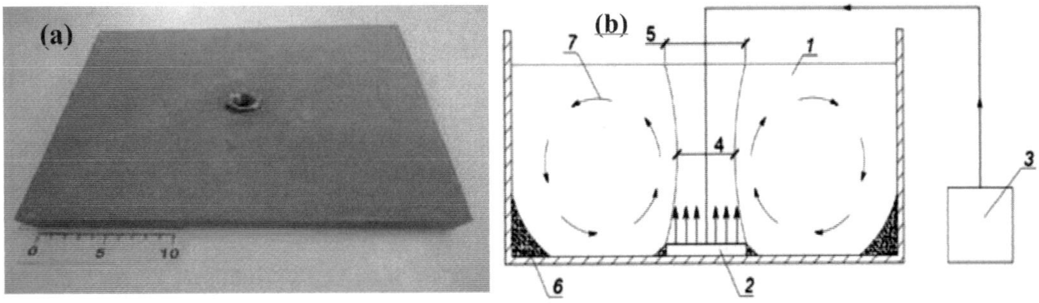

Figure 2. (a) Ceramic aerator of Bakor, and (b) Model of aeration tank
1 - reservoir; 2 - ceramic aerator of Bakor; 3 - compressor, 4 - aeration torch waist, 5 - aeration torch shoulders; 6 - zones of bottom sediments; 7 - direction of circulating flows.

Parameters of test mode for modified aerator with air-lift effect Fig. 3-(a): air flow rate Q = 50 l / min, oxygen content in water O2 = 3.02 mg / l, pressure P = 0.165 MPa. This type of aeration was accompanied by intense mixing of the contents of the model unit, the presence of vortices of air flows in the entire volume of the liquid was observed, and there were no areas and zones of stagnation of activated sludge at the bottom Fig. 3-(b).

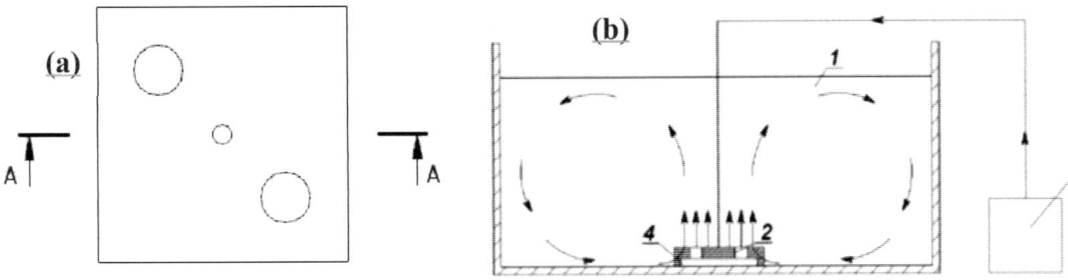

Figure 3. (a) Modified Bakor aerator with airlift effect, and (b) model of aeration tank
1 - tank; 2 - modified aerator of "Bakor" with airlift chambers; 3 - compressor; 4 - stands.

The defining criterion for the comparative assessment of the efficiency of the modified aerator was the sedimentation of suspended particles of wastewater. A bathometer was used to take water samples (three times repetitively to eliminate random errors) at six points in the volume of the model setup as illustrated in Fig. 4. After settling, the height of the sludge layer in cylinders with a volume of 100 ml was determined.

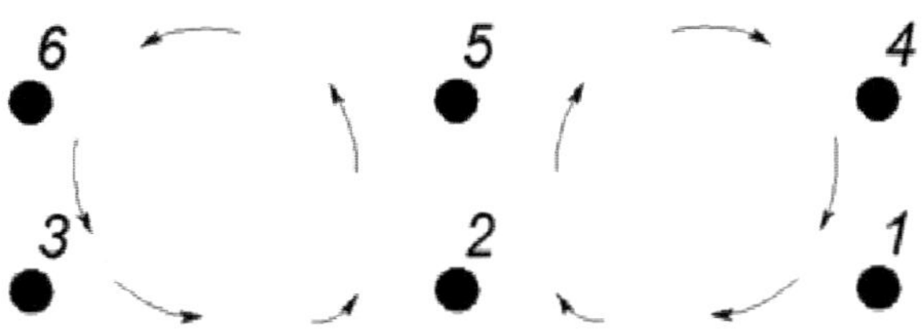

Figure 4. Water sampling points in the study of aerators: 1, 2, 3 at the bottom of the tank and 4, 5, 6 at the middle of the tank.

When using porous materials, the specific air consumption [4] per unit of the working surface of the aerators Ja, d depends on their individual properties and is assigned within the range = 30-100 m3 / (m2.h); for filter plates = 60-80 m3 / (m2.h).

2.1 Aerator of standard design:

Mode parameters: air consumption 48 and 96 l / min, air pressure P = 0.155 kPa. Technological indicators: fine-bubble aeration (intensity 35, 17 and 70.35 m3 / (m2.h)) does not provide proper roiling of particles at the bottom, and in remote areas of the model unit, there are zones of bottom sediments. Also, according to the indications of a portable oxygen meter (measurements were carried out directly during the aeration process), uniform oxygen distribution is not achieved in the entire volume of the aeration tank.

2.2 The aerator of modified design with airlift effect:

Parameters - air flow rate 48 and 96 l / min, air pressure P = 0.155 kPa. Technological indicators: intensity 35, 17 and 70.35 m3 / (m2.h), pronounced water-air eddies are formed in the volume, and at the bottom of the model, suspended substances in the imitation are distributed almost evenly throughout the entire volume, as noted earlier [5, 6]. The concentration of dissolved oxygen in the liquid at all points of the aeration tank volume is about 3.5 mg / l (+/- 0.05).

The results of the water samples taken with the analytical determination of the parameters of suspended solids demonstrate that the use of an aerator with an airlift effect makes it possible to achieve a more uniform distribution of concentrations over the volume.

3. Results and its discussion

The analysis of the results obtained shows that at equal air flow rates (48 and 96 l / min) an aerator with an airlift effect increases the concentration of sludge and suspended matter as illustrated in Figs. 3, 4 in the water of a modular installation, which indicates the involvement of bottom sediments in circulation flows. So, for the upper third of the tank, the concentration of sludge with aerator airlift is 184.08 mg / l, and with the typical one - 164.18 mg / l (+ 12%); the middle of the tank is characterized by values of 194.92 mg / l and 168.94 mg / l, respectively (+ 15%); the lower third of the tank is 184.08 mg / l and 169.14 mg / l (+ 8%), i.e. the participation of suspended solids in mass transfer reactions during aeration increases by an average of 11.8%. There is also an increase in the concentration of dissolved oxygen from 3.02 mg / L to 3.5 mg / L, i.e. by 15%, which indicates an improvement in mass transfer and, accordingly, a decrease in energy consumption for aeration

4. Conclusions

Based on the experimental data obtained, it can be concluded that the modified aerator is advantageous. An airlift effect over a typical Bakor aerator increases the concentration of suspended matter in the reactor volume due to the inclusion of bottom sediments in the process and in terms of dissolved oxygen. This, in turn, increases the coefficients of mass transfer, the use of the volume of the reactor, and a decrease in the specific energy costs for the aeration process. Therefore, a modified mode of pneumatic aeration through porous ceramic plates, with built-in airlifts, can be recommended for biological wastewater treatment plants and other aeration processes.

5. References

[1] K. Elmaadawy, A.W. Kandeal, A. Khalil, M.R. Elkadeem, B. Liu, S.W. Sharshir, Performance improvement of double slope solar still via combinations of low cost materials integrated with glass cooling, Desalination, 500 (2021) 114856.

[2] A.W. Kandeal, N.M. El-Shafai, M.R. Abdo, A.K. Thakur, I.M. El-Mehasseb, I. Maher, M. Rashad, A.E. Kabeel, N. Yang, S.W. Sharshir, Improved thermo-economic performance of solar desalination via copper chips, nanofluid, and nano-based phase change material, Solar Energy, 224 (2021) 1313-1325.

[3] G. Peng, S.W. Sharshir, Y. Wang, M. An, D. Ma, J. Zang, A.E. Kabeel, N. Yang, Potential and challenges of improving solar still by micro/nano-particles and porous materials - A review, Journal of Cleaner Production, 311 (2021) 127432.

[4] O.L.d.C. Maciel, I.Q. Willmer, T.D. Saint'Pierre, W. Machado, S. Siciliano, R.A. Hauser-Davis, Arsenic contamination in widely consumed Caribbean sharpnose sharks in southeastern Brazil:

Baseline data and concerns regarding fisheries resources, Marine Pollution Bulletin, 172 (2021) 112905.

[5] N. Alhaboubi, CO2 sequestration using a novel Belt Conveyor Reactor with rotating sieve trays compared with Airlift Bubble Column as photobioreactors, Journal of King Saud University - Engineering Sciences, (2021).

[6] L.-D. Shi, Z. Wang, T. Liu, M. Wu, C.-Y. Lai, B.E. Rittmann, J. Guo, H.-P. Zhao, Making good use of methane to remove oxidized contaminants from wastewater, Water Research, 197 (2021) 117082.

[7] T. Hu, L. Tang, H. Feng, J. Zhang, X. Li, Y. Zuo, Z. Lu, W. Tang, Metal-organic frameworks (MOFs) and their derivatives as emerging catalysts for electro-Fenton process in water purification, Coordination Chemistry Reviews, 451 (2022) 214277.

Experimental and process design of a treatment system to remove Aluminium ions from lean sludge in an Egyptian water treatment plant

Reham Samy[1], Amr Abdelghany[1], M. Hanafy[1], N. Abdel Monem[1], Mohamed M. El Bouraie[2]

[1]Cairo University, Faculty of Engineering, Chemical Eng. Department.
[2]Central Laboratory for Environmental Quality Monitoring (CLEQM), National Water Research Center (NWRC), El Qanater, Egypt.

Abstract. Aluminium ions traces in drinking water were found in El Khanka Water Treatment plant in Egypt because of the lean sludge generated from Alum's use in water treatment. This research studies using natural clay such as "bentonite" known as montmorillonite clay and combining bentonite – cationic polymer as coagulants to remove aluminium salts from lean sludge. Experimental studies were performed to determine the effect of adding bentonite on the removal of aluminium ions from lean sludge based on variation of pH. 98.8% removal of Al ions was achieved at pH 11, and 0.6 ppm concentration of bentonite in the sludge solution. Kinetic model of pseudo-second-order was found to be the best to describe the reaction and the rate constant of 11.922 $mol.^{-1}\ s^{-1}$ was calculated. An industrial process was suggested to the removal of Aluminium ions and brief economic estimate of applying it in Egypt was prepared as well.

Keywords: Lean Sludge, Alum, Settling, Bentonite, Cationic Polymer.

1. Introduction

Aluminum ion in water was one of the factors contributing to dialysis, dementia, and Alzheimer's disease by Shammas[1].In Egypt, alum $Al_2(SO_4)_3.12H_2O$ (aluminum salt compound) is continuously used in the coagulation-flocculation stage as a coagulant producing large amounts of lean sludge. Produced sludge containing a high concentration of aluminum salts is dumped by most of the water treatment plants in Egypt into the nearest waterways without further treatment and left to settle at the bottom of the water source. Aluminum ions are produced in waterways causing human diseases such as Alzheimer's and dementia, according to Rondeau et al. and Zubaidy et al.[2,3]

Because of the harmful effects of aluminum, many researchers investigated its removal. Many methods can remove aluminum, including reverse osmosis, cation exchange resin, and electro-dialysis. While coagulation, sedimentation, filtration, lime softening, and combined processes are moderately effective in Al^{+3} removal. Aeration, stripping chemical oxidation/disinfection, and anion exchange treatment methods were found to be ineffective for the removal of $Al^{+3,}$ according to Srinivasan et al. [4]

Othman et al.[5]used two chelating resins to remove aluminum from water. They investigated different concentrations, resin amounts, pH, and stirring time. They determined the optimum conditions from a batch system. Optimum results were concluded to vary according to pH values. They agreed with

Veres and Orolinova[6] that the efficient treatment of waste contaminated with metals should include chemical precipitation, ion exchange, membrane separations, and adsorption.

Wang et al. [7] suggested the addition of lime to the system to reach the required pH for coagulation and precipitation. The pH is reduced by the precipitation of aluminum and iron salts in their hydroxide forms.

Srinivasan et al. mentioned that Aluminium ion (Al^{+3}) was insoluble near pH values of 6.0 to 8.5.[4]. However, experimental analysis of water samples from the river Nile showed aluminum ions, which means that aluminum ions were soluble at pH 6.0 - 8.5.

On the other hand, bentonite clay was used as a traditional healing method to protect the body from disease. According to El Bouraie et al. [8] and Zhou et al. [9], Bentonite was used as an adsorbent for many toxic substances.

The bentonite contained two tetrahedral coordinated sheets of silicon ions between them octahedrally coordinated sheets of Aluminum ions. Exchanging silicon in Si-O tetrahedron and Aluminum in Al-O octahedron by cations produced excessive negative charge in the lattice that can be balanced by cations (Na^{+}, Mg^{2+}, Ca^{2+}) in the interlayer space. The exchange was unstable due to the loose binding and tended to exchange with other ions. The variety in exchangeable cations can affect the uptake of water and swelling. Swelling and water uptake were highest with sodium and most minor with potassium and magnesium. The most commonly used industrial bentonite was either sodium or calcium-based salts.

Sodium bentonite capacity for swelling was found to be high, and it formed gel-like masses with water. Al-Ani et al., Clem, et al., and Syafalni et al. explained this phenomenon by mentioning the origin of sodium bentonite as derived from volcanic ash deposited in marine environments. Calcium bentonite, was found to be a low swelling type evolved from volcanic ash deposited in freshwater environments. It could be improved after treatment with soda ash to form sodium-exchanged bentonite. However, these sodium-exchanged bentonites did not have a high swelling capacity compared to the natural sodium bentonites. [8][9–11].

Most industrial applications made use of the swelling property of bentonite to form water suspensions. These mixtures were used as suspending, plasticizing, and bonding agents. Bentonite breaks down into colloidal particles providing a large surface area per unit weight of clay. Thus worked as a stabilizing emulsion or as a medium that captures other chemicals as used by Zhou et al.[12]. Moreover, it is characterized by a large specific area, good adsorption, ion exchangeability, and non-toxicity. Thus, used in water treatment as a coagulant.

Modified bentonite as an adsorbent for anionic dyes by Deoclecio et al. [13]. Hua [14] used modified bentonite with manganese oxides and poly(di-methyl-diallyl ammonium chloride)in the adsorption of toxic material (Arsenic) from water. Carvalho et al., tested natural bentonite and two types of activated carbon to remove the harmful dyes such as orange 39 from water, and concluded that bentonite could be an alternative because of its low cost. [15].

This article is a comprehensive experimental study on the use of bentonite to remove Aluminium ions from lean alum sludge. Samples of lean sludge were collected from the "El Khanka" Water Treatment plant in Egypt. Cationic polyacrylamide was used as a coagulant with bentonite to test its combination effect as the cationic polymer did not change the pH of the water and was found to be more comfortable in handling as per Polasek and Mutl [16].In this study, the focus was given to using cheap natural adsorbents to remove Aluminium ions from lean sludge.

2. Experimental Work

The main objectives of the conducted experimental studies were to:

(i) Enhance settling of Aluminium from lean sludge resulting from the sedimentation tanks in the water treatment plant, where alum is used as a coagulant.

(ii) Dewater sludge (i.e., Increase its solids content), for easier handling and disposal.

The lean sludge and bentonite samples were characterized and analyzed using the XRD technique to achieve the previous objectives. In addition, lean sludge samples were characterized with inductively coupled plasma – optical emission spectrum (ICP-EOS) as well. Also, experimental jar tests were used to study parameters as bentonite dosage amounts, pH, and the mass of the added polymer. Finally, the treated samples were tested to measure the Aluminium concentration in water.

2.1. Samples of Lean Sludge

The lean sludge samples used in the experiments were collected from a water treatment plant in Abu Zaabal, El Khanka plant, Egypt. Samples used in this study were collected from the primary settling tank. Sample 1 was collected ten months before sample 2 from the exact location after the coagulation tank. In this plant, alum was used as a coagulant to treat water taken from the Ismailia canal. After treating with alum, filtration and chlorination were performed. Finally, the water was introduced for drinking, and the sludge was disposed of in the channel of the river Nile. The characteristics of Alum lean sludge are shown in Table 1.

Table 1. Characteristics of alum sludge from El Khanka plant, Abu Zaabal, Egypt

Parameter	Value	
	Sample 1	Sample 2
Time of collection	November	August
Al^{+3}filtrate, mg/L	0.90	0.83
Al^{+3}solids, mg/kg	7.00	8.85
TDS, ppm	557	530
pH	7.2	7.3

3. Materials

Chemicals used in this study were Sodium hydroxide, a cationic polymer, anionic polymer, and Hydrochloric acid (37%). A stock solution of Hydrochloric acid (1M) and sodium hydroxide (5%) were prepared using 1 liter of De-ionized water (Conductivity = 0.5 µS/m). Sodium Bentonite is the main adsorbent used in the jar test experiments. Properties and primary constituents of bentonite were determined using X-ray Diffraction (XRD) technique.

Bentonite was used as a coagulant for the experiments. All bentonite samples were dried at 105°C for 4 hours in a drying air oven, then crushed by a porcelain mill and screened to a diameter of 74 µm for the experiments. Bentonite chemical composition was analyzed in Kanater laboratory, and results are shown in Table 2.

Table 2. Chemical analysis of bentonite

Main constituents	Concentration (wt. %)
Na_2O	0.34
MgO	1.69
Al_2O_3	10.00
SiO_2	53.49
P_2O_5	0.02
SO_3	3.91
K_2O	0.46
CaO	2.42
TiO_2	0.60
MnO	0.01
Fe_2O_3	3.19
ZnO	0.01
SrO	0.13
ZrO_2	0.10
Nb_2O_5	0.02
F	0.06

Cl	0.03
L.O.I(Loss on ignition)	23.40

4. Methods

The jar test consisted of six beakers, 500 ml each, equipped with a mechanical stirrer; two pipettes were used, one for acid/alkali and the other for bentonite addition. The stirrers operated on the same motor to give the same rotation velocity for the six beakers.

Samples of lean sludge from the plant were investigated after adding different amounts and concentrations of bentonite. The pH was then adjusted to the desired values using HCl (1 M) or NaOH (5 %) and rapidly mixed at 150 rpm for one minute, then followed by slow mixing at 15 rpm for 1 minute.

The samples were left to settle for 20 minutes until all the particulates had fully settled. The treated samples were poured into the Büchner funnel for vacuum filtration to measure the moisture content.

The sludge samples were treated using a cationic polymer named polyacrylamide as a coagulant with bentonite. It was added as a solid (in grams) and subjected to rapid mixing (150 rpm) for 30 seconds, followed by slow mixing (15 rpm) for 30 seconds to enhance flocculation.

Standard bentonite solutions were prepared by dissolving different portions of bentonite in 1 liter of de-ionized water to form 10, 20, 30, 70, and 90 mg/L bentonite solutions. Then different volumes were taken from the solutions.

In the laboratory, other parameters of water were measured. Heraeus drier was used for the drying of the filtered cake. The total Aluminum content infiltrates of lean sludge before and after treatment were measured using the Inductively Coupled Plasma - Optical Emission Spectrum (ICP-OES) with Ultra Sonic Nebulizer (USN). The Nebulizer decreased the instrumental detection limits by 10%, and this ICP instrument is Perkin Elmer Optima 3000, USA origin.

Sludge samples were digested using microwave digestion technique, at which 0.1 gm of sludge sample is placed in Teflon vessel with 6 ml HNO_3 (65 %), and 2 ml HCl (95 %) to determine the heavy metal contents of sludge using Microwave digestion system, model MILESTONE MLS-1200 mega with microwave digestion rotor (MDR) technology.

5. Results and Discussion

Each bentonite sample was analyzed, and the composition of bentonite was investigated using X-ray diffraction analysis (XRD) presented in figure (1).

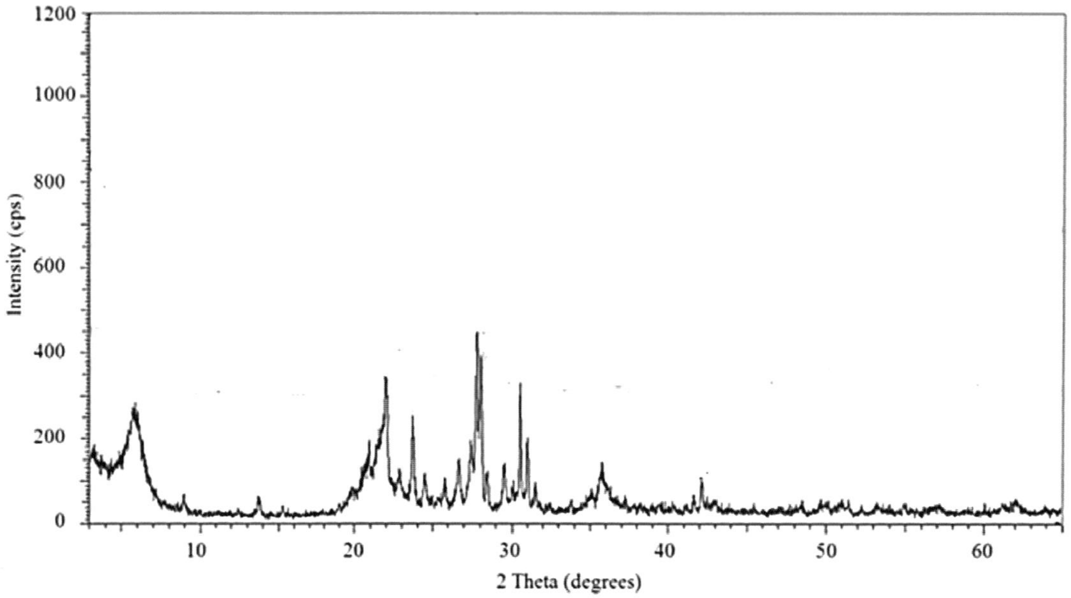

Figure 1. X-Ray pattern for bentonite

5.1. Effect of pH

Based on the performed jar tests, the optimum pH was determined in terms of Aluminum reduction in lean sludge filtrate with various pH values between 3 and 11 tested versus concentrations of bentonite from 0.6 to 8.8 mg/l, summarized in the following figure (2).
The highest removal efficiency of Al^{+3} at a pH of 11 was found to be 98.79%. due to the amphoteric nature of aluminum, it is soluble in acidic solutions, and its hydroxide form is insoluble at high pH

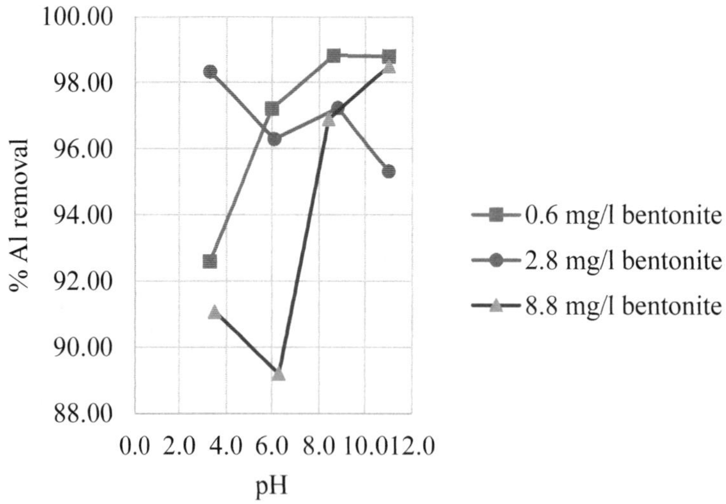

Figure 2. Effect of pH and bentonite concentration on lean sludge filtrate

5.2. *Effect of bentonite dosage*

Figure (3) showed that when the concentration of bentonite at pH 11 was increased, the Al^{+3} removal efficiency decreased then increased gradually in the lean sludge filtrate.

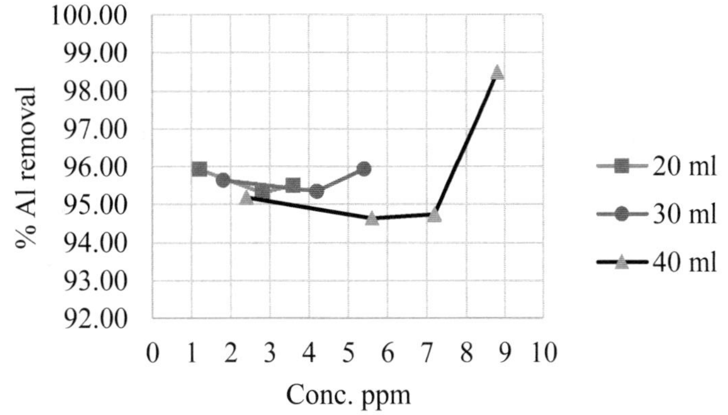

Figure 3. Effect of bentonite dosage on lean sludge filtrate

While for settled solids, dewatering that reaches 26% increases the bentonite dosage to 40 ml at pH 11

5.3. *Effect of polymer addition*

Cationic polyacrylamide polymers were recommended in the acidic media as aqueous suspensions at a pH of 3 -4 and above with a negative charge to settle solids quickly. Figure (4) showed the effect of adding masses of cationic polymer on a 3.6 mg/l bentonite at pH 3. The investigated polymer to bentonite mass ratios (P: B) was varied from (1.33:1) to (5.44:1). The concentration of Aluminium decreased gradually in the filtrate of lean sludge when the added mass of polymer increased. When inorganic and organic coagulants were used together, the coagulation was enhanced, but if there is an overdose with a cationic polymer concerning bentonite mass added, a phenomenon is known as re-stabilization may occur.

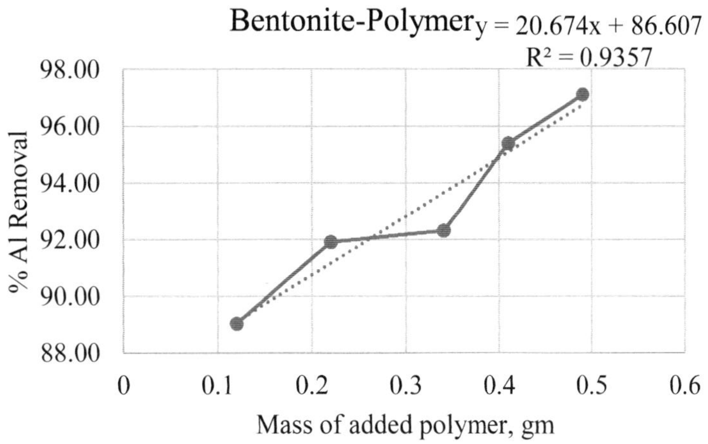

Figure 4. Effect of the mass of polymer on the lean sludge filtrate

From the above figure, the optimum added mass of cationic polymer was 0.49 gm to reach Aluminum removal of 97.1% in the lean sludge filtrate, and the solids content was increased to 22 % in a 3.6 ppm of bentonite solution and pH 3.

5.4. Effect of contact time

Time versus removal percentage is shown in figure (5). Bentonite concentration of 0.6 ppm at pH 11 gave the highest initial settling rate while aluminum removal percent remained constant after 20 min. When using a combination of bentonite and cationic polymer, the settling rate became constant after 12 min.

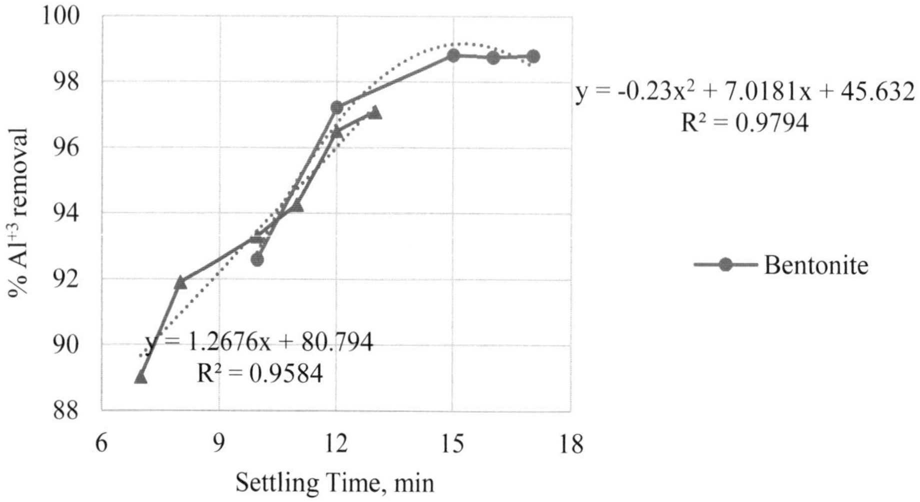

Figure 5. Effect of settling time of bentonite and polymer

5.5. Kinetics Model

The kinetics of Aluminium ions attraction on the interlayer of bentonite were investigated using pseudo-first-order and pseudo-second-order reactions. The pseudo-second-order equation did not require the specification of the adequate adsorption capacity or the rate constant of the second-order reaction. The initial rate could be determined through the equation without knowing or defining any parameter. Kinetics deal with the rate of adsorption, where equations are as follows,

For the first order,

$$\frac{d[C]}{[C]} = -k_1 dt \qquad (1)$$

For the second order,

$$\frac{d[C]}{dt} = -k_2 [C]^2 \qquad (2)$$

After linearization:

For the first order,

$$\ln C_f = \ln C_o - k_1 t \qquad (3)$$

And the second-order,

$$\frac{1}{C_f} = \frac{1}{C_o} + k_2 t \qquad (4)$$

, Where $[C_o]$: Initial concentration of reactant C

K_1: rate constant of the first order, K_2: rate constant of second-order

t: time, sec

Experiments were performed to identify the reaction order and mechanism. Then, first and second-order reaction kinetics were applied to the experiments, as shown in figures (6, 7).

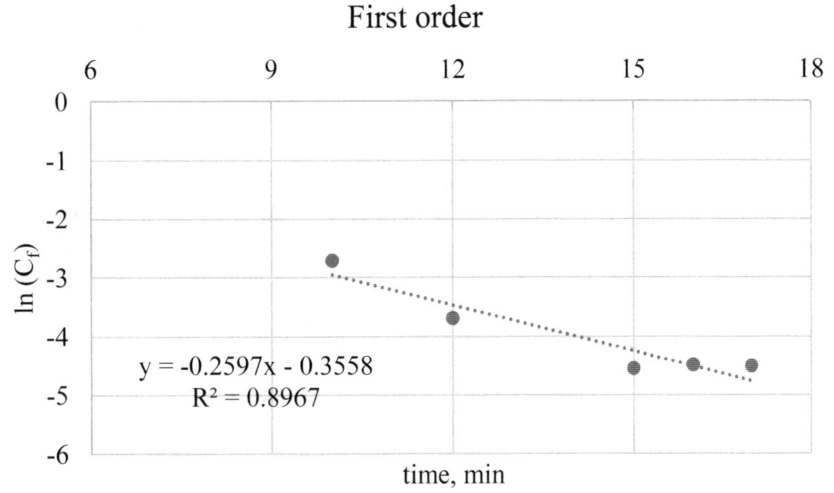

Figure 6. First order kinetics

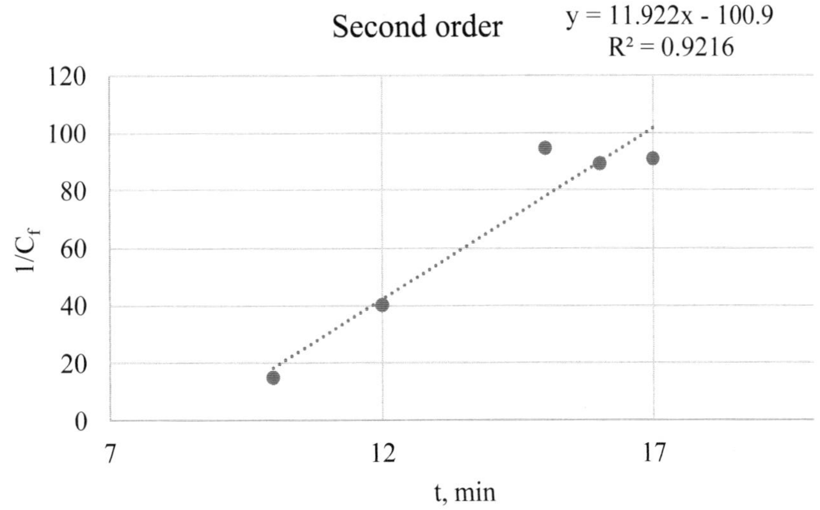

Figure 7. Second order kinetics

It can be noticed that the second-order is the actual kinetics as it has a higher determination coefficient (R^2), and it explains the attraction of Al^{+3} to the negative interlayer of bentonite as it becomes balanced by sodium, cations attached to the surface of flakes. Accurate matching of experimental results with the second-order kinetics suggests that the limiting step was chemisorption. Bentonite(Montmorillonite) adsorbed water resulting in moving flakes apart in a swelling mechanism. The following table 3 shows the values of constant kinetic parameters for both first and second-order reactions.

Table 3. Values of Kinetic constant parameters

First-order parameters			Second-order parameters		
C_o	K_1	R^2	C_o	K_2	R^2
0.7	0.2597	0.8967	0.01	11.922	0.9216

Based on the previous experimental results, a process for treating lean sludge was suggested as follows.

6. Process Description

A process was proposed to treat lean sludge in Figure 8, where the lean sludge passed through a screen to remove any solids. Sludge was moved to an equalization tank afterward for homogenization.

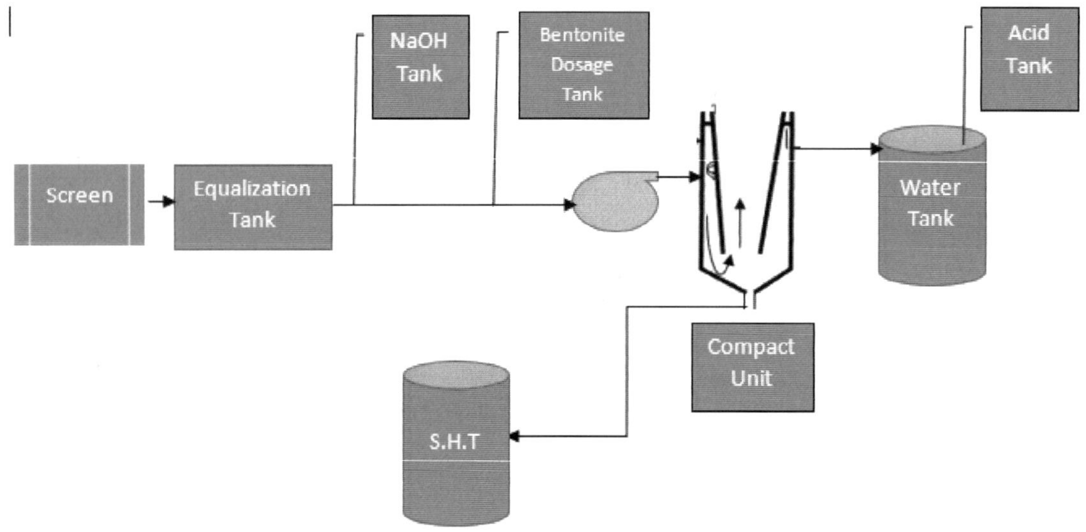

Figure 8. A proposed flow diagram for the lean sludge treatment unit to remove Aluminium ions

Sodium hydroxide was injected with the lean sludge to adjust pH; then, bentonite was added and pressurized into the compact treatment unit by a feed pump. The sludge was left to settle into the central cone at the compact unit. Next, the sludge was directed to the sludge-holding tank (S.H.T). Finally, the clarified supernatant was directed to a water tank to be used for irrigation purposes after being neutralized to pH 7 using HCl acid.

The compact unit consisted of a cylindrical tank with a conical bottom and an inner vessel, which may be cylindrical or conical. The inner vessel divided the basin into two separate zones: the outer zone is the annulus between the inner cylinder and the external body of the unit. Slow mixing was performed in this zone where the flocculation takes place. The internal zone is enclosed within the inner cylinder where the sedimentation occurs. The inlet feed pipeline was located on the outer cylindrical shell, with a slope angle (Θ), where (Θ) equals 60°.

7. Economic Consideration

The proposed system's capital investment and operating cost were estimated as follows.

It was found that the estimated fixed capital investment was around (EGP 659,500) in the case of using the compact concrete unit. Based on 20 m^3/h of lean sludge and was (EGP 669,500) in case of using epoxy-coated steel. The operating cost was calculated based on 365 working days/year and found to be

(EGP 884.44) in case of using the compact concrete unit. Therefore, the estimated cost for treating 1 m^3 of sludge was (EGP 1.84).

If this cost was added to the price of the drinking water produced from the station, then a small amount equal to (EGP 0.05) shall be added to the price of water from this 17,280 m^3/day (i.e., 200 L/sec.) station for treatment of the drinking water in Egypt. The following tables (4, 5) shows the itemized operating and fixed cost for the process.

Table 4. Items list for fixed cost

Item	No.	Cost
Screen	1	10,000.00
Equalization Tank	1	30,000.00
Feed Pump, 20 m3/hr	2	30,000.00
Dosing Pumps, 5 L/hr	4	35,000.00
Dosing Tanks	2	10,000.00
Coagulant tank	1	7,500.00
Compact treatment basin, Epoxy	1	110,000.00
Compact treatment basin, Concrete	1	100,000.00
Disinfection tank	1	10,000.00
Treated water tank	1	30,000.00
Sludge holding tank	1	25,000.00
Air Blowers	2	40,000.00
Sludge Pump	1	12,000.00
Sludge dewatering unit	1	250,000.00
Control Panel	1	30,000.00
Cables	Lot.	10,000.00
Piping, Valves and fittings	Lot.	20,000.00
Supernatant Pump, 10 L/hr.	2	10,000.00
Total, EGP	**Epoxy**	**669,500.00**
	Concrete	**659,500.00**
Total, EGP/day	**Epoxy**	**1,394.79**
	Concrete	**1,373.96**

Table 5. Items list for operating cost

Item	Cost/day
Labor: one technician,two operators	170.00
Capital charges	275.14
Insurance	18.34
Electrical energy	200.00
Maintenance	91.71
Coagulant (i.e. bentonite)	35.00
polymer	60.00
Sodium Hydroxide	20.00
Hydrochloric acid	20.00
Total EGP, for concrete	**884.44**

Where, Capital charges = (0.15 x fixed cost)/365
Insurance = (0.01 x fixed cost)/365, and maintenance = (0.05 x fixed cost)/365.

8. Concluding Remarks

Based on the previous results, discussion, and proposed flow sheet, the concluding remarks can be summarized as follows:

- Bentonite can be used to remove aluminum ions, and it was found to be a convenient option for developing countries like Egypt.
- Using a bentonite concentration of 0.6 mg/l gave the optimum removal at pH 11, generally removing 98.8% of aluminum ions with an Adsorption capacity of 1.48 mg/gm
- The estimated cost for the treatment of 1 m^3 of sludge is (EGP 1.84); this is an added cost (EGP 0.05 per liter) to the price of the drinking water produced from the station.

More research can be conducted on the dried sludge to be used in an economically friendly way instead of landfilling.

Acknowledgments
The authors of this article want to thank the organic and petrochemicals laboratory and the Central Laboratory for Environmental Quality Monitoring for their support in performing experiments and the analysis.

References

[1] Shammas N K 2005 *Coagulation and Flocculation Physicochemical Treatment Processes* **3**
[2] Rondeau V, Commenges D and Jacqmin-gadda, Hélène; Dartigues J 2000 Relation between aluminum concentrations in drinking water and Alzheimer ' s disease : an 8-year follow-up study *Am. J. Epidemiol.* **152** 59–66
[3] Zubaidy E A H Al, Mohammad F S and Bassioni G 2011 Effect of pH , Salinity and Temperature on Aluminum Cookware Leaching During Food Preparation *Int. J. Electrochem. Sci.* **6** 6424–41
[4] Srinivasan, P; Viraraghavan, T; Subramanian K 1999 Aluminium in drinking water: An overview *water SA* **25** 47–56
[5] Othman, Mohamad; Abdullah, Pauzi; Abd. Aziz Y 2010 Removal of Aluminium from drinking water *Sains Malaysiana* **39** 51–5
[6] Vereš J and Orolínová Z 2009 Study of the treated and magnetically modified bentonite as possible sorbents of heavy metals *Acta Montan. Slovaca* **14** 152–5
[7] Wang, Lawrence K.; Hung, Yung Tse; Shammas N K 2005 *Advanced Physicochemical Treatment Processes* (New Jersey: Humana Press, Totowa)
[8] Al-Ani T S and Olli 2008 Clay and Clay Mineralogy *Ind. Eng. Chem. Res.* **31** 1937–46
[9] Clem, Arthur G.; Doehler R W 1961 Industrial Applications of Bentonite *Clays Clay Miner.* **10** 272–83
[10] Syafalni, A.N., Abdullah, R., Abustan, I., Mohd Ibrahim A N . 2013 Wastewater Treatment Using Bentonite, The Combinations of Bentonite-Zeolite, Bentonite-Alum, and Bentonite-Limestone as Adsorbent and Coagulant *Int. J. Environ. Sci.* **4** 379–91
[11] Svarovsky L 2000 *Solid-Liquid Separation*
[12] Zhou F, Li J, Zhou L and Liu Y 2015 Preparation and Mechanism of a New Enhanced Flocculant Based on Bentonite for Drinking Water *Adv. Mater. Sci. Eng.* **2015** 1–8
[13] Brito D F, Da Silva Filho E C, Fonseca M G and Jaber M 2018 Organophilic bentonites obtained by microwave heating as adsorbents for anionic dyes *J. Environ. Chem. Eng.* **6** 7080–90
[14] Hua J 2018 Adsorption of low-concentration arsenic from water by co-modified bentonite with manganese oxides and poly(dimethyldiallylammonium chloride) *J. Environ. Chem. Eng.* **6** 156–68
[15] Carvalho L A S J, Konzen R A, Cunha A C M, Batista P R, Bassetti F J and Coral L A 2019 Efficiency of activated carbons and natural bentonite to remove direct orange 39 from water *J. Environ. Chem. Eng.*
[16] Polasek P and Mutl S 2002 Cationic polymers in water treatment Part 1: Treatability of water with cationic polymers *Water SA* **28** 69–82

Compendium data and preliminary results based on alum sludge and gravel continuous fixed-packed-bed column substrate as sustainable practice for constructed wetland development in northern Egypt

Maha A. Tony

Advanced Materials/Solar Energy and Environmental Sustainability (AMSEES) Laboratory, Basic Engineering Science Department, Faculty of Engineering, Menoufia University, Shebin El-Kom, Egypt

Corresponding author, Tel: +2 0482221549, Fax: +2 0482235695; E-mail: dr.maha.tony@gmail.com, ORCID ID: 0000-0003-3670-503X

Abstract: The present investigation represents the valuable reclaims of the aluminum-based waterworks sludge that is so called "alum sludge" to close the loop concerning the sludge waste production and the sustainable ecosystem. In this study the adsorption behaviors of aluminium based sludge, namely alum sludge (AS) and gravel as cost-efficient adsorbents regarding dye elimination from aqueous synthetic stream were analyzed in fixed bed column. The influence of variable boundaries such as column pack depth and influent concentration on the effluent concentrations via breakthrough curves were examined. It is estimated that the removal uptake and sorption capabilities elevated as the initial influent loading of the dye dismissed for the fixed bed height of the both adsorbents, namely, gravel and AS. Thereafter, a preliminarily results were checked to investigate the feasibility of the adsorption column in Reactive Blue 4 (RB4) adsorption. The maximum adsorption capacities were observed around 4 mg L-1 for sample with pH 7.5 that is signified as the natural pH of the synthetic stream. However, increasing the initial dye loading increasing the adsorption capacity which reached to 5.5 mg L-1 for the dye loading 27.8 mg L-1 of Reactive Blue 4 (RB4). Thus, the first results will be a valued scheme in the development of adsorption systems to build up a constructed wetland facility.

Keywords: Alum sludge; Bibliometric analysis; gravel; Fixed bed; Adsorption; Reactive Blue 4

1. Introduction:

To alleviate the increasingly strict environmental protection, searching for non-conventional tools for advanced and sustainable wastewater technologies remains attaining the scientists' attention [1, 2]. Constructed wetlands (CWs) could be the answer. CWs are engineered that could facilities the wastewater treatment facility to optimize naturally occurring processes, i.e. biological, chemical, and physical for the option of advanced wastewater treatment. Moreover, CWs provide a well appealing and recreational value of the landscape [3-5]. Globally, there is an interest and wide acceptance for their applications since CWs posses a lot of merits. Such advantages of CWs' system including the cost efficient for application in comparison to other treatment facilities, simple in building, operation and construction, easy for maintenance besides the very low energy needs are standing of its importance. According to the literature cited, CW systems verified to be an efficient technology for physical, chemical and biological treatment of wastewater [6-9].

Worldwide, several thousands of operating wetlands, CWs have been successfully adopted as a natural water purification technique. Recently, CWs have been applied in Egypt through the Ministry of Irrigation and The National Institute for Water Research. Their investigation is applied in certain locations, "Bahr Al-Baqar" drainage and "Al-Manzala" lake to treat the agricultural effluents to be reused in the irrigation of productive crops. However, in another location "Maryout" lake where used to treat municipal wastewater [10].

The main concepts of CWs sustainably combine landscape features and ecological functions. Therefore, they are simultaneously delivering reduced water pollution, and ecological functions, and fulfilling cultural and aesthetic, recreational needs.

2. Compendium data

2.1. Alum sludge production and quantities:

Drinking water treatment steps incorporating "coagulation, filtration and sedimentation" [11, 12] [13, 14] to produce a high and sufficient drinking water quality. The result is using coagulants aids. Aluminum and iron, are applied to precipitate the soluble "metal ions" in drinking waterworks treatment plants resulting in a huge amounts of sludge. Such sludge attaining a special concern since it including heavy metals residuals that is potentially toxic to the ecosystem.

The produced by-product sludge that is aluminum-rich by-product namely alum sludge and so called "alum sludge" should go through sludge dewaters technique prior to the final dumping. Alum sludge is signified as a two-phase combination consisting of both solids and water. The water content in such alum sludge is reached to 99% before thickening and could be reduced to 95% after thickening. According to European union (EU) classification "alum sludge" is signified as Code 19's list of waste. Globally, the data available in the last 20 years of the produced "alum sludge" in numerous selected countries is assessed and displayed in Fig. 1 [15-21].

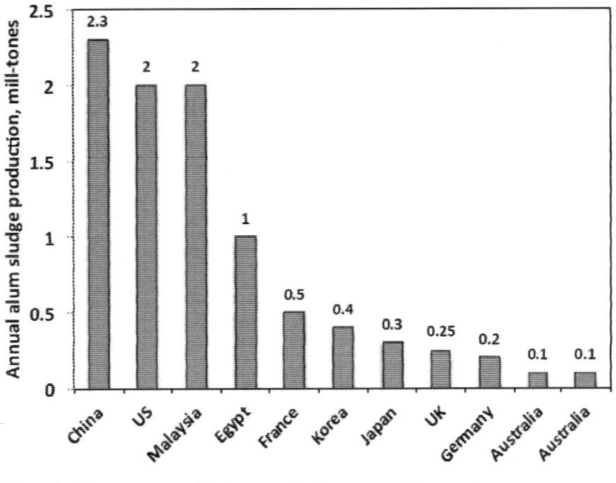

Fig. 1 Amount of alum sludge produced some areas

2.2. Bibliometric visualization analysis of CWs:

2.2.1. Key terms analysis

The term of "Constructed Wetland" or "Constructed Wetlands" has been emerged over the last decades and its numerous advantages related to its "Green" feature through treating wastewater and attaining a pleasant landscape in a reliable and economic value via low construction, operating and maintenance costs compared to traditional treatment technologies.

Recently, with the highly modernized search facilities, bibliometric visualization analysis were introduced as a valuable visualized tool for search items and analysis. Moreover, bibliometric visualization analysis could map key aspects in order to signify a certain research subject and achieve the most literature-cited items. Also, it could attain the relation between the search terms. From this regard, the state-of-the-art sketch of bibliometric analysis that include Constructed Wetland treating wastewater signified using the literature survey was conducted based on the "Web of Science" platform and the software of VOSviewer version 1.6.16.0 [18],. A search extracted from the Web of Science Core Collection using the search terms "alum sludge OR constructed wetland OR industrial wastewater treatment", the search is based till the year of 2021. The data given in Fig. 2 is designed related to mapping of incorporating network, overlay and density visualization for the total research articles as cumulative publication articles in the field of CWs based on alum sludge. The key techniques are incorporating some steps based on: (i) downloading the documents' data from "Web of Science" data; (ii) Incorporating the data into VOSviewer software according to its necessities; (iii) Choosing the type of analysis based on full counting method and the co-occurrence associated with authors' keywords; (iv) selecting the keyword to be located as its minimum occurrence number in the article. Through the current study, density visualization analyses were chosen and exhibited according to the arrangement of tags and rings. It is noteworthy to mention, the size of clusters signifies the significance of each occurrence. Fig. 1 displays the bibliometric network map generated in VOSviewer of 20 clusters and the most compactness visualization analysis terms that are presented in the system of labels and rings. The volume of the circles is indicating its significance. Among the research articles sources, the significant clusters of constructed wetland, constructed wetlands, industrial wastewater, adsorption [22], phosphorus, reuse, removal, performance, nitrogen and subsurface flow are considered the predominant terms used exploring target application. Research studies were focused on constructed wetlands uses and the target pollutants especially phosphorus and nitrogen as pollutants treated from wastewater. It is designated that these exploration articles established the varied guidelines in the field. It seems that studies are turning to wetlands wastewater treatments as an economic facility [23].

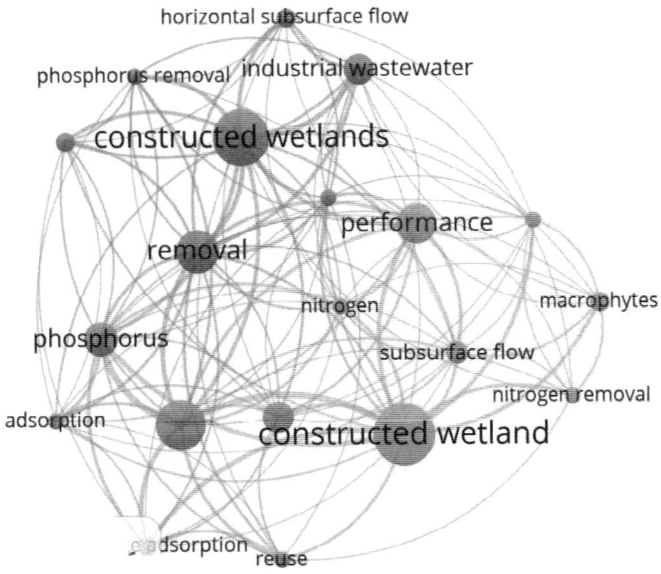

Fig. 2 Bibliometric network investigation mapping produced via "VOSviewer" based on CWs

2.3. Filter beds, reeds beds and CWs media for wastewater abetment:
Globally, pronounced attention has been established regarding the use of the alum sludge cake as a reactive filter substrate media in CWs beds. Since substrate is an important component in CW, its chosen should posses' multi-functions. Such functions including, supporter of biofilm formation, pollutants and contaminants adsorption and as a plants growth bed are the most pronounced function needed to a successful CWs medium. Regarding the applicability and real practical point of regard, drinking water treatment plants residues "alum sludge" is a non-hazard material [6, 24-26]. Significantly, alum sludge residue posses aluminum that is a predominant component in its composition has high adsorption affinity [27]. Thus, alum sludge cake own a good potential to be a good substrate in the CWs medium in order to endowing a double benefits of pleasant reeds and wastewater treatment. Various investigations are applied to evaluate the affinity of AS as an efficient adsorbent material for dyes [28, 29], phosphorous [27] and some of heavy metals adsorption [30-32] for introducing it as a filter bed or constructed wetland medium. The application of dewatered AS (cake) as the main CWs substrate to expand its scope for enhanced wastewater treatment facility is interesting as a new material substrates for CWs advancement. Thus, it could be mentioned that superior contaminants elimination efficiencies have been cited and globally could be also accepted through numerous constructed wetland research studies.

2.4. Impacts and concerns of Aluminium "Recover" from alum sludge
As a sustainable end-of-life waste criterion, alum sludge utilization is an interesting profit. However, aluminum metal presence in alum sludge is still remaining a barrier and stands towards its widening and uses. Environmental concerns are related to the probability of aluminum discharge to the biosphere and aquatic environment. Therefore, once Al- is claimed from such AS, a further increase in alum sludge applications could be increased. Scattered investigations have been studied and their research work have been cited to recover aluminum from aluminium based drinking water works sludge [33-35]. For example, some investigators reported that aluminum can be recovered via acid treatments (pH ranged 1.0-3.0) may achieve aluminum reclamation which could reach to 90% [34, 36]. However, other researchers encouraged alkaline treatments (pH 11) gave a reasonable aluminum recovery [37]. Also,

membrane separation methodologies [38] besides liquid ion exchange techniques enhanced aluminum recovery from such sludge [39].

2.5. Constructed Wetland in Egypt:

Although Egypt is well endowed with the river Nile as the source of water, according to recent studies, there is depletion in water resources. Hence, developing the research field wastewater treatments is gaining the priority of both academia and industry. Now, constructed wetland , CWs is a common used methodology for wastewater treatment. The multifunction landscape is a complicated harmony merging traditional landscape uses with recent human activities. Initiating a CWs parks is an application of such idea.

CWs is established in the late of eighteens in as an experimental studies. In 1990s, subsequently, full scale projects related CWs is pioneered in the last five years. Alexandria University applied a model for treating municipal wastewater prior to dumping at "Maryut Lake". Also, Suez Canal University had applied two CWs for both sewage wastewater at "Abou Atwa, Ismailia" and for industrial wastewater for "10th of Ramadan industrial city". Thereafter, their results were used as raw material to construct a full scale CW to treat municipal wastewater of Samaha village, Dakahlia. Moreover, another investigation is supported for the industrial and domestic pollution loads for the Lake Manzala environmental protection. Afterwards, two villages at Kantara, Ismailia where new municipal wastewater treatment plants are dealing with municipal wastewater treatment plants from the agricultural drains. Industrial or municipal pollution loads is being treated through using the drain for pollutants removal through water flow through these vegetation. The Egyptian constructed wetlands are as the following:

Abu Attwa, Ismailia:

In 1990, a subsurface gravel bet wetland was built to treat domestic sewage from overloaded septic tanks and effectively provided a secondary treatment stage. For 10 years, the wetland have been operated and monitored effectively in domestic sewage wastewater treatment [40].

10th of Ramadan City:

In 1992 at Sharkia Governorate, Egypt an experimental subsurface CW was used at 10th of Ramadan City for treating industrial effluents which showed a superior removal of hydrocarbons and was less effective in removing other compounds. Such trail concluded that the gravel bed wetlands could be applied for industrial wetlands but longer residence times may be needed to treat several industrial effluents [40].

Samaha Village, Dakahlia:

Wastewater Treatment Wetland of sewage wastewater from the village septic tanks and a collection of consist of a group of settling tanks collecting sewage from various houses is treated through the full scale wetland. Alum is added to the collected wastewater and the wastewater is then wastewater is passed through a number of horizontal flow subsurface wetland columns followed by vertical flow sand beds cells. The treated water is then go to disinfection tank before the final drain.

Lake Manzala CWs for Treatment of Bahr El Baqar drainage wastewater:

Bahr El Baqar drainage wastewater must be treated to reduced the pollution burden loading Manzala Lake and the Mediterranean Sea. Thus, CW is consists of a 3-stage treatment facility for treating a daily wastewater of 25 thousand of cubic meter from Bahr El-Baqr drain. Firstly, water is collected in a sedimentation tank for a couple of days followed by flowing through treatment beds where common reed plants are planted, i.e. cattail and water hyacinth to help in the treatment facility [41].

Environmental Assessment must be implemented for all CWs for their hydrological and ecological considerations besides their accumulative impacts to the environment and the socioeconomic issues. Thus, globally, numerous work were investigated to judge novel CWs. Form this regard, the volume and affinity of alum sludge as a cost-efficient adsorbent material is introduced as constructed wetland medium, however, it is not applied so far in Egypt, according to the author's knowledge. Alum sludge not only introducing the use of sludge waste as a lifespan end life but also, Al and P be restored

3. Preliminary results of AS filter bed:

3.1. Experimental section

3.1.1. Field sampling

Alum sludge:

Alum sludge by-product collected from the largest waterworks plant located at Southern part of Shebin El-Kom City that is the capital of Menoufia governorate in the Delta region in the north of Egypt. Then the sludge is stored and analysed in the laboratory according to the standard methods. Subsequently, the sludge will go through physical treatment then oven drying remove any water present followed by calcination. The obtained sludge cake is introduced as a porous adsorbent material. These treatment steps are the preliminary stage in this work as seen in Fig. 3. Thus the material will be a low-cost adsorbent material for removing organics.

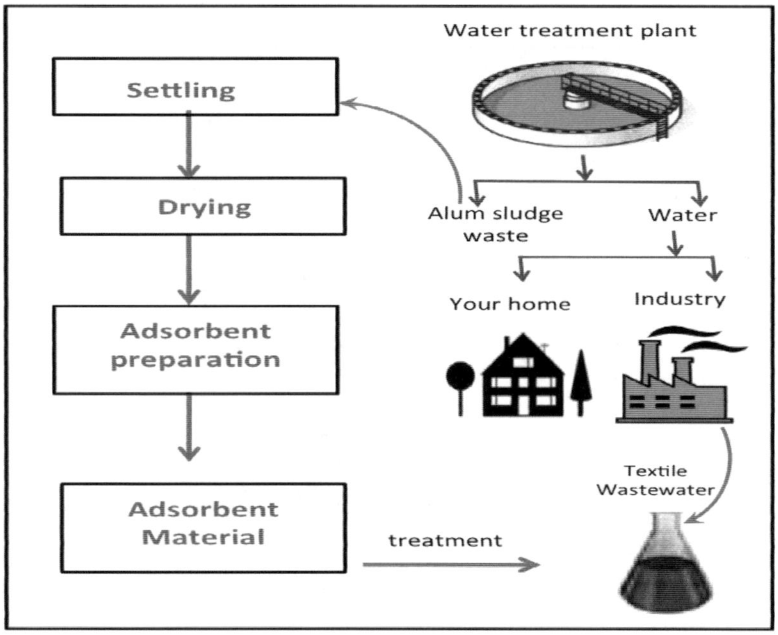

Fig. 3 Preliminary treatment steps of sludge to be an adsorbent material for wastewater treatment

Journal of Physics: Conference Series **2305** (2022) 012028 doi:10.1088/1742-6596/2305/1/012028

3.1.2. Alum sludge characterization:

The morphologies of the collected and prepared, dried alum sludge cake, samples were examined and imaged using field-emission scanning electron microscope (SEM) of model type FE-SEM, Quanta FEG 250. The used typical magnifications were x8000 and x60000. Field-emission scanning electron microscope is augmented with energy dispersive X-ray spectroscopy (EDX) for sludge composition analysis.

3.1.3. Adsorption column

Filter media preparation

The column of 100 mm was filled with 10-mm size of gravel to serve as support material was used as a support material, and then packed to 400 mm with the sludge cakes. Loaded onto the system at a hydraulic loading rate of 0.5m3/m2.d. The "tidal flow" operation was carried out applying a set of preprogramed digital electronic timers which ensured a cycle of fill and draw at a rate of 4 cycles/day.

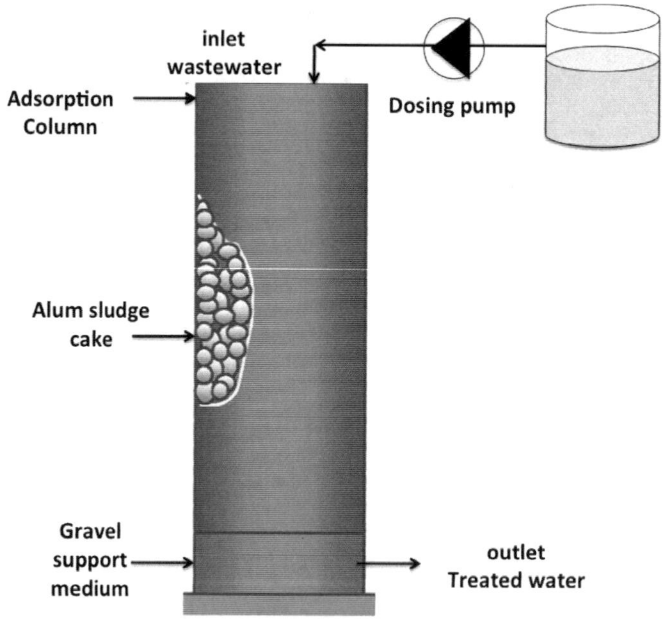

Fig. 4 Experimental set-up of the lab-scale adsorption column

Wastewater, synthetic effluent generation:

The synthetic wastewater was prepared following an established protocol, a stock solution is prepared then a diluted solutions are prepared. The pH of the original wastewater was monitored at 7.5.

Sampling and analytical determination

Procedure of the batch tests and sampling is followed the concentration of the dye elimination using UV–Vis analysis by using a model type Unico UV-2100 spectrophotometer, USA. The adsorption capacity of the materials (qe) was calculated by:

$$q_e = (C_o - C_t)\frac{V}{m} \qquad (1)$$

where Co and Ct are the initial and final concentrations of the aqueous dye stream, V is the solution volume (L) and m is adsorbent mass used (g).

3.1.4. Constructed Wetland System Design

Following the adsorption bench-scale test trails and optimizing the process parameters, the data will be applied to design the constructed wetland, CW methodology. Alum sludge substrate is an important essential in Constructed Wetland, thus, the best substrate was selected that posses multi-puroses, i.e. (i) supporter of biofilm formation; (ii) plants growth media; and (iii) adsorbent for pollutants. Construction of the wetland facility with the sludge cake will be obtained as shown in Fig. 5. The bottom of the tank will be filled with gravel as the supporting filter medium and dried alum sludge cake as the main wetland medium layer. Common reeds will be chosen and planted on the top of the column. The system will then be fed with a wastewater, which is collected from the textile factory in Quesna Industrial Zone as the industrial effluents source. Influent wastewater will be introduced into the column from the top while effluent will be drained from the bottom using a pump. Aeration in the column could be studied and supplied with an air diffuser. Therefore, Al-based resiude cake, as dewatered alum sludge posses a valuable possibility for recycle as raw environmental engineered substantial as the main substrate in Constructed Wetland for dual settlements of reed beds and wastewater treatment. Suggestions for enhancing strategies for constructed wetland:

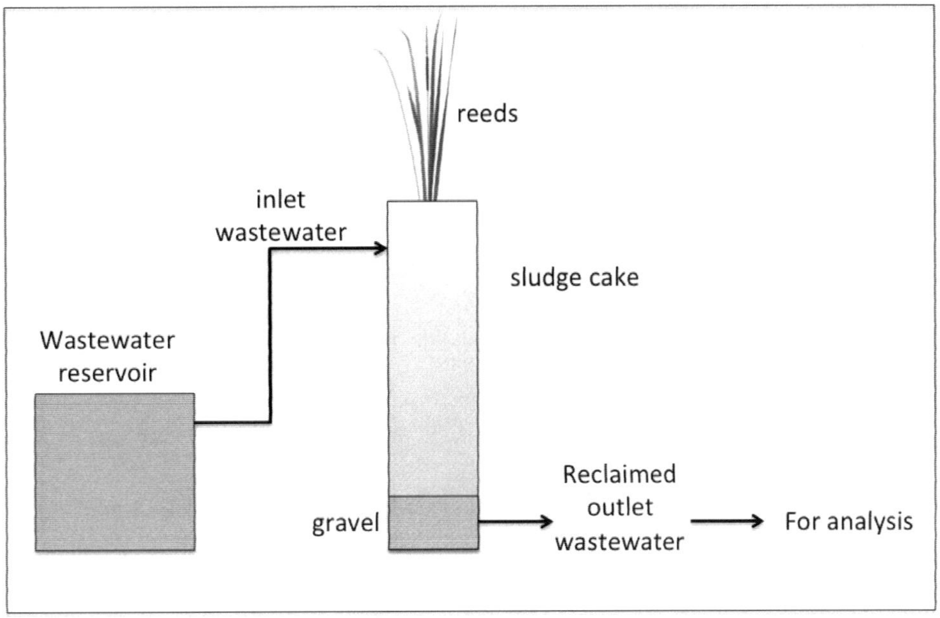

Fig. 5 Suggested constructed wet land lab-scale facility

Journal of Physics: Conference Series **2305** (2022) 012028 doi:10.1088/1742-6596/2305/1/012028

3.2. Results and Discussion:

3.2.1. Prepared AS adsorbent bed characterization:

The physical morphology of the alum sludge samples is displayed using SEM micrographs (Fig. 6a). The image explores that the material structure is porous. Such mixture could be signified as heterogeneous and non-uniform porous shape combination. Hence, the sludge could be considered had uneven structure nature. The dye molecules may be sorbet onto the AS material's surface penetrating the surface to the pores of the material and thus controlling the expansion of the surface reconstructing its structure.

EDX spectrum ascertained via EDX Energy dispersive X-ray spectroscopy, EDX analysis as illustrated in Fig. 6b. The results displayed in Fig. 6b clearly illustrates the peak of O, Si and Al as the main major components, that confirms the formation of alumina and silica, as well as, presence of S, Ca and Fe.

(a)

(b)

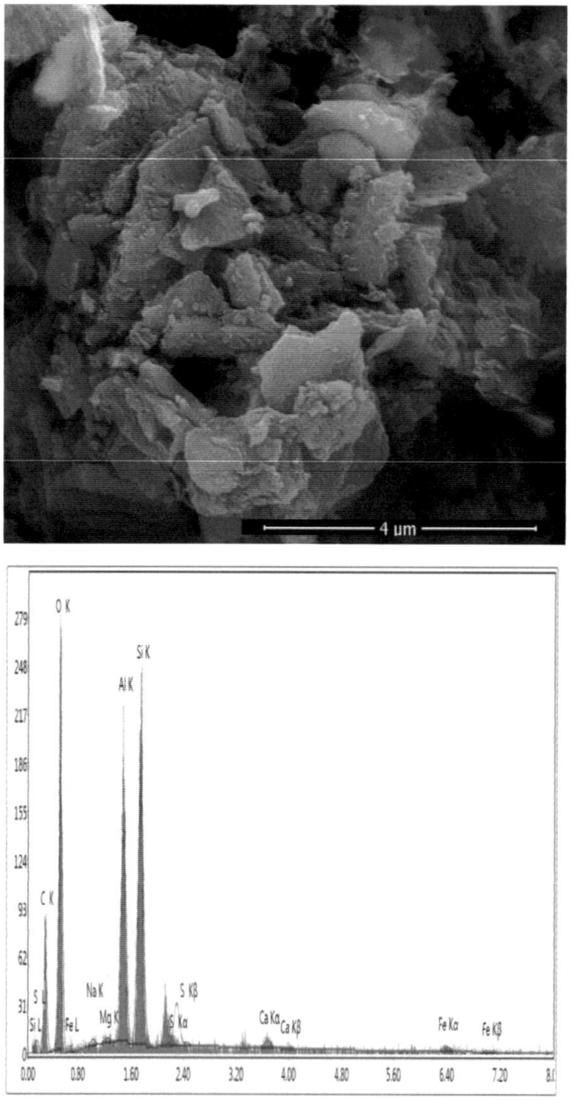

Fig. 6 (a) SEM images of alum sludge filter bed; (b) EDX composition of alum sludge

3.2.2. Adsorption Isotherm Time:

Initially, to establish the isotherm base, it is essential to inspect the sorption isotherm equilibrium time. The time-profile of Reactive Blue 4 dye adsorption using the alum sludge adsorbent material was explored at room temperature and displayed in Fig. 7. The majority of the dye is adsorbed within the first 120 min of contact time. This could be due to the existence of cavitation according to the SEM image result of AS; such cavities are effective in dye adsorption due to the ion exchange characteristics [3]. However, more time is not essential since the available active sites are occupied with the dye molecules so no more adsorbed dyes are adsorbed with the time increase.

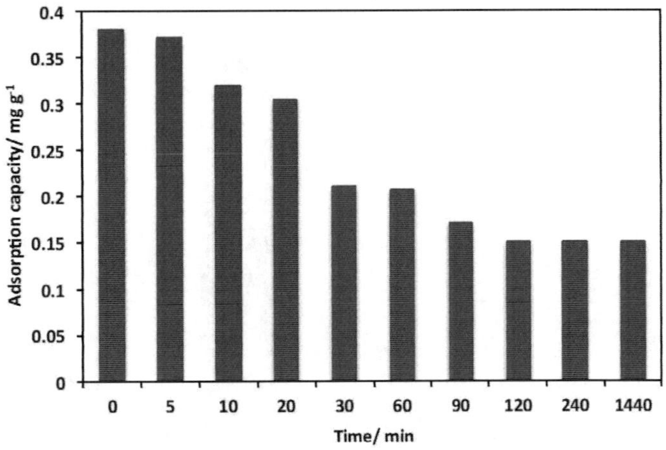

Fig. 7 Adsorption isotherm time

3.2.3. Effect of inlet RB4 dye Loading:

Shows the progressive mean treatment results obtained from the system over an experimental period of one cycle. Obvious improvement in adsorption uptake with elevating the initial RB4 concentration (Fig. 8). This may be illustrated by the improvement contact among the dye molecules and alum sludge. Porous nature and higher surface could facilitate the presence available adsorption sites to be utilized and thus increasing the adsorption uptake. These results of increasing the adsorption capacity with increasing the adsorbate are in agreement with that previously examined by Mittal and Gupta [42].

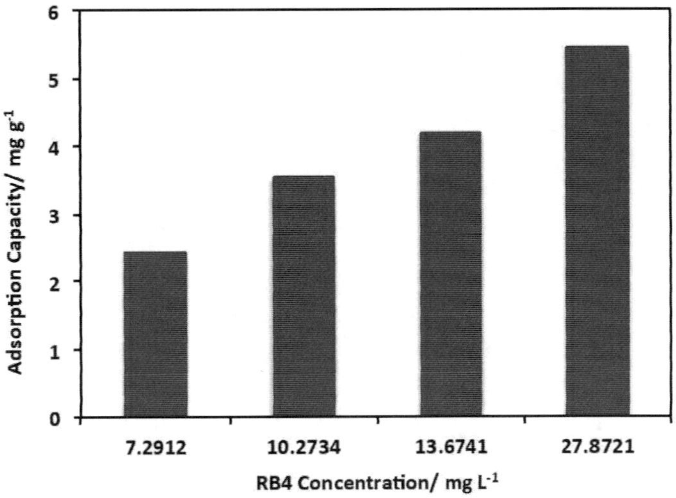

Fig. 8 Adsorption capacity plot of the treatment trend in the system

3.2.4. pH of RB4 dye solution effectiveness:

For exploring the maximal adsorption capacity, pH effect was tested at various initial pH values, i.e. 4, 6, 7.5 and 9. The experimental results are given in Fig. 9. The treatment efficiency enhanced with the elevation of the pH value and the neutral pH is favored. Although, more increase in the pH value attained no apparent enhancement. This is due to the loss in the adsorption capacity due to the decline of the active sites of alum sludge adsorbent material. Moreover, increasing the solution's pH weakening the attraction between the adsorbate and the alum sludge adsorbent substance [19, 29, 43, 44].

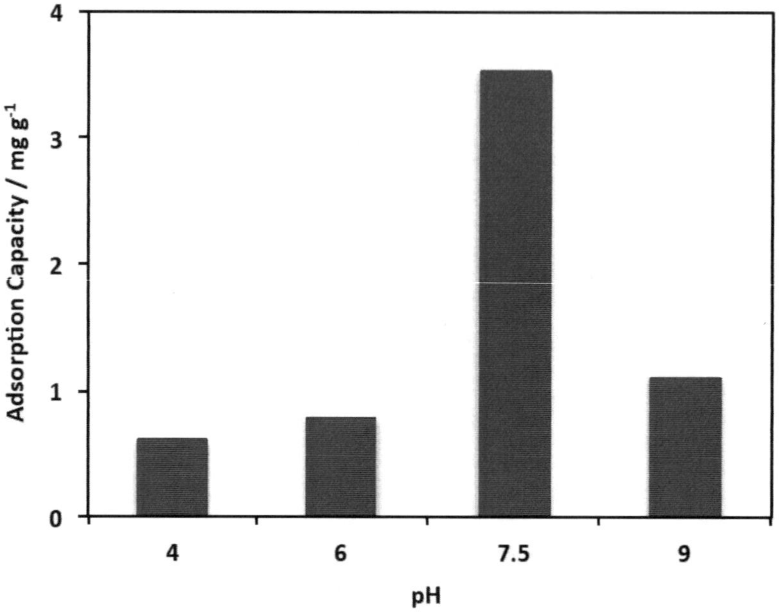

Fig. 9 pH effect related to adsorption capacity treatment trend in the system

3.2.5. Effect of temperature

Temperature behavior on RB4 adsorption capacity was studied. The reaction temperature was investigated over the range of 26 to 60 °C. The data displayed in Fig. 10 revealed that dye sorption capacity increased with temperature elevation. This may be attributed by the improvement of the pollutant molecules desorption stage in the adsorption development verifying the technique is exothermic [9]. Also, the temperature elevation weakening the sorptive forces between the adsorption active sites on alum sludge substance and the dye molecules.

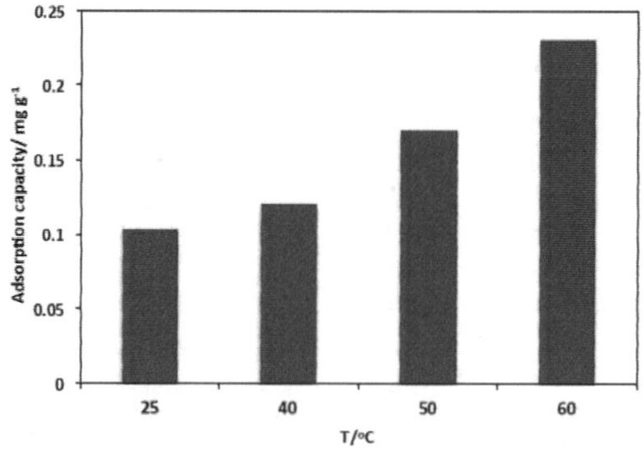

Fig. 10 Temperature effect on adsorption capacity

3.2.6. Isotherm Models

To investigate the adsorption mechanism and to control the adsorption uptake of alum sludge as adsorbent material, two isotherm models were applied namely Langmuir, and Freundlich models.
The Langmuir adsorption isotherm is founded on the basis of the maximal adsorption capacity is related to the monolayer adsorption on a homogeneous surface. Its linearized form as follows [43]:

$$\frac{C_e}{q_e} = \frac{1}{K_L} + \frac{a_L}{K_L} C_e \quad (2)$$

$$Q_o = \frac{K_L}{a_L} \quad (3)$$

where C_e is the dye concentration at equilibrium (mg L^{-1}), q_e is the equilibrium sorption uptake, and a_L and K_L are signified as Langmuir adsorption constants. Q_o is the monolayer adsorption capacity of the solid (the mass of dye adsorbed per unit mass of adsorbent) (mg g^{-1}). However, Freundlich model is appropriate for the highly heterogeneous surface systems. Such assumption is classified by the heterogeneity constant *1/n*. The linearized shape of the model equation could be categorized according to the ensuing form:

$$\ln(q_e) = lnK_F + \frac{1}{n} lnC_e \quad (4)$$

where K_F is Freundlich constant that relate to the adsorption capacity of the solid (L g^{-1}). *1/n* is signified as the heterogeneity constant that evaluates the adsorption intensity and gives a significance of how the adsorption is favourable. A (n) greater than unity represents favorable adsorption conditions [9]. Hence, the data were verified over the selected models.

The equilibrium data of alum sludge explored a respectable association to Langmuir isotherm (Fig. 11). It can be suggested, the adsorption surface is a homogenous one and RB4 dye adsorption on alum sludge is going to be monolayer coverage till the active sites saturation took place.

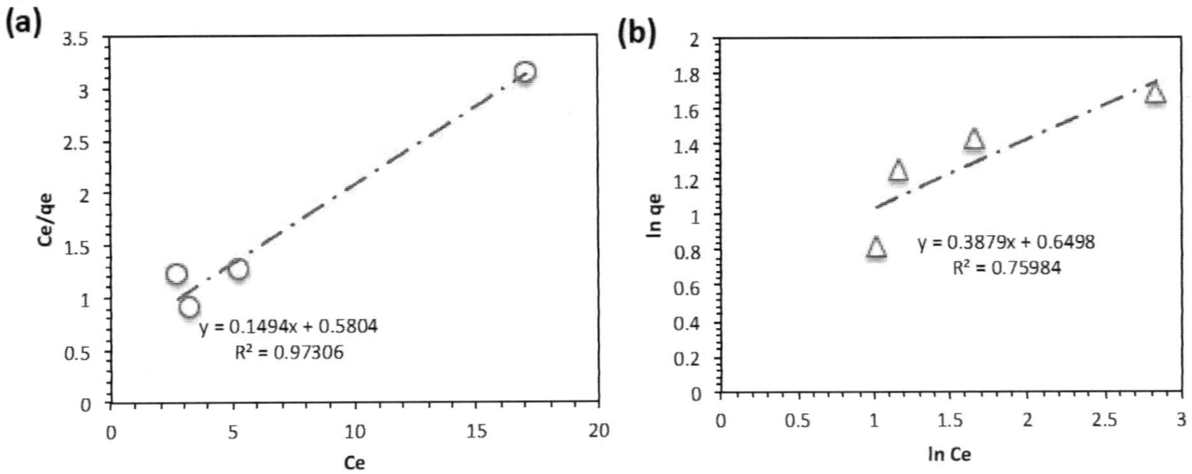

Fig. 11 Linearized adsorption isotherms of alum sludge for RB4 adsorption (a) Langmuir, (b) Freundlich,

3.2.7. Kinetics of dye adsorption

Kinetic models of adsorption that could assess the extent and mechanism of the proposed dye adsorption process via AS-Sorbents. The comparative estimation of the applicability of the Lagergren's pseudo-first order, pseudo-second order and Bangham's equation were analyzed and the parameters of each model are computed in Fig. 12.
Associating correlation coefficient values, R^2, for comparison of the pseudo-first-order model by Lagergren's and the pseudo-second-order model in Table 4 represents Lagergren's model is not as a well fit of the adsorption process as the pseudo-second-order model. For the alusme sludge adsorbent system, excellent R^2 of the pseudo-second-order model. Furthmore, a reasonable comparison of the suggested model with those in the cited research for similar adsorption methodologies [9].

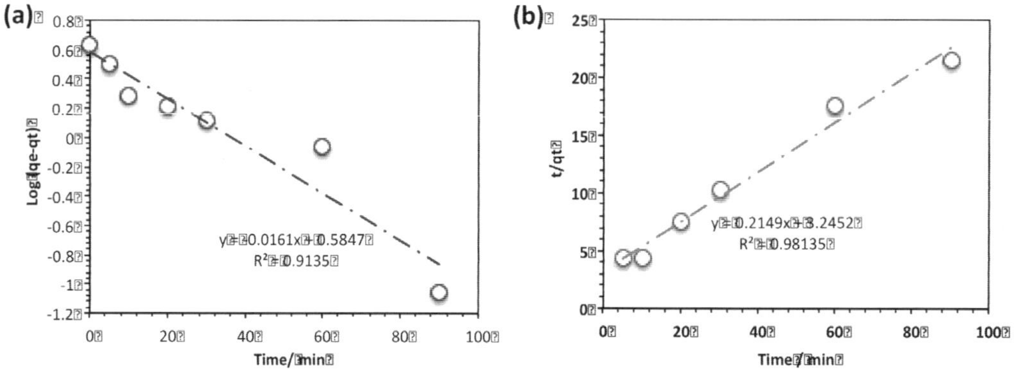

Fig. 12 kinetic models for dye adsorption

Numerous studies have been investigated the application of alum sludge as a low cost adsorbent material for pollutant remediation from aqueous streams. Such pollutants comprising dyes, phenol, phosphorus and lead as displayed in Table 1. But, detailed and more published studies are related to phosphorus removal from various wastewater effluents and types, their equilibrium isotherm, kinetic analysis and the effect of system parameters have been studied. The presence of amorphous aluminium nature in the alum sludge maximizing its affinity to adorb phosphorous anions from contaminated streams [51]. Alum sludge sorption capabilities are related to the particle size, surface area and surface charge of the alum sludge particles.

Table 1. Adsorption capacities for various adsorbate materials using alum sludge

Alum Sludge	Adsorbate	Working parameters	Adsorption capacity	Ref.
Southern Shebin El-Kom City, Menoufia governorate, Egypt	RB4 dye	pH 7.2; Co 7.5 mg/kg; 25 ◦C; 120min	3.5 mg/g.	Current study
Bradenton, USA	As(V), S(III)	pH 6-6.5; Co 1875 mg/kg; 23 ◦C; 2 d	7.68 mg/g 6.00 mg/g	[45]
Tampa, Florida. and Bradenton, Florida, USA	As(V)	pH 3-7; Co 200 mg/L	12 000 mg/kg	[46]
9th WTP, Beijing,	Co(II)	Co(II) 25 - 800 mg/L; pH 6.0; 48 h	17.31 mg/g	[47]

China				
WTP, Shebin El-Kom, Menoufia Governorate, Egypt	Phenol	AS 2 g; Co 100-1000 mg L-1; 298 K; 1 h	275 mg-phenol g-1	[48]
Sludge drying bed of Ming-Der WTP, Miaoli, Taiwan	Ammonium	AS 300 g; Co 17 g/L; 1000 mL wastewater; 60 min	11.3 mg/g	[49]
Southern Shebin El-Kom City, Menoufia governorate, Egypt	Procion Blue dye	AS 2 g; pH 7; Co 11.8 mg/L; 298 K; 1 h	6.5 mg/g.	[43]
Ballymore-Co. Kildare, Dublin	P	pH 4.3	22.4 mg/g	[50]
Leixlip WTW, Dublin	P	pH 4.3	20.1 mg/g	[50]

4. Conclusion

More efficient and cost viable alternatives for the high disposal costs especially of sludge landfilling, alum sludge beneficial application as a constructed wetland material substrate is discussed. But, still further work is needed especially for the field of aluminium recovery and release of alum sludge concerns. A lab-scale vertical flow adsorption filter bed system was operated with a tidal flow using dewatered alum sludge cakes as the main substrate supported with gravel. The technique is revealed a good adsorption capacity in each cycle. Although further experimental results and work is still required, it seems that the basis of application is sound, and that constructed wetland application is worthwhile.

Acknowledgment

This work was supported by the grant funded by Scientific Research Department, Graduate Studies and Research Sector, Menoufia University, Egypt.

References:

[1] Zhao Y Q, Babatunde A O, Hu Y S, Kumar J L G and Zhao X H 2011 Pilot field-scale demonstration of a novel alum sludge-based constructed wetland system for enhanced wastewater treatment Process Biochemistry 46 278-83

[2] Tony M A and Lin L-S 2020 Iron recovery from acid mine drainage sludge as Fenton source for municipal wastewater treatment International Journal of Environmental Analytical Chemistry 1-16

[3] Zhao Y, Liu R, Awe O W, Yang Y and Shen C 2018 Acceptability of land application of alum-based water treatment residuals–an explicit and comprehensive review Chemical Engineering Journal 353 717-26

[4] Thabet R H, Tony M A, El Sherbiny S A, Ali I A and Fouad M K 2020 Catalytic oxidation over nanostructured heterogeneous process as an effective tool for environmental remediation IOP Conference Series: Materials Science and Engineering 975 012004

[5] Tony M A Supports delivering sustainable environment using drinking water treatment plant waste for sullage treatment: Box–Behnken design model for process optimization. Military Technical College) pp 307-18
[6] Zhao Y Q, Keogh C and Tony M A 2009 On the necessity of sludge conditioning with non-organic polymer: AOP approach Journal of Residuals Science & Technology 6 151-5
[7] Tony M A, Tayeb A and Zhao Y Q 2016 An alternative arrangement for the alum sludge management: minimising waste with low-cost solar techniques Am J Chem Eng 4 30-7
[8] Tony M A and Ali I A 2021 Mechanistic implications of redox cycles solar reactions of recyclable layered double hydroxides nanoparticles for remazol brilliant abatement International Journal of Environmental Science and Technology 1-18
[9] Zou Y, Wang X, Khan A, Wang P, Liu Y, Alsaedi A, Hayat T and Wang X 2016 Environmental remediation and application of nanoscale zero-valent iron and its composites for the removal of heavy metal ions: a review Environmental science & technology 50 7290-304
[10] Haron A and Feisal Z 2020 Constructed wetland parks: a pathway to sustainability for Cairo Journal of Urban Research 36 61-81
[11] Konieczny K, Sąkol D, Płonka J, Rajca M and Bodzek M 2009 Coagulation—ultrafiltration system for river water treatment Desalination 240 151-9
[12] Yin C-Y 2010 Emerging usage of plant-based coagulants for water and wastewater treatment Process Biochemistry 45 1437-44
[13] De Sena R F, Tambosi J L, Genena A K, de Fpm Moreira R, Schröder H F and José H J 2009 Treatment of meat industry wastewater using dissolved air flotation and advanced oxidation processes monitored by GC–MS and LC–MS Chemical Engineering Journal 152 151-7
[14] Amaral Filho J, Azevedo A, Etchepare R and Rubio J 2016 Removal of sulfate ions by dissolved air flotation (DAF) following precipitation and flocculation International Journal of Mineral Processing 149 1-8
[15] Babatunde A O, Zhao Y Q, Yang Y and Kearney P 2007 From" fills" to filter: insights into the reuse of dewatered alum sludge as a filter media in a constructed wetland Journal of Residuals Science & Technology 4 147-52
[16] Xiong J B and Mahmood Q 2010 Adsorptive removal of phosphate from aqueous media by peat Desalination 259 59-64
[17] Hu Y, Zhao Y, Zhao X and Kumar J L G 2012 High rate nitrogen removal in an alum sludge-based intermittent aeration constructed wetland Environmental science & technology 46 4583-90
[18] Dassanayake K B, Jayasinghe G Y, Surapaneni A and Hetherington C 2015 A review on alum sludge reuse with special reference to agricultural applications and future challenges Waste Management 38 321-35
[19] Ahmad T, Ahmad K and Alam M 2016 Sustainable management of water treatment sludge through 3 'R'concept Journal of Cleaner Production 124 1-13
[20] Abd El-Razek T M, Elgendy A S and Fouad M M 2019 UTILIZATION OF DRINKING WATER TREATMENT SLUDGE TO PRODUCE ALUMINUM SULFATE Journal Archive 48
[21] Foss P J and O'Connell C A 1996 Irish peatland conservation plan 2000: Irish Peatland Conservation Council Dublin)

[22] Ashour E A and Tony M A 2020 Eco-friendly removal of hexavalent chromium from aqueous solution using natural clay mineral: activation and modification effects SN Applied Sciences 2 1-13
[23] Tony M A 2021 Paradigms of homo/heterogonous Fenton systems incorporating'Solar Energy'based on'Emerging Pollutants' removal–challenges, advancements and visualized bibliometric analysis International Journal of Environmental Analytical Chemistry 1-32
[24] Zhao W, Xie H, Li J, Zhang L and Zhao Y 2021 Application of Alum Sludge in Wastewater Treatment Processes:"Science" of Reuse and Reclamation Pathways Processes 9 612
[25] Turner T, Wheeler R, Stone A and Oliver I 2019 Potential alternative reuse pathways for water treatment residuals: Remaining barriers and questions—A review Water, Air, & Soil Pollution 230 1-30
[26] Maha A. Tony M M E 2021 End-of-life waste criteria: Synthesis and utilization of Mn-Zn ferrite nanoparticles as a super paramagnetic photocatalyst for synergistic wastewater remediation. Applied Water Science)
[27] Babatunde A O and Zhao Y Q 2007 Constructive approaches toward water treatment works sludge management: an international review of beneficial reuses Critical Reviews in Environmental Science and Technology 37 129-64
[28] Tony M A and Lin L-S 2020 Attenuation of organics contamination in polymers processing effluent using iron-based sludge: process optimization and oxidation mechanism Environmental Technology 1-10
[29] Thabet R H, Fouad M K, El Sherbiny S A and Tony M A 2021 Synthesis, characterization and potential application of magnetized nanoparticles for photocatalysis of Levafix CA Reactive Azo-Dye in aqueous effluent. Water and Environment Journal)
[30] Zhou Y-F and Haynes R J 2011 Removal of Pb (II), Cr (III) and Cr (VI) from aqueous solutions using alum-derived water treatment sludge Water, Air, & Soil Pollution 215 631-43
[31] Castaldi P, Mele E, Silvetti M, Garau G and Deiana S 2014 Water treatment residues as accumulators of oxoanions in soil. Sorption of arsenate and phosphate anions from an aqueous solution Journal of hazardous materials 264 144-52
[32] Silvetti M, Castaldi P, Garau G, Demurtas D and Deiana S 2015 Sorption of cadmium (II) and zinc (II) from aqueous solution by water treatment residuals at different pH values Water, Air, & Soil Pollution 226 1-13
[33] Dayton E A and Basta N T 2001 Characterization of drinking water treatment residuals for use as a soil substitute Water Environment Research 73 52-7
[34] Vaezi F and Batebi F 2001 Recovery of iron coagulants from Tehran water-treatment-plant sludge for reusing in textile wastewater treatment
[35] Stendahl K, Färm C, Fritzdorf I and Ulmert H 2006 The REAL process–a process for recycling sludge from water works Water science and technology 54 235-42
[36] Abdo M S E, Ewida K T and Youssef Y M 1993 Recovery of alum from wasted sludge produced from water treatment plants Journal of Environmental Science & Health Part A 28 1205-16
[37] Masschelein W J, Devleminck R and Genot J 1985 The feasibility of coagulant recycling by alkaline reaction of aluminium hydroxide sludges Water Research 19 1363-8

[38] Sengupta A K and Shi B 1992 Selective alum recovery from clarifier sludge Journal-American Water Works Association 84 96-103

[39] Cornwell D A 1982 Process for the extraction of metal ions. Google Patents)

[40] Stott R, Jenkins T, Bahgat M and Shalaby I 1999 Capacity of constructed wetlands to remove parasite eggs from wastewaters in Egypt Water Science and Technology 40 117-23

[41] El-Gamal A A 2017 Coastal Wetlands: Alteration and Remediation: Springer) pp 411-34

[42] Mittal A and Gupta V K 2010 Adsorptive removal and recovery of the azo dye Eriochrome Black T Toxicological and environ chemistry 92 1813-23

[43] Tony M A 2020 Zeolite-based adsorbent from alum sludge residue for textile wastewater treatment International Journal of Environmental Science and Technology 1-14

[44] Tony M A, M.M. Eltabey 2022, End-of-life waste criteria: Synthesis and utilization of Mn-Zn ferrite nanoparticles as a super paramagnetic photocatalyst for synergistic wastewater remediation, Applied Water Science, 12, 21, DOI: 10.1007/s13201-021-01555-6

[45] Makris, K. C., Sarkar, D., Parsons, J. G., Datta, R., & Gardea-Torresdey, J. L. (2009). X-ray absorption spectroscopy as a tool investigating arsenic (III) and arsenic (V) sorption by an aluminum-based drinking-water treatment residual. Journal of hazardous materials, 171(1-3), 980-986.

[46] Nagar, R., Sarkar, D., Makris, K. C., & Datta, R. 2010. Effect of solution chemistry on arsenic sorption by Fe-and Al-based drinking-water treatment residuals. Chemosphere, 78(8) 1028-1035.

[47] Jiao, J., Zhao, J., & Pei, Y. 2017. Adsorption of Co (II) from aqueous solutions by water treatment residuals. *Journal of Environmental Sciences*, 52, 232-239.

[48] Tony, M. (2019). Win-win wastewater treatment to sustain world: Porous adsorbents from waste waterworks sludge for phenol remediation. Paper presented at the Anaerobic Digestion Conference AD16, The International water Association, IWA.

[49] Cheng, W.-P., Chen, P.-H., Yu, R.-F., & Ho, W.-N. 2016. Treating ammonium-rich wastewater with sludge from water treatment plant to produce ammonium alum. *Sustainable Environment Research*, 26(2), 63-69.

[50] Zhao, Y., & Yang, Y. 2010. Extending the use of dewatered alum sludge as a P-trapping material in effluent purification: study on two separate water treatment sludges. Journal of Environmental Science and Health, Part A, 45, 1234–1239.

[51] Tony, M.A 2022. Valorization of undervalued aluminum-based waterworks sludge waste for the science of "The 5 Rs' criteria", *Applied Water Science*. **12**, 20, DOI: 10.1007/s13201-021-01554-7

Using biochar to control nitric oxide air pollution

Ghada Osama Mohamed [1]*, Maher Elsayed Saleh [2], Elsayed Ahmed Shalaby [1], Ahmed Samir Elsafty [1]

[1] Institute of Graduate Studies and Research, Alexandria University, Alexandria, Egypt
[2] Department of Soil & Water Science, Faculty of Agriculture, Alexandria University, Alexandria, Egypt

Abstract. This study deals with the ability of Rice Husk Biochar (RHB) to adsorb and reduce the concentration of 90 ppm Nitric Oxide (NO) gas which passed through it with a 1.2 liter/min flow rate within 8 minutes. The characteristics of biochar before and after adsorption were studied by SEM, EDAX, and FTIR analyses. To study the effect of different factors on adsorption the effect of manufacturing temperatures of biochar was studied by using two degrees 450 ^{O}C and 500 ^{O}C. It has also been measured the effect of biochar particles size using two types of size, coarse (C) and fine (F). The effect of the shape of the measurement system on the adsorption was also measured using two types of systems A and B. In general, the RHB450C A gives the best adsorption capacity for NO (95.7 mg/g) followed by RHB450F A (41.9 mg/g) followed by RHB500C A (24.8 mg/g). Both Langmuir and Freundlich isothermal models were applied to mathematical modelling of NO adsorption, and based on the coefficient of determination (r^2) Langmuir's model provided a perfect fit to the experimental data. It was also found that the pseudo-first-order model is suitable for studying adsorption kinetics.

Keywords: Rice husk Biochar, greenhouse gases, Nitric oxide adsorption, Air pollution.

1. Introduction

Air pollution is one of the most serious environmental problems that have direct and indirect negative effects on human health, the surrounding environment, and climate change [1]. Nitrogen oxides (NO_X) are one of the primary air pollutants which seriously affect atmospheric chemistry, including the destruction of ozone (O_3). Nitrogen oxides are mostly colorless and odorless gases and toxic and slightly soluble in water and they can stay in the atmosphere for several days [2]. Nitric oxide (NO) represents almost 90% of the compounds in the NO_X group. NO is highly reactive because of being a diatomic radical with one unpaired electron and being toxic. The numerous studies which are done after air pollution accidents show that the changes in air quality contribute to an increase in the number of deaths [3, 4]. For this reason, it was necessary to pay attention to the methods of controlling and reducing pollutants, studying the methods used in purifying, sequestering, and adsorbing pollutants, and knowing ways to treat them to reduce the concentrations higher than the permissible limit in the air. There are many methods used to control gases pollutants like NO_X gases which depend on the source of the pollution and its causes, for example (avoiding pollutant production by combustion control [5, 6]. Converting the product pollutant to another product with less polluting by introducing it into chemical reactions such as oxidation and reduction reactions, Selective catalytic reduction (SCR), Selective non-catalytic reduction (SNCR), and Catalytic decomposition [7]. All these methods have their limitations contrary to the situation in removing the pollutant from the source by filtration or adsorption techniques which are relatively simple, automatically easily controlled, and effectively capture low concentrations of hazardous substances making it a more important method [2, 8]. The materials which can adsorb gases pollutant and contributes to treating the gases produced from the flue (e.g., Nitrogen oxides, Sulfur, and volatile organic compounds (VOC)) are Zeolites, limestone powder, Polymers, molecular sieve adsorbents, and peat containing NH_3 and carbon materials such as activated carbon and charcoal [5]. Activated carbon with well-developed pore structures, large internal surface area, and a variety of active groups can be employed as an adsorbent for NO because the process of utilizing activated carbon to adsorption NO is simple and high efficiency. But most studies focused on the chemical adsorption mechanism and its effect on NO removal efficiency, not physical adsorption still rare [5]. Recently, some studies have reported using metal-loaded biochar for SCR of NO with ammonia at low temperatures [9]. But the catalytic oxidation of NO over biochar has been scarcely studied [10]. Biochar is a highly porous carbon-rich material that results as a by-product from pyrolysis of biomass, and its physicochemical properties depend on production process conditions (feedstock, temperature, oxygen limitation, and residence time) [11, 12]. Biochar is used in many environmental applications such as improving soil properties or purifying water such as sewage water and reducing greenhouse gases emissions from soil [13, 14]. The coal resulting from the thermal decomposition of the biomass is characterized by its high containment of organic carbon and a large surface area and porosity, which makes it applicable to more than one purpose. This has been demonstrated in many previous experiments. An experiment in which biochar was used in comparison with activated carbon in removing ammonia showed that non-activated biochar can replace commercial activated carbon in gaseous ammonia removal and that the removal efficiency increases when activated with phosphoric acid [15]. The results showed due to the influence of the chemical and physical properties of biochar that the adsorption of H_2S was better than that of carbon dioxide. It also showed a competitive removal of both H_2S and CO_2 gases for the biochar types used in the experiment [16, 17]. 30 million tons per year of biomass from agricultural waste has been monitored in Egypt [18, 19]. Since this biomass is the raw material for biochar production, this makes the production of biochar much less expensive compared to its counterparts from other carbon materials such as activated carbon [20].The selection of biomass for the development of biochar mainly depends on the abundance and availability of the lignocelluloses wastes [21]. Using agricultural wastes to produce biochar for gases pollutants would be a viable solution both to reduce the solid waste and, also, to reduce gases emission from combustion sources. Generally, biochar porosity properties and surface chemistry can be improved through some physical and/or chemical modifications [22]. In this study, the residues of rice husks (RH), which are produced in large quantities from rice production plants, were used as raw materials for the manufacture of biochar (RHB). It was used in its traditional form without chemical activation to study the possibility of offering it as an available and cheap material that can be manufactured in traditional ways and be able to compete with other commercial materials of high cost in absorbing gaseous pollutants. Nitric oxide has been used as one of the gases that cause environmental problems that must be controlled. Biochar has abundant oxygen-containing functional groups [23], which can significantly contribute to the oxidation of NO to NO_2 and its physical adsorption on the adsorbent is considerably higher than NO [24].

2. Materials and Methods

2.1. Adsorbent Materials
To prepare the Rice husk Biochar (RHB). Rice husk (RH) feedstock, was collected from a private sector rice mill located in Burg Al Arab city in Alexandria governorate. To remove extraneous materials such as dirt, sand, and other impurities from RH was washed with tap water and left to dry air, as shown in figure 1.

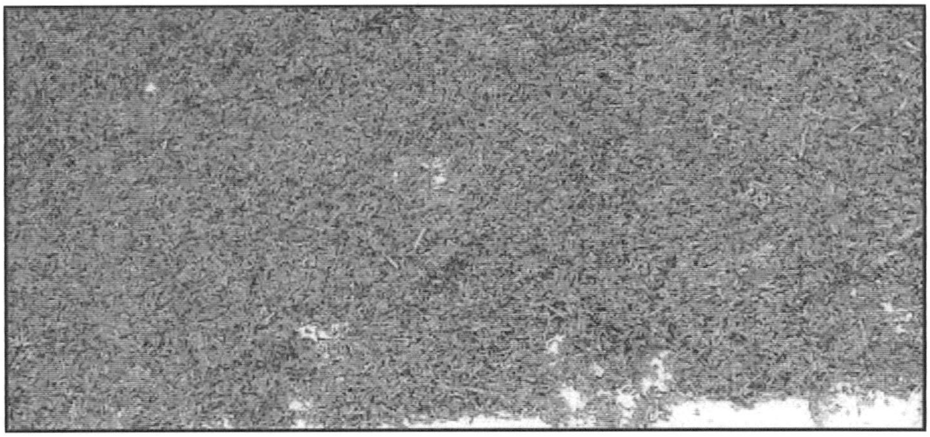

Figure 1. The feedstock of RH after washed and while leaving them to dry before using them in the pyrolysis process.

2.2. Biochar preparation

The air-dried RH feedstock was converted to biochar by slow pyrolysis at two temperature levels (450 °C and 500 °C) for 2 hours in a limited oxygen environment in a muffle furnace (VULCANE A-550). Quantities of the feedstock were placed in aluminum containers and covered to avoid contact with air by wrapping them tightly with aluminum foil and then piercing them from the top with a laboratory needle to create many pores to volatile gases exit. The muffle temperature was raised by 50°C until it reached 450°C or 500°C, then the pyrolysis process was left for 2 hours, then the muffle was closed to leave the biochar cooled inside it overnight. We had two parts of the quantities of produced biochar from each temperature, one of them was crushed and sieved by a stainless steel sieve with a diameter of 1 mm to obtain coarse particles (C) with a size (from 1 mm to 0.5 mm). The other part was crushed and sieved with a diameter of 0.250 mm to obtain fine particles (F) with a size (less than 0.250 mm) to be ready to use in the measurement experiment, as shown in Figure 2.

Figure 2. The RHB after pyrolysis process with two type of size ((C) from 1mm to 0.5 mm and (F) from 0.250 mm to 0.125 mm).

2.3. Adsorption System

2.3.1. System A. It is a cylinder made of stainless steel with 249.326 cm^3 of volume (14.4 cm in length and 4.7 cm in diameter), has two helical sides provided with a rubber bar and 400 mesh (37 μ) screen barrier, and the filter papers barrier added to it at inlet and outlet of gas.

2.3.2. System B. It is a canister made of reinforced plastic with 468.7 cm^3 of volume (19.7 cm in length and 5.8 cm in diameter). It has two metal and filter papers barriers with a metal screw spring to control the filling volume of the canister and install barriers inside. The inlet and outlet of gas enter and exit a free space (free of adsorption material) inside the canister from the same direction in the side to ensure that the gas is distributed well inside the canister.

2.4. Characterization of Biochar

2.4.1. Physiochemical properties of RHB. Some specific physicochemical parameters were determined in RHB as follow:

Biochar yield: The RHB yield was calculated according to the following equation [25]:

$$\text{Biochar yield } (\%) = (W_1/W_0)\ X\ 100 \tag{1}$$

Where W_0 is the weight of the air-dried feedstock of biochar (g), and W_1 is the weight of the produced biochar (g).

Ash content: The ash content was determined and calculated according to the following equation [26, 25]:

$$\text{Ash content } (\%) = (W_{ash} / W_{sample})\ X\ 100 \tag{2}$$

Where W_{ash} is the weight of ash (g) and W_{sample} is the weight of biochar (g).

Moisture content: it was determined according and calculated to the following equation [26]:

$$\text{Moisture content } (\%) = [(W_2 - W_3)/ (W_2 - W_1)]*\ 100 \tag{3}$$

Where W_1 is the weight of crucible (g), W_2 is the initial weight of crucible and biochar sample (g), W_3 is the final weight of crucible with biochar sample (g).

2.4.2. Scanning electron microscopy (SEM) and energy dispersive spectroscopy (EDAX) analysis. The morphology and elemental composition of RHB 450 (C and F) in their two cases before and after NO gas adsorption was studied using SEM (FEI Inspect S) for field emissions. During operation, the device's accelerating voltage was maintained at 25 kV.

2.4.3. Fourier transforms infrared (FTIR) spectroscopy analysis. To study the compositional properties and determine the functional groups of RHB 450 (C and F) before and after NO gas adsorption was used FTIR BRUKER M-IR spectrometer in the range 4000-500 cm^{-1}.

2.4.4. Iodine and methylene blue numbers. The analyses of iodine number and methylene blue number were used to infer the porosity of RHB by adsorbing both iodine dye and methylene blue dye on RHB. According to the American Standard test method for determination of iodine number [27] with using the following equation:

$$X/M= [A - (DF)\ (B)\ (S)]\ /M \tag{4}$$

Where A is (N_2*12693), B is (N_1*126.93), DF is ((I+H) /F), N_2 is Iodine normality, N_1 is Sodium Thiosulfate normality, DF is Dilution Factor, I is Iodine volume (ml), H is 5% hydrochloric acid used (ml), F is Filtrate (ml), S is Sodium Thiosulfate volume (ml), M is biochar used (g) and X/M is Iodine adsorbed per gram of biochar (mg/g).

And to determine of methylene blue number using the following equation [28]:

$$MB_{N\ (mg/g)} = \left[\frac{(Co - Ce) \times V}{M}\right] \quad (5)$$

Where C_0 is the concentration in the MB solution at the samples, C_e is the concentration in the MB solution at the blank, V is the volume of solution treated and M is the amount of adsorbent added (g).

2.5. NO gas adsorption experiment

2.5.1. First stage. Firstly, the concentration of NO standard gas was measured by connecting the rubber tubes from the cylinder to the FTIR gas analyzer (Gasmet 2002) directly, with a 1.2 (liter/minute) flow rate. The concentration was 90 ppm and used as a blank value for samples. Second, System A without biochar was connected with the standard NO gas cylinder and the measuring device, which gave the same readings for gas concentration as in the blank. This method was used to make sure that there is no affect of the system without the adsorption material on the measured gas concentrations readings and their time. Third, we filled system A with coarse sand (which is a non-adsorbent material) and used the same measurement method and no change was observed in the concentration reading, which also indicates that there is no effect of filling the system with the material on the gas flow, to make sure that the vacuum shape inside the system will not effect on gas movement inside it, that is to avoid any other effects on the adsorption process while using the system except the biochar.

2.5.2. Second stage. First, the same measurement technique mentioned earlier was used with system A filling with the first sample of RHB450 F biochar as the start of the experiment. A gradual increase was observed in the device readings of the gas concentration until it reached stability near the blank concentration during a certain period that was recorded. This process was repeated more than once to ensure the validity of the data. Then this method of measurement was adopted for all other biochar samples used in the experiment (RHB450 C, RHB500 F, and RHB500 C), with the filter paper replaced with new paper each time. Second, system A was replaced by system B and the same measurement process was repeated for system B. Figure 3 shows the path of gas starting from its exit from the cylinder, passing through the gas flow control unit, then the system containing the biochar (filter), and ending with the measuring unit gas concentration.

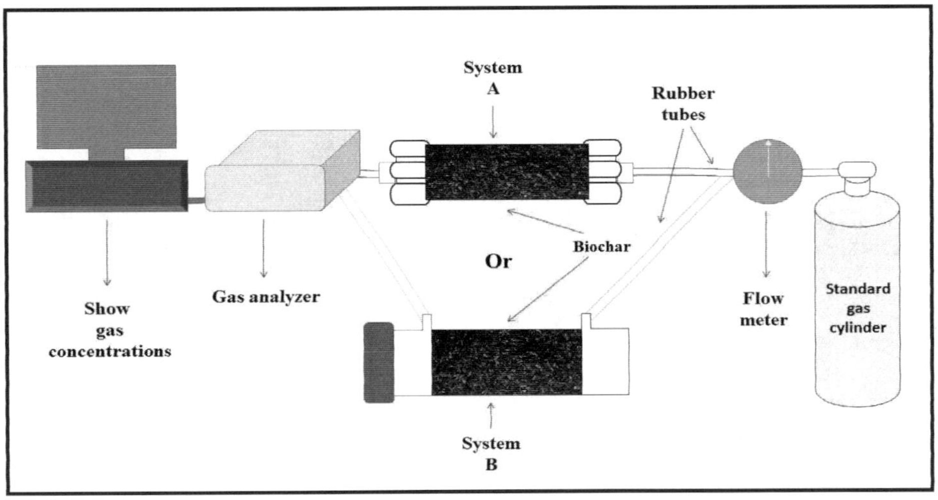

Figure 3. Diagram showing how the measurement process works.

2.6. Adsorption Isotherm

- Langmuir isotherm model assumes that adsorption takes place on a homogeneous surface (monolayer) and that it can only occur on a fixed active site and also can only hold one adsorbate molecule at a time without interaction between adsorbed molecules. The Langmuir adsorption isotherm can be represented by the following equation [29]:

$$\frac{C_e}{q_e} = [(\frac{1}{K_L}) + (\frac{b}{K_L})]C_e \quad (6)$$

Where, q_e is the amount of NO adsorbed per unit weight of RHB (mg/g), C_e is the NO concentration in effluent at equilibrium (mg/ L), b/K_L value is used to estimate the theoretical monolayer capacity of RHB and b (mg/g) is Langmuir constant.

- The Freundlich isotherm assumes that adsorption takes place on a heterogeneous surface with a non-uniform distribution of heat of adsorption through a multilayer adsorption mechanism. The Freundlich isotherm can be expressed by the following linearized form of equation [30]:

$$log\, q_e = [log\, k_F + (\frac{1}{n})log]C_e \quad (7)$$

Where q_e is q_e is the amount of NO adsorbed per unit weight of RHB (mg/g), C_e is the NO concentration at equilibrium, K_F are Freundlich constants (mg/g) and n is the Freundlich exponent

2.7. Kinetic adsorption

To determine an appropriate kinetic model it is necessary to analyze the experimental data to investigate the mechanism of adsorption process that may include mass transfer or chemical reaction. Many models have been applied to expound the adsorbate transfer onto the particles of adsorbent [31].The determination coefficient r^2 is used to examine the confirmation of the predicted values of models with experimental data. The validity of these models is evaluated by the determination coefficient r^2, which is within the range of 0–1, in which r^2 closer to unity implies the best fitting toward the particular kinetics model [32].

2.7.1. Simple-first-order model. The sorption kinetics is described by a simple order equation. The simple first-order describes by the following linear equation [32]:

$$log\, C_t = [(\frac{k_1}{2.303})\, t] + log\, C_0 \quad (8)$$

Where C_t and C_o are the concentration of adsorbate at time t and initially (mg/l), respectively, and k_1 is the first-order rate constant, (1/min).

2.7.2. pseudo-first-order model. The sorption kinetics is described by Lagergren pseudo first-order by the following linear equation [33, 34]:

$$log(q_e - q_t) = log\, q_e - [(\frac{k_1}{2.303})\, t] \quad (9)$$

Where q_e is the amount of NO adsorbed at equilibrium (mg/g), q is the amount of NO adsorbed at time t (mg/g), and k_1 is the pseudo-first-order constant (1/min).

2.7.3. pseudo-second-order model. The adsorption kinetics may also be described by pseudo-second-order by the following linear equation [35, 36]:

$$\frac{t}{q_t} = \left[\frac{1}{(K_2 q_e{}^2)}\right] + [\left(\frac{1}{q_e}\right) t] \quad (10)$$

Where K_2 is the equilibrium rate constant of pseudo-second-order adsorption (g/mg. min).

3. Results and discussion

3.1. Characterization of Biochar

3.1.1. Physiochemical properties. The Selected physicochemical properties of the RHB are shown in table 1 where RHB 500 has a higher value of ash content (4.61%) and moisture content (4.05%) than RHB450 which has ash content (3.51%) and moisture content (4.7%) but in yield content, RHB450 have a higher value (50.6%) than RHB500 (43.08%).

Table 1. Physical Parameters of RHB.

	RHB 450	**RHB 500**
Biochar yield %	50.6	43.08
Ash %	3.51	4.61
Moisture %	4.7	4.05

3.1.2. SEM. Scanning electron microscope (SEM) images were taken to study the shape, size, and nature of the pores in the particles of biochar before and after the adsorption process of NO gas with the scale of 50μ magnifications. As shown in figure 4 for the RHB450 C sample and figure 5 for the RHB450 F sample, the pores on RHB are not well shaped with the diminished structure of pores. There was a difference in the shape of the pores before and after the adsorption process, where the gas adsorption led to the close (or disappearance) of the pores that were visible before adsorption. Due to the lack of information on the use of biochar to remove nitrogen oxide (NO), activated carbon is the one that was used for these purposes [37]. The previous results of removing the NO by activated carbon indicated that the microspores around (10 Å) exhibited the best NO removal capacity [37]. Availability of microspores around 1-10 nanometers in the adsorbent material could help remove NO and enhance the efficiency of biochar.

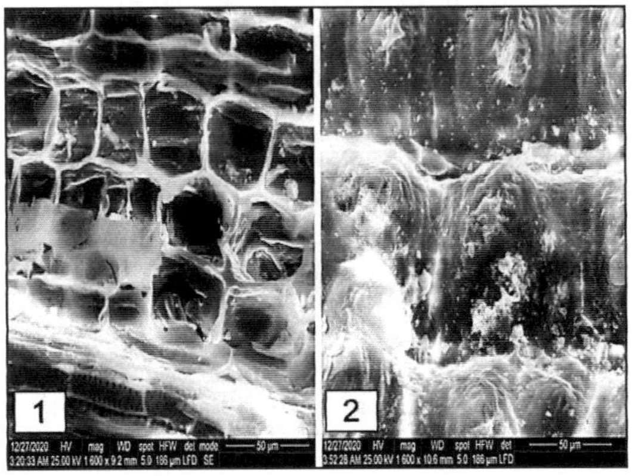

Figure 4. Scanning electron microscopy (SEM) of RHB 450 F before (1) and after (2) NO gas adsorption.

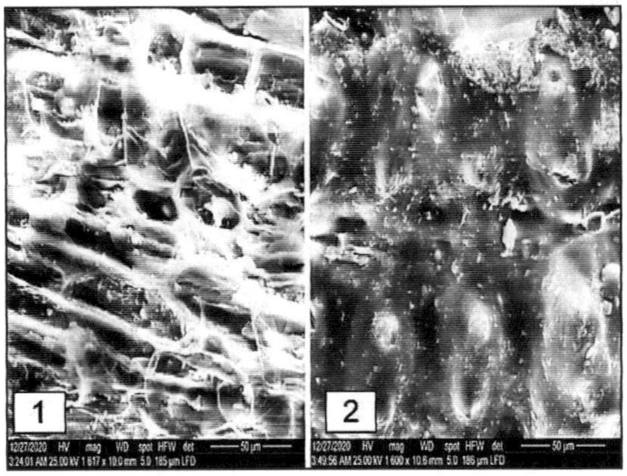

Figure 5. Scanning electron microscopy (SEM) of RHB 450 C before (1) and after (2) NO gas adsorption.

3.1.3. *EDAX.* The X-Ray (EDAX) data in table 1 show the elemental composition of biochar before and after nitrogen oxide removal, respectively. This analysis was done for RHB 450 only. It is clear from the content of the elements that silicon comes in third place after carbon and oxygen with the higher nitrogen content of N in biochar after the removal process of NO. This trend was observed in both coarse (C) and fine (F) particles of biochar. In general, the appearance of EDAX features of nitrogen (N) on the surface of biochar indicates the ability to bind gas. Through the results shown, it seems that the adsorption of nitrogen oxide on the biochar led to an increase in the content of nitrogen in the composition of the biochar, which may indicate that the adsorbed gases may react chemically and share the structure.

Table 2. Dispersive X-ray (EDX) of the RHB 450 C and RHB 450 F before and after NO gas adsorption.

	RHB 450 C				**RHB 450 F**			
	Before adsorption		**After adsorption**		**Before adsorption**		**After adsorption**	
Element	***Wt%***	***At%***	***Wt%***	**At%**	***Wt%***	**At%**	***Wt%***	**At%**
C	41.71	51.89	41.27	52.05	45.27	56.20	53.31	61.72
O	38.04	37.04	36.90	34.94	34.86	32.49	35.53	30.88
Si	19.25	10.68	19.54	10.54	18.43	09.78	07.38	03.66
N	01.00	00.40	02.29	02.47	01.44	01.53	03.78	03.75

3.1.4. FTIR. Fourier transform infrared spectroscopy (FTIR) is an important tool to identify the characteristic functional groups, which are instrumental in the adsorption of contaminants. The FTIR study provides the change in functional groups of biochar before and after the gas adsorption as shown in figures 6, 7. The FTIR spectrum of the two fractions of RHB 450 before NO adsorption revealed a few absorption peaks within the interval of 3500–500 cm-1, which is only a sign of the simple chemical nature of RHB. The peaks ranged from 3746 cm-1 to 2858 cm-1, indicating the existence of a free hydroxyl functional group due to the chemically absorbed water and the surface hydroxyl groups. Also, this range indicates the existence of free hydroxyl groups of alcohols and phenols functional groups [38]. The peak which is around 3206.24 cm-1 corresponding to the O-H due to carboxylic group and C-H stretching vibration indicates the presence of an alkane functional group. The peak which is around 2919.53 cm-1 corresponds to the O-H stretching, indicates the presence of carboxylic group and carbonyl. In both figures, the largest beak (1062-1048 cm-1) represents the most abundant active group, which is Si-O-Si, followed by the beak (795.23 to 791.5 cm-1), which represents the active group of Si-O in both cases before and after the absorption of nitrogen oxide. These active silicate groups play a major role in the surface reactivity of RHB towards the removal of most heavy metals from wastewater [39, 40], and in current work could play the same role for NO removal.

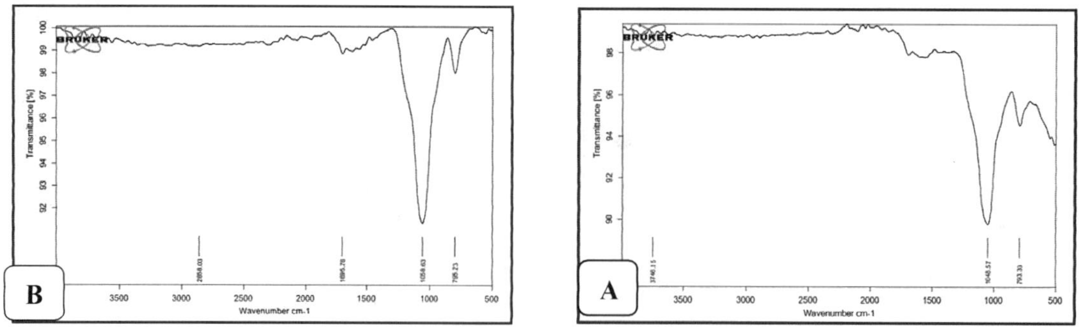

Figure 6. FTIR of the RHB 450 C before (B) and after (A) NO gas adsorption.

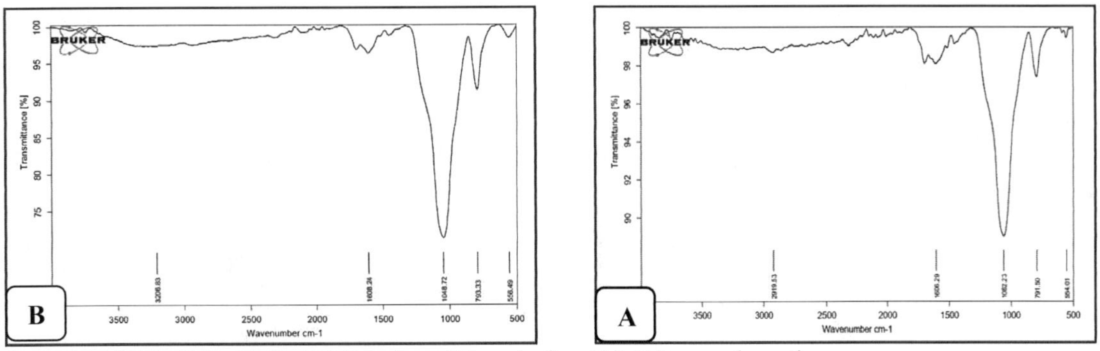
Figure 7. FTIR of the RHB 450 F before (B) and after (A) NO gas adsorption.

3.1.5. The iodine number and methylene blue number. Iodine adsorption is a simple and quick test, indicating the internal surface area of the carbon; in many types of activated carbons, the iodine number (expressed as milligrams of iodine per gram of carbon) is close to the Brunauer-Emmett-Teller (BET) surface area [41]. Table 3 shows the results of iodine and methylene blue adsorption on biochar samples (RHB 450 F, RHB 450 C, RHB 500 F, and RHB 500 C) used for nitrogen oxide removal. Iodine number (IN) is a parameter based on the capacity of iodine adsorption, providing a complementary idea of the microspore structure because iodine molecules are mainly adsorbed into microspores [42]. As shown in table 3 the fine samples of biochar had higher IN than the coarse samples for RHB. The iodine number resulted indicates that there is an effect of pyrolysis temperature on the surface area produced, as the IN of RHB fine produced with 450°C was higher than that produced with 500°C. This effect of temperature on the iodine number values does not have a specific direction so we cannot say that the increase in the temperature of the pyrolysis is accompanied by a decrease in the surface area. The biochar source and the associated thermochemical shifts (the transformation from aliphatic to aromatic structure and the degree of aromaticity) during pyrolysis has the greatest effect on both the total surface area and the porosity of the resulting biochar. While Methylene blue adsorption indicates the adsorption capacity for large molecules having similar dimensions to methylene blue; it is a quick test for medicinal and other carbons prepared to adsorb large molecules [41]. A similar trend was observed in the values of adsorbed methylene blue (methylene blue number, mg/g) by biochar samples table 3. The values of methylene blue number (MBN) expressed the surface reactivity and adsorption capacity of biochar towards the small organic molecules. The pyrolysis of rice husk feedstock leads to an increase in the microspores percentage and subsequent the specific surface area [43, 9]. Figure 8 summarizes the reactivity of biochar samples towards the adsorption of iodine and methylene blue. We can conclude that both the analyzes of the iodine number and the methylene blue concentration gave each an indication of the surface area and pores present on the samples used, which appears their ability to adsorb.

Table 3. The amount of adsorbed iodine and methylene blue (mg/g) on biochar samples used in the experiment.

Biochar samples	Iodine number (mg/g)	Methylene Blue number (mg/g)
RHB 450 F	1662.8	1036.5
RHB 450 C	896	300.2
RHB 500 F	1467.3	753.2
RHB 500 C	1188	580.2

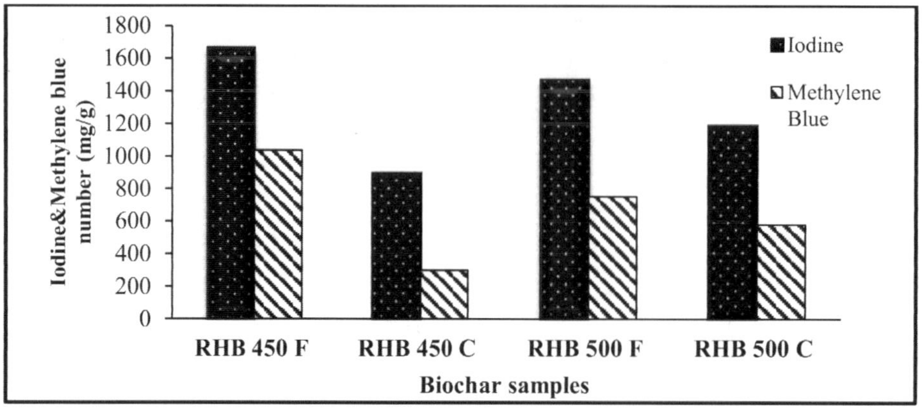

Figure 8. The amount of adsorbed methylene blue & iodine number of each type of biochar used in the experiment

3.2. Estimating and measuring the effects of different factors on the gas adsorption by biochar

The experiment results which are shown in table 4 were divided according to the factors that have effects on the measurement process, such as the preparation temperature of biochar, the particles size of biochar material, and also the kind of system which are used in the measurement process all of this was among the effects that have been studied by comparing the results and the graph of them and to come out with conclusions. The results in Table 4 indicated that the comparison was made between the results of NO gas concentration obtained from the experiment from two different preparation temperatures (450°C & 500°C) of Biochar that with the same sources of the feedstock material of Biochar (RHB) and that is with installed the comparison of each of the two cases of material particle size (F&C) and with the two systems which were used in the measurement (A&B) and that is in addition to the comparison with the blank of the experiment.

Table 4. Comparison of RHB in both of preparation temperature 450, 500 in each of his F, C particles size and both of A, B system with the readings of residual NO gas concentrations C_e (ppm) over time (second) and the NO adsorption capacity on RHB q_e (mg/g).

Time Sec	Blank C_e ppm	A RHB 450 F		A RHB 450 C		A RHB 500 F		A RHB 500 C		B RHB 450 F		B RHB 450 C		B RHB 500 F		B RHB 500 C	
		C_e	q_e	C_e	q_e	C_e	q_e	C_e	q_e	C_e	q_e	C_e	q_e	C_e	q_e	C_e	q_e
60	83.5	4.2	1.45	3	2.93	37	0.63	33.8	0.85	5.5	0.68	37.6	0.82	28.5	0.41	46.3	0.61
120	84.5	12.9	2.77	5.1	5.82	58	0.98	48.1	1.46	29.9	1.15	72.7	1.03	59	0.61	75.6	0.75
180	87.8	19.2	4.02	8.18	8.72	60	1.36	45.8	2.18	54.8	1.44	81.9	1.13	74	0.71	80.4	0.87
240	88.9	27.5	5.15	17.1	11.3	61.9	1.72	42.5	2.97	68.7	1.61	80.2	1.29	74.4	0.82	78.6	1.04
300	87.4	38.5	6.05	21.9	13.7	70.5	1.95	49.1	3.62	76.9	1.71	80.5	1.41	83.7	0.85	80.5	1.15
360	89.6	43.3	6.90	27.7	15.9	84	2.02	62.6	4.08	80	1.79	80.9	1.57	82.8	0.90	74.5	1.40
420	87.3	50.5	7.57	34.6	17.9	86	2.04	63.2	4.49	82.5	1.83	80.9	1.68	82.2	0.94	77.6	1.56
480	83.9	58.6	8.03	44.8	19.3	69	2.24	45	5.15	83.7	1.83	81.1	1.73	79	0.97	83	1.57
Total qe in 480 second			41.94		95.7		13		24.78		12.04		10.7		6.20		8.96

3.2.1. The effect of preparation temperature of RHB on the adsorption of NO gas. To study the effect of pyrolysis temperature on the NO adsorption capacity, the RHB prepared at different carbonization temperatures (450 and 500 ^{0}C) were subjected to the NO adsorption and the results are presented in figure 8 illustrates that in the same case of particles size fine (F) and system (A) the RHB 450 gave the highest adsorption capacity than the RHB500. After 8 minutes of passing the NO through RHB450, the adsorption capacity was 41.94 mg NO/g RHB while the adsorption capacity with RHB500 was 13 mg NO/g RHB. The same thing has happened with the system (B) in the same case with fine particles figure 9 RHB450 gave the highest adsorption capacity (12.04 mg NO/g RHB) than the RHB500 (6.20 mg NO/g RHB). Also, the same thing has happened with coarse particle (C) of RHB in the systems (A and B). According to the results of table 4 and figures 9,10,11,12, the pyrolysis temperature was effective to enhance the NO uptake capacity. The variations in the NO uptake capacity of biochar developed at different carbonization temperatures should be associated with the structural evolutions and transformations that take place during pyrolysis. And to confirm this premise one of the similar studies showed the adsorption of nitric oxide on biochar, where rubber seed shell biochar was used with different degrees of pyrolysis. Some of the spectroscopic analyzes that were conducted in the study showed the significant influence of carbonization temperature on the structural features of the biochar. This was proven in the alteration in the pore and microstructure of the biochar. Where the porosity development and enlarging of the pores along with the increase of the carbonization temperature are observed which are shown by the SEM analysis [44]. One of the important points that were observed in that study is that despite the development of porosity and creating more defects with increasing pyrolysis temperature, the adsorption capacity of NO for 900 was almost identical to that of 700 at all adsorption temperatures. High temperatures can cause new microspores to form due to biochar particles being broken down as they undergo the decellularization process. However, the higher temperature can cause the existing pores to widen and expand as indicated by the SEM images [44, 45]. It is also plausible that the surface chemical groups of biochar, which contribute to the gas uptake, were affected by high temperature. According to that, we can conclude that both the pore structure and surface chemical groups may be important for selective sorption of NO [44].

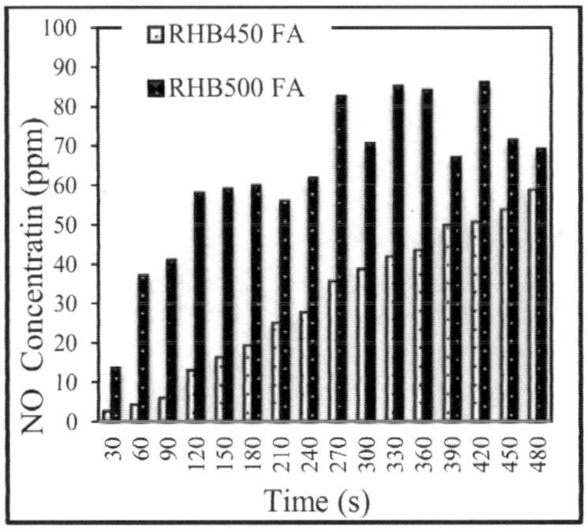

Figure 9. the adsorption of NO gas on both of RHB450 and RHB500 of the fine (F)particle using filtration system A.

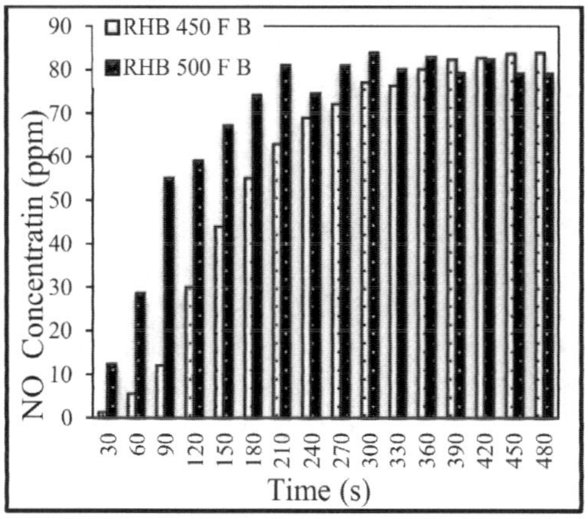

Figure 10. the adsorption of NO gas on both of RHB450 and RHB500 of the fine (F)particle using filtration system B.

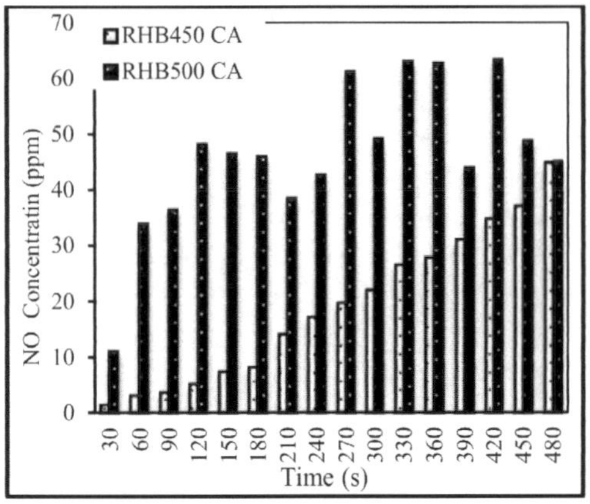

Figure 11. the adsorption of NO gas on both of RHB450 and RHB500 of the coarse (C)particle using filtration system A.

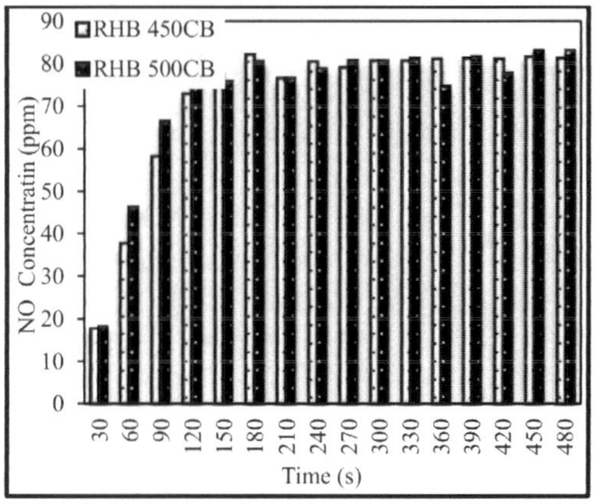

Figure 12. the adsorption of NO gas on both of RHB450 and RHB500 of the coarse (C)particle using filtration system B.

3.2.2. Effect of the particle size of RHB particles on the adsorption of NO gas. As shown in figure 13, in the same conditions of the manufacturing temperature of biochar450 and system (A), the RHB with coarse particles C gave the highest adsorption capacity (95.7 mg NO/g RHB) than the fine particles F (41.94 mg NO/g RHB) within the 8 minutes. In the same case of system A and time of adsorption with RHB500 as shown in figure 14 the RHB (C) gave the highest adsorption capacity (24.78mg NO/g RHB) than RHB (F)(13 mg NO/g RHB). While as shown in figure 15, in the case of the RHB450 type of biochar and with using measurement system (B) the RHB (F) gave a higher adsorption capacity (12.04 mg NO/g RHB) than RHB (C)(10.7mg NO/g RHB) with no significant difference between them. The opposite has happened with RHB500 in the system (B) as shown in figure 16 where the fine particles (C) gave the highest adsorption capacity (8.96 mg NO/g RHB)than the coarse size (F)(6.20 mg NO/g RHB). Since the kinetic diameters of N_2 (0.364 nm), NO (0.317), and NO_2 (0.340 nm) differ only slightly from any pore filling by these gases due to steric effects, the amount of adsorption is expected to be affected by pore size. Only for gases whose critical temperatures (TC) are higher than the applied adsorption temperature [46]. Accordingly, at atmospheric pressure, the adsorption of N2 and NO in the microspores is negligible compared to that of NO_2 adsorbed. It is also established that the extent of filling of the microspores in the carbon material is related to the van der Waals constant [46].there is a relationship between the pore size/structure and the uptake of NO_2 and the condensation of NO_2 is presumably favored in microspores of carbons. Therefore, it could be speculated that the excess heating of the biochar although contributing to the porosity development and creation of new pores, however, adversely affected the existing micro porous structure, which significantly contributed to the sorption and condensation of gas molecules, by enlarging the pores [44].

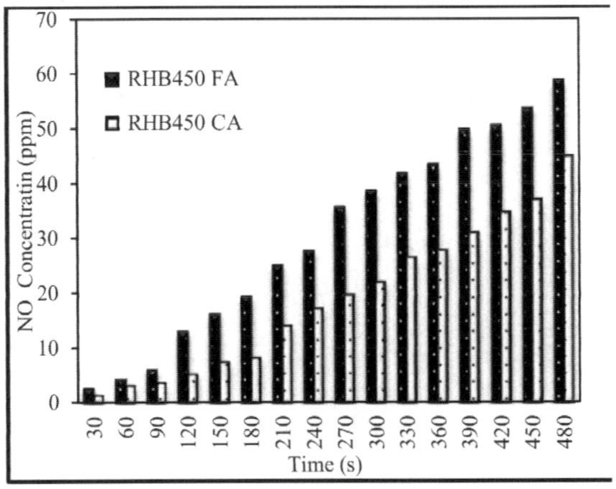

Figure 13. Effect of Particle size (F and C) of RHB 450 on adsorption of NO gas using filtration system A

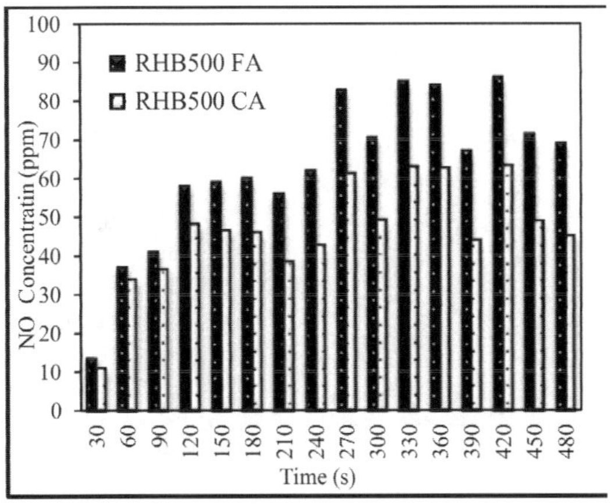

Figure 14. Effect of Particle size (F and C) of RHB 500 on adsorption of NO gas using filtration system A.

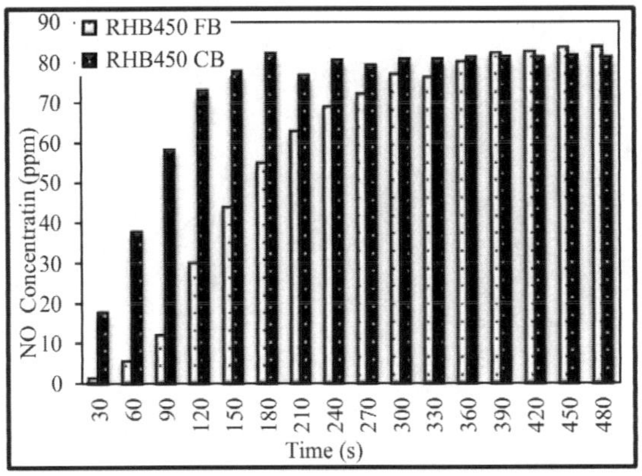

Figure 15. Effect of Particle size (F and C) of RHB 450 on adsorption of NO gas using filtration system B.

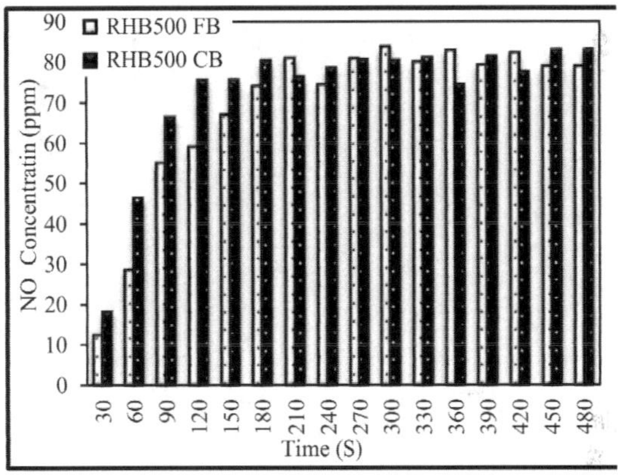

Figure 16. Effect of Particle size (F and C) of RHB 500 on adsorption of NO gas using filtration system B.

3.2.3. The effect of the filtration system on the Biochar adsorption of NO gas. Figure 17 illustrated that in the same case of the size of the fine particles (F) and manufacturing temperature of RHB450 system (A) gave the highest adsorption capacity than the system (B) after 8 minutes. In system (A) the adsorption capacity of NO gas was (41.94 mg NO/g RHB) while the adsorption capacity of NO gas at the same time with the system (B) was (12.04 mg NO/g RHB). As shown in figure 18, system (A) appeared the highest adsorption capacity than system (B). As shown in figure 19, the RHB450 and size (C) the system (A) gave the highest adsorption capacity (95.7 mg NO/g RHB) than system (B) (10.7 mg NO/g RHB). And the same thing happened in figure 20 with RHB500 and size (C) the adsorption capacity with the system (A) was (24.78mg NO/g RHB) while system (B) was (8.96 mg NO/g RHB). The filtration system A gave better gas adsorption than system B in both cases of the difference in the manufacturing temperature and the size of particles for RHB.

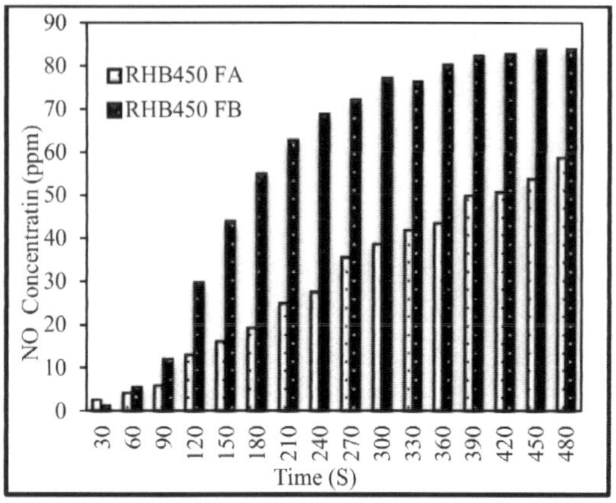

Figure 17. Effect of filtration system (A and B) of RHB 450 (F) on adsorption of NO gas.

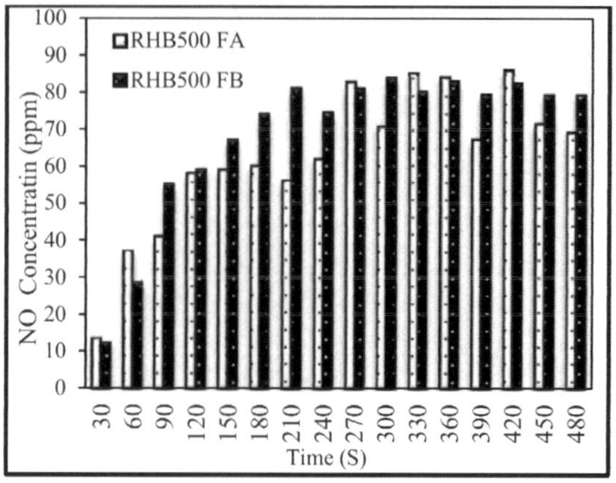

Figure 18. Effect of filtration system (A and B) of RHB 500 (F) on adsorption of NO gas.

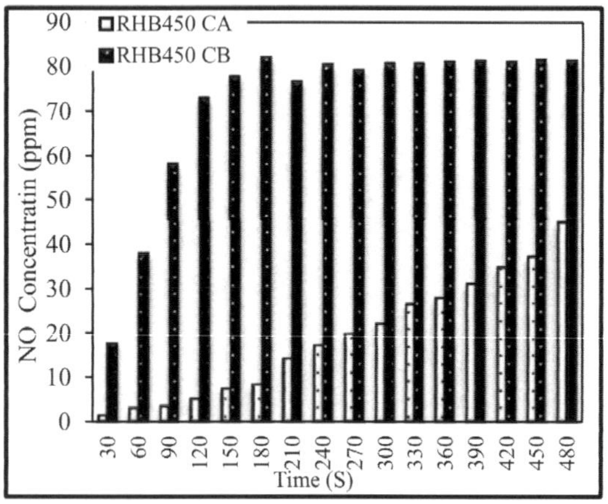

Figure 19. Effect of filtration system (A and B) of RHB 450 (C) on adsorption of NO gas.

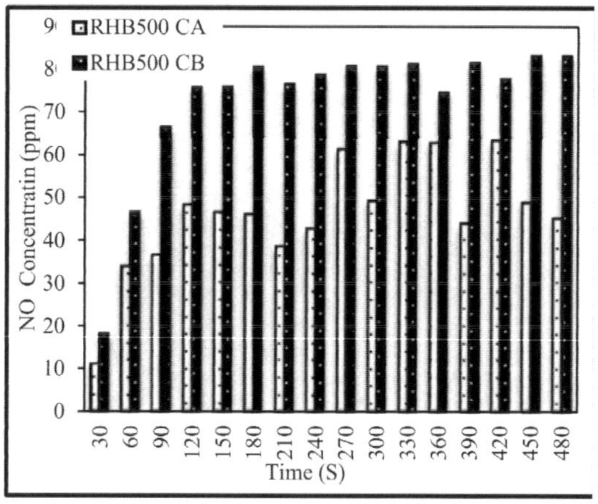

Figure 20. Effect of filtration system (A and B) of RHB 500 (C) on adsorption of NO gas.

3.3. Adsorption isotherm studies

The equilibrium data can be approximated using common and practical adsorption isotherms, which provide the basis for the design of adsorption systems. The amount of adsorbed NO onto all RHB as a function of its concentration at constant temperature can be described by the Langmuir equation (r^2 range 0.91 – 0.996) for filtration system B and (r^2 range 0.486 – 0.822) for filtration system A that is shown in table 5 where the Langmuir model gives the best fit toward the experimental data. This means that the surface of the RHB is heterogeneous and a restricted monolayer NO adsorption occurs as a result of adsorption NO, this explains what the SEM images showed after adsorption in the changing in the shape of the surface on the RHB450. While in the Freundlich isotherm model which is confirmed by the value of 1/n ranging between 0 and 1 which is a measure of adsorption intensity or surface heterogeneity and becomes more heterogeneous as its value gets closer to zero. That the adsorption is not favorable for RHB. And based on tabulated data, it was noted that the maximum capacity (q_e) of RHB for NO adsorption the q_{max} of the NO, derived from the Langmuir equation did not coincide well with the experimental (actual) values.

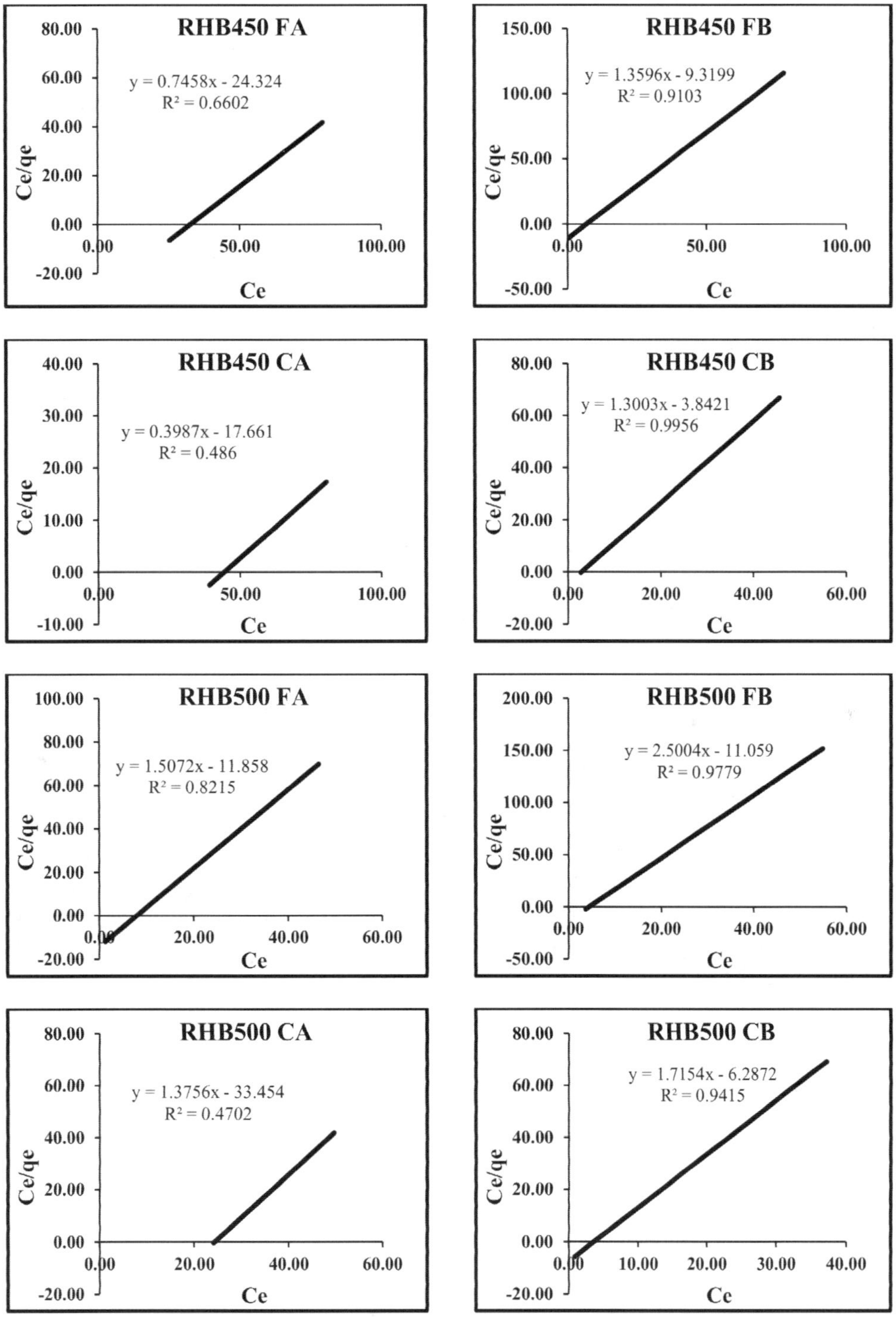

Figure 21. Langmuir isotherm of adsorption of NO gas on RHB samples.

Table 5. Parameters of Langmuir and Freundlich isotherm models at same temperature for adsorption of NO onto RHB samples.

			Langmuir constants					Freundlich constants			
			b l/mg	K_L	r^2	q_m mg/g	$q_{e\,L}$ mg/g	K_F	n	r^2	$q_{e\,F}$ mg/g
RHB 450	A	F	-0.03	-0.04	0.660	1.34	1.149	650.9	-0.8	0.68	0.265
		C	-0.02	-0.06	0.486	2.5	2.226	35334	-0.51	0.58	0.140
	B	F	-0.15	-0.11	0.910	0.7	0.630	1.928	-8.33	0.433	0.846
		C	-0.34	-0.26	0.996	0.76	0.660	2.33	-3.67	0.69	0.557
RHB 500	A	F	-0.13	-0.08	0.822	0.66	0.579	0.713	-4.16	0.83	0.694
		C	-0.04	-0.03	0.470	0.7	0.661	525.6	-0.68	0.35	0.309
	B	F	-0.23	-0.09	0.978	0.4	0.344	1.5	-3.53	0.85	0.103
		C	-0.27	-0.16	0.942	0.58	0.506	1.6	-5.08	0.35	0.547

3.4. Kinetic adsorption studies

The analysis of the kinetic adsorption process is a useful tool to estimate the time of residence for the adsorption process to complete and to determine the dynamics of adsorption. Thus, simple-first-order, pseudo-first-order, and pseudo-second-order of adsorption kinetic model were performed. In the simple first-order kinetic model for the adsorption kinetics of NO onto RHB the slope and intercept of plots of log C_t versus t were used to determine the simple first-order rate constants, k_1 and C_0 shown in table 6. Table 6 shows the simple first-order kinetic model for RHB did not fit well with the experimental data, with r^2 value found to be within the range of 0.37–0.95. The pseudo-first-order considers the rate of occupation of adsorption sites is directly proportional to the number of unoccupied sites. The slope and intercept of plots of log (q_e-q_t) versus t were used to determine the pseudo-first-order rate constants, k_1, and q_e shown in table 6 at different initial sorbent concentrations. Table 6 shows the pseudo-first-order kinetic model for RHB fit well with the experimental data of r^2 value found to be within the range of 0.827–0.99. The pseudo-second-order plot for the adsorption kinetics of NO onto RHB is given in table 6. The slope and intercept of plots of t/q_t versus t were used to calculate the pseudo-second-order rate constants k_2 and q_e. The second-order rate constants k_2, the maximum adsorption constant q_e values and correction coefficients (R^2) presented the pseudo-second-order model gives also fits with experimental data, with r^2 value within the range of 0.848– 0.99. the comparing the values of determination coefficients as stated in Table 6, both pseudo-first-order (r^2: 0.827 – 0.995) and pseudo-second-order (r^2: 0.848 – 0.998) equations can describe nitrogen oxide absorption in all prepared biochar (coarse and fine) produced at 450 and 500 degrees in both filtration systems.

Table 6. Kinetic parameters of NO adsorption onto RHB samples.

			Simple first-order			Pseudo-first-order			Pseudo-second-order		
			K_1	C_0	r^2	K_1	q_e	r^2	K_2	q_e	r^2
RHB 450	**A**	**F**	0.334	5.51	0.87	-0.026	41.12	0.984	$2.9*10^{-3}$	23.42	0.974
		C	0.386	2.6	0.955	-0.028	94.86	0.994	$3.3*10^{-4}$	98.04	0.903
	B	**F**	0.301	12.28	0.624	-0.014	11.22	0.827	0.19	2.41	0.992
		C	0.07	52.98	0.409	-0.014	9.92	0.989	0.195	2.21	0.92
RHB 500	**A**	**F**	0.09	42.9	0.676	-0.02	12.36	0.922	0.057	3.6	0.974
		C	0.053	37.8	0.405	-0.028	24.57	0.995	$1.8*10^{-3}$	21.6	0.848
	B	**F**	0.11	40.95	0.547	-0.013	5.75	0.909	0.42	1.2	0.998
		C	0.048	59.31	0.379	-0.019	8.53	0.984	0.1	2.32	0.903

3.5. Comparing the results of the study with previous studies

As shown in Table 7, the comparison between the adsorption capacity of different concentrations of nitric oxide on both RHB and different types of carbon has been tested under different conditions in previous studies. Where RHB showed a good adsorption capacity compared to the rest of the comparisons, and the type RSSB that was tested in one of the studies showed an increase in the ability to adsorption when activated chemically or with metal than its state without activation, which means that when applying the same method in the case of RHB it will give a higher adsorptive capacity than its current ability, which make it recommended to use it in more studies in this field.

Table 7. Comparison of the NO adsorption capacity of different carbon-based adsorbents from the literature.

Adsorbent	NO concentration in the gas	Adsorption condition	Time of adsorption	Flow rat	NO uptake capacity	References
Ordered mesoporous carbon (OMC)	500 ppm	P: atmospheric T: 50 °C			19.4 mg/g (OMC)	[47]
Coconut shell-based activated carbon	500 ppm	P: atmospheric T: 25 °C			23.0 mg/g	[48]
Cerium loaded palm shell activated carbon (Ce/PSAC)	500 ppm	P: atmospheric T: 150 °C			3.5 mg/g	[49]
Molecularly imprinted adsorbents	1500 ppm	P: atmospheric T: 40 °C			12.1 mg/g	[47]
Rubber seed shell biochar (RSSB)	2000 ppm	P: atmospheric T: 30 °C	180 minutes	75 ml/min	17.61 mg/g	[44]
Ce-loaded (RSSB)	2000 ppm	P: atmospheric T: 30 °C	18 minutes	0.1ml/min	142.53 mg/g	[44]
KOH-activated (RSSB)	2000 ppm	P: atmospheric T: 30 °C	23 minutes	0.1 l/min	470.59 mg/g	[44]
RHB 450 C	90 ppm	P: atmospheric T: 34 °C	8 minutes	1.2 l/min	95.7 mg/g	This study

4. Conclusions

According to Scanning Electron Microscopy (SEM) results the biochar has many pores which make it able to absorb pollutants. In general, the adsorption of biochar for Nitric Oxide (NO) gas in the case of C particles size (1.0 – 0.5 mm) was better than its case in F particles size (0.5 – 0.25 mm), in terms of the total amount of pollutants that was adsorbed on its surface. The experiment also proved that RHB prepared at 450 °C is more efficient than RHB prepared at 500 °C. The differences in the flow direction, amount of biochar, and system size are factors affecting the adsorption efficiency as shown in the two systems (A and B). The adsorption results were fitted on an Langmuir isotherm model, which gives a better fit to describe the NO gas adsorption process on RHB. Also, the adsorption kinetic studies matched with the pseudo-first-order kinetic model of NO gas adsorption on RHB.

5. Recommendations

We recommend that there is a possibility of using biochar produced from rice husks as an effective gaseous pollutant removal material by adsorption, but there are many factors affecting the adsorption efficiency which must be taken into consideration and need future studies, such as: Applying more studies on other raw materials of biochar types and gases which can be used in adsorption. Studying more factors that affect the adsorption process, such as changes in temperature, different gas flow rates, and different gas concentrations. Applying more than one type of measuring system with different sizes and shapes to achieve the best design for a high-performance measuring system. Using biochar after increasing its efficiency by preparing it through more advanced methods and chemically activated. Studying the possibility of reusing the biochar used in adsorption more than once to increase its economic advantages.

References

[1] Sumathi S, Bhatia S, Lee K and Mohamed A 2010 Cerium impregnated palm shell activated carbon (Ce/PSAC) sorbent for simultaneous removal of SO_2 and NO Process study *Chemical Engineering Journal* **162** 51-7

[2] Yang J, Zhou J, Tong W, Zhang T, Kong M and Ren S 2019 Low-temperature flue gas denitration with transition metal oxides supported on biomass char *Journal of the Energy Institute* **92** 1158-1166

[3] Abujarada S, Walton A S, Thomas A G, Chohan U K and Koehler S P 2019 Adsorption site, orientation and alignment of NO adsorbed on Au (100) using 3D-velocity map imaging, X-ray photoelectron spectroscopy and density functional theory *Physical Chemistry Chemical Physics* **21** 10939-10946

[4] Zhang P Y, Yang Q Y, Xu L S, Zeng H C and Zhang L 2004 Experimental study on adsorption and oxidation of activated carbon fiber to NO at low temperature *Electric Power Environmental Protection* **20** 25-28

[5] Bauer F W 1994 Fully engineered approach to acid rain control *Energy Engineering* United States p 91

[6] Lu W, Zhao X, Wang H and Xiao W 2000 Catalytic oxidation of NO *Chinese Journal of Catalysis* **21** 423-427

[7] Haywood J M and Cooper C D 1998 The economic feasibility of using hydrogen peroxide for the enhanced oxidation and removal of nitrogen oxides from coal-fired power plant flue gases *Journal of the Air & Waste Management Association* **48** 238-246

[8] Kong M, Liu Q, Guo F, Jiang L, Yao L, Ren S and Yang J 2018 Physicochemical properties of pine-derived bio-chars modified by metal oxides and their performance in the removal of NO *Journal of the Energy Institute* **91** 467-472

[9] Anthonysamy S I, Lahijani P, Mohammadi M and Mohamed A R 2020 Low temperature adsorption of nitric oxide on cerium impregnated biomass-derived biochar *Korean Journal of Chemical Engineering* **37** 130-140

[10] Le Yan S, Wang M, Zha J, Zhu L, Li W, Luo Q and Wang Z 2018 Environmentally relevant

concentrations of carbamazepine caused endocrine-disrupting effects on nontarget organisms, Chinese Rare Minnows (Gobiocypris rarus) *Environmental science & technology* **52** 886-894
[11] Li M, Liu Q, Guo L, Zhang Y, Lou Z, Wang Y and Qian G 2013 Cu (II) removal from aqueous solution by Spartina alterniflora derived biochar *Bioresource technology* **141** 83-88
[12] Novotny E H, Maia C M B D F, Carvalho M T D M and Madari B E 2015 Biochar: pyrogenic carbon for agricultural use-a critical review *Revista Brasileira de Ciência do Solo* **39** 321-344
[13] Trakal L, Komárek M, Száková J, Zemanová V and Tlustoš P 2011 Biochar application to metal-contaminated soil: Evaluating of Cd, Cu, Pb and Zn sorption behavior using single-and multi-element sorption experiment *Plant. soil and Environment* **57** 372-380
[14] Ro K S, Lima I M, Reddy G B, Jackson M A and Gao B 2015 Removing gaseous NH_3 using biochar as an adsorbent *Agriculture* **5** 991-1002
[15] Sethupathi S, Zhang M, Rajapaksha A U, Lee S R, Mohamad No N, Mohamed A R and Ok Y S 2017 Biochars as potential adsorbers of CH_4, CO_2 and H_2S *Sustainability* **9** 121
[16] Creamer A E, Gao B and Zhang M 2014 Carbon dioxide capture using biochar produced from sugarcane bagasse and hickory wood *Chemical Engineering Journal* **249** 174-179
[17] Gomaa H A, Steele P and Hamdy Y A 2011 Charcoal Industries Egypt *Food and Agriculture Organization of the United Nations Regional Office for the Near East* Cairo, Egypt
[18] El-Haggar S M, Mounir G and Gennaro L 2004 Agricultural waste as an energy source in developing countries, a case study in Egypt on the utilization of agricultural waste through complexes *International Centre for Science and High Technology (ICS) United Nations Industrial Development organization (UNODO)* 1-10
[19] Bamdad H, Hawboldt K and MacQuarrie S 2018 A review on common adsorbents for acid gases removal: Focus on biochar *Renewable and Sustainable Energy Reviews* **81** 1705-1720
[20] Fakayode O A, Aboagarib E A A, Zhou C and Ma H 2020 Co-pyrolysis of lignocellulosic and macroalgae biomasses for the production of biochar *A review Bioresource technology* **297** 122-408
[21] Do Minh T, Song J, Deb A, Cha L, Srivastava V, and Sillanpää M 2020 Biochar based catalysts for the abatement of emerging pollutants *a review Chemical Engineering Journal* **394** 124856
[22] Hassan M, Liu Y, Naidu R, Parikh S J, Du J, Qi F and Willett I R 2020 Influences of feedstock sources and pyrolysis temperature on the properties of biochar and functionality as adsorbents: A meta-analysis *Science of the Total Environment* 140-714
[23] Ghafari M and Atkinson J D 2016 Catalytic NO oxidation in the presence of moisture using porous polymers and activated carbon Environmental science & technology **50** 5189-5196
[24] Samsuri AW, Sadegh-Zadeh F, and Seh-Bardan B J 2014 Characterization of biochars produced from oil palm and rice husks and their adsorption capacities for heavy metals *Int. J. Environ. Sci. Technol.* **11** 967–976
[25] Lynch J and Joseph S 2010 Guidelines for the development and testing of pyrolysis plants to produce biochar *International Biochar Initiative* London UK 1 1 7
[26] ASTM International 2014 Standard Test Method for Determination of Iodine Number of Activated Carbon ASTM D4607-14
[27] Raposo F, De La Rubia M A and Borja R 2009 Methylene blue number as useful indicator to evaluate the adsorptive capacity of granular activated carbon in batch mode: Influence of adsorbate/adsorbent mass ratio and particle size *Journal of hazardousmaterials* **165** 291-299
[28] Ibrahim H G and Al-Meshragi M A 2020 Experimental Study of Adsorption on Capture Activated Carbon for CO2. CO2 Sequestration 111
[29] Olasehinde E F, Adegunloye AV, Adebayo M A and Oshodi A A 2018 Sequestration of aqueous lead (II) using modified and unmodified red onion skin *Analytical Letters* **51** 2710-2732
[30] Fadali O A, Magdy Y H, Daifullah A A M, Ebrahiem E E and Nassar M M 2004 Removal of chromium from tannery effluents by adsorption *Journal of Environmental Science and Health Part A* **39** 465-472
[31] Fang L, Liu J, Ju S, Zheng F, Dong W and Shen M 2010 Experimental and theoretical evidence

of enhanced ferromagnetism in sonochemical synthesized $BiFeO_3$ nanoparticles *Applied Physics Letters* **97** 242-501

[32] Hossain M A, Kumita M, Michigami Y and Mori S 2005 Optimization of parameters for Cr (VI) adsorption on used black tea leaves *Adsorption* 11 561-568

[33] Annadurai G, Juang R S and Lee D J 2002 Use of cellulose-based wastes for adsorption of dyes from aqueous solutions *Journal of hazardous materials* **92** 263-274

[34] Ho Y S and Chiang C C 2001 Sorption studies of acid dye by mixed sorbents *Adsorption* **7** 139-147

[35] Ho Y S and McKay G 1999 Pseudo-second order model for sorption processes *Process biochemistry* **34** 451-465

[36] Zhang W J, Rabiei S, Bagreev A, Zhuang M S and Rasouli F 2008 Study of NO adsorption on activated carbons *Applied Catalysis B: Environmental* **83** 63-71

[37] Khalil H A, Fizree H M, Jawaid M and Alattas O S 2011 Preparation and characterization of nano structured materials from oil palm ash: a bio-agricultural waste from oil palm mill *BioResources* **6** 4537-4546

[38] Saleh M E, El-Damarawy Y A, Assaad F F, Abdesalam A A and Youssef R A 2020 Removal of copper metal ions by sugarcane bagasse and rice husk biochars from contaminated aqueous solutions *Med. J. Soil Sci* **1** 1-17

[39] El-Damarawy Y A, Saleh M E, Assaad F F, Abel-Salam A and Youssef R A 2017 Adsorption of Lead onto a Waste Biomaterial-Biochar *Nature and science* **15** 154-1

[40] Marsh H and Rodríguez-Reinoso F 2006 Activation processes (chemical) *Activated carbon* 322-365

[41] Yenisoy-Karakaş S, Aygün A, Güneş M and Tahtasakal E 2004 Physical and chemical characteristics of polymer-based spherical activated carbon and its ability to adsorb organics *Carbon* **42** 477-484

[42] Saleh M E, Mahmoud A H and Rashad M 2013 April Biochar usage as a cost-effective bio-sorbent for removing NH4-N from wastewater *Int., Conf the Global Climate Change* (Biodiversity and Sustainability: Challenges and Opportunities in Arab MENA region and EuroMed) pp 15-18

[43] Mahmoud A H, Saleh M E and Abdel-Salam A A 2011 Effect of rice husk biochar on cadmium immobilization in soil and uptake by wheat plant grown on lacustrine soil *Alex. J. Agric. Res.* **56** 117-125

[44] Rubel A M, Stewart M L and Stencel J M 1995 Activated carbon for control of nitrogen oxide emissions *Journal of materials research* **10** 562-567

[45] Rubel A M and Stencel J M 1996 Effect of pressure on NOx adsorption by activated carbons *Energy & fuels* **10** 704-708

[46] Cao F, Chen J, Ni M, Song H, Xiao G, Wu W and Cen K 2014 Adsorption of NO on ordered mesoporous carbon and its improvement by cerium *RSC Advances* **4** 16281-16289

[47] Guo Y, Li Y, Zhu T and Ye M 2015 Investigation of SO_2 and NO adsorption species on activated carbon and the mechanism of NO promotion effect on SO_2 *Fuel* **143** 536-542

[48] Sumathi S, Bhatia S, Lee K T and Mohamed A R 2010 Selection of best impregnated palm shell activated carbon (PSAC) for simultaneous removal of SO_2 and NO_x *Journal of Hazardous Materials* **176** 1093-1096

[49] Li P, Ge B, Zhang S, Chen S, Zhang Q and Zhao Y 2008 CO_2 capture by polyethylenimine-modified fibrous adsorbent *Langmuir* **24** 6567-6574

Acknowledgments

All thanks and gratitude to the Main Chemical Laboratories in the Egyptian Armed Forces for their contributions with all possible capabilities and permanent support in many fields and scientific research in particular.

Radiological and mineralogical studies on the coastal area between Port-Said and Damietta

A. M. Osman[1*], R. A. Seif[1], R. M. Attia[2], A. EL-Azab[2], I. H. Khalifa[1]

[1] Geology Department, Faculty of Sciences, Suez Canal University, Egypt
[2] Authority of Nuclear Materials, Al-Katamyia, Egypt

* asmaa_3osman@yahoo.com

Abstract. The study area constitutes the Eastern flank of the Nile Delta between longitudes, 31° 39' 9.2″–32° 20' 9.6″ E and latitudes 31° 37′13″-31° 0′ 54″N, in the shoreline between Port-Said and Damietta. The radioelement contents were measured by gamma-spectrometry using NaI-detector and were studied mineralogically using binuclear and Environmental Scanning Electron microscopes after separation of heavy minerals by heavy liquids. eU, eTh, RaeU and K contents in the studied sediments and sabkhas are ranged from (1-2 ppm), (3-17 ppm), (1-4 ppm) and (0.45-1.25 wt.%) with averages 1.34, 7.18, 1.89 ppm and 0.77 wt.%, respectively. The average of radioelement contents is lower than the international averages, while the average of eTh/eU ratio is higher. High eTh/eU ratio reflects the poor weathering, rapid deposition of these sediments and the dominance of detrital radioactive minerals such as xenotime, monazite and zircon. eU/RaeU ratio average (0.9), is lower than unity and indicate the disequilibrium in the studied coastal sediments, suggesting disequilibrium and uranium migration out. The heavy minerals are represented meanly by magnetite, ilmenite, zircon, rutile, garnet, monazite, pyroxene and amphibole.

Keywords: Radiological studies, mineralogical studies, Port-Said, Damietta, Radioelement contents.

1. Introduction

Black sand deposits are deposits found in the Egyptian Coastal area between Alexandria and El-Arish on the Mediterranean Sea. They contain a lot of radioactive elements such as zircon, sphene and monazite. Detecting of radioactivity in the urbanized areas and marine environment has become very important and attract a lot of researchers, as they have bad effects on the human health. The study area constitutes the Eastern flank of the Nile Delta between longitudes, 31° 39' 9.2–32° 20' 9.6″ E and latitudes 31° 37′13″-31° 0′ 54″N in the shoreline between Port-Said and Damietta. Port-Said and Damietta Cities are considered of the main important cities in the Mediterranean coast because of the potential ports located there, the different activities which improve their economics such as: Fishing activities, agriculture, industry activities. These all causes made these cities, area of interest to many researchers and that's why, they are

worthy to be studied. The area is characterized by the presence of black sand deposits along its shoreline. Although the black sand has a great economic importance, it also, has bad effects on the human health due to the natural radioactivity emissions from the radioactive minerals they contain. Although the harmful effect, it has also economic importance, as example rutile mineral which found in the black sand is used in welding rods industry [1]. Many researchers globally studied regions of high radiation such as India by [2], in South Western Australia [3]. Locally [4] and [5] studied the distribution of radionuclides in Alexandria and Rosetta Coast of Egypt. The analysis of radioactivity of several fractions of heavy minerals sands illustrated that the monazite and zircon sands have high contents of uranium and thorium in comparison with other heavy minerals **(**[6]; [7]**)**.
The Egyptian black sands are coming from the Ethiopia and Central Africa Nile tributaries [8]**,** it deposited in the Nile Delta Coast and transported there by the Damietta and Rosetta branches, [9].According to [10], Ilmenite, Magnetite, Zircon, Garnet and Rutile besides trace amounts of cassiterite and gold are present in the beach placer deposits (black sands) that found along Egypt's Mediterranean Sea coast from El-Arish at the east to Abu-Qir at the west.
This study aims to evaluate the radioactivity levels in the study area and determine their economic minerals.

2. Methods and techniques:

Twenty-six sample were collected in the study area from different environments from the shoreline between Port-Said and Damietta and sabkha along the study area. The samples were collected manually by rotating spiral bar to a depth one meter and the weight of each sample is between five to seven kilograms. The site of each sample was detected by using GPS. The samples were distributed along two profiles from two different environments. 13 sediment samples were collected from beach sands along the shoreline of the Mediterranean coast from Port-Said city to Damietta and the other 13 sediment samples were collected from the Sabkha sediments in the same area **(Figure, 1)**.

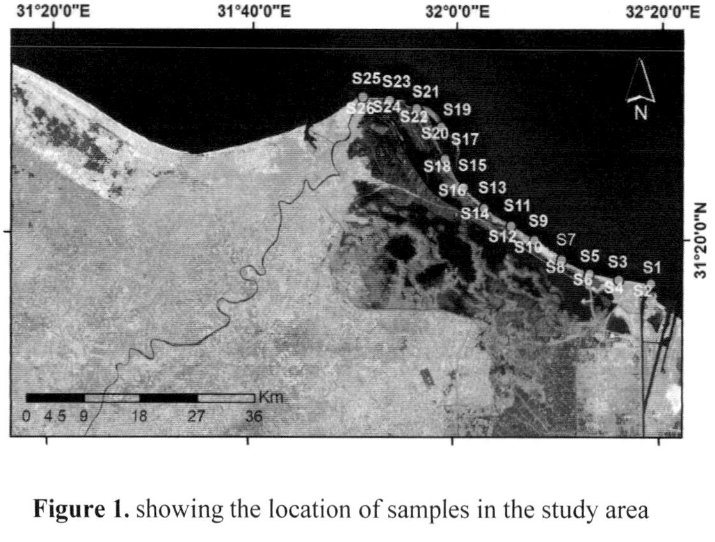

Figure 1. showing the location of samples in the study area

The samples were prepared for measuring their contents of 238U, 226Ra, 232Th and 40K and analyzed in the laboratory of Authority of Nuclear Materials, Al-Katamyia, Egypt by NaI (Tl)-detector. The NaI (Tl)-detector is characterized by high efficiency although it has a poor energy resolution. Its high efficiency allows fast and precise determinations of 40K, 238U, 226Ra and 232Th concentration in rock and soil samples. The results depend on the accuracy of the energy calibration procedure that takes into account the possible interference of each nuclide in each peak region ([11]; [12]; [13], [14] and [15]). The gamma ray spectrometry system consists of Bicron scintillation detector, NaI (Tl) crystal, 76x76 mm, hermetically sealed with a photomultiplier tube in aluminum housing. Stream sediment samples were packed in plastic containers, sealed and stored for at least 21 days to accumulate radon in order to attain radioactive equilibrium and be ready for the radiometric measurements. Uranium is estimated both as eU and Ra (eU) and thorium as eTh. The values of eU represent the concentration of U using Th234 energy peak (93 keV) and measure the first daughter isotope in the U238 decay series with very low possible loss. Radium is measured at the Pb214 energy peak (352 keV) which is considered as a measure for concentration of the U only in case of the secular equilibrium state between U238 and all its daughter isotopes. Thorium is measured at the Pb212 energy peak (238 keV).

3. Gravimetric and Magnetic Separation:

The gravimetric separation is used to separate the heavy minerals from the light minerals by using the heavy liquid as bromoform 2.88 specific gravity. This liquid separate between two minerals the first is lighter than these liquids and flow on it, and the second is heavier than these liquids and sink or fall or merged on it. So, this liquid is abundance and used for the gravimetric separation.

The original sand sample was dried and passed from sieve 1mm to remove the shell, shell fragments, and the gravel (concretion of sand). The sample is quartered until about 500gm, sub-sample was obtained, by John's splitters of different chutes as reference sample.

A representative sub-sample weighting about 60gm was taken from each prepared sample by quartering for mineral separation. Separation was conducted using bromoform (Sp. Gr. 2.86g/cm3) and magnetic fractionation using a Frantz Isodynamic Magnetic Separator (Model L-1).

Magnetic fractionation using a Frantz Isodynamic Magnetic Separator (Model L-1) were used to fractionate heavy minerals according to their magnetic susceptibilities. First, magnetite was collected by a hand magnet, and then the free-magnetite samples were subjected to the magnetic separation at 0.2, 0.5, 1.0, 1.5 magnetic and 1.5 nonmagnetic current amperes. The condition characterized a Frantz Isodynamic Magnetic Separator transvers slop 5, longitudinal slop 20. Each fraction obtained from the magnetic separation process was microscopically studied under binocular stereomicroscope and investigated to calculate the frequency distribution of the concerned minerals in the samples.

Semi quantitative EDX chemical analyses were carried out using a Phillips XL-30 Environmental Scanning Electron Microscope (ESEM).

The clean samples were subjected to heavy liquid separation. The basis of this method is that, mineral grains of specific gravity less than the liquid used will float, where those of higher specific gravity will sink. In these samples, light minerals are mainly quartz and feldspars. The heavy minerals are represented meanly by magnetite, ilmenite, zircon, rutile, garnet, monazite, pyroxene and amphibole, with specific gravity over 3.2. Bromoform of specific gravity 2.86 is very suitable for separation quartz and feldspars from the other heavy minerals.

4. Geology of the area:

Four geomorphic units were recognized on the coastal zone of the Nile Delta between Alexandria and Port-Said[15]. These are the beach and back shore flat, the coastal dunes, the coastal ridges and coastal lakes.

[17], studied the suspended material transported by the Nile; he mentioned that the suspended material varies quantitively, seasonally and annually. He found that the amount of suspended material decreases from one area to another due to the decrease in depth and the construction of artificial controls structures of the river.

The coastal zone area of the Nile Delta is exposed to extensive unplanned development projects which badly affected the land use and land cover [18], while [19] have warned since **1999** of erosion of the coastline intimidate to be the main environmental risk problem of the Delta as the result of this unplanned

strategy for the projects there. [20]demarcated four erosion stretches along the Nile Delta coast. These are Abu-Qir Bay, Rashid promontory and adjoining stretches, the area from El-Burullus outlet to Kitchener drain opening, the Damietta Promontory and the coastal stretches of Port-Said. The other parts of the Nile Delta coast are considered as accretion or stable areas. He supposes the boundaries between erosion and areas have moved from year to year.
After [21]analysis of the beach sand composition and grain size at the beach profile of the Nile Delta coastal line, they concluded that there is a general correspondence between sediment characteristic and patterns of shoreline erosion versus accretion. The high rates of erosion in the Nile Delta resulted in increased saline intrusion into rivers and fresh water aquifers [22].
Previous geological and remote sensing investigations indicate that the margins of the Nile Delta have experienced a land subsidence "submergence" during the mid- to late Holocene ([23]; [24]; [25].
At the end of the Eocene and during the Oligocene, Egypt was exposed to tectonic disturbance manifested by the rise of land and by the retreat of the sea shore to the north of Cairo. The Mediterranean fault system in the ENE-WSW, including the Ruzza Abu Hammad fault [26], besides the Gulf of Suez and Red Sea fault system in NW-SE direction, have been the result of that Palaeogene tectonic disturbance [27]
[28], concluded from his study of the coastal dunes on the coastal plains of the Mediterranean Sea that, the sand deposits show a stable drift from west to east as, it was controlled by the north-west (NW), west and south west (SW) wind direction. He recommended that the sands are actually resulted from the river rather than from the sea, and being a mixture of small angular colored grains of quartz and particles of magnetite.
[29], dealt with the geological studies of coastal area between Ras El-Bar and Port-Said. He concluded that the island in Lake Manzala can be interpreted as old shorelines or as banks of the former Nile branches.
Port-Said area dates back to Holocene Era which consisted of Nile deposits represented in three formations: Mit-Ghamer formation composed of about 115 m of sand, silt sand and mud with overlaps of limestone sand, especially in the upper part of the underground water bearing formations, this is due to its ecological age Pleistocene, Bilqas formation consists of 18 m of soft brown sand structures with overlaps of silt ([30]; [31].
According to [32], Rosetta and Damietta promontories stratigraphic interpretations of their subsurface sediments show three depositional phases, which are related to modern environments from base to top, the phases are transgression, regression and erosional transgression. The sediments of transgression phase include Holocene medium sand-sized particles which originated from re-working of the former Nile branches creating beach and coastal dunes. This sand was deposited under conditions of moderate to high rates of deposition associated with relatively rapidly rising sea level under stormy erosional conditions with slowly rising sea level and high enough rates of deposition, prodelta mud originated from the former branches.

5. Results and discussion:

The radiometric measurements in the studied stream sediments and sabkhas were carried out on twenty-six samples. eU, eTh, RaeU and K contents in the studied sediments and sabkhas are ranged from (1-2 ppm), (3-17 ppm), (1-4 ppm) and (0.45-1.25 wt.%) with averages 1.34, 7.18, 1.89 ppm and 0.77 wt.%, respectively.
It is evident that the average of uranium, thorium, radium and potassium contents are lower than the international averages ([33]; [34]; [35]), **(Table,1).** In contrast, eTh/eU ratio of the studied sediments and sabkhas (5.80) is higher than that of the Upper Continental Crust reported by [36] [37], and [38].The high Th/U ratio of the studied sediments may be related to preferential sorption of the ''immobile'' Th on clay minerals (mostly kaolin), the presence of thorium-bearing minerals like zircon, xenotime and monazite or from uranium mobilization (as $UO2^{2+}$ and/or complexes) relative to Th ([39] [40]; [41]).
The U-content and Th/U ratio in sedimentary rocks are generally used to deduce the conditions under which the highly anomalous mineralized or uraniferous types were formed [42]. Therefore, three types of sediments are differentiated according to their Th/U ratio;

i) The first type includes sediments of Th/U ratio ranging between 0.012 and 0.81. These sediments are developed under conditions where uranium was removed from its source and fixed in the sediments with continuous recharge.

ii) The second type of sediments has Th/U ratio varying between 1.47 and 1.49. They are characterized by their relatively high Th-content due to slightly more scavenging of U-content because of continuous leaching and recharging.

iii) The third type of sediments, exhibits Th/U ratio ranging between 1.49 and 5.47. These sediments reflect the poor weathering and rapid deposition of rock detritus. This type is dominant in detrital radioactive minerals such as xenotime, samarskite, thorite and euxenite usually dominate them.

The obtained data indicate that the eTh/eU ratio of the studied sediments and sabkhas are high and related to the third group reflecting the poor weathering and rapid deposition of these sediments and the dominance of detrital radioactive minerals such as xenotime, monazite and zircon, as will be mentioned later.

Inter-element relations:

eTh-eU variation diagram indicate that there is no any correlation between the two elements in these coastal sediments, suggesting their incorporation in discrete minerals or their separate mobilization. eTh/eU correlation with eU and eTh confirms this conclusion. eTh-K and eU-K relations illustrate thorium and uranium contents didn't relate to potassium minerals alteration processes to clay minerals. Low eTh/K ratios are considered an indication of potassium instability and lead to abrupt changes in this ratio in various sites with rising and lowering which may illustrate vigorous changes in physic-chemical conditions [43]. eTh/K- eTh/eU variation diagram clarify that some uranium was fixed in the accessory minerals while the rest migrated out.

Enrichment of such authigenic U can be calculated by the formula of [44]:

$$(Ua) = (U) - (Th)/3,$$

where the square brackets represent concentration expressed in ppm. The authigenic content, Ua, is extremely low. This U is interpreted as having been migrated in oxidized environment.

5.1 Radioactive equilibrium/disequilibrium:

The study of radioactive equilibrium/ disequilibrium sheds some lights on the surface weathering, alteration and movement of uranium by groundwater. It also helps to explain the isotopic variation of some radioactive nuclides and consequently can be used as a tool for U-exploration processes. The equilibrium state means a closed system (for physical and chemical exchanges of the radionuclides) with the surrounding environments whereas disequilibrium indicates an open system, i.e., the radioactive nuclides may migrate out from their parents.

The liberated radionuclides can be accumulated through several chemical and geological processes, including; chemical weathering, precipitation of minerals from aqueous solutions by biological and inorganic processes, adsorption on suspended clay minerals, etc.

Measurement of both the equivalent uranium (eU) and radium (RaeU) concentrations (in ppm) radiometrically. The equilibrium factor (P) is given as P=eU/RaeU [45]. The equilibrium is achieved if the activity ratio1between the parent/daughter or between daughter/daughter is equal to unity.

In the present study, the radioactive equilibrium/ disequilibrium states of the studied occurrences are discussed through the equilibrium factor (P), i.e., P= eU/RaeU. The obtained average (0.9), is lower than unity and indicate the disequilibrium in the studied coastal sediments **(Table, 1),** suggesting disequilibrium and uranium migration out.

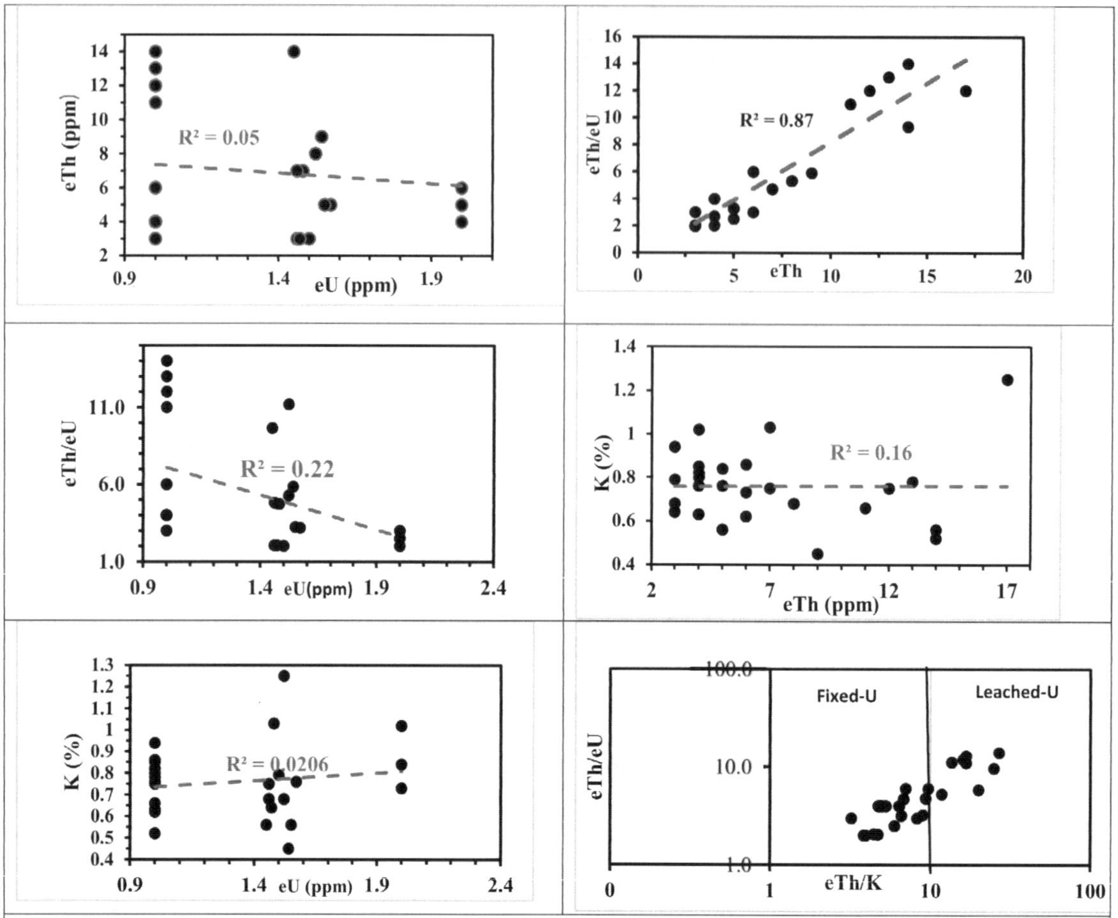

Figure 2. eU-eTh, eU-eTh/eU, eTh-eTh/eU, eU-K, eTh-K, and eTh/eU-eTh/K of the studied sediments

Table 1. radioelement contents (ppm and Bq/kg) of the studied sediments and their ratios.

Sample	eU	eTh	Ra	K	Th/U	eU/RaeU	eU-eTh/3	eTh/K	eU	eTh	Ra	K
.No	(ppm)	(ppm)	(ppm)	(%)	ppm	ppm	ppm	ppm	Bq/kg	Bq/kg	Bq/kg	Bq/kg
S1	1.52	17	4	1.25	11.18	0.38	-4.15	13.60	18.848	68.68	44.4	391.25
S2	1	13	3	0.78	13.00	0.33	-3.33	16.67	12.4	52.52	33.3	244.14
S3	1	11	3	0.66	11.00	0.33	-2.67	16.67	12.4	44.44	33.3	206.58
S4	1.45	14	2	0.56	9.66	0.73	-3.22	25.00	17.98	56.56	22.2	175.28
S5	1	14	3	0.52	14.00	0.33	-3.67	26.92	12.4	56.56	33.3	162.76
S6	1	12	2	0.75	12.00	0.50	-3.00	16.00	12.4	48.48	22.2	234.75
S7	1.54	9	1	0.45	5.84	1.54	-1.46	20.00	19.096	36.36	11.1	140.85
S8	1	6	2	0.62	6.00	0.50	-1.00	9.68	12.4	24.24	22.2	194.06
S9	1.5	3	1	0.79	2.00	1.50	0.50	3.80	18.6	12.12	11.1	247.27
S10	1	4	1	0.63	4.00	1.00	-0.33	6.35	12.4	16.16	11.1	197.19
S11	1.48	7	2	1.03	4.73	0.74	-0.85	6.80	18.352	28.28	22.2	322.39
S12	2	5	1	0.84	2.50	2.00	0.33	5.95	24.8	20.2	11.1	262.92
S13	1.46	3	1	0.68	2.05	1.46	0.46	4.41	18.104	12.12	11.1	212.84
S14	2	4	2	1.02	2.00	1.00	0.67	3.92	24.8	16.16	22.2	319.26
S15	1	3	2	0.94	3.00	0.50	0.00	3.19	12.4	12.12	22.2	294.22
S16	1.57	5	1	0.76	3.18	1.57	-0.10	6.58	19.468	20.2	11.1	237.88
S17	1	4	1	0.82	4.00	1.00	-0.33	4.88	12.4	16.16	11.1	256.66
S18	2	6	2	0.73	3.00	1.00	0.00	8.22	24.8	24.24	22.2	228.49
S19	1.47	3	1	0.64	2.04	1.47	0.47	4.69	18.228	12.12	11.1	200.32
S20	1.55	5	1	0.56	3.23	1.55	-0.12	8.93	19.22	20.2	11.1	175.28
S21	1	4	1	0.85	4.00	1.00	-0.33	4.71	12.4	16.16	11.1	266.05
S22	1	4	2	0.8	4.00	0.50	-0.33	5.00	12.4	16.16	22.2	250.4
S23	1.46	7	2	0.75	4.79	0.73	-0.87	9.33	18.104	28.28	22.2	234.75
S24	1	6	3	0.86	6.00	0.33	-1.00	6.98	12.4	24.24	33.3	269.18
S25	1	4	2	0.76	4.00	0.50	-0.33	5.26	12.4	16.16	22.2	237.88
S26	1.52	8	2	0.68	5.26	0.76	-1.15	11.76	18.848	32.32	22.2	212.84
Min.	1.00	3.00	1.00	0.45	2.00	0.33	-4.15	3.19	12.40	12.12	11.10	140.85
Max.	2.00	17.00	4.00	1.25	14.00	2.00	0.67	26.92	24.80	68.68	44.40	391.25
Av.	1.34	7.18	1.89	0.77	5.80	0.91	-1.05	10.19	16.62	29.00	21.01	239.56
Arenaceous sediments (([33]; [34]; [35])	1	3	-	1.4	3	-	-	-	33	45	32	412

5.2 Mineralogical studies:

5.2.1 Distribution of Total Heavy Minerals in the Study Area

The distribution of total economic heavy minerals (heavy Bromoform 2.86 Sp. gr.) **(Table 2).** The microscopic studies of heavy minerals separated from the different sand deposits show that the beach environment has total beach heavy minerals ranging from 3% to 7.1% with mean 3.8%.

Table 2. The percentage of Total Heavy Mineral in the studied samples:

S. No.	Total Heavy	S. No.	Total Heavy
S 1	4.2	S 14	4.3
S 2	5.5	S 15	3.5
S 3	3.95	S 16	3.9
S 4	4.2	S 17	3.7
S 5	3.2	S 18	3.5
S 6	4.1	S 19	3.84
S 7	3.5	S 20	3.57
S 8	3.75	S 21	3.5
S 9	7.1	S 22	4.3
S 10	4.1	S 23	3.8
S 11	3.9	S 24	3.1
S 12	3.85	S 25	3.0
S 13	3.3	S 26	3.6
Min.		3.0	
Max.		7.1	
Ave.		3.8	

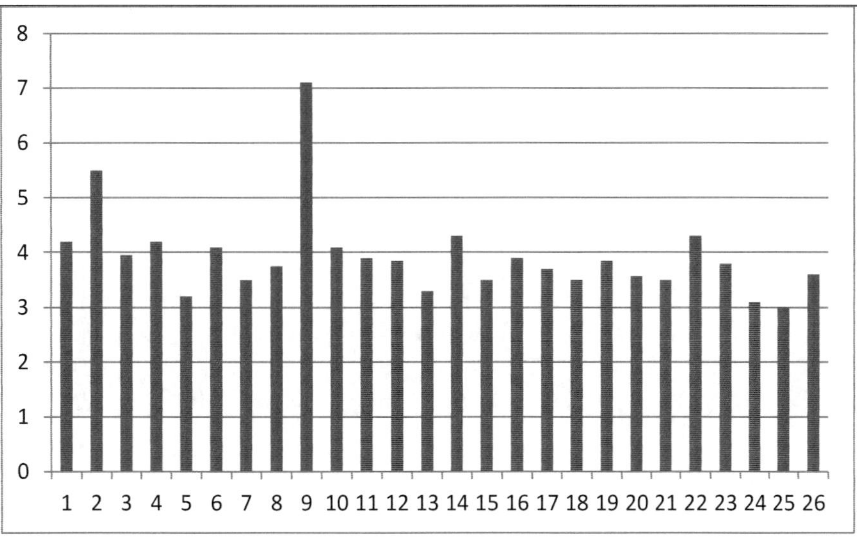

Figure 3. Histograms of percentages of the total heavy minerals for Beach environments.

The result revealed that the heavy minerals content in the beach is high at the western part of the coast and decrease eastward. Also, it shows that increasing of heavy minerals content is mainly related to the active Nile branches (Dumyat) which carrying sediments to the sea especially pre sixtieths of last century.
These sediments are related to the old extinct river branches; Canopic, Sebbenytic, and Bucoloic.
The productions of magnetic separation process were examined under binocular stereomicroscope to determine the proportions of mineral percentages in each fraction.
The counting was applied to a representative small sample (about 0.1gm) taken from each fraction and sprinkled as a layer of one grain thick on a glass plate. Counting of each mineral was carefully executed and the frequency was calculated according to equation given by [46]as follow:
The weight of each mineral in the fraction was calculated by multiplying the percentage value of the mineral by the weight of the fraction and divided by 100. In addition, weights of each economic mineral in the different fractions were summed to get the weight of mineral in the original sample. The obtained frequencies for the different minerals were recalculated again to estimate the percentage of each mineral in the original sample. This is does by multiplying the obtained frequency for each identified mineral by the frequency of the magnetic fraction and divided by 100. Then, the frequencies of each mineral in the different magnetic fractions were a cumulated and tabulated according to the environmental collection.

5.2.2 Physical Properties of the Heavy Economic Minerals

The heavy economic minerals in the Egyptian black sand deposits are mainly, magnetite, ilmenite, rutile, garnet, zircon, and monazite in addition to the alteration product of ilmenite "leucoxene". They differ greatly in optical properties and magnetic susceptibility from high magnetic to very low magnetic. Some of the properties of these minerals are study under binocular stereomicroscope. These minerals were counted and its percentages were estimated and tabulated in **(Table 3)**, according to environmental collection.

Table 3. The percentage of Heavy Economic Minerals in studied samples:

S. NO.	Mg.%	Ilm.%	Rut.%	Lux.%	Gar.%	Zr.%	Mz.%
S 1	0.35	0.58	0.06	0.03	0.01	0.01	0.001
S 2	2	0.5	0.04	0.02	0.05	0.01	0.005
S 3	0.6	0.78	0.06	0.04	0.02	0.02	0.01
S 4	0.7	1.1	0.12	0.01	0.03	0.01	0.01
S 5	0.16	1.02	0.15	0.05	0.06	0.01	0.002
S 6	0.1	0.5	0.11	0.03	0.08	0.2	0.01
S 7	0.4	1.1	0.09	0.01	0.04	0.01	0.03
S 8	0.14	1.26	0.22	0.04	0.06	0.07	0.01
S 9	1	2.1	0.1	0.02	0.05	0.05	0.01
S 10	0.8	0.8	0.12	0.1	0.04	0.03	0.01
S 11	0.2	0.61	0.02	0.15	0.01	0.01	0.004
S 12	0.4	1.03	0.13	0.03	0.05	0.2	0.02
S 13	0.17	1.1	0.02	0.02	0.02	0.1	0.01
S 14	1.1	0.7	0.05	0.1	0.04	0.07	0.006
S 15	1	0.5	0.03	0.06	0.01	0.01	0.02
S 16	0.3	0.61	0.04	0.02	0.03	0.01	0.01
S 17	0.2	0.36	0.1	0.15	0.06	0.02	0.03
S 18	0.35	0.58	0.16	0.03	0.04	0.01	0.001
S 19	2	0.5	0.14	0.02	0.06	0.01	0.005
S 20	0.6	0.8	0.16	0.04	0.03	0.02	0.01
S 21	0.4	1.23	0.13	0.03	0.06	0.2	0.02
S 22	0.17	1.13	0.12	0.02	0.03	0.1	0.01
S 23	0.75	0.25	0.04	1.1	0.07	0.006	0.1
S 24	0.58	0.33	0.02	1	0.01	0.02	0.06
S 25	0.69	0.4	0.03	0.3	0.01	0.01	0.02
S 26	0.42	0.21	0.06	0.2	0.02	0.03	0.15
Min.	0.1	0.21	0.02	0.14	0.04	0.05	0.02
Max.	2	2.1	0.22	1.1	0.08	0.2	0.15
Ave.	0.6	0.8	0.09	0.14	0.04	0.05	0.02

Mg.= Magnetite Ilm.= Ilmenite Rut.= Rutile Lux.= Leucoxene Gar.= Garnet Zr.= Zircon Mz.= Monazite
T. Eco. = Total Economic

Magnetite:

It is easily separated using a small hand magnet with suitable strength. Magnetite form small black, shiny euhedral to subhedral crystals, while partial alterations appear as yellowish stains on some grains **(Figure 4)**. Generally, magnetite as the heaviest mineral of the black sand minerals (Sp. Gr. 5.6). The magnetite in beach environment, shows very high to high concentration in some localities. The percentages of magnetite minerals ranging from 0.1% to 2% with mean 0.6% **(Table 3).**

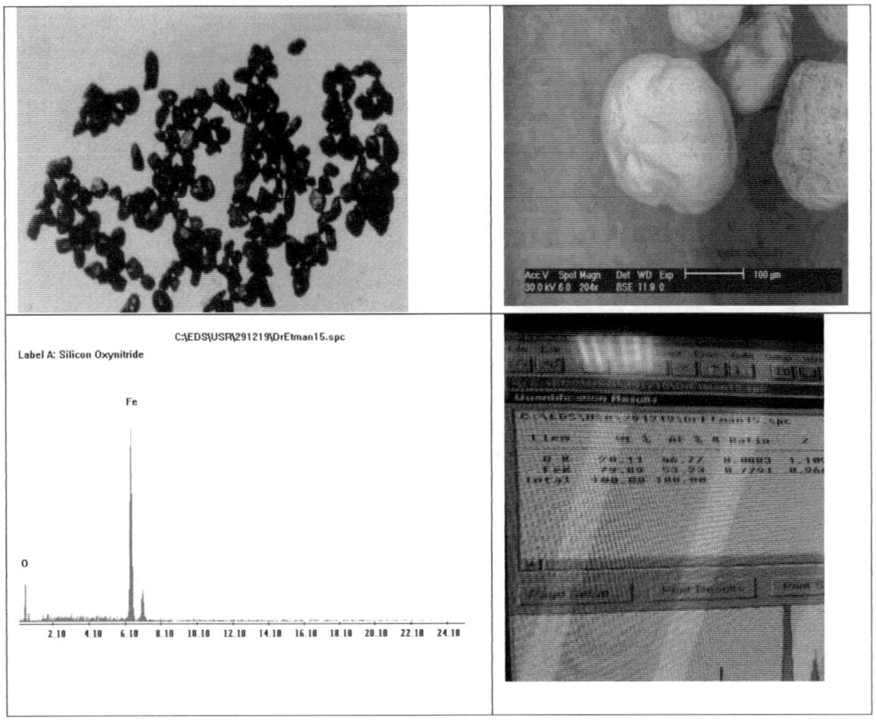

Figure 4. Photograph of **Magnetite** crystals.

Ilmenite:

Ilmenite is the most abundant heavy economic mineral in the Egyptian black sand deposits. It is found as well rounded to subrounded black grains with its characteristic purple gray submetalic luster, **(Figure. 5).** The percentages of ilmenite minerals ranging from 0.21% to 2.1% with mean 0.8% **(Table 3)**.

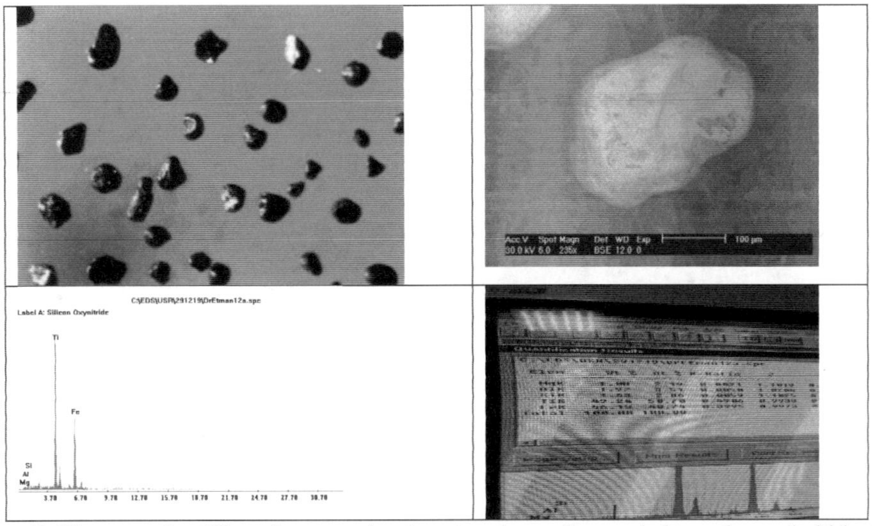

Figure 5. Photograph of ilmenite crystals, showing the well-rounded to subrounded black grains of submetalic luster, (X. 20).

Rutile:

Rutile has a wide range of magnetic susceptibilities. It concentrates in the low magnetic fields according to its iron content. Egyptian beach rutile is transparent, translucent and opaque. The Egyptian beach rutile particles appear under the binocular stereomicroscope as a small prismatic euhedral to subhedral grains. Some rounded grains have color varying from black, brown, red, and orange to yellowish, simple twinning was detected in some grains **(Figure. 6)**.

The percentages of rutile minerals ranging from 0.02% to 0.22% with mean 0.09% **(Table 3).**

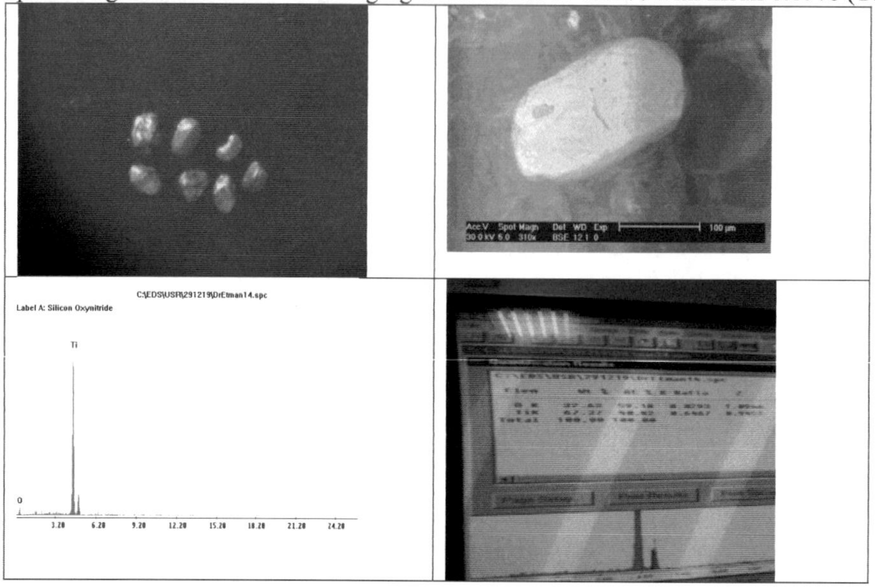

Figure 6. Photo of rutile crystals, showing the euhedral to subhedral, rounded grains, which varying in color from black to red, (X. 20).

Leucoxene:

Indeed, Leucoxene is an alteration product of ilmenite and is the transition stage between ilmenite and the secondary rutile. It is composed meanly from minute crystals of brookite and anatase.

The leucoxene commonly occurs as rounded grains, opaque in transmitted light. A rough pitted surface is characteristic of most grains". The color of leucoxene grains from the Egyptian black sands varies from white, yellowish white to brown, it depends mainly on the degree of alteration, **(Figure,7)**. [47]stated that, as the iron content decreases, the grain color becomes lighter brownish yellow, yellow, yellowish white and light creamy colors. Its specific gravity ranging from 3.5 to 4 according to the number of vugs formed in grains through iron withdrawal processes.

The percent of leucoxene minerals are ranging from 0.14% to 1.1% with mean 0.14% **(Figure, 7 and Table 3)**.

Figure 7. Photograph of leucoxene crystals, showing the rounded, rough pitted surface of the grains (X. 20).

Garnet:

Garnet crystallizes in the isometric system, at the Egyptian beach sand is mainly formed of angular to subrounded particles. Its hardness is about 7 (according to moho's scale of hardness) and its grain size is relatively coarser than the other economic minerals of the Egyptian beach sands.

[48] stated that, garnet occurs in all colors except blue. The reddish orange, red, violet, yellow, yellowish green, pale pink, green and colorless varieties are recorded in the Egyptian black sands. The more common color for Egyptian black sand garnet is the pale pink color **(Figure, 8)**. Garnet has moderate magnetic susceptibility which varies slightly according to variation in chemical composition [49]. The recorded garnet varieties in the study area are moderately magnetic.

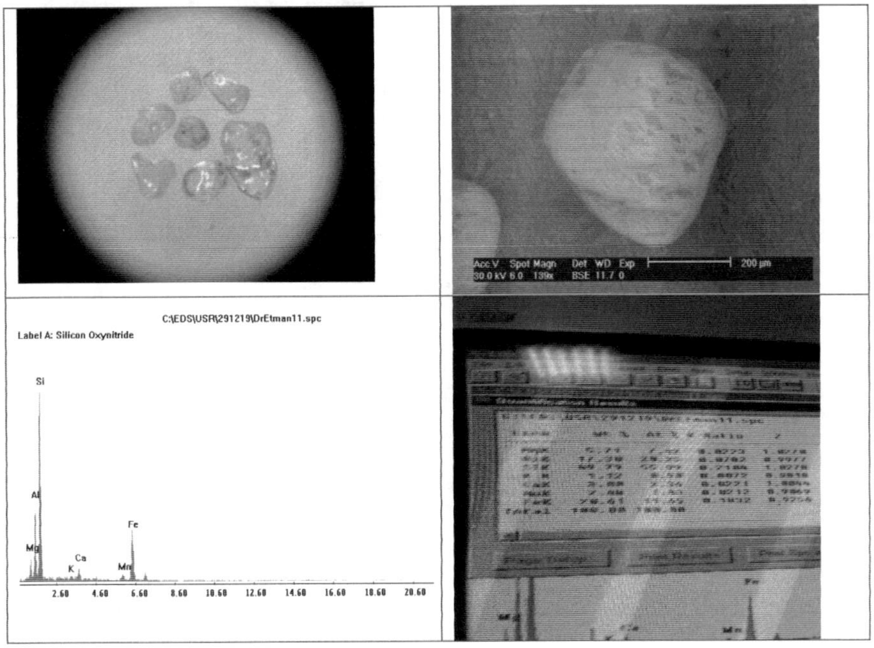

Figure 8. Photograph of garnet crystals, showing the angular to subrounded grains with variation of color from pale pink to reddish, (X. 20).

Zircon:

Zircon is one of the most important economic minerals of the Egyptian black sands due to its wide spread uses in engineering and atomic industries. The Egyptian beach zircon grains are mainly prismatic bipyramidal, and attains different degrees of roundness. It is mainly colorless and transparent with vitreous luster (Fig. 9), some zircon grains are colored in pale pink varieties. Inclusions are predominant in zircon grains, most of them are opaque with black and brown or red colors and irregularly arranged in the enclosing zircon grains. [50] showed that, in polished sections, the opaque inclusions in zircon were identified as mainly; magnetite, Ilmenite and hematite. Sometimes radioactive minerals inclusion as Uranothorite, Xenotime and Monazite may exist in Zircon [51].

The percentages of zircon minerals are ranging from 0.05% to 0.2% with mean 0.05% **(Figure 9 and Table 3)**.

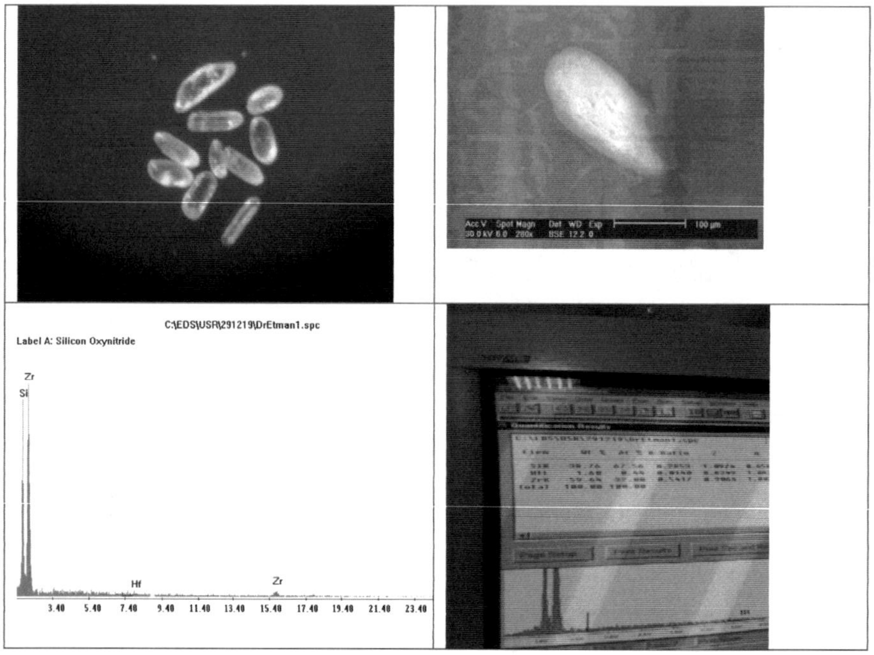

Figure 9. Photograph of zircon crystals, showing the prismatic bipyramidal grains of color less, and eroded faces in some crystals (X. 20).

Monazite:

The Egyptian beach monazite occurs as very well-rounded grains, mostly taking the oval shape with high sphericity **(Figure, 10)**. It is free from inclusions and has colors varying from colorless to deep brown. [52] identified six color varieties of the Egyptian beach monazite, they are brown, yellowish brown, reddish brown, yellow, honey yellow and lemon yellow. They concluded that, the two principal colors of this mineral are the lemon yellow (59%) and the honey-yellow ones (25%). According to **Milner (1962),** monazite is weakly magnetic. Monazite is essentially a phosphate of rare earth mainly cerium, thorium and uranium in which thorium and yttrium substitute for cerium.

The percentages of monazite minerals are ranging from 0.02% to 0.15% with mean 0.02% **(Figure10 and Table 3)**.

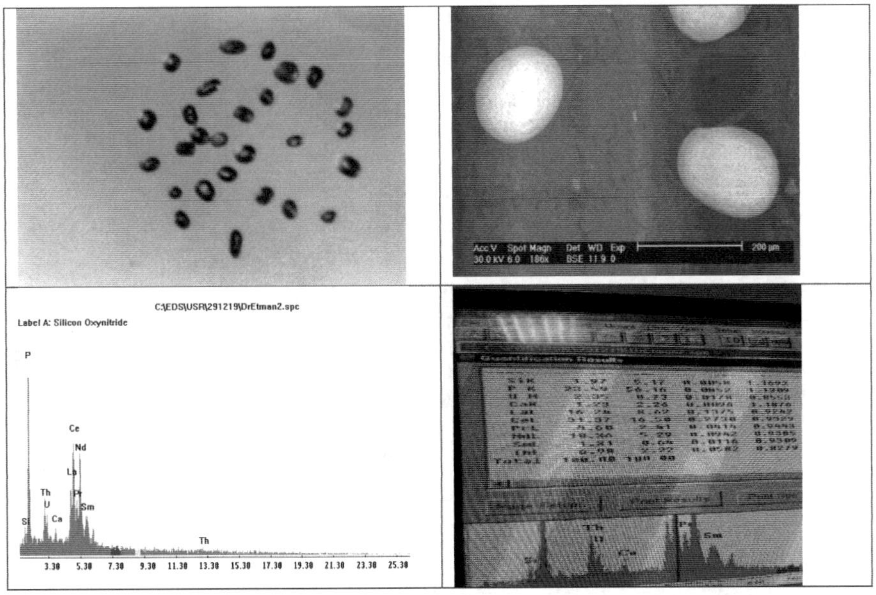

Figure 10. Photograph of monazite, showing the very well-rounded grains, oval shape, high sphericity, and honey color crystal (X. 20).

6. Conclusion

The average of radioelement contents is lower than the international averages, while the average of eTh/eU ratio is higher. High eTh/eU ratio reflects the poor weathering, rapid deposition of these sediments and the dominance of detrital radioactive minerals such as xenotime, monazite and zircon. eU/RaeU ratio average (0.9), is lower than unity and indicate the disequilibrium in the studied coastal sediments, suggesting disequilibrium and uranium migration out. The heavy minerals are represented meanly by magnetite, ilmenite, zircon, rutile, garnet, monazite, pyroxene and amphibole.

References

[1] Dabbour, G. A. (1991): Heavy minerals content in relation to the apparent specific gravity of the Egyptian black sands. Mansoura. Sci. Bull., Spec. Issue, pp. 212-220.

[2] Kannan, V., Rajan, M.P., Iyengar, M., Ramesh, R., (2002). Distribution of natural and anthropogenic radionuclides in soil and beach sand samples of Kalpakkam (India) using hyper pure germanium (HPGe) gamma ray spectrometry. Appl. Radiat. Isot. 57, 109–119.

[3] De Meijer RJ, James IR, Jennings PJ, Koeyers JE (2001): Cluster analysis of radionuclide concentrations in beach sand. Appl. Radiat. Isot. 54, pp.535-542.

[4] Abdel-Halim, A.A. and Saleh, I.H. (2015).Radiological Characterization of Beach Sediments along the Alexandria-Rosetta Coasts of Egypt, Journal of Taibah University for Science (2015), http://dx.doi.org/10.1016/j.jtusci.2015.02.016

[5] Kaiser, M.F. (2014). "Impact of the human activities on the local climate and environment of the Suez City in Egypt". International Journal of Geosciences, 5 (7), pp.700-709. http://dx.doi.org/10.4236/ijg.2014.57063 .ISSN Print: 2156-8359. ISSN Online: 2156-8367

[6] Alam, M.N., Chowdhury, M.I., Kamal, M., Ghose, S., Islam, M.N., Mustafa, M.N., Miah, N.M.H., Ansary, M.M.,(1999). The 226Ra, 232Th and 40K Activity in beach sand minerals and beach soils of Cox's Bazar, Bagladesh. J. Environ. Radioact. 46,pp.243–250.

[7] Mohanty, A.K., Sengupta, D., Das, S.K., Saha, S.K., Van, K.V., (2004). Natural radioactivity and radiation exposure in the high background area at Chhatrapur beach placer deposit of Orissa, India. J. Environ. Radioact. 75, 15–33

[8] Mowafi, W.A., El Tahawy, M.S.,(2009). Radiological investigation of the black sand region of the North-East of the Nile Delta. In: Proceedings of the 7th Conference on Nuclear and Particle Physics, 11e15 November, Sharm El Sheikh, Egypt.

[9] Dawood, Y.H. and Abdel-Naby, H.H (2007). Mineral chemistry of monazite from the black sand deposits, northern Sinai, Egypt: A provenance perspective. Mineralogical Magazine, August 2007,Vol. 71(4),pp. 389–406. DOI: 10.1180/minmag.2007.071.4.389.

[10] El Gemmizi, M. A. (1973): Physical and chemical proparties of monazite from some Egyptian deposits. Unpub. M.Sc. Thesis, Fac. Sci. Cairo Univ., Cairo, Egypt.

[11] Chiozzi,P, De Felice,P, Fazio,A, Pasquale,V, and Verdoya,M, (2000): Laboratory application of NaI (Tl) γ- ray spectrometry to studies of natural radioactivity in geophysics. Applied Radiation and Isotopes 53, pp. 127-132.

[12] Grasty, R.L., Mellander, H., and Parker, M., (1991): Airborne gammaray spectrometer surveying. IAEA Technical Reports Series No. 323.

[13] Anjos, A. M., Veiga, R., Soares, T., Santos, A. M. A., Aguiar, J. G., Frascá, M. H. B. O., Brage, J. A. P., Uzêda, D., Mangia, L., Facure, A., Mosquera, B., Carvalho, C. and Gomes, P. R.S., (2005): Natural radionuclide distribution in Brazilian commercial granites. Radiation measurements, 39, pp. 245-253.

[14] Anjos, R.M., Okuno, E., Gomes, P.R.S., Veiga, R., Estellita, L., Mangia, L., Uzêda, D., Soares, T., Facure, A., Brage, J.A.P., Mosquera, B., Carvalho, C., and Santos, A.M.A., (2004): Radioecology teaching: evaluation of the background radiation levels from areas with high concentrations of radionuclides in soil. Eur. J. Phys. 25 (2), pp. 133–144.

[15] Veiga, R.G., Sanches,N, Anjos,R,M, Macario,K, Bastos,J, Iguatemy,M, Auiar,J.G Santos,A,M,A, Mosquera,B Carvalho,C, BaptistaFilho,M and Umisedo,N,K, (2006): Measurement of natural radioactivity in Brazilian beach sands. Radiation measurement, 41, pp. 189-196.

[16] Frihy, O. E., El Fishawi, M. M. El Askary, M. A.(1988). Geomorphological features of the Nile Delta coastal plain: a review. Acta. Adriatica, V. 29: pp. 51-65.

[17] Ball, J.(1939). Contribution to the geography of Egypt. Geol. Surv. Egypt.

[18] Ali, E.M. and El-Maged, I.A. (2016). Impact of human interventions and coastal processes along the Nile Delta coast, Egypt during the past twenty-five years.Egyptian Journal of Aquatic Rese (2016) 42,pp. 1–10. http://dx.doi.org/10.1016/j.ejar.2016.01.002

[19] White, K. and El-Asmar, H.M., (1999). Monitoring changing position of coastlines using Thematic Mapper Imagery, an example from the Nile Delta. Geomorphology, 29, pp.93–105.

[20] Manohar, M.(1975a). Beach profiles. Proceeding of seminar on the Nile Delta sedimentology. UNDP/UNESCO project of coastal protection studies. Alexandria, pp.95-103.

[21] Frihy,O.E., Komar,P.D., (1993). Long term shoreline changes and the concentration of heavy minerals in beach sand of the Nile Delta, (Egypt), Mar. Geol. 115 (1993) 253.

[22] El-Nahry, A.H., Doluschitz, R., (2010). Climate change and its impacts on the coastal zone of the Nile Delta. Environ. Earth Sci. 59, pp.1497—1506.
[23] Sestini. G., (1989): Nile Delta Review of a Deposihonal Environments and Geological History. Egyptian Geological Society Special Publication. 41,pp.99—127.
[24] Stanley, D.J. and Warne, A.G. (1993a): Nile Delta: recent geological evolution and human impact. Science 260, 628–34. —— 1993b: Sea-level and initiation of predynastic culture in the Nile Delta. Nature 363, 435–38. —— 1998: Nile delta in its destruction phase. Journal of Coastal Research 14, 794–825.

[25] Becker, R. H., and Sultan, M. (2009). Land subsidence in the Nile Delta: Inferences from radar interferometry. The Holocene, 19(6), pp. 949– 954. https://doi.org/10.1177/0959683609336558
[26] Shata, A. and El Fayomy, I. F. (1970): Remarks on the hydrogeology of the Nile Delta. Proc. sysm. Hydrogeology of Deltas, UNESCO v. II.
[27] EL Shazly, E. M., Abdel Hady, M. A., El Ghawaby, M. A., El Kassas, I. A. and ElShazly, M. M. (1974): Geology of Sinai peninsula from Erts-I staellite images: publ. remote sensing research prog., Academy of Sc., and Tech., 7 p.
[28] Hume, W. F. (1925): Geology of Egypt. Surv. Dept., Cairo, (2 vol.).
[29] Lotfy, M.F., (1978). Geological studies of the coastal zone between Ras El Bar and Port Said. M.Sc. Thesis, Alexandria University, Egypt.
[30] SEAM Programme (2005). Damietta Governorate Environmental Profile, Ministry of State for Environmental Affairs, Egyptian Environmental Affairs Agency, Entec UK Ltd., ERM, UK Department for International Development, p. 189
[31] Environmental Description of Port Said Governorate (2007). Egyptian Environmental Affairs Agency. Arabic version.
[32] El Askary, M. A. and Frihy, O. M. (1986). Depositional phases of Rosetta and Damietta promontories on the Nile Delta coast. Jour. Of Afr., Earth Sci, 5(6), pp. 627-633.
[33] UNSCEAR, United Nations Scientific Committee on the Effects of Atomic Radiation, (2000). Sources and effects of ionizing radiation. United Nations Scientific Committee on the effect of atomic radiation. Report to the General Assembly, with Scientific Annexes. United Nations, New York
[34] IAEA (1979): International Atomic Energy Agency. "Gamma-Ray Survey in Uranium Exploration". IAEA Technical Report Series, No. 186, Vienna, Austria,pp. 90.
[35] Boyle R.W. (1982): Geochemical Prospecting for Thorium and Uranium Deposits [M]. pp.498, Elsevier, Amsterdam and New York.
[36] McLennan, S.M., and Taylor, S.R., (1980): Rare earth element–thorium correlations in sedimentary rocks and the composition of the continental crust. Geochimica et Cosmochimica Acta, 44,1833–1839.
[37] McLennan, S.M., and Taylor, S.R., (1991) :Sedimentary rocks and crustal evolution: tectonic setting and secular trends. J. Geol., 99, 1–21.
[38] Rudnick, R. L. and Gao, S. (2004): Composition of the Continental Crust. In: Treatise on Geochemistry. Holland, H.D. and Turekian, K.K. (Editors), Elsevier, Amsterdam. 3: 1-64.
[39] De Putter, T. h., Charlet, J.-M., Quinif, Y. (1999): REE, Y and U concentration at the fluid-iron oxide interface in Late Cenozoic cryptodolines from Southern Belgium. Chem. Geol. 153, 139–150.

[40] De Putter, T. H., Bernard, A., Perruchot, A., Yans, J., Verbrugghe, F. R., Dupuis, C. h. R. (2002): Neoformed mineral paragenesis in acid weathering systems: sedimentary vs volcanic environments. Proceedings of the 12th International Clay Conference, Elsevier Oxford.

[41] EL-KAMMAR A. MOHAMED ;EL-FEKY M. GALAL1; ABU ZIED H. TAHA. And OTHMAN D. AHMED., (2018):RADIOACTIVITY AND ENVIRONMENTAL IMPACTS OF SOME CARBONIFEROUS AND CRETACEOUS KAOLIN DEPOSITS IN SOUTHWESTERN - SINAI, EGYPT. Nuclear Sciences Scientific Journal, 103-116.

[42] Adams, J.A.S., Weaver, C.E., (1958). Thorium to uranium ratios as indicator for sedimentary processes; an example of geochemical facies. Am. Assoc. Pet. Geol. Bull. 42, 397-430.

[43] Shives R.B.K., Ford K.L., and Charoneau B.W., (1995). Application of gamma-ray spectrometric/magnetic/LF-EM surveys–Workshop Manual [J]. Geological Survey of Canada, Open File. 3061, 82.

[44] Wignall, P.B., Meyers, K.J.,(1988): Interpreting the benthic oxygen levels in mudrocks: a new approach. Geology 16, 452e455.

[45] El Galy, M. M., Nada, A., El Feky, M.G. and Roz, M. (2007): The radioactive disequilibrium state in U-series at G-II of U-mineralization, Gabal Gattar area, North Eastern Desert, Egypt". Arab Jour. Nuc. Sci. Appl., 40, (1), 17-26

[46] Muller, G. 1967. Sedimentary petrology. Part 1, Methods in sedimentary petrology. E. Schweizerbart, Sch Verlogsbuchhan dlung, (NAGELE u. OBERRMILLER) Stuttgart Hafner publishing Company New York, London.

[47] Ibrahim, M. I. M. 1995. Investigation of physical properties of zircon and rutile to prepare high purity mineral concentrates from black sand deposits. Unpub. M.Sc. Thesis, Mansoura Univ., Mansoura, Egypt.

[48] Rous, J. D. 1986. Garnet. Butterworth & Co. Ltd. Pp. 20-33.

[49] Milner, H. B. 1962. Sedimentary petrography. V, II, George Allen & Unwin, Ltd., London. pp. 15-205.

[50] Hammoud, N. S. 1973. Phisical and chemical properties of some Egyptian beach economic minerals in relation to their concentration problems. Unpub. Ph. D. Thesis, Fac. Sc. Cairo Univ., Cairo, Egypt.

[51] Dabbour, G. A. (1973). Physical properties and distribution of zircon in some Egyptian deposits. Unpub. M.Sc. Thesis, Fac. Sci. Cairo Univ., Cairo, Egypt.

[52] ZAGHLOUL, Z.M. and K. E. Kamel (1965): the minerlogical and petrographical features of monazite from the black sands of Rosetta, Journal of Geology, Egypt, V. 9, 1 , p. 17-30.

Air pollution remediation using synergetic coupling of AC/TiO₂ filter for benzene gas removal: regeneration and reuse of spent activated carbon after the treatment process

E. F. Mohamed

Air Pollution Department, Environment and Climate change Research Institute, National Research Centre, 33 EL Bohouth St., Dokki, Giza, P.O.12622, Egypt.

elham_farouk0000@yahoo.com

Abstract. Recently, climate change and air pollution are of the most important problems confronting the world. Hence, the present work is designed to provide an air purifier that can be re-utilized many times with high efficiency for several volatile pollutants. The air purifier fabricated from AC/TiO_2 was used to investigate the performance of this filter towards the removal of volatile organic compounds (VOCs) from the air. The filter was composed of a mixture of fiber layers (TiO_2, SiO2 and AC). The results indicated that the synergistic coupling of adsorption/oxidation technique is effective for removal of the air pollutants as well as for regeneration of the activated carbon to be extensional reused. The performance test of air purifier confirmed its high removal efficiency (98%). It was found that nearly up to 94% and 86% recovery of the filter activity could be achieved after the second and third cycles of treatment, respectively. From these results, AC/TiO_2 filter still has the capability to remove and control extra gas pollutants indicating the synergistic coupling effect of TiO_2 with the activated carbon. This filter concerns in developing an economic and feasible air purifier system to remove air contaminants for several times.

Keywords: Climate change; Air pollution; Remediation; Activated carbon; TiO2; Regeneration; Photodegradation.

1. Introduction

Each year, climate change and air contamination expand, resulting in harmful effects to living thinks. Our environment contains different pollutants such as volatile organic compounds (VOCs), greenhouse gases and hydrocarbons. In recent years, academic researchers and environmental scientists deal with of developing new technologies for monitoring and controlling these pollutants [1-6].

Adsorption onto activated carbons (ACs) is the most efficient method for removing the pollutants due to its high surface area [7-9]. However, the critical challenge is the recycling and reusing of the spent ACs. Consequently, many approaches have been studied to solve this drawback. Adsorption/Oxidation (ADOX) coupling is one of these approaches to improve the system performance used and to regenerate the spent ACs *in situ*. [10-12]. In this manner, the photocatalysis using TiO_2 can play a dual role; oxidation of the organic pollutants as well as regeneration of the spent ACs.

The main objectives of the present work are; to develop a new economic and effective ADOX filtration system by immobilization of TiO_2 onto ACs as a support sheet, to investigate the performance of the designed filter for the gaseous benzene treatment, to study the possible regeneration of ACs in this ADOX air purifier, and lastly to investigate its removal efficiency after the first cycle of utilization, which is a critical issue.

2. Experimental Section

2.1. *Materials*

The method of fixation of TiO_2 on ACs as a support was patented by the Ahlström Paper Group Company. The used catalyst is anatase TiO_2 PC 500 was supplied by Millennium. The colloidal silica used could play several roles as: inorganic binder, resistant to ultraviolet (UV) irradiation and photocatalysts, transparent to UV and TiO_2 protector. All the chemicals utilized were analytical grade reagents supplied by Sigma Aldrich.

2.2. *Fabrication of AC/ TiO_2 media*

A dry mixture of equal quantities of TiO_2 and SiO_2 was prepared. The silica particle size is ranged from 20 to 30 nm in diameter. Then a paste was formed by adding de-ionized water. Media of ACs were utilized as a support to fix this paste by using a cylinder rolling on the media. The total media were then dried to remove water. The auto-bound silica has the ability to form bonds during drying, resulting in a high resistant film.

2.3. *Characterization of the AC/TiO_2 media*

Full characterization of the AC/TiO_2 media was performed to examine its physical and chemical properties. Pore volume and specific surface area were evaluated by utilizing a Micrometric automated device ST-2000 based on the BET method (Brunauer–Emmett–Teller) with N_2 at 77K as an adsorbent. The structural morphology and composition were determined by energy dispersive x-ray analysis (EDX) and scanning electron microscopy (SEM).

2.4. *Application of the AC/TiO_2 filter for benzene gas control*

The air purifier AC/TiO_2 was synthesized to control the benzene pollutants. Schematic diagram of the air purifier device was depicted in Figure1. A fixed AC- TiO_2 sheet was used in the reactor. The sheet entirely covers the tore surface. A lamp, UV (Philips, λ = 365 nm) was situated in the reactor centre for illuminating the photocatalyst TiO_2. This lamp was surrounded by a double envelope and a water bath with circulating pump to cool the UV lamp. Storage bottle containing the liquid benzene mixture, 90% methanol solution, flow meter, and air pump were utilized to provide streaming of benzene polluted air in the ADOX process. Benzene concentration in the vapor state was obtained by adjusting the feed flow rate of the air stream. After the treatment process air stream was collected into a methanol–water mixture (90 : 10, v/v) for analysis using high performance liquid chromatography (HPLC, Agilent), with reversed phase column (Hypersil C-18, 5 μm, 250 × 4.6 mm) composed of 5 μm spherical silica particles transplanted with alkyl chains of C18.

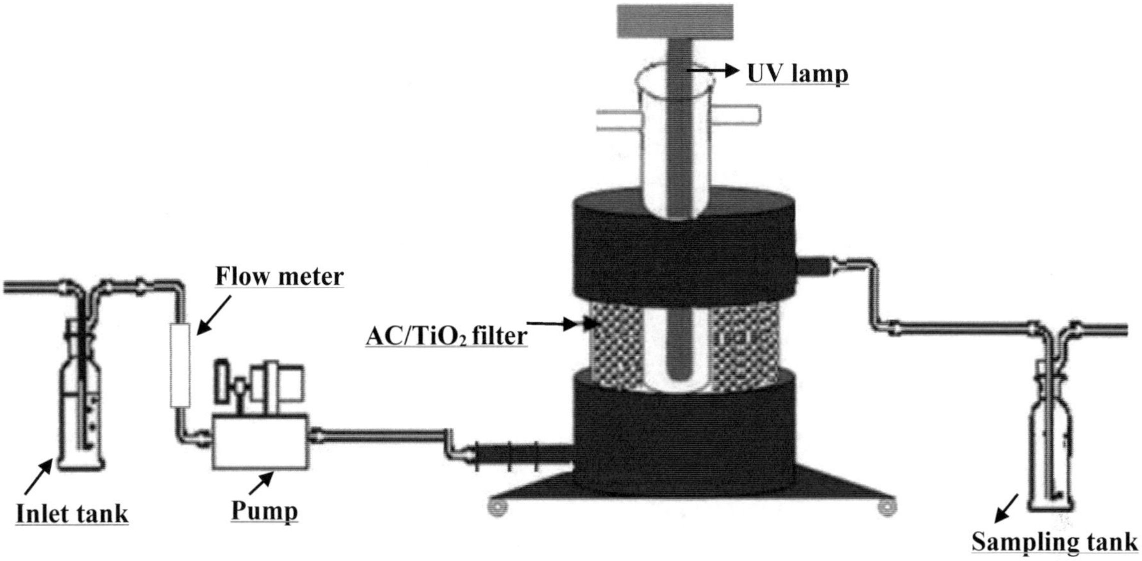

Figure 1. A schematic diagram of the air purifier device.

3. Results and discussion

3.1. Characterization of AC/ TiO_2 media

TiO_2 possesses elementary crystals ranged from 5 to 10 nm in diameter, anatase structure (> 99 %) with specific surface area (S_{BET}) of 320 $m^2.g^{-1}$. The granular AC has S_{BET} of 1065 $m^2.g^{-1}$, particle diameter of 0.25-0.60 mm and it is mostly microporous (0.46 $cm^3.g^{-1}$). The used colloidal silica has S_{BET} of almost 700 $m^2.g^{-1}$ with a diameter of 20-30 nm. The percentages of elemental composition of the used media are summarized in Table 1, indicating the existence of three main elements (Ti, Si, and O) in addition to C.

Table1. EDX elemental composition of AC in TiO_2 /AC filter

Media	Composition							
AC/TiO_2	C		O		Ti		Si	
filter	Wt%	*At%	Wt%	*At%	Wt%	*At%	Wt%	*At%
	79	84	21	16	18	7	22	15

*At%: atomic percentage

3.2. Application of AC/TiO_2 air purifier filter for benzene gas control

The treatment cycles of benzene gas using the new AC/ TiO_2 air purifier device are represented in Figure 2. It was noticed that the air purifier filter has excellent performance for removing the benzene gas pollutant in the first cycle of treatment. It may be interpreted on the hypothesis that the synergistic effect of ACs with TiO_2 leads to complete mineralization of the benzene gas. Hence, AC was used to remove benzene gas by the adsorption process followed by or in the same moment oxidation process using TiO_2 under UV irradiation, which plays a principal role in the benzene gas removal. The same trend is also observed when the air purifier system was reused to control the same concentration of benzene gas pollutant. Similarly, Zain et al. [13] reported that AC/TiO_2 composites revealed an enhanced removal of Ruthenium N_3 dye by coupling both the adsorption and photocatalysis, due to a synergistic influence.

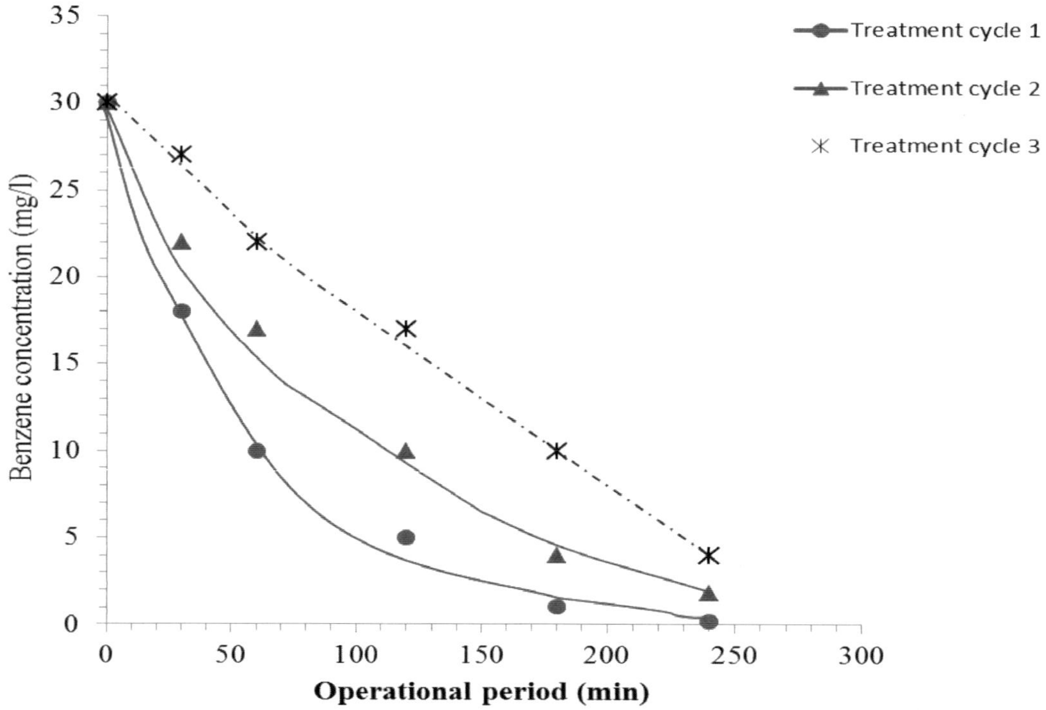

Figure 2. Benzene gas removal cycles using AC/TiO_2 air purifier device at 25 °C.

The air purifier life span is potentially important to select an air purifier that efficiently removes various pollutants from the air. Therefore, many filters can capture gaseous contaminants; however, they will eventually reach its maximum capacity and no longer be capable to remove extra gases from the air. Indeed, the air purifier life depends on several factors, including the filter type and the concentration and particle size of pollutants. Be attentive, however, all air purifiers have not the same filter components and specific characters. Air purifiers will require frequently more changes; hence the filter life challenge is a critical point of view that must be closely considered in fabricating the air purifier design. To study the system performance of the AC/TiO_2 purifier device, the removal efficiency of benzene waste gas was evaluated and represented in Figure 3.

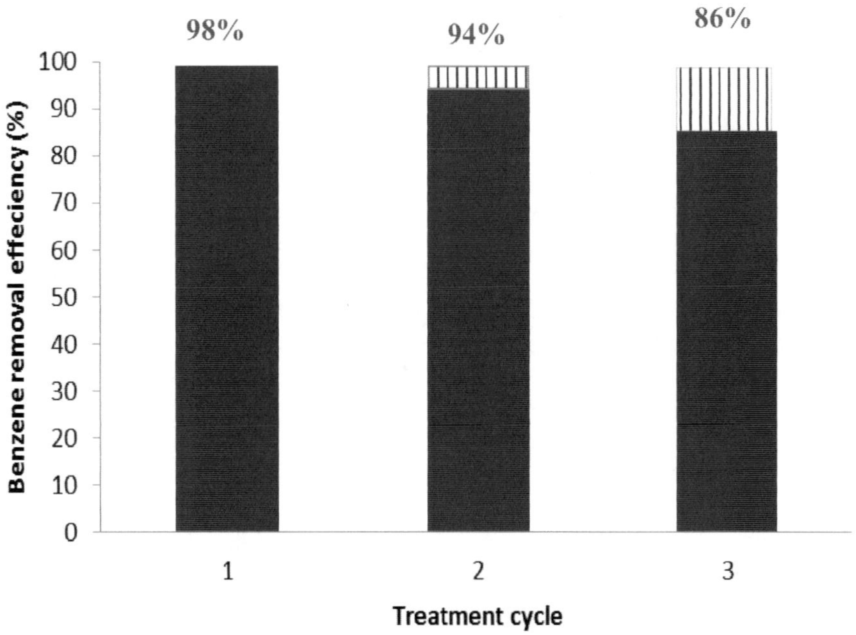

Figure 3. Benzene gas removal efficiency by using AC/TiO_2 air purifier device at 25 °C.

In the first treatment (cycle1), almost 98% of benzene could be removed. The results indicated that nearly 94% of the benzene gas was removed in cycle 2 of treatment. By taking into account only 5% of air purifier capacity was lost as compared to the cycle 1 of treatment. For the third treatment (cycle 3), it was observed that the efficiency of benzene removal using the air purifier system was reduced to be 86% with respect to that of cycle 1. At this stage, only 14 % of air purifier capacity was lost as compared to cycle 1. This performance seems to be due to the synergistic effect of ADOX coupling in which the AC serves to remove the gas pollutants by adsorption and TiO_2 degrades the adsorbed pollutants as well as it regenerates the activated carbon [9-11,14]. The results showed that there is a slight decrease in the air purifier performance during the reuse cycles; these are probably due to a decrease in the available site numbers required for adsorption owing to the blocking by by-products deposited in the filter. These results are in agreement with that obtained by Andriantsiferana et al. [10], which studied the tartrazine treatment with novel TiO_2 composite deposited on the AC porous surface, and the performance results indicated that the removal efficiency of phenolic pollutant was kept for several cycles of treatment. Rangkooy et al. [15] also used AC/TiO_2 in the filter to remove the pollutants from the air, with a removal efficiency reached 90%. Hence, the predominant conclusion is that the synergistic effect is due to the simultaneous occurrence of TiO_2 onto the AC support media, leading to carbon adsorption and finally, photo-oxidation. In comparison, Sun et al. [16] cited that the capacity of TiO_2/ACF adsorption was found to be 60 mg/g, which was more than that of the AC alone, in another work; Li et al. [17] reported that when TiO_2/AC system was continuously utilized for six times, the rate of formaldehyde removal was still at 92%.

The combination of AC and TiO_2 resulted in a synergistic mechanism that led to an increase in the benzene removal. Various possibilities could interpret this mechanism, including the following steps; (1) adsorption: AC was trapped and accumulated the benzene gas around the catalyst, which then transferred to the active site of TiO_2 for the photo-degradation process[18], (2) AC could acidify the OH groups in the TiO_2 surface, these could lead to extra degradation and complete removal of the pollutant, (3) AC could play a role in the oxidation step via the transfer of electron/hole and also the large AC pores can accelerate the flow of benzene to the photo-oxidative species [19], (4) in addition,

TiO_2 immobilized onto the AC not only augment the contaminant degradation, but also result in high AC regeneration performance [20], and (5) finally, AC can give a solution for the recovery problem of TiO_2 powder and enhance its catalytic performance.

4. Conclusion

Air purifiers TiO_2/AC can be synthesized simply by deposition of TiO_2 and SiO_2 particles as a paste on the AC sheet surface. High removal efficiency (98%) of benzene gas pollutant was obtained under UV light irradiation. These indicate that such combination of AC/ TiO_2 is attributed to the excellent adsorption influence of AC and high photocatalytic degradation of TiO_2 in the presence of UV irradiation. Benzene gas removal was only slightly decreased from 98% to 94% and 86% from the first to second and third cycles, respectively. Thus, coupling of AC and TiO_2 was significantly improved the performance of AC regeneration. The results demonstrated the high air purifier life span and reusability of AC/TiO_2 system for benzene removal as well as extending the ACs capacity, thus this could be a potential solution to reuse the ACs in the industrial application fields.

Acknowledgments

Funding and laboratory facilities were provided by the Air Pollution Department, National Research Centre, Egypt that are gratefully acknowledged.

References

[1] Essawy HA, El-Shakour A A, Tawfik M E, Mohamed EF, El-Sabbagh S H and El-Hashemy M A 2014 *RSC adv.* **4** 33555
[2] Zaitan H, Mohamed EF, Valdés H, Nawdali M, Rafqah S and Manero MH 2016 *Acta Chim. Slov.* **63** 798
[3] Mohamed EF, Awad G, Zaitan H, Andriantsiferana C andManero MH 2018 *Environ. Technol.* **39** 878
[4] Awad G and Mohamed EF 2019 *Appl. Catal. B: Environ.* **253** 88
[5] Mohamed EF and Awad G 2020 *Environ. Sci. Pollut. Res.* **27** 24507
[6] Mohamed EF and On-Do T 2020 *J. Taiwan Inst. Chem. Eng.* **111** 181
[7] Mohamed EF, El-Hashemy MA, Abdel-Latif NM and Shetaya WH 2015 *J. Air Waste Manag. Assoc.* **65** 1413
[8] Mohamed EF, Ahmed SAS, Abdel-Latif NM and EL-Mekawy A 2016 *RSC Adv.* **6** 57284
[9] Mohamed EF, Shetaya WH, El-Mekawy A, Shakour AA and Awad G 2017 *J. appl. biotechnol.* **5** 2
[10] Andriantsiferana C, Mohamed EF and Delmas H 2014 *Environ. Technol.* **35** 355
[11] Andriantsiferana C, Mohamed EF and Delmas H 2015 *Environ. Eng. Res.* **20** 181
[12] Mohamed EF, Awad G, Andriantsiferana C and El-Diwany A I 2016 *Environ.Technol.* **37** 1197
[13] Zain NM , Lim CM , Usman A, Keasberry N and Mahadi RTH 2021 *Environ. Nanotechnol. Monit. Manag.* **16** 100567
[14] Raso RA , Zeltner M and Stark WJ 2014 *Ind. Eng. Chem. Res.* **53** 19304
[15] Rangkooy HA, Jahani F and Ahangar AS 2020 *Environ. health eng. manag.* **7** 41
[16] Sun HF, Zhang GD and Zheng GY 2007 *J. Anhui Uni. Technol.* **24**, 4
[17] Li XH, Zheng XX, Yin ZY, Hou KS and Rui X 2009 *Chem. Res. Appl.* **21**, 2
[18] Guo X-P, Zang P, Li Y-M and Bi D-S 2021 *Water* **13**, 1622,1
[19] Xu W, Jin Y, Ren Y, Li J, Wei Z, Ban C, Cai H and Chen M 2021 *Can J Chem Eng.* 2021,1
[20] Zhu X, Wang D and Hui S 2021 *Adsorp. Sci. Technol.* **2021**, 1

Influence of Chemical Treatment on the Spectral and Thermal Behavior of Activated Carbon Obtained from Olive Oil Solid Waste

Wael A A elhelece*

*Radiation safety department, nuclear and radiological safety research center, Egyptian atomic energy authority

Abstract: Solid waste remained after the extraction of oil from olive oil fruits (OOSW), was used to obtain active carbon depending on chemical activation using H_3PO_4, KOH and $ZnCl_2$ as activating agents. The changes in composition due to chemical treatment and pyrolysis were investigated by chemical analysis, spectral (IR), and thermal analysis (DTA). Thermodynamic parameters of formation (activation energy (ΔEa^*), enthalpy change (ΔH^*), entropy change (ΔS^*) and free energy change (ΔG^*) related to the thermal decomposition process were calculated. Adsorption of both iodine and methylene blue at very low concentrations were done, the quantity of the iodine adsorbed (iodine number) indicated the micropores content of carbon by the uptake of iodine from the solution. It is equivalent to surface area of carbon between 900m²/g and 1100 m²/g. To characterize activated carbons as well as determining the adsorption capacity adsorption of methylene blue was done.

Keywords: Olive Oil Solid Waste (OOSW), Active Carbon Preparation, Spectral & Thermal Analysis, Thermodynamics Parameters Adsorption of Iodine and Methylene blue.

1. Introduction

Olive oil solid waste (OOSW) is one of the agricultural residues all over the world. Since the major by-products (skin, pulp, and pieces of stones) from oil extraction processes is the solid waste. The residue is a poor animal feed source as it consists mainly of ligno-cellulose and it is also poor in the energy content for so it is difficult to be used for energy production [1]. Thermal ignition in an inert atmosphere or law oxygen content of these residues gives a certain amount active carbon [2].

Olive oil represents approximately 20% of the olive fruit weight. The remaining biomass can be defined as waste and has to be treated according to specific rules and regulations [3]. Many solid and semi-solid wastes and effluents are obtained from olive planting and olive oil processing, which is a major challenge for most olive processing plants. The disposal of these residues is one of the biggest environmental problems that facing world today and the olive oil industry is considered as one of the most heavily polluting food industries [4].

The origin, environmental conditions, storage time as well as the oil extraction processes, these all parameters tend a variation in the constituents contents of the solid waste obtained [5, 6]. The olive oil production was based on the so-called three-phase centrifugal separation, which resulted in three different streams, i.e., olive oil, a watery liquid waste and a solid residue, with 40–45% water content [7, 8].

The control methods for pomace were to burn for energy however the combustion is inefficient and incomplete and emits dense smoke wealthy in carbon and carbon monoxide. In addition, the proposed use of incineration as an opportunity increases worries approximately its effect at the surroundings and public health. Waste in a sustainable and economically feasible way is rising undertaking and appropriate environmental practices need to be regulated. [9].

Pruning of this waste yield a go with the drift of approximately 75,000 tons of fronds in line with year. This has little monetary price and are every so often disposed of as waste or burnt and this is probably dangerous to the environment. The lots and clean availability of agricultural spinoff lead them to precise aid for Activate Carbon (AC) production. Lastly, those agricultural by-merchandise are generally cheaper and their powerful use is desirable [10].

Activate carbons (ACs) have been widely used in several fields such as; refining processes of vegetable and fatty oils, water purification and pollution treatment, air and gas purification (cigarette filters, vehicle exhaust gas control) and food and pharmaceutical industries [11]. We could prepare and activate carbon through two methods, indigenous activation or chemical activation. Indigenous activation technique includes carbonization of the precursor (uncooked material) in an inert environment and gasification of the ensuing char withinside the presence of steam, carbon dioxide or air. The uncooked material, in chemical activation is impregnated with a robust dehydrating agent after which accompanied with the aid of using pyrolysis at excessive temperature to put together activated carbon. ACs have been organized from many natural substances having a excessive content material of carbon like coal, wood, lignite, coconut shells and currently many agricultural with the aid of using-merchandise which include walnut shells, palm shells, pecan shells, date stones, almond shells, sugar cane bagasse, cotton stalks [12-14].

The intention of this research is to investigate the composition of the activated carbon intended through chemical activation the usage of H_3PO_4, KOH and $ZnCl_2$ of the (OOSW).

2. Experimental

All chemical used in this work were in High degree of purity.

HCl ; ACS reagent, 37% sigma-aldrich,
NaOH ; ACS reagent, ≥97.0%, pellets
H_3PO_4; sigma-aldrich85 wt. KOH: sigma-aldrich 85% KOH basis, pellets
$ZnCl_2$; sigma-aldrich 98% grade

2.1. Chemical characterization of OOSW

Samples of olive oil solid waste were collected from aljouf area, KSA, and allowed to different measures and analysis as;

2.1.1. Content of moisture [15];

It was determined by heating 1g of the OOSW in an oven for different time intervals.

2.1.2. Organic matters combustible [16];

It was determined by burning 1g of the OOSW in an oven at different temperatures, for different time intervals.

2.1.3. Water soluble matters [17];

It was determined by dissolving different amounts of the OOSW in cold as well as boiled water for different time intervals.

2.1.4. Ash content [17];

It was determined by complete burning amounts of the OOSW in an oven at different temperatures, for different time intervals.

2.1.5. Effects of HCl

It were determined by dissolving amounts of OOSW in different concentrations of HCl solutions at 25 ^{O}C and when boiling.

2.1.6. Effect of NaOH

It were determined by dissolving amounts of OOSW in different concentrations of NaOH solutions at 25 ^{O}C and when boiling.

2.2. Active carbon preparation

Active carbon was prepared from olive oil solid waste (OOSW), by chemical activation; The most common chemical agents are H_3PO_4, KOH and $ZnCl_2$.

2.2.1. H_3PO_4 activation [18]

50 gram OOSW was mixed with 50 ml H_3PO_4, left overnight then the burned in an oven at 300 ^{O}C for 2 hrs.

2.2.2. KOH activation [19]
50 gram OOSW was impregnated in 100 ml 10% KOH (wt/v), left overnight then the solid parat was burned in oven at 300 °C for 2hrs.and
2.2.3. $ZnCl_2$ activation [20]
50 gram OOSW was mixed well with $ZnCl_2$ (30 wt%) solution, left overnight then the solid part was burned in an oven at 300 °C for 2 hrs.

2.3. Active carbon characterization

2.3.1. Methylene Blue Number (MBN)

The extreme amount of dye adsorbed on (1g) of adsorbent, is the (MBN). Adsorption of methylene blue was done to characterize activated carbons as well as determining the adsorption capacity. It could suffixed as q_{eq} [23]. To calculate (MBN) form Langmuir model, a q_{eq} plot is made in function of C_e. The Langmuir parameters (q_{max} and K_L) are found by a least squares fitting regression. We could put 10 mg of activated carbon in contact with 10 mL of a methylene blue solution at different concentrations (10, 25, 50, 100, 250, 500 and 1000 mg/L) for 24 h at room temperature (approximately 25 °C). The remaining concentration of methylene blue is analyzed using a UV/Vis spectro. UV. Vis double beam PC 8 scanning auto cell (labomed UVD-3200) at 645 nm.

2.3.2. Iodine number (NI)

The ASTM D4607-94 method was used to determine the iodine number. It means the number of milligrams of iodine adsorbed on I g of carbon, the adsorption of iodine depends on surface area of the carbon and the activity of it. The higher the (NI) the higher the activation, adsorption of iodine from solution could be a measure of the porosity of the surface i.e. (0 to 20 Å, or up to 2 nm) is equivalent to surface area of carbon between 900 m^2/g and 1100 m^2/g [21]. Treatment of carbon samples were with 10 mL HCl (5%). The admixture was then boiled for 30 s and then cooled. Soon afterwards, 100 mL of 0.1 N (0.1 mol L^{-1}) iodine solution is added to the mixture and stirred for 30 s. Then the resulting solution is filtered and 50 mL of the filtrate is titrated with 0.1 N (0.1 mol L^{-1}) sodium thiosulfate, using starch as indicator [22].

2.3.3. Thermal studies;
Thermal studies were done using DTA-50, Shimadzu Thermogravimetric analyzer (Japan), with a rate of heating 10°C / min.
2.3.4. IR spectrum;
It was done using JASCO UV/VIS SPECTROMETER/DATA SYSTEM RESOLUTION
DELTAX 0.964233 X UNITS 1/CM, YUNITS ABSORBANCE,
FIRSTX 399.1927 LASTX 4000.6047

3. Results and discussions

3.1. Moisture content of olive oil solid waste (OOSW)

For determining the moisture content of the (OOSW), Samples were heated at 110 °C for different time periods. The moisture was pointed in tables 1and 2.

Table 1. Heating (OOSW) in an oven at 110°C for 1hr.

W_i g*	1	2	4
W_f g	0.63	1.1	2.03
ΔW g	0.36	0.9	1.97
Moisture%	36%	45%	49%

* g: means gram

Table 2. Heating Sun drayed (OOSW) in an oven at 110 °C for different time periods

	1hr.		
W_i g	1.0	2.0	4.0
W_f g	0.86	1.72	3.5
ΔW g	0.14	0.28	0.5
Moisture%	14%	14%	12.5%
	2hrs.		
W_i g	1.0	2.09	4.0
W_f g	0.90	1.89	3.63
ΔW g	0.10	0.20	0.37
Moisture%	10%	10%	9.25%
	3hrs.		
W_i g	1.0	2.0	4.0
W_f g	0.91	1.86	3.68
ΔW g	0.09	0.14	0.32
Moisture%	9%	7%	8%

3.2. Amount of water soluble matters from (OOSW)

1, 2 and 4 grams of (OOSW) samples were leached in 100ml distilled water for different time periods and the results were collected in table 3.

Table 3. Leaching with water

	1hr.		
W_i g	1	2	4
W_f g	0.69	1.55	2.93
ΔW g	0.31	0.45	1.07
Loss%	31%	23%	26.75%
	2hrs.		
W_i g	1	2	4
W_f g	0.69	1.34	2.8
ΔW g	0.32	0.66	1.2
Loss%	32%	33%	30%
	3hrs.		
W_i g	1	2	4
W_f g	0.69	1.34	2.72
ΔW g	0.33	0.66	1.28
Loss%	33%	33%	32%

The results in table 3, indicates that water dissolved about 30% by weight from (OOSW) samples.

3.3. Leaching with dil. HCl

Results in Table 4, indicated that (OOSW) was affected by loss about 3 % from its preliminary weight in line with the whole lack of alkali metals discovered withinside the OOSW While this percent rises to about 25% when boiling in HCl solution this may be due to degradation of part of lignin compounds.

Table 4. Effect of dil. HCl solution

at 25 ^{O}C.			
W_i g	1.0	3.0	5.0
W_f g	0.97	2.89	4.42
ΔW g	0.03	0.11	0.175
Loss%	3%	3.67%	3.5%
at 100 ^{O}C.			
W_i g	2.0	4.0	6.0
W_f g	1.50	2.99	4.94
ΔW g	0.50	1.01	1.06
Loss%	25%	25%	18%

3.4. Effect of sodium hydroxide solution on (OOSW)

Effect of sodium hydroxide solution concentration, as well as the time of treatment was studied; 1.0 g OOSW was dissolved in 100 ml sodium hydroxide solutions (2, 4, 6, 8 and 10% concentrations) for 2hrs. at 25°C. and when boiling. The obtained results are summarized in table 5.

Table 5. Leaching with sodium hydroxide solutions

For 2hrs. at 25°C.					
NaOH conc.	2%	4%	6%	8%	10%
W_i g	1.0	1.0	1.0	1.0	1.0
W_f g	0.70	0.74	0.72	0.71	0.71
ΔW g	0.30	0.26	0.28	0.29	0.29
loss%	30%	26%	28%	29%	29%
For 2hrs. at 100°C.					
NaOH conc.	2%	4%	6%	8%	10%
W_i g	1.0	1.0	1.0	1.0	1.0
W_f g	0.60	0.73	0.65	0.63	0.56
ΔW g	0.40	0.26	0.35	0.37	0.44
loss%	40%	26%	35%	37%	44%

Sodium hydroxide degrades lignin into sodium salts of phenolic compounds which soluble, the results indicates that the soluble percent was about 30-40% according to many effects as was mentioned above.

3.5. control firing

The results of control firing of (OOSW) samples as well as soda treated samples were collected in table 6., these results indicate that the OOSW samples contains at least 95% lingo-cellulosic part, also, treating with alkali rises thermal stability of the studied samples by decreasing the burned part into 40% instead of 97% at 600 ^{O}C for 4hours.

Table 6. Burning OOSW in the furnace

	Control firing at 200 °C for 2 hrs.		
W_i g	1.01	1.96	2.29
W_f g	0.703	1.23	1.38
ΔW g	0.307	0.73	0.91
Loss%	30%	37%	40%
	Complete burning in the furnace at 400 °C for 2 hrs.		
W_i g	1.0	2.0	3.0
W_f g	0.70	1.23	1.8
ΔW g	0.30	0.73	1.2
Loss%	30%	38.5%	40%
	Complete burning at 600°C for 4 hrs.		
	Untreated OOSW		Soda treated OOSW
W_i g	1.1	0.99	4.78
W_f g	0.06	0.07	2.72
ΔW g	0.9685	0.949	2.06
Loss%	97%	95 %	44%

3.6. Methylene blue number (MBN)

Equation 1 below was used to calculate the amount of methylene blue adsorbed from each solution:

$$q_{eq}\ (\text{mg g}^{-1}) = \{(C_0 - C_e) \times V\}/M \qquad (1)$$

Where:

C_0: MB Initial concentration (mgl^{-1}), C_e: MB concentration at equilibrium time (mgl^{-1}),

V: Solution volume (L) and M: Carbon mass (g).

Iodine number and the MBN increased proportionally with the surface area and pore volume, from this point of view they are interesting probes for the determination of these properties. While, the MB molecule has an area of 2.08 nm^2 and can only enter in large micropores and mesopores.

Table7. Amount of MB adsorbed on 1 g carbon

Initial Concentration C_0 mg/l	Absorbed amount		
	Phosphoric Acid Prepared	Potassium Hydroxide Prepared	Zinc Chloride Prepared
10	9.462	10	10
20	19.307	20	18
50	30.65	33.02	23.6
100	35.56	62.84	30.9
250	39.46	65.4	33.5
500	40.5	66.6	35.6
1000	43.5	70.3	37.3

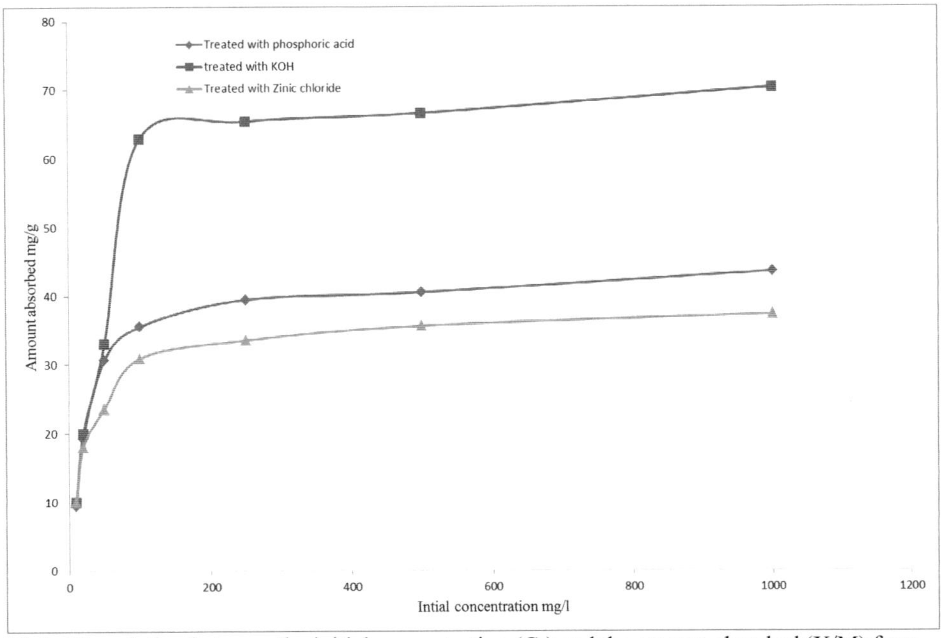

Fig1. The relation between the initial concentration (C_0) and the amount absorbed (X/M) from methylene blue solutions.

Determining the methylene number for three samples of carbon obtained from OOSR while treated with H_3PO_4, KOH and $ZnCl_2$ and comparing the results obtained, it was found that sample prepared from KOH has the most surface area and the more pore size.

3.7. Iodine number

Using the logarithmic axis, plot the amount of iodine per gram of carbon (X/m) against the iodine concentration of the filtrate (C). using equations 1 and 2 respectively.

$$X/M = \{(N1*126.93*V1) - [(V1+VHCl)/VF]*(NNa_2S_2O_3*126.93)*VNa_2S_2O_3\}/M \quad (2)$$

$$C = (N_{Na2S2O3} * V_{Na2S2O3}) \quad (3)$$

Where:

N_I: the normality of Iodine solution, V_I: the volume of Iodine solution.

V_{HCl}: the volume added of 5% HCl., V_F: the filtrate volume used in titration.

$N_{Na2S2O3}$: the normality of sodium thiosulfate solution.,

$V_{Na2S2O3}$: the volume of sodium thiosulfate solution.

M_C: is the mass of activated carbon.

As the iodine molecule is relatively small with an area of 0.4 nm^2 it enters in the smaller micropores. Determining the Iodine number for three samples of carbon obtained from OOSW while treated with H_3PO_4, KOH and $ZnCl_2$ and comparing the results obtained, it was found that sample prepared from KOH has the most surface area and the more pore size as we could see in fig 2.

Table8. Iodine amount adsorbed per gram of carbon

Initial Concentration C_0 mg/l	Absorbed amount		
	Potassium Hydroxide Prepared	Phosphoric Acid Prepared	Zinc Chloride Prepared
0.068	0.04	0.036	0.035
0.056	0.035	0.032	0.031
0.044	0.032	0.03	0.029
0.036	0.029	0.027	0.027
0.024	0.028	0.026	0.025

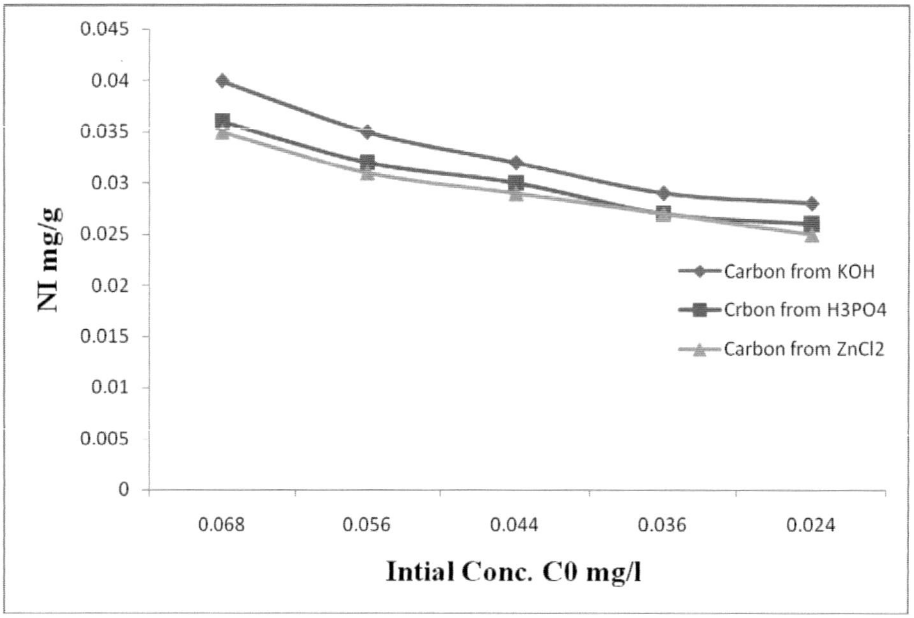

Figure 2. The relation between the initial concentrations (C_0) and the Iodine concentration absorbed.

Since (1mg) adsorbed Iodine, indicated that 1m^2 active carbon surface area, therefore IN is a very good index for activation process.

3.8. Differential Thermal Analysis (DTA) of (OOSW)

DTA analyses and investigations (Figure 3 and table 9), were found to comprise one major exothermic peak at 472 ^{O}C for the raw material, 481 ^{O}C for pyrolyzed (OOSW) at 300 ^{O}C, 522 ^{O}C for KOH activated carbon, 518 ^{O}C for H_3PO_4 activated carbon and 360°C for $ZnCl_2$ activated carbon. There were anther peaks, below 100 ^{O}C endothermic peak would be due to moisture loss, at higher temperatures there were anther exothermic peaks which were due to phase transformations occurring in the solid samples during the rise of temperature, this was only for samples prepared from $ZnCl_2$ activation. These results were collected and produced in table 9.

The DTA curves were used to determine the thermodynamic parameters related to the thermal changes occurring in the tested samples during firing. For this propose the following relations were used.

Journal of Physics: Conference Series **2305** (2022) 012032 doi:10.1088/1742-6596/2305/1/012032

The pyrolysis temperature of the uncooked substances at some stage in evaluation can decide the precise temperature for decomposition of natural substances, at some stage in the activation procedure for AC production.

Results acquired are summarized in table (9); the atypical values proven on this desk can be ascribed to a few facet reactions because the combustion of carbon and a few section changes of silica withinside the strong section.

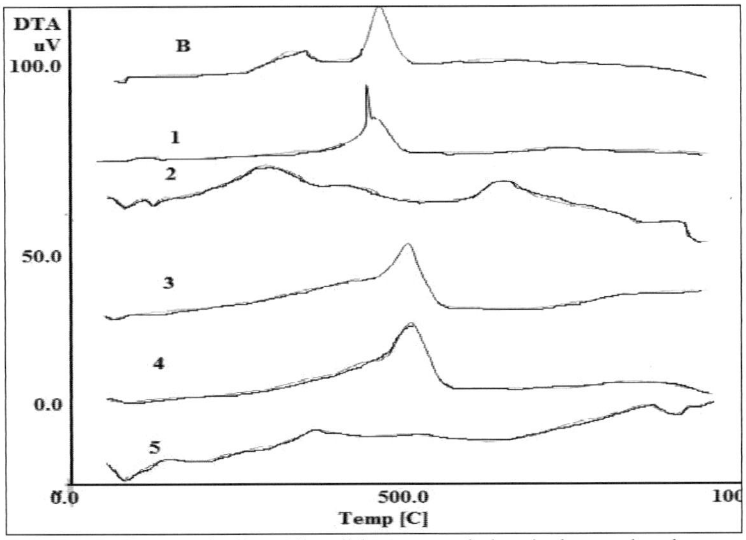

Figure 3. DTA of Olive Oil Solid Waste and chemical treated carbon.

Table9. DTA analysis of Olive Oil Solid Waste and chemical treated carbon

Sample no.	Peaks				
B		323-347 Exothermic	471 Exothermic	651 Exothermic	
1	53, 98 endothermic	283 Exothermic		667 exothermic	883 Endo.
2	46 endothermic			523 exothermic	
3			437 Exothermic	516 Exothermic	
4	52 endothermic	300 Exothermic		650 endothermic	938 Endo
5			481 Exotherm		

The thermodynamic parameters calculated from the differential thermal analysis and showed in table 10 were very high this may be due to the composition reactions occurring while measurements.

Table10. Differential thermal analysis of the tested samples

Sample Peak	OOSW			5	4				
	1	2	3	1	1	2	3	4	5
ΔH^* kj/mol	26	106	14	140	-12.5	-8	26	11.5	-25.5
ΔEa^* kj/mol	104	166	74.9	22.5	26	26.2	10	15	20
ΔS^* kj/mol	-1165	-6231	-1103	-1086	-8702	-302	-1028	-1394	-632
ΔG^* kj/mol	39632	239429	75495	487883	38488	155405	631252	915003	45272
Sample Peak	3		2			1			
	1	2	1	2	3	1	2	3	4
ΔH^* kj/mol	-1.4	25	53	111	21	-5.4	12	-7.8	-17
ΔEa^* kj/mol	234	16.6	22.5	350	109	2.6	26.2	10	15
ΔS^* kj/mol	-3245	-4321	-305.4	-1040	-630	-87.02	-302	-102.8	-139.4
ΔG^* kj/mol	46632	39429	78834	34560	131460	38488	155405	631252	915003

3.9. Thermal stability and hydrophilicity

Thermal stability and degree of hydrophilicity of carbon samples were investigated also from differential thermal analysis (DTA) and compared with the starting materials.

The first endothermic step less than or about 100 ^{O}C indicated that the degree of hydrophilicity increases from the raw material to the obtained carbon due to increasing in surface area and activity.

The thermal stability was studied from the decomposition in the second step of the thermograms as shown in Fig. 3. And table 9. The carbon is thermally stable up to 500◦C instead of 320◦C for the raw material. The decomposition of carbon treated with phosphoric acid started at 518◦C in one step this means that the carbon found in one phase, whereas the decomposition of carbon from KOH and pyrolyzed at 200◦C started at lower temperature (283◦C) and there was another peak at 667◦C this means that there were more than one phase of carbon, carbon prepared at the same conditions but at 400◦C was decomposed at 522◦C only one phase. Carbon from zinc chloride treatment showed less in thermal stability but it has higher in hydrophilicity it decomposed at 360◦C, and there was another endothermic peak at 963◦C which must be due to phase interchange of zinc oxide.

3.10. Analysis of IR spectra

The IR spectra of (OOSW) samples together with samples subjected to some chemical treatments were recorded in the solid state as KBr discs. The samples involved in the present study are as in scheme (1).The data reveal the existence of the bands characteristic of the vibration modes for the water molecules displayed as brood bands with strong intensity at 3445 –3425 cm^{-1} (γH_2O), 1630 – 1620 cm^{-1}(δH_2O) and 980 – 832 cm^{-1} (νH_2O). The various chemical treatments and pyrolysis affected much on the cellulosic materials, peaks between 2000 and 1000 cm^{-1} which were the characteristic peaks for C=C, C-OH completely disappeared or at least shortened as seen from fig.4 and table 11.

Table11. IR spectra analyses of OOSW and chemical treated samples

Sampl	IR spectral bands of diagnostic importance									
B		552	1020	1219	1461	1591	1712	2846	2913	3431
1		659	866	1370	1442	1555			2912	3233
2		506	1027			1569		2885		3383
3		522	1021	1369		1552		2844		3421
4	433		997			1600	1650	2381	2941	3400
	Si-O-Si	Si-O-Si str.,	SiO_2 lattice vib.	-	H_2O bending	-	-	Si-H str.,	-	H_2O str.,

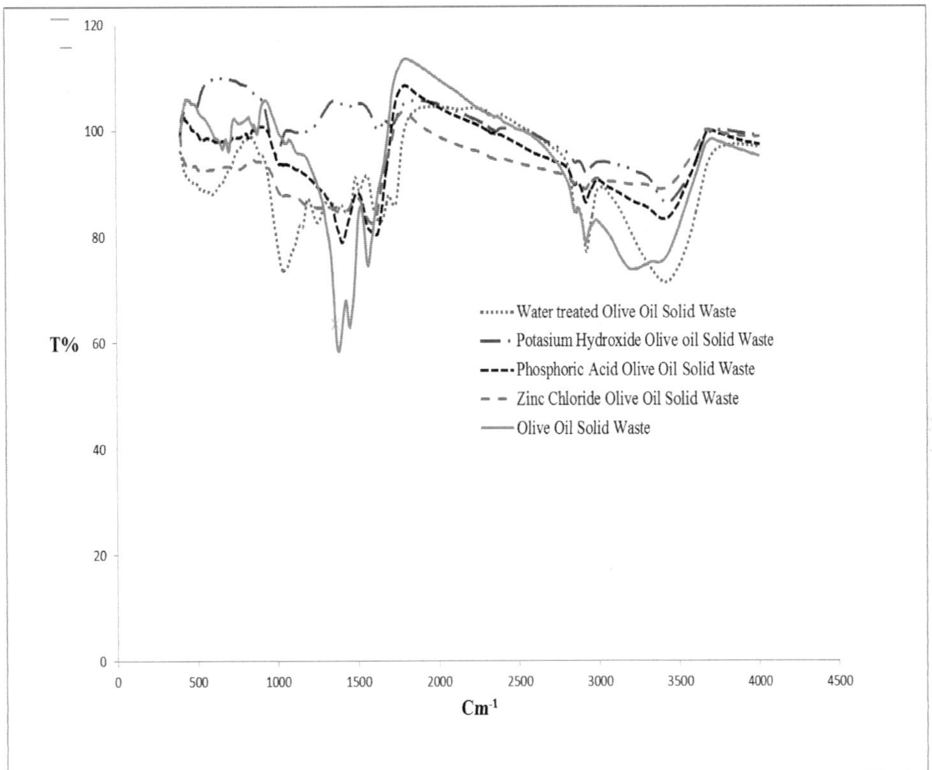

Figure 4. Infra-Red transmittance of OOSW, water treated OOSW, KOH treated OOSW, H_3PO_4 treated OOSW and $ZnCl_2$ treated OOSW samples

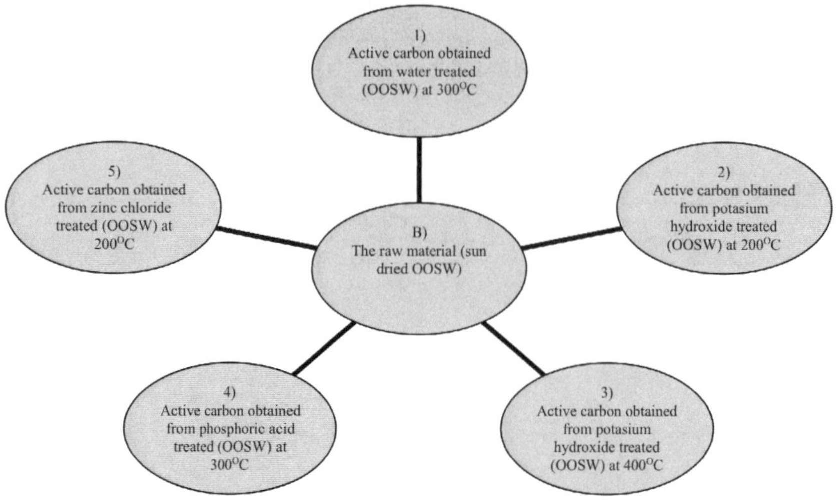

Scheme1; Indications of symbols and numbers used in the paper.

4. Conclusion

In the present study olive oil solid waste (OOSW) as well as the active carbon prepared by different methods were chemically studied and characterized with both spectral and thermal analysis. Samples were characterized through chemical treatment; this was leaching with water, hydrochloric acid, caustic soda and burning at different temperatures.

Characterization of OOSW and carbon from it was done using different methods such as experimental chemical treatment as well as thermal analysis (DTA) and spectral analysis (IR).

Comparing differential thermal analysis (DTA) for both (OOSW) samples as well as carbon prepared under different conditions was done.

The obtained results indicated that the carbon samples were all in the same phase but have different hydrophilicity and also differ in thermal stability according to the pre-treatment conditions as had been discussed before.

IR analysis has a great importance in detecting the modes of vibrations for various chemical bonds between atoms present in the tested samples. The obtained results showed that the obtained carbon under different conditions gathered from the band poor IR patterns; which means that the carbon present is in its most active state.

Acknowledgment

I would greatly thank all radiation safety department members especially Assis. Prof. Dr. Ashraf Ali and Prof. Dr. Eman Tharwat the head of the department for their hard review and kind support to introduce and write this paper.

References

[1] Sánchez-Gutiérrez M, Espinosa E, Bascón-Villegas I, Pérez-Rodríguez F, Carrasco E and Alejandro Rodríguez A 2020 Production of Cellulose Nanofibers from Olive Tree Harvest a Residue with Wide Applications Agronomy, 10, 696

[2] Tomczyk A, Sokołowska Z and Boguta P 2020 Biochar physicochemical properties: pyrolysis temperature and feedstock kind effects Rev Environ Sci Biotechnol 19, 191–215

[3] Al Naeb H, Caliandro L, Ozon M, Nakhleh K, Bourghoul A, Montel G L 2007 Olive oil by-product analysis and recovery in Syria In : Di Terlizzi B. (ed.), Dragotta A. (ed.), Jamal M. (ed.). Syrian national strategic plan for olive oil quality: final report. Bari: CIHEAM, p. 119-140

[4] Khdair A and Abu-Rumman G 2020 Sustainable Environmental Management and Valorization Options for Olive Mill Byproducts in the Middle East and North Africa (MENA) Region Processes, 8, 671

[5] Topi D, Beqiraj I, Seiti B and Halimi E 2014 Environmental Impact From Olive Mills Waste Disposal, Chemical Analysis Of Solid Wastes And Wastewaters Journal of Hygienic Engineering and Design UDC 628.312.5:634.63

[6] Uceda M, Beltrán G and Marquez A J 2006 Olive oil extraction and quality Grasas y Aceites 57, 1

[7] Borja R, Raposo F and Rincon B 2006 Treatment technologies of liquid and solid wastes from two-phase olive oil mills Grasas y Aceites 57, 1

[8] Kapellakis I E, Tsagarakis K P and Crowther J C 2007 Olive oil history, production and by-product management Reviews in Environmental Science and Bio/Technology 7,1,1-26

[9] https://www.ncbi.nlm.nih.gov/books/NBK233627/ National Research Council (US) Committee on Health Effects of Waste Incineration. Waste Incineration & Public Health. Washington (DC): National Academies Press (US); 2000. 3, Incineration Processes and Environmental Releases

[10] Al-swaidan H M and Ahmad A 2011 Synthesis and characterization of activated carbon from Saudi Arabian dates tree's fronds wastes IPCBEE vol.20 (2011) © (2011) IACSIT Press, Singapore

[11] Mistar E M, Alfatah T, Supardan M D 2020 Synthesis and characterization of activated carbon from Bambusa vulgaris striata using two-step KOH activation Journal of Materials Research and Technology, 9, 3,6278-6286

[12] Yakout S M, and Sharaf El-Deen G ٢٠١٦ Characterization of activated carbon prepared by phosphoric acid activation of olive stones" Arabian Journal of Chemistry, 9, 2, S1155-S1162

[13] Yang J and Qiu K 2010 Preparation of Activated Carbons from Walnut Shells via Vacuum Chemical Activation and Their Application for Methylene Blue Removal Chemical Engineering Journal 165, 1, 209-217

[14] Veli S, Arslan A, Zeybek S and Dimoglo A 2018 Activated Carbon Production from Walnut Shell by Application of Different Activating Agents Conference: 3rd International Conference on Civil and Environmental Engineering (ICOCEE) At: Izmir, Turkey

[15] Reeb J E and Milota M 1999 Moisture Content By the Oven-Dry Method for Industrial Testing WDKA

[16] Hossein M H 2018 Combustible Organic Materials: Determination and Prediction of Combustion Properties Berlin, Boston: De Gruyter

[17] ASTM D2876-00 (2020) Standard Test Method for Water-Soluble Matter of Vegetable-Tanned Leather

[18] Oginni O, Singh K, Oporto G, Dawson-Andoh B, McDonald L and Sabolsky E ٢٠١٩ Effect of one-step and two-step H3PO4 activation on activated carbon characteristics Bioresource Technology Reports, 8١٠٠٣٠٧,

[19] Yang H M Zhang D H, Chen Y, Ran M J and Gu J C 2017 Study on the application of KOH to produce activated carbon to realize the utilization of distiller's grains IOP Conf. Ser.: Earth Environ. Sci. 69 012051
[20] Hock P E and Zaini M A 2018 Activated carbons by zinc chloride activation for dye removal – a commentary Acta Chimica Slovaca 11, 2, 99-106
[21] Borsaru M, Charbucinski J, Eisler P L and Youl S F 1985 Determination of ash content in coal by borehole logging in dry boreholes using gamma-gamma methods Geoexploration, 23, 4, 503-518
[22] Vunain E, Kenneth D and Biswick T 2017 Synthesis and characterization of low-cost activated carbon prepared from Malawian baobab fruit shells by H3PO4 activation for removal of Cu(II) ions equilibrium and kinetics studies Appl Water Sci 7, 4301–4319
[23] Cleiton A, Nunes E, Mário and Guerreiro C 2011 Estimation Of Surface Area And Pore Volume Of Activated Carbons By Methylene Blue And Iodine Numbers Quim. Nova, 34, 3, 472-476

Radioactivity and Environmental Impacts of Granites from Um Ara, Southeastern Desert, Egypt

El-Afandy A. H[1], Mohamed G. El-Feky[1], Samia H. Taha[*2], S. M. El Minyawi[2], H. A. Sallam[2], A. F. Tawfic[3], A. Omar[3], M. G. El-Samrah[3]

[1] Nuclear Materials Authority, El Maadi P.O. Box 530, Cairo, Egypt.
[2] Physics Department, Faculty of Science / Al Azhar University, Girls' branch, Egypt
[3] Nuclear Engineering Department, Military Technical College, Cairo, Egypt

* tawficaf@mtc.edu.eg

Abstract. Granites from the Southeastern Desert of Egypt were investigated by using in situ γ-ray spectrometer. Consequently, estimating the radiological effects from these rocks were achieved through calculation of some radiological parameters; Annual effective dose equivalent AEDE, Internal hazard index H_{in}, External hazard index H_{ex}, Alpha activity index I_{α}, Excess lifetime cancer risk, ELCR, Annual gonadal dose equivalent AGDE, Gamma activity concentration index I_{γ} and Radium equivalent activity Ra_{eq}. The geological nature of the region under investigation was affected by the radioactivity measurements of the rocks under study. This was observed by the high U, Th, and K values, which exceed the average world value for safety used as building materials. Therefore, the granites of Om Ara may be inappropriate to choose as a building material or as an indoor and outdoor decorative. At the same time, the region under study represents valuable economic minerals sources that can be used in several industrial applications.

Keywords: Natural radioactivity; Southeastern desert; Excess lifetime cancer risk, Granites.

1. Introduction

The rugged natural stones are a more significant building material because they are usually suitable for interior and exterior building and decorative materials. Granites shows high concentrations of thorium (Th), potassium (K) and uranium (U).The attention of these radionuclides in granitic rocks is intimately related to their mineral compositions and general petrologic features [1-3].

Since granite under study can be used as a building material, radiation exposure to the population can be increased appreciably by using materials with moderate radioactivity levels. Therefore, it is crucial to calculate the radiation hazards assessment.

This study's objective is to present an assessment for the radiological nature at Um Ara area located at the southeastern desert of Egypt, the other objective of this study is to map the radioactivity background in this area and using the data collected from this map to assess any later radioactivity fluctuation.
For that reason, this research deals with natural radioactivity the mineralogical and chemical study of granite to understand and correlate the relationship of natural radioactivity with the radioactive mineral content and estimate possible implications from radiation exposure.

2. Geological Setting

Um, Ara area is located in the south Eastern Desert of Egypt, and it is bounded between latitudes 22° 38′–22° 41′ N and longitude 33° 47′ – 33° 53″. The geology of Um Ara area comprises Dokhan volcanics and metavolcanic .according to the petrographical background; these granites were classified as alkali-feldspar and monzogranite granite. The monzogranite generally has a homogeneous pinkish equigranular coarse-grained texture. [4-6]
Um, Ara granitic masses cover an area of about 300 km^2, characterized by moderate topography. The granitic rocks were classified as alkali feldspar and monzogranite granite. It also represented an excellent source of uranium in granitic rocks. [7, 8]

3. Materials and methods

A total number of 128 samples in Um Ara, Southeastern Desert, Egypt, were collected to measure their natural radioactivity levels according to ^{232}Th (ppm), ^{40}K and ^{238}U (ppm) (%).in situ γ-ray spectrometer, GS-256 with (NaI(Tl)) 3″×3″ crystal detector is used to study rocks in the field radiometrically.

4. Results

4.1 Radioactivity analysis

Measurements of elemental concentrations for eU, eTh, and K were taken expressed in ppm (**Table 1**). From the table, we can notice that the elemental concentrations of eU, eTh, and K have ranged from 11 to 430 with an average value of 48.8 ppm, from 36.5 to 280 with an average value of 69 ppm, and from 2.3 to 7.5 with average value 4.9 %, respectively. The remobilization factor ranged from -7.6 to 350, with an average value of 29.1. The frequency distributions of activity concentrations of the studied radionuclides in the examined granite samples are shown in figure 3.1, suggesting high radionuclides contents of these granites.
From the obtained results, we can deduce that the average values of eTh, eU and K were higher than the reported Earth's crust values, 10.5 ppm, 2.6 ppm and 2.4 % respectively, and also higher than the recommended safety values for using in construction materials, 12.1 ppm, 3.9 ppm and 1.5 %, respectively [9].
In general, the high contents of radioelements in Um Ara granites indicated that it's advisable to monitor the radioactivity levels of granites before adopting them for use as a building material.
The Egyptian younger granites are a good target for uranium exploration in the Eastern Desert of Egypt, which belong to the third epoch of uranium mineralization (0.70-0.50 Ga) during the upper Proterozoic-Cambrian times [10-13]. Um, Ara granites have a ^{206}Pb/^{238}U age of 603±14 Ma of zircons, which differs from the alkali feldspar micro granites [14]. Uranium deposits related to granitic nature can be divided into secondary and primary uranium deposits [15]. The first is formed in the rocks during magma crystallization. While the second is formed by later geological processes starting from dissolving and transportation of uranium. Then uranium can be divided into three types; (1) interstitial uranium at the grain boundaries, formed by hydrothermal solution migration and granites, (2) uranium in microfractures. and (3) absorbed uranium in chlorite, limonite and montmorillonite. The secondary uranium mineralization in Um Ara alkali feldspar microgranite formed at a two-stage of the metallogenic model as proposed by [16]; the first period was controlled by hydrothermal alteration accompanied by silicification, chlorination, K-feldspathization, and albitization .while the second period occurred near-surface, where water played an important role in moving the early, formed U-bearing minerals. The mineralogical

Journal of Physics: Conference Series **2305** (2022) 012033 doi:10.1088/1742-6596/2305/1/012033

composition of uranium deposits is dominated by uraninite and uranophane [17, 18] in Um Ara granite, attributed to the syngenetic hydrothermal origin. At the same time, accessories are represented by zircon, apatite, fluorite, sphene, topaz, and monazite. Uranium can be oxidized as U^{6+} in a near-surface area and is insoluble as a U^{4+} state.

4.2 Geochemistry of U and Th in the studied granites

The geochemistry of U and Th during magmatic differentiation has been studied in many types of granites from different areas. Generally, during magmatic differentiation, the Th/U ratio remains constant. [19] Suggesting 3.5 to 4 for the Th/U ratios in the granites. These ratios can be increased [20, 21] or decreased depending on the redox conditions, the volatile contents, or alterations by endogen or supergene solutions [22, 23]. Typical eTh/eU ratios in igneous rocks were 3:1 or 4:1. In contrast, the Th/U ratio of the studied granitic rocks was 2.2, indicating uranium enrichment. This was also confirmed by the average value of the remobilization parameter, which has a positive value of 29.1.
The relationships between ^{238}U, ^{232}Th, and ^{40}K (Table 1), were presented in figure 2. eU-eTh variation diagram of the studied granitic rocks reflected a strong positive correlation (R= 0.8), which indicated that magmatic processes played an essential role in the concentration of radioelements (Fig. a).
Thorium displayed a significant correlation with potassium (R= 0.7), and there was a positive correlation between ^{238}U and ^{40}K (R = 0.5), confirming the previous result (Fig. 1b & c).
The variation of eTh and eU against the eTh/eU ratio showed a negative correlation which has confirmed the idea that U and Th enrichment in these anomalies could be controlled by magmatic processes (Fig. 2d & e) [24,25].
Ratios of eTh/eU < 2 suggested relative uranium enrichment and implicates reducing conditions, but ratios >7 indicated preferential removal of uranium, possibly by leaching. The cross plot of eTh/eU and eTh/K ratios presented in figure 2f exhibit that altered granite samples lie in the leached-U sector.
The average value of eTh/K (14.3) is >2×10-4, which denoted thorium enrichment in the studied rocks. In general, low eTh/K ratios were an excellent indicator of potassium alterations [26].

Table 1. Values of ^{238}U (ppm), ^{232}Th (ppm), and K (%), as well as eTh/eU and eTh/K ratios and (eU-eTh/3.5) parameter for Um Ara granite.

Rock	eU (ppm)	eTh (ppm)	eK (%)	eTh/eU	eTh/eK	eU-eTh/3.5
granite	29	46	2.9	1.6	15.9	15.9
	140.1	78.9	6.4	0.6	12.3	117.6
	41.5	51.2	2.8	1.2	18.3	26.9
	117.8	82.7	6.7	0.7	12.3	94.2
	21.6	51.3	2.5	2.4	20.5	6.9
	206.6	160.2	6.9	0.8	23.2	160.8
	22.2	46.4	2.8	2.1	16.6	8.9
	153.9	87.6	**7.5**	0.6	11.7	128.9
	34.3	56.4	4.3	1.6	13.1	18.2
	121.5	88.9	6.6	0.7	13.5	96.1
	40.1	46.6	3.3	1.2	14.1	26.8
	108.9	78.8	6.3	0.7	12.5	86.4
	16.4	46.8	3.2	2.9	14.6	3
	226.1	91.3	7.4	0.4	12.3	200
	38.3	47.8	3.2	1.2	14.9	24.6
	279.9	108.1	6.8	**0.4**	15.9	249
	23.4	41	3.6	1.8	11.4	11.7
	46.3	90.9	6.7	2	13.6	20.3

22.5	**36.5**	3.2	1.6	11.4	12.1
76.9	82.7	6.8	1.1	12.2	53.3
23	45.1	**2.3**	2	19.6	10.1
141.8	97.5	6.5	0.7	15	113.9
22.8	42.1	2.3	1.8	18.3	10.8
76.7	77	7.2	1	10.7	54.7
28.2	50.8	4.4	1.8	11.6	13.7
72.8	74.9	6	1	12.5	51.4
17	50.1	3.8	2.9	13.2	2.7
56.8	95.6	4.6	1.7	20. 8	29.5
24.9	56.6	4.1	2.3	13.8	8.7
85.7	98.3	6.3	1.1	15.6	57.6
17.9	48.5	4	2.7	12.1	4
80.6	84.7	6.8	1.1	12.5	56.4
24.2	50.9	4.3	2.1	11.8	9.7
80.9	84.8	5.6	1	15.1	56.7
36.4	40.8	3.5	1.1	11.7	24.7
56.4	70.1	5.1	1.2	13.8	36.4
33.5	44.2	3.1	1.3	14.3	20.9
52.3	55.6	5.1	1.1	10.9	36.4
21	40.5	3.2	1.9	12.7	9.4
58	80.1	6.3	1.4	12.7	35.1
11	65	4.8	**5.9**	13.5	-7.6
22.5	76.6	6.1	3.4	12. 6	0.6
16.5	54.7	4.4	3.3	12.4	0.9
32.6	76.5	5.5	2.3	13.9	10.7
12.3	54.2	4.4	4.4	12.3	-3.2
32.4	75.6	6	2.3	12.6	10.8

Table continues

granite	14.3	44.1	3.4	3.1	12.9	1.7
	60.5	72.3	6.3	1.2	11.5	39.8
	20.4	59.3	4.2	2.9	14.1	3.5
	46.2	88.4	5.6	1.9	15.8	20.9
	18.9	41.8	3.7	2.2	11.3	7
	31.9	81.9	6.5	2.6	12.6	8.5
	13	54.6	4.1	4.2	13.3	-2.6
	30.1	83.1	5.7	2.8	14.6	6.4
	16	49.1	4.3	3.1	11.4	2
	70.4	77.8	6.8	1.1	11.4	48.2
	20.2	58.2	4	2.9	14.6	3.6
	35.7	76.9	5.7	2.2	13.5	13.7
	17.1	50	3.6	2.9	13. 9	2.8
	30.1	65.9	5.3	2.2	12.4	11.3

	21.3	52.1	3.7	2.4	14.1	6.4
	42	83.1	6.3	2	13.2	18.3
	18.9	53.6	3.7	2.8	14.5	3.6
	33	92.4	5.8	2.8	15.9	6.6
	29	61	3.8	2.1	16.1	11.6
	126	77	5.6	0.6	13.8	104
	40	43	2.6	1.1	16.5	27.7
	70	72	4.8	1	15	49.4
	29	49	3.1	1.7	15.8	15
	430	**280**	6.7	0.7	**41.8**	**350**
	38	54	2.4	1.4	22.5	22.6
	140	91	5.5	0.7	16.6	114
	26	65.7	4.8	2.5	13.7	7.2
	31.6	77	6.6	2.4	11.7	9.6
	32.1	72.5	4.1	2.3	17.7	11.4
	55.8	85.6	5.8	1.5	14.8	31.3
	24.4	58.7	4.1	2.4	14.3	7.6
	56.9	85.1	5.5	1.5	15.5	32.6
	30.1	51.9	4.1	1.7	12.7	15.3
	50.9	76.5	5.9	1.5	12.9	29
	12	59.1	3.4	4.9	17.4	-4.9
	81.9	91.2	6.2	1.1	14.7	55.8
	18	50.1	4.2	2.8	11.9	3.7
	24	70.2	5.5	2.9	12.8	3.9
	11.1	50.1	4.1	4.5	12.2	-3.2
	29.6	72.4	5.6	2.4	12.9	8.9
	18.5	51.3	4.1	2.8	12.5	3.8
	27.8	83.7	6.1	3	13.7	3.9
	14.8	59.3	4	4	14.8	-2.1
	47.9	97.6	6.3	2	15.5	20
	19.4	60.7	4	3.1	15.2	2.1
	33.1	89.9	5.8	2.7	15.5	7.4
	16.9	65.6	4.3	3.9	15.3	-1.8
	38.1	97.3	6.5	2.6	14.9	10.3
	16.3	58.9	4.5	3.6	13.1	-0.5

Table continues

granite	41.6	82.4	5.8	2	14.2	18.1
	18.9	50.4	3.5	2.7	14.4	4.5
	27.3	59.6	5.1	2.2	11.7	10.3
	17.2	51.2	3.5	3	14.6	2.6
	30.2	67.3	5.6	2.2	12	11
	20.6	42.3	3.6	2.1	11.8	8.5
	99.1	90.2	6	0.9	15	73.3

	17.7	44.8	3.5	2.5	12.8	4.9
	89.6	84.2	7.1	0.9	11.9	65.5
	35.2	42.3	4.2	1.2	**10.1**	23.1
	53.6	61.5	5.8	1.1	10.6	36
	53	51.8	2.9	1	17.9	38.2
	70.2	66	4.8	0.9	13.8	51.3
	30.1	50.1	3.2	1.7	15.7	15.8
	78.1	67.9	5.6	0.9	12.1	58.7
	25.7	51.6	3.5	2	14.7	11
	79.9	88.7	6.2	1.1	14.3	54.6
	18.4	53.9	4.4	2.9	12.3	3
	40.1	71.2	5.7	1.8	12.5	19.8
	16.1	66.4	3.8	4.1	17.5	-2.9
	33.2	76.6	6.3	2.3	12.2	11.3
	15	52.7	3.5	3.5	15.1	-0.1
	74.4	77	6.3	1	12.2	52.4
	13.9	53.6	3.4	3.9	15.8	-1.4
	32.5	80.1	6	2.5	13.4	9.6
	13.5	73.8	4.4	5.5	16.8	**-7.6**
	24.7	87.2	6.4	3.5	13.6	-0.2
	12.5	55	4.2	4.4	13.1	-3.2
	24.7	80.2	5.6	3.2	14.3	1.8
	13.6	57.5	3.2	4.2	17.9	-2.8
	26.7	90	6.1	3.4	14.8	1
	12.5	65.4	4.1	5.2	15.9	-6.2
	22.5	79.8	5.8	3.5	13.8	-0.3
Aver.	48.8	69	4.9	2.2	14.3	29.1
Crust	2.9	10.8	2.7			
Safety for building	4	12.3	1.6			

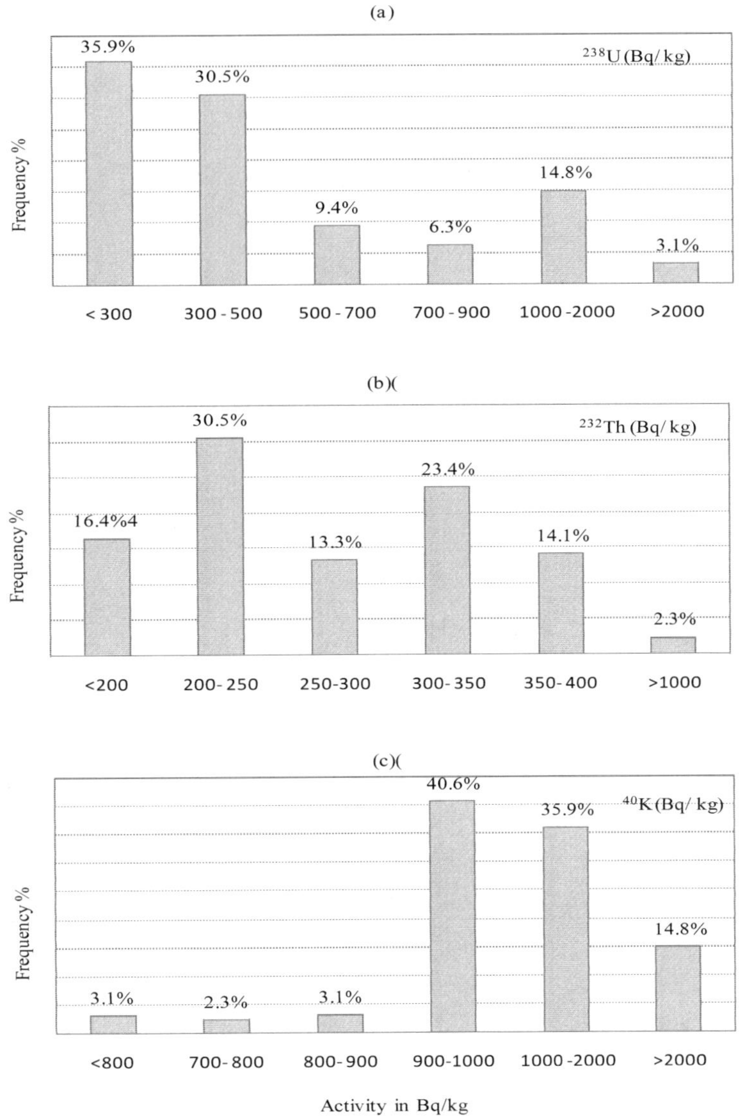

Figure 1. Frequency distributions of activity concentrations (Bq/ kg) for Um- Ara granite rocks (a) ^{238}U (b) ^{232}Th and (c) ^{40}K

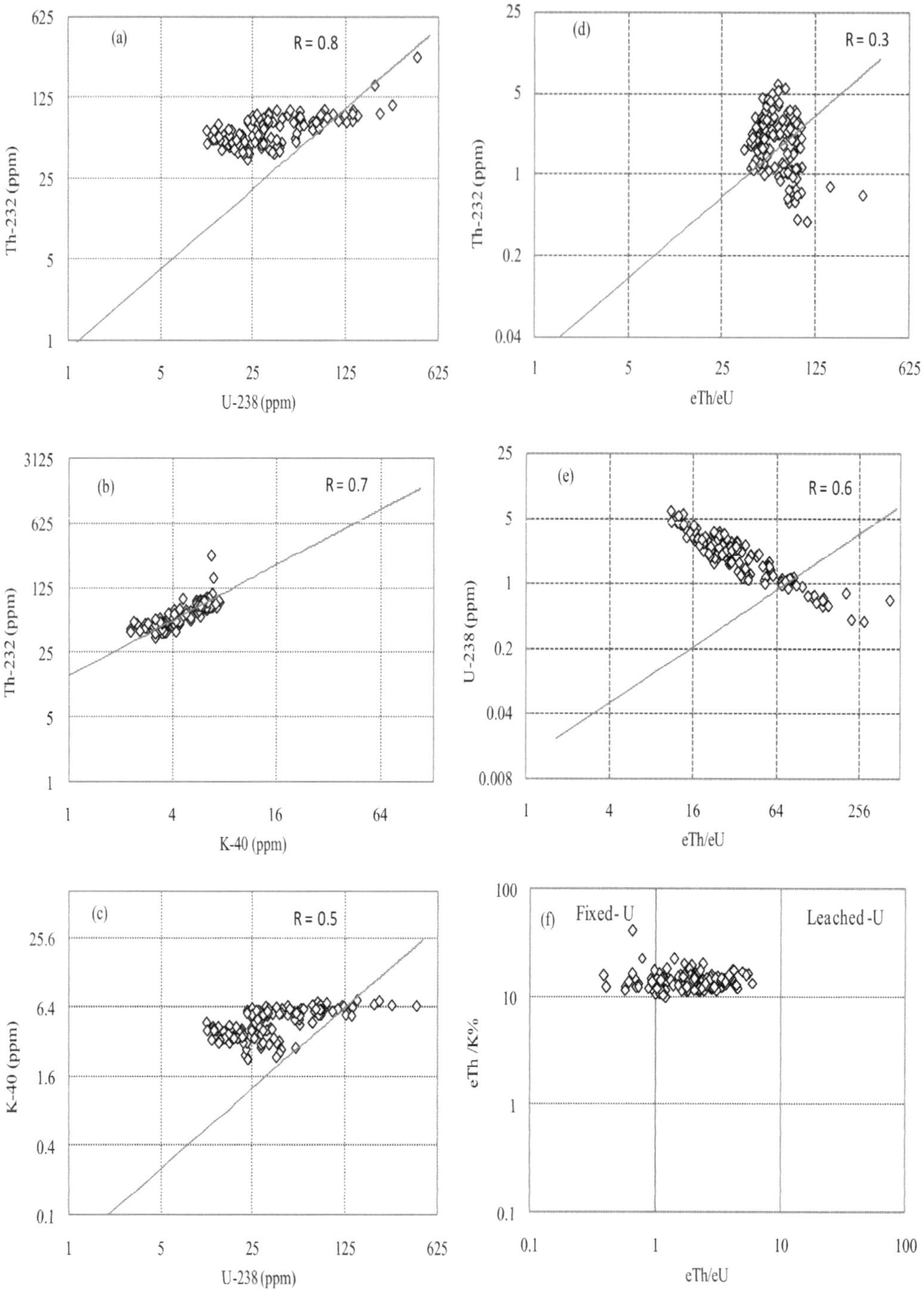

Figure 2. (a) U vs. Th, (b) K vs. U and (c) K vsTh (d) eTh/eU vs eTh, (e) eTh/eU vs eU and (f) eTh/eU vs eTh/K% for granite.

4.3 Evaluation of radiological hazard effects

ETh and eU in ppm and K, in %, were converted to activity concentration, Bq/kg, using the conversion equations given by IAEA [27-29] and Polish Central Laboratory for Radiological Protection [28]. The specific parent activity of a sample containing one ppm, 1.1% of ^{40}K is 310 Bq/ kg by weight, one ppm of Ra is 11.3 Bq/kg, of U, is 12.43 Bq/kg, and one ppm of Th 4.26 Bq/kg. Therefore, it is possible to predicate the radiation effect by measuring the following values.

4.3.1 Radium equivalent activity, Ra_{eq}

It is the activity sum of the, ^{232}Th, ^{238}U and ^{40}K radionuclides based on the assumption that 4710 Bq/ kg of ^{40}K, 261 Bq/ kg of ^{32}Th, and 372 Bq/ kg of ^{238}U produce the same gamma-ray dose rate [30-32]

$$Ra_{eq} = A_U + 1.43\ A_{Th} + 0.077\ A_K$$

where A_K, A_{Th}, and A_U are the specific activities of ^{40}K, ^{232}Th, and ^{238}U in Bq/ kg. Ra_{eq} were estimated from the collected rocks and is presented in Table 2. From the recorded results, we can observe that the calculated values of Ra_{eq} ranged between 514.6 and 7097.6, with an average value of 1120.4 Bq/kg. The obtained values were more than the recorded world value (369 Bq/kg).

4.3.2 Gamma activity index, I_γ

The restriction on building materials for gamma radiation is based on a dose range of 0.3 –1 mSv/y. To examine whether a building material meets these limits of dose criteria, the gamma activity concentration index I_γ is calculated from the following equation [33,34]:

$$I_\gamma = \frac{A_U}{300} + \frac{A_{Th}}{200} + \frac{A_K}{3000}$$

The presented equation depend on the radionuclides contribution to the external irradiation, it sums the three specific activities for gamma radiation index to form unit sum. $I_\gamma \leq 2$ related to a dose criterion of 0.3 mSv y^{-1}, while $I_\gamma \leq 6$ related to 1 mSv y^{-1} [35] .

Thus, the activity concentration index should be used only as a screening tool for identifying materials that might be of concern to be used as construction or covering materials. According to this dose criterion, materials with $I_\gamma \geq$ six should be avoided. In contrast, these values correspond to dose rates higher than one mSv/y, the highest dose rate value in air recommended for the population [9, 33].

Values of gamma activity index (I_γ) for Um Ara granite ranged between 1.8 and 24.1 with an average value of 3.9, higher than the acceptable value 2.

The value showed $1.9 \leq I_\gamma \leq 5.9$, where the gamma dose for these samples is higher than 0.29 mSvy^{-1}, but this value is less than the recommended dose rate in air value 0.9 mSv/y .

4.3.3 Alpha activity index, I_α

Alpha particles indices are related to radon inhalation produced from building materials. The alpha index (I_α) is defined as follows:

$$I_\alpha = \frac{A_U}{200} \leq 1$$

The ^{238}U activity concentrations in contracting materials are 199 and 99 Bq/kg, respectively, as suggested by ICRP, 1994. Therefore, I_α = 0.9 when the maximum concentration value of ^{238}U is 199 Bq/kg, the indoor radon concentration could exceed 199 Bq/kg.

I_α ranged between 0.69 and 26.4, with an average value of 3. It indicates that the average value of the alpha activity index is higher than the acceptable value.

4.3.4 Hazard indices (H_{ex} and H_{in})

External hazard index, H_{ex}

The external hazard index (H_{ex}) measures the external hazard due to the emitted gamma radiation. The equation calculated it.

$$H_{ex} = A_K/4810 + A_{Th}/259 + A_U/370 \leq 1$$

To reduce the radiation risk from constructing materials, this index value must be less than one. For that, the Ra_{eq} value must be less than 370 $Bqkg^{-1}$.

Values of H_{ex} ranged between 1.4 and 19.2, with an average weight of 3. This result indicates that this value is higher than the acceptable value. This suggests that Um Ara granites' radiation risk can't be negligible.

Internal hazard index, H_{in}

The internal hazard index (H_{in}) controls the internal exposure to ^{222}Rn and its radioactive levels. The following equation gives it.

$$H_{in} = A.U./185 + A_{Th}/259A_K/4810 \leq 1$$

If the uranium concentration values in samples under investigation are half that of the acceptable limit, then the value of H_{in} should be less than 1. Values of H_{in} ranged from 1.8 to 33.5, with an average value of 4.7. The results are more than the required values, this indicates that the radiation risks from these samples increases, so it can't be used as a building material.

4.3.5 External dose from naturally occurring radionuclides

The external absorbed dose rate in outdoor air, D_{out}

The external absorbed dose rates D can be calculated according to the following formula;

$$D = A_{Ei} \times C_F$$

A_{Ei} is the activity concentration (Bq kg^{-1}), and $C_{.F.}$ is the dose conversion factor (absorbed dose rate in the air per unit activity per unit mass in units of nGy h^{-1} per Bq kg^{-1}). the considered dose rate conversion factors for ^{40}K, ^{232}Th, and ^{238}U were determined where C_K= 0.0429, C_{Th}= 0.5835 and C_U= 0.4551.

The external absorbed dose rate in indoor air, D_{in}

D_{in} is calculated as follows[36]:

$$D_{in} = D_{out} \times 1.4$$

Values of D_{in} ranged from 320.6 to 4321.6, with an average value of 704.3 nGy/h. It is observed from these values that the average values for granite samples under study were much higher than the acceptable world values and higher than fair Egypt values (Table 1).

4.3.6 Annual effective dose, AEDE

The AEDE is determined using the following equations:

$$\text{AEDE (outdoor)} = 0.7 \text{ Sv/Gy} \times 0.2 \times D_{out} \text{ nGy/h} \times 8760 \text{ h}$$

$$\text{AEDE (indoor)} = 0.7 \text{ Sv/Gy} \times 0.8 \times D_{in} \text{ nGy/h} \times 8760 \text{ h}$$

Using these equations, the annual effective doses for children and infants are calculated with the numerical values of 0.8 and 0.9 of conversion factor for infants and children.

The annual outdoor effective dose of Um Ara granitic rocks varied from 0.3 to 3.9 with an average value of 0.6 mSv, from 0.3 to 4.4 with an average value of 0.7 mSv, and from 0.4 to 5 with an average value of 0.8 mSv for adults, children, and infants, respectively. At the same time, the annual indoor effective doses were ranged from 1.6 to 21.8 with an average value of 3.5 mSv, from 1.8 to 24.9 with an average value of 3.9 mSv, and from 2 to 28 with an average value of 4.4 mSv for adults, children, and infants, respectively. The total annual effective dose equivalent of Um Ara rocks is 4.1 mSv for adults, 4.7 mSv for children, and 5.2 mSv for infants. These values were greater than the world's value (0.49 mSv).

4.3.7 Annual gonadal dose equivalent, AGDE
(AGDE) was calculated using the following formula.

$$AGDE = (0.314 \times A_K + 4.18 \times A_{Th} + 3.09 \times A_U) \times 10^{-3}$$

Values of AGDE ranged between 1.6 and21.8, with an average value of 3.5 mSv. The results are higher than the world value.

4.3.8 Excess lifetime cancer risk (ELCR)
Excess lifetime cancer risk (ELCR) is shown as

$$ELCR = AEDE \times DL \times RF$$

AEDR is the Annual Equivalent Dose Equivalent, DL is the Average Duration of Life, and R.F. is the Risk Factor (Sv^{-1}), i.e., fatal cancer risk per Sievert. Values ranged from 6.5×10^{-3} to 89.8×10^{-3}. This value is higher than the world value (0.29×10^{-3}).

5. Conclusion

The geological structure of the region under investigation was affected by the radioactivity levels of the rocks under study. This was observed by the high U, Th, and K values, which exceed the average world value for safety used as building materials.

Therefore, the granites of Om Ara may be inappropriate to choose as a building material or as an indoor and outdoor decorative. At the same time, the region under investigation shows high economic values of minerals concentration.

References

[1] A. H. El-Farrash, M. G. El-Feky, S. F. Hassan, Soliman A. Abu Elatta, and M. Refaat (2020): Radiological and environmental studies on the metamorphosed sandstones at Wadi Sikait area, south eastern desert, Egypt. IOP Conf. Series: Materials Science and Engineering 975 (2020) 012017 IOP Publishing doi:10.1088/1757-899X/975/1/012017.

[2] Ahmed M. EL-Kammar; Mohamed G. EL-Feky 1; Hassan T. Abu zied and Dina A. Othman (2018): RAdioactivity and environmental impacts of some Carboniferous and Cretaceous kaolin deposits in southwestern-Sinai, Egypt. Nuclear Sciences Scientific Journal, 7, 87- 100.

[3] A. Chandrasekaran, R. Ravisankar, G. Senthilkumar, K. Thillaivelavan, B. Dhinakaran, P. Vijayagopal, S.N. Bramha, B. Venkatraman.2014 Spatial distribution, and lifetime cancer risk due to gamma radioactivity in Yelagiri Hills, Tamilnadu, India. Egyptian Journal of Basic and Applied Sciences I, 38- 48.

[4] Ajibode, M. O., Avwiri, G. O and Agbalagba, E. O. 2013. Evaluation of radiation hazard indices in an oil mineral lease (oil block) in Delta State, Nigeria. International Journal of Engineering and Applied Sciences. Vol. 4, No. 2.

[5] A. F. Tawfic, Hesham M. H. Zakaly, Hamdy A. Awad, Hesham R. Tantawy, Akbar. Abbasi,Neveen S. Abed, Mostafa Mostafa (2020) "Natural radioactivity levels and radiological implications in the high natural radiation area of Wadi El Reddah, Egypt" Journal of Radioanalytical and Nuclear Chemistry. doi.org/10.1007/s10967-020-07554-2

[6] Avwiri, G.O., Osimobi, J.C and Agbalagba, E.O.2012. Evaluation of Radiation Hazard Indices and Excess Lifetime Cancer Risk Due to Natural Radioactivity in the soil profile of Udi and Ezeagu Local Government Areas of Enugu State, Nigeria. Comprehensive Journal of Environmental and Earth Sciences Vol. 1(1), pp. 1 – 10.

[7] Ahmed N. Shabaka, Ahmed Omar, Sayed A. El-Mongy & A. F Tawfic (2020) "Analysis of natural radionuclides and 137Cs using HPGe spectrometer and radiological hazards assessment for Al-Nigella site, Egypt" International Journal of Environmental Analytical Chemistry, DOI: 10.1080/03067319.2020.1724985

[8] D.I.J wanbot M.M.Izam, G.G. Nyam, M.Yusuf, 2014. Indoor and Outdoor Gamma Dose Rate Exposure Levels in Major Commercial Building Materials Distribution Outlets in Jos, Plateau State-Nigeria. Asian Review of Environmental and Earth Sciences Vol. 1, No. 1, 5-7

[9] UNSCEAR, 1993. Exposure from natural sources of radiation. Forty-second session of United Nations Scientific Committee on the Effect of Atomic Radiation, Vienna 12-28 May.

[10] Elham Bavarnegin, Masoud Vahabi Moghaddam, and Nasrin Fathabadi, 2013. Natural radionuclide and radiological assessment of building materials in high background radiation areas of Ramsar, Iran. J Med Phys. 2013 Apr-Jun; 38(2): 93–97.

[11] El-Afandy, A.H., Mohamed G. El-Feky, Samia Taha, S.M. El Minyawi, H.A. Sallam (2016): Distribution of Radioelements and Evaluation of Radiological Hazard effects on Stream Sediments and Cataclastic Rocks of Wadi Abu Rusheid, Southeastern Desert, Egypt. Greener Journal of Geology and Earth Sciences, 4, (3), 056-069.

[12] A. F. Tawfic, Hamed I. Mira & Ahmed M. Omar (2019) " Chemical alteration processes for uranium migration and their significance on uranium isotopic ratio changes" International Journal of Environmental Analytical Chemistry, doi.org/10.1080/03067319.2019.1702172

[13] Harpy, N. M., El Dabour, S. E., Sallam, A. M., Nada, A. A., El Aassy, A. E. & El Feky, M. G. (2019): Radiometric and environmental impacts of mill tailings at experimental plant processing unit, Allouga, Egypt. ISSN: 1527-5922 (Print) 1527-5930 (Online) Journal homepage: https://www.tandfonline.com/loi/uenf20

[14] Samia H. Taha, Osama R. Sallam, Abd Elhadi A. Abbas & Neveen S. Abed, 2020. Radioactivity and environmental impacts of ferruginous sandstone and its associating soil. International Journal of Environmental Analytical Chemistry

[15] A. F. Tawfic, A. Omar,, Neveen S. Abed, and Hesham R. Tantawy. (2020) "Investigation of Natural Radioactivity in Wadi El Reddah Stream Sediments and Its Radiological Implications" Radiochemistry Vol. 63, No. 2, pp. 245–252 DOI:10.1134/S1066362221020156

[16] Hamed I. Mira , Neveen S. Abed , Hesham R. Tantawy & A. F. Tawfic (2020) "Mineralogical and natural radioactivity investigations of Wadi El Reddah stream sediments" International Journal of Environmental AnalyticalChemistry, doi.org/10.1080/03067319.2020.1796991

[17] Iqbal, M., Tufail, M. and Mirza, S. M. 2000. Measurement of natural radioactivity in marble found in Pakistan using a NaI(Tl) gamma-ray spectrometer. Journal of Environmental Radioactivity, 51, 255-265.

[18] Koshi, C., Takao, I. And Hideo, S. 2001. Terrestrial gamma radiation in Koshi prefecture, Japan. Journal of Health Science, 47 (4), 362-372.

[19] Mahmoud R. Khattab, Ahmed F. Tawfic & Ahmed M. Omar (2019) "Uranium-series disequilibrium as a tool for tracing uranium accumulation zone in altered granite rocks" International Journal of Environmental Analytical Chemistry, doi.org/10.1080/03067319.2019.1686495

[20] Mohamed Galal El-Fekya, Hesham Samir Mohammed, Ahmed Mohammed El-Shabasy, Mohammed Ramadan Ahmed, Yasser Kamal Abdel-Monem & Hamed Ibrahim Mira (2021): Mobilisation of radionuclides during uranium and gold processing of granitic rock at El-Missikat area, central Eastern Desert, Egypt. International Journal of Environmental Analytical Chemistry ISSN: (Print) (Online) Journal homepage: https://www.tandfonline.com/loi/geac20

[21] M.M. El Galy, A.M. El Mezayn, A.F. Said, A.A. El Mowafy, M.S. Mohamed, 2008. Distribution and environmental impacts of some Radionuclides in sedimentary rocks at Wadi Naseib area, southwest Sinai, Egypt, Journal of Environmental Radioactivity 99 (2008) 1075-1082.

[22] Malczewski, D., Taper, L., Dorda, J., (2004). Assessment of natural and anthropogenic radioactivity levels in rocks and soils in the environs of Swieradow Zdroj in Sudetes, Poland by in situ gamma-ray spectrometry. J. Environ. Radioact. 73, 233-245.

[23] Moamen G. El-Samrah , A.F. Tawfic , Samir E. Chidiac (2021) "Spent nuclear fuel interim dry storage; Design requirements, most common methods, and evolution: A review" Annals of Nuclear Energy , doi.org/10.1016/j.anucene.2021.108408

[24] Mohd Zubair, Deepak Verma, Ameer Azam, Sukanta Roy, 2013 Natural radioactivity and radiological hazard assessment of soil using gamma-ray spectrometry. Radiat. Prot. Dosimetry, 155 (4): 467-473

[25] Quindos, L. S., Fernández, P.L., Ródenas, C., Gómez-Arozamena, J. and Arteche, J. 2004. Conversion factors for external gamma dose derived from natural radionuclides in soils. Journal of Environmental Radioactivity, 71, 139- 145.
[26] Ravisankar, R., Vanasundari, K., Chandrasekaran, A., Rajalakshmi, A., Suganya, M., Vijayagopal, P. and Meenakshisundaram, V., 2012. "Measurement of Natural Radioactivity in Building Materials of Namakkal, Tamil Nadu, India Using Gamma-Ray Spectrometry", Applied Radiation and Isotopes, 70, 699-704.
[27] (IAEA) International Atomic Energy, Agency, 1989. Measurement of radionuclides in food and the environment, a guidebook, Technical Reports Series No. 229, Vienna.
[28] Serena Righi and Luigi Bruzzi, 2006. Natural radioactivity and radon exhalation in building materials used in Italian dwellings. Journal of Environmental Radioactivity 88 (2006) 158-170.
[29] Khaled Elewa, Adel Belal, Ola El Monayeri ,A.F. Tawfic.2022" Application of metal-organic framework (Zn-Ph-D CP) for copper ion removal from aqueous solution" Ain Shams Engineering Journal 13 (2022) 101670 doi.org/10.1016/j.asej.2021.101670
[30] Taskin, H., Karavus, M., Ay, P., Topuzoglu, A., Hidiroglu, S. &Karahan, G., 2009. "Radionuclide concentration in soil and lifetime cancer risk due to the gamma radioactivity in Kirklareli, Turkey". Journal of environmental radioactivity, 100, 49-53.
[31] Mohamed A.E.M. Ali , A.F. Tawfic, Mostafa A. Abdelgawad , Mohamed Mahdy, A. Omar. 2022" Gamma and neutrons shielding using innovative fiber reinforced concrete" Progress in Nuclear Energy 145 (2022) 104133. doi.org/10.1016/j.pnucene.2022.104133
[32] Y. Orgun, N. Altinsoy, S. Y. Sahin, Y. Gungor, A. H. Gultekin, G. Karaham and Z. Karaak, 2007. Natural and anthropogenic radionuclides in rocks and beach sands from Ezine region (Canakkale), Western Anatolia, Turkey. Applied Radiation and Isotopes, 739–747: 65.
[33] A. F. Tawfic & S. E. Dickson & Y. Kim & W. Mekky & Mohamed Gobara ,Ahmad Baraka. 2017 "Preparation and characterization of nickel hexacyanoferrate films for the removal of cesium ion by electrically switched ion exchange (ESIX)" J Solid State Electrochem DOI 10.1007/s10008-017-3635-y
[34] Shams ISSA, Mohamed UOSIF, Reda ELSAMAN, 2013. Gamma radioactivity measurements in Nile River sediment samples. Turkish Journal of Engineering and Environmental Sciences. 37: 109- 122
[35] Tzortzis, M., Tsertos, H., Christofides, S. and Christodoulides, G. 2003. Gamma radiation measurements and dose rates in commercially-used natural tiling rocks (granites). Journal of Environmental Radioactivity, 70, 223–235.
[36] UNSCEAR, 2000. United Nations Scientific Committee on the Effects of Atomic Radiation. Report to the General Assembly, with scientific annexes. Sources and Effects of Ionizing Radiation. United Nations sales publications No. E.00.IX.3 Volume I: Sources) and No. E.00.IX.4 (Volume II: Effects). United Nations, New York, 1220 pp.

Comparative Studies on Hydrometallurgy and Pyrometallurgy to the Extraction of Beryllium from Decrepitated Beryl from Wadi El Gemal Region, Eastern Desert, Egypt

Mohamed Shabaan Atrees[1], Hamed I. E. Mira[1], Yasser M Kawassik[1], And Mohamed E. M. Ali[2]

[1] Nuclear Materials Authority.
[2] Environmental Research and Climate Changes Institute, National Research Centre

ABSTRACT. Comparative Studies on hydrometallurgy and pyrometallurgy to the extraction of beryllium from decrepitated beryl from Wadi El Gemal Region, Eastern Desert, Egypt are carried out in this work in short, useful summary. Beryl is one of the important ore mineral constituents of the mineralized zones at Wadi El Gemal area, Eastern Desert, Egypt. Extraction of beryllium from beryl ore involves both (a) pyrometallurgical techniques through roasting of beryl with Na_2SiF_6 followed by leaching of roasting beryl with commercial (dilute) H_2SO_4 (b) hydrometallurgical processing techniques of beryl through direct leaching with H_2SO_4 to produce directly sulfate liquor. 88% of beryllium was extracted by roasted beryl with Na_2SiF_6 at 850°C for 1.5 hours and leaching the roasted product with 25% H_2SO_4 AT 85 °C for 3/4 hour. On the other hand, 95 % of beryllium was extracted directly after 20 minutes by direct leaching at 280°c with 10 mol/liter H_2SO_4. Therefore, this work sheds light on the study of the factors affecting both forms of beryllium extraction.

Keywords: Beryl, extraction, hydrometallurgical pyrometallurgical, leaching, roasting

1, INTRODUCTION

The demand for beryllium and its alloys is steadily increasing owing to widening the circle of nuclear and non-nuclear application. Beryllium is a very attractive metal for cladding purposes since it joins to its very weak thermal neutron absorption good thermal conductivity and good corrosion resistance. Its low thermal neutron-absorption confined to low atomic weight, give it suitable moderating properties, and hence, beryllium metal and oxide are suitable reflectors and moderators Asmaa Mansour Abo El Souad et al (2019). Beryl is the most important beryllium mineral and has the chemical composition3BeO.Al_2O_3.6SiO_2. Among beryllium-bearing minerals, one can mention Helvite, phenacite, chryso-beryl, herderite, bertrandite, euclasse... etc [1]. Chryso-beryl and bertrandite have been marketed as by-products, but non of the former five minerals are known in workable deposits. The remaining minerals (about 30 minerals containing beryllium) are listed as rare on very rare Demerdash, [2-3] have briefly reported the occurrence of beryl in the area around Zabara, Sikait, Nugrus, Um Kabu, and Um Debaa in the Eastern Desert, Starting in Wadi El-Gemal Area. Hassan and El-Shatoury, 1976 [4] found that beryl occurs in metamorphic schists and occurs as green hexagonal crystals in several localities in wadi Nugrus, Wadi Abu Rucheid, Wadi Um Harbe, Wadi- Um El-Keiron, Wadi Um Kabu, and Wadi Um Debaa. They found that beryl is mostly present either in lenticular passes of pegmatitic quartz, which occurs in mica-rich schist's or the schist's themselves. Hughes, 1965 [5] reported on the prospering of beryllium at the wadi El-Gemal area. He concluded that beryl grade in the prospected localities of Wadi El-Gemal, in the south Eastern Desert, is commonly low and should be concentrated in various ways particularly flotation. X-ray data for beryl were demonstrated. Beryl ore is very inert in that it is not attacked by the mineral acids under the normal conditions of temperature and pressure. There are two methods for extraction of beryllium from beryl ore:

(a) pyrometallurgical techniques(fluoride technique) through roasting of beryl with a fluorinating agent followed by (hydrometallurgical technique) leaching of roasting beryl with water and commercial (dilute) H_2SO_4

(b) Direct hydrometallurgical processing techniques (sulfate technique) of beryl through direct leaching with H_2SO_4 to produce directly sulfite liquor. This study is concerned with the light consideration of two methods for extraction of beryllium from Egyptian beryl ore [6].

2. EXPERIMENTAL

2.1. Materials

Beryl samples were provided by the Abo Rushied Project of Nuclear Materials Authority. Chemical analysis of the samples is given in Table (1)

Table (1): The Chemical Analysis of the Two Beryl Samples

Chemical composition of rock type	Quartz vein	Mica schist
Na_2O	1.87	1.85
MgO	3.22	3.21
Al_2O_3	16.88	16.57
SiO_2	63.62	63.35
Fe_2O_3	0.95	1.01
BeO	13.46	13.85
Total	100	100

X-ray diffraction analysis of studied samples indicates the presence of the following minerals: beryl, mica (muscovite and biotite), quartz, and feldspar

Apparatus

- Silicofluoride route tests were carried out by using platinum crucible for sintering in a muffle furnace, and leaching of the roasted product in a 250 cm^3 round bottom flask placed in a thermostat. The reaction mixture was agitated at a rate of 500 r.p.m and then filtered under a vacuum in a Buchner –type filter.
- Direct (pressure) leaching of beryl with H_2SO_4 was performed using an autoclave type, "has manometer, temperature regulator and valves "to take samples during leaching process which previously described [7]. The leaching container is made of titanium and the speed of its stirrer reaches about 1200 r.p.m.

3. RESULTS AND DISCUSSION

3.1. Silicofluoride process or pyro –hydrometallurgical techniques

This process for extracting beryllium from beryl ore is more elegant than the sulfate process since by its use only the beryllium content of the ore need be attacked, the alumina and silica contents being largely unaffected.

3.1.1. Effect of fluorinating agent. CaF_2, Al F_3, and Na_2SiF_6 as fluorinating agents were firstly examined as a source of fluorine. This series of experiments were performed at a temperature ranging from 500 to 900 °C for 1.5 hours using beryl: fluorinating agent1:2 weight ratio [6]. After roasting the product was leaching with water and 25 H_2SO_4 at 95 °C for 3/4 hour separately. As shown in Fig. (1), the obtained results reveal that the effect of the fluorinating agent increases in the following order:

$$Na_2SiF_6 > Al\ F_3 > CaF_2 \qquad (1)$$

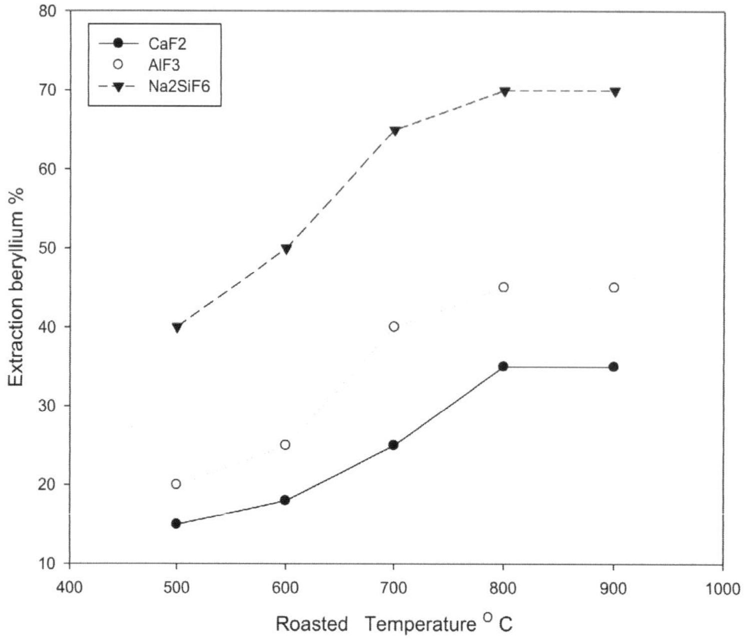

Fig. (1): Effect of fluorinating agent at different roasting Temperature on beryllium extraction

The process, as invented by Copuaux (9), consisted of heating beryl with sodium silicofluoride in the proportion 2 by weight (1:6 molar) at a temperature of 750°C. The reaction was alleged to take place and,

$3\ BeO.Al_2O_3.6SiO_2 + 6\ Na_2SiF_6 = 3\ Na_2BeF_4 + 2\ Na_3AlF_6 + 9\ SiO_2 + 3\ SiF4$ (2)

of the reaction products, only Na_2BeF_4 was water-soluble so that a fairly pure beryllium solution could be obtained by simple aqueous leaching. Therefore, Na_2SiF_6 is the most convenient fluorinating agent. Morana et al presented a simplified breakdown of the reaction steps with Na_2SiF_6

$Na_2SiF_6 = 2\ NaF + SiF_4$ (3)

$2\ BeO + SiF_4 = 2\ BeF_2 + SiO_2$ (4)

$BeF_2 + 2\ NaF = Na_2BeF_4$ (5)

It is noticed that leaching the roasted product with 25 % H_2SO_4 raises beryllium extraction

3.1.2. Effect of a beryl: Na_2SiF_6 Ratio. To determine the effect of the beryl: Na_2SiF_6 ratio, a series of experiments was carried out at a temperature ranging from 400 to 900 °C for 1.5 hours using beryl: Na_2SiF_6 ratio from 1:1 to1:3 and leaching the roasted product with 25 % H_2SO_4, separately at 90 °C for 3/4 hour. Results in Fig.(2) show that the extent of beryllium extraction is directly proportional to the amount of Na_2SiF_6 reaching 90 % at 1:3 beryl: Na_2SiF_6 weight ratio by leaching 25%H_2SO_4

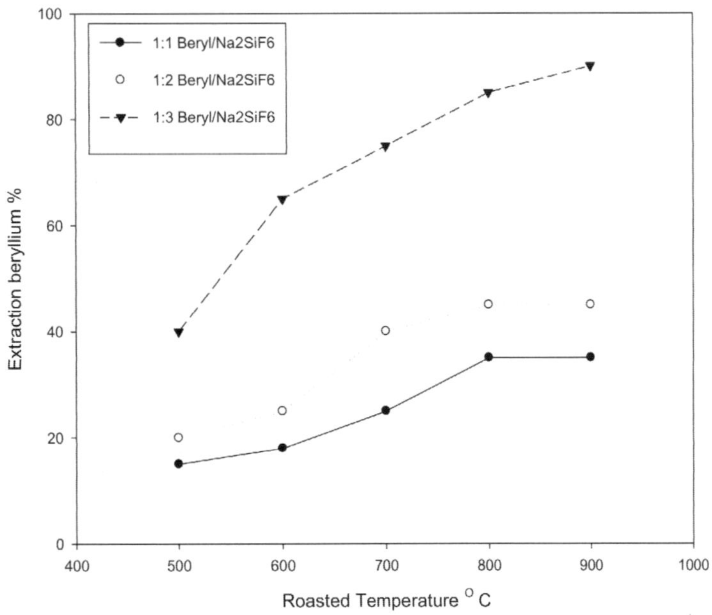

Fig. (2) Effect of beryl : Na_2SiF_6 ratio at different roasting Temperature on beryllium extraction

3.1.3. Effect of leaching time of roasted product. To study the effect of leaching time ranging from 15-90 minutes of roasted product, a series of experiments was carried out at a temperature ranging from 50 to 90 °C with 25 % H_2SO_4 on the leachability of beryllium product at 800 °C with 1:3 beryls: Na_2SiF_6 weight ratio for 1.5 roasting times. Results in Fig. (3) Show that the leachability of beryllium from the roasted product is increased by increasing temperature and leaching time, reaching a maximum value at 85 % at 90 ° C after 60 minutes of leaching time. It can be concluded that 85 % of beryllium value can be extracted by fluoride route using 1:3 beryl: Na_2SiF_6 weight ratio at 800 ° C for 1.5 roasting times and leaching the product with 25% H_2SO_4 at 90 ° C for 3/4 hour

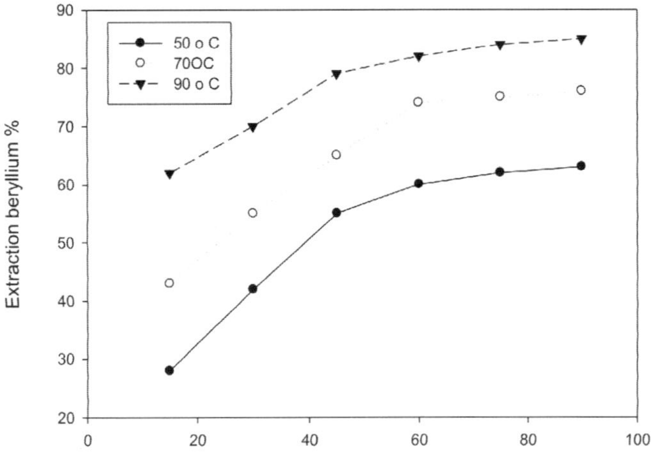

Fig. (3) Effect of leaching time of roasted product at different Temperature on beryllium extraction

3.2. Direct hydrometallurgical techniques or sulfate techniques

The fused or sintered beryl dissolved in sulphuric acid. Heat-treated beryl will only react with hot, concentrated acid (unless pressure is used), but alkali–decomposed beryl will also be dissolved. The alkali–decomposed beryl consumes much greater amounts of acid since alkali itself must also be converted to sulfate. The respective reaction is (12):

$$3\ BeO.Al_2O_3.6SiO_2 + 6H_2SO_4 = 3BeSO_4 + Al_2(SO_4)_3 + 9SiO_2 + 3H_2O \quad (6)$$

$$\text{Beryl} + \text{alkali product} + (6+n)\ H_2SO_4 =$$

$$3\ BeSO4 + Al_2(SO_4) + nM''SO_4 + 9\ SiO_2 + (6+n)H_2O \quad (7)$$

This technique is achieved through direct leaching of beryl with sulphuric acid. The main factors affecting leaching rate are particle size, reaction temperature, reaction time, and acid concentration. These factors were systematically studied.

3.2.1 Effect of grain size. A series of experiments were carried out using different grain sizes from -250 to +64 microns. The experiments were conducted at 260 °C using H_2SO_4 10 mol/l at 10% solid content for a different times ranging from 15 to 120 minutes. As shown in Fig. (4), results reveal that decrease in the particle size leads to an increase in beryllium extraction value reaching its maximum value of 85%at minus 64 microns, this is due to the increase in surface area which promotes the reactivity.

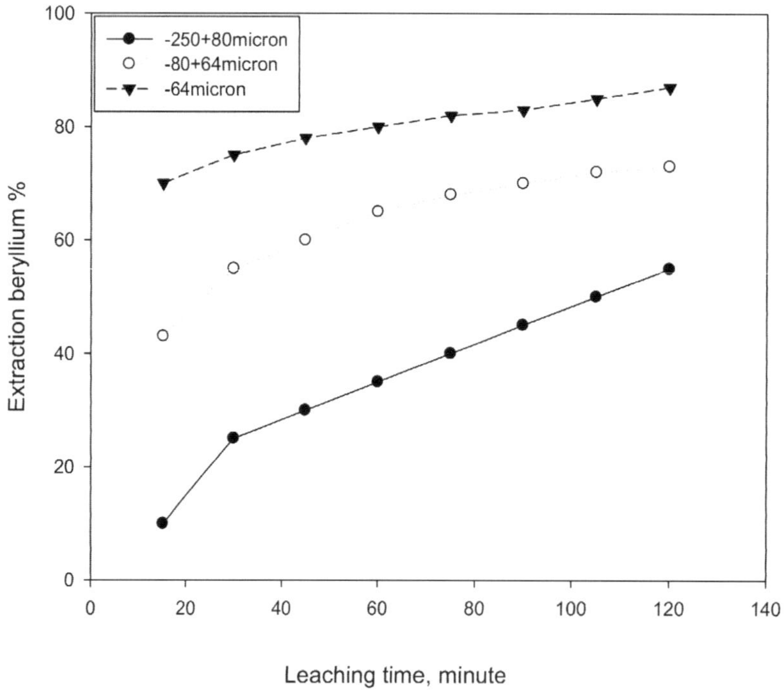

Fig. (4) Effect of grain size on beryllium extraction by direct leaching

3.2.2 Effect of temperature. Experiments were carried out at temperatures from 230 to 270 °C. The leaching was achieved using -64 microns grain size and 10 mol /l at 10% solid content for a different times ranging from 15 to 120 minutes. Results which, are shown in Fig. (5) reveal that, leaching rate increases with the increasing temperature reaching its maximum value of 90% after 30 minutes. An increase of time reaction over 30 minutes gives no appreciable increase in beryllium extraction; this is due to the reaction being temperature dependent.

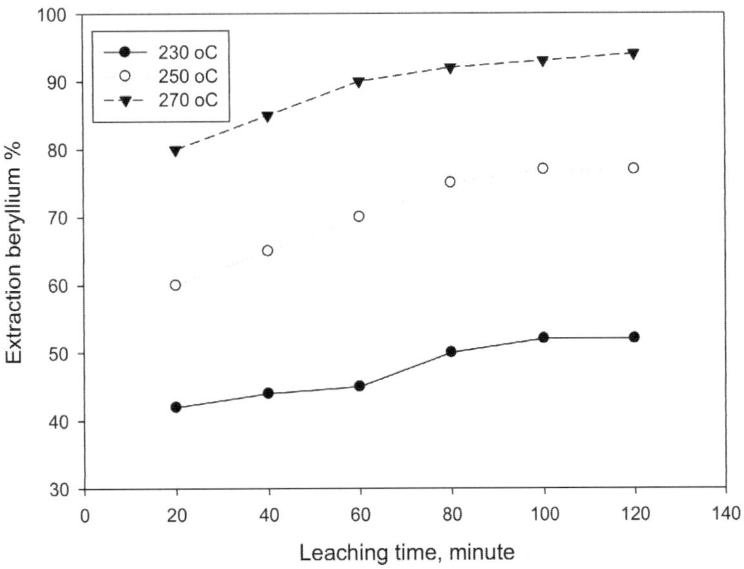

Fig. (5) Effect of temperature on beryllium extraction by direct leaching

3.2.3 Effect of sulphuric acid concentration. A series of experiments were conducted at 270 °C for 30 minutes at 10% solid content with grain size -64 micron using H_2SO_4 concentration ranging from 4 to 12 mol/L. As shown in Fig. (6), Results reveal that the reaction is dependent on acid concentration till 10 mol/L. More increase of acid concentration to 11 mol/L leads to beryllium extraction from 92 to 97 %. It can be concluded that 92 % of the beryl can be extracted by direct leaching with 10 mol/l sulphuric acids at 10% solid content at 270 °C for 30 minutes.

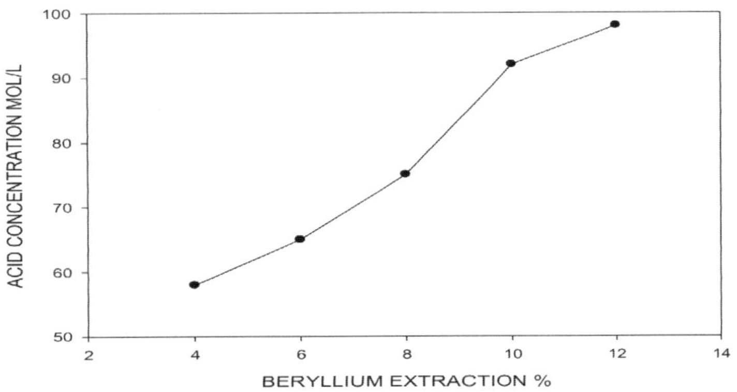

Fig. (6) Effect of sulphuric acid concentration on beryllium extraction by direct leaching

4. CONCLUSION

The optimum conditions of fluoride techniques are 1:3 beryl: fluorinating agent as Na_2SiF_6 ratio, 800 °C, 1.5 hours, and leaching the roasted product with 25% H_2SO_4 at 90°C for 3/4 hour. At these conditions, 85% of beryllium was extracted. The optimum conditions of sulphuric acid techniques by direct leaching are: 270 °C temperature, 30 minutes leaching time, 10% solid content with grain size -64 micron using H_2SO_4 concentration 10 mol/L. At these conditions, 92 % of the beryl can be extracted. Comparing the two evaluates techniques reveal that direct leaching has the following advantages: It takes place in only one stage while the fluoride technique takes place in two stages namely roasting then leaching treatment. Short time is needed for the reaction 30 minutes and low chemical consumption compared to fluoride technique.

References

1. Atrees, Sh.M., 2001; "Physicochemical studies of some Egyptian beryl ores for producing high-grade beryllium concentrate" Ph.D. Thesis, Faculty of Science, Helwan University.
2. Demerdash, M. Hashim 2006; "Hydrometallurgical comparative study between ion exchange and solvent extraction techniques on beryllium compounds" Master Science Thesis. Cairo Univ. Egypt.
3. El-Assay, E. Botros, N. H. Ibrahim, M. E. Hammd, M. S. and Hassan, M. A. 1993: "A new beryl occurrence in Sinai" Egyptian Mineralogist, Vol. 5, pp. 1-10.
4. Hassan, M. A., and El-Shatoury, H. M. (1976); "Beryl occurrences in Egypt" Min. Geol., Vol. 26, pp. 253-262
5. Huges, C.J. 1965:Report on the prospecting of beryllium in Wadi El Gemal Area, during field season (1964-1965), Inshass
6. Ibrahim, A.I., and Amer, A.M. 1997: Development of Beryllium Extraction In Egypt, Proceedings of the XXIMPC-Aachen,21-26 September pp141- 148.
7. Mormon, S.I., and Simon, G.F. 1962: beryllium extracted by the fluoride process, J.metals,14, 571.

Waste Cooking Oil Management in Egypt: Production of Biodiesel-Development of Rapid Test Method

Marwa Mohamed[1], Nourhan Sherif *[1], Omar Aboelazayem [2], Hany A. Elazab[1,3], Mamdouh Gadalla[1,4], Basudeb Saha[5]

[1] The British University in Egypt, Department of Chemical Engineering, 11837, Egypt
[2] School of Chemical and Process Engineering, University of Leeds, LS2 9JT, UK
[3] Jacobs School of Engineering, University of California San Diego, La Jolla, CA 92093-0403, USA.
[4] Port Said University, Department of Chemical Engineering, 42526, Egypt
[5] Department of Engineering, Lancaster University, Lancaster, LA1 4YW, UK

* nourhan.sherif@bue.edu.eg

Abstract. According to the United Nations Office for South-South Cooperation (UNOSSC) report in 2017, Egypt produced 500,000 tonnes of WCO from various resources including food industries, restaurants and hotels. Based on our previous funded project, we have reported unique properties for Egyptian WCO as it has very high range acidity ($\approx$18 mg KOH/ g oil) due to the extensive usage of oil in the cooking process. Further, the repeated heating of cooking oil for long time produces carcinogenic compounds that have significant impact on people health. On the other hand, high acidity WCO could be valorised into soap, biodiesel and value added chemicals. However, most of these industries requires costly pre-treatment for the WCO to reduce the acidity prior processing. Accordingly, the WCO should be produced within a guideline of specified properties that could detect if it has been extensively used. Low-acidity WCO will prevent several health consequences for Egyptian people and allow industries to easily valorise it into value added chemicals and fuels. The waste cooking oil was subjected to filtration before being fed to the reactor the produced biodiesel was treated to remove any methanol and the characteristics of the final product was compared with the international standards. This work develops a guideline for WCO quality testing to ensure that it has not been extensively used in cooking process. The work also applies supercritical conditions to convert WCO into biodiesel. The response surface methodology was used to conclude the optimum conditions to produce biodiesel.

Keywords: Biodiesel, Waste management, WCO, Cooking Oil.

1. Introduction

Cooking oil as well as other oils used in eating, along with fats that are produced from households are currently disposed with other public waste or are dispensed into drainage. For a long time, the recycling of this type of waste had been essential from the food industry companies along Europe. Recycling of this type of waste on a household level is still not the great matter even if it occupies a significant part of the overall production, around 19% of the overall production comes from households in South Korea. In some countries, the idea of fat waste recycling is grabbing the interest of the government; in Czech Republic municipalities are forced to enable the separate collection of fats and Waste Cooking Oil (WCO) in 2020. Due to the physical properties and structure of fats, fat waste is a branch of waste management which gets the interest of many. However, there are some challenges as the animal fats and vegetable oils becomes solid at low temperatures and have high melting temperatures. Simultaneously, the previous tend to appear in a semi-solid or as paste at room temperature, while the other has higher viscosity [1]. Unlike the handling of glass, paper and plastic, the handling of fats requires some work. When fat is dumped off to sewage, it combines with other solid wastes in a sewer which causes clogging in the waste system. However, fats and waste cooking oils show a valuable feedstock and commodity which can be used for energy generation [2].

Waste cooking oil is an inland waste that is produced as a result of frying and cooking food using vegetable oil. Waste cooking oil results mainly from frying at elevated temperatures which is dumped to sewage systems after being used, it is municipal solid waste (MSW). Even though the usage of acronym MSW carries some error and is solid evidence that this type of waste is given poor attention in terms of management and collection. In Europe the statistics of the level of WCO collection is between 100,000 and 700,000 tons/year, which is equal to 1 kg/inhabitant/year. Negative environmental impacts and economic loss are created from the uncontrolled dumping of WCO [3]. Altering the improper ways of disposal of WCO increases the lifecycle of the product and the avoids the groundwater's contamination with this liquid waste. Not only for the environment, the collection of the unwanted WCO has an economic point of view too, it is a promising and a cost-effective raw material for the production of synthesized biofuel as well as recovering valuable energy content. Ever since the 70s, the European Commission suggested that it is essential to collect recognizable amounts of waste oils to avoid environmental contamination [4].

When WCO is thrown into water systems, the oil forms a layer on the surface which stops the oxygen from dissolving, therefore causing a drastic extinction in the marine life. Moreover, the mixture of water and oil elevates the chemical oxygen demand (COD) of water and causes it to be poisonous as a result of the production of by-products formed from oil degradation. The sea creatures absorb carcinogenic compounds and are returned back to humans through the food chain [5]. When the temperature reaches above 180 °C in the absence of oxygen, by-products of alkenes, alkanes, lower fatty acids, symmetric ketones, CO, CO_2 and alkanes are formed from the non-oxidative breakdown of saturated fatty acid [6].

An average of 500,000 tons of WCO is produced yearly from Egypt from different sources. It is produced from food factories, restaurants, hotels and houses, but it is not managed efficiently by the previously mentioned entities to be reused and to benefit from it. Conventionally, it is sold to some collectors of waste. The waste collectors filter it and sell it back to the food industry. Not only food industry, but to soap companies too and are sometimes exported to be used in biofuel industry abroad. The idea of using WCO in the production of biodiesel started in 2013 in Egypt, the conversion process takes around 3 hours and results in two products, biodiesel and glycerin. One of the leading tourism companies is currently using the produced biodiesel in its hotel's vehicles, generators and machinery

[7]. The work of this research aims at 2 objectives, first is to propose a systematic method to measure the quality of WCO, second is to convert WCO into useful fuel such as biodiesel.

2. Methodology

2.1 WCO Quality Testing Method

The acid value of cooking oil is measured as a very important parameter determining the quality of cooking oil. Oil is the most essential and widely used ingredient in Egyptian households. It enhances the taste and flavour of the food. But do you know how poisonous it can be to your health if you keep on reusing the same cooking oil for various purposes and for several times. In order to avoid the wastage of oil, many people reuse it. It is very important to understand that this practice is extremely unsafe for families and loved ones and can lead to a number of health problems. Nutritionists explain that the risks in reusing cooking oil depends on different categories. "When someone does not have a deep fryer or uses a normal fryer, it is recommended to use the oil for frying chips only once. When somebody uses a deep fryer such, as common in restaurants, they can reuse the oil three or four times." Reusing cooking oil without using a deep fryer is extremely harmful to your health. The harmful effects of reusing cooking oil:

1- A high amount of toxins/chemicals is released such as aldehydes/4-hydroxy-trans-2-nonnal associated with many health conditions such as heart-related disorders, dementia cancer, Alzheimer's, which alters the function of DNA, RNA and protein in our body.
2- Gives rise to trans fatty level of oil: it increases levels of bad cholesterol.
3- Oils become partially or completely oxidised and thus chemical composition changes.
4- Blood pressure levels can increase when repeatedly heated cooking oil.
5- Can generate compounds including polycyclic aromatic hydrocarbons that may lead to carcinogenic risks.

Acid value of the feedstock is one of the most important parameters which affects biodiesel production and yield. Based on the acid value of the feedstock, the production process as well as the catalyst to be used are determined or selected. Waste cooking oil WCO which contains triglycerides is converted through transesterification with methanol and at the presence of base catalyst into biodiesel. Typically, 1 mole of WCO reacts with 3 moles of methanol to produce 1 mole of methyl ester or biodiesel and 1 moles of glycerol as by product. The acid value test is used twice in biodiesel production. The first time determines how much catalyst to use in the biodiesel subject to the properties of the oil. The second time determines how long the fuel's shelf-life will be subject to the amount of free fatty acids FFAs in the biodiesel. Oils and fats tend to break down through oxidation over time to form free fatty acids. When these feedstocks are used to make biodiesel by transesterification, the FFAs quickly consume some of the catalysts to form sodium and potassium soaps. Too much soap can complicate the separation process and reduce the yield by carrying some of the fuel into the glycerol phase. Soaps must be removed from the biodiesel by water washing or some other purification processes. Residual soap in the fuel can lead to sodium/potassium deposits when burned that are abrasive to the engine. The acid value is used to determine the amount of free fatty acids in the oil/fat, which in turn is used to calculate the exact amount of catalyst needed to produce biodiesel. Having too little catalyst can lead to an incomplete reaction resulting in fuel with too high levels of total glycerides. According to ASTM specification, biodiesel can have no more than 5 ppm of sodium and potassium and no more than 0.24 % total glycerides in the fuel sample. The ASTM D6751 biodiesel acid-number limit was harmonized with the European biodiesel value of 0.50 mg KOH/g fuel. Titration is a chemical procedure that can be used to obtain this number, i.e. acid value for waste cooking oil and biodiesel product. The acid value is the quantity of a known concentration of potassium hydroxide KOH consumed by titrating an oil/fat sample (i.e. neutralise the fatty acids) and is expressed in the units of milligrams of KOH per gram of sample.

In this activity six different samples were collected to investigate the effect of frying oil reuse on its acid value from Egyptian market/location. The First sample was pure oil as reference. While the second, third, fourth and fifth ones are representative samples of oil reused once, twice, three times and four times, respectively. Last sample was collected from an Egyptian fast food restaurant where they reused the oil several times (more than 10). After collecting the samples, the acid values were measured using titration method.

Manual titration requires a colour indicator such as phenolphthalein and organic solvent such as toluene or ethyl alcohol to dissolve the sample in order to obtain a clean end point. The standard method for acid and base number using colour indicator titration (ASTM D974) was implemented to measure the acid value of the samples (waste cooking oil/biodiesel). Titration solvent, colour indicator and KOH are the main chemicals employed in titration. Instructions for standardising the titrant (KOH) are found in ASTM D664. Acid value test of oil samples was carried out by titrating 20 grams of oil in 100 mL ethanol by KOH 0.1 N using phenolphthalein as indicator. Every sample was subject to 3 experiments with equal volumes to measure the acid value and an average number is considered to reduce the uncertainty in values. The procedure is as follows:

1) Add 100 mL of the titrating solvent (ethanol) to a 250 mL flask.
2) Add approximately 10 drops of phenolphthalein indicator solution.
3) Pour weighed sample of approximately 20 grams into the solvent.
4) Titrate while stirring with KOH titrant to the end point which is a definite colour change to orange or green brown colour that persists for at least 15 seconds.
5) Report the millilitres of titrant consumed.
6) Calculate the Acid value from the below equation:

$$Acid\ Value = \frac{(V - V_b) \times N \times Eq\_wt}{W_{oil}} \qquad (1)$$

where V and V_b are the volumes of 0.1 N KOH needed for titration in mL for both the oil sample and blank sample, N is the normality of KOH which equals about 0.1 geqL^{-1}, Eq_wt is the equivalent weight of KOH = 56.1 and W_{oil} is the oil sample mass in g. This works because the KOH is a basic solution, which reacts with an acidic solution until all of the acid is consumed and the solution is neutral. Any KOH added to the sample after it is neutral causes the solution to become basic and change colour.

7) To express the acid value as a percent of free fatty acids (% FFA), divide the acid value by 2 if the fatty acid profile is mostly oleic, palmitic or stearic acids. Oils with high percentages of lauric acids such as coconut, should be divided by 2.8.

$$\%\ FFA = \frac{Acid\ Value}{2.0} \qquad (2)$$

8) Use this same method to measure the acid value of the biodiesel product. Sample size will be larger and a weaker normality of KOH should be used to achieve accurate results.

When oil feedstock acid value is greater than 2, extra catalyst should be added to account for what will be lost in forming soap. It takes 1 mole of catalyst to neutralise one mole of soap. The amount of additional catalyst can be calculated using the following equation:

$$NaOH = (\%\ FFA) \times (0.144) + 1\% \qquad (3)$$

$$KOH = \frac{(\%\ FFA) \times (0.197)}{0.86} + 1\% \qquad (4)$$

2.2 Biodiesel Production: Materials

Waste cooking oil is supplied which originates from sunflower oil by Uptown biodiesel Ltd. Moreover, methanol bought from Fisher scientific in the United Kingdom. The calibration curves were prepared by using standard methyl esters, and the heptadecanoic acid methyl ester was used as an internal standard and bought from Sigma-Aldrich, UK. The liquid CO_2 cylinder (99.9%) equipped with a dip tube was bought from BOC Ltd., UK.

2.3 Biodiesel Production: Experimental Setup

The waste cooking oil was filtered in order to get rid of all of the impurities that originated from cooking/frying. The reaction was carried out in supercritical conditions in a reactor made of stainless steel and has a volume of 100 ml and could withstand high pressure. The reactor's type is 49590, Parr Instrument Company, USA. The reactor is supplemented with a heating mantle, model 4848 controller, J-type thermocouple and a mechanical stirrer. Figure 1 shows a schematic for the experimental setup. The oil and methanol with a specific molar ratio were added to the reactor then heated. After the oil and methanol are mixed with a specific molar ratio, the mixture was fed to the reactor. In order to reach the required temperature, stirring at 300 rpm and heating were continuously applied to the reactor. The needed pressure was achieved by pumping CO_2 using a supercritical fluid pump. Before any pressure is applied by the CO_2, the reactor is left to heat as the heated methanol produces vapor which causes a build-up pressure. The reaction conditions were achieved nearly after 15 minutes; this is when the reaction time starts. After the reaction time is over the reaction is stopped by quenching, which is using an ice bath for the reactor to stop the reaction. The pressure of the reactor is released and a centrifuge is used to separate the reaction's product to glycerol and biodiesel. Furthermore, the biodiesel is heated again to 80 °C to recuperate any methanol that didn't react in the distillation process. Finally, the pure biodiesel is analyzed and its properties is then compared to the European standard (EN14214). The biodiesel's product yield is calculated using the following equation [8].

Equation 1:yield of the produced biodiesel equation

$$BD\ Yield\ (\%) = \frac{Total\ weight\ of\ methyl\ esters}{Total\ weight\ of\ waste\ oil\ used} \times 100 \qquad (5)$$

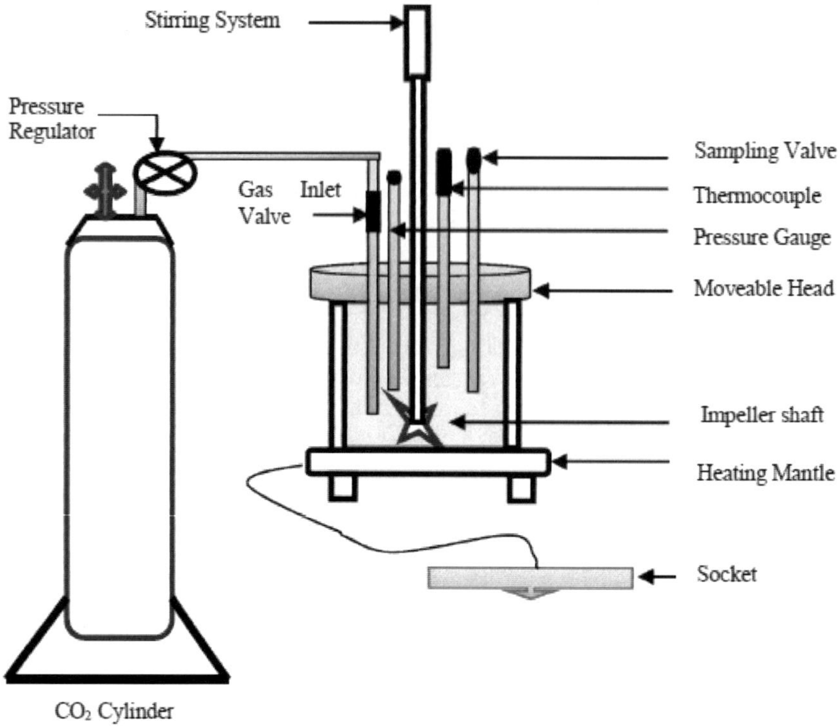

Figure 1. Schematic of the experimental setup

2.4 Experimental Design

Response surface methodology (RSM) is a multi-variable method which is capable of developing a model that shows the response function which depends on the reaction in the experimental studied independent variables [9].To reach the optimum reaction conditions that are needed for the production of biodiesel RSM is used, this is done by learning the relationship of each of the variables with the response yield. There are four variables on which the experimental runs have been operated; methanol-to-oil molar ratio, temperature, pressure and time, labelled as A, B, C, D respectively. Three levels of each of these variables are coded as -1,0,1 as shown in the table below. BBD is one of the RSM techniques used to study the effect of the process variable on the response. Moreover, it studies the interactions of the variables on the response [10]. The yield of biodiesel is selected as the response for this study. In order to lower the effect of unexplained consistency in the responses, the experiments are completed in a random order [11]. Twenty-nine runs are performed in a random way and their response are calculated from the experimental results using the yield equation.

Table 1. Experimental design variables and their coded levels

Factor	Code	Levels		
		-1	0	+1
M:O (molar ratio)	A	20	31	42
Temperature (°C)	B	240	260	280
Pressure (bar)	C	180	220	260
Time (min)	D	12	22	32

3. Results and Discussion

Results are reported for the main activities of this research, i.e. results of quality test method of WCO, and results of biodiesel production.

3.1 Results of WCO Quality Testing Method

Table 2 reports the measured acid values for the waste cooking oil samples collected previously.

Table 2. Acid value data for WCO samples

WCO Sample	Reuse Cycle	Acid Value (mg KOH/g Oil)
1	0	0.1
2	1	2,5
3	2	4
4	3	7.5
5	4	10
6	>10	25

As indicated from table 2, the acid value increases with the increase of the number of reuse cycles. A possible reason for this increase is the hydrolysis and oxidation of oil due to the introduction of wet food and high temperature. It should be noted that there are another important factors that affect the acidity of the reused oil, which is the type of food being fried, time, frying temperature and the type of used oil (palm, soybean). Also, the number of double and triple bonds already exist in the structure of the oil molecule have a significant influence on its hydrolysis and oxidation in addition to their decomposition to other polar components.

The data obtained were plotted so, the acid value on the ordinate (y-axis) and the number of reuse cycles of WCO on the abscissa (x-axis). Then data are fitted to a proper model and correlation is extracted. The data are shown in Figure 2 below which fit well to a linear interpolation. The correlation derived from the data plot is reported to be:

$$A_V = m \times N_c + c$$

Where A_v is the predicted acid value by the model, N_c is the number of reuse cycles of WCO, and m and c are the model constants (m=2.4, c=.1). The error of the predicted values by the model with respect to experimental data is calculated from the following equation (N_s=total number of WCO samples), and shown in Table 3:

$$\% \, Error = \frac{Predicted - Experimental}{Experimental} \quad (6)$$

$$\% \, Max \, Error = \frac{\sum_1^{N_s}(Predicted - Experimental)}{N_s} \quad (7)$$

The maximum error of the predicted acid values is calculated to be 5.3. The correlation can be employed to predict the acid value corresponding to a number of reuse cycles, e.g. for a number of 7 times reuse cycles, the predicted acid value will be 16.9 mg KOH/g oil.

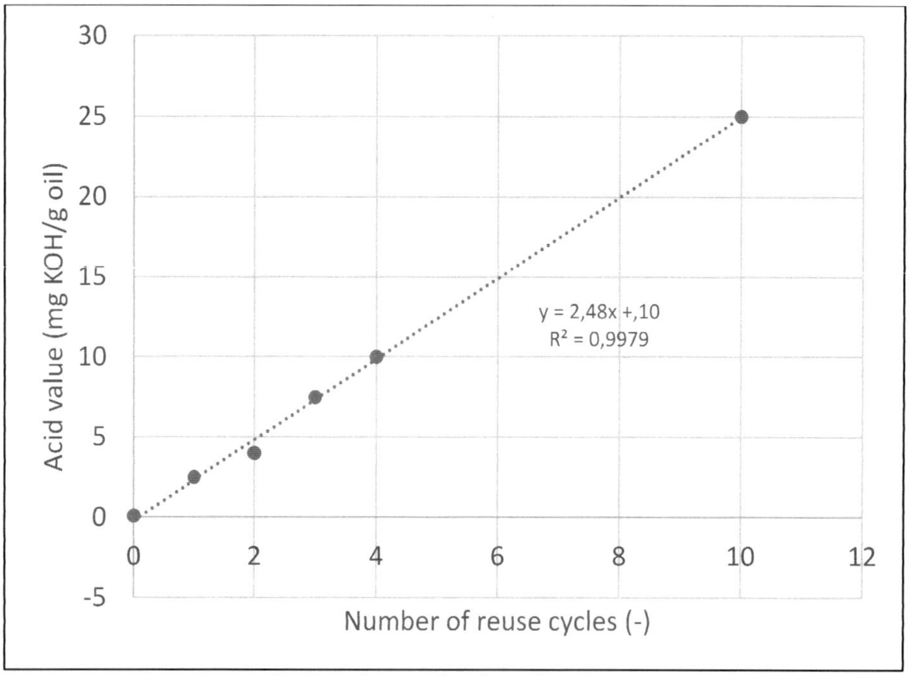

Figure 2. Correlation plot for acid values

Table 3. Experimental/predicted data for acid value data of WCO samples

WCO Sample	Reuse Cycle	Experimental (mg KOH/g Oil)	Predicted (mg KOH/g Oil)	Deviation (mg KOH/g Oil)	% Error
1	0	0.1	0,10	0,00	0
2	1	2.5	2,50	0,00	0
3	2	4	4,90	0,90	22,5
4	3	7.5	7,30	0,20	2,7
5	4	10	9,70	0,30	3
6	>10	25	24,10	0,90	3,6
			Average Error %		5,3

The calculations of the extra catalyst concentration and the % FFA for the measured data are shown in Table 4 due to higher values of acid numbers.

Table 4. Experimental/predicted data for acid value data of WCO samples

WCO Sample	Reuse Cycle	Experimental (mg KOH/g Oil)	% FFA	KOH Catalyst amount (%)	Extra KOH catalyst (%)
1	0	0,1	0,05	1,01	0,01
2	1	2,5	1,25	1,29	0,29
3	2	4	2	1,46	0,46
4	3	7,5	3,75	1,86	0,86
5	4	10	5	2,15	1,15
6	>10	25	12,5	3,86	2,86

3.2 Physiochemical Properties

Waste cooking oil feedstock properties are calibrated as shown in the table 5. The purified biodiesel produced from the supercritical trans-esterification reaction using the optimum conditions that were analyzed for evaluating the chemical and physical properties. The obtained results were obtained and compared with the European standards of biodiesel, EN14214. The analysed properties have been repeated two times and the final result have been obtained as an average of the two results. ASTM D4052 method was used to determine the standard density, while ASTM D445 method was used to determine the kinematic viscosity. ASTM D974 method was used to calibrate the produced biodiesel total acid number (TAN) of the produced biodiesel.

Table 5. Properties of WCO feedstock

Test	Calibration Method	Result
Kinematic viscosity at 40 °C	ASTM D-445	54.2 cSt
Density at 15°C	ASTM D-4052	0.883 g/cm^3
TAN	ASTM D-974	0.8 mg KOH/g oil

3.3 Physiochemical Properties of the Produced Biodiesel

The standard density of the produced biodiesel is 887 kg/m^3. The produced biodiesel at methanol-to-oil molar ratio 37:1, temperature 253°C, pressure 199 bar and reaction time of 15 minutes has been subjected to examination ensuring that its properties abide by the European Biodiesel Standards, EN14214 [12]. Table 6 shows the properties of the produced biodiesel and the European biodiesel standard acceptable range. The other byproducts produced were discarded.

Table 6. Comparison between produced biodiesel properties and European biodiesel standard EN14214

Test	Unit	Produced biodiesel	Biodiesel (EN14214)
Density at 15°C	kg/m^3	887	860 - 900
Kinematic viscosity at 40°C	cSt	4.63	3.5 - 5
TAN	mg KOH/ g oil	0.09	< 0.5
Pour point	°C	-6	N/A
Flash point	°C	161	> 101
Cetane number		59	> 51

4. Conclusion

Based on our previous projects results, the collected WCOs from restaurants and industries usually have very high acidity due the extensive usage in cooking process. It has been recently reported that the repeatedly usage of cooking oil at elevated temperatures produce polycyclic aromatic hydrocarbons (PAH) that are recognized as carcinogenic compounds. Further, the high acidity obtained from literature review creates several problems for WCO valorization feedstocks, i.e. biodiesel. One biodiesel company (Biodiesel Misr), who are the only operating biodiesel plant in Egypt working to convert WCO into biodiesel, have confirmed the difficulties of processing high acidity WCO where costly pre-treatment steps are vital. Accordingly, the extensive usage of cooking oil from industries and restaurants are affecting the Egyptian people's health and causing problems for WCO valorization processes. The project is aiming to develop guidelines for WCO management including disposal standard regulations to be implemented by households, restaurants, and industries. The guidelines will be submitted to the Egyptian government and a practical implementation method will be also proposed. The project will develop a rapid testing method for onsite inspection of WCO prior collection. The optimum reaction conditions are at methanol-to-oil molar ratio 37:1, reaction temperature of 253°C, reaction pressure of 199 bar and reaction time of 15 minutes.

Acknowledgment

The authors would like to acknowledge Newton Mosharafa Fund Impact Scheme through STDF Egypt and The British Council UK (Project ID 43941) for funding this research.

5. References

[1] L. Chuah, A. Aziz, Y. S, J. J. Klemes and A. Bokari, "Waste cooking oil biodiesel via hydrodynamic cavitation on a diesel engine performance and greenhouse gas footprint reduction," *Chemical Engineering transactions,* no. 50, pp. 301-306, 2016\.

[2] J. Matušinec, D. Hrabec, R. Šomplák, V. Nevrlý, J. Pecha, V. Smejkalová and Y. Redutskiy, "Cooking Oil and Fat Waste Management: A Review of the Current State," *The Italian Association of Chemical Engineers,* pp. 763-768, 2020.

[3] T. Ramos, M. Gomes and A. P. Barbosa-Póvoa, "Planning waste cooking oil collection systems," *waste management,* no. 33, p. 1691–1703, 2013.

[4] G. De Feo, A. Di Domenico, C. Ferrara, S. Abate and L. S. Osseo, "Evolution of Waste Cooking Oil Collection in an Area with Long-Standing Waste Management Problems," *MDPI,* vol. 12, no. 8578, pp. 2-16, 2020.

[5] M. G. Kulkarni and A. K. Dalai, "Waste Cooking OilAn Economical Source for Biodiesel: A Review," *Industrial & Engineering Chemistry Research,* no. 45, pp. 2901-2913, 2006.

[6] H. Kamila, K. Sudesh and T. Yang, "The Management of Waste Cooking Oil: A Preliminary Survey," *Health and the Environment Journal,* vol. 4, no. 1, pp. 76-81, 2013.

[7] OUDA, "Operational Unit for Development Assistance," 2019. [Online]. Available: https://www.ouda.org.eg/sustainable-development-solutions/ssda-solutions-demo/converting-cooking-oil-to-biodiesel/. [Accessed 10 December 2021].

[8] J. V. Gerpen, "Biodiesel processing and production," *Fuel Process. Technology,* vol. 86, pp. 1097-1107, 2005.

[9] J. Yuan, J. Huang, G. Wu, J. Tong, G. Xie, J. Duan and M. Quin, "Multiple responses optimization of ultrasonic-assisted extraction by response surface methodology (RSM) for rapid analysis of bioactive compounds in the flower head of Chrysanthemum morifolium Ramat.," *Industrial Crops and Products,* vol. 74, pp. 192-199, 2015.

[10] M. Khajeh, "ptimization of microwave-assisted extraction procedure for zinc and copper determination in food samples by Box-Behnken design," *Journal of Food Composition and Analysis ,* vol. 22, pp. 343-346, 2009.

[11] H. Jaliliannosrati, N. S. Amin, A. Talebian-Kiakalaieh and I. Noshadi, "Microwave assisted biodiesel production from Jatropha curcas L. seed by two-step in situ process: Optimization using response surface methodology," *Bioresource Technology,* vol. 136, pp. 565-573, 2013.

[12] B. Sajjadi, A. Raman and H. Arandiyan, "A comprehensive review on properties of edible and non-edible vegetable oil-based biodiesel: Composition, specifications and prediction models," *Renewable and Sustainable Energy Reviews,* vol. 63, pp. 62-92, 2016.

[13] G. commodities, "GF commodities," 30 April 2020. [Online]. Available: https://gfcommodities.com/blog/a-greasy-problem-whats-the-proper-way-to-discard-used-cooking-oil/. [Accessed 10 December 2021].

Multivariate-Parameters Effect of Dewatered Aluminium-Based Sludge in Catalytic Oxidation into Wastewater Treatment with A Special Strategy for Solar Energy Application

Rahma H. Thabet[1], Mai K. Fouad[2], Shakinaz A. El Sherbiny[2], Maha A. Tony[1*]

[1]Advanced Materials/Solar Energy and Environmental Sustainability (AMSEES) Laboratory, Basic Engineering Science Department, Faculty of Engineering, Menoufia University, Shebin El-Kom, Egypt.
[2]Chemical Engineering Department, Faculty of Engineering, Cairo University, Egypt

* corrosponding author: Tel: +2048-2221549, Fax: +20482235695, Egypt, E-mail: dr.maha.tony@gmail.com

Abstract. In this study, the Fenton/adsorption of dye effluent containing Levafix blue dye has been carried out via a cost-efficient and eco-friendly heterogeneous Fenton and solar-Fenton methods. Alum sludge, the water treatment by-product were successfully collected and calcined to 400 °C (AS400) were tested for use as a catalyst in the Fenton process. X-ray diffraction, Scanned electron microscopy techniques were applied to investigate the structural and morphological properties of such catalyst. The results showed that the material had a better photocatalytic activity under sunlight irradiation. The process of solar photo-Fenton reagent including AS400 was found to be competent for the oxidation of Levafix dye compared with the process of using Fenton process. The oxidation was improved to a great extent from 60 to 99% removal in the treatment time of 40-min irradiation. The key factor is containing 1.5 g/L and 800 mg/L of AS400 and H_2O_2, respectively, at pH 3.0. As a result, the practical application of such a by-product as a catalyst on the Fenton system employing natural solar radiation on the treating dye effluent indicates that technique is extremely efficient for a safer environment.

Keywords: Fenton's reagent, Levafix blue, Wastewater, Alum sludge, Solar

1. Introduction

Textile dye wastewater has been one of the health and environmental issues in developing economical countries due to its high COD and colour with non-biodegradable organics from textile dyestuffs. As a result, numerous methods for treating textile wastewater have been developed, including biological chemical, physico-chemical and advanced oxidation processes [1-4]. In contrast, Fenton is often regarded as the most appropriate technique of destruction not only for ternary treatment but also for enhancing the biodegradability of dye wastewater for subsequent conventional treatment [5, 6].

Alum Sludge (AS) is often released as waste from a water supply plant's coagulation operation. It is regarded as a low-cost, easily accessible, and ecologically acceptable method of removing contaminants from water [7]. AS was initially used to remove colors from dyeing effluent [8]. Furthermore, a substantial amount of study has been conducted on the use of AS to remove phosphate in water [9]. Also, AS was effectively employed for the adsorption of various inorganic compounds in water, such as fluoride ions [10, 11] and hexavalent chromium [12], and organic compounds like removal of humic acid in water [13]. Most research on adsorption by AS focus solely on the removal of inorganic chemicals in water, with little attention paid to the removal of organic materials. Given because adsorption is merely a simple phase transfer of the substance and cannot achieve the degradation of pollutants [14, 15], particularly organic pollutants in water, degradation can provide a more comprehensive removal of pollutants. Furthermore, because saturated adsorbents are often difficult to renew, adsorption capacity will be reduced. However, the use of recent AOP techniques have recognized as an effective method for treatment of wastewater [16].
Advanced oxidation processes are efficient methods for removing resistant chemicals from industrial effluent. The objective of any advanced oxidation process (AOP) is to generate hydroxyl radicals (OH•) in solution as a powerful oxidant ($E_0 = 2.80$ V) in order to remove pollutants that are difficult to oxidise using ordinary oxidants [17, 18]. Hydroxyl radicals degrade contaminants by dehydrogenation or hydroxylation. Recently, several AOP (Ozonation, Fenton and Photo Fenton processes etc.) have been developed for the treatment of dyes in wastewater effluent [19-21]. Although the Fenton and Fenton-like processes are less affordable and do not create harmful by-products when compared to other AOPs, most Fenton-like processes documented in the literature have been shown to be efficient at acidic pH levels [18]. Among these AOPs, the Photo Fenton method showed to be particularly successful for dye and organic pollutant removal even at low wastewater pH [22].
Levafix CA reactive dyes were chosen for their excellent fastness profile and ability to fulfil the majority of textile producer criteria. They offer a variety of advantageous dyeing qualities, including a high fixation yield and wash off properties [23].
Wastewater treatment systems consume a significant amount of energy [24]. As seen in Fig. 1, 14% of such energy is exclusively used for water treatment plants [25, 26]. Solar-powered photocatalysis for water treatment has been studied in recent decades as an environmentally acceptable method of decontaminating water. In this regard, the photo-Fenton technique has emerged as a potential approach for the elimination of persistent organic pollutants especially that based on the solar energy activation reaction [27].

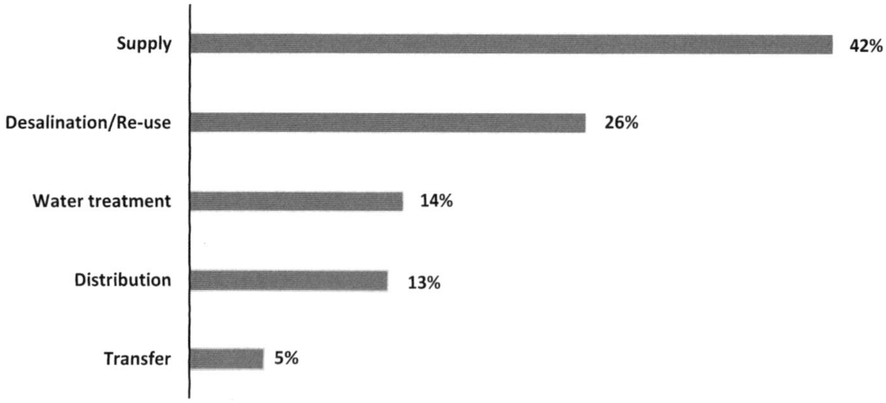

Figure 1. Electricity consumption in the water sector by process. Source [IEA 2009]

The purpose of this investigation is to assess the calcined alum sludge residue created as a by-product from a potable water treatment facility in Giza Governorate, Egypt in order to determine its potential as a photocatalytic material for oxidation Levafix blue CA dye under sunlight. The calcination process yields important compounds, such as graphite and zeolite, which are used in wastewater treatment sector. Maximum dye oxidation of alum sludge materials was examined in both dark and sunlight conditions with a dual Fenton/adsorption processes. The optimal operating conditions such as alum sludge concentration, hydrogen peroxide concentration and pH were evaluated. The eco-sustainable utilization of solar energy with AS by-product to treat contaminated wastewater effluent is regarded as a green and sustainable solution.

2. Materials and Methods

2.1. Chemicals and reagents

Levafix Blue CA was obtained from Dystar Co, Egypt branch and its structure is not disclosed yet. For the pH adjustment to the desired value, laboratory grade sulfuric acid and sodium hydroxide were provided. The initiator of the AS-Fenton reaction was 30 percent w/v hydrogen peroxide (H_2O_2). All chemicals were of analytical grade and were used without any further purification. Distilled water was used throughout this study.

2.2. Preparation of alum sludge

Alum sludge (AS) was collected at the Gezirat El Dahb treatment facility in Giza governorate, Egypt, as a by-product residual from the drinking water treatment plant. As a result of the plant's usage of $Al_2(SO_4)_3$ as a principal coagulant for flocculating water, aluminum-based sludge is generated as a byproduct. The alum sludge utilised in this investigation was collected straight from the sedimentation tank's underflow channel for further treatment and reuse according to the previously mentioned process [8]. Following that, the collected alum sludge is brought to the laboratory and gravity settled to remove the surplus water. The produced sludge cake is then scrubbed with distilled water to remove any pollutants before being dried in an oven overnight (105°C) to remove any water content before being stored for future use. The dried alum sludge is subjected grinding. Following that, AS is calcined in an electrical furnace for 2 hours at 400°C at a rate of 10°C/min (AS400).

2.3. Experimental procedure

2.3.1. Fenton methodology

The oxidation of Levafix Blue CA was examined using 100 mL of aqueous solution with 100 ppm concentration with AS powder in dark condition, as shown in Figure 2.

Using a digital pH-meter, the pH of the solution was first corrected if necessary by adding sulfuric acid and/or sodium hydroxide by (AD1030, Adwa instrument, Hungary). Before the oxidation reaction, the necessary quantity of AS was disseminated in the dye aqueous solution for 5 minutes using sonication, followed by the addition of various dosages of hydrogen peroxide. The samples are then removed from the reaction mixture and filtered using a micro filter before being submitted to a spectrophotometer (UV-1601, Shimadzu, Model TCC-240A, Japan) to monitor the dye concentrations remained in solution.

2.3.2. Solar-Fenton methodology

Advanced Materials/Solar Energy and Environmental Sustainability (AMSEES) Lab., Faculty of Engineering, Menoufia University, Shebin El-Kom Cit, Egypt is site to the pilot-scale experimental setup. The city of Shebin El-Kom is located 75 kilometres north of Cairo. The climate in the region is often moderate, sunny, and dry, especially during the summer months [28]. During the experimental research, the temperature rises to more than 40 °C. A pilot-scale solar reactor is intended to harness solar energy. It is made up of a tubular silica glass tube positioned on the focal line of a parabolic trough-concentrating collector made of stainless steel sheet. A peristaltic pump with a maximum flow rate of 2280 mL min^{-1} provides circulation (BT300-M, Shenchen pump Yz1515X, China). Rubber tubes 12-inches in diameter are utilized to link a feed/discharge reservoir to the tubular reactor.

2.3.3 Characterization study

The morphologies of the AS400 sample was investigated and photographed using a field-emission scanning electron microscope (SEM) (FE-SEM, Quanta FEG 250). Typical magnifications utilised were x8000 and x60000. The crystal structure of the calcined alum sludge sample, AS400, was determined using an XRPhillips X'pert (MPD3040) diffractometer equipped with a Cu Ka radiation source (= 1.5406) and operating at 40 kV and 40 mA.

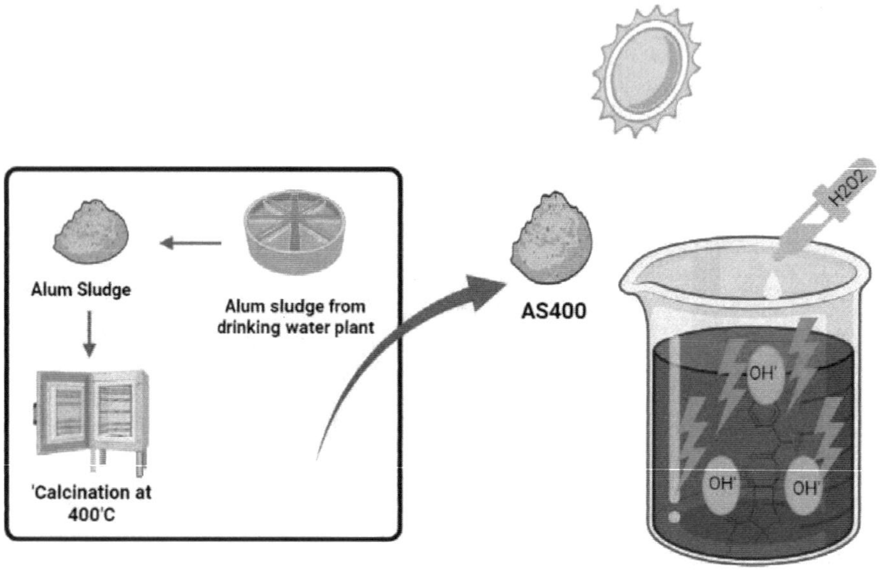

Figure 2. Schematic diagram of solar-Fenton for Levafix dye removal

***3. Results and discussions*:**
3.1. XRD analysis and Morphological characterization of the AS400

Fig. 3 shows XRD patterns used to knowing the mineral composition of AS400. The crystalline rich-silica zeolite, ZSM-12 (sodium aluminium silicate, Na1.16Al2Si77.4O158) with a monoclinic crystal structure appears in the sludge calcined at 400 °C (AS400) with calcium aluminosilicate (CaAl2Si2O8), quartz and graphite as well.

Scanning electron microscopy (SEM) was used to describe the morphologies of AS400, as seen in Figure 4. The morphology of calcined sludge surfaces were relatively similar, with a smooth surface because the surface was dominated by an aluminosilicate glass phase with average particle size around 1µm. The calcined sludge was uneven and diverse in form, with a porous texture.

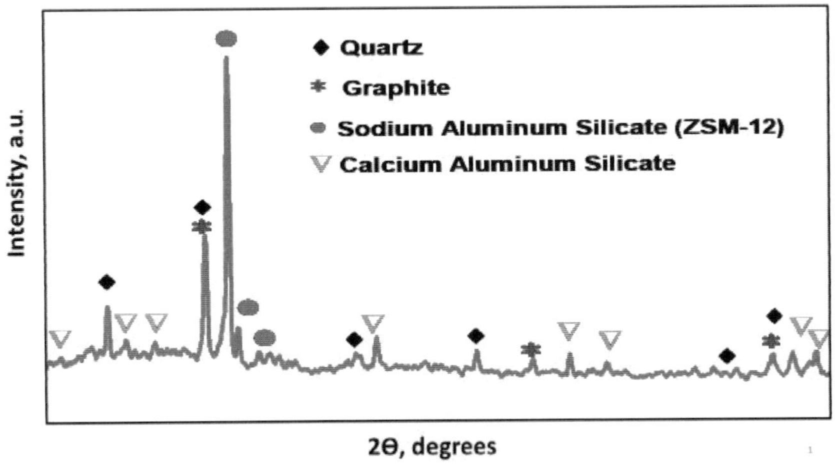

Figure 3. XRD patterns for AS400

Figure 4. SEM Image of AS400

3.2. Effect of wastewater oxidation

3.2.1. Effect of Solo-H_2O_2:

In addition, variety of experiments was carried out with the aim of proving the relevance of the Fenton reaction. The oxidation with H_2O_2 in the absence of AS400 is not significant, particularly at low pH, as is the case. Different H_2O_2 dosages on Levafix oxidation was studied using a wide range of H_2O_2 reagents (100 to 1600 mg/L) in dark condition with 100 ppm initial dye concentration. The data in Figure 5 reveal that the dye oxidation rate was low due to low H_2O_2 concentrations, while increasing such reagent up to 1600 mg/L increases the dye oxidation rate. The dye removal rate was raised from 5% to 21% within 40 minutes oxidation time and pH 3 by increasing the reagent concentration from 400 to 1600 mg/L, respectively. Because of the low H_2O_2 dosage, the concentration of OH• radicals, the major oxidising species, is minimal. Furthermore, by increasing the concentration of peroxide reagent, more • OH radicals were produced, resulting in a higher yield of hydroxyl radicals in the reaction medium that could react with dye molecules. Nonetheless, increasing the H_2O_2 concentration leads in a decrease in dye removal because the large quantity of • OH produced combines with the excess H_2O_2 rather than the dye, producing the HO_2 radical. These radicals have a lower reactivity than • OH radicals. As a result, the total response is reduced. Consequently, in the absence of AS400, oxidation with H_2O_2 is insignificant, especially at low pH, as is the case here [29, 30].

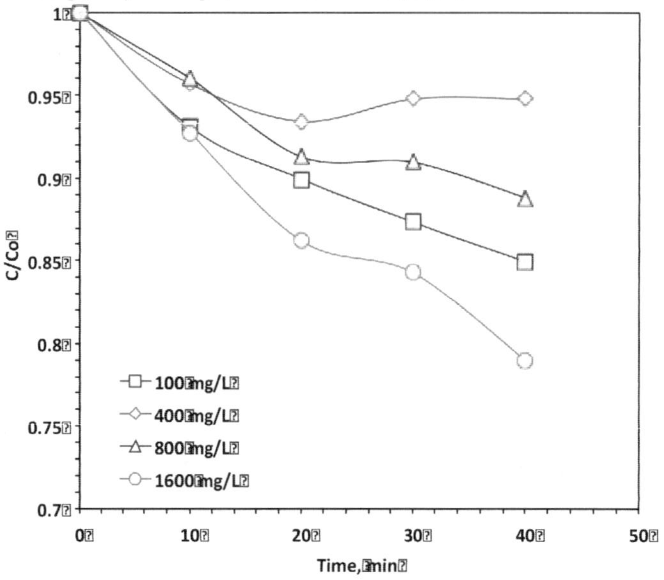

Figure 5. Effect of solo H_2O_2 on dye oxidation

3.2.2. Effect of aluminium-based sludge catalytic oxidation system:

3.2.2.1. Effect of H_2O_2

The impact of increasing the hydrogen peroxide reagent on the Levafix dye oxidation rate while maintaining the other parameters constant at pH 3.0 and AS400 concentration of 1 g/L is shown in Fig. 6. Various initial H_2O_2 concentrations ranged from 400 to 3000 mg/L was studied. The determined % degradation is established, which corresponds to the simultaneous elimination of dye via surface adsorption and Fenton reaction. The results show that by increasing the hydrogen peroxide concentration from 400 to 1600 mg/L, Levafix dye oxidation increased from 58 to 63% after 40 minutes of illumination time. Levafix dye oxidation increased from 58 to 63% after 40 minutes of oxidation reaction when exposed to high levels of hydrogen peroxide (3000 mg/L). It might be owing to the increase in •OH caused by the addition of hydrogen peroxide according to equations (1-2). Several writers have previously reported on the effect of a high dosage of H_2O_2 on Fenton's reaction [31-33].

$$AS400 + H_2O_2 \rightarrow AS400^+ + HO^- + HO^\cdot \quad (1)$$
$$AS400 + H_2O_2 \rightarrow AS400^+ + HO_2^- + H^+ \quad (2)$$

where AS400 and AS400$^+$ represent both the reduced and oxidised states of the catalyst, respectively.

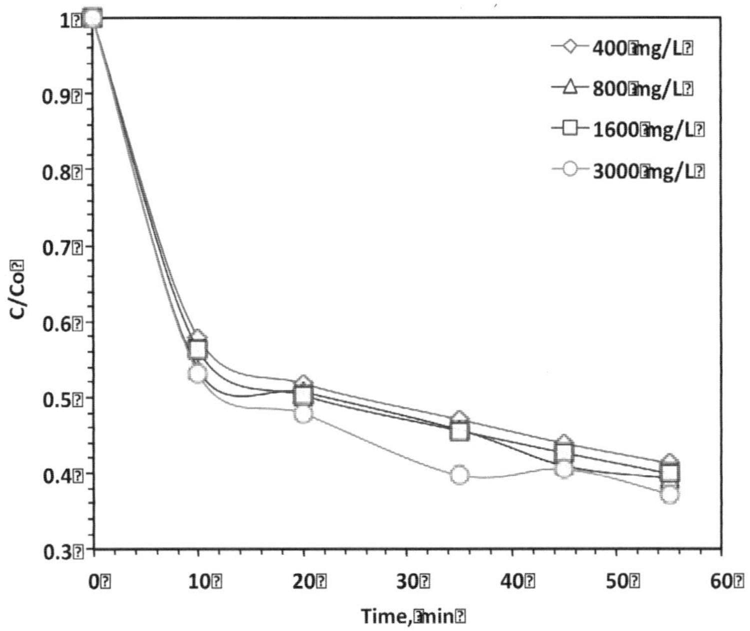

Figure 6. Effect of the concentration of H_2O_2 on the removal of Levafix dye, [Dye]o= 100 mg/L, AS400= 1g/L, pH 3.0.

3.2.2.2. Effect of dewatered aluminium-based sludge:

A number of experiments were carried out with different concentrations of AS400 ranging from 0.5 to 2 g/l in order to investigate the influence of the initial concentration of AS400 on the oxidation of Levafix dye . The results in Fig.7, revealed that the Fenton/adsorption of Levafix dye is remarkably dependent on the concentration of AS400 with constant values of hydrogen peroxide 800 mg/l and pH 3.0.
The results revealed that 1.5 g/L of AS400 in the oxidation of Levafix dye is the optimal value. When the AS400 load is raised from 0.5 to 1.5 g/L, the oxidation efficiency is maximised. The oxidation of dye is reduced after the ideal value (2 g/L AS400). As a result, raising the catalyst concentration improves the simultaneous removal of dye by surface adsorption and Fenton reaction
Hence, the adsorption rate of the dye on the surface of AS400 may influence the Fenton reaction's effectiveness. This might be demonstrated by the presence of more active sites at high concentrations of AS400. However, higher amounts of the reagent reduces the adsorption rate on the surface of AS400 [34, 35]. However, increasing the catalyst load led in a reduction in Fenton's oxidation effectiveness owing to particle aggregation, which decreases the surface area of the catalyst susceptible to oxidation. this is in accordance with the previous investigation cited in the literature [36].

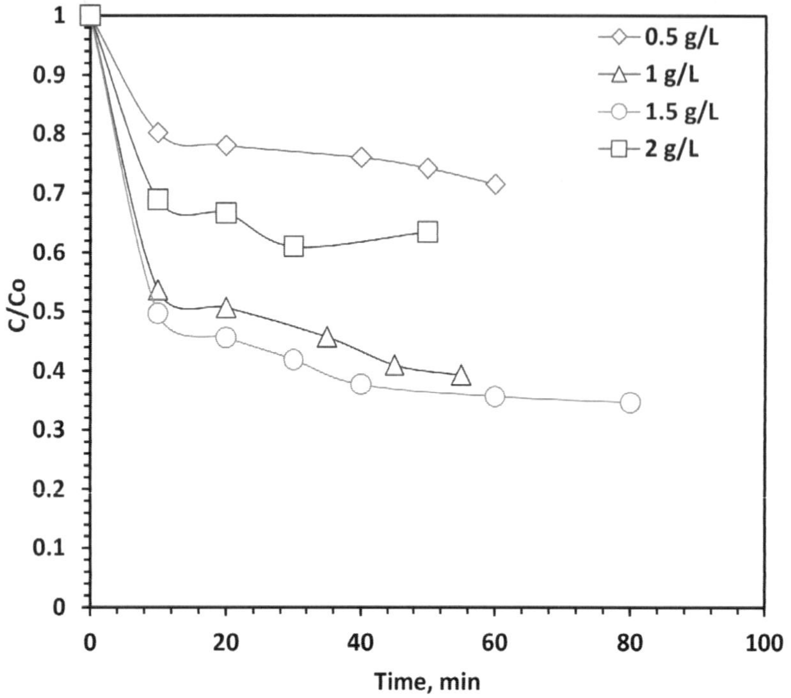

Figure 7. Effect of the concentration of AS400 on the removal of Levafix dye, [Dye]o= 100 mg/L, H_2O_2= 800 mg/L, pH 3.0.

3.2.2.3. Effect of pH:

The influence of starting pH may be described using the data in Fig. 8. The pH of solutions is a primary parameter of the treatment efficiency of Fenton's reagent, since it influences the production rate of hydroxyl radical [41]. The effect of initial pH value of solutions on the adsorption/oxidation of Levafix dye was obtained in the pH range 3.0 to 8.0.

The maximum removal of Levafix dye on the AS400 surface occurs at beginning pH 2.0 (final decolorisation efficiency 99.7%), while the second largest removal occurs at initial pH 3. (final decolorisation efficiency 60 %). At solution pH 7 and 8, no significant change in dye concentrations was detected within the first 60 minutes. Nonetheless, the concentration of dye reduced somewhat.

In comparison to the other examined factors, dye removal efficiency is particularly sensitive to pH variation. Thus, the optimal pH-value is 2.0. This result is most likely owing to the acidic medium producing the most $OH^{\bullet}$ radicals. In addition, an organometallic complex is present in the acidic media, which might result in H_2O_2 regeneration. According to the literature, low activity of Fenton reagent is supplied at high pH because undesirable radicals were created rather than OH radicals, scavenging the reaction rate [37-39].

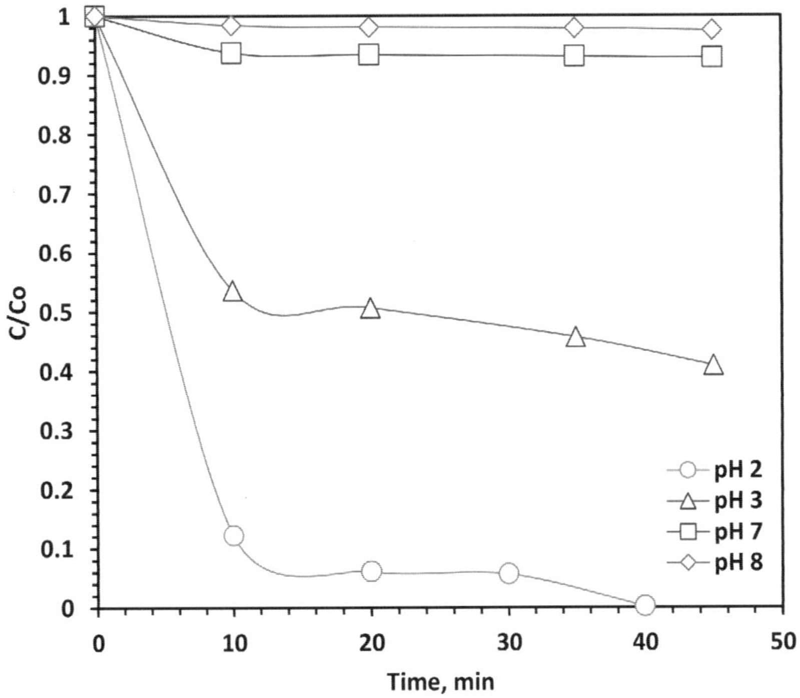

Figure 8. Effect of pH on the removal of Levafix dye, [Dye]o= 100 mg/L, H_2O_2= 800 mg/L, AS400=1 g/L.

3.2.3. Solar energy/aluminium-based sludge catalytic system:

Shebin El-Kom city, Menoufia government is one of the most plentiful solar radiation towns in Egypt. Such a region is well-endowe with solar energy and has significant solar radiation, particularly during the summer months [28]. The intensity of sun radiation reached to (1014 W/m^2) in the noon during Augest, the time of the study. Fig. shows the solar photo-catalysis process at conditions pH 3.0 and 800 and 1.5 g/L of H_2O_2 and, AS400, respectively using pilot scale reactor. With sun the green, free and cheapest energy source on the earth that has a various electromagnetic spectrum wavelenghs which is a key reason for the almost compelet removal of the 100 ppm Levafix dye effluent (99%) within only 30 minutes. However, when a visible light was used at the sae condition to illuminate the reaction, the percentage elimination reported was just 70% after around 55 minutes of reaction time. This demonstrat

es that photons taken from sunlight are more efficient than visible light at producing reaction complexes when using high light solar intensity radiation. This result suggests that the reaction is more likely to be employed around solar noon using a natural UV source when the sun radiation is between 1014 and 995 W /m^2, at operating daytime on 11:00 a.m., 12:00 p.m. These findings are consistent with those of Neppolian et al. [40] and Hussein [41] in the treatment of textile wastewater effluents with photocatalysts and solar energy. In the treatment of synthetic coloured wastewater, Akbal [42] discovered that UV light had a higher detoxification rate than sun irradiation.

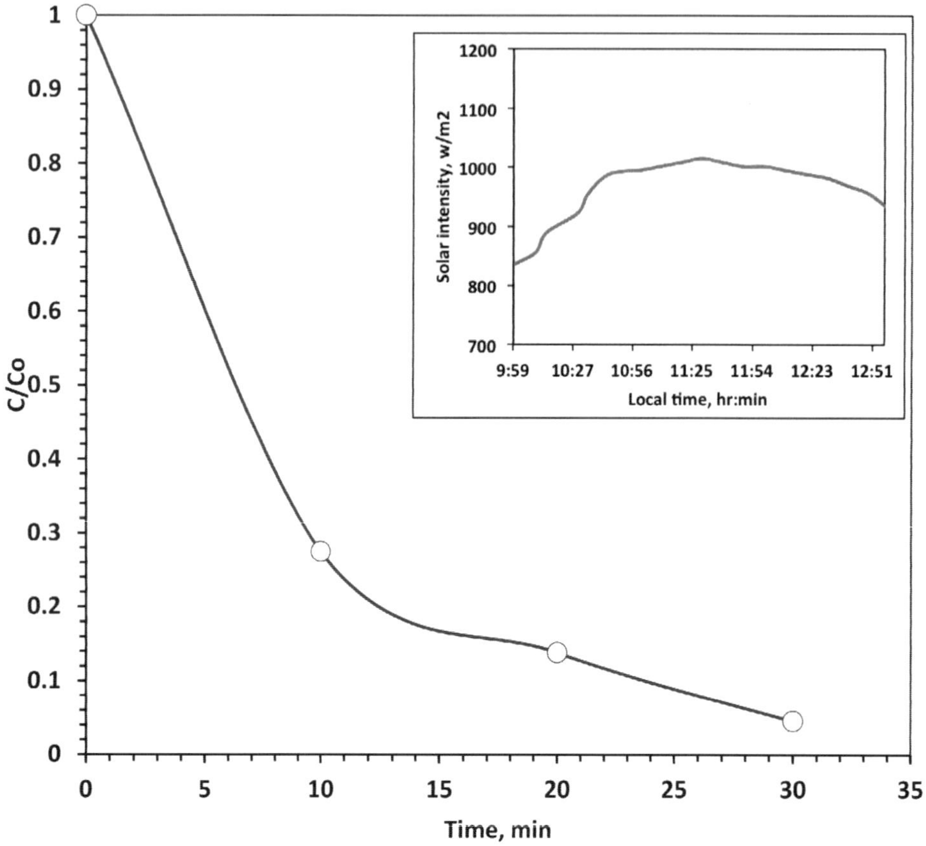

Figure 9. Solar experiment of the removal of Levafix dye, [Dye]o= 100 mg/L, H_2O_2= 800 mg/L, AS400=1 g/L, pH 3

4. Conclusion

Calcined alum sludge powder can be effectively used as a photo-catalyst for oxidation of Levafix blue dye. The dual adsorption/Fenton of Levafix dye was studied with and without solar radiation. It was found that the solution pH, the initial H_2O_2 concentration and the initial iron concentration, are the main factors that have strong influences on the oxidation of levafix blue dye. The optimal operation parameters for the Fenton oxidation were 800 mg/L for H_2O_2 and 1.5 g/L for AS400 at an initial pH of 2.0. Under this condition, 99% oxidation efficiency of Levafix aqueous solution within 40 minutes of reaction time. Using a pilot-scale photo-reactor, solar photo-Fenton demonstrated an effective treatment solution for dying effluent.

It was proposed that the process be carried out around solar noon since raising the sun intensity enhances the response rate to 99% compared to just 60% at dark condition. As a result, the procedure offers a potential technology for treating textile dying effluent wastewater in a cost-effective manner utilising renewable energy.

References

[1] Kuleyin A, Gök A and Akbal F 2021 Treatment of textile industry wastewater by electro-Fenton process using graphite electrodes in batch and continuous mode *Journal of Environmental Chemical Engineering* **9** 104782

[2] Dalari B L S K, Giroletti C L, Dalri-Cecato L, Domingos D G and Hassemer M E N 2020 Application of heterogeneous photo-fenton process using chitosan beads for textile wastewater treatment *Journal of Environmental Chemical Engineering* **8** 103893

[3] Patil A D and Raut P D 2014 Treatment of textile wastewater by Fenton's process as a Advanced Oxidation Process *IOSR Journal of Environmental Science, Toxicology and Food Technology (IOSR-JESTFT) e-ISSN* 2319-402

[4] Tony M 2021 Valorization of undervalued aluminum-based waterworks sludge waste for the science of "The 5 Rs' criteria". Applied Water Science)

[5] Tony M A and Mansour S A 2019 Removal of the commercial reactive dye Procion Blue MX-7RX from real textile wastewater using the synthesized Fe 2 O 3 nanoparticles at different particle sizes as a source of Fenton's reagent *Nanoscale Advances* **1** 1362-71

[6] Maha A. Tony M M E 2021 End-of-life waste criteria: Synthesis and utilization of Mn-Zn ferrite nanoparticles as a super paramagnetic photocatalyst for synergistic wastewater remediation. Applied Water Science)

[7] Tony M A 2021 Nexus approach: ZSM-12 derived from industrial waste into microencapsulated in organic wax for solar energy storage system *Energy Sources, Part A: Recovery, Utilization, and Environmental Effects* 1-19

[8] Tony M A 2020 Zeolite-based adsorbent from alum sludge residue for textile wastewater treatment *International Journal of Environmental Science and Technology* 1-14

[9] Van Truong T and Kim D-J 2021 Phosphate removal using thermally regenerated Al adsorbent from drinking water treatment sludge *Environmental Research* **196** 110877

[10] Sujana M G, Thakur R S and Rao S B 1998 Removal of fluoride from aqueous solution by using alum sludge *Journal of Colloid and Interface Science* **206** 94-101

[11] Tony M A 2021 Low-cost adsorbents for environmental pollution control: a concise systematic review from the prospective of principles, mechanism and their applications *Journal of Dispersion Science and Technology* 1-23

[12] Pradhan J, Das S N and Thakur R S 1999 Adsorption of hexavalent chromium from aqueous solution by using activated red mud *Journal of Colloid and Interface Science* **217** 137-41

[13] Xu Y, Chen T, Cui F and Shi W 2016 Effect of reused alum-humic-flocs on coagulation performance and floc characteristics formed by aluminum salt coagulants in humic-acid water *Chemical Engineering Journal* **287** 225-32

[14] Geng Y, Zhang J, Zhou J and Lei J 2018 Study on adsorption of methylene blue by a novel composite material of TiO 2 and alum sludge *RSC advances* **8** 32799-807

[15] Tony M A 2021 Paradigms of homo/heterogonous Fenton systems incorporating'Solar Energy'based on'Emerging Pollutants' removal–challenges, advancements and visualized bibliometric analysis *International Journal of Environmental Analytical Chemistry* 1-32

[16] Thabet R H, Tony M A, El Sherbiny S A, Ali I A and Fouad M K 2020 Catalytic oxidation over nanostructured heterogeneous process as an effective tool for environmental remediation *IOP Conference Series: Materials Science and Engineering* **975** 012004

[17] Tony M A 2020 Central composite design optimization of Bismarck Dye oxidation from textile effluent with Fenton's reagent *Applied Water Science* **10** 1-9

[18] Tony M A and Lin L-S 2021 Iron Coated-Sand from Acid Mine Drainage Waste for Being a Catalytic Oxidant Towards Municipal Wastewater Remediation *International Journal of Environmental Research* 1-11

[19] Santos V P, Pereira M F R, Faria P C C and Órfão J J M 2009 Decolourisation of dye solutions by oxidation with H2O2 in the presence of modified activated carbons *Journal of hazardous materials* **162** 736-42

[20] Thabet R H, Fouad M K, El Sherbiny S A and Tony M A 2021 Synthesis, characterization and potential application of magnetized nanoparticles for photocatalysis of Levafix CA Reactive Azo-Dye in aqueous effluent. *Water and Environment Journal*

[21] Chang C-C, Trinh C, Chiu C-Y, Chang C-Y, Chiang S-W, Ji D-R, Tseng J-Y, Chang C-F and Chen Y-H 2015 UV-C irradiation enhanced ozonation for the treatment of hazardous insecticide methomyl *Journal of the Taiwan Institute of Chemical Engineers* **49** 100-4

[22] Thabet R H, Fouad M K, Ali I A, El Sherbiney S A and Tony M A 2021 Magnetite-based nanoparticles as an efficient hybrid heterogeneous adsorption/oxidation process for reactive textile dye removal from wastewater matrix *International Journal of Environmental Analytical Chemistry* 1-23

[23] El-Desoky H S, Ghoneim M M, El-Sheikh R and Zidan N M 2010 Oxidation of Levafix CA reactive azo-dyes in industrial wastewater of textile dyeing by electro-generated Fenton's reagent *Journal of hazardous materials* **175** 858-65

[24] Ashour E A and Tony M A 2020 Eco-friendly removal of hexavalent chromium from aqueous solution using natural clay mineral: activation and modification effects *SN Applied Sciences* **2** 1-13

[25] Murena A, Borea L, Zarra T, Boguniewicz-Zablocka J, Belgiorno V and Naddeo V 2020 *Frontiers in Water-Energy-Nexus—Nature-Based Solutions, Advanced Technologies and Best Practices for Environmental Sustainability*: Springer) pp 33-5

[26] Colgan J D 2009 The international energy agency *Challenges for the 21st Century. GPPi Energy Policy Paper* **6**

[27] Tony M A 2021 Solar concentration for green environmental remediation opportunity—international review: advances, constraints and their practice in wastewater treatment *International Journal of Environmental Analytical Chemistry* 1-33

[28] Farahat M A, Mousa M M and Mahmoud N H Solar distiller with flat plate collector and thermal storage. Military Technical College) pp 1-11

[29] Feng H E and Le-Cheng L E I 2004 Degradation kinetics and mechanisms of phenol in photo-Fenton process *Journal of Zhejiang University-SCIENCE A* **5** 198-205

[30] Georgiou D, Melidis P, Aivasidis A and Gimouhopoulos K 2002 Degradation of azo-reactive dyes by ultraviolet radiation in the presence of hydrogen peroxide *Dyes and pigments* **52** 69-78

[31] Kavitha V and Palanivelu K 2005 Destruction of cresols by Fenton oxidation process *Water research* **39** 3062-72

[32] Gulkaya I, Surucu G A and Dilek F B 2006 Importance of H2O2/Fe2+ ratio in Fenton's treatment of a carpet dyeing wastewater *Journal of Hazardous Materials* **136** 763-9

[33] Tony M A and Lin L S 2020 Performance of acid mine drainage sludge as an innovative catalytic oxidation source for treating vehicle-washing wastewater *Journal of Dispersion Science and Technology*

[34] Modirshahla N, Behnajady M A and Ghanbary F 2007 Decolorization and mineralization of CI Acid Yellow 23 by Fenton and photo-Fenton processes *Dyes and pigments* **73** 305-10

[35] Hamoud H I, Finqueneisel G and Azambre B 2017 Removal of binary dyes mixtures with opposite and similar charges by adsorption, coagulation/flocculation and catalytic oxidation in the presence of CeO2/H2O2 Fenton-like system *Journal of environmental management* **195** 195-207

[36] Tony M A and Ali I A 2021 Mechanistic implications of redox cycles solar reactions of recyclable layered double hydroxides nanoparticles for remazol brilliant abatement *International Journal of Environmental Science and Technology* 1-18

[37] Clarizia L, Russo D, Di Somma I, Marotta R and Andreozzi R 2017 Homogeneous photo-Fenton processes at near neutral pH: a review *Applied Catalysis B: Environmental* **209** 358-71

[38] Lu M C, Lin C J, Liao C H, Ting W P and Huang R Y 2001 Influence of pH on the dewatering of activated sludge by Fenton's reagent *Water Science and Technology* **44** 327-32

[39] Tony M A and Lin L S 2020 Attenuation of organics contamination in polymers processing effluent using iron-based sludge: process optimization and oxidation mechanism *Environmental Technology* 10

[40] Sakthivel S, Neppolian B, Shankar M V, Arabindoo B, Palanichamy M and Murugesan V 2003 Solar photocatalytic degradation of azo dye: comparison of photocatalytic efficiency of ZnO and TiO2 *Solar energy materials and solar cells* **77** 65-82

[41] Hussein T A T and Ahmed T 2012 Estimation of hourly global solar radiation in Egypt using mathematical model *Int. J Latest Trends Agr. Food Sci* **2** 74-82

[42] Akbal F and Kuleyin A 2011 Decolorization of levafix brilliant blue E-B by electrocoagulation method *Environmental Progress & Sustainable Energy* **30** 29-36

Study of Elastomeric Heat Shielding Materials for Solid Rocket Motor Insulation

M Mosa[1], M Mokhtar Kotb[1], H Fouda[2] and Mohamed Gobara[1]

[1]Chemical Engineering Department, Military Technical College, Kobry Elkobbah, Cairo, Egypt
[2]Defense Systems Studies Center, Cairo, Egypt

Abstract. Thermal insulation of solid rocket motor casing is necessary to overcome the catastrophic breakdown during the rocket propellant combustion. In our study, a comprehensive review on different types elastomeric heat shielding materials (EHSMs) such as polyurethane, silicon rubber, ethylene propylene diene monomer (EPDM) and nitrile rubber was conducted. Various types of fillers are added to the EHSMs to enhance their thermal, mechanical and ablative performances. The charring phenomenon, which offers a means of protection to the virgin polymer layer from direct flame, as well as the synergetic effect, that may occur between fillers and polymer matrices, were studied. All different methods and machines used for manufacturing EHSMs are described in detail. Moreover, the different techniques used to characterize the EHSMs are discussed. Last but not least, the empirical models that relate the thermal insulation performance to the filler concentration are presented.

Keywords: Thermal insulation; Solid rocket motor; Elastomeric heat shielding materials; Ablation test; EPDM; Synergetic effect; Charring phenomenon; Theoretical models.

1. Introduction

Rocket motor is one of the primary types of jet propulsion engines. The solid rocket motor (SRM) is a chemical propulsion system in which the propellant grain is composed of both the oxidizer and the fuel in a single solid phase inside the motor case, the insulator is an important component of an SRM. Internal insulation shields the outer case from disintegration during rocket fuel combustion and is defined as a sheet of heat retardant material bonded, coated, or painted within the motor case to prevent the motor case from reaching temperatures that could destroy its structure during combustion. In order to protect the spacecraft from failure, designers have established some components and designs to protect the spacecraft from failure[1].

The heat shielding materials (HSMs) are produced from the reinforcement of elastomers with different types of fillers, are used to protect the metallic case of the rocket motor [2]. EHSM are the most efficient materials used for the protection of the interior metal case of rocket motor, as they have high thermal stability, low thermal conductivity, low density and good mechanical properties[3]. Also, convenient mixing of particulate fillers and fibers to the EHSM delivers better thermal, mechanical and ablative performance [4]. The addition of nanoparticle materials to the polymer that is used in thermal insulation enhances properties of the composite. nanoparticles of silica are added to polymers to enhance the thermal stability and mechanical properties [5]. Addition of two or more fillers to form a composite material may have a synergetic effect where properties enhancement occurs from the combination of these fillers specially in the halogenated polymers which is used as flame retardants[6,7]. EHSMs are prepared by using the two rolls calendar where the composite material is produced by mixing the different types of materials and the produced sheet is placed in a mold under the high pressure and temperature of hydraulic press for vulcanization and obtaining the desired shape [8]. The produced EHSM must be characterized to confirm that it can be used in thermal insulation of SRM, so all methods of thermal analysis, mechanical analysis and ablation test will be mentioned. The ablation process is the thermal decomposition of the organic constitutes of a material under high temperature. The char layer formed during the combustion process acts as a barrier and protects the virgin layer of the composite material from being ablated by the direct flame[9]. Most of prepared composite materials are based on using empirical equations and theoretical models in order to get the best results depending on types, thermal conductivities and compositions of the used fillers[10].

2. Thermal insulation types:

Thermal insulation composites and materials are divided according to the ranges of temperature to different classes as the following: low range temperature (-75 to 15) °C, intermediate range temperature up to 300 °C, high range temperature up to (800) °C and very high range temperature (above 800) °C. examples of different classes are given in Table (1)[11].

Table 1. Types of thermal insulators according to temperature range[11].

Low temperature range (-75 °C to 15 °C)	Intermediate temperature range (15 °C to 300 °C)	High Temperature range (300 °C to 800 °C)	Very high temperature range (above 800 °C)
• Elastomeric foamed plastic • Glass fiber • Mineral fiber • Polyethylene • Polyurethane • Polystyrene	• Elastomeric foamed plastic • Glass fiber • Mineral fiber • Phenolic • Polystyrene • Polyurethane	• Calcium silicate • Cements • Ceramic fibers • Glass fibers • Polymeric composites	• Polymeric composites • Ceramic composites

To reach the demanding specifications and field conditions, thermal insulation systems for rocket motors must be designed using current techniques and materials. Rigid material, such as steel or composites, is used to make the rocket motor housing. Within the case, the propellant is placed, and the thermal insulation composite forms a liner between them. There are two classifications of rocket motor insulation: internal Insulation, which is a bonded layer between the case and the propellant, which prevents the case from overheating and rising in temperature during combustion process of the rocket. The second classification is external insulation which is used to coat the outside of rocket motor bodies, especially in tactical missiles and launch rockets with high-acceleration[12].

3. Main types of elastomeric heat shielding materials:

EHSMs are classified into four different categories which are ethylene propylene diene monomer (EPDM), silicone (Si) rubber, nitrile rubber (NBR) based EHSM and polyurethane (PU) based EHSM [13,14].

3.1. EPDM:

In EPDM, compositions of ethylene, propylene and diene ratio can be varied to meet different types of applications, 'figure 1' represents the structure of EPDM, table (2) shows different types of EPDM. Nowadays, EPDM is one of the most important EHSM used as a thermal insulator for the SRM[15]. EPDM has great advantages as thermal insulator such as good mechanical, thermal and ablative properties, high thermal stability, low density, and good aging resistance[16]. It has a significant resistance to ozonation and oxidation, lowest density among the elastomers (about 0.85 g/cm^3) and very good properties at low temperature as its Tg is -50 °C[17]. the low density plays an important role in lowering the weight of the

HSM so it is recommended to be used in the lining of the SRM casing[18,19]. table (2) shows different densities of some elastomeric ablatives[20].
The aramid fibers, such as Kevlar and Twaron, are most commonly used for the reinforcement of EPDM in the rocket motor case is due to their high mechanical properties, thermal stability and low density[21,22]. Also when aramid fibers added to EPDM, they enhance the ablative properties where a layer of char is formed at the top surface of the polymer when exposed to the flame protecting the virgin layer from the direct flame[22].

Table 2. different types of EPDM.

Type	Ethylene content	Propylene content	Diene content
EPDM 4640	55%	40%	5%
EPDM 4570	50%	35%	5%
EPDM 4760	67%	28%	5%
EPDM 4725	70%	25%	5%

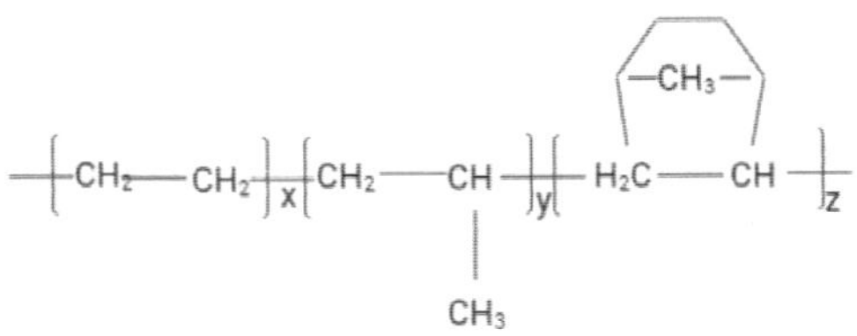

Figure 1. Structure of EPDM[15].

3.2. Silicone rubber based EHSM:
In the SRM and thermal protective systems(TPS), EHSMs based on Si rubber are employed as ablative materials[23]. The good thermal stability of Si rubber composites makes them a good selection for aerospace applications[24]. Si rubbers have good chemical resistance and thermal stability. Adding methyl phenyl siloxane as a copolymer with polymethyl siloxane proves to increase the degradation temperature up to 400°C [25,26].

The char layer, consisting of residues of inorganic silica, released when silicones are exposed to flames acts as a protective layer protecting the internal layer of the rubber Mixing Si rubber with EPDM at a ratio of (1:1) by weight enhances the ablative properties of the elastomeric material; the ablation rate of EPDM by oxy-acetylene for (20) seconds is (0.215) mm/s while the ablation rate of the mixture is (0.06) mm/s, [28]. Si rubber has a low tensile strength value of (2.4) MPa, to obtain highly effective EHSM, Si rubber must be reinforced with some additives as fillers that increase its thermal stability, mechanical properties and ablative properties, flame retardant materials and vulcanizing agents[22]. while the poor mechanical properties of Si rubbers limits its use as thermal insulations in SRM[27].

3.3. Nitrile rubber based EHSM:
NBR is a copolymer of butadiene and acrylonitrile, 'figure 2' represents the structure of NBR, it is used as thermal insulator for SRM casing. The chemical and physical properties of NBR relies on the nitrile percent in the copolymer.

In the development of HSM for rocket motor casing, NBR, a synthetic copolymer of acrylonitrile and butadiene, was used[29]. The chemical structure of EPDM elastomer is represented in Figure (6)[30]. EPDM rubber is a good insulator because of its stable mechanical properties such as tear strength, great thermal resistance, and good property retention after age due to the saturated main chain. Because EPDM has a low specific gravity, it has a novel application in case-bonded insulation for rocket engines. It has also been shown to have improved ageing qualities and a high filler loading capacity[31].

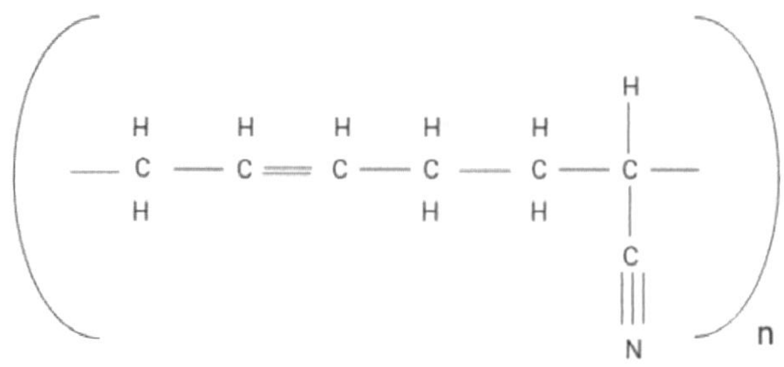

Figure 2. The structure of NBR[31].

NBR processing is difficult because of the strong intermolecular force between molecules[13]. SRM still uses NBR as an insulation liner. NBR heat shielding material containing SiO_2 powder and asbestos fibers for reinforcing and hence improves thermal stability. Because of their stronger affinity with higher char yield resins like phenolic resin, which are added to improve ablation resistance[32]. When NBR is reinforced with aramid fibers as Kevlar fibers by means of mixing in the two rolls calendar, the product produced is characterized by excellent ablative and mechanical properties, excellent thermal stability will be achieved and the ablation rate will be decreased as the aramid fibers composition increased[4]. EHSM based on NBR are used in the Indian aerospace system[33].

3.4. Polyurethane based EHSM:

PU is used as an EHSM due to its high thermal stability and high specific surface area. 'figure 3' shows PU as EHSM and the different types of fillers used for reinforcement[34,35].

PU composites used as EHSM reinforced with alumina, silica, carbon black, and zirconia have good ablation rate and thermal conductivity values. The lowest erosion rate is (0.324 mm/s) for 10% carbon black[36]. Elastomeric PU nanocomposite materials are one of the best choice of thermal insulation liner made by Air Forces Research Laboratory which have low density, good ablation, but more cost of fabricating if compared with Kevlar / EPDM composites[37].

The performance of PU nanocomposite as EHSM was investigated and it was concluded that they could be used as EHSM for SRM with the addition of different fillers and reinforcements[38]. Melamine phosphate and PU encapsulated with melamine phosphate/polyoxymethylene were examined in terms of flame retardancy and mechanical properties. Thermal resistance was shown to be better in polyoxymethylene/PU reinforced with melamine phosphate[39]. Table (3) lists the main advantages and disadvantages of EHSM[22]:

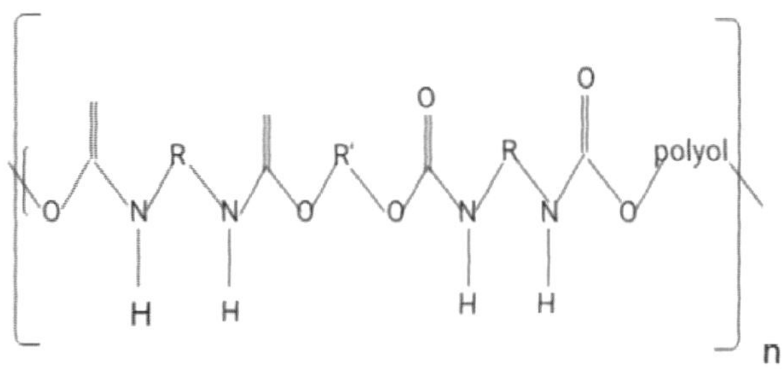

Figure 3. PU as EHSM [39].

Table 3. Advantages and disadvantages of EHSM.

Elastomeric heat shielding material	Advantages	disadvantages
EPDM based	Thermal stability, low density, low ablation rate, resistance for oxidation and ozonation	Premature vulcanization
Nitrile rubber based	Low rate of ablation	Low shelf-life time, high density
Silicone rubber based	Convenient with proper fillers for aerospace applications	Bad mechanical properties
Polyurethane based	Low ablation rate when adding proper filler, good bonding and compatibility to the propellant	Short range isocyanate and polyols mixture is not always useful

4. Different fillers and additives to EHSM:

Mixing of suitable fillers and fibers with rubber enhances the thermal insulation by increasing thermal, mechanical, flame retardancy and ablative properties, especially if the fillers on the nano scale, good bonding or adhesion between rubber and fillers such as aramid fiber, carbon fiber, and powder silica is a must to be sure that the composite is stable, characteristics of different types of fillers are shown in table (4). The homogeneous mixing of the fillers and additives with the rubber matrix plays a very important role on enhancing thermal, mechanical and ablative properties of the produced composite material. The concentration of the filler in the rubber matrix usually does not exceed 20 % to maintain the primary properties of the matrix[40]. There are some factors affecting the properties of the produced insulation composite as filler and matrix interaction, mixing method and time of mixing between fillers and the matrix, filler-filler interactions and the surface area of the used fillers[40]. Different matrices without fillers or fibers as (polyesters silicone, and PU) as shown in table (4), are tested and ensured that ablation rates have high values to be used as EHSM[40]. So, fillers are added to various types of matrices to achieve better ablation performance. Microspheres are used in the EPDM as EHSM, the composite contains 20 phr microspheres showing best properties. Perlite is also tested but fails in the manufacturing process[41]. Reinforcement of elastomeric materials with suitable fillers and fibers confirmed that ablation properties and thermal behavior properties are enhanced[42].

Historically, asbestos fibers have been used for reinforcement of various elastomers and materials. however, asbestos is banned for environmental and health reasons. So, other types of fillers and fibers, of the same or even better mechanical and thermal properties, are being used as alternatives for asbestos [43].

Rubbery composite materials are modified and reinforced with Kevlar fiber that have a relatively good thermal resistance, it can be used as an EHSM. Different types of fibers as polyimide, carbon fibers and polyetherimide can give polymers same properties as Kevlar. but they may be more expensive. Conventional fibers or fillers used for EHSM are expanded graphite, fumed silica, and perlite. These different types of fillers and fibers have the ability to improve the mechanical and thermal properties of the thermal insulation used in the SRM. Thermal conductivity of the fumed silica is very low, this is explained by nanoscale structure and high porosity. Fumed silica is available at low cost and can be mixed homogeneously with high simplicity to the different types of rubbers[44]. Characteristics of different types of fillers are shown in table (5).

Table 4. Characteristics of different types of fillers.

property	Asbestos fiber	Kevlar fiber	Fumed silica	Fiber glass	Chopped carbon fiber
Thermal conductivity (W/m. K)	0.08	0.04	0.03	0.2	10.4
Tensile strength (MPa)	1100(MPa)	2700 (MPa)	152.2(MPa)	160 (MPa)	3400 (MPa)
Elongation %	1.54	3.4	------	2.2	2.1
Hardness (shore A)	60	85.7	30	45-50	88-100
Density (g/cm^3)	2.53	1.44	2.2	1.7-1.85	1.8
shape	fiber	fiber	powder	fiber	fiber

Table 5. Addition of different types of fillers on EHSM[22].

sample	Ablation rate (mm/s)	Thermal conductivity (W/m K)	Tensile strength (MPa)	Elongation (%)	reference
EPDM matrix					
0–50% phr KF/EPDM	0.13–0.18	0.013–0.018	2–13	175–300	[14]
MWCNT/EPDM	0.078–0.092	--	6.28–7.84	6.28–7.84	[45]
Aramid fiber-silica-ZnO/EPDM	0.11/0.12	--	--	--	[46]
15 phr Kevlar/EPDM	0.03	--	--	--	[34]
Virgin EPDM	0.29 to 0.72	--	--	--	[47]
20% phr KP + 5% alumina + 5% silica + 40% dechlorane + 20% Sb_2O_3/EPDM	0.10	0.21	8.5	16	[48]
Aramid fiber-PR-mineral oxide/EPDM	--	--	4.15	45	[49]
30 phr KP/EPDM	0.015	0.171	9.35	11.7	[50]
25 phr CF/25 phr KP/EPDM	--	0.198	11	7.4	[51]
SiO2/EPDM	--	0.202	5.2	350	[22]
Polyurethane matrix					
20–50 phr KF/EPDM	--	0.363	--	--	[52]
5% nano clay/HTPB-PU	0.42–0.69	--	1.07/1.47	700/650	[53]
5% phr perlite/HTPB	0.70	--	1.22	130	[54]

23% POSS/HTPB-based PU	0.298	--	--	--	[55]
		Nitrile rubber matrix			
0% MMT/NBR//10% MMT/NBR	0.087/0.075	0.156/0.151	2.58/2.54	157/151	[56]
Carbon + SiC/NBR	--	0.52–1.61	85–125	0.4–0.47	[57]
Fumed silica-nano clay-EG/HNBR	0.063/0.047/0.067	--	--	--	[58]
6 phr Kevlar fiber/NBR	0.137	--	15.4	666.6	[59]
5% SCF/NBR	0.06	0.08	4.75	22	[60]
SiO2/NBR	----	0.224	7.7–10	300-450	[61]

5. The synergetic effect:

Polymeric materials have the ability to accept two or more types of fillers and additives to their matrices and the effect obtained from the composite is higher than the sum of effects of each filler. One example is the combination of ammonium sulphate and antimony tri-oxide as flame retardants and the obtained composite has high ablative performance[6,7]. Several different types of chemical compounds including (molybdenum oxide, bismuth oxide, boron and tin oxide) are added to halogenated polymers and form a synergetic effect, the commonly used is antimony tri-oxide which have low flame retardancy properties when used alone in a polymeric material, but once it is used in a halogenated resin the flame retardancy properties become extremely high[62].

The combination of aramid fibers with carbon fibers to the EPDM as a thermal insulator in SRM has a synergetic effect due to the formation of a char layer of very high strength which is affecting on the ablation rate of the insulation and protects the virgin layer of the polymer composite[63]. Synergetic effect occurs by interaction of nano silica with fillers in polymeric composites affecting the erosion rate of the polymer and reduce the ablation rate, improving the char strength and reduce the rate of oxidation of EHSM [30].

6.The charring phenomenon:

When ablation process takes place on an ablative material, pyrolytic gases are produced at the location of the ablation forming a char layer at elevated temperatures, the availability of this char layer have the ability to control the diffusion of heat to enter into the surface, the pyrolytic gases are then trapped as an under layer of the charring layer, ablation process is explained by the following scheme in 'figure 4'[80].

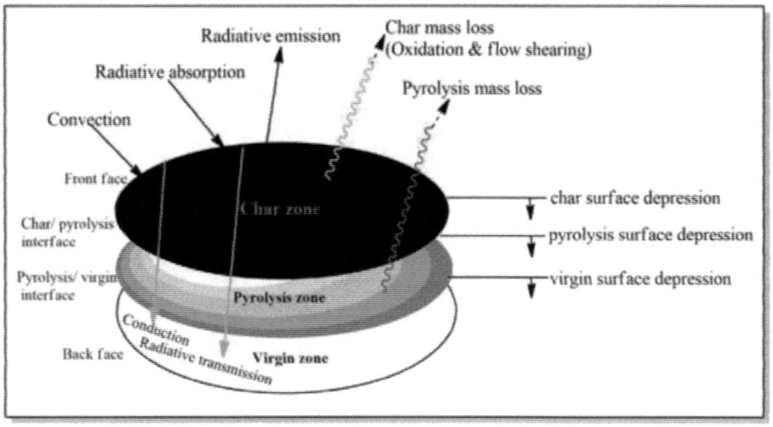

Figure 4. The ablation process mechanism[80].

When the surface of the material is exposed to extremely high temperature from the heat source the char layer decomposes, new char layers will be formed with the reactant at short time and the pyrolysis process of virgin materials will take place, the char layer is responsible for the protection of the protection of the virgin layer of the polymer by acting as a barrier for the heat flux[81].

7. Methods of manufacturing EHSM:

7.1 preparation of EPDM composite:

The EPDM composite is prepared by means of two rolls calendar, where EPDM is added into the calendar, mixing the EPDM for 10 minutes, then antimony tri-oxide and ammonia sulfates are added and mixed for 2 minutes, then paraffin oil and Kevlar fiber are added to the calendar rolls and mixed for 10 minutes, and nano silica is added with the DCP and time of mixing is 30 minutes. The time of mixing is for ensuring the good homogeneity of the rubber composite materials. The produced composite is left to cool for 24 hours, then placed in molds to get the desired formations and vulcanization of the composite is under the hydraulic press at 180 °C and 50 bar for 12 minutes[64].

Another method for preparing, is that EPDM rubber is heated at 70 °C in an oven, then 3 (part per hundred) phr of nano silica and fibers percentage are added to the matrix at two rolls milling calendar at 40 rpm at 70 °C. in addition, antioxidants and curing agents are introduced with good mixing into the matrix. All previous added constituents' compositions are shown in table (6). Finally, a sheet of the prepared EPDM composite is produced and analyzed to ensure its performance[65].

Table 6. Recipes used for production of the EPDM composites [65].

components	EPDM	EPDM/Kevlar	EPDM/Kevlar/nano silica
EPDM	100	100	100
Dicumyl peroxide	1	1	1
antioxidant	1	1	1
Kevlar fiber	0	10	10
Nano silica	0	0	3

7.2 preparation of NBR composites:

NBR is an acrylonitrile-butadiene copolymer with a density of 1–1.2 g/cm^3. The intermolecular interaction between the polymer chains, as well as the density and glass transition temperature of the elastomer, increases as the acrylonitrile content increases. Because of the elastomer's high intrinsic stiffness, which is directly related to the strong intermolecular contact, processing NBR in open or closed mixers is relatively challenging. Sulfur or peroxides can crosslink NBR, which is vulcanized at temperatures between 140 °C and 190 °C[59].

7.3 preparation of silicone rubber composites:

The sheet of silicone rubber is produced by means of vulcanization at high temperature. Carbon fiber and silicon carbide powder are prepared by casting methods. The composite is completely dissolved in toluene with 1 wt. % peroxide. The produced solution is dried in an oven for 48 hours at 60°C, Carbon fibers is then introduced to the solution. produced solution is stirred until homogeneity takes place, the composite is cured under a hydraulic press at 175 °C for 10 min [66].

Silicone rubber composite can be prepared by adding fillers to the silicone rubber matrix, the fillers are dried for 12 hours in a vacuum oven at 80 °C and gradually added to the matrix. The silicone rubber composites were mixed for 20 minutes at room temperature in a rheometer. The composites were then combined in a two-roll open mill and a vulcanizing agent is added., the mixing was done at room temperature for 20 minutes. The samples were then vulcanized in a mold at 180 °C for 10 minutes at a pressure of 15 MPa. Finally, silicone compounds were secondary vulcanized in an oven for 2 hours at 200 °C [66].

7.4 calendar mixer

A mill is made up of two horizontal hollow metal cylinders that rotate in opposite directions is shown in 'figure 5'. Distance between the two cylinders can be adjusted, ranging from 0.25 to 2 mm in most cases. A nip is the space between the rolls, raw elastomer is placed in the nip. The back roll rotates quicker than the front one, and the speeds of the two rolls are frequently varied. The friction ratio is the difference in speed between the rolls that allows a shearing action at the nip to scatter the components while forcing the polymer to stay on one roll. Other additives are then added at the nip, and the mixing process generates heat. This heat must be dissipated by flooding the inside of the roll with cooling water or by pouring water via drilled channels in the roll's wall. after the completion of the mixing operation Sheets of the polymer composite are withdrawn from the mill[67].

Figure 5. Double rolls calendar[67].

7.5 hydraulic press:

Hydraulic press represented on 'figure 6' is specifically designed to mold test samples of materials. It is capable to exert a pressure of about 300 bar, Fully guided, heated and insulated platens are operating temperature to 300 °C. The hydraulic press was set to vulcanize composite samples at 50 bar and 180 °C, the samples were vulcanized in a range of 12 minutes.

Figure 6. Hydraulic press.

8. characterization of thermal insulation composites:

8.1 Hardness measurement:

Although hardness is not a significant feature of a composite, it is regarded a good quality control tool. When paired with other data, such as composition, hardness levels can predict a variety of qualities. Hardness is measured by a hardness tester shore (A) durometer, shown in 'figure 7', which is used for the measurement of hardness at 25 °C. Hardness shore (A) is used to measure the soft materials; where the hardness test depends on determination of the penetration of the computing pin through the material at definite conditions of force and time. Hardness measurements of different samples are shown in table (7), There are no units for hardness values[68].

Figure 7. Hardness tester shore (A) durometer.

Table 7. Hardness measurements of different samples[68].

Sample No.	Composition (Phr)	Hardness (Shore A)	Specific gravity
1	100 EPDM + 60 Ammonium poly phosphate + 5 peroxide +5 Kevlar	74.5	1.0898
2	100 EPDM + 60 Ammonium poly phosphate + 5 Peroxide+10 Kevlar	78	1.0952
3	100 EPDM + 60 Ammonium poly phosphate + 5 Peroxide+20 Kevlar	87	1.1082
4	100 EPDM + 60 Ammonium poly phosphate + 5 Peroxide +30 Kevlar	90.7	1.1221

8.2 Mechanical Properties measurements:

Tensile strength can be measured by using the test machine (ZWICK Z050) 'figure 8'[68]. The test is carried out on at least three different specimens and an average is reported for the sample. A remote control is equipped by the test machine which controls the software and enables obtaining results and print produced data with low effort. There are some equations used for the measurement of tensile strength and young's modulus specimen have some specifications 'as in equations ((1),(2),(3),(4),(5),(6))'[70].

$$\text{Tensile strength} = \frac{\text{force (load)}}{\text{cross-section area}} \tag{1}$$

$$\text{Tensile strength at yield (psi)} = \frac{\text{maximum load recorded}}{\text{cross-section area}} \tag{2}$$

$$\text{Tensile strength (psi)} = \frac{\text{load recorded at break}}{\text{cross-section area}} \tag{3}$$

$$\text{Young's Modulus} = \frac{\text{Stress}}{\text{Strain}} \tag{4}$$

$$\text{Stress} = \frac{\text{force}}{\text{cross-section area}} \tag{5}$$

$$\text{Strain} = \frac{\text{change in length}}{\text{orignal length}} \tag{6}$$

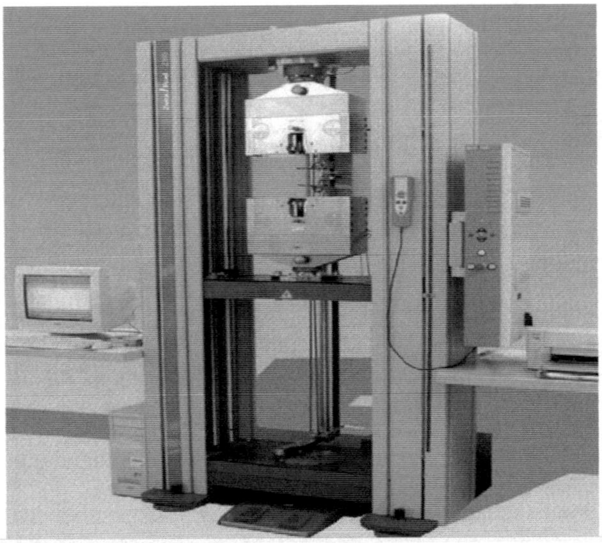

Figure 8. Universal test machine (ZWICK Z050)[69].

Kesiya George showed that the mechanical properties of EPDM composites are shown in table (8) [70]:

Table 8. Mechanical properties of EPDM composites.

Sample	Tensile strength (MPa)	Elongation at break (%)	Tensile modulus (MPa)	Hardness (Shore A)	Density (g/cm^3)
EPDM 4640	2.33 ± 0.2	280 ± 11	2.9 ± 0.5	67	0.930
Kevlar /EPDM	3.93 ± 0.1	228 ± 8	51.65 ± 6	80	0.959
EPDM /Kevlar /silica	4.88 ± 0.3	376 ± 13	64.09 ± 4	93	0.962

8.3 Specific gravity measurements:

Specific gravity is defined as ratio of the substance density to that of the reference substance. The principle of measuring the specific gravity is that when a solid is weighed in air, then in water; the difference between the weights is the weight of the water displaced by the volume of the solid, according to Archimedes' principle. If the solid is less dense than water, some method of entirely submerging it must be used. The ratio of the solid's weight in air to the difference between its weight in air and its weight immersed in water called its specific gravity[69].

An analytical balance is being used for determining the specific gravity of the produced samples with respect to standard techniques as in 'figure 9'. The specific gravity is determined 'as in equation (7)'[69].

$$\text{Specific gravity} = \frac{w_1 \times \rho_w}{w_1 - w_2} \quad (7)$$

Where: W_1.... weight of sample in air (g)

ρ_w.... density of water

W_2.... Weight of Sample in the water (g)

Figure 9. The used method to determine the density[69].

Kesiya George showed that the densities of different compositions of Kevlar and Nano silica with EPDM is as shown in table (9) and 'figure 10' [71].

Table 9. Different compositions of EPDM composites[70].

No.	Sample code	Kevlar fiber	Nano silica
1	S1	0	0
2	S2	5	0
3	S3	5	1
4	S4	5	3
5	S5	5	5
6	S6	10	0
7	S7	10	1
8	S8	10	3
9	S9	10	5

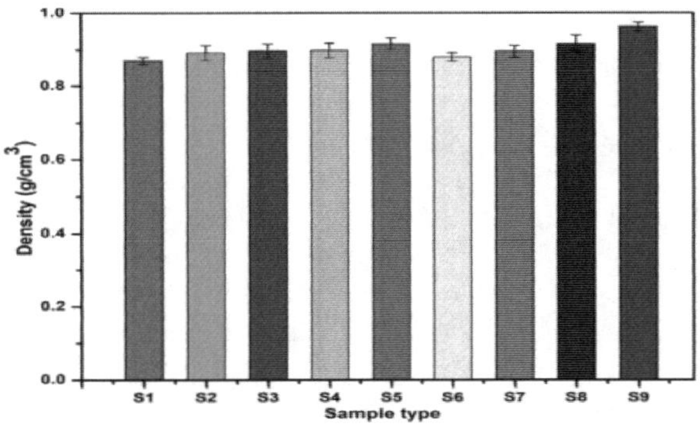

Figure 10. Different densities of the samples[71].

8.4 Thermal Measurements:

8.4.1 Thermal Conductivity test:

Thermal conductivity can be defined as the rate of heat energy which is flowing through a material. samples are measured by TCFCM-N20 Thermal Conductivity device, which is used for measuring high to medium thermal conductivity samples as ceramics metals, plastic, metal alloys, glass, epoxies, geological and composite materials. thermal conductivity may be measured for unknown sample by making a comparison with a known thermal conductivity of known sample. The Best obtained outcomes are gained by via a reference sample which is matched to very closely to the predictable thermal conductivity of the strange sample. Two thermocouples are used to measure the gradient of the temperature of the sample while performing the test. Two materials of known thermal conductivities are located above and under the sample, the reference materials are equipped by thermocouples. Heaters are located at the both ends and temperature of both heaters are controlled, the thermal conductivity of the sample 'as in equation (8)'[72].

$$K_{TestSample} = \frac{1}{2}\left(\frac{\Delta x}{\Delta T}\right)_{TestSample}\left[\left(K\frac{\Delta T}{\Delta x}\right)_{Top\ Re\,frence} + \left(K\frac{\Delta T}{\Delta x}\right)_{Bottom\ Re\,frence}\right] \quad (8)$$

K = thermal conductivity of reference material.

ΔX= distance between thermocouples in each sample.

ΔT=temperature difference.

Kesiya George showed that the thermal conductivity of different compositions of Kevlar and Nano silica on EPDM is as shown in table (10)[71].

Table 10. Thermal conductivity values[71].

Sample	Thermal conductivity (W/m K)
EPDM	0.188
EPDM/Kevlar	0.178
EPDM/Kevlar/silica	0.159

8.4.2 Differential Scanning Calorimeter (DSC):

DSC model Q2000 as shown in 'figure 11' is used to measure the temperature flow and heat correlated to the changes of material as a function of time and temperature. Quantitative and qualitative analysis takes place on endothermic and exothermic steps of materials at physical variations which results from change in phases, melting and oxidation. The DSC operation relies on measuring the flux of heat between the reference and the sample while changing the temperature for an area when the flux is varying. DSC is the most used method for thermal analysis, it has two aluminum pans, the first is used for putting the sample and the second is an empty pan for putting the reference. There is a heater used to save the temperature unchanged for both reference and sample. Heating is controlled by means of a software program which controls heating rates to adjust the heat transfer between reference and sample. A thermogram graph is obtained from DSC device which is the difference heat amount supplied to the sample and reference with respect to sample temperature. DSC is used for the determination of specific heat capacity (Cp) of a sample[73]. I Kingstone Lesley Jabez shows the curves of DSC of the nitrile rubber, EPDM and EPDM composite in 'figure 12'.

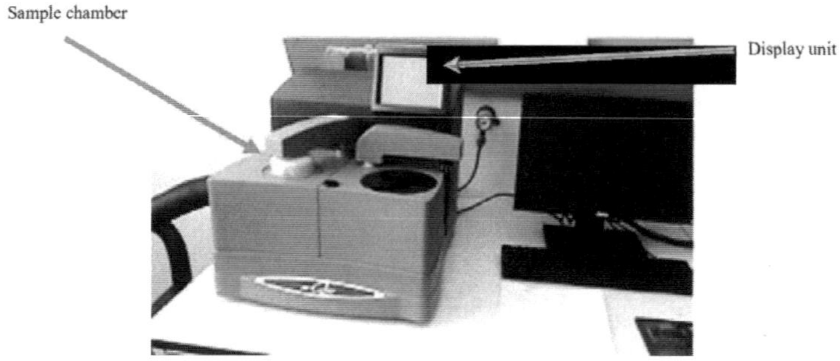

Figure 11. DSC device[73].

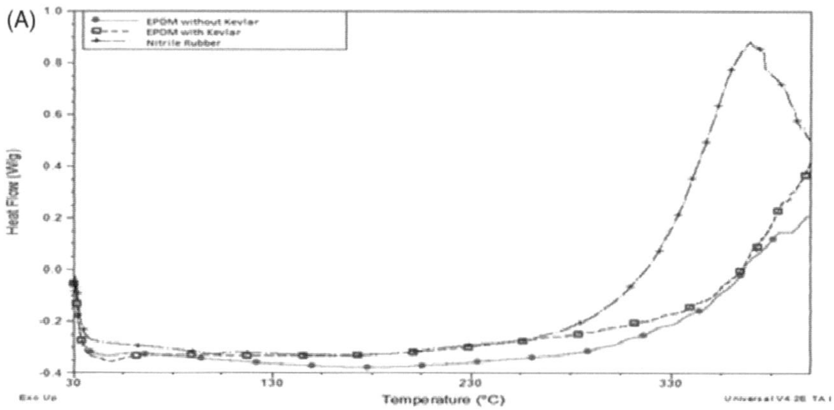

Figure 12. DSC curves for nitrile rubber, EPDM and EPDM composite.

8.4.3 Thermogravimetric Analyzer (TGA):

The TGA model Q5001 device is a weight varying analysis device as shown in 'figure 13' which is used to determine the amount and rate that the weight varies of the sample and determine the thermal stability of the samples, while temperature increasing with time. software program controls the TGA device which gives the calculation needed for thermal analysis. TGA is also used to know changes in phases as result of decomposition and dehydration. samples weights vary for a range from (2 to 5) ±1 mg. Analysis takes place in presence of nitrogen where the flow rate of nitrogen is 20 ml/min[74]. The main parts of the TGA device[74]:

- A balance used to measure the weight of the sample.
- A furnace which is used to control the atmosphere and the temperature of the sample.
- A case where the electronics systems and mechanics are located.
- Two controllers of mass flow to control the purge gas to the balance and the furnace.

I Kingstone Lesley Jabez shows the curves of TGA of the nitrile rubber, EPDM and EPDM composite in 'figure 14' where the decomposition of the nitrile rubber and the EPDM composite is less than that of the EPDM without Kevlar[75]. Chengzhi Jiang shows the TGA curve of silicone rubber composition 'figure 15'[76].

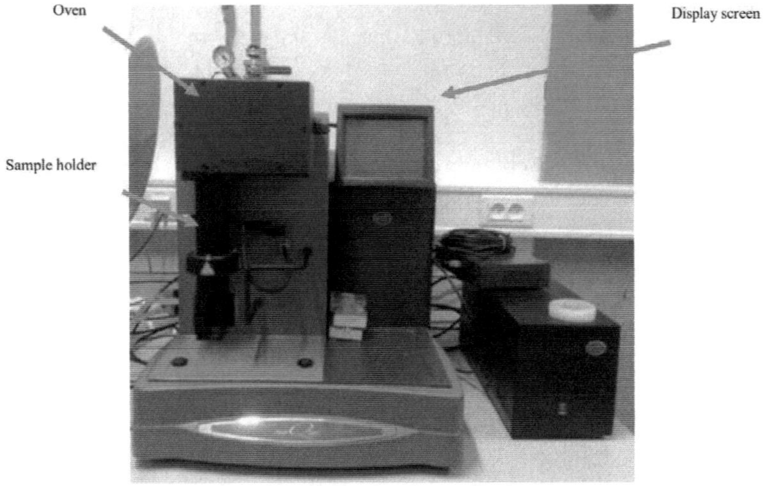

Figure 13. Thermogravimetric analyzer[74].

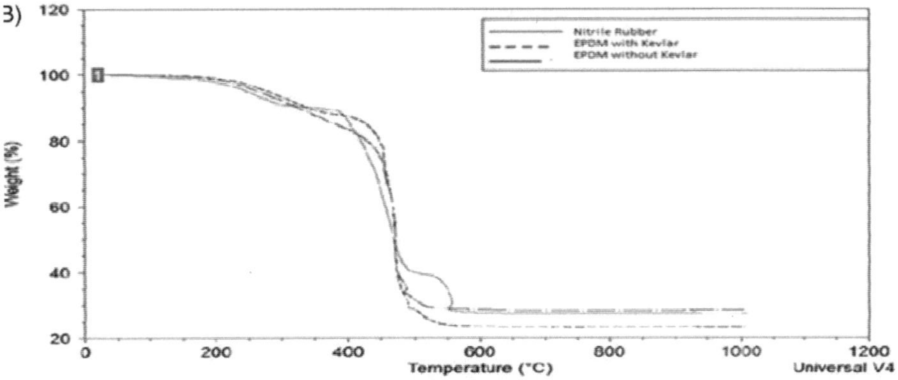

Figure 14. TGA curves of nitrile rubber, EPDM and EPDM composite[75].

Journal of Physics: Conference Series **2305** (2022) 012037 doi:10.1088/1742-6596/2305/1/012037

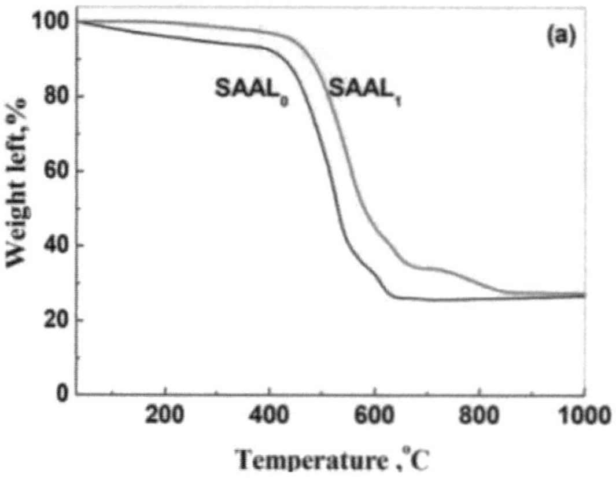

Figure 15. TGA curve of silicone rubber composition.

8.4.4. Thermomechanical Analyzer (TMA):
TMA instrument as shown in 'figure 16'is used to measure the deformation of materials varies at controlled conditions of time, pressure, force and temperature. The used force can be applied in compression, flexure or tensile modes of distortion using specially designed probes.

TMA measures important properties of materials as young's modulus, coefficient of thermal expansion, glass transition temperatures and softening points. There are two modes (modes or models, not important anyway) which are Q400EM or Q400. TMA patterns for samples containing carbon nanotubes (CN) are shown in 'figure 17'. TMA can be used also for measurement of viscoelastic properties by usage of creep tests. In creep test we fix the stress while measuring the strain[77].

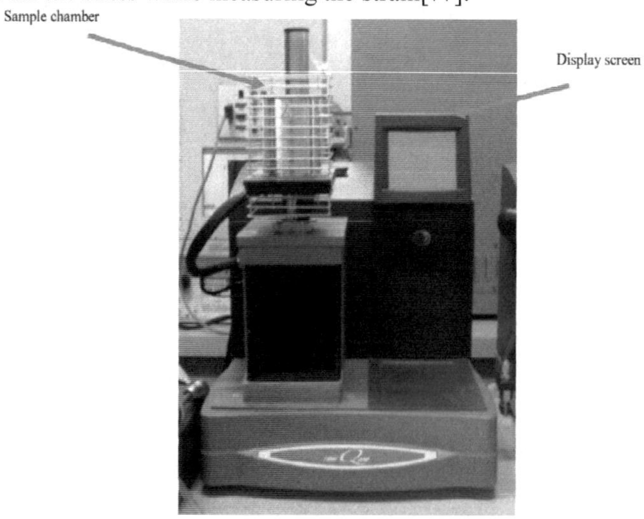

Figure 16. Thermomechanical analyzer[77].

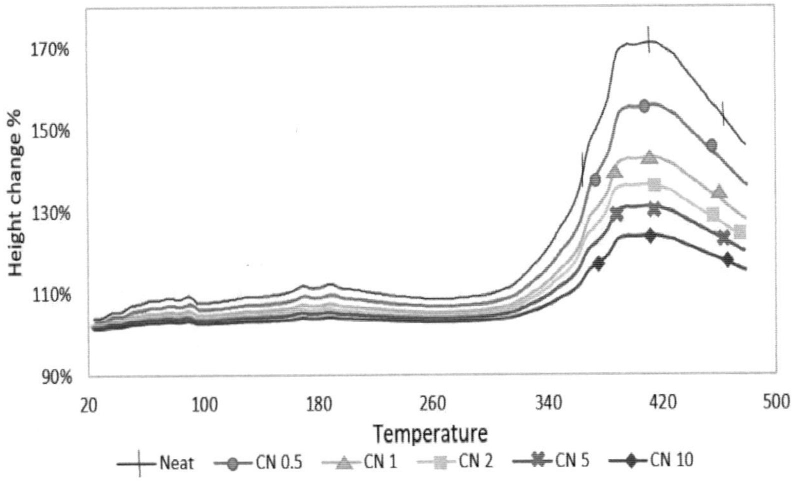

Figure 17. TMA patterns for samples containing carbon nanotubes.

8.4.5 Ablation Test:

Ablation standard test is a method used to evaluate the performance of thermal insulation of rubber ablative materials at which the performance is determined by the usage of an oxyacetylene flame torch of a temperature 2000 °C on the surface of the sample as shown in ‘figure 18’ [78].

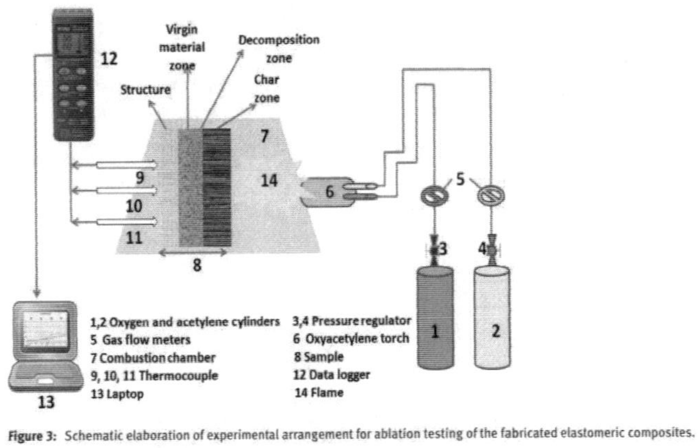

Figure 18. Schematic diagram of ablation test setup[78].

Ablation test measures the ablation rate and properties of the samples and the efficiency of the thermal insulation will be showed and examined.

The sample is exposed to the flame for approximately 60 seconds, and the rate of ablation is calculated by dividing the difference between the final and initial thicknesses by the burning time. A thermocouple attached to the back of the sample is used to measure the temperature. The sample is 3 mm thick, with a length and breadth of 15 cm, the sample's weight is determined, and the sample is adhered to the surface of a steel plate with adhesives such as Epon and Versamide, as illustrated in ‘figure 19’[79].

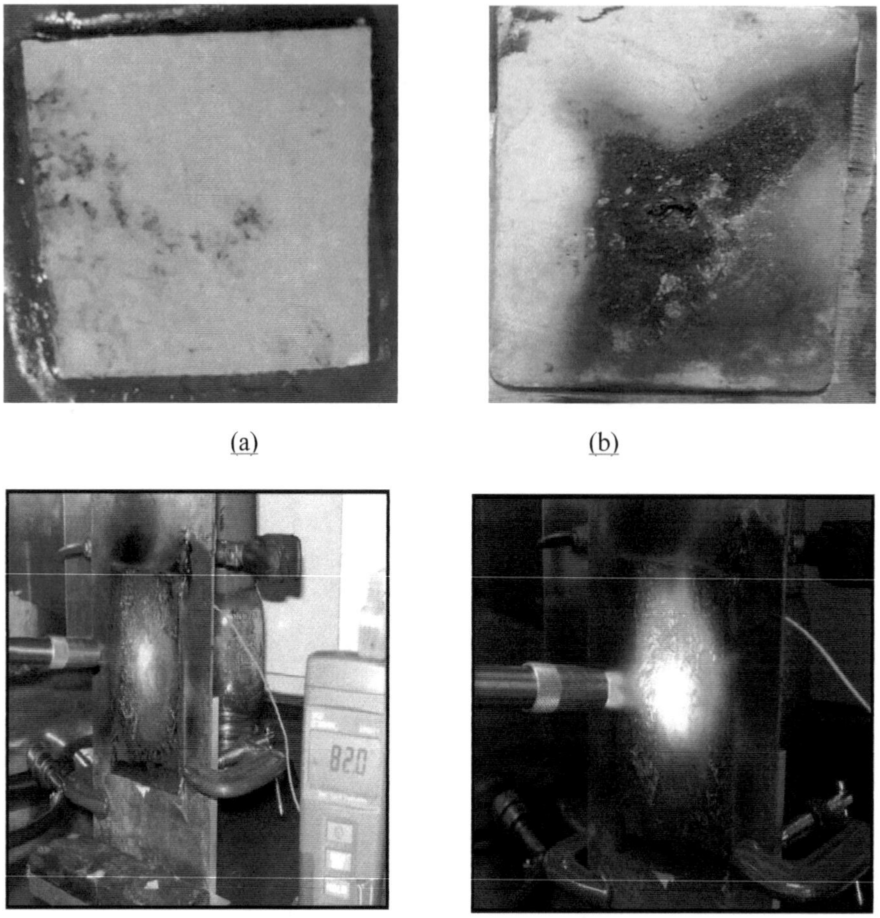

(a) (b)

(c)

Figure 19. (a) Sample image before applying the ablation test.
(b) Sample image after applying ablation test. (c) Samples during ablation test[79].

8.5 Morphological analysis:

Scanning Electron Microscope (SEM) is used to image or photograph the surface and indicates the homogeneity of distribution of fillers of the insulating sample by scanning it in a raster scan pattern with a high-powered electron beam as shown in 'figure 20'. The electrons interact with the sample's atoms to produce signals that contain information on the sample's composition, surface topography, and other properties such as electrical conductivity. SEM can be particularly useful for examining sample morphology since its magnification ranges from 25 x to 250000 x, which is 250 times greater than any other light magnification microscope. The uniformity of the fiber dispersion inside the polymeric matrix was examined using a SEM. For a high-quality image, the insulation samples are covered with a thin film layer of an electrically conducting material during electron radiation to prevent the formation of static charges on the surface of the samples[82].

Figure 20. ZEISS scanning electron microscope[82].

I Kingstone Lesley Jabez shows the curves of SEM of the nitrile rubber, EPDM and EPDM composite in 'figure 21'[75].

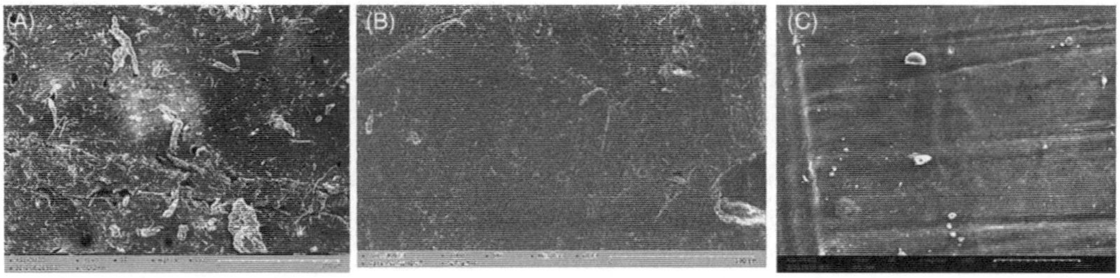

Figure 21. SEM images of (a) EPDM, (b) EPDM composite and (c) Nitrile rubber[75].

9. Empirical equations and theoretical models:

Nowadays, the possibility of prediction of different properties as thermal conductivity (TC) by using the empirical and/or theoretical models is available[10]. A lot of models are used in these days to predict the TC of composite systems based on their compositions. [83, 84].

The major goal of the models is to choose the convenient forms which contains the different agents for fillers and matrix that we need to be in count. In our study, empirical models and theoretical models are the different methods that are used for the prediction of the appropriate properties of the composite matrix[10].

9.1. Theoretical models:

9.1.1. Russell's model

Russell's model indicates that the TC of composites is accompanied by volume fraction of the filler in the matrix. Russell's model interferes the TC of the filler in the matrix, the combination of parallel - series networks are also used for TC calculations are 'as in equation (9)'.

$$\lambda = \lambda_1 \frac{v^{2/3} + \frac{\lambda_1}{\lambda_2}(1 - v^{\frac{2}{3}})}{v^{2/3} - v + \frac{\lambda_1}{\lambda_2}(1 - v^{\frac{2}{3}})} \quad (9)$$

Where λ, λ_1 and λ_2 are the thermal conductivities of the composites, polymer matrix and filler respectively. v (%) is the volume fraction of particles. Extensive estimates of heat conductivity of porous materials are probably not advisable at this point in our knowledge. The formulas will be useful if they explain some well-known facts and points to direction of better insulation manufacture. In the absence of direct experimental data, these or other formulas may function in the absence of more data on the conductivity of refractory materials in the nonporous condition, and when the nature of the pores in a certain brick can be more precisely defined[85, 86].

9.1.2. Maxwell-Eucken's model

Maxwell-Eucken model assumed that particles of filler are diluted and spread in the matrix with no interactions between them [87, 88]. analytical expressions shows that particles have a minimum rate in matrices, matrix will be covered with particles of the filler, the formula 'as in equation (10)' [89].

$$\lambda = \lambda_1 \frac{2\lambda_1+\lambda_2+2v(\lambda_2-\lambda_1)}{2\lambda_1+\lambda_2-v(\lambda_2-\lambda_1)} \tag{10}$$

The TC of composites was found to be near to the Maxwell-Eucken theoretical value when the volume fractions of fillers in the matrix were less than 20%, under the condition of a reduced volume proportion of fillers, Maxwell's model may accurately predict the TC of polymer composites. Because in Maxwell's model, there is no mutual contact of particles, and the experimental TC deviates from Maxwell's model at a high volume fraction[83, 90].

9.1.3. Bruggeman's model:

Prediction model of TC is showed by Bruggeman[91]. and is derivatized by Landauer [92] who can evaluate a lot of heterogeneous materials that is combined by TC, electrical conductivity, thermal diffusivity, magnetic permeability or electric permittivity [83]. The differential effective medium hypothesis emerges from the Bruggeman approach. This theory is applicable to a wide range of materials, including composites, porous materials, nano fluids, aerosols, and so on. It also allows for two- and multi-component systems[89]. The formula is 'as in equation (11)'.

$$1 - v = \frac{\lambda_2-\lambda}{\lambda_2-\lambda_1}\left(\frac{\lambda_1}{\lambda}\right)^{\frac{1}{3}} \tag{11}$$

When the volume fraction of the filler is less than 22.5 percent of the matrix, the TC values fit well, but when the volume fraction is larger, the theoretical model value differs from the experimental results. In other words, the modified Bruggeman's model can forecast the TC of the composite at a higher volume % of fillers than Bruggeman's model[89].

9.1.4. Hamilton-Crosser's model:

The most common issue encountered while developing statistical techniques for TC models is the need for more knowledge about the system's microstructures. The lack of information on the layered structures of the material locally causes the TC values of the materials to be incorrect. To solve this problem, equations based on upper and lower limit information of processes that may exist without comprehensive knowledge of scattered phases must be developed. On the basis of this method, several second-order models have been built [93].The following lower bound 'as in equation (12)' was developed by Hamilton-Crosser [94].

$$\lambda = \lambda_1 \frac{\lambda_2+(n-1)\lambda_1+(n-1)v(\lambda_2-\lambda_1)}{\lambda_2+(n-1)\lambda_1-v(\lambda_2-\lambda_1)} \tag{12}$$

Particles of Fillers with smaller sphericity ψ can increase thermal conduction for the filled polymer, however the specific surface area of filler particles is difficult to achieve by experiment, and sphericity is also difficult to obtain by calculation, thus this model's use is restricted[94].

9.2. Empirical models:

9.2.1. Nielsen's model:

Nielsen et al. pointed out that data from particulate-filled composites over their whole concentration range can be better fitted using Kerner [95] or Halpin-Tsai models [96]. Modified equations 'as in equations ((13),(14))' are used for TC prediction can be expressed in the following way according to the model[97].

$$\lambda = \lambda_1 \frac{AB\mathrm{v}}{1-B\phi\mathrm{v}} \tag{13}$$

$$A = K_e - 1,\ B = \frac{\frac{\lambda_2}{\lambda_1}-1}{\frac{\lambda_2}{\lambda_1}+A}\ ,\quad \phi = 1 + \frac{1-\mathrm{v}_m}{\mathrm{v}_m{}^2}\mathrm{v} \tag{14}$$

A is a constant connected to the generalized Einstein coefficient K_e [98] in these equations, and B is a constant related to the relative conductivity of the components in these equations. The function is connected to the volume fraction of fillers, and V_m is the maximum compressed volume fractions of fillers. The v_m value was introduced by Nielsen et al. to the mutual interaction between particles [89]. It should be emphasized that the model considers the effects of both filler size and shape, but the interface is not considered. Between the matrix and the particles, there is heat resistance [98]. The Lewis-Nielsen model has the advantage of being simple and covering a large range of particle forms and patterns. When v reached a value of more than 40% and was nearing to v_m, The difference between experimental and theoretical values would be larger when v increased[98].

9.2.2. Mixed empirical model:

The parallel and vertical models were used to create the mixed empirical model of polymer composites[98] 'as in equations ((15),(16),(17))'.

$$\lambda = \mathrm{v}_1\lambda_1 + \mathrm{v}_2\lambda_2 + \mathrm{v}_3\lambda_3 + \ldots\ldots \tag{15}$$

$$\frac{1}{\lambda} = \frac{\mathrm{v}_1}{\lambda_1} + \frac{\mathrm{v}_2}{\lambda_2} + \frac{\mathrm{v}_3}{\lambda_3} + \cdots \tag{16}$$

$$\ln\lambda = \mathrm{v}C_2 \ln\lambda_2 + (1-\mathrm{v})\ln(C_1\lambda_1) \tag{17}$$

Where ξ is a constant according to the values in between (+1 and -1). The lower limit of the theoretical value according to the vertical model is -1, and the top limit of the theoretical value according to the parallel model is +1 [89]. Due to the difficulties in determining ξ value via experiment, only a few validated tests for this model were conducted[89].

10. Conclusion:

In this review paper, EPDM is the most used elastomer as it has the lowest density of all the elastomers, allowing it to add a large amount of fillers for the casing of the SRM. EPDM based EHSM had the lowest ablation rate (0.015 mm/s), compared to (0.13 and 0.06 mm/s) for NBR based EHSM and PU based EHSM respectively. According to a comparison of research, nitrile rubber is still employed as an insulation liner for SRM, but not to the extent that EPDM based EHSM is. Because of advantages of PU based EHSM, we may infer that it is a good alternative to EPDM based EHSM, because of their easier processability and reinforcing options and composite propellant compatibility at the moment EPDM reinforced with Kevlar fiber have the best TC values. The best mechanical and thermal properties is when reinforcing the EPDM with Kevlar and nano silica. When the polymeric materials are exposed to the flame the char layer is produced which protects the virgin layers of the polymer, the ablation is also used to evaluate the performance of the prepared EHSM. The empirical equations and theoretical models afford a prediction of different properties as TC and the compositions of materials used in preparation different composites to give the best performance.

References:

[1] S. K. Bapanapalli 2007 Design of an integral thermal protection system for future space vehicles, *University of Florida*

[2] M. Natali and L. Torre 2011 Ablative materials *Encyclopedia of Composites Wiley New York*

[3] A. Donskoy 1996 Elastomeric heat-shielding materials for internal surfaces of missile engines *International journal of polymeric materials* vol. **31** pp. 215-236

[4] B. Boonstra 1979 Role of particulate fillers in elastomer reinforcement: a review *Polymer* vol. **20** pp. 691-704

[5] A. Bouty, L. Petitjean, C. Degrandcourt, J. Gummel, P. Kwaśniewski, F. Meneau, *et al.* 2014 Nanofiller structure and reinforcement in model silica/rubber composites: a quantitative correlation driven by interfacial agents *Macromolecules* vol. **47** pp. 5365-5378

[6] C. Bhuvaneswari, S. Kakade, V. Deuskar, A. Dange, and M. Gupta 2008 Filled Ethylene-propylene Diene Terpolymer Elastomer as ThermalInsulator for Case-bonded Solid Rocket Motors *Defence science journal* vol. **58**, p. 94

[7] D. G. Guillot and A. R. Harvey 2008 EPDM rocket motor insulation *Polymer*

[8] A. F. Ahmed and S. Van Hoa 2010 Optmization and Thermal Conduction Modeling for a Hybrid Reinforced Composite Material for Solid Rocket Motor Insulation *Journal of Materials Science and Engineering* vol. **4**

[9] Q. Zhu, C. Zhang, and W. Yang 2019 The Influence of permeability on the ablation process for an ablative material *Fractals* vol. **27** p. 1950105

[10] N. Kucukdogan, L. Aydin, and M. Sutcu 2018 Theoretical and empirical thermal conductivity models of red mud filled polymer composites *Thermochimica Acta* vol. **665** pp. 76-84

[11] C. M. Blow 1971 *Rubber technology and manufacture edited by C. M. Blow*. London: Butterworths for the Institution of the Rubber Industry

[12] S. Singh, P. Guchhait, G. Bandyopadhyay, and T. Chaki 2013 Development of polyimide–nanosilica filled EPDM based light rocket motor insulator compound: influence of polyimide–nanosilica loading on thermal, ablation, and mechanical properties *Composites Part A: Applied Science and Manufacturing*, vol. **44** pp. 8-15

[13] M. Natali, J. M. Kenny, and L. Torre 2016 Science and technology of polymeric ablative materials for thermal protection systems and propulsion devices: A review *Progress in Materials Science*, vol. **84** pp. 192-275

[14] J. H. Koo and T. Mensah 2017 Novel Polymer Nanocomposite Ablative Technologies for Thermal Protection of Propulsion and Reentry System for Space Application *Nanotechnology Commercialization: Manufacturing Processes and Products* pp. 177-244

[15] C. Bhuvaneswari, M. Sureshkumar, S. Kakade, and M. Gupta 2006 Ethylene-propylene diene rubber as a futuristic elastomer for insulation of solid rocket motors *Defence science journal* vol. **56** p. 309

[16] R. Gallager 1962 Low-density parity-check codes *IRE Transactions on information theory vol. 45* pp. 21-28

[17] M. Natali, M. Rallini, D. Puglia, J. Kenny, and L. Torre 2013 EPDM based heat shielding materials for solid rocket motors: a comparative study of different fibrous reinforcements *Polymer degradation and stability* vol. **98** pp. 2131-2139

[18] S. Kakade, S. Navale, and V. Narsimhan 2003 Studies on interface properties of propellant liner for case-bonded composite propellants *Journal of energetic materials* vol. **21** pp. 73-85

[19] A. Donskoi 1998 Physico-chemistry of elastomer heat-shielding materials *Nova Science Pub Incorporated*

[20] C. L. Whipple and J. A. Thorne 1966 Performance of elastomeric silicones in ablative and space environments *Rubber Chemistry and Technology* vol. **39** pp. 1247-1257

[21] A. S. Deuri, A. K. Bhowmick, R. Ghosh, B. John, T. Sriram, and S. De 1988 Thermal and ablative properties of rocket insulator compound based on EPDM *Polymer degradation and stability* vol. **21** pp. 21-28

[22] J. C. Q. Amado, P. G. Ross, N. B. Sanches, J. R. A. Pinto, and J. C. N. Dutra 2020 Evaluation of elastomeric heat shielding materials as insulators for solid propellant rocket motors: A short review *Open Chemistry* vol. **18** pp. 1452-1467
[23] V. Fitch, N. Eddy, V. Fitch, and N. Eddy 1997 Space Shuttle solid propellant rocket motors, asbestos filled insulation replacement in *33rd Joint Propulsion Conference and Exhibit* p. 2992.
[24] G. Deshpande and M. E. Rezac 2002 Kinetic aspects of the thermal degradation of poly (dimethyl siloxane) and poly (dimethyl diphenyl siloxane) *Polymer Degradation and Stability* vol. **76** pp. 17-24
[25] M. Amin, M. Akbar, and S. Amin 2007 Hydrophobicity of silicone rubber used for outdoor insulation (an overview) *Rev. Adv. Mater. Sci* vol. **16** pp. 10-26
[26] N. Grassie, K. Francey, and I. Macfarlane, "The thermal degradation of polysiloxanes—Part 4: Poly (dimethyl/diphenyl siloxane) 1980 *Polymer Degradation and Stability* vol. **2** pp. 67-83
[27] A. S. Alex, S. Bhuvaneswari, N. Sreenivas, V. Sekkar, and C. Gouri 2019 Short silica fibre-reinforced polymethylsilsesquioxane–phenolic interpenetrating networks: Exploration for use as ablative thermal protection system in aerospace *Polymer Bulletin* vol. **76** pp. 3941-3956
[28] S. Wu, S. Zhang, R. Akram, A. Yasir, B. Wang, Z. Han, *et al* 2019 EPDM-based heat-shielding materials modified by hybrid elastomers of silicone or polyphosphazene *High Performance Polymers* vol. **31** pp. 1112-1121
[29] S. Banaee and S. S. Que Hee 2020 Glove permeation of chemicals: The state of the art of current practice—Part 2. Research emphases on high boiling point compounds and simulating the donned glove environment *Journal of occupational and environmental hygiene* vol. **17** pp. 135-164
[30] D. Schmidt 1968 Chemical propulsion exhaust *Environmental effects on polymeric materials environments* vol. **1** pp. 413-486
[31] H. Bertram 1981 Developments in Acrylonitrile—Butadiene Rubber (NBR) and Future Prospects *Developments in Rubber Technology—2*, ed: Springer pp. 51-85
[32] N. Iqbal, S. S. Iqbal, A. Anwar, A. Sarwar, F. Jabeen, and S. Jabeen 2014 Acrylonitrile Butadiene Rubber/Phenolic Resin Blended Ablative Composites for High Temperature Applications *International Journal of Advanced Technology in Engineering and Science* vol. **2**
[33] R. Muraleekrishnan, S. Bhagawan, A. Raghavan, and S. Rao 1997 IRC. Rocket motor lowdensity insulation based on NBR *Proceedings of the 3rd International Conference* pp. 12-14
[34] K. George, B. P. Panda, S. Mohanty, and S. K. Nayak 2018 Recent developments in elastomeric heat shielding materials for solid rocket motor casing application for future perspective *Polymers for Advanced Technologies* vol. **29** pp. 8-21
[35] F. Xie, T. Zhang, P. Bryant, V. Kurusingal, J. M. Colwell, and B. Laycock 2019 Degradation and stabilization of polyurethane elastomers *Progress in Polymer Science* vol. **90** pp. 211-268
[36] S. Benli, Ü. Yilmazer, F. Pekel, and S. Özkar 1998 Effect of fillers on thermal and mechanical properties of polyurethane elastomer *Journal of applied polymer science* vol. **68** pp. 1057-1065
[37] Q. Jing, J. Y. Law, L. P. Tan, V. V. Silberschmidt, L. Li, and Z. Dong 2015 Preparation, characterization and properties of polycaprolactone diol-functionalized multi-walled carbon nanotube/thermoplastic polyurethane composite *Composites Part A: Applied Science and Manufacturing* vol. **70** pp. 8-15
[38] D. Puglia, L. Valentini, I. Armentano, and J. Kenny 2003 Effects of single-walled carbon nanotube incorporation on the cure reaction of epoxy resin and its detection by Raman spectroscopy *Diamond and related materials* vol. **12** pp. 827-832
[39] Y. Liu, M. Liu, D. Xie, and Q. Wang 2008 Thermoplastic polyurethane-encapsulated melamine phosphate flame retardant polyoxymethylene *Polymer-Plastics Technology and Engineering* vol. **47** pp. 330-334
[40] G. P. Milhomem, A. P. C. de Paula Nunes, A. Andrianov, O. Shynkarenko, and J. Lee 2017 Experimental studies of heat insulation materials for hyprid propellant rocket motors *Cobem*

[41] M. Rallini, I. Puri, L. Torre, and M. Natali 2018 Thermal and ablation properties of EPDM based heat shielding materials modified with density reducer fillers *Composites Part A: Applied Science and Manufacturing* vol. **112** pp. 71-80

[42] B. Serge and F. Xavier 2002 Heat resistance and flammability of high performance fibers: A review *Fire Mater* vol. **26** pp. 155-168

[43] M. Jaramillo, J. Koo, and M. Natali 2014 Compressive char strength of thermoplastic polyurethane elastomer nanocomposites *Polymers for advanced technologies* vol. **25** pp. 742-751

[44] S. S. Iqbal, A. Sabir, A. Islam, S. Z. U. A. Bukhari, M. Yasir, M. A. Bashir, *et al.* 2018 Effect of graphene for ablation study of advanced composite materials for aerospace applications *Key Engineering Materials* pp. 118-125

[45] M. Guo, J. Li, K. Xi, Y. Liu, and J. Ji 2019 Effect of multi-walled carbon nanotubes on thermal stability and ablation properties of EPDM insulation materials for solid rocket motors *Acta Astronautica* vol. **159** pp. 508-516

[46] J. Li, K. Xi, X. Lv, Q. Li, and S.-x. Wang 2018 Characteristics and formation mechanism of compact/porous structures in char layers of EPDM insulation materials *Carbon,* vol. **127** pp. 498-509

[47] L. Yang, Y. Xiao-Jing, M. Jian-Ming, G. Yi-Wen, L. Jiang, L. Qiang, *et al.*, "A volumetric ablation model of EPDM considering complex physicochemical process in porous structure of char layer," *Open Physics,* vol. **15**, pp. 344-353, 2017.

[48] A. El-Dakhakhny, A. Ahmed, M. Rutkevičius, S. El-Marsafy, and E. Abadeer 2018 Effects of aramid fibers and colloidal particle fillers in composite ethylene propylene diene monomer rubber thermal insulators for rocket motor insulation *Journal of Composite Materials* vol. **52** pp. 1989-1995

[49] Z.-S. Zhang, J.-S. Xu, X. Chen, and J. Jiang 2016 Study on thermal and mechanical properties of EPDM insulation *Mechanical Engineering and Control Systems: Proceedings of 2015 International Conference on Mechanical Engineering and Control Systems (MECS2015)* pp. 66-70

[50] A. F. Ahmed and S. V. Hoa, "Thermal insulation by heat resistant polymers for solid rocket motor insulation," *Journal of composite materials,* vol. **46**, pp. 1549-1559, 2012.

[51] S.-M. Sheu, H.-T. Chu, and J.-H. Chou 2011 Experimental study of ablative materials by oxyacetylene flame flow system *Journal of Propulsion and Power* vol. **27** pp. 1108-1113

[52] C. Yezzi and B. Moore 1986 Characterization of Kevlar/EPDM rubbers for use as rocket motor caseinsulators *22nd Joint Propulsion Conference* p. 1489

[53] J. A. Bocchio, M. M. Escobar, and J. C. Quagliano Amado 2020 Ablative properties of polyurethanes reinforced with organoclay *Polymer Engineering & Science* vol. **60** pp. 630-635

[54] J. R. Pinto, N. B. Sanches, M. F. Diniz, R. S. Santos, J. I. D. Oliveira, and R. D. C. L. Dutra 2018 Expanded perlite/cork fillers applied to aerospace insulation materials *Anais da Academia Brasileira de Ciências* vol. **90** pp. 3197-3206

[55] H.-J. Kim, C. K. Kim, and Y. Kwon 2015 Ablation and fire-retardant properties of hydroxyl-terminated polybutadiene-based polyurethane-g-polyhedral oligomeric silsesquioxane composites *High Performance Polymers* vol. **27** pp. 749-757

[56] S. S. Iqbal, T. J. Nadeem Iqbal, and A. Bashir 2018 Exfoliated sodium montmorillonite reinforced elastomeric nanocomposites: Ablation, thermal transport/decomposition/transitions and mechanical aspects *Int J Mater Mech Manuf* vol. **6** pp. 19-25

[57] D. Chukov, A. Stepashkin, A. Salimon, and S. Kaloshkin 2018 Highly filled elastomeric matrix composites: Structure and property evolution at low temperature carbonization *Materials & Design* vol. **156** pp. 22-31

[58] Y. Guan, L.-X. Zhang, L.-Q. Zhang, and Y.-L. Lu 2011 Study on ablative properties and mechanisms of hydrogenated nitrile butadiene rubber (HNBR) composites containing different fillers *Polymer Degradation and Stability* vol. **96** pp. 808-817

[59] Y. Zhao, S. Hu, W. Liu, G. An, Z. Wu, D. Wu, *et al.* 2015 Nitrile butadiene rubber-based heat-shielding insulations for solid rocket motors: Effect of polyimide fibrous reinforcement on the morphology and properties *High Performance Polymers* vol. **27** pp. 153-160

[60] M. Bassyouni, N. Iqbal, S. S. Iqbal, S.-S. Abdel-Hamid, M. Abdel-Aziz, U. Javaid, *et al.* 2014 Ablation and thermo-mechanical investigation of short carbon fiber impregnated elastomeric ablatives for ultrahigh temperature applications *Polymer degradation and stability* vol. **110** pp. 195-202

[61] A. Donskoy 1995 Silicon organic heat-shielding materials for aircraft external surfaces *New Approaches to Polymer Materials; Zaikov, GE, Ed.; Nova Publishers: Hauppauge, NY, USA* p. 5392

[62] M. Biron 2003 *Thermosets and composites*: Elsevier

[63] Y. Ji, S. Han, L. Xia, C. Li, H. Wu, S. Guo, *et al.* 2021 Synergetic effect of aramid fiber and carbon fiber to enhance ablative resistance of EPDM-based insulators via constructing high-strength char layer *Composites Science and Technology* vol. **201** p. 108494

[64] M. H. Al-Saleh and U. Sundararaj, "Review of the mechanical properties of carbon nanofiber/polymer composites 2011 *Composites Part A: Applied Science and Manufacturing* vol. **42** pp. 2126-2142

[65] K. George, M. Biswal, S. Mohanty, S. K. Nayak, and B. P. Panda 2021 Nanosilica filled EPDM/Kevlar fiber hybrid nanocomposites: Mechanical and thermal properties *Materials Today: Proceedings* vol. **41** pp. 983-986

[66] E. Soo Kim, T. H. Lee, S. H. Shin, and J. S. Yoon 2011 Effect of incorporation of carbon fiber and silicon carbide powder into silicone rubber on the ablation and mechanical properties of the silicone rubber-based ablation material *Journal of Applied Polymer Science* vol. **120** pp. 831-838

[67] L. E. Nielsen 1973 Thermal conductivity of particulate-filled polymers *Journal of applied polymer science* vol. **17** pp. 3819-3820

[68] S. K. Jindal, M. Sherriff, M. G. Waters, and T. J. Coward 2016 Development of a 3D printable maxillofacial silicone: Part I. Optimization of polydimethylsiloxane chains and cross-linker concentration *The Journal of prosthetic dentistry* vol. **116** pp. 617-622

[69] B. R. Swarner 2016 Material Evaluation of an Elastomer, Epoxy and Lightweight Concrete Rail Attachment System for Direct Fixation Light Rail Applications *Applied Science and Manufacturing*

[70] A. M. Enew, M. Abd Elfattah, S. R. Fouda, and S. A. Hawash 2021 Effect of aramid and carbon fibers with nano carbon particles on the mechanical properties of EPDM rubber thermal insulators for solid rocket motors application *Polymer Testing* vol. **103** p. 107341

[71] K. George, S. Mohanty, M. Biswal, and S. K. Nayak 2021 Thermal insulation behaviour of Ethylene propylene diene monomer rubber/kevlar fiber based hybrid composites containing Nanosilica for solid rocket motor insulation *Journal of Applied Polymer Science* vol. **138** p. 49934

[72] H.-D. Vosteen and R. Schellschmidt 2003 Influence of temperature on thermal conductivity, thermal capacity and thermal diffusivity for different types of rock *Physics and Chemistry of the Earth, Parts A/B/C* vol. **28** pp. 499-509

[73] D. MacNeil, S. Trussler, H. Fortier, and J. Dahn 2002 A novel hermetic differential scanning calorimeter (DSC) sample crucible *Thermochimica acta,* vol. **386** pp. 153-160

[74] S. A. Kumar and T. S. Narayanan 2002 Thermal properties of siliconized epoxy interpenetrating coatings *Progress in organic coatings* vol. **45** pp. 323-330

[75] I. K. L. Jabez, U. Das, M. Ramalingam, and S. Anne 2021 Evaluation of Ethylene Propylene Diene Terpolymer based insulation against Nitrile rubber based insulation for large Composite Rocket Motor Casing *Polymers for Advanced Technologies* vol. **32** pp. 111-122

[76] C. Jiang, Y. Jin, and J. Gao 2021 Ablation and thermal insulation properties of silicone rubber-polyarylacetylene-carbonwoven laminates for solid rocket motor *Plastics, Rubber and Composites* pp. 1-8
[77] Y. Gao, N. R. Choudhury, J. Matisons, U. Schubert, and B. Moraru 2002 Part 2: inorganic–organic hybrid polymers by polymerization of methacrylate-substituted oxotitanium clusters with methyl methacrylate: thermomechanical and morphological properties *Chemistry of materials* vol. **14** pp. 4522-4529
[78] M. Asghar, N. Iqbal, S. S. Iqbal, M. Farooq, and T. Jamil 2016 Ablation and thermo-mechanical tailoring of EPDM rubber using carbon fibers *Journal of Polymer Engineering* vol. **36** pp. 713-722
[79] E. ASTM 2008 285-80. Standard test method for oxyacetylene ablation testing of thermal insulation materials *Annu. B. ASTM Stan,* vol. **15**
[80] R. A. Vaia, G. Price, P. N. Ruth, H. T. Nguyen, and J. Lichtenhan 1999 Polymer/layered silicate nanocomposites as high performance ablative materials *Applied Clay Science* vol. **15** pp. 67-92
[81] D. L. Schmidt 1969 Ablative polymers in aerospace technology *Journal of Macromolecular Science—Chemistry* vol. **3** pp. 327-365
[82] D. McMullan 1995 Scanning electron microscopy 1928–1965 *Scanning* vol. **17** pp. 175-185
[83] K. Pietrak and T. S. Wisniewski 2015 A review of models for effective thermal conductivity of composite materials *Journal of Power Technologies* vol. **95** p. 14
[84] J. Xu, B. Gao, and F. Kang 2016 A reconstruction of Maxwell model for effective thermal conductivity of composite materials *Applied Thermal Engineering* vol. **102** pp. 972-979
[85] H. Russell 1935 Principles of heat flow in porous insulators *Journal of the American Ceramic Society* vol. **18** pp. 1-5
[86] R. Progelhof, J. Throne, and R. Ruetsch 1976 Methods for predicting the thermal conductivity of composite systems: a review *Polymer Engineering & Science* vol. **16** pp. 615-625
[87] B. Zhu, H. Zheng, J. Wang, J. Ma, J. Wu, and R. Wu 2014 Tailoring of thermal and dielectric properties of LDPE-matrix composites by the volume fraction, density, and surface modification of hollow glass microsphere filler *Composites Part B: Engineering* vol. **58** pp. 91-102
[88] L. Qian, X. Pang, J. Zhou, J. Yang, S. Lin, and D. Hui 2017 Theoretical model and finite element simulation on the effective thermal conductivity of particulate composite materials *Composites Part B: Engineering* vol. **116** pp. 291-297
[89] M.-x. Shen, Y.-x. Cui, J. He, and Y.-m. Zhang 2011 Thermal conductivity model of filled polymer composites *International Journal of Minerals, Metallurgy, and Materials* vol. **18** pp. 623-631
[90] B. Gao, J. Xu, J. Peng, F. Kang, H. Du, J. Li, *et al.* 2015 Experimental and theoretical studies of effective thermal conductivity of composites made of silicone rubber and Al2O3 particles *Thermochimica Acta* vol. **614** pp. 1-8
[91] C. Subramaniam 2019 On the Effective Thermal Conductivity of Nanofluids With Fractal Aggregation *Journal of Heat Transfer* vol. **141**
[92] R. Landauer 1952 The electrical resistance of binary metallic mixtures *Journal of Applied Physics* vol. **23** pp. 779-784
[93] D. Bigg 1995 Thermal conductivity of heterophase polymer compositions *Thermal and electrical conductivity of polymer materials* pp. 1-30
[94] R. L. Hamilton and O. Crosser 1962 Thermal conductivity of heterogeneous two-component systems *Industrial & Engineering chemistry fundamentals* vol. **1** pp. 187-191
[95] E. Kerner 1956 The elastic and thermo-elastic properties of composite media *Proceedings of the physical society. Section B* vol. **69** p. 808
[96] J. H. Affdl and J. Kardos 1976 The Halpin-Tsai equations: a review *Polymer Engineering & Science* vol. **16** pp. 344-352
[97] A. Einstein 1906 Zur theorie der brownschen bewegung *Annalen der physik* vol. **324** pp. 371-381

[98] L. Ren, K. Pashayi, H. R. Fard, S. P. Kotha, T. Borca-Tasciuc, and R. Ozisik 2014 Engineering the coefficient of thermal expansion and thermal conductivity of polymers filled with high aspect ratio silica nanofibers *Composites Part B: Engineering* vol. 58 pp. 228-234

Implementation of Mass Integration Design Concept for Material Preservation in Acrylonitrile Industry

Nessren M Farrag[1] **and Ayat O Ghallab**[2]

[1]The British University in Egypt, Department of Chemical Engineering, Egypt,11837
[2]Cairo University, Department of Chemical Engineering, Egypt, 12613
E-mail Address: nessren.farrag@bue.edu.eg and ayatossama@hotmail.com

Abstract. The industrial production has suffered from huge amounts of waste discharges that contain valuable unreacted components as well as major contaminants that affect the environment. Just like heat integration, mass integration evolves around recycling the discharge streams back into the process to increase the overall plant performance and decrease the fresh feedstock resources used as well as extracting the valuable by-products whether in product streams or waste streams; thus, decreasing the overall cost and increasing the production value of outcomes. The study has tackled mass integration graphically on an existing production process, acrylonitrile, and evaluated the performance of mass-exchange networks. It also suggested improvements can be recommended in terms of composition driving forces inside mass exchangers. The mass network developed improved both the production yield and the product quality. In this research, the plant, located in Baroda complex, India, on a 500 m^2 area, has been studied for mass integration using mass separating agents. The mass separating agents used were water and benzene. A mass exchange network design has been conducted in the feasible design regions to achieve maximum mass recovery and mass targets of each mass exchanger. Three extra units were added for Acetonitrile removal, Hydrogen cyanide removal, and Acrylonitrile extraction.

Keywords: Acrylonitrile, Mass Exchange Network, Pinch Analysis, Mass Transfer

1. Introduction

Acrylonitrile (ACN) is one of the key monomers used in many production processes to produce day-to-day use products such as fibers, resins, intermediate chemical for certain products, synthetic rubber etc. It has been produced in many different methods. The "SOHIO Acrylonitrile Process" in the 1950's by the Standard Oil Company, now part of British Petroleum has witnessed promising production route for acrylonitrile. The process was deemed a great success as for the first time it provides an economical one-step conversion from raw materials to products through gas phase ammoxidation process over a catalyst [1].

Acrylonitrile is also used in fatty acid and vegetable oils extraction, unsaturated oil hydrocarbon extraction, plastic molding, removal of tars, phenols, and petroleum coloring substances, wool resin purification, steroid recrystallizing, DNA synthesis and sequence of peptides solvents, as reaction promotion medium, copper removal and refinement, chlorinated solvent stabilizer. It is also used in chromatographic examination of high-pressure fluids, transition-metal component and catalyst component, production of perfumes and solvents [2].

Acrylonitrile can be produced either by the ammoxidation of propylene, heating of ethylene cyanohydrin in the presence of a dehydrating catalyst, by reacting acetylene and hydrogen cyanide in the presence of an aqueous solution of cuprous chloride catalyst or by the reaction of acetaldehyde-hydrogen cyanide reaction. Although a variety of chemical pathways have been shown and several techniques have been devised to acrylonitrile, the current paper focuses mainly on propylene ammoxidation. Standard Oil of Ohio (SOHIO) process fluid bed method is used in the vast majority of situations. This method is followed by approximately 90% of total acrylonitrile production plants. Reaction is too highly selective and fast for acrylonitrile production without the need for excessive recycling efforts [3].

$$2CH_2{=}CH{-}CH_3 + 2NH_3 + 3O_2 \rightarrow 2CH_2{=}CH{-}C{\equiv}N + 6H_2O \quad (1)$$

In this paper, it is desired to implement the graphical mass integration method developed by Farrag et al. [4] on the SOHIO plant, as designed by Hansora in order to efficiently and economically separate acrylonitrile, and extract the useful side products as well as handle the toxic ones. Then the obtained results would be used to design mass exchanger units in order to be implemented in Hansora's process for separation. The new renovated plant will be evaluated accordingly and compared to Hansora's initially designed plant performance.

A strong motive for considering Mass Exchange network (MEN) was to deal with environmental concerns through using mass separating agents (MSAs) to remove the environment endangering species. It uses Pinch Analysis depending on the target and supply compositions of different streams. The streams are classified as lean or rich according to whether the composition of considered material in the stream is low or high, accordingly. Using these data, the cost of MSAs needed can be determined. Even though HI and MI is very similar, yet MI is far more complicated as it counts for more factors that affect the integration, such as the direct interactions between the streams in the mass exchangers and the MSA regeneration of considerations [5].

Mass exchanger networks (MEN) work on matching the rich streams with lean streams and MSAs, accounting for their varying target and supply compositions, for the required load of mass exchange as per the targeted values using the equilibrium relations. Very often when using MSAs the flow of MSAs is determined according the requirements of the network to remove the predetermined mass load from the system [6].

Gadalla has developed the first of its kind graphical tool that can be used to assess existing as well as new network designs against the targets determined. The tool has PA principles integrated in a graphical tool that represents the compositions of lean streams versus it equivalent rich streams, as per equilibrium data, on the diagram. This is suggested to be used when revamping a plant with an already existing MEN to further improve the performances [7].

Using the composition driving forces, Farrag et al. developed another graphical tool to assess diminish losses, both environmental and economic. This method illustrates the lean streams compositions against the driving force for each individually considered exchanger, which is shown in forms of straight lines on the graphical tool. Since the costs are somewhat related to the driving forces as it affects the design considerations for the number of plates within the unit, the chosen method of assessing the performance was through the use of the composition driving force, within the mass exchange unit. The plot could be divided into different sections in order to determine the optimum section to place a mass exchanging unit and accordingly achieve the best possible solution. This method could also be used to asses existing plant in order to increase the resources sustainability and integration efficiency [8].

Farrag et al. further developed a graphical approach to evaluate the efficiencies of existing plants, and design new ones through using composition driving forces. This tool, using pinch concepts, plots the lean stream compositions, between the inlets and outlets, against the corresponding composition driving force for each point in the mass exchanging unit. This method can also be used on new developing MEN or already existing ones to assess the performance. Moreover, it was uncovered that this method could be openly and efficiently used to evaluate linear systems with dilute streams The decision of matches was taken as to pick the parameters that allow the most efficient target achievement such as low flows, highest driving forces and minimum costs [4].

In this project, the aim will be to implement the graphical mass integration method developed by Farrag et al. on the plant outlined in the previous section, as designed by Hansora in order to efficiently and economically separate ACN, and extract the useful side products as well as handle the toxic ones. Then the obtained results would be used to design mass exchanger units in order to be implemented in Hansora's process for separation. The new renovated plant will be evaluated accordingly and compared to Hansora's initially designed plant performance.

2. Synthesis of mass exchanger network

In the standard SOHIO process, air, ammonia, and propylene are introduced into a fluidized- bed catalytic (bismuth phosphomolybdate in silica) reactor operating at 0.3-2 bar pressure and 400-510°C. Ammonia and air are fed to the reactor in slight extra of stoichiometric proportions because extra ammonia drives the reaction closer to integration and air continually regenerates the catalyst. An important feature of the process is the high conversion of reactants on a once-through basis with only just a few seconds habitation time. The heat generated from the exothermic reaction is recovered via a waste-heat-recovery boiler. Figure 1 shows the standard SOHIO process.

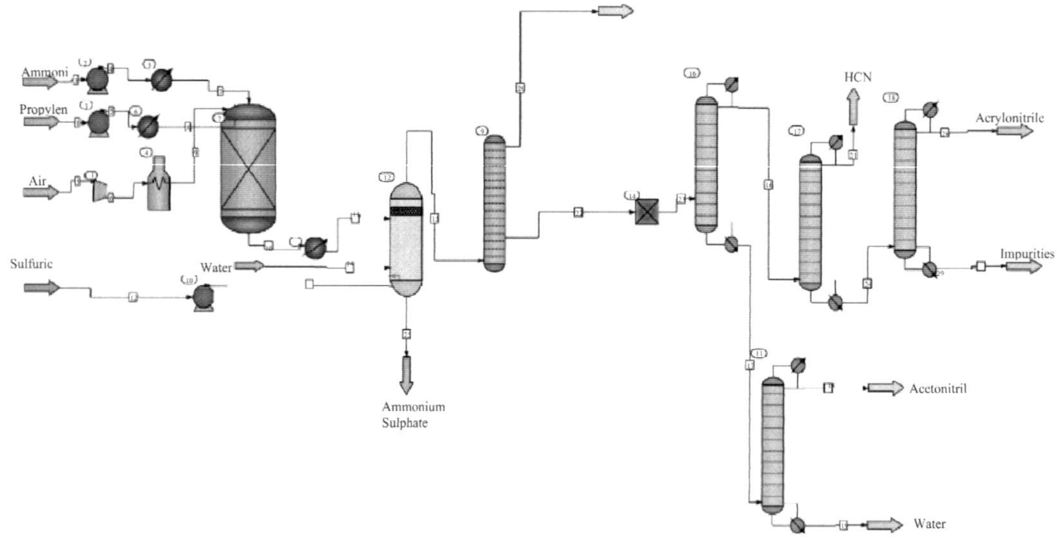

Figure 1. The standard SOHIO process.

Water acidified with sulphuric acid, which may retrieve by crystallization, removes unreacted ammonia, as aqueous ammonia sulphate. The product stream then flows through opposite current water absorber-stripper to reject inert gases and recover reaction products. The reaction products mixture of acrylonitrile, acetonitrile, and water and then is sent to a fractionator to remove hydrogen cyanide. The final two steps involve the drying of the acrylonitrile stream and the final distillation to route out heavy ends. The fiber-grade acrylonitrile obtained from the process is 99+% pure. Several fluid-bed catalysts have been used since the inception of the SOHIO ammoxidation process [9].

Hansora has used the developed process in order to construct a plant that processes 70,000 tons per year of acrylonitrile product on a basis of 24 hours per day, and 330 days per year. The reactor processes ammonia, air and propylene yielding high conversion of propylene on a catalyst based on bismuth molybdenum. This in turn produces some side products such as toxic HCN, acetonitrile, acrylic acid, etc. [10].

2.1. Introduction

The idea and concept of the mass exchanger networks (MENs) was established by El- Halwagi (2017) [6]. The mass exchanger applies a direct interaction that occurs between two streams (rich stream and lean stream) to separate the components or species which are targeted to be removed from the rich streams. All strategies of the mass transfer which are adsorption, absorption, stripping, extraction, leaching, and ion exchange can be utilized in the mass exchange method. The leans streams (mass separating agents) could be found in the process itself or it could be achieved externally and thus they will be considered as external utilities, where these lean streams can be an adsorbent, a stripping agent, a solvent, an ion exchange resin. The rich streams are products streams or any type of wastes.

2.2. Composition driving force graphical design

The transfer of the targeted components needed to be removed from the rich streams and sending them to the lean streams, are ruled by the composition driving force, moreover, to achieve the theoretical plates number and the composition driving force. The total cost estimation is related to the composition driving force. For sizing the equipment's of mass exchanger, the graphical approach of the composition driving force is considered.

A straight line that has a slope which is linked with the rich and lean streams mass flows, refer to the unit of mass exchanging. The length of this straight line is proportional to the transferred mass capacity that occurred inside the exchanger unit. The mass exchangers are analyzed and located according to the pinch analysis rules. [4, 8].

2.3. Mass integration analysis and network design

The compositions of the rich stream and the lean stream are shown in figure 2. The lines of the lean streams are constructed as vertical straight lines which begins from the (x_{eqv}) point that rely on the x-axis. The lines of the rich streams are straight lines that intercepts at the points (y) where they are constructed by an inclination of 45° and they carry slopes equal to (−1) as shown in figure 2. The construction of the rich stream-lines begins from the y-axis where the $\Delta C = y$ that corresponds to $x_{eqv} = 0$ on the x-axis, and the end of the rich stream-lines lies on the x-axis on the points which is $y = x_{eqv}$ and $\Delta C = 0$. The composition of the lean pinch is constructed parallel to the lean streams and the same for the composition of the rich pinch where it is constructed parallel to the parallel to the rich streams.

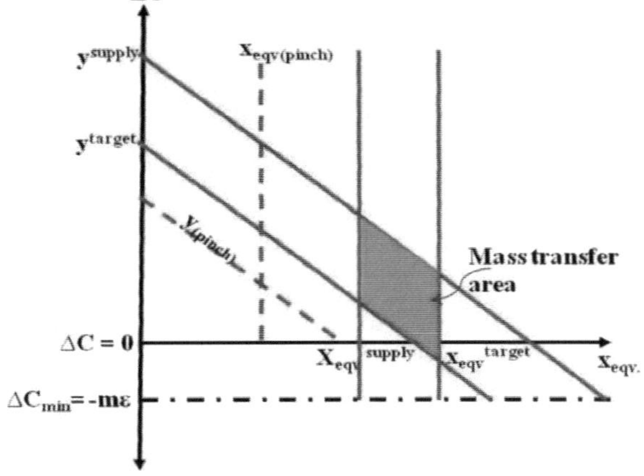

Figure 2. Streams representation for network design.

Journal of Physics: Conference Series 2305 (2022) 012038 doi:10.1088/1742-6596/2305/1/012038

The mass networks are described in the (CDF) graphical representation by means of three dimensions. Across the inlet or exit of the exchanger the composition driving force is displayed by the first coordinate, the second coordinate indicates the equivalent compositions of the lean streams, and the third one shows the compositions of the rich streams that passes through the exchanger [4, 8].

2.4. *Process description and rigorous modelling*

The process described by Hansora is for the purpose of increasing an industrial capacity of an Indian acrylonitrile plant by the ammoxidation of propylene mechanism. The initial process achieves the product separation through the consecutive distillation units in order to obtain the required product specifications as well as remove the contaminants in the product mixture [10].

In order to achieve the required processes efficiently, the efficient use of streams within the system to remove the components desired, in the main aim of reducing the process use of external resources. The limit is achieved within the materialistic and utilities use; in other words, this limits the reagents as well as heating system uses.

The project is going to do so by first removing the key contaminants from the product stream which is rich in some components that must be removed; acetonitrile and hydrogen cyanide. Acetonitrile is easily removed by the use of water, which demonstrates highly miscible characteristics [11]. Also as water demonstrates limited solubility of acrylonitrile, this shows that the water may be used as a solvent [12].

The simulation, illustrated in figure 3, was done in order to extract the main data to be used in mass exchanger network design.

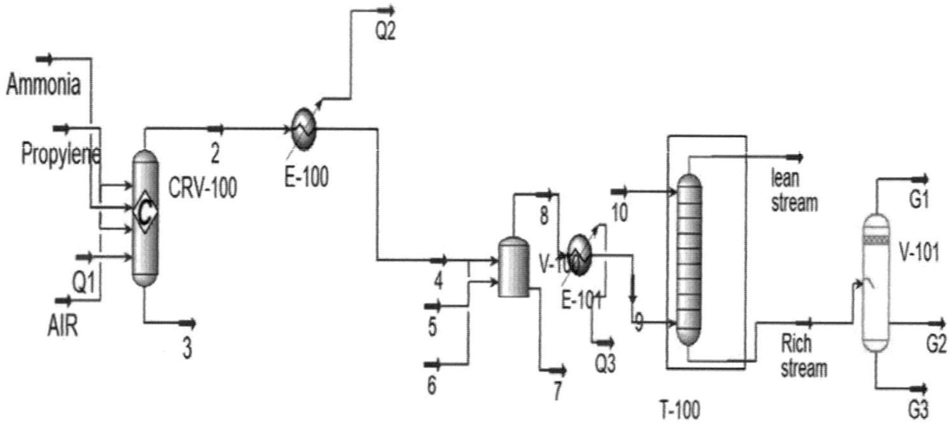

Figure 3. Acrylonitrile plant simulation.

2.5. *Network Design Calculations*

It is assumed in the separation system that $\varepsilon=0.001$, and that the product stream is at 1 atm and 40°C. The product (Rich) stream is separated into three different streams using three phase separators. The gas stream (G_1) flows at 896.4 kg/h, the liquid stream (G_2) flows at 5243 kg/h, while the aqueous stream (G_3) flows at 35540 kg/h. The key contaminants will be removed using separating agent to reduce the duty of the used separators.

2.6. *Choice of Separating agents*

The plant provides a stream of waste water that has been used to remove unreacted ammonia with sulphuric acid dissolved within the mixture. Due to the miscibility of acetonitrile and HCN in water, the stream will be used to target these key concentration components in the streams. Looking at the mass separation coefficient dependent on which the separation quality isdetermined, it was chosen to carry out the processes at 40°C and 1 atm for water as a separating agent.

Although HCN is toxic and would contaminate water, water has been chosen as a separating agent to remove the component from the mixture in order to allow for better extraction of the desired product from all other by-products and unreacted materials.

Also due to the miscibility of acrylonitrile in benzene, this is the separating agent decided to be used for the extraction of acrylonitrile from the streams for which key contaminants have been removed before the process of acrylonitrile extraction.

3. Results for the integrated process

3.1. Acetonitrile removal

3.1.1. Pinch determination

The mass exchange from the rich stream to the lean stream is determined across the different ranges of the mass fraction of the acetonitrile content. This depends on the total flow rate of the rich streams for each range. The mass exchanged from each of the streams is calculated for each range using the following equation.

$$\Delta M_i = \Delta y_i \times G_i \tag{2}$$

For the range y = 0.0005 to 0.0040

$\Delta M_{y=0.0005 \text{ to } 0.0040} = (0.0040 - 0.0005) \times 5243 = 18.3505$ kg/h

For the range y = 0.0040 to 0.0048

$\Delta M_{y=0.0040 \text{ to } 0.0048} = (0.0048 - 0.0040) \times (896.4 + 5243) = 4.9115$ kg/h

For the range y = 0.0048 to 0.0191

$\Delta M_{y=0.0048 \text{ to } 0.0191} = (0.0191 - 0.0048) \times 5243 = 74.9749$ kg/h

The pinch plot will be plotted using the data found in table 1.

Table 1. Pinch stream plot data.

y (wt/wt)	ΔM (kg/h)
0	0
0.0005	0
0.004	18.3505
0.0048	23.2620
0.0191	98.2369

For T = 40°C and pressure = 1 atm, $m = 0.07715$

$$y = m(x + \varepsilon) + b \tag{3}$$

b is neglected as it only affects the position not the behavior of mass transfer in thenetwork.

$$x_{eq} = y - m\varepsilon \tag{4}$$

Table 2 shows the lean stream compositions on different scales.

Table 2. Lean stream composition on different scales.

Point	Mass fraction of acrylonitrile in lean stream, x (wt/wt)	Corresponding mass fraction in rich stream, y (wt/wt)	x_{eq} (wt/wt)
Supply	0.0007	1.3116x10^{-6}	5.4005x10^{-5}

Target	0.3000	0.0232	0.0232

According to figure 4, the pinch points upon which the network is classified were determined to illustrate the acrylonitrile pinch diagram.

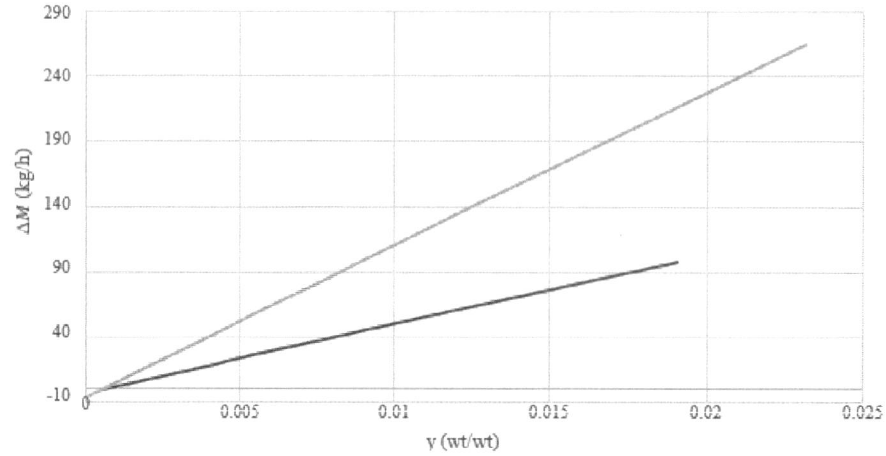

Figure 4. Acrylonitrile pinch diagram.

The point of intersection which also the pinch point is at $y = 0.0005$ and $\Delta M = 0$ kg/h, on the equivalent scale this represents an acetonitrile composition of $x = 0.005$ anda composition of $x_{eq} = 0.0004285$.

3.1.2. Streams compositions

The acetonitrile is present in significant concentrations in order to affect the quality of extraction of acrylonitrile using benzene, it is also very highly soluble. The acrylonitrile composition (y_{in}) is 0.0048, 0.0191 and 0.0054 in (G_1), (G_2) and (G_3), respectively.

Acetonitrile is infinitely soluble in water. The outlet mass fraction will be limited to 0.3 in order to dodge the azeotrope effect during extraction of acetonitrile. The target mass fraction of acrylonitrile (y_{out}) is 0.004 and 0.0005 in (G_1) and (G_2), respectively. The concentrations of acetonitrile are going to be reduced in the rich streams using the process lean stream.

3.1.3. Mass exchanger network design

The acetonitrile mass exchanger network includes two streams, gas phase stream (G_1) and liquid phase stream (G_2). The gas phase (G_1) has a supply composition of $y_{in} = 0.0048$ and the liquid phase (G_2) has a supply composition of $y_{in} = 0.0191$. The gas phase (G_1) has a targeted composition of $y_{out} = 0.004$ and the liquid phase (G_2) has a targeted composition of $y_{out} = 0.0005$. This is presented as follows on the mass exchanger network in figure 5. The lean stream is represented by two vertical lines at the equivalent x mass fraction.

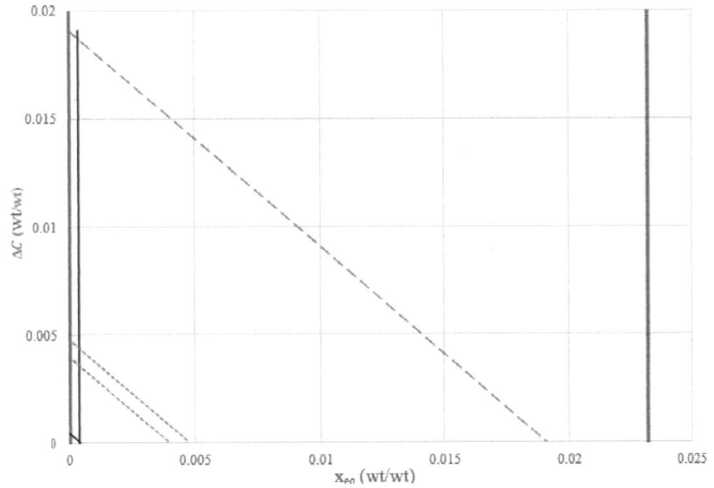

Figure 5. Acetonitrile mass exchanger network before applying mass exchangers.

Acetonitrile mass exchanger network above allows for two mass exchangers one for each of the rich streams G_1 = 896.4 kg/h and G_2 = 5243 kg/h.

$$L' = \frac{L}{m} = \frac{900.7}{0.07715} = 11674.65 \text{ kg/h} \tag{5}$$

Using the following equation to get the slope of each exchanger line in the network

$$S = \frac{L'}{G} - 1 \tag{6}$$

For G_1: $S_{G1} = \frac{11674.65}{896.4} - 1 = 12.0394$

For G_2: $S_{G2} = \frac{11674.65}{5243} - 1 = 1.2267$

Using S_{G1} = 12.0394, starting at the intersection of the pinch line with the line representing the outlet stream $\Delta C = 0.004 - x_{eq.}$ The exchanger line stops at ΔC = 0.0043 and x_{eq} = 0.00048. For the second stream, using S_{G2} = 1.2267, starting at the intersection of the pinch line with the line representing the outlet stream $\Delta C = 0.0005 - x_{eq.}$ The exchanger line stops at ΔC = 0.0103 and x_{eq} = 0.0087760. Figure 6 illustrates the acetonitrile mass exchanger network with mass exchangers.

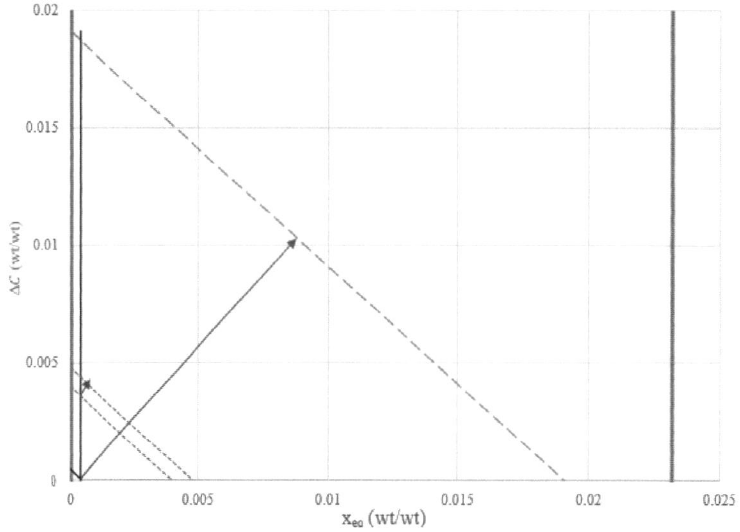

Figure 6. Acetonitrile Mass Exchanger Network with mass exchangers.

Using the equation derived for the number of theoretical plates in terms of the slope, the mass exchanger can be designed

$$NTP = \frac{\ln\left[\frac{1}{S+1}\left(\frac{S\Delta y}{\Delta C_{lean}+m\varepsilon}\right)+S+1\right]}{\ln(S+1)} \quad (7)$$

For Gas Stream the NTP is 1.006, with 60% plate efficiency, thus the actual number of plates is almost 2.

From Hysys, the gas stream is evaluated to have a volumetric flow rate of 2 m^3/h and a lean stream flowrate of 0.7379 m^3/h, based on which, the unit will be designed. The velocity is taken as 1.2 m/s in a vessel with diameter 1 m. For a residence time of 20 min, height will be 2.16 m.

For Liquid Stream: NTP is 8, with 60% plate efficiency, thus the actual number of plates is almost 14.

From Hysys, the gas stream is evaluated to have a volumetric flow rate of 6.4257 m^3/h and a lean stream flowrate of 0.7379 m^3/h, based on which, the unit will be designed. The velocity is taken as 1.2 m/s in a vessel with diameter 1 m. For a residence time of 20 min, height will be 10 m.

3.2. Hydrogen cyanide removal

A process for the removal of hydrogen cyanide from gas streams which comprises contacting the gas stream in the presence of water in an amount at least sufficient to hydrolyze the contained hydrogen cyanide with a catalyst consisting of at least one alkali metal hydroxide deposited on a support selected from the group consisting of alumina, silica, alumina silica and the zeolites in which the alkali metal hydroxide deposited on the support as alkali metal is present in an amount of from about 0.1 to about 10 percent in on the total weight of the alkali metal and the support at a temperature from about 200°F to about 800°F for a time sufficient to hydrolyze a substantial quantity of the hydrogen cyanide present in the gas stream to ammonia.

3.2.1. Pinch determination

The mass exchange from the rich stream to the lean stream is determined across the different ranges of the mass fraction of the hydrogen cyanide content. This depends on the total flow rate of the rich streams for each range. The mass exchanged from each of the streams is calculated for each range using equation (1).

For the range y = 0.01 to 0.02

$\Delta M_{y=0.01\ to\ 0.02} = (0.02 - 0.01) \times 5243 = 52.43$ kg/h

For the range y = 0.02 to 0.0417
$\Delta M_{y=0.0040 \text{ to } 0.0048} = (0.0417 - 0.02) \times (896.4 + 5243) = 133.225$ kg/h
For the range y = 0.0417 to 0.0787
$\Delta M_{y=0.0048 \text{ to } 0.0191} = (0.0787 - 0.0417) \times 5243 = 193.991$ kg/h
The pinch plot will be plotted using the data found in table 3.

Table 3. Hydrogen Cyanide pinch streams plot data.

y (wt/wt)	ΔM (kg/h)
0	0
0.01	0
0.02	52.43
0.0417	185.655
0.0787	379.646

For T = 40°C and pressure = 1 atm, $m = 0.07715$

$$y = m\,(x + \varepsilon) + b \tag{8}$$

b is neglected as it only affects the position not the behavior of mass transfer in the network.

$$x_{eq} = y - m\varepsilon \tag{9}$$

Table 4 shows the lean stream composition on different scales.

Table 4. Lean stream compositions on different scales.

Point	Mass fraction of HCN in lean stream, x (wt/wt)	Corresponding mass fraction in rich stream, y (wt/wt)	x_{eq} (wt/wt)
Supply	0.0007	$7.5023\text{x}10^{-5}$	$3.0892\text{x}10^{-5}$
Target	0.01	$4.8544\text{x}10^{-4}$	$1.31\text{x}10^{-6}$

The range at which the lean stream will fall on the pinch diagram does not intersect with the rich composite curve thus, the pinch point cannot be determined. Also, since the richstreams have equivalent compositions higher than that the water stream has, mass transfer is impractical. Thus, an external lean stream, or an external mass separating agent will be used.

The water supply will be assumed at a negligible concentration of hydrogen cyanide of $x_{in} = 0.001$ which is equivalent to $x_{eq} = 0.0004$. The target composition would be limited to $x_{out} = 0.025$ which is equivalent to $x_{eq} = 0.0114$.

3.2.2. Streams compositions

The hydrogen cyanide is also present in considerable concentrations that would affect the quality of extraction of acrylonitrile from the mixture. The hydrogen cyanide has thefollowing compositions in the three streams as listed below in table 5.

Table 5. Hydrogen Cyanide mass fractions.

Stream	Symbol	Flow rate (kg/h)	Mass fraction of HCN, y_{in}
Gas steam	G_1	896.4	0.0787
Liquid stream	G_2	5243	0.0471
Aqueous stream	G_3	35540	0.0201

Hydrogen cyanide is soluble in water. The outlet mass fraction will be reduced in orderto allow for the better extraction of acrylonitrile from the product streams. Water used to removethe hydrogen cyanide must be treated later on as hydrogen cyanide is toxic, so it should not be thrown out directly. Table 6 illustrates the supply and target mass fraction of Hydrogen Cyanide.

Table 6. Supply and target mass fraction of hydrogen cyanide.

Stream	Symbol	Supply mass fraction of HCN, y_{in}	Target mass fraction of HCN,y_{in}
Gas steam	G_1	0.0787	0.01
Liquid stream	G_2	0.0471	0.02

The concentrations of hydrogen cyanide are targeted to be removed from the rich streams (0.004 and 0.0005 in Gas and Liquid streams respectively) using the process lean stream.

The hydrogen cyanide mass exchanger network includes two streams, gas phase stream (G_1) and liquid phase stream (G_2). The gas phase (G_1) has a supply composition of $y_{in} = 0.0787$ and the liquid phase (G_2) has a supply composition of $y_{in} = 0.0417$. The gas phase (G_1) has a targeted composition of $y_{out} = 0.01$ and the liquid phase (G_2) has a targeted composition of $y_{out} = 0.02$. This is presented as follows on the mass exchanger network in figure 7.

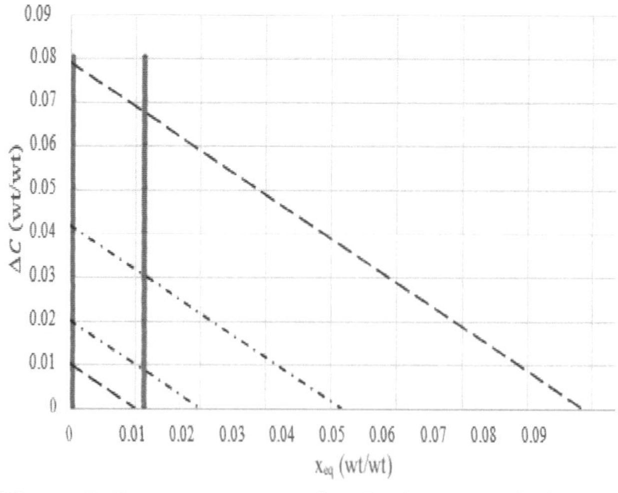

Figure 7. Streams representation for the removal of hydrogen cyanide mass exchanger network.

3.2.3. Mass exchanger network design

Since no pinch conditions can be determined due to the unknown flowrate of external mass separating agent, the mass exchangers will be placed on the network as to remove all the required contaminant, by using the constraints set on the concentration of the streams flowing into and out of the mass exchanger. Hydrogen cyanide mass exchanger network is developed for two mass exchangers one for each of the rich streams G_1 = 896.4 kg/h and G_2 = 5243 kg/h.

Using the following equation to get the slope of each exchanger line in the network:

$$S = \frac{\Delta y}{\Delta x_{eq}} - 1 \quad (10)$$

For G_1: $S_{G1} = \frac{0.0687}{0.011} - 1 = 6.2455$

For G_2: $S_{G2} = \frac{0.0217}{0.011} - 1 = 0.9727$

Matching the two points through which the mass exchange happens for each stream thefollowing network is obtained as a result. Figure 8 illustrates the hydrogen cyanide mass exchanger network.

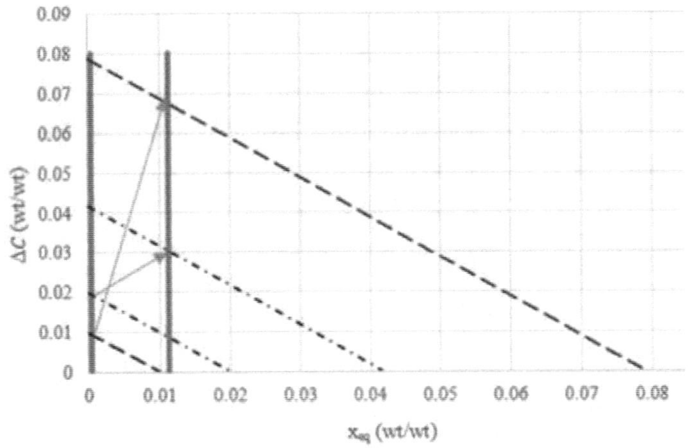

Figure 8. Hydrogen cyanide mass exchanger network.

For gas stream: NTP is 1.3, with 60% plate efficiency, thus the actual number of plates is almost 3.

From Hysys, the gas stream is evaluated to have a volumetric flow rate of 2 m^3/h and an external lean stream can be calculated to be 808.3735 kg/h or 0.9308 m^3/h (assuming benzene density of 868.5). The velocity is taken as 1.2 m/s in a vessel with diameter 1 m. For a residence time of 20 min, height will be 2.24 m. For liquid stream: NTP is 3, with 60% plate efficiency, thus the actual number of plates is almost 6.

From Hysys, the gas stream is evaluated to have a volumetric flow rate of 6.4257 m^3/h and an external lean stream can be calculated to be 34737.15 kg/h or 39.9967 m^3/h (assuming benzene density of 868.5). The velocity is taken as 1.2 m/s in a vessel with diameter 1 m. For a residence time of 20 min, height will be 11.75 m. For aqueous Stream: NTP is 1.12, with 60% plate efficiency, thus the actual number of plates is almost 2.

From Hysys, the gas stream is evaluated to have a volumetric flow rate of 37.0089 m^3/h and an external lean stream can be calculated to be 32049.972 kg/h or 36.9027 m^3/h (assuming benzene density of 868.5). The velocity is taken as 1.2 m/s in a vessel with diameter 1 m. For a residence time of 20 min, height will be 8 m.

3.3. Acrylonitrile extraction

3.3.1. Pinch determination

The mass exchange from the rich stream to the lean stream is determined across the different ranges of the mass fraction of the acrylonitrile content. This depends on the total flow rate of the rich streams for each range. The mass exchanged from each of the streams is calculated for each range using equation (1).

For the range y = 0.0005 to 0.0048

$\Delta M_{y=0.0005\text{ to }0.0048} = (0.0048 - 0.0005) \times (5243 + 896.4) = 26.39942$ kg/h

For the range y = 0.0048 to 0.077

$\Delta M_{y=0.0048\text{ to }0.077} = (0.077 - 0.0048) \times 5243 = 378.544$ kg/h

For the range y = 0.077 to 0.1209

$\Delta M_{y=0.077\text{ to }0.1209} = (0.1209 - 0.077) \times (5243 + 35540) = 1790.3737$ kg/h

For the range y = 0.1209 to 0.8285

$\Delta M_{y=0.1209\text{ to}0.8285} = (0.8285 - 0.1209) \times 5243 = 3709.9468$ kg/h

For T= 30°C and pressure = 1 atm, $m = 0.9018$

$$y = m\,(x + \varepsilon) + b \tag{11}$$

b is neglected as it only affects the position not the behavior of mass transfer in thenetwork.

$$x_{eq} = y - m\varepsilon \tag{12}$$

Table 7 shows the lean stream compositions on different scales.

Table 7. Pinch stream plot data.

Point	Mass fraction of HCN in lean stream, x (wt/wt)	Corresponding mass fraction in rich stream, y (wt/wt)	x_{eq} (wt/wt)
Supply	0.0000	0.0009	0
Target	0.1250	0.1136	0.1127

3.3.2. Mass exchanger network

The acrylonitrile mass exchanger network is to include three streams, gas phase stream (G_1), liquid phase stream (G_2) and aqueous phase stream (G_3). The gas phase (G_1) has a supply composition of $y_{in} = 0.0048$, the liquid phase (G_2) has a supply composition of $y_{in} = 0.8285$ and the aqueous phase (G_3) has a supply composition of $y_{in} = 0.1209$. The gas phase (G_1) has a targeted composition of $y_{out} = 0.0005$ and the liquid phase (G_2) has a targeted composition of $y_{out} = 0.0005$ and the aqueous phase (G_3) has a supply composition of $y_{in} = 0.0770$.

Acrylonitrile mass exchanger network above allows for three mass exchangers one for each of the rich streams $G_1 = 896.4$ kg/h, $G_2 = 5243$ kg/h and $G_3 = 35540$ kg/h.

Using the following equation to get the slope of each exchanger line in the network:

$$S = \frac{\Delta y}{\Delta x_{eq}} - 1 \tag{13}$$

After calculating the slope of each exchanger line, figure 9 is developed to represent the acrylonitrile mass exchanger network after implementing mass integration.

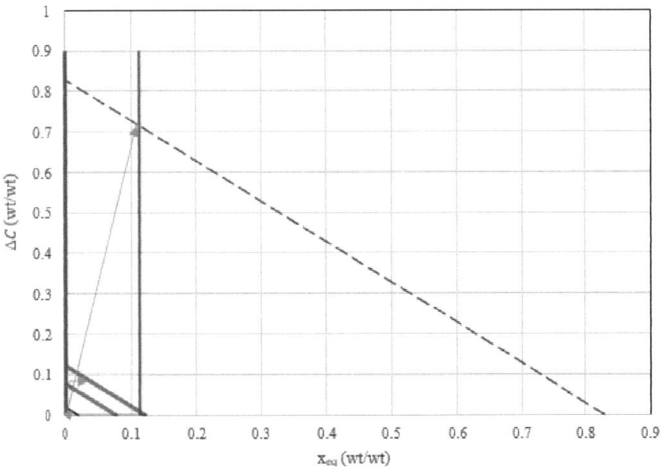

Figure 9. Acrylonitrile mass exchanger network.

3.3.3. *Mass exchanger design*

For gas stream: NTP is 1.12, with 60% plate efficiency, thus the actual number of plates is almost 2.

From Hysys, the gas stream is evaluated to have a volumetric flow rate of 2 m^3/h and an external lean stream can be calculated to be 808.3735 kg/h or 0.9308 m3/h (assuming benzene density of 868.5). The velocity is taken as 1.2 m/s in a vessel with diameter 1 m. For a residence time of 20 min, height will be 2.24 m. For liquid stream: NTP is 3, with 60% plate efficiency, thus the actual number of plates is almost 6.

For liquid stream: NTP is 3, with 60% plate efficiency, thus the actual number of plates is almost 6.

From Hysys, the gas stream is evaluated to have a volumetric flow rate of 6.4257 m^3/h and an external lean stream can be calculated to be 34737.15 kg/h or 39.9967 m^3/h. The vessel with diameter 1.5 m, for a residence time of 20 min, height will be 11.75 m.

For Aqueous stream: NTP is 1.12, with 60% plate efficiency, thus the actual number of plates is almost 2.

From Hysys, the gas stream is evaluated to have a volumetric flow rate of 37.0089 m^3/h and an external lean stream can be calculated to be 32049.972 kg/h or 36.9027 m^3/h. The vessel with diameter 1.5 m, for a residence time of 20 min, height will be 8 m.

4. *Conclusion*

This study addressed the implementation of mass exchange design concept for the acrylonitrile production plant based in India located on a 500 m^2 land. The mass exchanger networks were designed using graphical design approach. The product stream was split into three different streams; gas, liquid, and aqueous phase streams. Two mass separating agents were utilized, which were benzene and water. The mass separating agents were used to remove the contaminants attached with the main production stream. These contaminants were mainly acetonitrile and hydrogen cyanide. The acetonitrile was in the gas stream, liquid stream, and aqueous stream. The concentrations of acetonitrile and hydrogen cyanide was reduced in the rich stream (main product stream) using the process mass separating agents (water and benzene). For the acetonitrile removal, the mass load integrated was 0.72 kg/h and 97.52 kg/h for gas and liquid streams, respectively. For its removal no surplus loads of external mass separating agents were needed, the process lean stream was sufficient. For the hydrogen cyanide removal, the mass load integrated was 61.58 kg/h and 113.77 kg/h for gas and liquid streams, respectively. For its removal surplus loads of external mass separating agents were added for more efficient mass separation. Finally, the extraction of acrylonitrile took place over three cascaded units where the maximum mass transfer loads reached were 3.85 kg/h, 434.12 kg/h and 156 kg/h.

References

[1] Rezaie F, Pirouzfar V and Alihosseini A 2020 Technical and economic analysis of acrylonitrile production from polypropylene. *Thermal Sci. and Eng. Prog.* **16** 1-9.

[2] Dittrich E 2015 *The Sustainable Laboratory Handbook, Design, Equipment and Operation.* (John Wiley & Sons).

[3] Cespi D, Passarini F, Neri E, Vassura I, Ciacci L and Cavani F 2014 Life Cycle Assessment comparison of two ways for acrylonitrile production: the SOHIO process and an alternative route using propane. *J. Cleaner Prod.* **69** 17-25

[4] Farrag N, Kamel D, Ghallab A., Gadalla M. and Fouad M 2021 Graphical Design and Analysis of Mass Exchange Networks Using Composition Driving Forces. *South Afr. J. Chem. Eng.* **36** 94-104.

[5] El-Halwagi M and Spriggs D 1998 Solve design puzzles with mass integration. *Chem. Eng. Prog.* **94** 25–43.

[6] El-Halwagi M 2017 Synthesis of Mass-Exchange Networks. *Sus. Design through Process Integ.* (Elsevier: IChem, United States) p.163–97.

[7] Gadalla M 2015 A new graphical-based approach for mass integration and exchange network design. *Chem. Eng. Sci.* **127** 239–52.

[8] Farrag N, Kamel D, Ghallab A, Gadalla M and Fouad M 2018 A novel graphical approach for mass exchange networks using composition driving force. *Chem. Eng. Trans.* **70** 259–64.

[9] Grover A, Sharma M, Patel D and Mitra S, Manufacture of Acrylonitrile 2012 *SRM University Report.*

[10] Hansora D 2014 Industrial manufacturing process of Acrylonitrile. *LAMBERT Academic Publishing.*

[11] Creager S 2007 Solvents and supporting electrolytes. Handbook of Electrochemistry. (Elsevier) Cynthia Zoski p 57-72.

[12] Lu Q, Weng Z, Zhou S, Huang Z and Pan Z 2002 Effects of acrylonitrile water solubilityon limiting conversion and copolymer composition in suspension polymerization. *European Polymer Journal* **38** 1337–42.

Spectrophotometric and biological studies for the interaction of rifaximin with Co, Cu, Fe and vitamin C

A. S. El-Kholany[*1], H. A. Hafith[1] and N. M. Gaber[1]

[1] Medical Laboratory Department, Faculty of Applied Health Sciences Technology, Pharos University in Alexandria, Alexandria, Egypt

* dr.ahmed.salah@pua.edu.eg

Abstract. The interaction of Rifaximin with Iron, Copper, Cobalt and Vitamin C was studied by Uv-Vis. spectrophotometric technique at different pH and time intervals. The maximum absorbance bands for Co, Fe and Cu rifaximin complexes were 446 nm, 397 nm and 241 nm respectively. The molar ratio method was used to the most stable stoichiometries for the prepared complexes. The yellow color that formed from the interaction of rifaximin with vitamin C, allowed its determination spectrophotometry. The antimicrobial activities of Iron, Copper and Cobalt complexes with Rifaximin against standard human pathogenic Staphylococcus aureus (Gram-positive) and Escherichia coli (Gram negative) using agar diffusion method were studied. Iron and Cobalt complexes were found to exhibit higher zone of inhibition against the tested pathogens while copper inhibit completely the effect of Rifaximin. The biological study for Rifaximin - Vitamin C mixture, vitamin C decreased the effect of the antibiotic on staphylococcus, and this inhibition effect increased on E-coli.

Keywords: rifaximin, biological activity, Spectrophotometric, vitamin C

1. Introduction:

Rifaximin is a semi-synthetic non-absorbable antibiotic derived from rifamycin and with a broad-spectrum activity against Gram-positive, Gram-negative, aerobic and anaerobic microorganisms proposed to act on the gut microenvironment [1]. Virtually unabsorbable, which minimizes systemic exposure and adverse events is the main advantage of rifaximin over similar antibiotics [2]. Animal and human studies demonstrate that the systemic absorption of rifaximin, after oral administration, is negligible, being less than 0.4% of the administered dose. Rifaximin is a topical antibiotic which exerts its antibacterial activity against microorganisms that cause gastrointestinal infections, however now no longer systemic infections; indeed, rifaximin, launched withinside the gastrointestinal. tract, is excreted, primarily, in feces as

unchanged drug [3]. Due to its localized activity, rifaximin has a favorable side-effect and a low ability for drug interactions [4]. As less than 0.1% of its oral dose is absorbed, rifaximin administration is associated to a very low side-effect incidence [5].
Rifaximin, as the other rifamycins, exerts its antibacterial activity by binding to the β-subunit of bacterial DNA-dependent RNA poly-merase, thus inhibiting bacterial RNA synthesis [6]. Beyond its antibacterial activity, such as alteration of virulence, prevention of gut mucosal adherence and bacterial translocation, modulates microbial virulence [7] and epithelial cell function[8], it's been proven to have both bactericidal and bacteriostatic properties [9]. Rifaximin reveals a vast spectrum of in vitro and in vivo activity [10]. It is very effective for a variety of clinical uses and initially approved for the treatment of traveler's diarrhea caused by noninvasive strains of E. coli and hepatic encephalopathy (due to its inhibition of ammonia-producing enteric bacteria and consequent reduction of circulating ammonia in patients with cirrhosis) [11]. Further medical proof brought about the approval of using rifaximin for the treatment of diarrhea-fundamental irritable bowel syndrome (IBS)[12]. Moreover, withinside the medical practice, rifaximin is regularly prescribed for different gastrointestinal disorders, along with inflammatory bowel disease (IBD), small intestinal bacterial overgrowth (SIBO), and diverticular disease, due to its theoretical functionality to modulate the intestinal microbiota [13]. It also can be considered as an optional treatment for recurrent C. difficile infection [14]. It's a potential treatment for Crohn's disease too [15].
The kind of the rifaximin actions, further to its antimicrobial consequences, might also additionally probably provide an explanation for the consequences of the drug in kind of diseases and syndromes. All those residences make rifaximin a terrific candidate to deal with numerous gastrointestinal diseases. Rifaximin exerts a few anti-inflammatory consequences with best a minimum impact on the general composition of the intestine microbiota. One of the most intriguing aspects of this drug is that it can shorten the duration of infection with minimal effects on the colonic bacterial flora [16]. In fact, the most important characteristic of rifaximin, is its efficacy without significant changes in gut microbiota composition [17]
Some observations recommended that rifaximin can be greater powerful in the remedy of infections withinside the small intestine, due to the better awareness of bile on this vicinity than in the colon, as bile acids appear to solubilize rifaximin increasing the drug's antimicrobial effect[18].

Iron is an essential element for almost all living organisms as it participates in a wide variety of metabolic processes, including oxygen transport, deoxyribonucleic acid (DNA) synthesis, and electron transport. However, iron can form free radicals that concentration in body tissues needed to be tightly regulated because excessive amount may cause tissue damage. Disorders of iron metabolism are encompassing a broad spectrum of diseases ranging from anemia to iron overload, and possibly to neurodegenerative diseases. For many years, nutritional interest focused on the role of iron in oxygen transport. and hemoglobin formation. Nowadays, although low iron intake and/or bioavailability are responsible for most anemia [19]. The total content of iron in transferrin (≈3 mg) [20].

Cobalt is an essential trace element for the human body. The organic form is a necessary component of vitamin B12 and plays a very important role in forming amino acids and some proteins in nerve cells, and in creating neurotransmitters that are indispensable for correct functioning of the organism. Its excess or deficiency will influence it unfavorably. Salts of cobalt have been applied in medicine in the treatment of anemia [21]. The daily intake of Cobalt in adults is 3 µg its content is 0.012 µg [22]

Copper has very important role in metabolism, because it allows many critical enzymes to function properly [23]. Copper is essential for maintaining the skin, epithelial, blood vessels and connective tissue throughout the body. Cu has an important role in the manufacturing of hemoglobin, myelin, melanin and it additionally keeps thyroid gland functioning normally [24]. The overall quantity of copper in the body ranged from75 to 100 milligrams [25].

2. Experimental:

2.1. Preparation of stock solutions:

0.005M of vitamin c with high purity which obtained from IEC International Egyptian Company. 0.005 M of Rifaximin solution was prepared by using a mixture of ethanol and double distilled water with ratio 1:3. All metallic ions (Iron, Copper and Cobalt) were prepared by dissolving the ultra pure metal salts 99%) ($FeSO_4.7H_2O$, $CuCl_2.2H_2O$ and $CoCl_2.6H_2O$) with concentration 0.01 M using double distilled water.

2.2. Spectrophotometric studies:

The absorbance of mixtures of different concentration ranged from (0.5 to 5 x 10^{-3} M) of metallic ions (Fe^{2+}, Co^{2+} and Cu^{2+}) interacted with 0.005 M of rifaximin ($C_{43}H_{51}N_3O_{11}$, M. Wt. 785.9) and 0.0025 M of vitamin C with different concentration of rifaximin were measured by UV-1800 spectrophotometer (SHIMADZU) Japan

2.3. The biological activity:

The experiments at the antimicrobial interest had been achieved as defined through the Clinical and Laboratory Standards Institute [26]. Antimicrobial activity of the synthesized antibiotic metal complex become examined towards the human pathogenic bacteria Staphylococcus aureus and Escherichia coli through using agar diffusion method. The wells had been made through the use of sterile cork borer (5mm). Wells become created into every Petri-plates. A different volume of the synthesized (25-50 μl) become used to evaluate it as antibacterial efficiency [27].

Mueller Hinton Agar (MHA) become hired for bacterial pathogens, MHA medium containing (g/l): Beef (dehydrated infusion from 300) 2, Casein hydro lysate 17.5, Starch 1.5, agar 17 and dissolved in 1L distilled water with very last pH 7.3 ± 0.1 at 25°C and sterilized through autoclaving at (121°C) for 15 mins for sterilization [28]. The inoculated plates had been incubated at 37°C (24 hr).

3. Results and Discussion:

3.1. Spectrophotometric studies: -

The importance of Rifaximin for spectrophotometric and visual colorimetric is the determination of Fe^{2+}, Co^{2+} and Cu^{2+}. The colored complexes are formed between Fe^{2+}, Co^{2+} and Cu^{2+}. Rifaximin by preparing serial concentrations for Fe^{2+}, Co^{2+}, Cu^{2+} and Rifaximin. This leads to investigate suitable sensing materials and immobilization techniques for these low-cost commercially available indicators in optical sensing. The absorption spectra of pure rifaximin, pure iron, copper and cobalt salts were given, Figures (1-4).

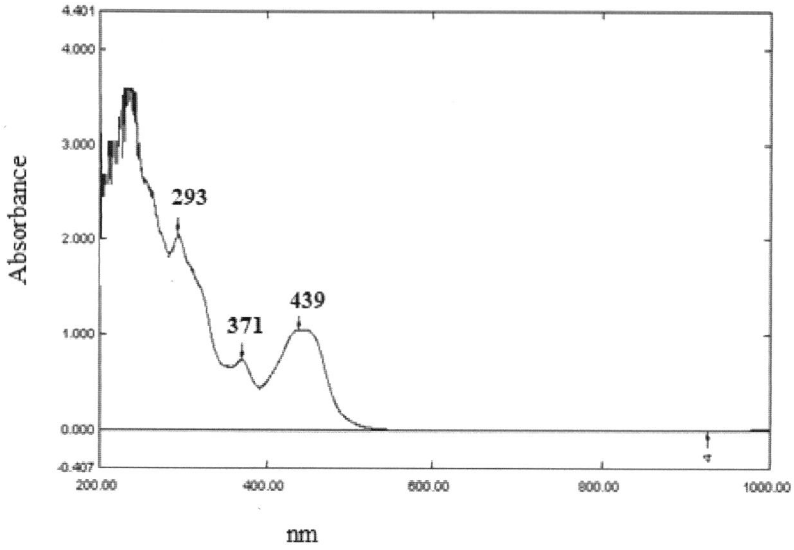

Figure 1. The adsorption spectrum for pure Rifaximin.

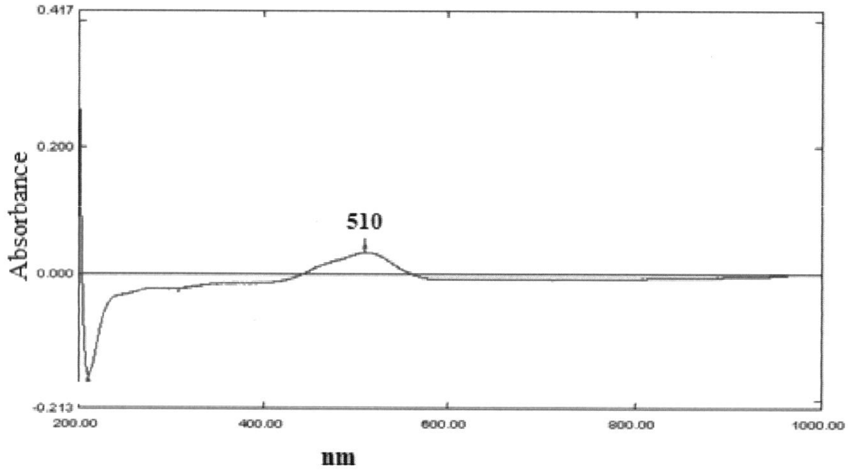

Figure 2. The adsorption spectrum for Co^{+3} ions.

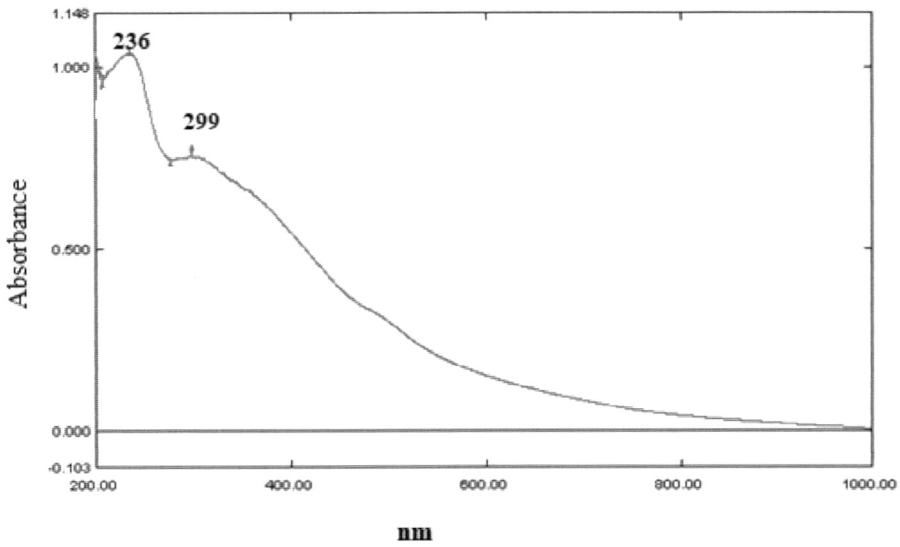

Figure 3. The adsorption spectrum for Fe^{+2} ions.

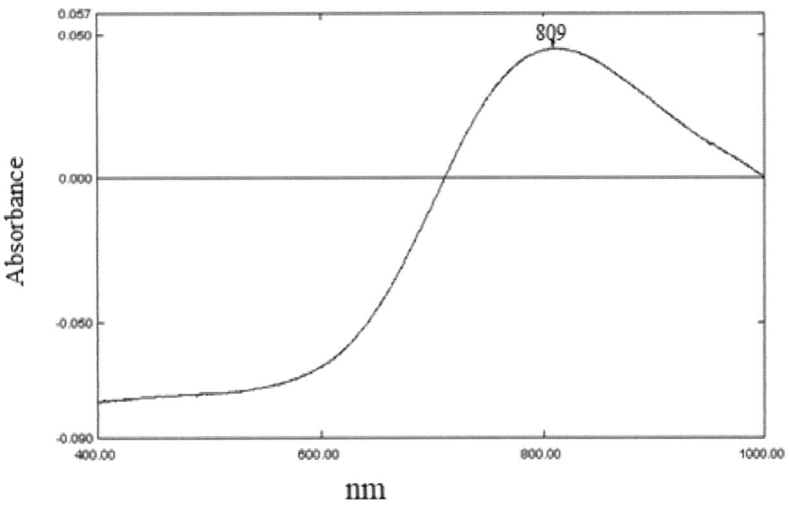

Figure 4. The adsorption spectrum for Cu^{+2} ions.

The complexation of metal ions with ligand, molar ratio, effect of time and pH have been studied.

3.1.1. Fe-Rifaximin complex:

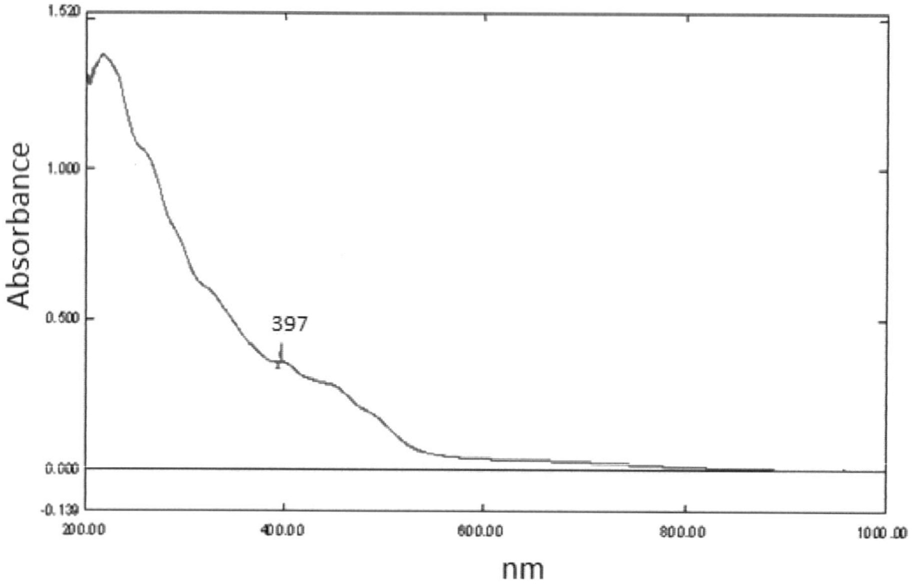

Figure 5. Absorption spectrum for the reaction products of Fe^{+2}- Rifaximin in the UV-Vis range.

The absorption spectrum for the reaction of Fe^{+2}- Rifaximin with different wavelengths, figure (5), showed maximum absorbance at 299 nm.

Table 1. Different concentration of Fe ion with $5*10^{-3}$M Of Rifaximin at 397 nm

Concentration of Fe^{2+} ion $*10^{-3}$ M	0.500	1.000	1.500	2.000	2.500	3.500	4.000	4.500	5.000
Absorbance	0.261	0.191	0.160	0.266	0.372	0.304	0.301	0.417	0.362

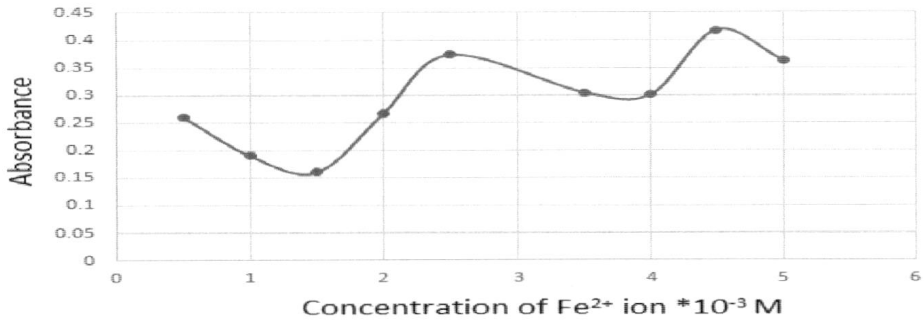

Figure 6. Absorption spectrum for the reaction products of different concentrations of Fe^{+2} with 0.005M of Rifaximin.

From this curve the maximum absorbance showed at 4.5 concentration is 0.417M m**olar ratio**

Table 2. Different absorption for molar ratio between different concentration of Fe ion and $2.5*10^{-3}$M Rifaximin.

Molar ratio	0.4	0.6	0.8	1	1.4	1.6	1.8	2
Absorbance	0.19	0.16	0.27	0.37	0.30	0.30	0.41	0.36

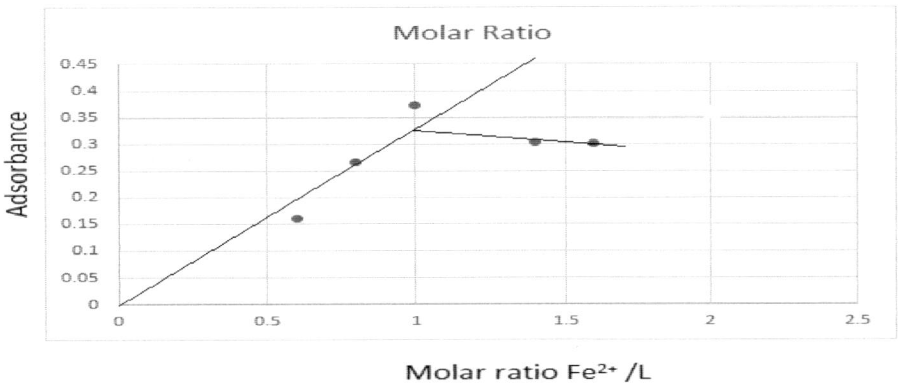

Figure 7. The absorbance spectrum for molar ratio between different concentrations of Fe^{+2} ion and $2.5*10^{-3}$M of Rifaximin.

-From the mole ratio method, it's found that the most stable formation of complex at ratio (1:1).

Effect of time on Fe (II)-Rifaximin Complex at Wavelength: 397 nm

Table 3. effect of time on absorbance for $2.5*10^{(-3)}$ Molar concentration of Fe (II) – rifaximin Complex

Time(min)	5	10	15	20
Absorbance	0.141	0.146	0.146	0.146

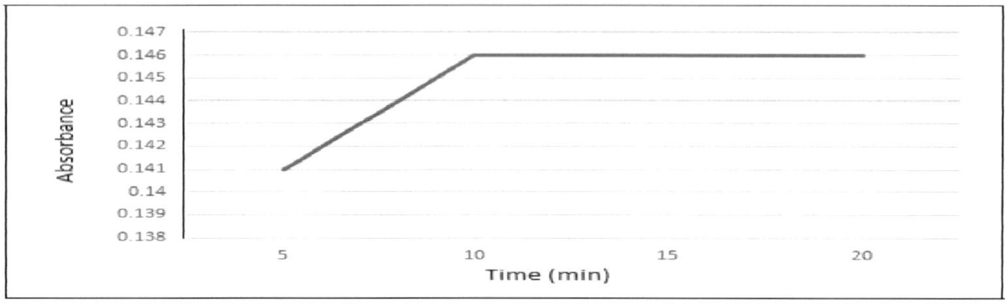

Figure 8. Spectra of effect of time on absorbance of Fe (II)- Rifaximin Complex at Wavelength= 397 nm

The absorbance increased by time then became constant after 10 min.

Table 4. Effect of pH on Fe (II)-Rifaximin Complex at 397 nm

pH	2.5	5.6	11
Absorbance	0.112	0.139	0.114

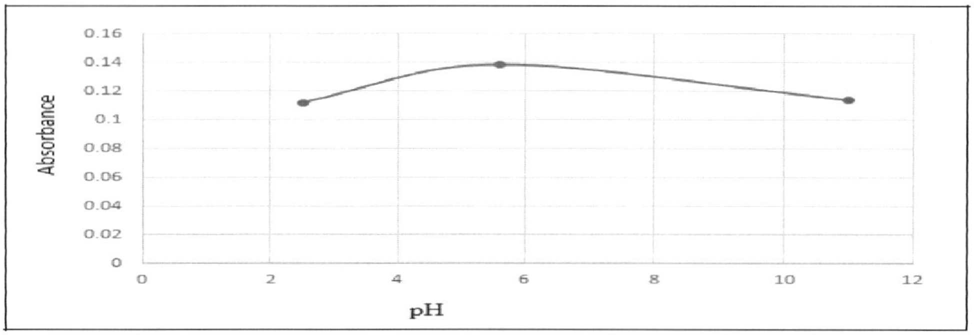

Figure 9. Spectra of effect of pH on absorbance of Fe(II)-Rifaximin Complex at wavelength= 397 nm.

From Table The maximum absorbance is at pH= 5.6.

From studding the effect of time and pH its recommended to proceeds this analysis at pH 5.6 after 10 minutes.

3.1.2. Cu-Rifaximin complex:

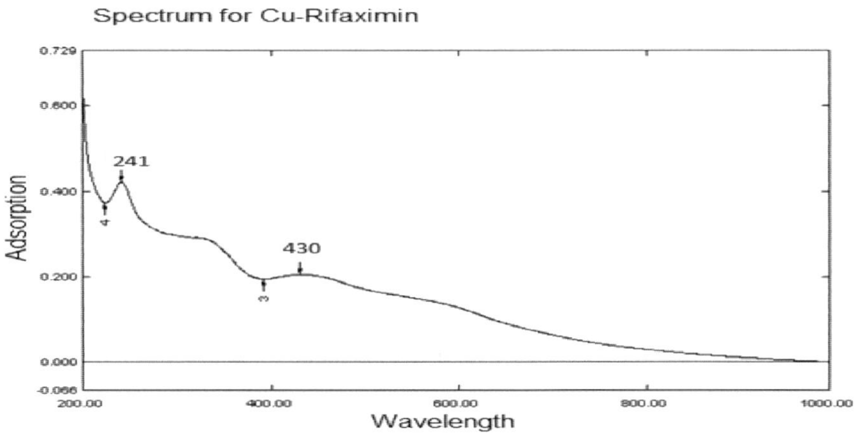

Figure 10. The absorption spectrum for the reaction of Cu^{+2}- Rifaximin with different wavelengths.

The spectrophotometric study for Cu-Rifaximin complex at different wavelengths, Figure (10), it is noticed that the maximum absorbance at 241 nm.

Absorption for different concentrations of Cu-Rifaximin complex at wavelength (241) nm

Table 5. Different absorption for molar ratio between different concentration of Cu^{2+} ion and $2.5*10^{-3}$M Rifaximin.

Concentration of copper ion $*10^{-3}$M	0.500	1.000	2.000	2.500	3.500	4.000	5.000
Absorption	0.466	0.312	0.443	0.117	0.273	0.196	0.500

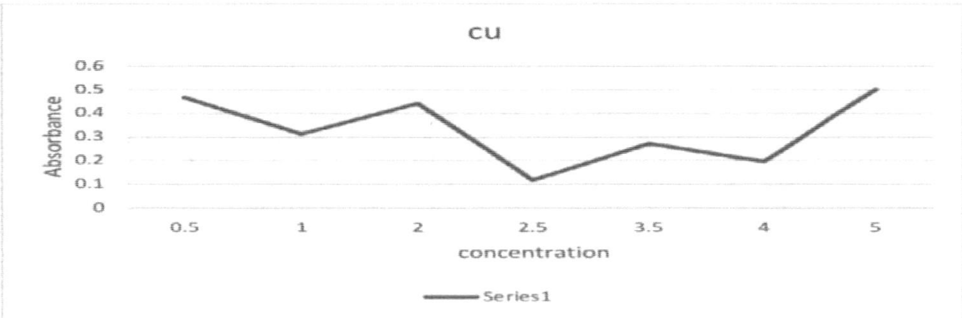

Figure 11. The absorption spectrum for the reaction of different concentrations of Cu^{+2} with 0.0025M of Rifaximin at wavelength= 241 nm.

-Molar Ratio

Table 6. Different absorption for molar ratio between different concentrations of copper ion and $2.5*10^{-3}$M of Rifaximin.

Molar ratio	0.2	0.4	0.8	1	1.4	1.6	2
Absorption	0.47	0.31	0.44	0.18	0.27	0.196	0.5

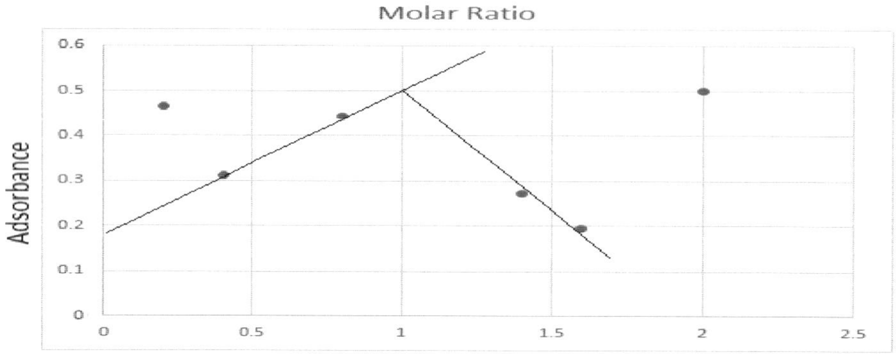

Figure 12. The absorption spectrum for molar ratio between different concentrations of copper ion and $2.5*10^{-3}$M of Rifaximin.

From figure 12 and table 4, it's noticed that the maximum absorption was (0.5) at molar ratio (1:1) at wavelength 240 nm.

-Effect of pH with Copper-Rifaximin Complex

Table 7. Effect of pH on Cu (II)-Rifaximin Complex at 241 nm

pH	2.5	5.6	11
Absorbance	0.187	0.288	0.424

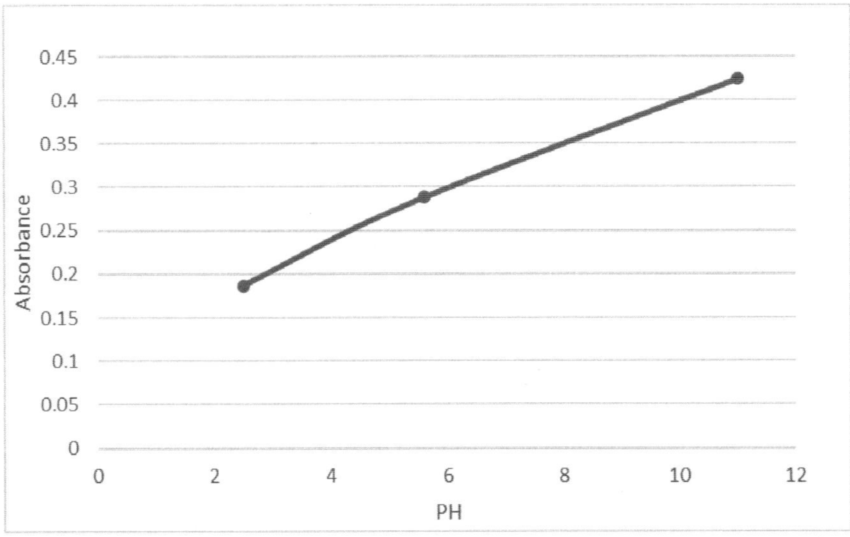

Figure 13. Spectra of effect of pH on absorbance of Cu-Rifaximin Complex at wavelength= 241nm.

From Table 7. The maximum absorbance is at pH= 11, were the reversible rection shifted toward the complex formation.

-Effect of time on Cu (II)-Rifaximin complex at 241 nm

Table 8. effect of time on absorbance for 2.5*10$^{(-3)}$ Molar concentration of Cu(II) – rifaximin Complex

Time(min)	0	5	10	15	20
Absorbance	0.439	0.470	0.446	0.447	0.447

Figure

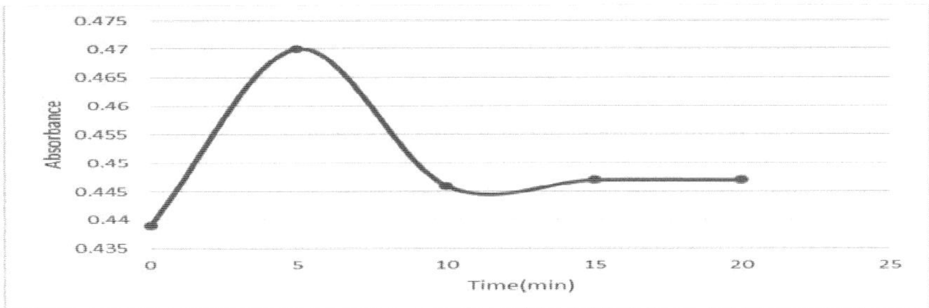

Figure 14. Absorbance of Cu(II)-Rifaximin complex at different times at 241nm.

The rate of absorbance started higher, by the time the rate of absorbance becomes stable.

3.1.3. Co-Rifaximin complex:

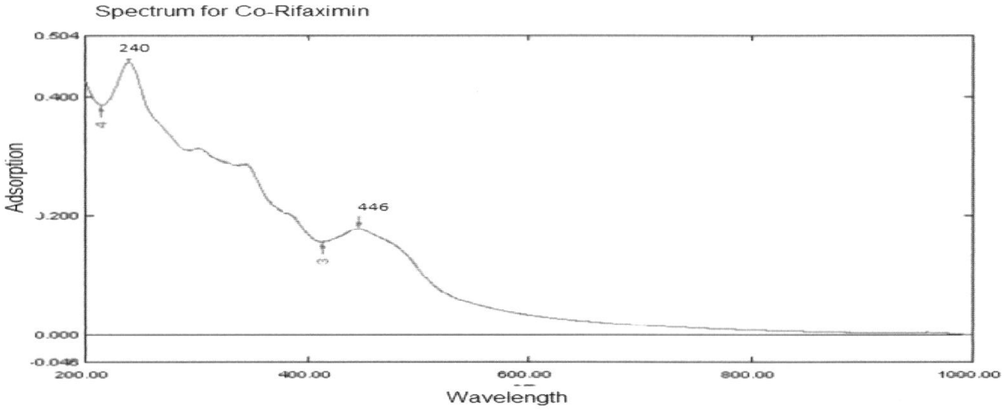

Figure 15. The adsorption spectrum for the reaction of Co^{+2}-L complex with different wavelengths.

From the spectrum of Co^{2+} -Rifaximin at different wavelengths, Figure 15, We noticed that the best wavelength of the cobalt ion with Rifaximin is 446 nm.

Absorption for different concentrations of Co-Rifaximin complex at Wavelength: 446 nm

Table 9. Different concentrations of Cobalt ion with $2.5*10^{-3}$M of Rifaximin

Concentration of cobalt ion*10^{-3}M	1.000	1.500	2.000	2.500	3.000	3.500	4.000	4.500	5.000
Absorption	0.721	0.605	0.513	0.569	0.229	0.335	0.557	0.160	0.238

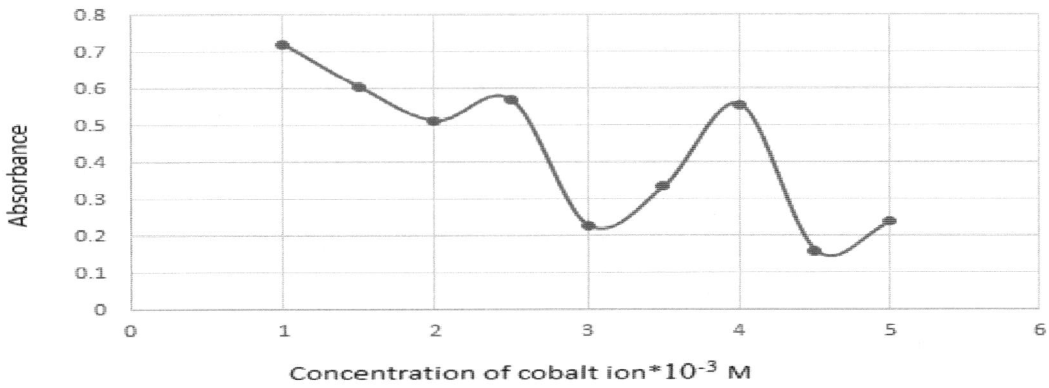

Figure 16. The absorption spectrum for the reaction of different concentrations of Co^{+2} with 0.0025M of Rifaximin at wavelength 446 nm.

The importance of Rifaximin for spectrophotometric and visual colorimetric is the determination of Co^{+2}. A Dark orange color is formed between Co^{+2} and Rifaximin by preparing serial concentrations for Cobalt and Rifaximin. This leads to investigate suitable sensing materials and immobilization techniques for these low-cost commercially available indicators in optical sensing. The complexation of a metal ion with ligand have been investigated leads to study the complex linearity, the formation constant and molar absorptivity table 9, figure 16.

-Molar Ratio

Table 10. The adsorption for molar ratio between different concentrations of Co^{2+} and $2.5*10^{-3}$M of Rifaximin.

Molar ratio	0.4	0.6	0.8	1	1.2	1.4	1.6	1.8	2
Absorbance	0.721	0.6	0.51	0.56	0.2	0.34	0.56	0.16	0.24

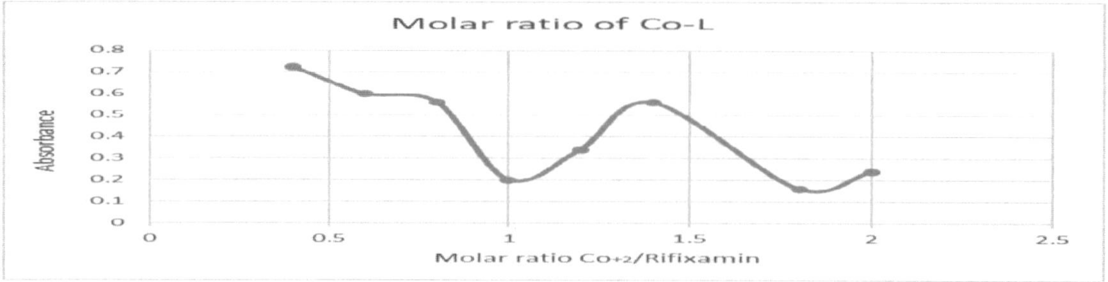

Figure 17. The absorption spectrum for molar ratio between different concentration of cobalt ion and $2.5*10^{-3}$M of Rifaximin.

From figure 17, and table 10 . it's noticed that the maximum absorption was (0.721) at molar ratio (0.4) at wavelength 446 nm.

Effect of time on Cobalt-Rifaximin complex at 446 nm

Table 11. effect of time on absorbance for $2.5*10^{(-3)}$ Molar concentration of Co (II) – rifaximin Complex

Time (min)	0	5	10	15	20	25
Absorbance	0.415	0.405	0.401	0.397	0.397	0.397

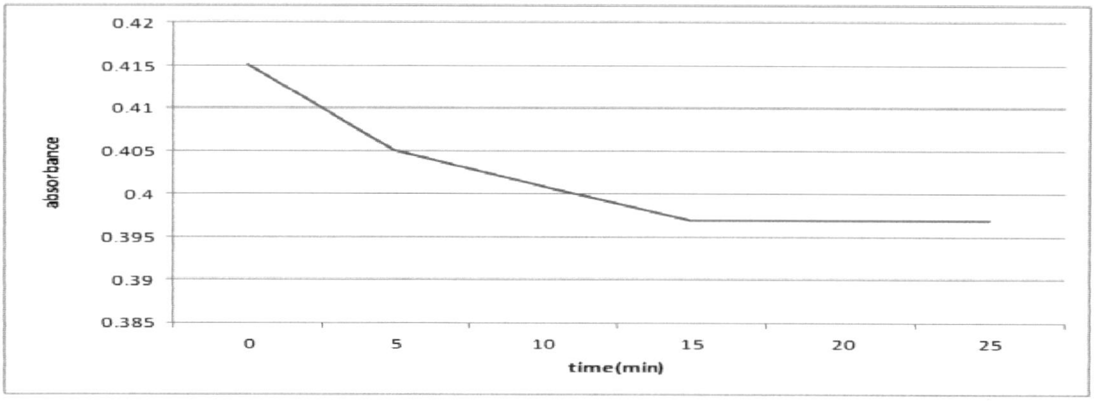

Figure 18. Spectra of effect of time on absorbance of Co-Rifaximin complex at 446 nm.

The reaction is reversible which the rate of the reaction at beginning is the highest and by the time the rate of reaction becomes the lowest, then it becomes stable after 10 min.

Effect of PH on Cobalt-Rifaximin complex at 446 nm

Table 12. Effect of pH on Cu (II)-Rifaximin Complex at 241 nm

pH	2.5	5.6	11
Absorbance	0.209	0.136	0.271

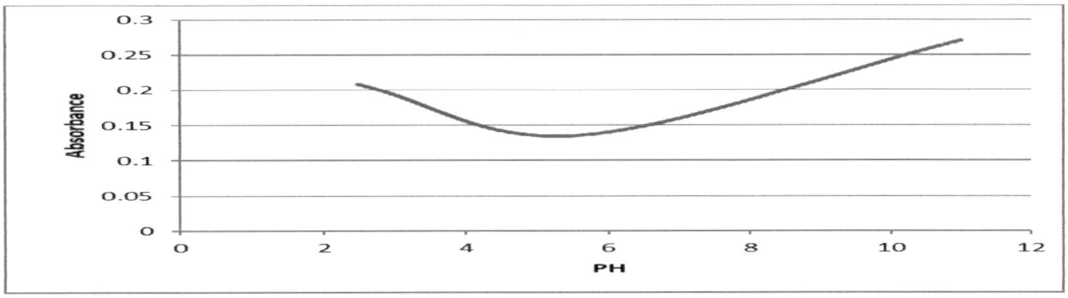

Figure 19. Spectra of effect of pH on absorbance of Co-Rifaximin complex at wavelength= 446

We measure absorbance at different pH (acidic, neutral and basic), the maximum absorbance is at pH =11.

The best recommendation for reaction of Co-Rifaximin complex absorbance is at basic pH.

3.1.4.The reaction of vitamin C with Rifaximim

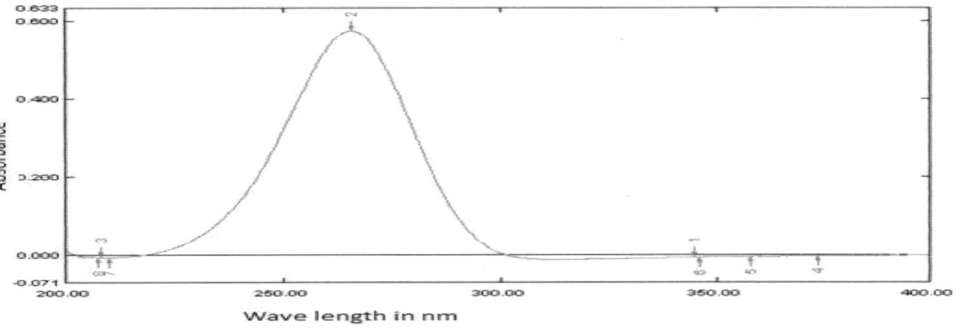

Figure 20. The absorption spectra of the vitamin C solution at (PH=7)

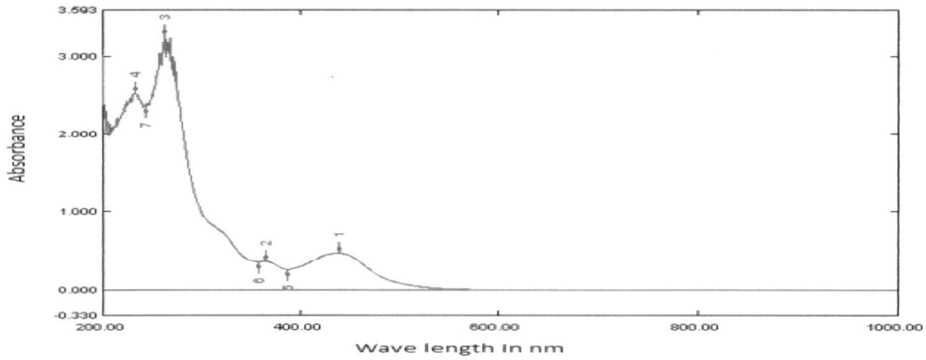

Figure 21. The absorption spectra of the rifaximin-vitamin C solution at (pH=7)

Table 13. Different concentrations of Rifaximin with $2.5*10^{-3}$M of Vitamin C.

Conc of Rifaximin * 10^{-3}	0.50	1.0	1.500	2.00	2.50	3.00	3.50	4.00	4.50	5.00
Absorbance	0.079	**0.084**	**0.116**	**0.135**	**0.185**	**0.252**	**0.254**	**0.318**	**0.408**	0.417

Absorption for different concentrations of vitamin C-Rifaximin complex at 438nm

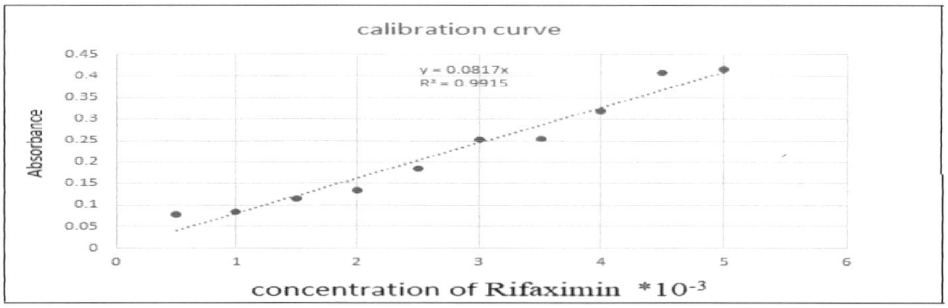

Figure 22. The absorption spectrum for the reaction of different concentrations of Rifaximin with 0.0025M vitamin C.

When adding vitamin C to Rifaximin its color changes from brown to yellow. A series of different concentrations of rifaximin and vitamin C were prepared and the absorbance were measured after 24 hours from the preparation. The relation between the different concentrations of Rifaximin with 0.0025M vitamin C were studied and a straight line was obtained which represent the calibration curve for this reaction.

The Molar absorption coefficient (ε) were determined to be 817 L $mole^{-1}cm^{-1}$ and the concentration of rifaximin could be calculating from be Beer–Lambert law A = ε C L.

Effect of time on the reaction of vitamin C-Rifaximin complex at 438 nm

Table 14. effect of time on absorbance for $2.5*10^{(-3)}$ Molar concentration of **vitamin C**—rifaximin.

Time (min)	5	10	15	20	25	30
Absorbance	0.34	0.41	0.4	0.43	0.43	043

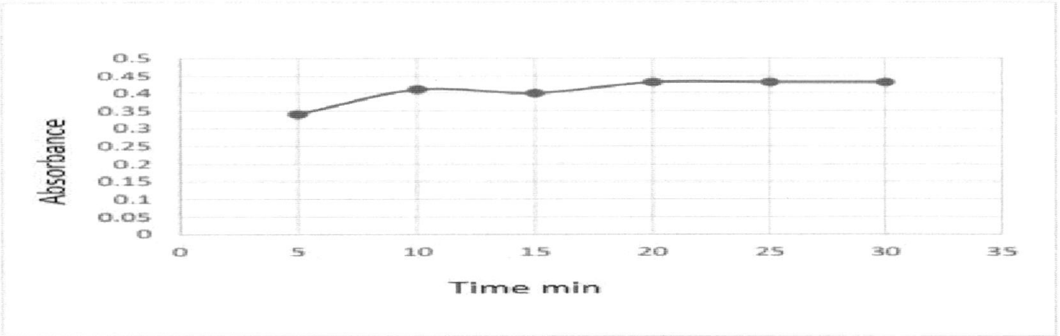

Figure 23. Spectra of effect of time on absorbance of vitamin C-Rifaximin mixture at 438

3.2. The biological activity:

3.2.1. The biological effect of Metals-Rifaximin complexes

The antimicrobial activity of the prepared Copper (Cu), Cobalt (Co) and iron (Fe) complexes of Rifaximin was evaluated using the well diffusion method, against tested multidrug-resistant human pathogens. Briefly, MHA ager plates were prepared and inoculated with 0.5 McFarland standard bacterial inoculums of 1×10^8 CFU/ml and uniformly spread onto the plates [29]. Then, different amount of Copper (Cu), Cobalt (Co) and iron (Fe) complexes of Rifaximin (L:M, 1:1). (25- 50 μl) aliquot solution was loaded into wells. The plates were then incubated at 37°C for 24 hr. Finally, the inhibition zone was measured. The suppression of bacterial growth was determined in Petri after 24 h at 37°C, Diameter of inhibition zone is measured.

Among the Gram-positive bacteria tested, Staphylococcus aureus shows strong inhibition by Rifaximin 25μl (28 mm), Rifaximin 50μl (29mm) Besides, Gram negative bacteria tested Escherichia coli (24 mm) at 25μl and (25mm) at 50μl of Rifaximin, it could be due to the release of Rifaximin ions that increased membrane permeability and ROS generation leading to cell death.

- Antibiotic-Copper (Cu) complex (25μl and 50 μl) give negative result it stops the effect of antibiotic against both the Gram-positive bacteria tested Staphylococcus aureus and Gram-negative bacteria tested Escherichia coli no inhibition zone is formed. So, it is not recommended to take Cu with Rifaximin as it yelled no effect against bacteria[30].

- Iron (Fe) (25μl and 50μl) give large inhibition zone (26 mm) on Staphylococcus aureus (G+) and inhibition zone ranged from (15 to 20 mm) on (G-) Escherichia coli.
- Cobalt (Co) (25μl and 50μl) give inhibition zone ranged from (14 to 15mm) with Gram-positive bacteria tested Staphylococcus aureus and lead to stop the effect of antibiotic against Gram negative bacteria tested Escherichia coli no inhibition zone is generated so it is not recommended to take cobalt with Rifaximin.
- Rifaximin-Vitamin C complex (25μl and 50μl) give inhibition zone ranged from (19 to 22mm) on tested G- Escherichia coli and (26 to 28 mm) on Staphylococcus aureus.

Staphylococcus aureus (G+)

Table 15. Diameter of inhibition zone against **Staphylococcus aureus** (mm) with 25 and 50μl of different concentrations of Rifaximin and its complexes, Rifaximin with vitamin c

SampleH	L25	L50	Fe25	Fe26	Cu25	Cu50	Co25	Co50	V.c25	V.c50	L.v.c25	L.v.c50
Diameter of inhibition zone	28	29	26	26	-Ve	-Ve	14	15	-Ve	-Ve	26	28

(L= Rifaximin Cu, Co, Fe complexes with florfenicol, v= vitamin C, L-V Rifaximin with vitamin C)

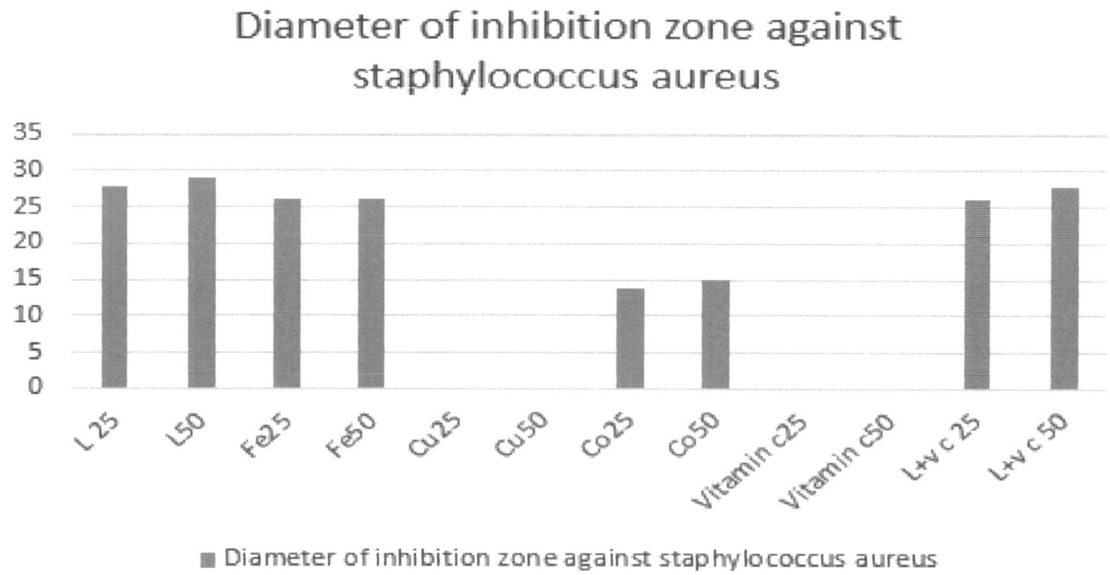

Figure 24. Diameter of inhibition zone against Staphylococcus aureus with 25 and 50μl of different concentrations of Rifaximin, its complexes, Vitamin C and Rifaximin with vitamin C

Escherichia coli (G-)

Table 16. Diameter of inhibition zone against Escherichia coli (mm) with 25 and 50 μl of different concentrations of Rifaximin and its complexes, Rifaximin with vitamin c

Sample	L25	L50	Fe25	Fe50	Cu 25	Cu50	Co 25	Co50	V. c 25	V. c 50	V. c - L 25	V. c -L 50
Diameter of inhibition zone against E. coli	24	25	15	20	-ve	-ve	-ve	-ve	-ve	-ve	19	22

(L= Rifaximin Cu, Co, Fe complexes with florfenicol, v= vitamin C, L-V Rifaximin with vitamin C)

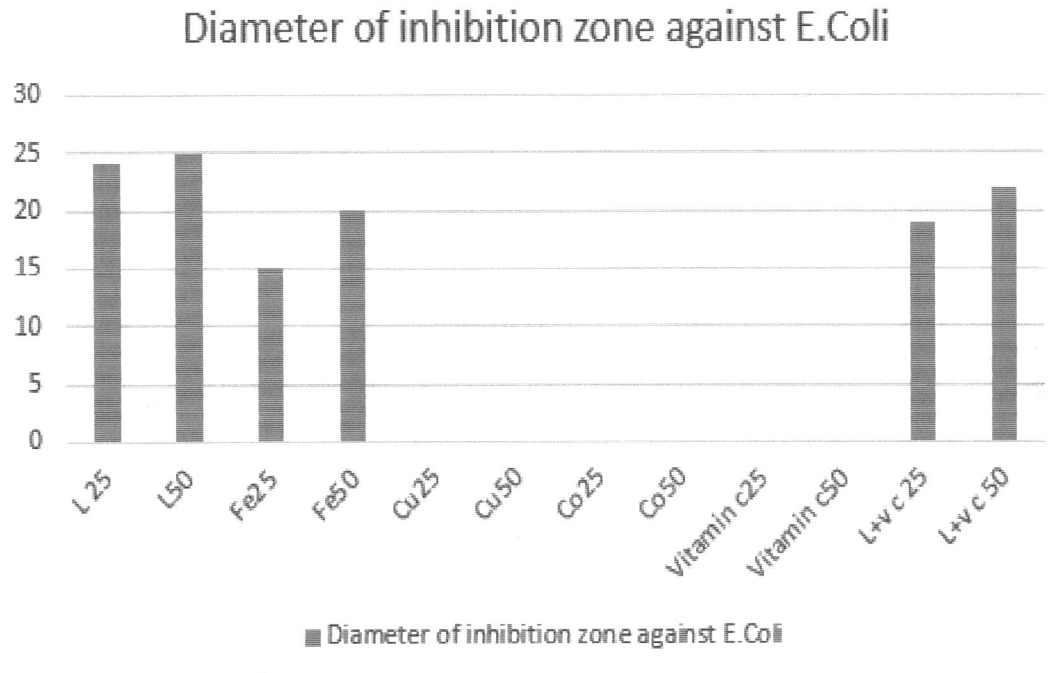

Figure 25. Diameter of inhibition zone against Escherichia coli with 25 and 50μl of different concentrations of Rifaximin, its complexes, Vitamin C and Rifaximin with vitamin C

Cobalt has antibacterial effect against the tested bacterial pathogens, it was confirmed that cobalt shown antibacterial potential against different bacterial strains [31-32-33]. The possible reason for the antimicrobial activity of Cobalt is the electrostatic interaction amid cobalt with Staphylococcus aureus bacteria which interrupt the cell wall of bacteria [34-35]. Another reason could be the release of cobalt ions leading to disruption in the DNA replication and cellular protein inactivation as it interacts with the cell wall of bacteria and interrupt the membrane structure causing cell death and leading to inhibition of bacterial growth [36].

4. Conclusion

From the mole ratio method, it's found that the most stable rifaximin complexes with Cu, Co and Fe formed at ratio (L:M, 1:1). The yellow color that formed from the interaction of rifaximin with vitamin C, allowed the determination of rifaximin spectrophotometry. The antibacterial effect of rifaximin on Escherichia coli and staphylococcus aureus decreases on complexation with Fe (20, 7.2 %) ,Co (100, 50 %) respectively and Cu (100% for both Escherichia coli and staphylococcus aureus). The bacteria resistant to rifaximin complexes for Escherichia coli is higher than that for staphylococcus aureus. The bacteria resistant to rifaximin complexes for Escherichia coli is higher than that for staphylococcus aureus.

References

[1] Maccaferri S, Vitali B, Klinder A, Kolida, S, Ndagijimana M, Laghi L, Calanni F, Brigidi P, Gibson G & Costabile, A. 2010 *J. Antimicrob. Chemother.* **65**, 2556–2565.

[2] Ferrer M, Aguilera M, Martinez V.2021 *Int. J. Mol. Sci.* **22** 500.

[3] Jiang Z, Ke S Palazzini E, Riopel L.& Dupont H. 2000 *Antimicrob Agents Chemother.* **44** 2205–2206.

[4] Calanni F, Renzulli C, Barbanti M, Viscomi G. 2014 *The Journal of Antibiotics* **67** 667–670.

[5] Scarpellini E, Gabrielli M, Lauritano C, Lupascu A, Merra G, Cammarota G, Cazzato I, Gasbarrini G.& Gasbarrini A. 2007 *Alimentary Pharmacology & Therapeutics* 782-785.

[6] Jiang Z. & DuPont H. 2005 *Chemotherapy* **51** 67–72.

[7] Jiang Z, Ke S. & DuPont H. Int 2010 *J. Antimicrob Agents* **5** 278–281.

[8] Brown E. 2010 Antimicrob Agents Chemother 54 388–396.
[9] Taylor D. 2006 Am. J. Trop. Med. Hyg. 74 1060–1066.
[10] Gomi H.2001 Antimicrob Agents Chemother. 45 212–216.
[11] Mullen K, Sanyal A, Bass N, Poordad F, Sheikh M, Frederick R, Bortey E. & Forbes W. 2014 Clin. Gastroenterol. Hepatol. 12 1390–1397.
[12] Kane J. & Ford A. 2016 Expert Rev. Gastroenterol. Hepatol. 10 431–442.
[13] Lopetuso L, Napoli M., Rizzatti G. & Gasbarrini A. 2018 Expert Opin. Investig. Drugs 27 543–551.
[14] Mattila E., Arkkila P, Tarkka P, Tissari P. & Anttila V.2013 Alimentary Pharmacology and Therapeutics 37 122–128.
[15] Prantera C. 2012 Gastroenterology 142 473–481.

[16] Taylor D. 2006 Am. J. Trop. Med. Hyg. 74 1060–1066.
[17] Maccaferri S. 2010 J. Antimicrob. Chemother 65 2556–2565.
[18] Darkoh C. 2010 Antimicrob Agents Chemother 54 3618–3624.
[19] Abbaspour N., Hurrell R. & Kelishadi R. 2014 J of Research in Medical Sciences 19 164-74.
[20] Theil E., Chen H., Miranda C., Janser H., Elsenhans B. & Núñez M. 2012 J. Nutr. 142 478-83
[21] Czarnek K., Terpiłowska S. & Siwicki A. 2015 Central-European Journal of Immunology 40 2 236-242.
[22] Karovic O., Tonazzini I. & Robola N.2007 Biochem Pharmacol. 73, 694-708.
[23] Harris E. 2001 Nutr Rev 59 281-285.
[24] Osredkar J., Sustar N. 2011 Journal of Clinical Toxicology 40 6310
[25] Willis M., Monaghan S., Miller M., McKenna R. & Perkins W. 2005 Am J Clin Pathol. 123 125-131.
[26] Clinical and Laboratory Standards Institute, CLSI 2006.
[27] Khan Z, Sadiq H, Shah N, Khan A, Muhammad N., Hassan S. & Ullah F. 2019 Journal of Photochemistry and Photobiology B: Biology 192 147-157.
[28] El-Moslamy S, Shehata N, Ahmed W. & El-Naggar M. 2018 Journal of Pure and applied Microbiology 12 2 439-458.
[29] Oves M., Ansari M, Darwesh R, Hussian A, Alajmi M. & Qari H. 2020 Coatings 10 10 950.
[30] Poole K. 2017 Trends in Microbiology 25 10 820–832.
[31] Wen-Ru L, Xiao-Bao X, Qing-Shan S. & Shun-Shan D. 2011 Biometals 24 135–141.
[32] Dogra V., Kaur G., Jindal S., Kumar R., Kumar S. & Singhal N. 2019 Sci Total Environ 1 681 350-364.
[33] Raheem H., Al-Ghazali L., & Al-Marzook F. 2019 International Journal of Pharmaceutical and Clinical Research 10 138-141.
[34] Kadiyala U, Turali-Emre E, Bahng J, Kotov N. & VanEpps 2018 J. Nanoscale 10 4927- 4939.
[35] Saratale R, Saratale G, Shin H, Jacob J, Pugazhendhi A, Bhaisare M. & Kumar G. 2018 Environ. Sci. Pollut. Res. 25 11 10164–10183.
[36] Chattopadhyay S, Dash, S, Tripathy S, Das B, Mandal D, Pramanik P. & Roy S. 2015 Chem. Biol. Interact 226 58–71.

Solar-driven Photocatalyzed Hydrogen Production from Water cleavage: Evaluation and Process Life Cycle Assessment

Mohamed E.M. Ali[1,*], Mohammad M. Almutairi[2], Mohamed S. Mahmoud[3,4], Ebrahim Ismail[4], Yasser M. Khawassek[5], Mohamed S. Atrees[5]

[1] Water Pollution Research Department, Environmental Research and Climate Changes Institute, National Research Centre (NRC), Egypt, P.O. 12622.
[2] Ministry of Electricity and Water, Kuwait, Kuwait, P.O.12010
[3] Collage of Applied Science, Department of Engineering, Suhar, 331, Sultanate of Oman
[4] Chemical Engineering Department, Menia University, El-Menia, Egypt.
5 Nuclear Materials Authority, Maadi Road, Elqattamia, Cairo, Egypt. P.O. 530.

*Corresponding author: alienv81@yahoo.com

Abstract. Hydrogen is proper chemical storage for solar energy and is considered a green and non-pollution energy. The feasibility of solar-driven hydrogen generation from water cleavage is investigated using commercial TiO_2 nanoparticles, and the process of life cycle assessment (LCA) is estimated. TiO_2 is fully characterized. The results confirm the efficacy of TiO_2 for solar-driven photocatalytic water splitting with the same production rate of H_2. The maximum calculated photocatalytic rate of hydrogen over PC was 4.76 μmole/h under sunlight. As well the rate of produced H_2 is dependent PC dose. There is a signification corelation between Kapp with PC dose with r^2 of 0.96. Reaction rate constant for hydrogen production is significantly dependent concentration of TiO_2. The outcome emphasized the practical application of solar-driven, cost-effective technology for production renewable energy. Also, LCA study of solar-driven H_2 production is showed that higher carbon dioxide and as well as global warming potential with lower climatic change impact originated from operation of system.

Keywords: Hydrogen, solar-driven photocatalytic, water splitting, LCA, Renewable Energy

1.Introduction

Hydrogen had been reckoned a green energy source without negative environmental impact or toxic emission to human health. As well deployment of hydrogen in power generation did not generate GHGs. Hydrogen is a guaranteed energy carrier that could replace oil in the transport or domestic energy supply. Amongst the challenges of the future is to replace traditional fuels with renewable energy sources [1-2]. These energy resources have limited quantity, are non-renewable, and bring severe environmental pollution. Hence, the questing for non-polluted alternate energy has is a crucial primacy mission. Photocatalysis-based H_2 production via water splitting is considered modern fashion technology to supply non-polluted energy. Upon irradiation of photocatalyst (PC) material surface with energy (photons) identical to or greater than PC energy gap, charge carriers on PC surface were efficiently separated; in conduction and valence bands

of PC. The produced electron and hole are capable for reduction and oxidization water, producing hydrogen and oxygen [1]. The major shortcoming of the approach is employing artificial UV/visible light lamps as irradiation sources. Thus, search for an alternative source of light is a great priority, and solar light was considered a renewable and costless source of irradiation.
Solar energy is an unlimited principal energy source [3, 4]. Conversely, the direct capability for solar energy storing or incessant provision is not available. Therefore, the adaptation of solar energy to be stored is of essential magnitude. The Solar Atlas was issued, Egypt is provided with the elevated intensity of direct solar irradiation varying between 5.5 – 8.8 kWh/m^2/day. The availability of sunlight time varied between 9-11 h/day [5]. LCA illustrates life cycle phases of process/product while evaluating the performance of a system, including the construction phase, operation, use, and disposal. The assimilation of LCA is necessary for assessing sustainability to attain consistent outcomes [6]. LCA relies on setting up an analytical model that attempts to illustrate the system under assessment. More information concerning LCA can be presented somewhere else [7]. LCA methods spotlight the environmental feature of sustainability [6].

One of the important approaches for storing solar energy is H_2 production via photocatalyzed water splitting. One of the major problems in photocatalytic hydrogen production via water splitting was the high cost of irradiation source; UV lamps. Herein, UV lamp as irradiation source is replaced by solar energy. Moreover, the LCA of the used system for H_2 production will be studied to estimate the environmental impact of the process. The current work investigates the feasibility of utilizing solar energy for a chemical energy source, H_2.

2. Experimental

2.1. Materials
Titanium dioxide ($TiCl_4$) with 99.8 % purity, ammonium hydroxide solution, and absolute ethanol were purchased from Sigma-Aldrich (Germany). High purity Sulphuric acid and sodium hydroxide pellets were purchased from Fluka (Germany). These chemicals were used without further purification.

2.2. Characterization of Photocatalyst
The prepared powder was prepared as previously described [1]. and it was characterized by different tools that are previously described [1].

2.3. Solar Photocatalytic Experiments
Fig.1 shows used solar reactor used for photocatalytic hydrogen production. The photoreactor consisted of a cylindrical glass reactor vessel with an effective volume of 2000 mL and a gas collector at the top of system. Definite PC amount are suspended in 2000 mL of water at ambient temperature that is spareged with N_2 gas for an hour, and the system was exposed to sunlight. At various time periods, the gas was collected in gas samplers which are previously sparaged ith N_2 gas and was detected by a hydrogen gas sensor.

2.4. Life Cycle Assessment Methodology
LCA methodology consists of main steps, including reviewing the available literature of the studies and collecting all the required data.

2.4.1. Goal and Scope Definition. The study targets to assess the environmental impacts of the construction stage and solar-driven photocatalytic hydrogen production, specifically the principles and framework methodology of life cycle assessment of solar-driven hydrogen production, and then develop an assessment approach for the process. The data are gathered from the experimental used data.

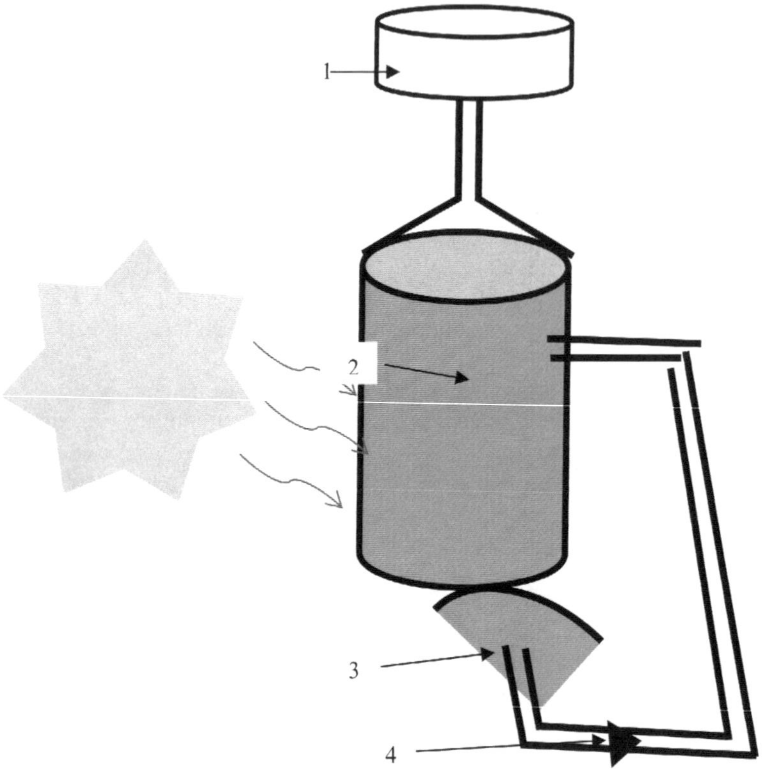

Figure 1 Solar Reactor for photocatalytic Hydrogen production, 1: gas collecting container, 2: Solar photoreactor; 3: suspension of photocatalyst Water, 4: Circulation pump

2.4.2. Framework for LCA analysis. The boundary of this work includes the construction phase and the production process of hydrogen. Consumption of materials and energy involves construction stage and the process of hydrogen. The output flows to the environment involves gaseous emissions like CO_2 from the operation phase. The functional unit is a one-day operation and 200 days operation.

2.4.3. Life cycle inventory. The report confronts troubles associated with availability and quality of database. Data for the inventory is collected from a laboratory experiment, evaluation from experts, relevant literature, and LCA databases. LCI comprises the materials and energy consumption in the construction and operation phases. In addition, some factors and parameters like retention time, gas production rate, and composition of collected gases will take from experimental measurements (See Table 1).

2.4.4. Life cycle impact assessment. The life cycle impact assessment (LCIA) was conducted for global warming potential (GWP, kg CO2 eq.

3. Results and discussion

3.1. Summary of Results

The XRD patterns exhibited strong diffraction peaks at 25° and 48° indicating the major phase of TiO_2 particles is anatase. The occurrence of additional peaks at 36° and 55° is indicating to small fractions of the rutile phase of TiO_2. TiO_2 has a mixture of anatase/rutile. Figure 3 shows the spectra of FT-IR for commercial TiO_2 sample diluted with KBr and referenced against KBr in IR range up to 4000 cm^{-1}. The large broadband at 3310–3606 cm^{-1} centered at 3451.95 cm^{-1} that ascribed to OH stretching mode. The hydroxyl stretching region bands correspond to the O–H vibration of the Ti–OH groups and H_2O molecules. The rather narrow bands at 1644.02 cm^{-1} can be assigned to OH bending mode of hydroxyl (OH) groups, representing the water as moisture . The peaks at 529.4 cm^{-1} show stretching vibration of Ti-O, and peaks at 1473.34 cm^{-1} show stretching vibrations of Ti-O-Ti. TiO_2 surface area was successfully measured by BET adsorption apparatus, and it is found to be 30.5 m^2/g. Figure 4 showed that the PC sample has an IUPAC isotherm IV pattern with a mesoporous hysteresis loop. The pore volume and diameter are 0.08 cm^3/g and 4.9 nm, respectively, as shown in Figure (4b). The porous diameter of the catalyst indicates the occurrence of mesopores. TEM image of TiO_2 nanocrystals presented in Fig. 5a that exhibited the spherical shape and homogeneously distributed nanocrystals. As well SEM image of TiO_2 material exhibited in Fig. 5b, it was observed a homogeneous structure of TiO_2. The energy of the bandgap (Eg) is a characteristic feature of PC, and it determines their application potential. Figure 6a shows the UV–visible spectrum of TiO_2 photocatalyst, while Figure (6b) illustrates the application of the Kubelka-munk model, where the bandgap energy of TiO_2 photocatalyst is found by calculating the absorption spectra from the reflection spectrum. From Figures 6 (a,b), It is obvious that TiO_2 has an absorption edge at 390-400 nm with a corresponding Eg of 3.1 eV according to the Kubleka-Munk model. This moderate energy bandgap indicated that TiO_2 could only be activated by UV–visible spectrum of the sunlight irradiation (**Hashimoto, et al., 2005**).

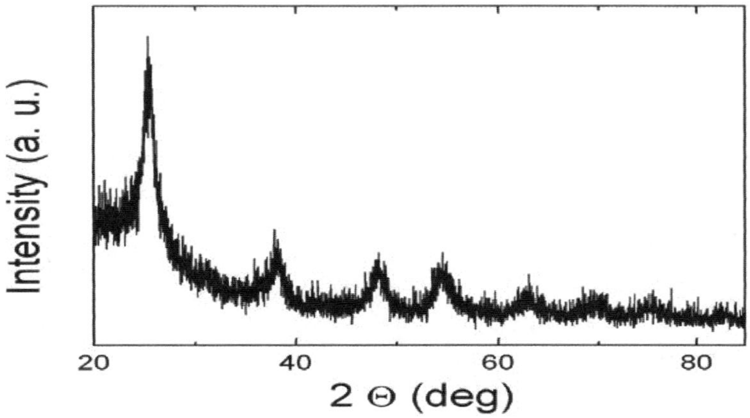

Figure 2 XRD pattern of TiO_2 photocatalyst

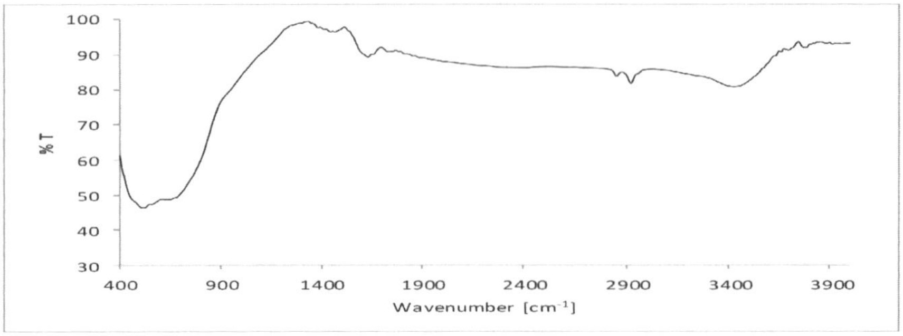

Figure 3 FT-IR spectrum of TiO_2 photocatalyst

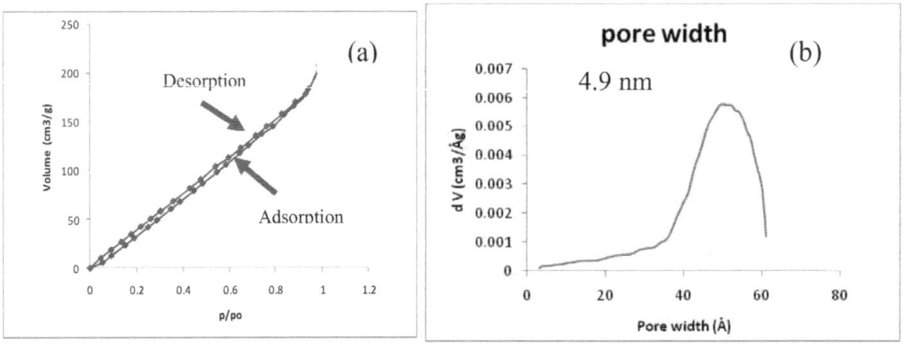

Figures 4 N_2-adsorption-desorption of TiO_2 nanoparticles (a), pore distribution (b).

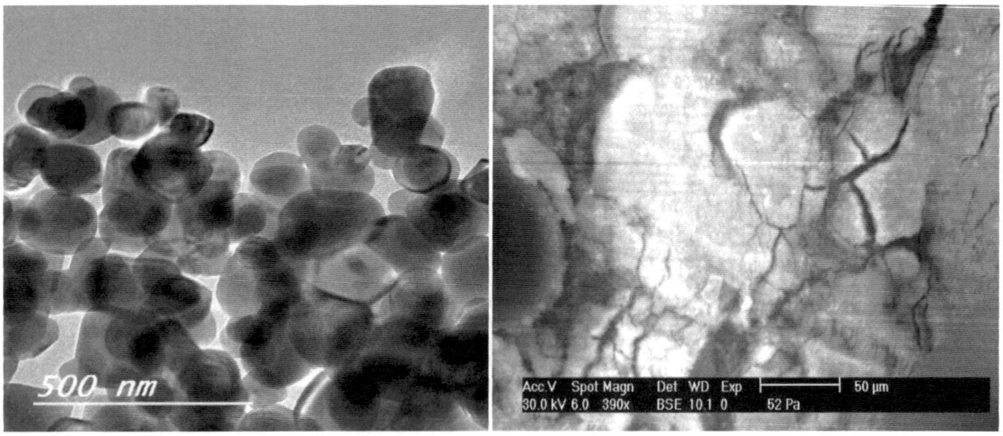

(a) (b)

Figure 5 TEM (a) and SEM (b) of TiO_2 nanoparticles

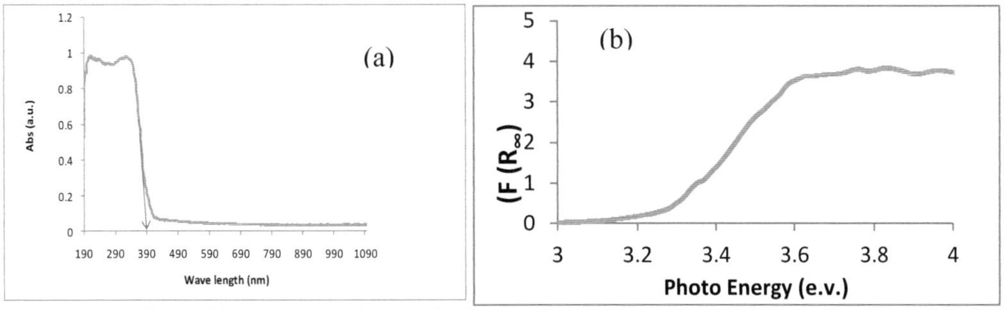

Figure 6 DRS of TiO_2 photocatalyst; (a) absorption spectrum and (b) application of Kubelka-Munk model

3.2. Photocatalyzed cleavage of water

First, the effectiveness of TiO_2 was investigated for photocatalyzed water cleaving for H_2 production under sunlight as an irradiation source. Blank trials are carried out in the absence of irradiation and/or photocatalyst (PC) under similar conditions of hydrogen production. No H_2 was observed, demonstrating the prominence of light energy and PC for H_2 generation from photocatalyzed water cleaving. Figure 7 shows the photocatalytic produced H_2 amount and the rate of produced H_2 over 1 g of PC and under sunlight. It was observed that amount of produced H_2 and rate of hydrogen from photo-splitting of water increased with increasing reaction period up to 140 min under sunlight; after that, it declined. It was assigned that the increasing irradiation time leads to more photogenerated hole-electron pairs, and as well as increasing reduction of water to H_2 molecules [4]. Meanwhile, the prolonged irradiation time directs to the accumulation of photogenerated holes on PC surface that enhances the recombining rate of hole-electron pairs. The maximum calculated rate of hydrogen over PC was 4.76 µmole/h. While the solar photocatalytic was operated for 200 days using the same PC without further treatment (data (not shown). The results confirm the efficacy of TiO_2 for photocatalytic water splitting with the same production rate of H_2; that outcome emphasized the feasible application of the solar system for renewable energy with cost-effective technology.

Figure 7 Photocatalytic Hydrogen production from water splitting

3.2.1. Effect of photocatalyst dose on H_2 generation over TiO_2. The photocatalytic production of H_2 has influenced photocatalyst concentration in suspension. To quantify the dependency of H_2 production on PC concentration, the curves of H2 generation with irradiation time were plotted at various concentrations of TiO2 and shown in Figure 8. With photocatalyst concentration increasing from 0.5 to 2 g/L, the photocatalytic H2 production is varied in the order of 2 g/L> 1.5 g/L > 1 g/L > 0.5 g/L [4, 8] stated the same trends for H2 generation. The photocatalysis on the surface of TiO_2 is an intricate approach and controlled with transmitting of irradiation in the slurry system and energetic positions on the Surface of the PC.

At a lower dose of PC concentration, the interaction is essentially directed by the available energetic locations for harvesting light and reactant [1, 9, 10]. The active sites are increased with the increment of photocatalyst concentration. However, when the photocatalyst concentration is above the optimum level, the solution become turbid and light is greatly scattered by the suspended PC. The observed phenomenon can be previously rationalized in terms of the accessibility of energetic positions on the TiO_2 surfaces, the light absorption, and the penetration depths into the suspension. But to select the most favorable dosage of PC, other calculations were investigated. Figure 9 showed the time profile of specific photocatalytic H_2 production rates over different doses of TiO_2. As noticed, the specific production rate of H_2.

The obtained results showed that the higher dose of titania does not improve the photocatalytic rate of H_2. Figure (10) shows the apparent rate constant (k_{app}) of H_2 production from water splitting as a function of TiO_2 loading. There is a highly fitted relation between K_{app} with PC dose with r^2 of 0.96. It is obvious that k_{app} is the highly dependent concentration of TiO_2. This is characteristics of heterogeneous photocatalyst and agrees with earlier reports [4, 10]. Thus, from an economic point of view, 1 g/L was selected as the optimal dose of PC for solar photocatalytic H_2 production from water splitting.

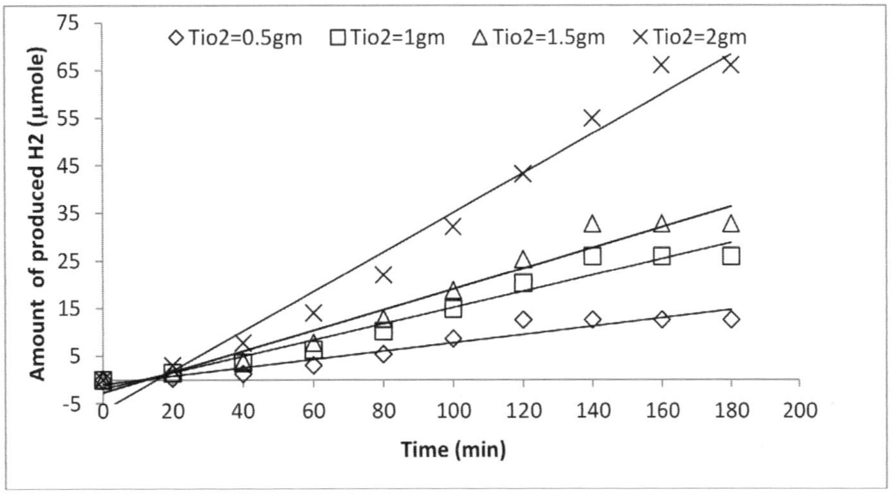

Figure 8 Effect of the concentration of photocatalyst on photocatalytic hydrogen production from water splitting using TiO_2 under sunlight irradiation

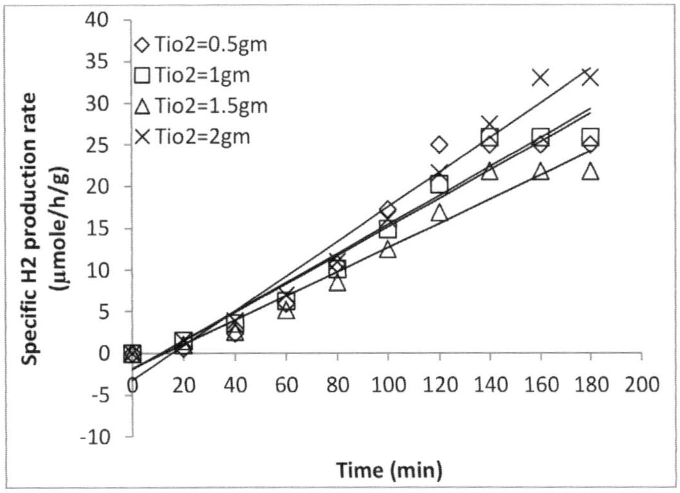

Figure 9 Effect of PC concentration on rate constant of photocatalytic hydrogen production from water splitting over pure TiO_2

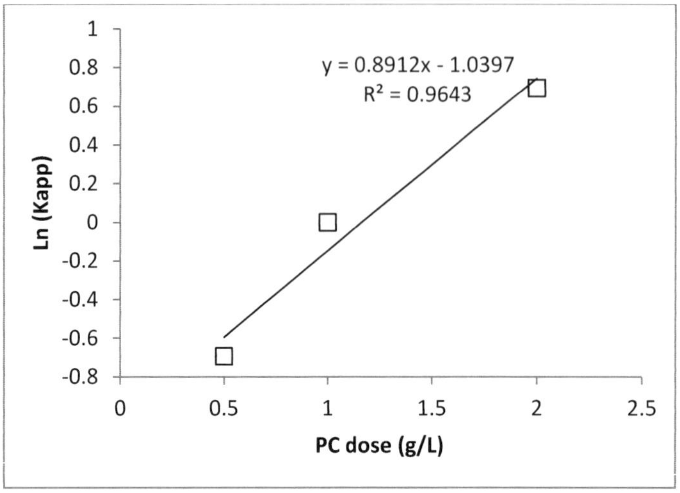

Figure 10 Dependence of rate constant of H_2 production on PC dose

3.3. Life cycle assessment (LCA)

LCA was applied to confirm the environmental and economic sustainability of H_2 production from photocatalytic water splitting using sunlight. LCA methodology is employed in accordance with ISO 14040 and ISO 14044 [11, 12]. The assessment was made using life cycle inventory (LCI) data collected from the database. The outcomes ought to afford different enterprisers with a better understanding of the environmental sustainability of solar-driven hydrogen production, which consecutively should assist progress the technology so that it becomes ready for industrial-scale application. This is the first study to date dealing with LCA of photocatalytic water splitting processes to the preeminent of the researchers' information. Table 2 shows the design parameter of a system for the production of 15 mmoles of hydrogen for production of 1 kWh of electricity. For limited data, only CO_2 emission for different system components was calculated. It was shown the most of CO_2 emission originate from system construction. For long term operation of the system, it was noted ecological impact of system construction is declined. Otherwise, the impact saved due to the reuse of produced hydrogen for power generation is conferred in Fig. 11. To envisage LCIA, the impact of equivalent quantity of electricity is saved. It is observed from the results that the produced hydrogen for power generation significantly reduce the environmental impact by 22 kg CO_2 e/200 Kwh, i.e. 0.11 kg CO_2 e/Kwh. Fig. 12 showed the contribution of different component of system in global warming potential. It was noted that construction phase is have the dominant part of global warming during short term operation.

Table 1 Life cycle inventory of system

Item	**Description**	**Carbon dioxide emission-kg/unit**
Construction of photoreactor unit	Capacity = 12.5 L, 3.12 kg glass	2.65
Circulation Pump with power of 0.05 Kwh	for 12 h, 0.6 Kwh	0.275
Titanium dioxide (TiO_2)	0.002 kg	0.0467
Total Global warming Potential		2.9717

Table 2 Design parameters of system for production 15 mmole equivalent one kWh of electricity

Item	Description
Flow rate of water	0.05 m^3/d
Circulation rate	0.02 m^3/min
Amount of PC	0.002 g

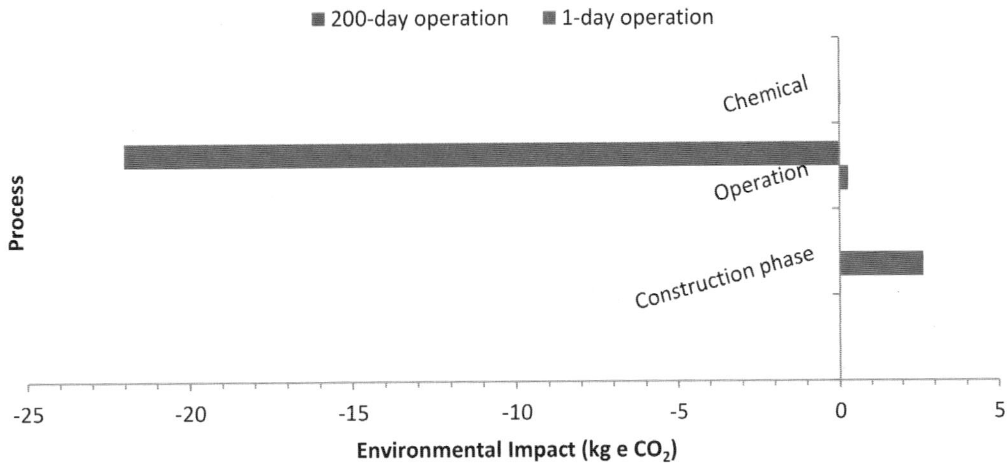

Figure 11 Environmental impact of photocatalytic hydrogen production before and after H_2 utilization

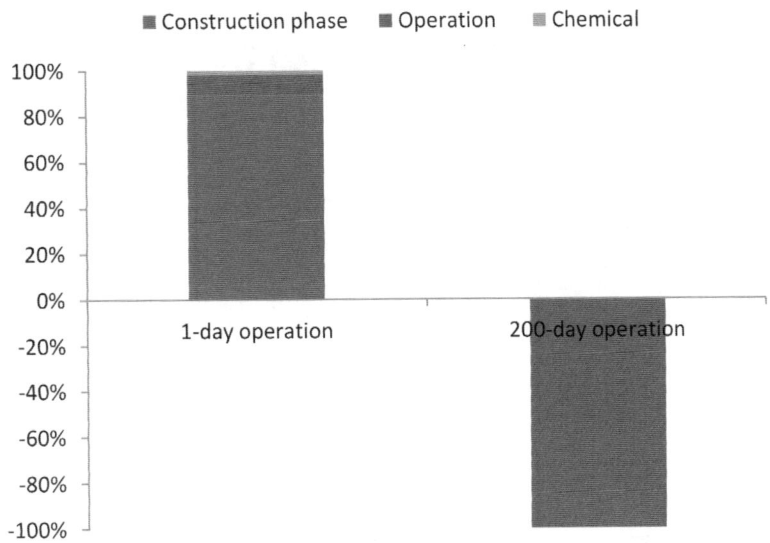

Figure 12 Environmental impact Contribution from different components of process

4. Conclusions

Prepared TiO_2 nanocomposite photocatalyst for generating H_2 was fully characterized. The solar photocatalytic system was operated for 200 days using the same PC without further treatment. The results confirm the efficiency of TiO_2 for generating H_2 with the same rate of H_2 that outcome emphasized the feasible application and utilization of solar systems for renewable energy with cost-effective technology. It was noted the reliance of solar photocatalytic rate of H_2 production on PC dose. LCA of the system confirmed that higher carbon dioxide and global warming potential originated from system construction. The produced hydrogen for power generation significantly reduces the environmental impact by 22 kg CO_2 e/200 Kwh, i.e., 0.11 kg CO_2 e/Kwh. Finally, LCA is a useful tool to determine the application of new environmentally friendly processes. Solar-driven water splitting is considered promising green hydrogen source as energy emission-free.

5. References

1. Badawy M.I., Ali M.E.M., Ghaly M.Y., El-Missiry M. A.. Mesoporous simonkolleite–TiO_2 nanostructured composite for simultaneous photocatalytic hydrogen production and dye decontamination, Process Safety and Environmental Protection 94 (2015) 11–17.
2. Ali, M.E.M., Assirey, E.A., Abdel-Moniem, S.M., Hanan S. Ibrahim. Low temperature-calcined TiO2 for visible light assisted decontamination of 4-nitrophenol and hexavalent chromium from wastewater. Sci Rep (9) 19354 (2019) doi:10.1038/s41598-019-55912-2
3. Zhu Y, Li J. Solar energy and renewable energy. Solar Energy (1995) 4: 2–6 (in Chinese).
4. Atrees, M.S., Ebraheim, E.E., Ali, M.E.M., Khawassek, Y.M, Mahmoud, M.S., Almutairi, M.M., Synergetic effect of metal-doped GO and TiO_2 on enhancing visible-light-driven photocatalytic hydrogen production from water splitting, Energy Sources, Part A: Recovery, Utilization and Environmental Effects, In press 2019.
5. Global Solar Atlas, World bank, 2016. http://globalsolaratlas.info.
6. Environmental Protection Agency (EPA). 2006. Life Cycle Assessment: Principles and Practice. Scientific Applications International Corporation (SAIC). Contract No. 68 C02-067. Work Assignment 3-15, EPA/600/R- 06/060.
7. Finkbeiner, M., Schau, E.M., Lehmann, A., Traverso, M. Towards Life Cycle Sustainability Assessment. Sustainability. (2010) 2, 3309 - 3322.
8. Zheng X., Wei L., Zhang Z., Jiang Q., Wei Y., Xie B., Wei M., Research on photocatalytic H_2 production from acetic acid solution by Pt/TiO_2 nanoparticles under UV irradiation International Journal of hydrogen energy (2009) 34, 9033 - 9041.
9. Abdel Moniem, S.M., Ali, M.E.M., Gad-Allah, T.A., Khalil, A.S.G., Ulbricht, M., El-Shahat, M.F., Ashmawy, A.M, Ibrahim, H.S. Detoxification of hexavalent chromium in wastewater containing organic substances using simonkolleite-TiO_2 photocatalyst. Process Safety and Environmental Protection, 95, 2015, Pages 247-254.
10. Ali, M.E.M, Alanezi A. A., Azeez F.A, Ghaly M.Y, Photoassisted mineralization of remazole red F3B over NiO/TiO_2 and CdO/TiO_2 nanoparticles under simulated sunlight, Separation Science and Technology (2018) 53 (1), 170-180.
11. International Standardization Organization (ISO) 2006a. ISO 14044: 2006. Environmental management - Life cycle assessment - Requirements and guidelines.
12. International Standardization Organization (ISO), 2006b. ISO 14040: 2006. Environmental Management-Life Cycle Assessment - Principles and Framework.

Implementing Nanocomposite Shear Thickening Fluid to Improve The Impact Resistance of Polyamide Fabrics Against Intermediate Velocity Projectiles

Remon A Mankarious[1]* and Mostafa A Radwan[2] and M H Roushdy[3]

[1]Assistant Lecturer, Department of Chemical Engineering, The British University in Egypt, El-Sherouk City, Egypt.
[2]Professor, Department of Chemical Engineering, The British University in Egypt, El-Sherouk City, Egypt.
3Lecturer, Department of Chemical Engineering, The British University in Egypt, El-Sherouk City, Egypt.

*Corresponding author:
Email: remon.ashraf@bue.edu.eg

Abstract

The novelty of this research is the synthesis and comparison of a soft liquid composite shield made from two different Shear Thickening Fluids (STFs) / Polyamide fabric composite material and choosing the best one for impact resistance based on characterization and performance. The STFs microstructural properties are characterized, and STF / Polyamide composites are prepared while testing their ballistic resistance using an air gun with intermediate velocities (<100 m/s) for energy absorption. Materials used for this purpose are fumed silica with polyethylene glycol (PEG) STF and starch with water STF based dispersions. The results confirmed the superior performance of STF to absorb energy, especially for starch and water system. The rheological properties for starch/water STF showed higher viscosity than silica/PEG reaching 523.6 Pa.s and 178.9 Pa.s respectively. Other characterization tests proved the superior performance of starch with water STF leading to more dispersion stability and obvious thickening effect at higher particle concentration leading to a greater jump in viscosity at sudden shear rate. The highest concentration of silica/PEG STF/fabric showed improvement in indentation depth by 26.9%, while the starch/water STF by 43.6%. These results show the viability of the STF/Polyamide composite material if integrated for personal protection applications.

Keywords: Impact Resistance; Intermediate Velocity Projectiles; Composite Material; Nanomaterials; Shear Thickening Fluid; Polyamide Fabrics

1. Introduction

Personal bodily defense against threats and damage has been a necessity for people since prehistoric times. As a result of this necessity, protective clothes, today known as armor, have always been under development. Modern military operations and high-complexity weaponry need a continuous enhancement of improved ballistic protection armor systems that are adaptable, lightweight, comfortable, and have great energy absorption capabilities [1]. The great strength, durability, incombustibility, and capacity to protect Aramid-based fibers such as Kevlar and Twaron are most commonly utilized for safety without significantly limiting durability and user comfort [2], [3]. Several pieces of study have cited literature on the influencing elements on cloth ballistic performance [4]. Wonjin et al. [5] investigated fiber characteristics, fabric structure, number of fiber layers, areal density, project parameters, and impact parameters as the most critical aspects in the energy absorption of textiles. Impregnation of these textiles with shear thickening fluids (STF) results in what is commonly referred to as liquid armors [6]. Shear thickening fluids exhibit non-Newtonian fluid behavior in that apparent viscosity increases abruptly at a given shear rate. Shear causes fluid thickness in tightly packed colloids. When mixed at a low shear rate, such suspensions seem like thin fluids, but when agitated at a high shear rate, they feel quite thick [7]. Research also includes scientists who primarily analyze their chemical and structural origins, as well as the causes for their thickening behavior. Osborne Reynolds was the first to speak about dilatant condensed solid suspensions in 1885 [8]. The "Order-Disorder Transition" (ODT) refers to the transition from ordered to fragmented particles [9]. The hydrodynamic clustering theory provides a hypothesis for shear densification, which is caused by particle cluster formation. When the energy of a rapid hit surpasses the repulsive forces between the particles, there is a sudden critical shear rate. The STF particles link together to create hydroclusters [10]. A recent study proposes another hypothesis, the contact rheology hypothesis, as well as the hydro-clustering idea. Friction forces increase during the thickening stage when colloidal particles come into contact with each other at high shear frequencies [11]. Contact networks have the potential to withstand applied shear pressures, and as a result, suspension introduces viscosity increase and yields stress development. It should be noted at this point that the main contributors to the increase in viscosity are inter-particle contacts rather than hydrodynamic interactions, and the process operates in such a way that particle attraction only induces contact network formation; these expanded branch-like structures are resistant to shear [12]. STF is used in sports equipment and liquid armoring to increase the efficacy of energy resistance in materials. Shear thickening of fluids has lately gained more interest in the energy absorption management field, notably in ballistic and soft armor applications [13]. M. Bajya et al. [14] were among the researchers that investigated the ballistic output of multifunctional Kevlar impregnated cloth impregnated with 40% silicon in STF. They discovered that STF impregnation Kevlar fabric had a substantial resistance to ballistic penetration. Park et al. [15] in the laboratory, has combined STF with Kevlar materials to create soft body armors. The findings of the experiments reveal that the impregnation of Kevlar with a thickening fluid (silica particles dispersed in polyethylene glycol) enhances ballistic penetration resistance. The full investigation of the influence of STF impregnation on the mechanism for energy absorption of aramid fabric will result in the creation of a ballistic-resistant configuration for high-speed impact applications with optimum flexibility and comfort while boosting structural impact resistance [16]. A. Haris et al. and S. Gürgen et al. used different quantities of carbide particles incorporated in a shear stiffening polymer, and the effects of additives were studied using rheological tests. According to rheological studies, carbide particles result in substantially stiffer polymer characteristics [17,18]. Multi-phase suspensions were created by incorporating aluminum oxide particles into a nano-silica/PEG-based STF. However, it performed poorly in predicting the shear-thinning zone before the commencement of thickening [19].

This research discusses the enhanced properties of STF impregnated fibers to obtain an efficient personal shield capable of increasing impact resistance of ballistic fabrics towards intermediate speed projectiles (<100 m/s) [20,21]. The overall aim of this analysis is to study and present a novel comparison clarifying the effect of the impact resistance of two colloids developed STFs, one composed of silica and polyethylene glycol and the other one is starch and water. The research will provide a primary comparison between the properties of the prepared STFs while comparing their impact resistance capabilities as STF impregnated fabrics.

2. Materials and Experimental Approach

2.1 Materials:

The solid particles utilized in the formulation of one of the shear thickening fluids are nano-sized hydrophilic fumed silica (Aerosil 200) particles acquired from EVONIK Industries Corporation. The average particle size of used silica was determined experimentally using the dynamic light scattering technique to be 16.49 nm, with a specific surface area of 200 m2/g [22]. SIGMA-ALDRICH Company provides polyethylene glycol (BioUltra 400). It is the carrier liquid used in the manufacture of silica-based shear thickening fluid. PEG was chosen as a solvent due to its non-toxicity, low volatility, thermal stability, and H-bond efficiency. Polyethylene glycol (PEG) has an average molecular weight of 380-420 g mol-1 and a viscosity of 120 mPa.s at 20 degrees Celsius [23]. Commercial food-grade corn starch is used in the manufacture of a shear thickening fluid made of distilled water and starch. The particle size range of the used starch was determined experimentally using the dynamic light scattering method to be 303.7 nm. Starch, in general, is a massive macromolecule having the formula C27H48O20 with a molecular weight of 692.6 g mol-1 [24]. Twaron (CT 714) is a polyamide fabric purchased from Teijin Aramid Company that is commonly used in the manufacturing of body armors and is characterized by its high strength and tensile properties with 1100 dtex, Areal Density of 190 (g/m2), Strength at Break of 267 (N), plain weave, and 2 mm thick/layer [25].

2.2 Experimental Work:

The tests involve the fabrication of multi-system STFs made of silica, polyethylene glycol, and starch, each with different filler content. To provide a fair comparison, the STF samples will be processed under identical circumstances. Characterization of the prepared STFs, as well as STF impregnated fabrics, is carried out, including particle size detection and Zeta potential tests using the Malvern Panalytical Company's Zetasizer Nano ZS device to investigate the average particle size of the used silica nanoparticles and starch, as well as the stability of the dispersed particles in their carrier mediums [26].

To determine the overall performance of the samples under shear rate, rheological characterization is done using a concentric cylinder viscometer Model 35A acquired from Fann Instrument CompanyRheology is a field described by the American Society of Rheology in 1929 as the study of matter deformation and flow. Depending on their intrinsic micro and molecular structures, all materials exhibit distinct behaviors. As a result, regulating some parameters is critical in industrial applications. A viscometer is a device that uses an electric motor or a manual crank to move fluid between two cylinders. The outer sleeve, also known as the rotor sleeve, is pushed at a certain rotating velocity based on the applied force. Torque is produced on the inner cylinder by the spinning of the rotor sleeve in the fluid. The movement is restrained by a torsion spring. Usually, a dial linked to the bob reveals bob displacement through an angle known as fan reading. The final findings show the viscosity range vs the change in shear rate applied, producing a thickening of the tested fluid in logarithmic graphs.

All prepared samples will be subjected to real impact tests using an air gun and an impactor at intermediate speed simulating the geometry of caliber 9mm FMJ RN bullets to simulate the impact mechanism of fabrics towards real ballistic projectiles and clarify the role of STF in enhancing impact resistance. The impactor diameter = 9 mm and the base diameter was extruded to fit into the dimensions of the barrel = 19.3 mm. The indenter length is 1.6 cm. The results are evaluated by measuring the indentation depth (trauma) that occurred in a modeling clay fixed after the target to simulate injuries in the human body after impact. As much as the indentation depth decreases, this indicates the increase of STF enhancement in absorbing impact energy.

The mixing of fumed silica in PEG and starch in water was done with a Konmixchina Company Lab homogenizer (KRH-II) set at 7000 rpm for 3 hours to distribute the particles homogeneously into their carrier fluids. For the rheological testing, prepared samples with varying weight ratios of fillers were fixed at concentrations of 10, 20, and 30% weight to weight for both systems. Aramid fabric / STF composites were made using the same formulated STFs with filler concentrations of 10, 30, and 60% to investigate the potential of STFs for the fabrication of liquid armors and to assess the effect of increased particle concentration on fluid performance under shear. Aramid Twaron fabrics were cut to squares of 20 cm x 20 cm. For each impact test, 5 layers of fabrics have been used and impregnated for several minutes each layer into the STF solution to ensure complete impregnation. A ballistic test is performed using a lab-scale air gun represented schematically in Figure 1. The real setup of the gun using gas under pressure to accelerate a projectile with a particular speed that simulates the real one in ballistic threat is shown in Figure 2. A laser speed detector is used to calibrate the real speed of the tested projectile. Since the driving force for launching the projectile is gas pressure, a speed detector is needed to translate the applied pressure into velocity. It consists of two laser systems, two emitters, and two receivers as shown in Figure 3. The impact test is assessed through the modeling clay behind the target which shows the dent/trauma caused due to impact and its depth is measured using Vernier Caliper as seen in Figure 4.

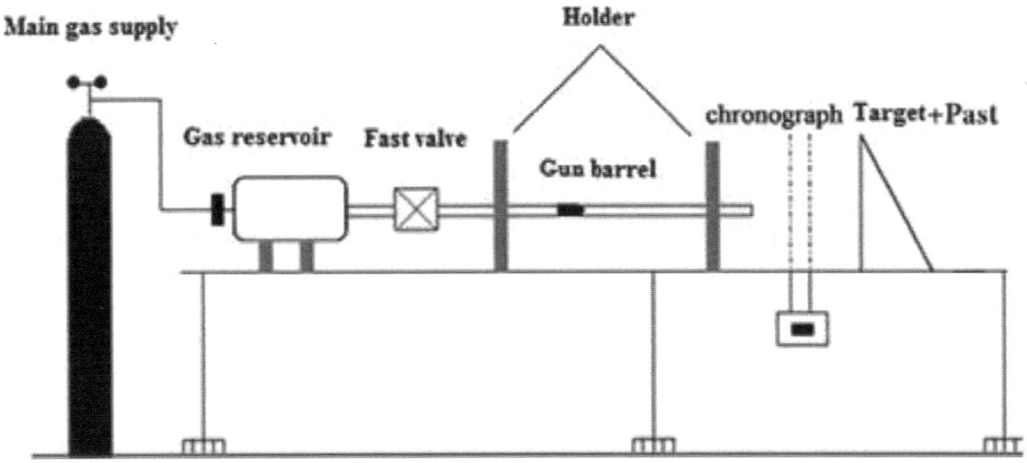

Figure 1: Schematic Representation of a Lab Scale Impact Test Procedure [20]

Figure 2: Real Setup for Impact Test

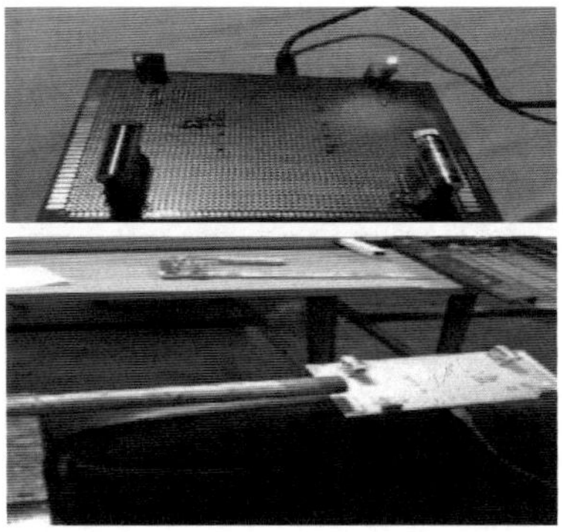

Figure 3: Laser Speed Detector System

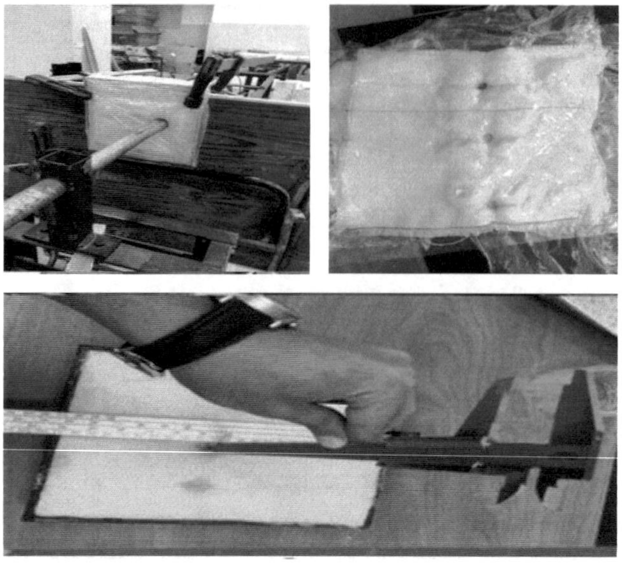

Figure 4: Real Impact Test Procedure on Impregnated Fabric with STF

3. Results and Discussion

The output of the characterization tests will be pointed out briefly while the detailed information is found in a previous publication [2]. The impact test results are shown in detail and discussed to clarify the significance of adding STF to Polyamide fabrics in impact resistance. The indentation depth of the back face signature is employed as an indicator of the ballistic performance of fabrics.

3.1 Characterization Tests

The average particle size of 84.5% of the fumed silica particles has an average size of 16.49 nm, while the average size of 89% starch particles is determined as 303.7 nm.

Results also show that starch has a higher zeta potential value (22.6 mV) than silica (-12.8 mV). Rheological characterization was performed for the two STFs synthesized from different fumed silica concentrations dispersed in PEG 400 g mol-1 and commercial starch in water. Results show that shear thickening happens for Silica prepared with PEG 400 g mol-1 with all particle weight fractions having a concentration range from 10 to 30 wt% ratio silica nanoparticles. The viscosity with the highest value was for Silica 30% as 178.9 Pa.s. The same procedure has been made for starch/water STF. The viscosity range against the change in the shear rate of 10 to 30 wt% ratio of starch particles in distilled water. The highest viscosity achieved for starch 30% is 523.6 Pa.s [2].

3.2 Impact Test

The impact test applied is identified as a non-penetration method where the projectile does not penetrate the target and remains in it. Accordingly, the indentation depth of the backface signature is employed as an indicator of the ballistic performance of fabrics. The velocity of the projectile has been tested several times and recording its average value to be 40 m/s at air pressure equals 2 bar produced from the gas compressor attached. The kinetic energy of the projectile is determined from equation (1) as follows:

$$\text{Kinetic Energy} = 0.5mv^2 \quad (1)$$

$= 0.5 \times 0.044\ (kg) \times 40^2\ (m/sec) = 35.2\ J.$

Where (m) is the mass of the projectile used in the test and (v) is the velocity of the projectile which is equal to 40 m/s.

3.2.1 Impregnation of Fabrics with STFs. Sample 1 is a neat fabric without impregnation taken as a reference to compare the impregnated fabrics with. The weight of the 5 neat layers of fabric without impregnation is 36 (gm). Description of samples of impregnated fabric with STF composed of (Silica/PEG) is shown in Table 1. While samples of impregnated fabric with STF composed of (Starch/Water) is shown in Table 2. The maximum percentage of the addon filler materials could not exceed 60 % because this would negatively affect fluid mobility.

Table 1: Description of Samples of Impregnated Fabric with STF Composed of (Silica/PEG).

Sample	**Weight Ratio of Silica**	**Composition (gm)**			**Weight of fabric after impregnation (gm)**	**Weight of STF absorbed by the fabric (gm)**	**Weight of STF absorbed without ethanol (gm)**
		PEG	Silica	Ethanol			
1	0	0	0	0	--------	---------	---------
2	10	200	20	0	75	75-36 = 39	39
3	30	200	60	400	155	155-36 = 119	33
4	60	200	120	600	223	223-36 = 187	30

Table 2: Description of Samples of Impregnated Fabric with STF Composed of (Starch/Water).

Sample	Weight Ratio of Starch	Composition (gm)		Weight of fabric after impregnation (gm)	Weight of STF absorbed by the fabric (gm)
		Water	Starch		
5	10	200	20	76	76-36 = 40
6	30	200	60	89	89-36 = 53
7	60	200	120	85	85-36 = 49

The process of impregnating fabrics with starch and water is much easier and more sufficient amounts of STF are absorbed that exceeded the add-on weight ratio of 1:1 of fabric to STF. The ease of impregnation might be due to the low viscosity of water compared to PEG which facilitates the wetting of the fabric.

3.2.2 Samples After Impact. Figure 5 shows the samples after the impact test. All samples are prepared from 5 layers of Twaron CT (714) fabric cut into 20x20cm and placed directly in front of the modeling clay. Sample number 1 shows the result of shooting a neat fabric without impregnation with STF. Samples 2, 3, and 4 are all impregnated with STF (silica/PEG) at different concentrations of silica. Samples 5, 6, and 7 are all impregnated with STF (starch/water) at different concentrations of starch.

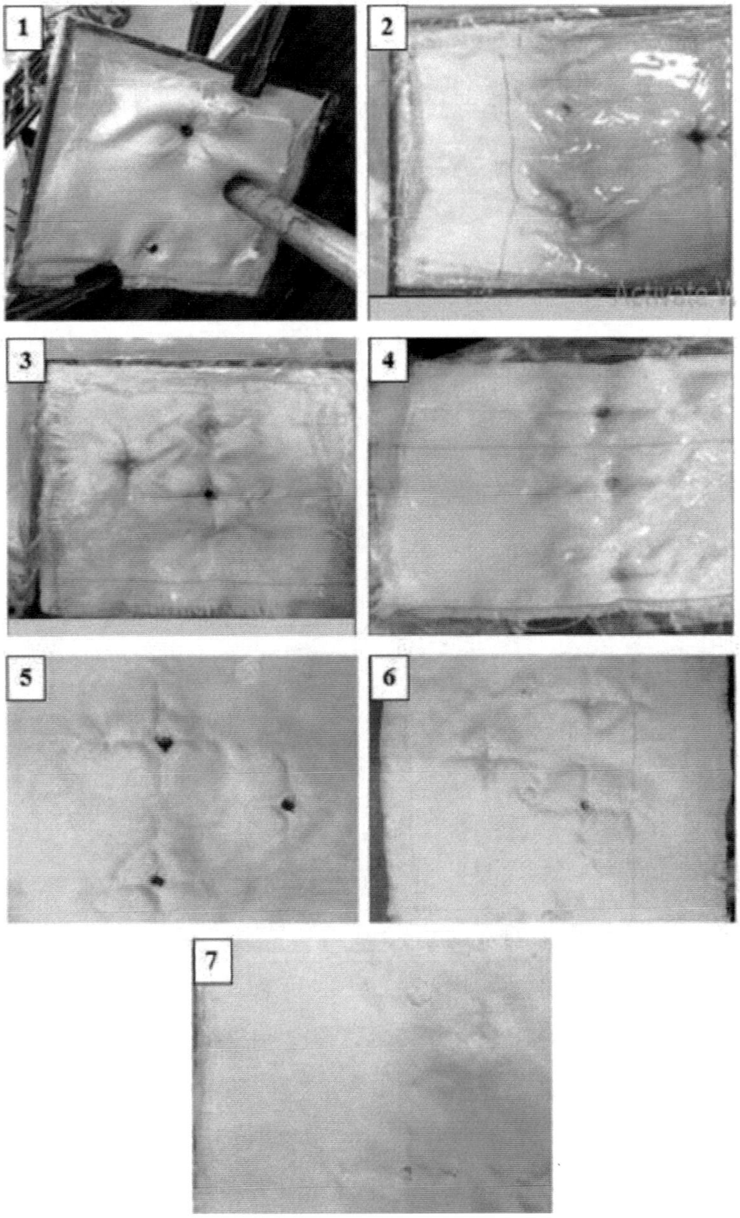

Figure 5: Impact Test on Impregnated Twaron Fabrics (1) Plain Sample. Samples (2, 3 & 4) of (10,30,60 %) Silica/PEG respectively. Samples (5, 6 & 7) of (10,30,60 %) Starch/Water respectively.

Figure 5 shows the damaged area of neat Twaron five-layered fabrics and STF-Twaron fabrics after the impact testing. Yarn that was directly affected by the projectile in untreated Twaron fabrics has been drawn out of the harmed area substantially, while yarn pull-out has declined significantly, as the filler rate in the area of fabric treated with STF has been increased. This can be explained by the additional restriction between yarns, which contribute to a less transverse deformation, of the impregnated STF. In addition, the thickening phenomena play a significant role in absorbing energy as clarified before in theory due to the collision of particles and hydrocluster formation at the point of impact. STF impregnated fabrics with starch/water system show superior results compared to silica/PEG STF at the same concentration of the filling materials.

3.2.3 Average Indentation Depth of Each Sample. Figure 6 shows the average indentation values of each sample after being shot multiple times. The figure shows that both systems show a decreasing trend in the indentation depth meaning better energy absorption as much as the STF concentration increases. The results show less indentation depth in starch/water STF-Twaron fabrics for the same concentration of filler than silica/PEG system as shown in samples 4 and 7. The indentation depth in sample 7 shows less value than sample 4 although both have the same concentration of filler which is 60 % of the whole fluid. The improvement percentage in indentation depth of each sample is a direct indication of the percentage of energy absorption improvement.

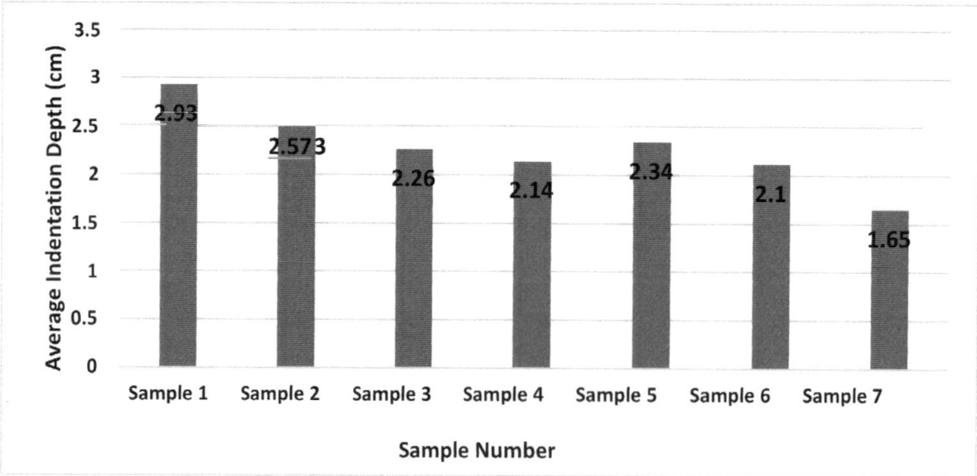

Figure 6: Average Indentation Depth of Each Sample

3.2.4 The Improvement Percentage in the Indentation Depth for Each Sample. This can be calculated relative to the indentation depth in the plain Twaron fabric sample as shown in Figure 7; in which the indentation depth improvement percentage is calculated from equation (2) as follows:

$$\frac{\text{Average Deapth in Plain Sample} - \text{Average Deapth in STF Sample}}{\text{Average Deapth in Plain Sample}} \text{x}100\% \quad (2)$$

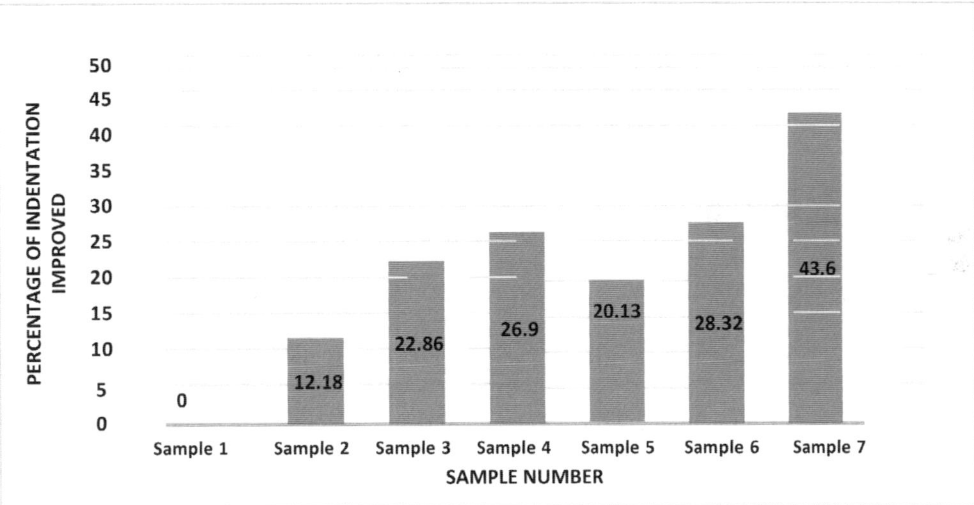

Figure 7: The Improvement Percentage in the Indentation Depth for Each Sample

Table 3: The Improvement Percentage in the Indentation Depth for Each Sample.

Sample	Weight Ratio of Silica	Weight Ratio of Starch	Average Depth of Indentation Marks (cm)	Percentage Improved
1	0	0	2.93	-
2	10	-	2.573	12.18 %
3	30	-	2.26	22.86 %
4	60	-	2.14	26.9 %
5	-	10	2.34	20.13 %
6	-	30	2.1	28.32 %
7	-	60	1.65	43.6 %

Results shown in Table 3 highlight the improved performance in higher Silica concentrations in (PEG) impregnated fabrics as demonstrated in 60 wt% ratio compared to 30 and 10 wt% ratio. This is because, at higher concentrations of silica, more friction between fabric filaments resists transverse deformation besides the shear thickening effect in which more collisions between particles occur forming clusters at sudden shear rates resisting the projectile movement and dissipating energy as stated in theory. The real promising results appeared in samples 5-7 where a starch/water system is used for fabric impregnation. The indentation depth shows smaller values than silica/PEG in an obvious way although both had a similar concentration range. The improvement rate for all samples is determined in Figure 7 which is the percentage at which shear thickening fluid (STF) will improve impact resistance against projectiles. Such changes are strong evidence of a process similar to the spontaneous solid phase produced by the shear thickening fluids when maximum load rates are reached. It was proposed that the STF be used in soft ballistic body amours.

4. Conclusion

In conclusion, this research aims to develop an approach to improving the impact resistance of ballistic fabrics for applications like protecting clothing and energy dissipation. The Shear Thickening Fluid (STF) has shown new opportunities for improvement of impact resistance by impregnating it with ballistic fabrics while reducing the majority of disadvantages associated with commonly used armor materials, including rigidity, wide thicknesses, and high costs. Improved friction in fabric elements at various levels may tend partly to boost the behavior of energy dispersion in fabrics treated with STF. Increased friction can prevent mobility between the yarns at higher stress levels and lead to increased yarn pull-out energy. The findings also verify rheological test results for both STF systems. The superior performance of the starch/water system over the silica/PEG shear thickening fluid can be attributed to many reasons; one of the most significant is the higher viscosity values obtained in the rheological tests of the starch/water system. The similar shear rates that both STF systems were subjected to, reveal that the maximum viscosity rise in starch and water system was 523.6 Pa.s while silica/PEG was only 179.8 Pa.s. Another realistic reason might be due to the higher amount of STF absorbed by the fabric in the case of starch and water system as shown in the impregnation procedure. The larger amount of STF absorbed is directly translated to a more obvious thickening effect at the impact test leading to more impact resistance. Finally, the starch molecule structure had a vital role in the abundance of hydrogen bonds and increased Zeta-potential values and polydispersity of starch particles greatly affect the stability of the particles in the fluid and fill all the inter-yarn spaces of the fabric. Therefore, results showed superior shear thickening effect of starch and water system over silica and PEG system as expected from the characterization and rheological tests performed earlier. Larger starch particle size over silica and Zeta Potential recorded respectively lead to more dispersion stability and obvious thickening effect at higher particle concentration leading to a greater jump in viscosity at sudden shear rate. The highest concentration of silica/PEG STF/fabric showed improvement in indentation depth by 26.9%, while the starch/water STF by 43.6%. This opens a route in the future to try various hybrid STFs and test the performance of the intercalated nanoparticles on the rheological behavior which is directly translated into the ballistic performance as well. Results demonstrate that STF enhanced fabrics have taken advantage of the STF properties to fully improve the impact response of the composite material.

Acknowledgment

We are grateful for the support of The Center of Advanced Material (CAM) in The British University in Egypt (BUE).

Disclosure Statement

No potential competing interest was reported by the authors.

References

[1] R. A. Mankarious, M. A. Radwan, M. Shazly, and H. A. Elazab, "*Bulletproof vests/shields prepared from composite material based on strong polyamide fibers and epoxy resin*," J. Eng. Appl. Sci., **vol. 12**, no. 10, 2017.

[2] R. A. Mankarious and M. A. Radwan, "*Shear Thickening Fluids Comparative Analysis Composed of Silica Nanoparticles in Polyethylene Glycol and Starch in Water*," J. Nanotechnol., **vol. 2020**, 2020.

[3] O. Hamdy, R. A. Mankarious, M. A. Radwan, M. A. Sadek, and H. A. Elazab, "*Composite materials based on twaron and nano materials*," Int. J. Innov. Technol. Explore. Eng., **vol. 9**, no. 1, 2019.

[4] R. A. Mankarious and M. A. Radwan, "*Preparation and Characterization of Novel Composite Material Based on Strong Polyamide Fibers and Epoxy Resin Applied in Bulletproof Vests and Shields*," Int. Undergrad. Res. Conf., no. July 2016, 2016.

[5] W. Na et al., "*Shear behavior of a shear thickening fluid-impregnated aramid fabrics at high shear rate*," Compos. Part B Eng., **vol. 97**, pp. 162–175, 2016.

[6] N. Asija, H. Chouhan, and S. Amare, "*Impact Response of Shear Thickening Fluid (STF) Treated High Strength Polymer Composites – Effect of STF Intercalation Method*," Procedia Eng., **vol. 173**, pp. 655–662, 2017.

[7] Erdoğan T., "*Development of Liquid Armor Materials and Rheological Behavior of Shear Thickening Fluids (STFs)*," Izmir Inst. ofTechnology, November, p. 147, 2011.

[8] S. R. Raghavan and S. A. Khan, "*Shear-thickening response of fumed silica suspensions under steady and oscillatory shear,*" J. Colloid Interface Sci., **vol. 185**, no. 1, pp. 57–67, 1997.

[9] B. Yang, S. Wang, G. Z. Xu, and F. Xin, "*Preparation of SiO2/PEG Shear Thickening System by Centrifugal Dispersion,*" Adv. Mater. Res., **vol. 560–561**, no. April, pp. 586–590, 2012.

[10] B. J. F. Bossis G., "*The rheology of Brownian suspensions*," Br. J. Psychiatry, **vol. 111**, no. 479, pp. 1009–1010, 1989.

[11] S. Gürgen, M. A. Sofuoğlu, and M. C. Kuşhan, "*Rheological compatibility of multi-phase shear thickening fluid with a phenomenological model,*" Smart Mater. Struct., **vol. 28**, no. 3, 2019.

[12] Gürgen, Selim; Kushan, Melih Cemal; and Li, Weihua, "*Shear thickening fluids in protective applications: A review*" (2017). Faculty of Engineering and Information Sciences - Papers: Part B. 1015.

[13] P. V. Cavallaro, "*Soft Body Armor: An Overview of Materials, Manufacturing, Testing, and Ballistic Impact Dynamics,*" NUWC-NPT Tech. Rep. 12,057, no. August, pp. 1–22, 2011.

[14] M. Bajya, A. Majumdar, B. S. Butola, S. K. Verma, and D. Bhattacharjee, "*Design strategy for optimizing weight and ballistic performance of soft body armour reinforced with shear thickening fluid,*" Compos. Part B Eng., **vol. 183**, no. December 2019, 2020.

[15] J. L. Park, B. il Yoon, J. G. Paik, and T. J. Kang, "*Ballistic performance of p-aramid fabrics impregnated with shear thickening fluid; Part I – Effect of laminating sequence,*" Text. Res. J., **vol. 82**,

no. 6, pp. 527–541, 2012.
[16] Y. Salim, M. Y. Yahya, H. A. Israr, and R. Mohamed, *"A Review on Stab and Spike Resistance Performance of Shear Thickening Fluids (STF) Impregnated Fabrics,"* pp. 22–27, 2017.
[17] A. Haris, H. P. Lee, T. E. Tay, and V. B. C. Tan, *"Shear thickening fluid impregnated ballistic fabric composites for shock wave mitigation,"* Int. J. Impact Eng., **vol. 80**, pp. 143–151, 2015.
[18] S. Gürgen and T. Yıldız, *"Stab resistance of smart polymer coated textiles reinforced with particle additives,"* Compos. Struct., **vol. 235**, p. 111812, 2020.
[19] Y. Park, A. H. Baluch, Y. Kim, and C.-G. Kim, "*High Velocity Impact Characteristics of Shear Thickening Fluid Impregnated Kevlar Fabric,"* Int. J. Aeronaut. Sp. Sci., **vol. 14**, no. 2, pp. 140– 145, 2013.
[20] T. A. Sanders, *"Penetration of Composite Laminates by Conical Indenters And Projectiles,"* p. 136, 1997.
[21] Y. Gong and G. Yang, *"Manufacturing and physical properties of all-polyamide composites,"* J. Mater. Sci., **vol. 44**, no. 17, pp. 4639–4644, 2009.
[22] Evonik Industries, *"AEROSIL ® 200 Hydrophilic fumed silica,"* 2015.
[23] Sigma-Aldrich, *"Poly(ethylene glycol) BioUltra, 400,"* pp. 4–6, 2019.
[24] C. Ding, *"Experiments on the Viscosity and Freezing Point of a Starch Water Mixture,"* **vol.81**, no. June, pp. 1–5, 2013.
[25] Teijin Aramid, *"Ballistics material handbook,"* pp. 1–11, 2016.
[26] M. Fahool and A. R. Sabet, *"Parametric study of energy absorption mechanism in Twaron fabric impregnated with a shear thickening fluid,"* Int. J. Impact Eng., **vol. 90**, pp. 61–71, 2016.

AUTHOR INDEX